# Selected Titles in This Series

21 **Shigui Ruan, Gail S. K. Wolkowicz, and Jianhong Wu, Editors,** Differential equations with applications to biology, 1999

20 **Peter A. Fillmore and James A. Mingo, Editors,** Operator algebras and their applications II, 1998

19 **William G. Dwyer, Stephen Halperin, Richard Kane, Stanley O. Kochman, Mark E. Mahowald, and Paul S. Selick (Editor-in-Chief), Editors,** Stable and unstable homotopy, 1998

18 **Panos M. Pardalos and Henry Wolkowicz, Editors,** Topics in semidefinite and interior-point methods, 1998

17 **Joachim J. R. Cuntz and Masoud Khalkhali, Editors,** Cyclic cohomology and noncommutative geometry, 1997

16 **Victor P. Snaith, Editor,** Algebraic K-theory, 1997

15 **Stephen P. Braham, Jack D. Gegenberg, and Robert J. McKellar, Editors,** Sixth Canadian conference on general relativity and relativistic astrophysics, 1997

14 **Mourad E. H. Ismail, David R. Masson, and Mizan Rahman, Editors,** Special functions, $q$-series and related topics, 1997

13 **Peter A. Fillmore and James A. Mingo, Editors,** Operator algebras and their applications, 1997

12 **Dan-Virgil Voiculescu, Editor,** Free probability theory, 1997

11 **Colleen D. Cutler and Daniel T. Kaplan, Editors,** Nonlinear dynamics and time series: Building a bridge between the natural and statistical sciences, 1997

10 **Jerrold E. Marsden, George W. Patrick, and William F. Shadwick, Editors,** Integration algorithms and classical mechanics, 1996

9 **W. H. Kliemann, W. F. Langford, and N. S. Namachchivaya, Editors,** Nonlinear dynamics and stochastic mechanics, 1996

8 **Larry M. Bates and David L. Rod, Editors,** Conservative systems and quantum chaos, 1996

7 **William F. Shadwick, Perinkulam Sambamurthy Krishnaprasad, and Tudor Stefan Ratiu, Editors,** Mechanics day, 1996

6 **Anna T. Lawniczak and Raymond Kapral, Editors,** Pattern formation and lattice gas automata, 1996

5 **John Chadam, Martin Golubitsky, William Langford, and Brian Wetton, Editors,** Pattern formation: Symmetry methods and applications, 1996

4 **William F. Langford and Wayne Nagata, Editors,** Normal forms and homoclinic chaos, 1995

3 **Anthony Bloch, Editor,** Hamiltonian and gradient flows, algorithms and control, 1994

2 **K. A. Morris, Editor,** Control of flexible structures, 1993

1 **Michael J. Enos, Editor,** Dynamics and control of mechanical systems: The falling cat and related problems, 1993

FIELDS INSTITUTE COMMUNICATIONS

THE FIELDS INSTITUTE FOR RESEARCH IN MATHEMATICAL SCIENCES

# Differential Equations with Applications to Biology

Shigui Ruan
Gail S. K. Wolkowicz
Jianhong Wu
Editors

American Mathematical Society
Providence, Rhode Island

## The Fields Institute for Research in Mathematical Sciences

The Fields Institute is named in honour of the Canadian mathematician John Charles Fields (1863–1932). Fields was a visionary who received many honours for his scientific work, including election to the Royal Society of Canada in 1909 and to the Royal Society of London in 1913. Among other accomplishments in the service of the international mathematics community, Fields was responsible for establishing the world's most prestigious prize for mathematics research—the Fields Medal.

The Fields Institute for Research in Mathematical Sciences is supported by grants from the Ontario Ministry of Education and Training and the Natural Sciences and Engineering Research Council of Canada. The Institute is sponsored by McMaster University, the University of Toronto, the University of Waterloo, and York University and has affiliated universities in Ontario and across Canada.

1991 *Mathematics Subject Classification.* Primary 34-xx; Secondary 92-xx, 35-xx, 58-xx.

---

**Library of Congress Cataloging-in-Publication Data**

Differential equations with applications to biology / Shigui Ruan, Gail S. K. Wolkowicz, Jianhong Wu (editors).
p. cm. — (Fields Institute Communications, ISSN 1069-5265 ; v. 21)
"Proceedings of the International Conference on Differential Equations with Applications to Biology, held at Dalhousie University, July 25–29, 1997, in Halifax, Nova Scotia, Canada"—Pref.
Includes bibliographical references.
ISBN 0-8218-0944-X (alk. paper)
1. Biomathematics—Congresses. 2. Differential equations—Congresses. I. Ruan, Shigui, 1963– . II. Wolkowicz, Gail S. K. (Gail Susan Kohl), 1950– . III. Wu, Jianhong. IV. International Conference on Differential Equations with Applications to Biology (1997 : Nova Scotia) V. Series.
QH323.5.D56 1998
570′.1′51535—dc21 98-33721
CIP

---



∞ The paper used in this book is acid-free and falls within the guidelines
established to ensure permanence and durability.
This publication was prepared by The Fields Institute.
Visit the AMS home page at URL: http://www.ams.org/

10 9 8 7 6 5 4 3 2 1 04 03 02 01 00 99

# Contents

# Preface

This volume contains the proceedings of the International Conference on Differential Equations with Applications to Biology, held at Dalhousie University, July 25 - 29, 1997, in Halifax, Nova Scotia, Canada. The conference was held in honor of the retirement of Professor Lynn Erbe and Professor Herb I. Freedman from the University of Alberta. Professor Erbe and Professor Freedman have made significant contributions in both research and the training of graduate and undergraduate students. In particular, they helped organize the very successful G. J. Butler Memorial Conferences on Differential Equations and Population Dynamics that have been among the most influential and most important conferences in this area.

In the last few decades, the theory of differential equations and dynamical systems has advanced significantly and various applications have been found in biology, ecology, epidemiology, medicine, neurobiology, physiology, zoology, and many other fields. The subject is of current interest and of increasing importance, as witnessed by the recent contributions in such topics as population dynamics, neural networks, resource management, and disease transmission.

The objective of the conference was to bring together mathematicians, physical scientists, and biologists to present their current research. There were 120 participants from 25 countries. There were 94 talks, 12 of which were one-hour plenary lectures. The plenary speakers were Leon Glass (McGill University), Karl P. Hadeler (Universität Tübingen), Morris Hirsch (University of California at Berkeley), Simon Levin (Princeton University), Michael Mackey (McGill University), John Mallet-Paret (Brown University), Klaus Schmitt (University of Utah), Lee Segel (Weizmann Institute), Hal Smith (Arizona State University), Pauline van den Driessche (University of Victoria), Paul Waltman (Emory University), and Glenn Webb (Vanderbilt University).

The topics of the conference included the qualitative and numerical analysis of ordinary, partial, functional, and stochastic differential equations and their applications to biology and other related fields. The main areas of emphasis were: bifurcation theory, chaos, stability theory, boundary value problems, persistence theory, neural networks, disease transmission, predator-prey, competition, pattern formation, etc. This proceedings contains 41 papers by 66 authors and covers almost all of the topics discussed at the conference.

The conference was supported by The Fields Institute for Research in Mathematical Sciences in Toronto, le centre de recherches mathématiques (CRM) in Montreal, the Applied Mathematics Institute at the University of Alberta, and Dalhousie University.

We wish to thank the Department of Mathematics, Statistics and Computing Science at Dalhousie University for their hospitality. We also wish to thank Mrs. Erna Unrau, the Publications Assistant of The Fields Institute, for her professional assistance. Finally we would like to thank all of the participants for making the conference a success and all of the referees for their thorough and timely review of the manuscripts submitted to this volume.

Shigui Ruan
Dalhousie University

Gail S. K. Wolkowicz
McMaster University

Jianhong Wu
York University

Editors

Fields Institute Communications
Volume **21**, 1999

# On a Class of Nonlocal Problems with Applications to Mathematical Biology

**W. Allegretto**
Department of Mathematical Sciences
University of Alberta
Edmonton, Alberta, Canada T6G 2G1
retl@retl.math.ualberta.ca

**P. Nistri**
Dipartimento di Sistemi ed Informatica
Università di Firenze
Via di S. Marta 3
50139 Firenze, Italy
pnistri@ingfi1.ing.unifi.it

*To our friends Lynn Erbe and Herb Freedman*

**Abstract.** We consider the problem of finding positive solutions to a family of nonlocal partial differential equations, with mixed boundary conditions. It is a feature of these problems that maximum principle methods are not applicable, and we employ topological methods in the analysis. A key role is played by related eigenvalue estimates.

## 1 Introduction

Let $B_\delta(x,y)$ denote a mollifier in $R^N$, i.e., $B_\delta(x,y) = B_\delta(|x-y|) \in C_0^\infty$, $\int_{R^N} B_\delta(x,y)dy = 1$ for any $x$, $B_\delta(|x-y|) = 0$ if $|x-y| \geq \delta$, $B_\delta(|x-y|)$ bounded away from zero if $|x-y| < \mu < \delta$. Consider a smooth bounded domain $\Omega \subset R^N$ and define for any $u \in L^1(\Omega)$,

$$\overline{u}(x) = \int_\Omega B_\delta(x,y)u(y)dy.$$

In this paper we consider the problem of finding positive solutions to the nonlocal partial differential equation:

$$-\nabla[a\nabla u + \vec{b}u] = [h - g\overline{u}]u \tag{1}$$

1991 *Mathematics Subject Classification.* Primary 35J56; Secondary 35J25.
Research supported in part by NSERC (Canada), CNR (Italy).

subject to the mixed boundary conditions:

$$\begin{cases} u = 0 \quad \text{on} \quad \partial\Omega_D; \\ \text{natural boundary conditions on } \partial\Omega_N, \text{ i.e., in this case :} \\ a\,\frac{\partial u}{\partial n} + (\vec{b}\cdot\vec{n})u = 0 \end{cases} \tag{2}$$

where $\partial\Omega_D$, $\partial\Omega_N$ are disjoint parts of $\partial\Omega$, $\partial\Omega = \partial\Omega_D \cup \partial\Omega_N$, $\partial\Omega_D$ closed. We shall consider explicitly only cases where $\partial\Omega_D$ and $\partial\Omega_N$ are not trivial but simpler versions of our proofs work in the other cases. We also require that regularity conditions be obtained at points of $\partial\Omega$ (resp. $\partial\Omega_D \cap \overline{\partial\Omega_N}$) by means of biLipschitz maps to hemispheres (resp. quarter spheres).

Furthermore, given any constant $K$ we ask that there exists a neighbourhood $\mathcal{N}$ of $\partial\Omega$ such that

$$\int_{\mathcal{N}} a|\nabla\omega|^2 + (\vec{b}\cdot\nabla\omega)\omega > K\int_{\mathcal{N}} \omega^2 \tag{3}$$

for any nontrivial $\omega \in H_0^{1,2}(\mathcal{N} \cap (\Omega \cup \partial\Omega_N))$. The reader interested in detailed descriptions of such conditions may find some of them given in, e.g., [**18**], [**19**], [**22**], [**24**] and elsewhere. As for the functions $a, h, g, \vec{b}$ in (1), we recall their possible physical meanings: $a(x)$ represents the diffusion process; $\vec{b}(x) = (b_1(x), \ldots, b_N(x))$ is a possible drift; $h(x)$ the growth condition and $g(x)$ a factor to describe the magnitude of the "crowding effects". We ask that $a, h, g$ be nonnegative in $L^\infty(\Omega)$ and piecewise smooth with $\alpha < a(x) < \beta$ for some constants $0 < \alpha < \beta$, $a$ smooth near $\partial\Omega_N$ and that $\vec{b}$ be $C^1$. We thus allow for some abrupt environmental changes in the physical problem being described as was earlier done by H. I. Freedman and his coworkers, [**10**], [**11**], for local problems. Unlike such earlier approaches, we do not consider the problem in separate "patches", with suitable continuity conditions at the patch boundaries. Instead we formulate the problem globally in weak form and recover the patch continuity conditions as a consequence. This is indicated at the end of the paper. Consequently, we feel that this approach has also some advantages for purely local problems, although we do not pursue this topic here.

Observe that we require more smoothness for $\vec{b}$ than for the other coefficients. This has to do with our method of proof, but we conjecture that the results are true for $\vec{b}$ piecewise smooth, say.

Nonlocal equations similar to (1) have been previously introduced to model biological processes by [**13**], [**8, 9**], [**5**]. In these papers, further significant references may be found.

Some differences between the present results and earlier work with which we are familiar are as follows: as mentioned above, we consider the case of mixed boundary conditions, with piecewise smooth $L^\infty$ coefficients in which $g$ is also allowed to vanish somewhere in $\Omega$. Furthermore, the kernel $B_\delta(x,y)$ vanishes for $|x-y| > \delta$. We do not require that the left-hand side of (1) be coercive. We feel that these conditions are more realistic than many often used, and certainly more general. They do introduce new difficulties in the proofs.

We note that here $\overline{u}$ is calculated using only values of $y$ inside $\Omega$, i.e., the effects on $u(x)$, for $x$ with dist $(x, \partial\Omega) < \delta$, of values $u(y)$ with $y \notin \Omega$ are disregarded except through the boundary conditions. If such interaction effects were significant, then $\Omega$ could be enlarged. It may also be of interest to assume that $u$ is given outside $\Omega$. This is actually what we do here, with the choice $u(y) \equiv 0$ if $y \notin \Omega$, for most of the paper, but the small $\delta$ case is treated differently.

We emphasize to the reader that order arguments (i.e., upper/lower solutions) are in general not applicable to this family of problems, as was illustrated by explicit example in [**5**], even if $g(x) = \varepsilon \simeq 0$. The problem we consider is thus different from those considered in [**3**], [**9**], [**12**], [**15**], where if $g(x)$ was small enough then order procedures were indeed applicable, and from those considered in, e.g., [**20**], [**21**] where order arguments can always be employed. We thus proceed by Topological Methods (Degree Theory).

The paper is in final form and structured as follows: We shall first consider the case where $g$ is bounded (essentially) away from zero. Next, we will allow $g$ to be zero at some points of $\Omega$. The case of $\delta$ small will then be considered. Some comments, related results and open problems will conclude the paper.

Finally, we point out that our proofs work without any changes in the case that the right-hand side of (1) is replaced by the more general expression $f\big(x, u, T(u)\big)u$ under suitable assumptions on $f$ and the operator $T$. More comments on this will be made in the last section.

## 2 Existence for general $\delta$

Note first that equation (1) and the maximum principle show that if $0 \leq u \in L^\infty$, nontrivial, then $u > 0$. Since most of the coefficients are only in $L^\infty$, to see this note that we must still have $u \in C^{\alpha_0}(\overline{\Omega})$ for some $\alpha_0$ determined by the problem data but not $u$, [**24**]. If $u(P) = 0$ at some $P \in \Omega$, then $\|u\|_{L^1(S_r(P))} = 0$ for small $r$ by the generalized Harnack's Inequality, [**14**, p. 194], where $S_r(P)$ denotes the ball of radius $r$, center $P$. It follows that $u \equiv 0$ in $S_r(P)$ and thus $u > 0$ in $\Omega$ since $\Omega$ is connected.

As may be expected, we shall formulate results in the sequel based on suitable eigenvalue estimates. Specifically here we consider the two parameter local problem:

$$\ell_0 \omega = -\nabla[a\nabla\omega + \vec{b}\omega] + \eta g\omega = \mu_1(\eta)h\omega \tag{4}$$

subject to boundary conditions (2), where $\mu_1(\eta)$ denotes the eigenvalue with a positive eigenvector. Problem (4) is classical, it and related problems have been studied by many authors, see, e.g., [**4**], [**6**]. It is well known that $\omega > 0$ exists, and $\mu_1(\eta)$ is monotone increasing in $\eta$.

Our specific assumption is:

$$\mu_1(0) < 1 < \mu_1(K) \tag{5}$$

for large $K$, i.e., "eigenvalue crossing" occurs for this related problem. It is fairly easy to see that if this condition is dropped, then the nonlocal problem (1) - (2) may not have a positive solution.

We collect such preliminary facts as well as other results for the reader's convenience in Lemma 0, and give brief proofs.

**Lemma 1** *For* $\eta \geq 0$ *:*

(a) $\mu_1(\eta)$ *exists and* $\omega > 0$ *in* $\Omega \cup \partial\Omega_N$. *Moreover,* $\mu_1(\eta)$ *and* $\omega$ *are unique.*
(b) $\mu_1(\eta) \uparrow \eta$. *If* $g > \alpha_0 > 0$ *then* $\mu_1(\eta) \to \infty$ *as* $\eta \to \infty$

(c) $\theta_1(\eta) \le \mu_1(\eta) \le \theta_2(\eta)$
*where* $\theta_1(\eta)$, $\theta_2(\eta)$ *are the first eigenvalues of the formally self-adjoint problems:*

$$-\nabla\left[a\nabla\omega_1 + \frac{\vec{b}}{2}\omega_1\right] + \frac{\vec{b}\cdot\nabla\omega_1}{2} + \eta g\omega_1 = \theta_1(\eta)h\omega_1$$

*and*

$$-\nabla\left[a\nabla\omega_2 + \frac{\vec{b}}{2}\omega_2\right] + \frac{\vec{b}\cdot\nabla\omega_2}{2} + \left[\frac{|\vec{b}|^2}{4a} + \eta g\right]\omega_2 = \theta_2(\eta)h\omega_2$$

*respectively, with the associated conditions* (2). *Here the existence of* $\theta_1(\eta)$ *is assumed.*

(d) *If problem* (1) – (2) *has a positive solution then condition* (5) *holds.*

(e) *Let* $0 \le u \in C^\alpha$ *satisfy*

$$\begin{cases} -\nabla[a\nabla u + \vec{b}u] \le hu \\ \text{Boundary conditions (2)} \end{cases}$$

*and assume* $P \in \Omega$. *Let* $\delta^* > 0$ *be given, and let* $S_{\delta^*}(P)$ *denote the sphere centered at* $P$ *of radius* $\delta^*$. *Then there exists* $K$, $\varepsilon_0 > 0$ *independent of* $u, P$ *such that*

$$\sup_{S_{\varepsilon_0}(P)\cap\Omega} u(x) \le K\Big(\int_{S_{\delta^*}(P)\cap\Omega} u^2\Big)^{1/2}.$$

(f) *Let* $z \in C^\alpha(\overline{\Omega})$ *satisfy* $-\nabla[a\nabla z+\vec{b}z]-\lambda hz = f$ *with boundary conditions (2) and* $0 \ge f \in L^\infty$. *There exists* $\lambda_0 > 0$ *independent of* $f$, $z$ *such that if* $0 < \lambda < \lambda_0$ *and* $z \ge 0$ *then* $z \equiv 0$.

**Proof** (a) That $\mu_1(\eta)$ exists and $\omega > 0$ is an immediate consequence of the fact $\ell_0^{-1}$ is a positive map for $\eta \ge 0$ (by the maximum principle applied to the formal $L^2$ adjoint, ([**23**], p. 97)) from $C^{\alpha_0}$ to itself if the coefficients are smooth. See, e.g., ([**16**], p. 67), and note that $(\ell_0)^{-1}(hz) \ge cz$ for some $c > 0$, $0 \le z \in C_0^\infty$ and if $\omega \ge 0$ then $\omega > 0$ in $\Omega$ by our earlier comments at the start of this section. To obtain the result in our case we pass to the limit over a normalized (in $L^2$) sequence, employing the estimates of part (c) to show that the sequence of approximating eigenvalues is bounded. That $\omega > 0$ in $\partial\Omega_N$ follows from the extension process of part (e).

To see that there is only one such $\mu_1(\eta)$, note that by the Fredholm Alternative we also have

$$\ell_0^*(\omega^*) \triangleq \mu_1^*(\eta)h\omega^*$$

for some $\mu_1^*(\eta)$, $\omega^* > 0$ where $\ell_0^*$ denotes the formal adjoint, i.e.,

$$\mu_1(\eta)(h\omega^*,\omega) = (\omega^*,\ell_0\omega) = \mu_1^*(\eta)(h\omega^*,\omega).$$

Since $(h\omega^*,\omega) > 0$, this shows $\mu_1(\eta)$ is unique. Finally to see that $\omega > 0$ is also unique (up to scalar multiples), note that if there were two such eigenfunctions: $\omega$, $v$ then $\ell_0(\omega - \varepsilon v) - \mu_1(\eta)(\omega - \varepsilon v) = 0$. Set $E = \{\varepsilon \mid \omega - \varepsilon v \ge 0 \text{ in } \Omega\}$. Clearly $0 \in E$, $E$ is closed and if $\varepsilon \in E$ then $(\omega - (\varepsilon + \delta)v)^-$ must have support near $\partial\Omega$ for small $\delta > 0$, since given any connected compact set $K \subset \Omega$ then, again by Harnack's Inequality, either $\omega - \varepsilon v \equiv 0$ in $K$ or it is bounded away from zero in $K$. But the form associated with $\ell_0$ is coercive over such functions, by inequality (3), whence $\varepsilon + \delta \in E$ for $\delta$ small, i.e., $E$ is also open, and the result follows.

(b) Again by the Fredholm Alternative (see also [**2**] for related concepts) we have for any $0 < u \in \text{Domain}(\ell_0)$, $\ell_0 u - \mu_1(\eta) hu \in C^\alpha$, and assuming at first smooth coefficients with $h > 0$ that:

$$(\omega^*, \ell_0 u - \mu_1(\eta) hu) = 0,$$

i.e., for some $x_0 \in \Omega$,

$$\ell_0 u(x_0) = \mu_1(\eta) h(x_0) u(x_0)$$

whence

$$\inf_{x\in\Omega} \left(\frac{\ell_0 u}{hu}\right) \leq \mu_1(\eta) \leq \sup_{x\in\Omega} \left(\frac{\ell_0 u}{hu}\right).$$

For example, if $-\nabla[a\nabla u_i + \vec{b}u_i] + \eta_i g u_i = \mu(\eta_i) h u_i$ with $i = 1, 2$ and $\eta_2 > \eta_1$, then

$$\begin{aligned}\mu_1(\eta_1) &\leq \sup_{x\in\Omega} \left[\frac{-\nabla[a\nabla u_2 + \vec{b}u_2] + \eta_1 g u_2}{hu_2}\right] \\ &= \sup_{x\in\Omega} \left[\frac{\mu_2(\eta_2) hu_2 + (\eta_1 - \eta_2) g u_2}{hu_2}\right] \leq \mu_2(\eta_2).\end{aligned}$$

Passing to a limit then gives the result for our case. That $\mu_1(\eta) \to \infty$ as $\eta \to \infty$ for $g > \alpha_0 > 0$ follows from $\mu_1(\eta) \geq \theta_1(\eta)$ (shown in part (c)) and the Courant Min-Max Principle.

(c) Observe $\theta_1(\eta) \leq \mu_1(\eta)$ follows from:

$$\mu_1(\eta) = \frac{\int_\Omega a|\nabla\omega|^2 + (\vec{b}\cdot\nabla\omega)\omega + \eta g\omega^2}{\int_\Omega h\omega^2}$$

while $\theta_1(\eta)$ is the infimum of the right-hand side over $\omega \in H_0^{1,2}(\Omega \cup \partial\Omega_N)$, $h\omega \not\equiv 0$. Next we obtain $\theta_2(\eta) \geq \mu_1(\eta)$ by modifying the arguments in [**1**]. Specifically, let $\varphi \in C_0^\infty(\Omega \cup \partial\Omega_N)$, and observe the identity

$$\begin{aligned}&\int_\Omega (\omega^*)^2 a \left|\nabla\left(\frac{\varphi}{\omega^*}\right)\right|^2 + \omega^*\varphi\vec{b}\cdot\nabla\left(\frac{\varphi}{\omega^*}\right) + \frac{|\vec{b}|^2}{4a}\varphi^2 \\ &\qquad = \int_\Omega a|\nabla\varphi|^2 + (\vec{b}\cdot\nabla\varphi)\varphi + \left(\frac{|\vec{b}|^2}{4a} + \eta g\right)\varphi^2 - \mu_1(\eta)\int_\Omega h\varphi^2.\end{aligned}$$

As in [**1**], we observe that the left-hand side is nonnegative definite and obtain the result by taking the infimum over $\varphi$.

(d) Suppose (1) – (2) has a solution $0 < u \in L^\infty$. Then

$$-\nabla[a\nabla u + \vec{b}u] - hu = -g\overline{u}\, u \leq 0 \quad (\not\equiv 0).$$

We thus have (by the arguments in (b)) that there exists $0 < \theta < 1$ and $\omega > 0$ such that

$$-\nabla[a\nabla\omega + \vec{b}\omega] - \theta h\omega = 0$$

and $\omega$ satisfies condition (2). In fact, $\mu_1(0) = \theta < 1$ since $\theta = 1$ implies $g\overline{u}\, u = 0$. Furthermore, if we choose $K \geq \overline{u}$ in $\overline{\Omega}$ then

$$-\nabla[a\nabla u + \vec{b}u] + gKu \geq hu.$$

For example, $\mu_1(K) \geq 1$. We can only have equality if $u$ is actually the eigenvector, i.e., $K = \overline{u}$. But in this case, $\mu_1(K+1) > 1$.

(e) Note that for each point $Q \in \overline{\partial\Omega_N}$ we have a neighbourhood $\mathcal{N}$ such that $\Omega\cap\mathcal{N}$ can be mapped either to a hemisphere or a quarter sphere if $Q \in \overline{\partial\Omega_N}\cap\partial\Omega_D$ of radius $r = r(Q)$ by biLipschitz maps $L$. Without loss of generality, $\mathcal{N} \subset S_{\delta^*}(Q)$. Set $\mathcal{M} = L^{-1}[S_{\varepsilon r}(L(Q))]$, $0 < \varepsilon < 1$. First cover $\overline{\partial\Omega_N} \cap \partial\Omega_D$ in this way and then the remainder of $\partial\Omega_N$ so that for each $\mathcal{N}$ we have either $\mathcal{N} \cap \partial\Omega_D = \emptyset$ or $\mathcal{N}$ is a neighbourhood of some $Q \in \overline{\partial\Omega_N} \cap \partial\Omega_D$. By compactness of $\overline{\partial\Omega_N}$ we have

$$\overline{\partial\Omega_N} \subset \bigcup_1^m \mathcal{N}_i$$

and thus

$$\overline{\partial\Omega_N} \subset \bigcup_1^m \mathcal{M}_i$$

for some $\varepsilon$, $0 < \varepsilon < 1$. Clearly we have dist $\left(\partial(\overset{m}{\underset{i}{\cup}}\mathcal{M}_i),\ \overline{\partial\Omega_N}\right) \geq \mu > 0$ for some $\mu$.

Now if dist $(P, \overline{\partial\Omega_N}) \geq \mu$ then the result is immediate by [**14**] for some $\varepsilon_0 < \mu/2$. On the other hand, if dist $(P, \overline{\partial\Omega_N}) < \mu$ then $P \in \mathcal{M}_i$ for some $i$. Let $L : \mathcal{N}_i \cap \Omega \to H_i$ where $H_i$ is a hemisphere of radius $r_i$ (and thus $L(\mathcal{M}_i \cap \Omega)$ is a hemisphere of radius $\varepsilon r_i$) if $\mathcal{N}_i \cap \Omega_D = \emptyset$. $H_i$, $L(\mathcal{M}_i \cap \Omega)$ are quarter spheres otherwise. We extend $u$ as an even function to the whole sphere of radius $r_i$ (resp. to the upper hemisphere) if $H_i$ is a hemisphere (resp. quarter sphere) and extend the coefficients as outlined in [**23**]. In the latter case where $H_i$ is a quarter sphere, we extend $u$ to the whole sphere by taking $u \equiv 0$ on the bottom hemisphere. Note that this is different from what was done in [**23**] where the extension across the Dirichlet boundary was as an odd function, not as zero. Since $P \in \mathcal{M}_i$, $\varepsilon_i < 1$, we now apply the interior results of [**14**] to the extended problem in each sphere and conclude the result for $\varepsilon_0$ small enough. Since $L^{-1}$ is also Lipschitz, the result holds for $P$ and $\Omega$. Finally, since the number of $\mathcal{N}_i$, $\mathcal{M}_i$ is finite, we need only take the smallest $\varepsilon_0$ and largest $K$.

(f) Again by part (a) we have the existence of an eigenvalue $\xi$ and positive eigenvector $\tau$ to the problem

$$\mathcal{L}^*\tau = \xi\tau$$

with boundary conditions (2), where

$$\mathcal{L}(\omega) = -\nabla[a\nabla\omega + \vec{b}\omega].$$

We then note:

$$0 \geq (\tau, f) = (\tau, \mathcal{L}z - \lambda hz) \geq (\xi - \lambda|h|_{L^\infty})(\tau, z)$$

and the conclusion if $\lambda_0 < \xi/|h|_{L^\infty}$.

□

Observe that Lemma 1(c) makes it possible to estimate $\mu_1$ by first estimating $\theta_1, \theta_2$ via the Courant Min-Max principle. We shall use this later.

**Lemma 2** *Let* $g > \alpha_0 > 0$ a.e. $\Omega$, *and let* $a_n, g_n, h_n$ *be smooth pointwise approximations to* $a, g, h$ *with* $\alpha < a_n < \beta$, $\alpha_0 \leq g_n \uparrow g$, $0 \leq h_n \uparrow h$. *Consider the regularized problem:*

$$-\nabla[a_n\nabla u_n + \vec{b}u_n] = [h_n - g_n\overline{u}_n]u_n \tag{1*}$$

*subject to conditions* (2). *If assumption* (5) *holds, then problem* $(1^*) - (2)$ *has a solution* $0 < u_n \in C^\alpha(\overline{\Omega})$ *with* $\|u_n\|_{C^\alpha} \leq K_1$, $\|u_n\|_{C^\alpha} \geq K_0$, *for some positive constants* $\alpha, K_1, K_0$, *independent of* $u_n, n$.

**Remark** We recall that upper/lower solutions procedures cannot be applied to $(1^*)$ due to the presence of $\overline{u}$. Note also that by Lemma 1(b), for assumption (5) to hold it suffices to have $\mu_1(0) < 1$.

**Proof** We may assume that $\vec{b} \cdot \vec{n} > 0$ on $\partial\Omega_N$, for if not we need only make the variable change $u_n = e^\omega v_n$ with $\nabla\omega \cdot \vec{n} \gg 0$ with $\omega$ chosen suitably. This will also change the coefficients and $B_\delta$ but it will not affect their relevant properties. In particular, we will still have $\mu_1(0) < 1$. Next, add a large positive constant $C_0$ to both sides of $(1^*)$ so that the left-hand side becomes coercive over $H_0^{1,2}(\Omega \cup \partial\Omega_N)$. We thus obtain the equation:

$$\mathcal{L}u_n \equiv -\nabla[a_n \nabla u_n + \vec{b}u_n] + C_0 u_n = [h_n + C_0 - g_n \overline{u}_n]u_n.$$

With which we associate family of problems:

$$\mathcal{L}u_n = \lambda[h_n + C_0 - g_n \overline{u}_n]u_n^+, \tag{1**}$$

for $0 \leq \lambda \leq 1$. A routine bootstrap argument shows that any nontrivial solution $u_n \in L^\infty$ of $(1^{**})$ must be classical except at $\lambda_D : \partial\Omega_D \cap \overline{\partial\Omega_N}$, and furthermore $u_n \geq 0$ by choice of $C_0$. The earlier arguments then show $u_n > 0$. Observe that there exists a $\lambda_0 > 0$ independent of $n$ such that if $\lambda \leq \lambda_0$ then $u_n \equiv 0$ by Lemma 1(e) and $u_n$ is thus bounded in this case. While, if $\lambda_0 < \lambda \leq 1$ then $u_n$ cannot have a positive maximum on $\partial\Omega$ since on $\partial\Omega_N$ we would have $\frac{a\partial u}{\partial n} = -(\vec{b}\cdot\vec{n})u < 0$ at such a point. It follows that if $P_n \in \Omega$ is a maximum of $u_n$ then from the equation we have at $P_n$ :

$$-\operatorname{div}(\vec{b}) + C_0 \leq \lambda[h_n + C_0 - g_n \overline{u}_n]$$

and we conclude $\overline{u}_n(P_n)$ is bounded independently of $n$. By the definition of $\overline{u}$ and by the assumed smoothness of $\partial\Omega$ we have $\int_{S_p(P_n)\cap\Omega} u_n \leq C$ for some sphere $S_p(P_n)$. We need only apply Lemma 1(e) to obtain

$$|u_n(P_n)| \leq K\left[\int_{S_p(P_n)\cap\Omega} u^2\right]^{1/2}$$

and thus

$$\|u_n\|_{L^\infty} = |u_n(P_n)| \leq K\left[\|u_n\|_{L^\infty} \int_{S_p(P_n)\cap\Omega} u\right]^{1/2}$$

i.e.,

$$\|u_n\|_{L^\infty} \leq K' \int_{S_p(P_n)\cap\Omega} u$$

for some $K'$.

We conclude $\|u_n\|_{L^\infty(\Omega)} < K$ for some $K$ independent of $n$, $\lambda$. But then, $u_n$ satisfies a differential equation with bounded coefficients and we conclude by, e.g., [**24**], $\|u_n\|_{C^\alpha(\overline{\Omega})} \leq K_1$ for some $\alpha, K_1$ independent of $u_n$, $\lambda$. On the other hand, note that if the right-hand side of $(1^{**})$ is replaced by:

$$[h_n + C_0 - g_n \overline{u}_n]u_n^+ + t\theta$$

with $t \geq 0$ and some smooth $\theta \geq 0$, by (5) we must have for any nontrivial solution $u$ some point $P \in \Omega$ such that

$$\overline{u}_n(P) \geq c > 0 \tag{6}$$

with $c$ independent of $t$, and thus $\|u_n\|_{C^\alpha(\overline{\Omega})} \geq K_0$ with $K_0$ independent of $n, t$. Since $C^\alpha$ is compactly embedded in $C^{\alpha_0}$ for $\alpha > \alpha_0$, we then have by a routine but lengthy argument using Degree Theory, [**25**], that $(1^{**}) - (2)$ has a positive solution in some annulus $B_R - B_r$ where $B_p$ denotes the sphere of radius $p$ in $C^\alpha$ for some $\alpha > 0$.

□

**Theorem 1** *Problem* $(1) - (2)$ *has a positive solution* $u$.

**Proof** Note that $\{u_n\}$ are bounded in $H^{1,2} \cap L^\infty$ and thus we may assume $u_n \to u$ weakly in $H^{1,2}$ and strongly in $L^\xi$ for any large $\xi$. By (6), $\overline{u}_n(P_n) \geq c$ and since we also may assume $P_n \to P$ and thus $\overline{u}_n(P_n) \to \overline{u}(P)$ by the mollifier properties, we must have $0 \leq u$ nontrivial. Finally note that $\int_\Omega a\nabla u \cdot \nabla\varphi$ is an equivalent inner product to the usual one in $H^{1,2} \cap \{v | v = 0 \text{ on } \partial\Omega_D\}$ and thus

$$\int_\Omega a_n \nabla u_n \cdot \nabla\varphi - \int_\Omega a\nabla u \cdot \nabla\varphi = \int_\Omega [a_n - a]\nabla u_n \cdot \nabla\varphi + \int_\Omega a\nabla(u_n - u) \cdot \nabla\varphi.$$

The first term converges to zero by Holder's Inequality since $0 < \alpha < a_n \uparrow a$ in $L^\xi$ for any large $\xi$, while the second term tends to zero by inner product equivalence. Since convergence of the other terms is immediate, we conclude that $u$ is a positive (weak) solution of $(1) - (2)$ in $L^\infty$, and thus in $C^\alpha(\overline{\Omega})$ for some $\alpha$ and classical where the coefficients are smooth, except on $\partial\Omega_D \cap \overline{\partial\Omega_N}$.

□

We now pass to the situation where $g \geq 0$ may actually vanish on parts of $\Omega$. We present the results assuming all coefficients to be smooth, since arguments identical to Theorem 1 can then be used to extend the results to more general cases. Elementary examples show that now the existence of positive solutions depends on the relation between the various coefficients in $(1) - (2)$. We shall assume the following condition:

Let $K > 0$ be given and consider the eigenvalue problem:

$$-\nabla[a\nabla\omega + \vec{b}\omega] + gKI\big(S_{\delta'}(x)\big)\omega = e(K)h\omega \tag{7}$$

with boundary conditions (2) and some $\delta' < \delta$ where $\delta$ is as given in the definition of $\overline{u}$. We recall that $S_{\delta'}(x)$ is the ball centered at $x$ of radius $\delta'$, $I(S_{\delta'}(x))(y) = 1$ if $y \in S_{\delta'}(x)$, $= 0$ otherwise. We then assume that there exists a $K$ such that given any $x \in \Omega$ at which $g(x) = 0$, $e(\lambda)$ satisfies:

$$e(0) < 1 < e(K). \tag{8}$$

Clearly condition (8) excludes problems with very small $\delta$, all other quantities being fixed. For these situations an analysis can however be made by perturbation arguments as we show in the next section.

**Theorem 2** *If condition* (8) *holds, problem* $(1) - (2)$ *has a positive solution.*

**Proof** The arguments are similar to those of Lemma 1. We thus only make the following comments where some differences do arise. If (8) holds, then any nonnegative solution $u \in L^\infty$ of

$$-\nabla[a\nabla u + \vec{b}u] = \lambda[h - g\,\overline{u}]u$$

and conditions (2) must vanish for $\lambda \leq e(0)$ and is thus clearly bounded. Note that $e(0)$ depends only on the coefficients.

To show that any solution must be bounded in $L^\infty$, it suffices by our earlier arguments to show there exist numbers $\xi > 0,\ K_1 > 0$ independent of $u$ such that $\int_{S_\xi(P)\cap\Omega} u \leq K$, at any maximum point $P$.

Set $Z = \{x | g(x) = 0\}$, $Q = \{x | \operatorname{dist}(x, Z) \geq \delta_0\}$ with $\delta_0, \delta''$ such that $0 < \delta_0 < \delta' < \delta' + \delta_0 < \delta'' < \delta$ and choose $\theta > 0$ such that $g(x) \geq \theta$ for $x$ in $Q$. Now if $P \in Q$ then we proceed exactly as before. If $P \in \Omega - Q$ then there exists $y \in Z$ such that $\|P - y\| < \delta_0$. Condition (8) implies the existence of some $x \in S_{\delta'}(y)$ such that $K \geq \overline{u}(x)$. Since $\|x - P\| < \delta' + \delta_0 < \delta'' < \delta$ we obtain the result with $\xi = \delta'' - (\delta' + \delta_0)$. The rest of the proof is identical to the earlier one.

□

## 3 The small $\delta$ case

We consider here the case of $\delta$ small with $g$ still allowed to vanish somewhere in $\Omega$, and smooth coefficients. Our starting points are the local problem:

$$-\nabla[a\nabla u + \vec{b}u] = [h - gu]u \tag{9}$$

subject to conditions (2), and the fact that the degree is invariant under perturbation. We basically exploit the observation that $\overline{u}(x)$ and $u(x)$ are near each other if $\delta$ is sufficiently small. The existence of a solution to (9) will then imply the existence of a solution to (1).

We modify the definition of $\overline{u}$ as follows: If $u \in C^\alpha(\overline{\Omega})$ then there exists $\widehat{u}$, $C^\alpha$ extension of $u$, with $\|\widehat{u}\|_{C^\alpha(R^N)} \leq K\|u\|_{C^\alpha(\overline{\Omega})}$. There are clearly infinitely many possible extensions, but here we select one such and then define $\overline{u}$ by

$$\overline{u}(x) = \int_\Omega B_\delta(x,y)\widehat{u}\,(y)dy$$

in this section. Note that this is different from the earlier $\overline{u}$, unless Dirichlet conditions are chosen on the whole $\partial\Omega$ and we set $u^* \equiv 0$ in $\mathbb{R}^n - \Omega$.

Unlike the earlier results, problem (9) - (2) can be considered by upper/lower solution arguments, but it is advantageous for our purposes to use only half of these arguments. Specifically, consider once again the related linear eigenvalue problem (4) with conditions (2), and again assume that condition (5) holds if $K$ is large enough. It follows that we can find a value of $K$ and a $\theta > 0$ such that for any constant $z > 0$ the problem:

$$\begin{cases} -\nabla[a\nabla\omega_1 + \vec{b}\omega_1] + Kg\omega_1 = (1+\theta)h\omega_1 & \\ a\,\frac{\partial\omega_1}{\partial n} + (\vec{b}\cdot\vec{n})\omega_1 = 0 & \text{on} \quad \partial\Omega_N \\ \omega_1 = z & \text{on} \quad \partial\Omega_D \end{cases} \tag{10}$$

has a positive solution $\omega_1$. This follows from the observation that for $C_0$ sufficiently large the problem

$$\begin{cases} -\nabla[a\nabla\omega_2 + \vec{b}\omega_2] + [\lambda_0 g - (1+\theta)h + C_0]\omega_2 = 0 \\ \omega_2 = 1 & \text{on} \quad \partial\Omega_D \\ a\,\frac{\partial\omega_2}{\partial n} + (\vec{b}\cdot\vec{n})\omega_2 = 0 & \text{on} \quad \partial\Omega_N \end{cases}$$

has a solution $\omega_2 \geq 0$ by coercivity of the left-hand side. Indeed $\omega_2 > 0$ in $\overline{\Omega}$ since it cannot have a minimum of zero in the interior by the Generalized Harnack's Inequality while on the boundary either $\omega_2 = 1$ or without loss of generality $\vec{b}{\cdot}\vec{n} > 0$. On the other hand the equation:

$$\ell_2 z_0 \equiv -\nabla[a\nabla z_0 + \vec{b}z_0] + [\lambda_0 g - (1+\theta)h]z_0 = 0$$

with the same boundary conditions has a solution $z_0$ by our eigenvalue assumptions. But then we have $\ell_2(z_0 - \omega_2) \geq 0$ and thus $z_0 - \omega_2 \geq 0$ by the same arguments as in Lemma 0(a). We then need only note that $\omega_1 = zz_0$. By the maximum principle, if $z$ is large enough we have for the solution $\omega_1$ of (10):

$$\begin{cases} -\nabla[a\nabla\omega_1 + \vec{b}\omega_1] + g\omega_1^2 \geq h\omega_1 \\ a\,\frac{\partial\omega_1}{\partial n} + (\vec{b}\cdot\vec{n})\omega_1 = 0 & \text{on} \quad \partial\Omega_N \\ \omega_1 > 0 & \text{on} \quad \partial\Omega_D \end{cases}$$

i.e., $\omega_1$ is an upper solution to (9) – (2).

**Theorem 3** *If* (5) *holds then problem* (1) - (2) *has for* $\delta$ *small enough a positive solution.*

**Proof** Note that (9) may be rewritten as

$$-\nabla[a\nabla u + \vec{b}u] + cu = f(x,u) \tag{11}$$

with a $c > 0$ chosen large enough so that the left-hand side is coercive and $f$ is monotone in $u$ for $u$ in the interval $0 \leq u \leq \omega_1$. Let $\widetilde{u} = \min(\omega_1, u)$ and replace $f(x,u)$ in (11) by $f(x,\widetilde{u})$. We now apply the same Degree Theory arguments as before, setting

$$\widetilde{T} = [-\nabla[a\nabla\cdot + \vec{b}\cdot] + c\cdot]^{-1}(f(x,\widetilde{u}^+)),$$

we again find that $\mathrm{Deg}\,[I - \widetilde{T},\, B_R - B_r, 0] = 1$ for some $0 < r < R$ with $B_p$ the ball of radius $p$ in $C^\alpha(\overline{\Omega})$.

For example, there is a positive solution of (9) – (2) since $\omega_1$ is an upper solution and furthermore this solution is unique since if $u_1, u_2$ are two solutions then we may construct a supersolution $\omega^*$, by solving (10) with $z \gg 0$, such that $\omega^* > \max(u_1, u_2)$. By the usual local problem iteration procedure we construct a sequence $\{\omega_n\}$ such that $\omega_0 = \omega^*$ and $\omega_n \geq \omega_{n+1} \geq \max(u_1, u_2)$. Furthermore, $\omega_n$ converges to a solution pointwise, so that we may conclude the existence of three solutions $u_1, u_2, u_3$ with $u_3 \geq u_1$. But then we have from (9) that $gu_3 \geq gu_1$ and by the same arguments as in the Proof of Lemma 1(b) it follows that $gu_3 = gu_1$ somewhere in $\Omega$ and thus it follows that $u_3 \equiv u_1$ and hence $u_2 \equiv u_1$. If need be we replace $\omega_1$ by $\omega^*$ and thus conclude without loss of generality that the (unique) positive solution $u$ satisfies $u < \omega_1$ in $\overline{\Omega}$. If we now choose a small ball in $C^\alpha$ about the unique local problem solution just found, then the perturbed problem:

$$-\nabla[a\nabla u + \vec{b}u] + cu = f(x, \widetilde{u}^+) + g\widetilde{u}\,\widetilde{u}^+ - g\,\overline{\widetilde{u}}\,\widetilde{u}^+$$

must also have a solution in this ball for $\delta$ small enough, since $\|\widetilde{u} - \overline{\widetilde{u}}\|_{L^\infty(\Omega)} \le K\delta^\alpha$ by the (present) definition of $\overline{u}$. In summary, we have found a solution $u^*$ of

$$-\nabla[a\nabla u^* + \vec{b}u^*] + cu^* = [h + c - g\overline{(\widetilde{u}^*)}](\widetilde{u}^*)^+$$

near $u$. We conclude for $\delta$ small enough, that $u^* \le \omega_1$, i.e., $\widetilde{u}^* = u^*$ and we have thus found a solution to the original nonlocal problem.

□

## 4 Examples

In this section we illustrate the above results with a pair of concrete examples.

**Example 1** Suppose that in Equation (1) we have $a \equiv 1$, $\vec{b} \equiv \vec{0}$, $h = 1$ and $\Omega = (0, L)$. Finally assume Dirichlet conditions: $u(0) = u(L) = 0$. We find that $e(0) = \frac{\pi^2}{L^2}$. Suppose $g$ is given by:

$$g = \begin{cases} 0 & \text{for} \quad x \in (\frac{L}{2} - \varepsilon,\ \frac{L}{2} + \varepsilon) \\ g_0 \text{ (constant)} & \text{otherwise} \end{cases}$$

with $0 < \frac{L}{2} - \varepsilon < \frac{L}{2} + \varepsilon < L$. Now if $\delta > 2\varepsilon$ as $K \to \infty$ in (7), $e(K)$ converges to a value at least equal to the smaller of the three eigenvalues obtained by considering the three segments: $(0, x - \delta)$, $(\frac{L}{2} - \varepsilon,\ \frac{L}{2} + \varepsilon)$, $(x + \delta,\ L)$. By symmetry and eigenvalue monotonicity, this will be least for $\Omega^* = (0, \frac{L}{2} + \varepsilon - \delta)$. If we choose $\frac{L}{2} > \delta + \varepsilon \ge 3\varepsilon$ then the least eigenvalue will be no less than $\frac{\pi^2}{(\frac{L}{2} + \varepsilon - \delta)^2}$. A sufficient requirement for the existence of a positive solution to (1) - (2) thus becomes:

$$\pi^2 < L^2 < \frac{L^2\pi^2}{(\frac{L}{2} + \varepsilon - \delta)^2}.$$

Observe that $g_0$ does not appear in this condition. Its magnitude will have an effect on the size/shape of the solution $u$ but not on its existence.

**Example 2** Assume now $g > 0$ in $\overline{\Omega}$, that the coefficients are smooth and the boundary conditions are purely Dirichlet. To apply our results we need only show that the eigenvalue problem

$$-\nabla[a\nabla\omega + \vec{b}\omega] = \mu_1(0)h\omega$$

with conditions (2) has least eigenvalue $\mu_1(0) < 1$, since $\mu_1(K) > 1$ for $K$ large by Lemma 1(b).

We apply Lemma 1(c) by comparing $\mu_1(0)$ with $\theta_2(0)$ where this is given by:

$$-\nabla[a\nabla\omega_2] + \Big[-\frac{1}{2}\,\mathrm{div}\,(\vec{b}) + \frac{|\vec{b}|^2}{4a}\,\omega_2\Big] = \theta_2(0)h\omega_2.$$

That $\mu_1(0) < 1$ is thus ensured by showing the existence of a $\omega_3$ such that

$$\int_\Omega \beta|\nabla\omega_3|^2 + (\vec{b}\cdot\nabla\omega_3)\omega_3 + \frac{|\vec{b}|^2}{4\alpha}\,\omega_3^2 - h\omega_3^2 < 0,$$

where we recall $0 < \alpha < a < \beta$. This becomes:

$$\int_\Omega \beta|\nabla\omega_3|^2 + \Big[-\frac{1}{2}\operatorname{div}(\vec{b}) + \frac{|\vec{b}|^2}{4\alpha} - h\Big]\omega_3^2 < 0.$$

It is easy to obtain sufficient conditions for this to be the case. Suppose, for example, that $\Omega$ contains a cube $C = \prod_{i=1}^{N} \{0 < x_i < L\}$. The least eigenvalue for the Dirichlet Problem associated with the cube is $(\frac{\pi}{L})^{2N}$ with associated eigenvector $v = \prod_1^N \sin(\frac{\pi x_i}{L})$. We pick $\omega_3$ by setting $v = 0$ in $\Omega - C$ and thus obtain our result if:

$$\frac{\beta\pi^{2N}}{2^N L^N} + \int_C \Big[-\frac{1}{2}\operatorname{div}(\vec{b}) + \frac{|\vec{b}|^2}{4\alpha} - h\Big]\prod_1^N \sin^2\Big(\frac{\pi x_i}{L}\Big) < 0.$$

## 5 Open questions and related problems

A natural question concerns the uniqueness of the solutions to (1) - (2) found in the preceding sections. In general, no uniqueness results are known to the authors, but in some specific cases, uniqueness can be established. We note, see also [1], that if $B_\delta$ is constant, i.e., if we consider

$$-\nabla[a\nabla u + \vec{b}u] = \Big[h - g\int_\Omega u\Big]u,$$

then the solution must be unique: $u$ is the least eigenvector of (10) with $e(\mu) = 1$ and $\int_\Omega u = \mu$. We conjecture that the solution is unique for (1) - (2), but observe that if we allow the kernel $B_\delta$ to be smooth but otherwise arbitrary, then this is not the case. To see this, consider, e.g., the least eigenvalue Dirichlet problem for:

$$-\Delta u + g_1(1+\varepsilon_1)u + g_2(1+\varepsilon_2)u = \mu_1(\varepsilon_1,\varepsilon_2)u,$$

with $g_1, g_2$ smooth nonnegative functions with disjoint supports. We may choose $\varepsilon_1, \varepsilon_2$ such that $\mu_1(\varepsilon_1, 0) = \mu_1(0, \varepsilon_2)$. The choice of any two functions $f_1, f_2$ such that:

$$\int_\Omega f_1u_1 = (1+\varepsilon_1), \quad \int_\Omega f_1u_2 = 1, \quad \int_\Omega f_2u_1 = 1, \quad \int_\Omega f_2u_2 = (1+\varepsilon_2)$$

yields a nonlocal problem with kernel $g_1(x)f_1(y) + g_2(x)f_2(y)$, $g = 1$, $h = \mu_1(\varepsilon_1, 0)$ and two positive solutions: $u_1$, $u_2$.

Finally, as mentioned earlier, $[h - g\overline{u}]u$ can be replaced by a more general $f$ and $\overline{u}$ can be replaced by a more general operator $Tu$. One case arises from a kernel $B_\delta(|x-y|)$ which satisfies similar properties to the ones postulated before, except $B_\delta(0) = \infty$. One possible such situation comes from the system

$$\begin{cases} -\Delta u = [h - gv]u \\ -\Delta v = u \end{cases}$$

with, for simplicity, Dirichlet conditions. We then obviously have:

$$-\Delta u = [h - gT(u)]u$$

$$\text{where} \quad T(u) = (-\Delta)^{-1}(u) = \int_{\Omega} G(x,y)u(y)dy$$

and $G(x,y)$ represents the Green's Function for $-\Delta$. The earlier procedures are applicable as follows: Let $u > 0$ be a solution of

$$-\Delta u = \lambda[h - gT(u)]u$$

with $0 < \lambda_0 < \lambda < 1$. Choose $\Omega_\varepsilon = \{x | x \in \Omega, \text{ dist}\,(x, \partial\Omega) < \varepsilon\}$ and let $\omega > 0$ solve in $\Omega_\varepsilon$ :

$$\begin{aligned} -\Delta\omega - h\omega &= 0 \\ \omega = 1 \quad &\text{on} \quad (\partial\Omega_\varepsilon) \cap \Omega \\ \omega = 0 \quad &\text{on} \quad \partial\Omega. \end{aligned}$$

This is possible if $\partial\Omega$ is regular enough and $\varepsilon$ is small. Since $-\Delta u - hu \leq 0$ in $\Omega_\varepsilon$ we then have: $u \leq \omega \|u\|_{L^\infty(\overline{\Omega})}$ and thus the maximum of $u$ must occur inside some compact set $K \ll \Omega$. But if $P \in K$ then $v \geq c \int_{B_{\delta/2}(P)} u$ for some constants $c, \delta > 0$. Indeed, $v$ exceeds $-\Delta z = u$ in $B_\delta(P)$, $z = 0$ on $\partial B_\delta(P)$, for some small $\delta$ independent of $u, P$, i.e.,

$$\begin{aligned} v(P) &\geq c_1 \int_{|P-y|<\delta} \left[\frac{1}{|P-y|^{n-2}} - \frac{1}{\delta^{n-2}}\right] u \\ &\geq c_2 \int_{B_{\delta/2}(P)} u\,. \end{aligned}$$

We can thus bound $\|u\|_{L^\infty}$ as before.

Note also that if the coefficients are assumed regular near $\partial\Omega$, then the neighbourhood $\mathcal{N}$ of the introduction need only be a neighbourhood of $\partial\Omega_D \cap \overline{\partial\Omega}_N$. If $\vec{b}/a$ is a gradient field, then this condition can be dropped altogether.

As a final observation, we point out that if the coefficients in (1) are subject to a finite jump at some point $x_0 \in \Gamma$, where $\Gamma$ is a smooth surface in $\Omega$ then besides the $C^\alpha$ continuity of $u$ we may also conclude by [**17**, p. 222] that, the "flux matching" condition holds. For example, if $a(x)$ has a jump at $x_0$ then $a\,\frac{\partial u}{\partial n}\,|_{x_0}$ is continuous, where $n$ here denotes the normal to $\Gamma$ in a chosen direction.

## References


[1] Allegretto, W. [1971] *A comparison theorem for nonlinear operators,* Annali Scuola Normale Superiore Pisa **XXV**, 41-46.

[2] Allegretto, W. [1987] *On the principal eigenvalues of indefinite elliptic problems,* Math. Z. **195**, 29-35.

[3] Allegretto, W. and Barabanova, A. [1996] *Positivity of solutions of elliptic equations with nonlocal terms,* Proc. Royal Soc. Edin. **126A**, 643-663.

[4] Allegretto, W. and Mingarelli, A. [1989] *Boundary problems of the second order with an indefinite weight-function,* J. Reine Agnew. Math. **398**, 1-24.

[5] Allegretto, W. and Nistri, P. [to appear] *Existence and optimal control for periodic parabolic equations with nonlocal terms,* I.M.A. Journal of Mathematical Control and Information.

[6] Binding, P. and Browne, P. V. [1987] *Spectral properties of two parameter eigenvalue problems,* Proc. Royal Soc. Edin. **106A**, 39-51.

[7] Calsina, A., Perello, C. and Saldana, J. [1994] *Non-local reaction-diffusion equations modelling predator-prey coevolution,* Publicacions Matematiques **38**, 315-325.

[8] Calsina, A. and Perello, C. [1995] *Equations for biological evolution,* Proc. Royal Soc. Edin. **125A**, 939-958.

[9] DeFigueiredo, D. G. and Mitidieri, E. [1986] *A maximum principle for an elliptic system and applications to semilinear problems,* SIAM Journal Math. Anal. **17**, 836-849.

[10] Freedman, H. I. and Takeuchi, Y. [1989] *Global stability and predator dynamics in a model of prey dispersal in a patchy environment,* Nonl. Anal. TMA **13**, 993-1002.

[11] Freedman, H. I. and Wu, J. [1992] *Steady state analysis in a model for population diffusion in a multi-patch environment,* Nonl. Anal. TMA **18**, 517-542.

[12] Freitas, P. and Sweers, G. [1994] *Positivity results for a nonlocal elliptic equation,* preprint.

[13] Furter, J. and Grinfeld, M. [1989] *Local vs. nonlocal interactions in population dynamics,* J. Math. Biol. **27**, 65-80.

[14] Gilbarg, D. and Trudinger, N. S. [1983] *Elliptic Partial Differential Equations of Second Order*, $2^{nd}$ Edition, Springer-Verlag, New York.

[15] Hernandez, J. [1986] *Continuation and comparison methods for some nonlinear elliptic systems,* in: Contributions to nonlinear partial differential equations, Vol II *(J. I. Diaz and P. L. Lions, eds.)*, Longman, Essex, 137-147.

[16] Krasnoselskii, M. A. [1964] *Positive Solutions of Operator Equations*, Noordhoff, Groningen.

[17] Ladyzhenskaya, O. A. and Uraltseva, N. N. [1968] *Linear and Quasilinear Elliptic Equations*, Academic Press, New York and London.

[18] Lieberman, G. M. [1992] *Intermediate Schauder Theory for second order parabolic equations* IV: *time irregularity and regularity,* Diff. Int. Eq. **5**, 1219-1236.

[19] Murthy, M. K. V. and Stampacchia, G. [1972] *A variational inequality with mixed boundary conditions,* Israel J. Math. **13**, 189-224.

[20] Pao, C. V. [1992] *Nonlinear Parabolic and Elliptic Equations*, Plenum Press, New York.

[21] Lenhart, S. and Protopopescu, V. [1994] *Optimal control for parabolic systems with competitive interactions,* Math. Meth. Appl. Sci. **17**, 501-524.

[22] Shamir, E. [1968] *Regularization of mixed second order boundary value problems,* Israel J. Math. **6**, 151-168.

[23] Troianiello, G. M. [1987] *Elliptic Differential Equations and Obstacle Problems*, Plenum Press, New York.

[24] Xie, H. [1995] $L^{2,\mu}$*-estimate to the mixed boundary value problem for second order elliptic equations and applications in thermistor problems,* Nonl. Anal. TMA **24**, 9-28.

[25] Zeidler, E. [1986] *Nonlinear Functional Analysis and its Applications*, Vol. I, Springer-Verlag, New York.

Fields Institute Communications
Volume **21**, 1999

# Integrodifference Equations Applied to Plant Dispersal, Competition, and Control

**Linda J. S. Allen**
Department of Mathematics and Statistics
Texas Tech University
Lubbock, TX 79409-1042, USA
lallen@math.ttu.edu

**Edward J. Allen**
Department of Mathematics and Statistics
Texas Tech University
Lubbock, TX 79409-1042, USA
eallen@math.ttu.edu

**David N. Atkinson**
Department of Biology
Case Western Reserve University
Cleveland, OH 44106-7080, USA
dna@po.cwru.edu

**Abstract.** A system of integrodifference equations is formulated for $n$ competing annual plants that spread by dispersing their seeds. The competition model, a system of difference equations, is analyzed and it is demonstrated that asymptotic behavior mimics that of the continuous-time Lotka-Volterra models in some cases. The system of integrodifference equations is analyzed in terms of the amount of control required to prevent the spread of one of the competing species. The amount of control depends on the dispersal kernel and the population parameters.

## 1 Introduction

A system of difference equations is derived and analyzed for $n$ competing annual plants, extending a model of [**13**] for two competing annual plants to $n$ species. The solution behavior is compared to the continuous-time Lotka-Volterra equations in the case of two and three competing species. Surprisingly, the solution behavior is similar to the continuous-time models. For $n$-species competition, it is shown that solutions are bounded and in the case of weak competition, i.e., small competition coefficients, the system is uniformly persistent.

The model is modified to realistically model the growth rate for small population sizes. This leads to an alternate model for $n$-species competition and is

1991 *Mathematics Subject Classification.* Primary 39A10, 92D26.
The first two authors were partially supported by NSF Grant DMS–9626417.

particularly applicable when control and dispersal are included. The model with competition and dispersal is a system of integrodifference equations, where the time variable is in discrete units, $t = 0, 1, 2, \dots$, and spatial variable is continuous, $x \in \mathbb{R}$. Control of an invasive plant species is investigated in the modified model. It is shown for certain dispersal kernels and population parameters that one of the competing species can be controlled in the region $x > 0$, that is, $\int_0^\infty A(t,x)dx < C < \infty$ for $t = 1, \dots$, where $A(t,x)$ is the density of adult plants at time $t$ and position $x$. This result generalizes a previous result of [**2**], where the control was applied uniformly over the spatial region. The requirement on the dispersal kernel is that the tail of the distribution is sufficiently small. The kind of dispersal kernels, for which the results are applicable, are discussed in relation to other integrodifference equation models ([**6**], [**7**], [**8**], [**15**]). Numerical and analytical results demonstrate that even for solutions with an increasing rate of spread the control conditions are sufficient to prevent dispersal.

## 2 Competition

**2.1 Model development.** The basic competition model for annual plants is developed from first principles. For two-species, our competition model simplifies to one developed in [**13**]. The variables and parameters for the model are defined below.

$y_i(t)$ = new seed density produced by species $i$ in year $t$
$r_i(t)$ = residual seed density produced by species $i$ in year $t$
$A_i(t)$ = density of adult plants of species $i$ in year $t$
$m_i$ = maximum seed density of species $i$
$P_A^i$ = probability that plants become adults
$P_E^i$ = probability that plants become established
$P_G^i$ = probability that seeds germinate
$P_W^i$ = probability that seeds survive winter
$k_{ij}$ = competition coefficients.

A difference equation for $n$-species competition is formulated in terms of the density of adult plants $A_i(t)$. New seeds and residual seeds in generation $t$ that survive the winter, germinate, become established and survive to adulthood form the new adults in generation $t+1$, i.e.,

$$A_i(t+1) = P_A^i P_E^i P_G^i P_W^i \left[y_i(t) + r_i(t)\right] = a_i \left[y_i(t) + r_i(t)\right], \tag{2.1}$$

where $a_i = P_A^i P_E^i P_G^i P_W^i$. The residual seeds or seeds in the seedbank are those that don't germinate but survive the winter. Application of equation (2.1) allows residual seeds to be expressed in terms of adult plants. Assuming that the parameters are time-independent,

$$\begin{aligned} r_i(t) &= P_W^i(1 - P_G^i)[y_i(t-1) + r_i(t-1)] \\ &= b_i \frac{A_i(t)}{a_i}, \end{aligned}$$

where $b_i = P_W^i(1 - P_G^i)$. Thus,

$$A_i(t+1) = a_i y_i(t) + b_i A_i(t).$$

The competitive relationship between adult plants $A_i$ and new seeds $y_i$ takes the following form ([**1**], [**2**], [**3**], [**13**]):

$$y_i(t) = \frac{m_i A_i(t)}{A_i(t) + \sum_{j=1,j\neq i} k_{ij} A_j(t)}, \quad i = 1, \ldots, n.$$

Thus, a system of autonomous first order difference equations for competition between $n$ adult plants is given by:

$$A_i(t+1) = \frac{a_i m_i A_i(t)}{A_i(t) + \sum_{j=1,j\neq i} k_{ij} A_j(t)} + b_i A_i(t), \; i = 1, \ldots, n.$$

In deriving this system, it was assumed that all parameters were independent of time. In practice, this is not the case; the parameters depend on environmental conditions such as temperature and soil moisture which vary with time and with spatial location. Thus, an implicit assumption in these models is that environmental changes have a negligible effect on the parameters. However, some of the analytical results are applicable to the system with time-dependent parameters and, of course, numerical simulations can be generalized to include parameters that vary with time.

For analysis purposes, the $n$-species competition model is expressed in a dimensionless form. Let

$$\tilde{A}_i = \frac{A_i(1-b_i)}{a_i m_i} \quad \text{and} \quad \tilde{k}_{ij} = k_{ij}\frac{a_j m_j (1-b_i)}{a_i m_i (1-b_j)},$$

$i, j = 1, \ldots, n$. Then, the dimensionless system has the form:

$$\tilde{A}_i(t+1) = \frac{(1-b_i)\tilde{A}_i(t)}{\tilde{A}_i(t) + \sum_{j=1,j\neq i}^{n} \tilde{k}_{ij} \tilde{A}_j(t)} + b_i \tilde{A}_i(t) \tag{2.2}$$

where $b_i = P_W^i(1 - P_G^i)$ and $i = 1, \ldots, n$. We shall refer to the parameters $b_i$ as the seedbank parameters. For notational simplicity, the dimensionless system will be written without the tilde notation.

It is important to consider the model behavior in the case of a single species. In this case, the dimensionless model reduces to

$$A(t+1) = (1-b) + bA(t). \tag{2.3}$$

The solution to model (2.3) is $A(t) = b^t A(0) + 1 - b^t$ and $\lim_{t\to\infty} A(t) = 1$. The model is unrealistic for small population sizes since when $A(0) = 0$ the population density still approaches one. A simple modification includes an additional factor in the term $(1 - b)$:

$$A(t+1) = (1-b)\min\{1, \; cA(t)\} + bA(t), \quad c > 1. \tag{2.4}$$

For this latter model, $A(0) = 0$ implies $A(t) = 0$ and the limiting behavior is the same as (2.3). This modification is especially important when the impact of control is considered. For example, by reducing the adult plants to an amount $\rho A(t)$, so that in equation (2.4), $cA(t)$ is replaced by $c\rho A(t)$ and $c\rho < 1$, then the population is driven to extinction, $\lim_{t\to\infty} A(t) = 0$. The factor $\rho$ is not included in the term $bA(t)$ in equation (2.4) because, for example, with control such as a herbicide it is assumed that the seedbank is not affected.

In the next two sections, we consider the model (2.2) for two and three-species and demonstrate its close relationship to the continuous-time Lotka-Volterra equations. However, in later sections, we include the factor $\min\{1, cA(t)\}$. Then, the dimensionless model has the following form:

$$A_i(t+1) = \frac{(1-b_i)\min\{1, c_i A_i(t)\} A_i(t)}{A_i(t) + \sum_{j=1, j\neq i}^{n} k_{ij} A_j(t)} + b_i A_i(t), \ i = 1, \dots, n. \tag{2.5}$$

**2.2 Two and three species competition.** It is shown here that the two and three species discrete-time competition models of the form (2.2) have, in many cases, asymptotic behavior similar to that of the continuous-time Lotka-Volterra competition equations. First, note that the isosurfaces of the dimensionless system (2.2), $\Delta A_i = A_i(t+1) - A_i(t) = 0$, are linear equations:

$$A_i + \sum_{j=1, j\neq i}^{n} k_{ij} A_j = 1.$$

Through a detailed analysis of the phase plane in two dimensions, it was shown in [**13**] that the asymptotic behavior of the two-species discrete competition model (2.2) is the same as that of Lotka-Volterra two-species competition. This relationship is rather surprising given the diverse behavior (e.g., period-doubling and chaos) exhibited by discrete versions of logistic growth [**9**]. The results of [**13**] stated in terms of the dimensionless model (2.2) depend only on the magnitude of the competition coefficients $k_{ij}$ and not on the seedbank parameters $b_i$; the competition coefficients determine how the isoclines cross. When $k_{12},\ k_{21} < 1$, then $\lim_{t\to\infty}(A_1(t), A_2(t)) = (A_1^*, A_2^*) > 0$. When $k_{12}, k_{21} > 1$, then either $\lim_{t\to\infty}(A_1(t), A_2(t)) = (1,0)$ or $(0,1)$, depending on the initial conditions. When $k_{12} < 1$ and $k_{21} > 1$, then $\lim_{t\to\infty}(A_1(t), A_2(t)) = (1,0)$ but when the inequalities are reversed, then $\lim_{t\to\infty}(A_1(t), A_2(t)) = (0,1)$.

For three species, the stability results depend on the magnitude of the seedbank parameters $b_i$. First, we state the local stability conditions for each of the equilibria. Because the isosurfaces are linear, there are at most seven equilibria in the interior or on the boundary of the region $\mathbb{R}_+^3$. Note that the origin cannot be an equilibrium of system (2.2); all equilibria are nondegenerate. There are three single-species equilibria given by (1,0,0), (0,1,0), and (0,0,1), three two-species equilibria, $(A_1^*, A_2^*, 0)$, $(A_1^*, 0, A_3^*)$, $(0, A_2^*, A_3^*)$, and one three-species equilibrium, $(A_1^{**}, A_2^{**}, A_3^{**})$. The single-species equilibrium, where the $i$th population is one and the others are zero is locally asymptotically stable if $k_{ji},\ k_{li} > 1,\ j, l \neq i$ and $j \neq l$. The two-species equilibria, where

$$A_i^* = \frac{1-k_{ij}}{1-k_{ij}k_{ji}} \quad A_j^* = \frac{1-k_{ji}}{1-k_{ij}k_{ji}}, \ i \neq j,$$

is locally asymptotically stable if $k_{ij},\ k_{ji} < 1$ and

$$k_{mi}A_i^* + k_{mj}A_j^* > 1, \quad m \neq i, j.$$

These conditions agree with the local stability conditions for the continuous-time Lotka-Volterra competition equations:

$$\frac{dA_i(t)}{dt} = r_i A_i(t) \left[1 - A_i(t) - \sum_{j=1, k\neq j} A_j(t)\right], \ i = 1, 2, \dots, n. \tag{2.6}$$

These stability conditions do not depend on the seedbank parameters.

The stability conditions for the three-species equilibrium are too unwieldy to write down. However, the Jacobian matrix $J$ evaluated at the three-species equilibrium is given below; the magnitude of the eigenvalues must be less than one. Note that the local stability conditions depend on the seedbank parameters.

$$J = I - \text{diag}([1-b_i]A_i^{**})\begin{pmatrix} 1 & k_{12} & k_{13} \\ k_{21} & 1 & k_{23} \\ k_{31} & k_{32} & 1 \end{pmatrix}, \tag{2.7}$$

or simply $J = I - \text{diag}([1-b_i]A_i^{**})K$, where $I$ is the $3 \times 3$ identity matrix and $K = (k_{ij})$ is the matrix of competition coefficients, $k_{ii} = 1$. The three-species model is complicated to analyze, such is the case even for the continuous-time Lotka-Volterra competition equations. However, a simple case of three-species competition is considered, where the results of the discrete model (2.2) are compared to that of the analogous continuous-time Lotka-Volterra competition model.

2.2.1 *May-Leonard competition model.* Consider the following three-species competition model:

$$\begin{aligned} A_1(t+1) &= \frac{(1-b_1)A_1(t)}{A_1(t)+\alpha A_2(t)+\beta A_3(t)} + b_1 A_1(t) \\ A_2(t+1) &= \frac{(1-b_2)A_2(t)}{A_2(t)+\beta A_1(t)+\alpha A_3(t)} + b_2 A_2(t) \\ A_3(t+1) &= \frac{(1-b_3)A_3(t)}{A_3(t)+\alpha A_1(t)+\beta A_2(t)} + b_3 A_3(t). \end{aligned} \tag{2.8}$$

This model is a discrete version of the well-known Lotka-Volterra competition model analyzed by [**10**] and others (e.g., [**11**] and [**14**]). For model (2.8), the two-species equilibria are given by

$$A_i^* = \frac{1-\alpha}{1-\alpha\beta} \quad \text{and} \quad A_j^* = \frac{1-\beta}{1-\alpha\beta}, \; i \neq j$$

and the three-species equilibrium by

$$A_i^{**} = \frac{1}{1+\alpha+\beta}, \; i = 1, 2, 3. \tag{2.9}$$

If $\alpha < 1 < \beta$, system (2.8) exhibits nontransitive competition, that is, in pairwise competition, $A_1$ dominates $A_2$, $A_2$ dominates $A_3$, and $A_3$ dominates $A_1$. It was shown in [**10**] in the continuous Lotka-Volterra model that the condition for local asymptotic stability of the three-species equilibrium is $\alpha + \beta < 2$. We show that this condition differs from that of the discrete model (2.8). The stability region for the three-species equilibrium (2.9) for the case that the seedbank parameters are equal, $b_1 = b_2 = b_3 = b = 0.75$, is graphed in Figure 1. It can be seen from Figure 1 that the stability region for model (2.8) is a subset of the stability region in the May and Leonard model ($\alpha + \beta < 2$). In addition, the stability region for the discrete model increases as the seedbank parameter increases; as $b \to 1$ the stability region approaches that of the May and Leonard model.

The dynamics of (2.8) are illustrated for the case $b = 0.75$ when the parameters $\alpha$ and $\beta$ lie outside of the stable region given in Figure 1. Figure 2 is a graph of a numerical solution to (2.8) when $(\alpha, \beta) = (0.1, 1.85)$. These parameter values lie outside of the stable region for the discrete model (2.8) but inside the stable region for the continuous-time Lotka-Volterra model ($\alpha + \beta < 2$). The positive equilibrium of model (2.8) in this case is unstable. However, it appears that there exists a stable limit cycle. As the parameters $\alpha$ and $\beta$ increase, the amplitude of the cycle increases. See Figure 3 for $(\alpha, \beta) = (0.5, 1.7)$. For the parameters $(\alpha, \beta) = (0.5, 5.0)$, a numerical solution is graphed in Figures 4 and 5. As can be seen from the graph of the solution as a function of time (Figure 5), the solution spends an increasing amount of time near the single-species equilibria and cycles ever closer to these equilibria in the direction: (1,0,0)→ $(0,0,1) \rightarrow (0,1,0) \rightarrow (1,0,0)$. This type of behavior has been demonstrated in the Lotka-Volterra continuous-time model. Although none of the state variables approach zero, the persistence behavior is very weak; it has been referred to as "weak-weak persistence", where solutions satisfy the condition $\limsup_{t\to\infty} A_i(t) > 0,\ i = 1,2,3$ but not the condition $\limsup_{t\to\infty} \min_{1\le i\le 3}\{A_i(t)\} > 0$ [**4**].

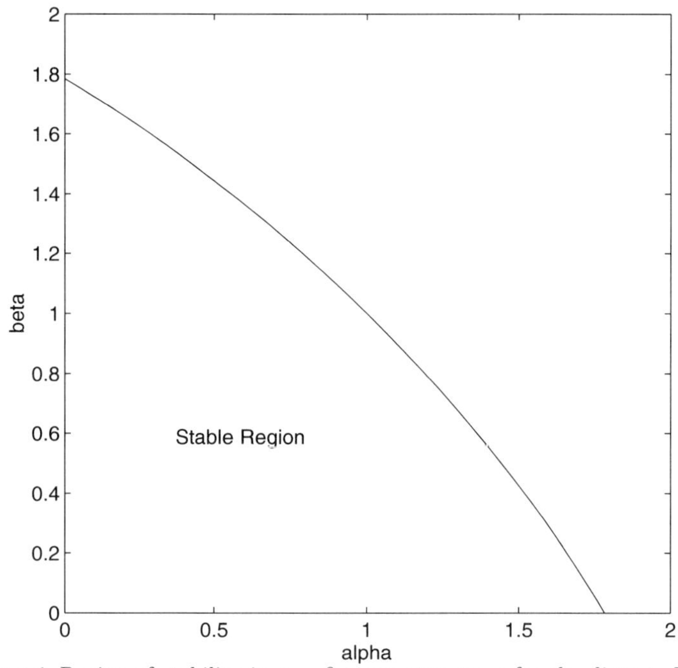


**Figure 1** Region of stability in $\alpha - \beta$ parameter space for the discrete May-Leonard model (2.8), $b = 0.75$.

The behavior of the discrete three-species competition model (2.2) differs in some respects from the continuous-time Lotka-Volterra model. These differences depend in part on the values of the seedbank parameters $b_i$. However, the discrete competition model appears to be much better behaved than is often the case for difference equations.

In the next section, a persistence result for the $n$-species competition model (2.2) is stated.

**2.3 N-species competition.** First, it is shown that solutions are bounded.

**Lemma 2.1** *For systems (2.2) and (2.5), if* $A_i(0) \in \mathbb{R}_+$, *then*

$$0 \le \liminf_{t\to\infty} A_i(t) \le \limsup_{t\to\infty} A_i(t) \le 1.$$

*In addition, if* $0 < A_i(0) \le 1$, *then*

$$0 < A_i(t) \le 1.$$

**Proof** Note that

$$b_i A_i(t) \le A_i(t+1) \le 1 - b_i + b_i A_i(t).$$

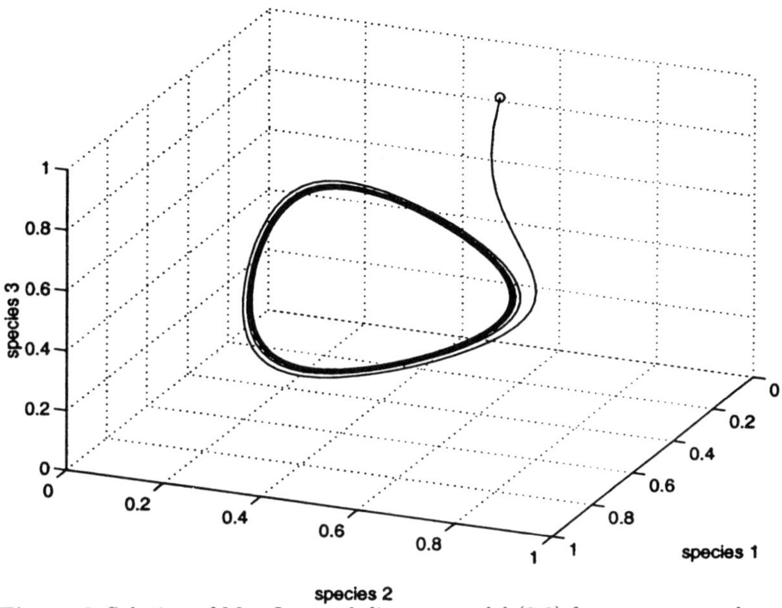

**Figure 2** Solution of May-Leonard discrete model (2.8) for parameter values outside of the stable region, $\alpha = 0.1$, $\beta = 1.85$, and $b = 0.75$.

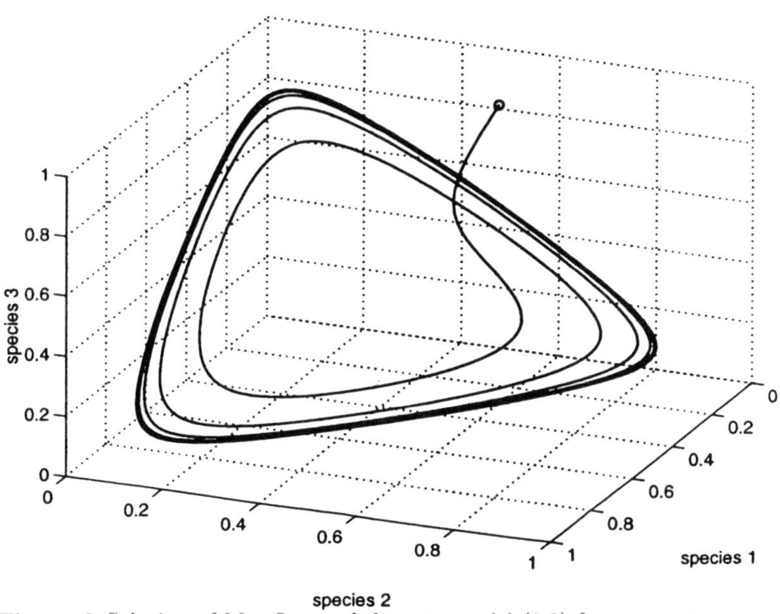

**Figure 3** Solution of May-Leonard discrete model (2.8) for parameters $\alpha = 0.5$, $\beta = 1.7$, and $b = 0.75$.

The lemma follows immediately from iteration of the above inequalities:

$$b_i^t A_i(0) \leq A_i(t) \leq 1 - b_i^t + b_i^t A_i(0).$$

□

It is shown that if the competition coefficients satisfy $\sum_{j \neq i} k_{ij} < 1$, then solutions to (2.2) are uniformly persistent. Such behavior can be demonstrated in the

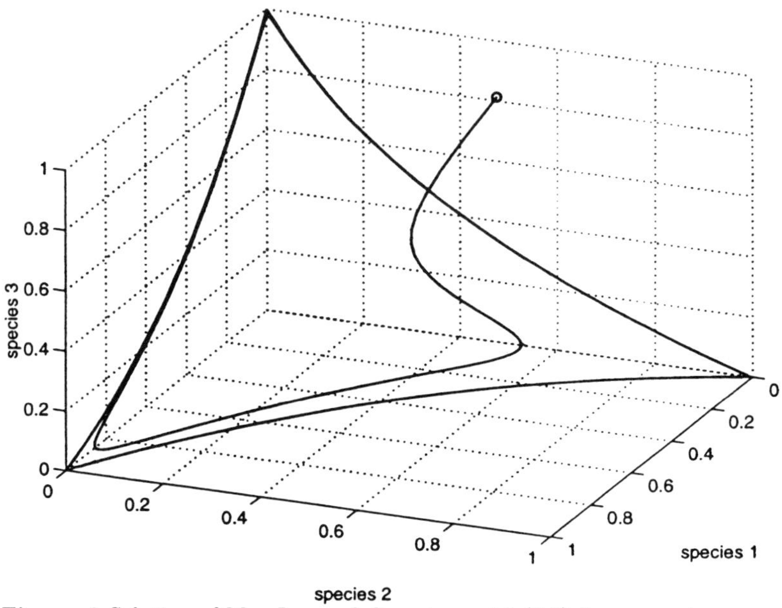


**Figure 4** Solution of May-Leonard discrete model (2.8) for parameters $\alpha = 0.5$, $\beta = 5.0$, and $b = 0.75$.

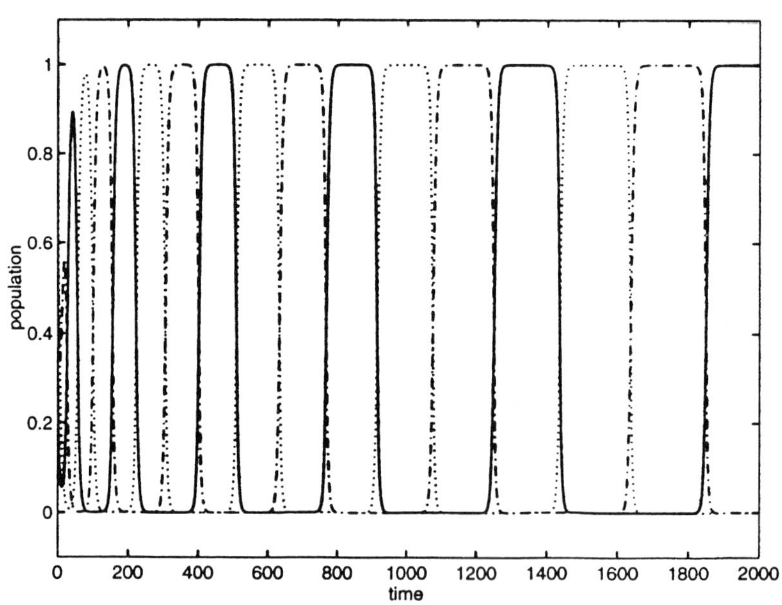


**Figure 5** Solution of May-Leonard discrete model (2.8) for $t = 0, \ldots, 2000$, $\alpha = 0.5$, $\beta = 5.0$, and $b = 0.75$.

normalized continuous-time Lotka-Volterra competition equations (2.6). Persistence behavior in the case of three-species Lotka-Volterra competition and in more general three-species, continuous-time competition models has been investigated in [**5**].

The asymptotic behavior of the following scalar difference equation is needed in the proof of the following theorem. Consider the difference equation:

$$x(t+1) = \frac{(1-b_i)x(t)}{x(t)+c_i} + b_i x(t) = f_i(x). \tag{2.10}$$

The function $f_i$ is increasing on $(0,\infty)$ and for $0 < c_i < 1$ it follows that the difference equation (2.10) has a unique positive fixed point $1 - c_i$. If the initial condition satisfies $0 < x(0) < 1 - c_i$, then the sequence $\{x(t)\}_{t=0}^{\infty}$ generated by (2.10) is monotonically increasing and $\lim_{t\to\infty} x(t) = 1 - c_i$.

**Theorem 2.2** *If the competition coefficients of system (2.2) satisfy*

$$\sum_{j=1, j\neq i}^{n} k_{ij} < 1, \; i = 1, \ldots, n,$$

*then system (2.2) is uniformly persistent, i.e., there exists $\delta > 0$ such that*

$$\liminf_{t\to\infty} A_i(t) > \delta, \; i = 1, \ldots, n.$$

**Proof** Choose a $c > 1$ and $l < 1$ such that $c\sum_{j=1,j\neq i}^{n} k_{ij} < l$ for all $i$. By Lemma 2.1 there exists $T_i$ such that $t \geq T_i$ implies $A_i(t) \leq c$. Let $T = \max_i T_i$. Then for $t \geq T$,

$$A_i(t+1) \geq \frac{(1-b_i)A_i(t)}{A_i(t) + c\sum_{j=1,j\neq i}^{n} k_{ij}} + b_i A_i(t).$$

Let $c_i = c\sum_{j=1,j\neq i}^{n} k_{ij}$, $0 < c_i < 1$. Now we show that $\liminf_{t\to\infty} A_i(t) \geq 1 - c_i$. If for $t > T$, $A_i(t) \geq 1 - c_i$ we are done. Suppose there exists $t_j$ such that $A_i(t_j) < 1 - c_i$ and $t_j > T$. Consider the difference equation given by (2.10) with $t = t_j$, where $x(t_j) = A_i(t_j)$, then

$$\begin{aligned} A_i(t_j+1) &\geq \frac{(1-b_i)A_i(t_j)}{A_i(t_j)+c_i} + b_i A_i(t_j) = f_i(A_i(t_j)) \\ &= f_i(x(t_j)) \\ &= x(t_j+1) \geq x(t_j). \end{aligned}$$

In general, it follows that $A_i(t_j + k) \geq x(t_j + k)$ because $f_i$ is monotonically increasing. Thus, $\liminf_{t\to\infty} A_i(t) \geq \lim_{t\to\infty} x(t) = 1 - c_i$. □

The restriction on the competition coefficients given by Theorem 2.2 implies that the $n$-species equilibrium of (2.2) is locally asymptotically stable. The Jacobian matrix (2.7) of the three-species discrete system can be extended to $n$ species. Let the $n$-species equilibrium be denoted by $A_i^{**}$, $i = 1, \ldots, n$. Then the Jacobian matrix at this equilibrium is given by $J = I - C$, where $C = diag((1-b_i)A_i^{**})K$ and $K$ is the matrix of competition coefficients, $K = (k_{ij})$, $k_{ii} = 1$. It is shown in the following corollary that the $n$-species equilibrium of (2.2) is locally asymptotically stable under the restriction imposed in Theorem 2.2.

**Corollary 2.3** *Suppose the $n$-species equilibrium of system (2.2) is positive. If*

$$\sum_{j=1, j\neq i}^{n} k_{ij} < 1, \; i = 1, \ldots, n,$$

*then the $n$-species equilibrium is locally asymptotically stable.*

**Proof** It follows from Lemma 2.1 that the elements of matrix $C$ are positive and less than one. The assumption on the competition coefficients guarantees that $C = (c_{ij})$ is strictly diagonally dominant:

$$c_{ii} = (1 - b_i)A_i^{**} > (1 - b_i)A_i^{**} \sum_{j \neq i} k_{ij} = c_{ij}.$$

It follows from Gershgorin's circle theorem [**12**] that the eigenvalues $\lambda_C$ of $C$ satisfy

$$0 < (1 - b_i)A_i^{**} \left[1 - \sum_{j \neq i} k_{ij}\right] \leq \lambda_C \leq (1 - b_i)A_i^{**} \left[1 + \sum_{j \neq i} k_{ij}\right] < 2.$$

However, the eigenvalues $\lambda_J$ of $J = I - C$ satisfy $\lambda_J = 1 - \lambda_C$ and hence, $-1 < \lambda_J < 1$. Thus, the $n$-species equilibrium is locally asymptotically stable. □

Inclusion of the factor in (2.5) changes the dynamics near the origin. It is possible to drive the entire system to extinction. If $c_i < 1$ and $0 < A_i(0) \leq 1$, then it follows directly from Lemma 2.1 and (2.5) that

$$A_i(t + 1) \leq [(1 - b_i)c_i + b_i]A_i(t).$$

Since $[(1 - b_i)c_i + b_i] < 1$, $\lim_{t \to \infty} A_i(t) = 0$. This extinction behavior will be especially important when dispersal and control are included.

In the next section, the $n$-species competition model is extended to a competition and dispersal model. The effect of a control on the system dynamics is investigated.

## 3 Dispersal, competition, and control

**3.1 Model development.** The $n$-species competition model can be generalized to include dispersal and control. The model has the form:

$$A_i(t + 1, x) = \frac{a_i m_i(t, x) N_i(t, x)}{N_i(t, x) + \sum_{j \neq i} k_{ij} N_j(t, x)} + b_i A_i(t, x), \tag{3.1}$$

where

$$N_i(t, x) = \int_{\mathbb{R}} \rho_i(t, z) p_i(x, z) A_i(t, z) dz$$

and

$$m_i(t, x) = \min\{m_i, \ d_i N_i(t, x)\}, \ i = 1, \ldots, n.$$

The additional functions and parameters are defined below:

$d_i$ = average number of seeds dispersed per adult plant $i$
$p_i(x, z)$ = probability of a seed spreading from position $z$ to position $x$; $p_i$ is the dispersal kernel
$\rho_i(t, z)$ = proportion of plants remaining after a control is applied in year $t$ to adult plants $A_i(z)$.

The additional factor is included as in the competition model (2.5) so that the growth of a single species is limited at low population densities.

The dimensionless model, using the same variable changes as described earlier, takes the form:

$$A_i(t + 1, x) = \frac{(1 - b_i) m_i(t, x) N_i(t, x)}{N_i(t, x) + \sum_{j \neq i} k_{ij} N_j(t, x)} + b_i A_i(t, x), \tag{3.2}$$

where

$$N_i(t,x) = \int_{\mathbb{R}} \rho_i(t,z) p_i(x,z) A_i(t,z) dz$$

and

$$m_i(t,x) = \min\left\{1, \ \frac{a_i d_i}{1-b_i} N_i(t,x)\right\}, \ i = 1, \ldots, n.$$

The functions $\rho_i(t,z)$ and $p_i(x,z)$ satisfy $0 \leq \rho_i(t,z), \ p_i(x,z) \leq 1$. It is assumed that dispersal depends on distance from a source so that $p_i(x,z) \equiv p_i(u)$, where $u = |x-z|$. Typical forms for $p_i(u)$ are the normal or Weibull density functions:

$$\frac{\exp(-u^2/D)}{\sqrt{D\pi}}, \quad \frac{\exp(-u/D)}{2D}, \quad \text{or} \quad \frac{u\exp(-u^2/D^2)}{D^2}. \tag{3.3}$$

Each of the functions in (3.3) have exponentially bounded tails. However, more general dispersal kernels are possible.

**3.2 Control of one species.** The impact of a control, $\rho_i(t,z)$, is considered. It is shown that application of a control to adult plants such as a herbicide can prevent invasion of a particular species provided that the dispersal kernel and model parameters satisfy certain restrictions.

Consider the following situation. Let species one, $A_1$, represent a weed species occupying primarily the region $x < 0$ and let the region $x > 0$ represent an agricultural field occupied by the remaining species $A_j$, $j = 2, \ldots, n$. Species $A_1$ may be considered an introduced species that will invade the native plant community consisting of species $A_j$, $j = 2, \ldots, n$, unless control methods are applied. In addition, if $A_1$ is a dominant competitor, it will replace some of the native plants over time. Therefore, the objective is to prevent species $A_1$ from invading the region $x > 0$.

It is possible to prevent the spread of $A_1$ into the region $x > 0$ if the dispersal kernel and model parameters of species $A_1$ satisfy certain restrictions. The restriction on the dispersal kernel $p_1(x,z)$ is

$$\int_0^\infty \int_{-\infty}^0 p_1(x,z) dx dz = c_1 < \infty. \tag{3.4}$$

Such a restriction was required for control in a one-species model developed in [**2**]. The inequality (3.4) puts a restriction on the tails of the dispersal kernel. For example, density functions with exponentially bounded tails satisfy this restriction, i.e., if $p(u) \leq K_1 \exp(-K_2 u)$). Kernels such as the normal and Weibull density functions given in (3.3) satisfy this restriction. In addition, this restriction is satisfied by some "fat-tailed" dispersal kernels as discussed in [**7**]. Fat-tailed kernels do not have a moment generating function ($\int_{\mathbb{R}} p(u)\exp(su)du = \infty$ for all $s > 0$) but all of the moments of the function exist ($\int_{\mathbb{R}} |p(u)u^n| du < \infty$, $n = 1, \ldots$). One such fat-tailed kernel satisfying the restriction (3.4) is the dispersal kernel

$$p(u) = \frac{D^2}{4}\exp(-D\sqrt{u}). \tag{3.5}$$

For a simple integrodifference equation of the form:

$$A(t+1,x) = \int_{\mathbb{R}} p(x-z) f(A(t,z)) dz, \tag{3.6}$$

Kot, Lewis and van den Driessche in [**7**] showed that a traveling wave solution, with $p$ defined as in (3.5), spreads with increasing speed. Other investigators (e.g., [**6**], [**8**], [**15**]) have studied integrodifference equations of the form (3.6); however, many of these investigations concentrated on the existence, stability, and shape of traveling wave solutions with constant speed, where the dispersal kernel has an exponentially bounded tail.

There are other types of fat-tailed dispersal kernels satisfying condition (3.4) that do not have finite moments. For example, the function

$$p(u) = \frac{D}{(1+u^m)}, \quad m \geq 3,$$

where $D^{-1} = \int_0^\infty (1+u^m)^{-1} du$, has moments only for $n = 1, \dots, m-2$; this kernel satisfies the restriction (3.4). A fat-tailed dispersal kernel not satisfying the restriction (3.4) is the Cauchy distribution,

$$p(u) = \frac{\pi^{-1}}{1+u^2}.$$

In addition to the restriction on the dispersal kernel, another restriction on the model parameters is required to prevent invasion of species $A_1$ into the region $x > 0$. The following theorem is a slight generalization of a theorem given in [**2**]; competition and a nonuniform control were not considered in the models of [**2**].

**Theorem 3.1** *Let $A_i(t,x)$, $i = 1, \dots, n$ denote the solution to (3.2), where $\int_0^\infty A_1(0,x)dx \leq \hat{A}_1 < \infty$ and $0 \leq A_1(0,x) \leq 1$. Suppose that the dispersal kernel satisfies condition (3.4) and*

$$\rho_1(t,x) < \hat{\rho}_1 < \frac{1-b_1}{a_1 d_1}, \text{ for } x > 0,$$

*and $t = 0, \dots$ , then*

$$\int_0^\infty A_1(t,x)dx \leq \hat{r}_1^t \hat{A}_1 + a_1 d_1 c_1 \frac{(1-\hat{r}_1^t)}{1-\hat{r}_1},$$

*where $\hat{r}_1 = \hat{\rho}_1 a_1 d_1 + b_1 < 1$.*

**Proof** From (3.2), it follows that

$$\begin{aligned} A_1(t+1,x) &\leq a_1 d_1 N_1(t,x) + b_1 A_1(t,x) \\ &\leq \hat{\rho}_1 a_1 d_1 \int_0^\infty p_1(x,z) A_1(t,z)dz + b_1 A_1(t,x) \\ &\quad + a_1 d_1 \int_{-\infty}^0 p_1(x,z) A_1(t,z) dz. \end{aligned}$$

Integrate both sides of the above inequality and apply Lemma 2.1 for a fixed $x$, where $0 < A_1(0,x) \leq 1$,

$$\begin{aligned} \int_0^\infty A_1(t+1,x)dx &\leq \hat{\rho}_1 a_1 d_1 \int_0^\infty \int_0^\infty p_1(x,z)dx A_1(t,z)dz + a_1 d_1 c_1 \\ &\quad + b_1 \int_0^\infty A_1(t,x)dx \\ &\leq (\hat{\rho}_1 a_1 d_1 + b_1) \int_0^\infty A_1(t,x)dx + a_1 d_1 c_1. \end{aligned}$$

The above inequality can be iterated to obtain the conclusion of the theorem:

$$\int_0^\infty A_1(t,x)dx \leq \hat{r}_1^t \int_0^\infty A_1(0,x)dx + a_1 d_1 c_1 \frac{(1-\hat{r}_1^t)}{1-\hat{r}_1}.$$

□

If a control is applied every $m$ years, $m > 1$, instead of every year, then it is still possible to prevent the invasion of $A_1$ if

$$b_1(a_1d_1 + b_1)^{m-1} < 1.$$

The proof of the corollary follows in a manner similar to the proof of Theorem 3.1.

**Corollary 3.2** *Let $A_i(t, x)$, $i = 1, \dots, n$ denote the solution to (3.2), where $\int_0^\infty A_1(0, x)dx \leq \hat{A}_1 < \infty$ and $0 \leq A_1(0, x) \leq 1$. Suppose that the dispersal kernel satisfies condition (3.4) and*

$$\rho_1(t, x) < \hat{\rho}_1 < \frac{1 - b_1(a_1d_1 + b_1)^{m-1}}{a_1d_1(ad + b)^{m-1}} \text{ for } x > 0$$

*for $t = jm$, $m > 1$, $j = 1, 2, \dots$, where $b_1(a_1d_1 + b_1)^{m-1} < 1$, then*

$$\begin{aligned}\int_0^\infty A_1(mt, x)dx &\leq [\hat{r}_1 r_1^{m-1}]^t \hat{A}_1 \\ &\quad + a_1d_1c_1\left(\frac{\hat{r}_1[1 - r_1^{m-1}]}{1 - r_1} + 1\right)\left(\frac{[1 - (\hat{r}_1 r_1^{m-1})^t]}{1 - \hat{r}_1 r_1^{m-1}}\right),\end{aligned}$$

*where $\hat{r}_1 = \hat{\rho}_1 a_1 d_1 + b_1$ and $r_1 = a_1d_1 + b_1$.*

**Proof** As in the proof of Theorem 3.1 it follows that for years $t = 1, \dots, m-1$,

$$\int_0^\infty A_1(m-1, x)dx \leq r_1^{m-1}\int_0^\infty A_1(0, x)dx + a_1d_1c_1\frac{[1 - r_1^{m-1}]}{1 - r_1}.$$

In year $t = m$,

$$\begin{aligned}\int_0^\infty A_1(m, x)dx &\leq (\hat{\rho}_1 a_1 d_1 + b_1)\int_0^\infty A_1(m-1, x)dx + a_1d_1c_1 \\ &\leq \hat{r}_1 r_1^{m-1}\int_0^\infty A_1(0, x)dx \\ &\quad + a_1d_1c_1\left(\frac{\hat{r}_1[1 - r_1^{m-1}]}{1 - r_1} + 1\right).\end{aligned}$$

By continued iteration, the conclusion of the corollary follows. □

**3.3 Numerical examples.** Several numerical examples are presented for two-species competition with and without control. Species one, $A_1$, is the dominant competitor. Without control, species $A_1$ will eliminate species $A_2$. Species $A_2$ has the normal dispersal kernel $p_2(u) = \exp(-u^2)/\sqrt{\pi}$, whereas species $A_1$ has one of the following dispersal kernels:

$$(1)\quad \frac{\sqrt{2}\pi^{-1}}{1 + u^4}, \qquad (2)\quad \frac{\pi^{-1}}{1 + u^2}. \tag{3.7}$$

Both kernels have fat tails. The first dispersal kernel satisfies condition (3.4) but the second dispersal kernel does not satisfy this condition.

The results of Theorem 3.1 are illustrated in the following examples. Simulations are performed for model (3.1). In Figure 6, no control is applied to species $A_1$ with dispersal kernel (1) given in (3.7). It is evident in Figure 6 that without control, species $A_1$ has a traveling wave solution which moves to the right eliminating species $A_2$. It is not evident from Figure 6 whether the wave moves at an increasing speed as demonstrated by other types of integrodifference equations with fat-tailed dispersal kernels [**7**]. However, since the dispersal of the competitor $A_2$

is governed by the normal dispersal kernel, where waves travel at a constant speed [**1**], the competition with $A_2$ may affect the rate of spread of $A_1$.

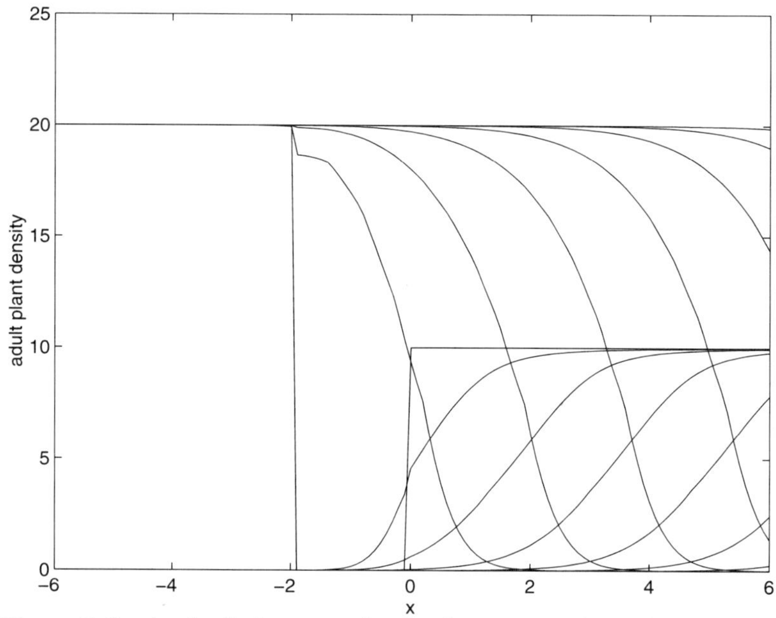


**Figure 6** Species $A_1$ eliminates species $A_2$ when no control is applied. Species $A_1$ travels to the right pushing $A_2$ to the right also. The parameters for both species are: $a_1 = a_2 = 0.5$, $b_1 = b_2 = 0.5$, $k_{12} = k_{21} = 1.0$, $d_1 = 4.0$, $d_2 = 2.0$, $m_1 = 20.0$, and $m_2 = 10.0$. The initial densities are zero except on part of the domain: $A_1(0, x) = 20.0$ for $x \leq -2$ and $A_2(0, x) = 10.0$ for $x \geq 0$. Solutions are graphed for $t = 0, 4, 8, \ldots, 28$.

In Figure 7, the same parameters and dispersal kernels as in Figure 6 are used, except that control is applied to species $A_1$. The control is $\rho_1(t, x) = 0.2 < \hat{\rho}_1 = (1 - b_1)/(a_1 d_1) = 0.25$ for $x > 0$ and $t = 1, 2, \ldots$ so that Theorem 3.1 is applicable. In this case, after approximately $t = 28$ years the traveling wave has slowed down and simulations for longer periods of time show that the wave has stopped moving to the right. For this example, $\hat{A}_1 = 0$ and $c_1 \approx 0.35355$. It follows from Theorem 3.1 for the dimensionless model (3.2) that $\int_0^\infty \tilde{A}_1(t, x)dx < a_1 d_1 c_1/(1 - \hat{r}_1) \approx 7.07$. However, in terms of the dimensional model (3.1), which is used in this example,

$$\int_0^\infty A_1(t, x)dx < \frac{a_1 m_1}{1 - b_1} \frac{a_1 d_1 c_1}{1 - \hat{r}_1} \approx 141.42.$$

In the final example, the dispersal kernel (2) in (3.7) is used for species $A_1$. As this dispersal kernel does not satisfy (3.4), Theorem 3.1 cannot be applied to determine the level of control required. However, the same amount of control as in Figure 7 is applied, $\rho_1(t, x) = 0.2$ for $x > 0$ and $t = 1, \ldots$. In the computational results for Figure 8, solution $A_1$ travels further into the region $x > 0$ than in Figure 7 but the solution eventually stops moving to the right; the level of control is sufficient to stop the spread of $A_1$.

## 4 Conclusion

A general $n$-species competition model is developed for annual plants. The model is a system of first order difference equations for the density of adult plants

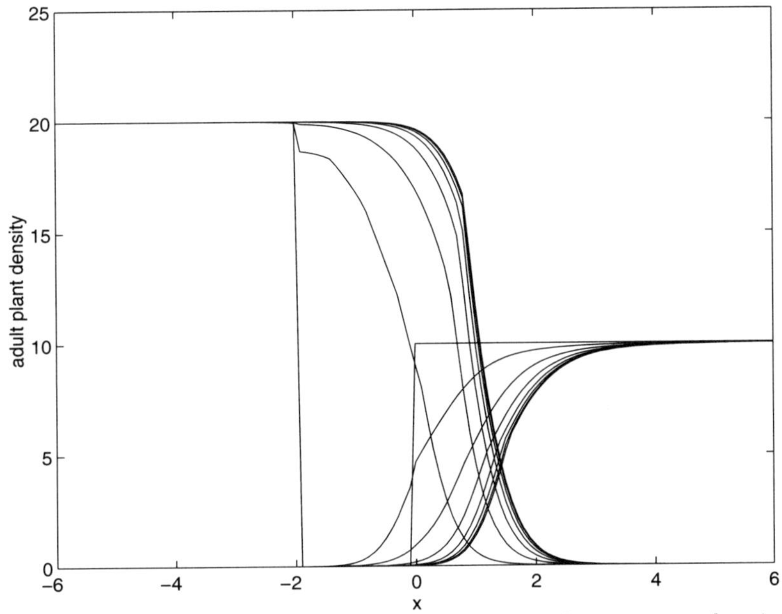


**Figure 7** The same parameter values and initial conditions are used as in Figure 6, except that $\rho_1(t,x) = 0.2 < \hat{\rho}_1 = 0.25$ for $x > 0$ and $t = 1, 2, \ldots$ . Species $A_1$ moves to the right but has slowed down by $t = 28$. Solutions are graphed for $t = 0, 4, 8, \ldots, 28$.

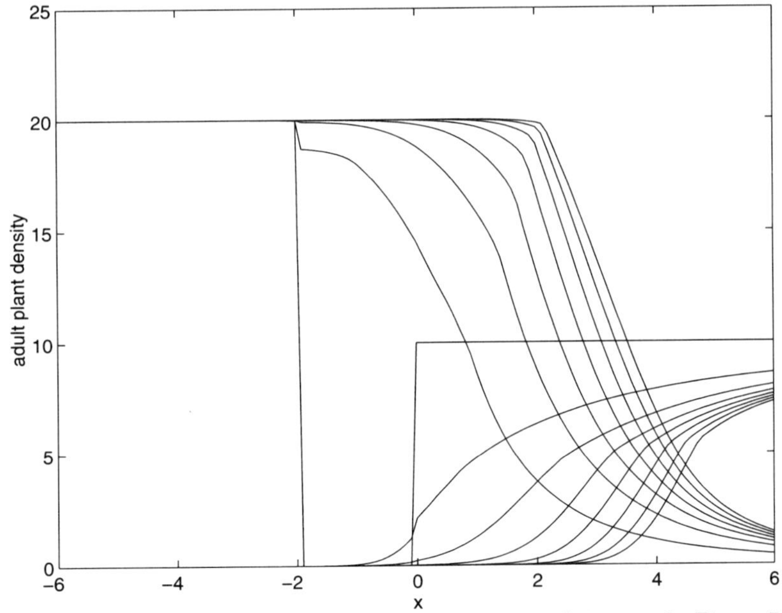


**Figure 8** The same parameter values and initial conditions as in Figure 7 are used, except that the dispersal kernel (2) in (3.7) is used for species $A_1$. A control, $\rho_1(t,x) = 0.2 < \hat{\rho}_1 = 0.25$, $x > 0$ and $t = 1, 2, \ldots$ , is applied to species $A_1$. The control is sufficient to eventually stop the spread of $A_1$ into the region $x > 0$. Solutions are graphed for $t = 0, 4, \ldots, 28$.

in year $t$. The model has asymptotic behavior similar to the continuous-time Lotka-Volterra competition model and does not exhibit the chaotic behavior inherent in many discrete analogues of continuous-time models.

A modified model with more realistic behavior for small population sizes is formulated for plant dispersal and control. The impact of control on one of the competing plants is investigated analytically and numerically. Sufficient conditions are derived that prevent plant dispersal in the region $x > 0$, i.e., the integral of plant density is bounded for all time. The conditions depend on the amount of reduction $\rho(t, z)$ of adult plants and on the magnitude of the dispersal kernel in the tail of the distribution. However, the conditions are less restrictive than the requirements on the dispersal kernel for existence of traveling wave solutions with constant speed (e.g., **[1]**, **[6]**, **[7]**, and **[15]**). Thus, the stated conditions in Theorem 3.1 prevent dispersal of adult plants into the region $x > 0$ even for traveling waves with increasing speed. However, it is important to note that the conditions in Theorem 3.1 are sufficient but not necessary. The conditions in Theorem 3.1 do not include the effects of competition which can be an important factor in preventing the spread of an invasive plant species. Future work is to investigate the impact of competition and spatial dimension on the rate of spread.

## References


[1] Allen, E. J., Allen, L. J. S., and Gilliam, X. [1996], *Dispersal and competition models for plants,* J. Math. Biol. **34**, 455-481.

[2] Allen, L. J. S., Allen, E. J., and Ponweera, S. [1996], *A mathematical model for weed dispersal and control,* Bull. Math. Biol. **58**, 815-834.

[3] Atkinson, D. N. [1997], *Mathematical models for plant competition and dispersal,* Master's Thesis, Texas Tech Univ., Lubbock, TX.

[4] Freedman, H. I. and Moson, P. [1990], *Persistence definitions and their connections*, Proc. Amer. Math. Soc. **109**, 1025-1033.

[5] Freedman, H. I. and Waltman, P. [1985], *Persistence in a model of three competitive populations,* Math. Biosci. **73**, 89-101.

[6] Kot, M. [1992], *Discrete-time travelling waves: ecological examples*, J. Math. Biol. **30**, 413-436.

[7] Kot, M., Lewis, M. A., and van den Driessche, P. [1996], *Dispersal data and the spread of organisms*, Ecology **77**, 2027-2042.

[8] Kot, M. and Schaffer, W. M. [1986], *Discrete-time growth-dispersal models,* Math. Biosci. **80**, 109-136.

[9] May, R. M. [1976], *Simple mathematical models with very complicated dynamics*, Nature **261**, 459-467.

[10] May, R. M. and Leonard, W. J. [1975], *Nonlinear aspects of competition between three species,* SIAM J. Appl. Math. **29**, 243-253.

[11] Merino, S. [1996], *Cyclic competition of three species in the time periodic and diffusive case,* J. Math. Biol. **34**, 789-809.

[12] Ortega, J. M. [1987], *Matrix Theory*, Plenum Press, New York.

[13] Pakes, A. G. and Maller, R. A. [1990], *Mathematical Ecology of Plant Species Competition,* Cambridge University Press, Cambridge.

[14] Schuster, P., Sigmund, K., and Wolff, R. [1979], *On $\omega$-limit sets for competition between three species*, SIAM J. Appl. Math. **37** 49-54.

[15] Weinberger, H. F. [1982], *Long-time behavior of a class of biological models,* SIAM J. Math. Anal. **13**, 353-396.

Fields Institute Communications
Volume **21**, 1999

# Differential and Stochastic Epidemic Models

**Michel Benaïm**
Département de Mathématiques
Université Paul Sabatier
118 Route de Narbonne 31062
Toulouse Cedex, France
benaim@corail.cict.fr

**Morris W. Hirsch**
Department of Mathematics
University of California
Berkeley, CA 94720-3840, USA
hirsch@math.berkeley.edu

*This paper is dedicated to Lynn Erbe and Herb Freedman.*

**Abstract.** Both dynamical systems and stochastic processes are frequently used to model biological processes involving interacting populations. In this article we apply recent results on monotone dynamical systems and discrete Markov processes to reveal close mathematical connections between these two ways of modeling the spread of certain types of infectious disease. A stationary Markov chain $\mathcal{C}^N$ is proposed as a model for an SIS epidemic in a population of size $N$ having constant or state-dependent infection probabilities between certain fixed population groups. The deterministic vector field $G$ giving the expected value of the change of state is a differential equations model for the same epidemic, and is independent of $N$ up to a scalar multiple. Any weak - * limit of a sequence $\{\mu^N\}$ of invariant measures for the processes $\mathcal{C}^N$ is an invariant measure $\mu$ for the semiflow of $G$. Because $G$ is cooperative and irreducible, the support of $\mu$ lies in the set of stable equilibria. When the stable equilibrium $e$ is unique, as in the Lajmonovich-Yorke model, sample paths tend to cluster near $e$ as $N \to \infty$. In this way the differential equations model is rigorously connected to a more basic stochastic model.

---

1991 *Mathematics Subject Classification.* Primary 58F40, 60J20, 60G17, 92C60; Secondary 58F11, 60B10.

This research was partially supported by National Science Foundation Grant DMS-9424382 and North Atlantic Treaty Organization Grant CRG 950857.

## 1 Introduction

Lajmanovich and Yorke in [**24**] consider a disease that does not confer immunity (e.g., gonorrhea) in a population divided into $n$ groups $\mathcal{G}_i$ having constant contact rates with each other. They study an autonomous differential equation that models the evolution of the infection levels in the $n$ groups. Using a Liapunov function, they prove a striking dichotomy: Either the disease dies out, or else there is a unique positive equilibrium state at which the disease stabilizes regardless of initial infection levels (assuming some infection is present to start with). This result was the basis for profound investigations into the transmission and control of gonorrhea [**14**].

Their differential equation ((2.4) below) is of the type called *cooperative*. The infection levels in different groups cooperate in the following sense: If the level in one group $\mathcal{G}_i$ is exogenously raised, the growth rates of levels in all other groups increase. The theory of cooperative differential equations is used below to investigate more general models.

In presenting their model, Lajmanovich and Yorke use the language of probability theory, but only heuristically. The goal of this paper is to use recent results in dynamics and stochastic approximation to draw a rigorous connection between the dynamical behavior of equations similar to theirs and precise (albeit simplistic) probabilistic models. Along the way we take the opportunity to compare the two kinds of models, in both their foundations and conclusions.

The Lajmanovich and Yorke model, like every model, is fictitious. Actual populations are finite, yet we model their dynamics with differential equations. The philosopical issue of justifying this common procedure is rarely discussed. This problem of relating the finite to the continuum occurs frequently in science, for example in passing from particle mechanics to continuum mechanics, and from statistical mechanics to thermodynamics. In these subjects the physics of the continuum is the goal. In population dynamics, however, the integer variables themselves, the population sizes, are the primary focus. We tend to forget that the use of differential equations is merely a tool.

Behind this problem is an even more fundamental one: the relation between the properties of individual entities – molecules, neurons, animals, consumers – and the behavior of large finite ensembles – fluids, brains, species, economies. Here models based on probability seem more basic than those given as differential equations. In many cases, as in this article, a differential equation can be rigorously derived from the probabilistic models, rather than merely being postulated, and its dynamical properties rigorously related to the probabilistic model.

These issues are not merely metaphysical, but are reflected in the mathematics. The Lajmanovich-Yorke differential equation, like many population models, makes biological sense only if the total population $N$ is "arbitrarily large" or "goes to infinity", because hidden in its derivation is the implicit use of the Law of Large Numbers. This means that inferences from differential equations should be independent of $N$, or should at least stabilize as $N \to \infty$. It turns out that this sharply constrains the way infection rates can depend on $N$.

Probability theory is a useful tool for deriving ensemble laws. Markov processes, the stochastic counterparts of dynamical systems, have the notable virtue of making sense for discrete and even finite state spaces. To a stationary, discrete-time Markov process in a convex subset $K \subset \mathbf{R}^n$ we associate a deterministic map $G : K \to \mathbf{R}^n$, called the mean vector field, giving the conditional expected value of the change in state as a function of the current state. The theme of this article is the mathematical connection between sample paths of Markov processes modeling an epidemic and trajectories of their mean vector fields.

A key feature of the stochastic models presented below is that the mean field is independent of the total population $N$, up to a constant scalar multiple. Its dynamics is thus independent of $N$; consequently we are free to use consequences of the Law of Large Numbers in deriving stochastic interpretations of its dynamical features.

Trajectories of the mean field are best viewed as caricatures of sample paths of the Markov process. (In game theory, the dynamics of the mean field is termed "fictitious play"; (see, e.g., [**13**] and [**8**].) But the mean field also has rigorous connections to the Markov process. In many settings, it is highly probable that as a parameter such as total population becomes infinite, sample paths of the Markov process closely track trajectories of the mean field. Rigorous development of this theme of stochastic approximation goes back to [**26**], [**25**] and [**23**]. Recent work includes [**1**], [**2**], [**9**], [**3**], [**5**], [**6**], and [**11**].

We examine variants of the Lajmanovich-Yorke epidemic model Equation (2.4) from this point of view. In Section 2 we go through a heuristic derivation of their system, Equation (2.4). Pointing out some problems with its interpretation, we replace it in Section 3 with Equation (3.1), which describes the evolution of the vector of *per capita* infection densities. This equation, based on slightly altered informal probabilistic assumptions, has the same algebraic form as the Lajmanovich-Yorke equation, and their convergence theorem holds. We describe several dynamical theorems that apply to more general epidemic models, and in particular to the *generalized Lajmanovich-Yorke equation* (3.4). The dynamics of these systems is shown to be sharply restricted, because the generating vector fields are cooperative and irreducible, and sometimes subhomogeneous. These properties are exploited in later sections.

In Section 4 we build stochastic models. We define a sequence of discrete time stochastic processes $\{\mathcal{C}^N\}_{N\in\mathbf{N}^+}$ in $I^n$, based on data similar to that motivating the generalized Lajmanovich-Yorke differential equation. Each process $\mathcal{C}^N = \{X_k^N\}_{k\in\mathbf{N}}$ is an irreducible Markov chain in the common state space $I^n$; its state $X_k^N$ is the vector of infection densities at time $k$.

It turns out that the mean vector field of $\mathcal{C}^N$ is $N^{-1}$ times the vector field $G$ of a generalized Lajmanovich-Yorke model:

$$\mathsf{E}\big(X_{k+1}^N \mid X_k^N = y\big) = y + \frac{1}{N}G(y). \tag{1.1}$$

This gives a different interpretation to the differential equation $dy/dt = G(y)$. Rather than governing the evolution in continuous time of infection densities, it describes the evolution in discrete time of the expected value of the changes in these densities.

Equation (1.1) leads to the important recursion relation

$$X_{k+1}^N = X_k^N + \frac{1}{N}\big[G(X_k^N) + U_{k+1}\big], \tag{1.2}$$

where the $U_{k+1}$ are mean 0 random variables. Stochastic processes satisfying such recursions are the subject of the broad field of stochastic approximation.

In Section 5 we use results on Equation (1.2) and find further relations between the family of processes $\{\mathcal{C}^N\}$ and the vector field $G$. Our chief mathematical result

(Theorem 5.1), based on the special nature of the epidemic vector field, is the following.

*For each $N$ let $\mu^N$ denote the unique probability measure on $I^n$ that is invariant under the Markov process $\mathcal{C}^N$. Then every weak-* limit $\mu$ of the sequence $\{\mu^N\}$ is a measure invariant under the semiflow generated by $G$, and the support of $\mu$ lies in the set of stable equilibria of $G$.*

This implies our basic result about the epidemic (Corollary 5.2):

*As $N \to \infty$, sample paths tend to cluster at stable equilibria of $G$.*

These results are briefly discussed in the final section.

## 2 Heuristic derivation of the Lajmanovich-Yorke equation

We use the following notation. Indices $i$ and $j$ will always run from 1 to $n$. The *nonnegative orthant* in Euclidean $n$-space $\mathbf{R}^n$ is

$$\mathbf{R}^n{}_+ = \{y \in \mathbf{R}^n : y^i \geq 0\},$$

and its interior is the set of *strongly positive* vectors.

$$\mathbf{R}^n{}_{++} = \{y \in \mathbf{R}^n : y^i > 0\}.$$

For $u, v \in \mathbf{R}^n$ we write:

$u \geq v$ (or $v \leq u$) if $u - v \in \mathbf{R}^n{}_+$,

$u > v$ if $u \geq v$ but $u \neq v$,

$u \gg v$ if $u - v \in \mathbf{R}^n{}_{++}$.

The closed unit interval in $\mathbf{R}$ is denoted by $I = [0, 1]$. The Cartesian product of $n$ unit intervals is the closed unit $n$-cube $I^n \subset \mathbf{R}^n$; the open unit $n$-cube is $\text{Int}I^n = \{x \in I^n : 0 < x^i < 1\}$.

Consider now a disease which spreads through a population by contacts between individuals. We always assume the disease confers no immunity, that is anyone not infected is susceptible. The population is static; there are no births or deaths.

Following [**24**], we assume a constant population of $N$ individuals, subdivided into $n \geq 1$ unchanging disjoint groups $\mathcal{G}^i$. The population of $\mathcal{G}_i$ is $N_i$, so that $N = \sum N_i$.

Let $M^i = M^i(t) \in \mathbf{N}$ (the set of natural numbers) denote the number of members of group $i$ who are infected at time $t$. The number of susceptible individuals in $\mathcal{G}_i$ is therefore $N_i - M^i$. The proportion of infecteds in $\mathcal{G}_i$ is the *infection density* $X^i = M^i/N_i \in I$.

In any time interval, and for each $i, j$, some susceptibles in $\mathcal{G}_i$ contract the disease from infecteds in group $j$. We make the following heuristic assumptions. There are constants $A_{ij} \in I = [0, 1]$ such that for each pair of individuals $\alpha \in \mathcal{G}_i, \beta \in \mathcal{G}_j$, and for each time interval $J = [t, t + \Delta t]$, the probability that $\alpha$ and $\beta$ meet at some time in $J$ is

$$A_{ij}\Delta t + o(\Delta t).$$

We assume there is sufficient probabilistic independence in these events to justify the following estimates.

The expected number of meetings at which $\alpha$ is susceptible and $\beta$ is infected is

$$(N_i - M^i)M^j A_{ij}\Delta t + o(\Delta t).$$

Given such a meeting, the probability that $\alpha$ is thereupon infected is a constant $B_{ij} \in I$. The expected number of new infections in $\mathcal{G}_i$ is

$$(N_i - M^i)M^j \sum_j C_{ij}\Delta t + o(\Delta t) \tag{2.1}$$

where $C_{ij} = A_{ij}B_{ij}$.

There is a constant cure rate $K_i$, $0 < K_i < 1$, such that the expected number of infecteds in $\mathcal{G}_i$ that are cured is

$$K_i M^i(t)\Delta t + o(\Delta t). \tag{2.2}$$

The change in $M^i$, namely $\Delta M^i = M^i(t + \Delta t) - M^i(t)$, equals the number of new infections in $\mathcal{G}_i$ minus the number of cures, so the conditional expected value of $\Delta M^i$ is

$$\mathsf{E}\Delta M^i = \big[(N_i - M^i(t))\sum_j C_{ij}X^j(t) - K_i M^i(t)\big]\Delta t + o(\Delta t), \tag{2.3}$$

where $C_{ij} = A_{ij}B_{ij}$. Dividing by $\Delta t$ and taking limits as $\Delta t$ goes to zero, we are led to the system of differential equations proposed in [**24**]:

$$\frac{dM^i}{dt} = (N_i - M^i)\sum_j C_{ij}M^j \; - K^i M^i. \tag{2.4}$$

In this equation the $M^i$, of course, denotes nonnegative real numbers.

Lajmanovich and Yorke proved, as we discuss in more detail below, that under mild hypotheses there is an equilibrium $e \in I^n$ that attracts all positive solutions.

The dynamics of Equation ((2.4)) must be viewed as a continuum approximation, as total population $N$ goes to $\infty$, to the integer dynamics of the actual populations $M^i$. For in deriving (2.4) from (2.3) we have implicitly invoked the equality

$$\lim_{N\to\infty}\lim_{\Delta t\to 0}\frac{\mathsf{E}\Delta M^i}{\Delta t} = \lim_{\Delta t\to 0}\frac{\Delta M^i}{\Delta t},$$

which, if it is justifiable, must be based on the Law of Large Numbers.

But letting the $N_i$ become arbitrarily large obscures the interpretation of (2.4). An obvious resort is to assume the ratios $N_i/N$ converge to positive limits $p_i$ as $N \to \infty$, and to rewrite (2.4) in terms of the *infection densities* $X^i = M^i/N_i$. Dividing (2.4) by $N_i$, after some manipulation we obtain

$$\frac{dX^i}{dt} = (1 - X^i)N\sum_j C_{ij}p_j X^j \; - K^i X^i. \tag{2.5}$$

In this equation, however, $N$ is a parameter, and its unique nonnegative equilibrium $e_N$ depends on $N$.

It is easy to see that $e^i_N \to 1$ as $N \to \infty$. In other words, for large $N$ almost the whole population becomes infected in the long run. Equations (2.1) and (2.2) provide a Malthusian explanation as follows: Infections increase quadratically with $N$, while cures increase only linearly. This conclusion is not only unduly pessimistic, but it is not even interesting, because it is independent on the contact and cure rates.

Revising the model, we look for a differential equation for infection densities that is independent of $N$. To prevent new infections from dominating cures, *we assume that the contact rates $A_{ij}$ are inversely proportional to $N$.* We absorb this proportionality factor and the population ratio $p_j$ into $C_{ij}$, and Equation (2.5) now takes the form:

$$\frac{dX^i}{dt} = (1 - X^i) \sum_j C_{ij} X^j - K^i X^i, \ 0 \le X^i \le 1. \tag{2.6}$$

This equation (but not its interpretation) is independent of $N$.

## 3 ODE epidemic models and cooperative irreducible systems

The variables $X^i$ in (2.6) are of course real numbers, but in their biological interpretation they are rational numbers. Later we treat them as random variables. To avoid further confusion we rewrite (2.6) in terms of real variables $y^i$:

$$\frac{dy^i}{dt} = (1 - y^i) \sum_j C_{ij} y^j - K^i y^i, \ 0 \le y^i \le 1. \tag{3.1}$$

As this has the same form as (2.4), we refer to it as the Lajmanovich-Yorke equation.

Lajmanovich and Yorke make the simplifying assumption of *mutual infectibility*: each group $\mathcal{G}_i$ can directly or indirectly transmit infection to every other group. The mathematical formulation of this, which we assume from now on, is that the matrix $[C_{ij}]$ is *irreducible*: For each $i, j$ there is a finite sequence of indices $i = l_0, \dots, l_r = l$ such that $C_{l_{m-1}, l_m}(t) > 0$ for all $m \in \{1, \dots, r\}$, $t \ge 0$. Equivalently, in the directed graph with vertices $\{1, \dots, n\}$ and an arrow from $j$ to $i$ if and only if $C_{ij} > 0$, every ordered pair of vertices can be joined by a directed path.

Lajmanovich and Yorke prove the following striking result, which applies to Equations (3.1) and (2.4):

**Theorem 3.1** ([**24**]) *Either every solution $y(t)$ converges to $0$; or else there is a unique equilibrium $e \gg 0$ such that $y(t) \to e$ if $y(0) > 0$.*

It turns out that the same conclusion holds under more general circumstances. Let $F : I^n \to \mathbf{R}^n$ be the vector field obtained from the right-hand side of Equation (3.1):

$$F^i(y) = (1 - y^i) \sum_j C_{ij} y^j - K_i y^i. \tag{3.2}$$

Let $\Phi = \{\Phi_t : I^n \to I^n\}_{t \in [0, \infty)}$ denote the solution semiflow generated by $F$, so that the solution to the initial value problem

$$\frac{dy}{dt} = F(y), \quad y(0) = v, \ v \in I^n, \ t \ge 0 \tag{3.3}$$

is given by $y(t) = \Phi_t v$. The following facts are easily verified:

(a) *Invariance*: $\Phi_t v$ is defined and in $I^n$ for all $t \ge 0, v \in I^n$. Moreover $\Phi_t v > 0$ if $t > 0, v > 0$.

(b) *Cooperativity*: $\dfrac{\partial F^i}{\partial y^j} \ge 0$ for $i \ne j$.

(c) *Irreducibility*: The Jacobian matrix $DF(y) = \left[\dfrac{\partial F^i}{\partial y^j}(y)\right]$ is irreducible provided $x \in \mathrm{Int} I^n$.

(d) *Strong subhomogeneity*: $F(\lambda y) < \lambda F(y)$ if $\lambda > 1$, $v \in \mathrm{Int} I^n$.

Invariance is readily verified from Equation (3.2).

Cooperativity means that the rate of change in the infection density in $\mathcal{G}_i$, which is $F^i(y^1, \dots, y^n)$, increases, or at least does not decrease, if the infection density $y^j$, $j \neq i$ in any other group increases exogenously while other densities are held fixed. This has the crucial dynamical consequence that $\Phi$ is *monotone*: $\Phi_t v \geq \Phi_t u$ provided $t \geq 0$ and $v \geq u$. Biologically this is plausible: greater initial infection densities in all groups will result in greater densities at all future times.

Irreducibility, a mathematically convenient way of modeling mutual infectibility, implies that $\Phi$ has the important property of being *strongly monotone*: $\Phi_t v \gg \Phi_t u$ provided $t > 0$ and $v > u$ ([**15**], as corrected in [**22**], [**28**] or [**31**]).

Strong subhomogeneity is what economists call decreasing returns to scale, if the "product" is infection. In the epidemic context it means this: When all infection densities are increased exogenously by the same factor $\lambda > 1$, the rate of change in each infection density is multiplied by a factor less than $\lambda$. Examination of Equation (3.1) shows this is due to the decline in the number of susceptibles in $\mathcal{G}_i$ when the infection density in $\mathcal{G}_i$ is raised.

**Theorem 3.2** *Suppose a $C^1$ vector field $F : I^n \to \mathbf{R}^n$ satisfies (a), (b), (c) and (d). Then the differential equation $dy/dt = F(y)$ satisfies the conclusion of the Lajmanovich-Yorke Theorem 3.1.*

**Proof** Follows from ([**18**], Theorem 5.6). □

There are analogous results under the more realistic assumption that the infection and cure coefficients are periodic functions of time [**30**].

Next we discuss more general dynamical systems, without subhomogeneity.

An equilibrium $e$ is *stable* if it has arbitrarily small neighborhoods invariant under the solution semiflow in positive time. The set of stable equilibria of a vector field $F$ is denoted by $\mathcal{E}_{\text{stab}}(F)$. We call $e$ *asymptotically stable* if it has a neighborhood in which all trajectories converge uniformly to $e$ as $t \to \infty$. If $\operatorname{Det} DF(e) \neq 0$ then $e$ is *simple*. Asymptotically stable equilibria, and simple equilibria, are isolated in the set of all equilibria.

It seems likely that every closed set is the set of stable equilibria of some $C^\infty$ vector field. But for cooperative irreducible systems the possibilities are quite limited:

**Proposition 3.3** *Let $F : I^n \to \mathbf{R}^n$ be a $C^1$ vector field in $\mathbf{R}^n$ satisfying (a), (b) and (c). Then*

(i) *Each connected component of $\mathcal{E}_{\text{stab}}(F)$ is a point or a totally ordered arc.*
(ii) *Let $J, K \subset \mathcal{E}_{\text{stab}}(X)$ be distinct components. If there exist $u \in J, v \in K$ such that $u < v$ then $J \ll K$.*
(iii) *$\mathcal{E}_{\text{stab}}(F)$ is a nonempty finite set of asymptotically stable equilibria provided $F(0) > 0$, and either $F$ is real analytic, or every equilibrium is simple.*

**Proof** Follows from [**19**] and [**21**] for the analytic case of (iii). □

The dynamics of cooperative irreducible vector fields are also very restricted.

**Theorem 3.4** *Suppose a $C^1$ vector field $F : I^n \to \mathbf{R}^n$ satisfies (a), (b), (c). Then the solution semiflow to $dy/dt = F(y)$ has the following properties,*

(i) *The trajectory of almost every initial state (in the sense of Lebesgue measure) converges as $t \to \infty$ to an equilibrium, and to an asymptotically stable equilibrium under assumption (iii) of Theorem 3.3.*

(ii) *If there are at most two equilibria, then exactly one of them is asymptotically stable and every trajectory converges.*

**Proof** (i) follows from Smith and Thieme [**29**]. (ii) follows from strong monotonicity of the semiflow and a well known argument based on fixed point indices (compare [**10**], [**17, 20**]). □

The Lajmanovich-Yorke theorem can be generalized to state dependent data in various ways. For example, replace the constants $C_{ij}, K_i$ by $C^1$ functions

$$C_{ij} : I^n \to I, \quad K_i : I \to I,\ 0 < K_i < 1,$$

taking $C_{ij}(y)$ to be a function only of $y^i$ and $y^j$. For later use we add a constant term $Q_i \in I$ representing an exogenous source of infection (see Section 4). Analogous heuristic reasoning leads to a *generalized Lajmanovich-Yorke epidemic model*:

$$\begin{aligned} \frac{dy^i}{dt} &= Q^i + (1-y^i)\sum_j (1-Q_i)C_{ij}(y)y^j - K_i(y^i)y^i \\ &\equiv G_i(y),\ y \in I^n. \end{aligned} \tag{3.4}$$

There is the following extension to the Lajmanovich-Yorke theorem:

**Theorem 3.5** *Let $G = (G_1, \dots, G_n) : I^n \to \mathbf{R}^n$ be the vector field defined by (3.4), and assume*

(i) $K_i(0) = 0 < K_i(1), \quad 0 < Q_i < K_i(1),$ *and* $C_{ij}(y) > 0$ *if* $y^j > 0$,

(ii) $\dfrac{\partial C_{ij}}{\partial y^j}(y)y^j + C_{ij}(y) \ge 0$ *for* $i \ne j$,

(iii) *the matrices* $\left[\dfrac{\partial C_{ij}}{\partial y^j}(y)y^j + C_{ij}(y)\right]$ *are irreducible for* $y \in \mathrm{Int} I^n \setminus \{0\}$.

*Then $G$ and its semiflow satisfy (a), (b), (c) above, and therefore the conclusion of Theorem 3.5 is valid. Moreover there are equilibria $u, v$ such that $0 < u^i \le v^i < 1$ and for every solution $y(t)$:*

$$u^i \le \liminf_{t\to\infty} y^i(t) \le \limsup_{t\to\infty} y^i(t) \le v^i.$$

*If, in addition,*

(iv) $\displaystyle\sum_{k=1}^n y_k \frac{\partial C_{ij}}{\partial y^k} \le 0$ *and* $\dfrac{dK_i}{dy_i} \ge 0$,

*then there is a unique equilibrium $e$, $0 < e^i < 1$, that attracts all solutions.*

**Proof** Assumption (i) implies the invariance condition (a); (ii) implies cooperativity (b); (iii) implies irreducibility (c); and (iv) implies strong subhomogeneity (d). Therefore the conclusion follows from Theorems 3.4 and 3.2, with $u$ and $v$ obtained as follows. Equation (3.4) and the assumptions $0 < Q_i < K_i(1)$ imply $0 < G(0) < G(q)$ where $q = (1, \dots, 1)$, whence strong monotonicity and a result of [**27**] imply the trajectory of 0 and $q$ converges respectively upward and downward to equilibria $u$ and $v$ as required. □

The biological interpretations of the assumptions in Theorem 3.5, and their plausibility, are left to the reader.

Lajmanovich and Yorke interpret their result as a "bio-theorem": infection levels ultimately stabilize at a common equilibrium. Similarly, Theorems 3.4(a) and 3.5 seem to mean that for these more general epidemics, almost all infection densities ultimately stabilize. But these results reveal nothing about any particular nonequilibrium state, and the biological interpretation of "almost all" is far from clear. In the real world the current infection density is simply a fact of nature; it is unique, and for all we know it belongs to the measure zero set comprising states having nonconvergent trajectories. All "generic" theorems in applied mathematics have this same difficulty.

The Lajmanovich-Yorke Theorem 3.1, like its generalizations 3.4(ii) and 3.5 (iv), does not have this particular problem of interpretation, as *all* trajectories converge. But it is hard to see how real data can guarantee that gonorrhea, for example, satisfies the assumptions of such theorems.

These results, like many in mathematical biology, are not intended to closely model reality. But they may provide useful insights and suggest interesting research (e.g., [**14**]).

In the next section we consider precise stochastic models that are rigorously linked to the Lajmanovich-Yorke equation. We present results that hold for all initial states, but only with probability 1. This shifts the philosophical issue to the interpretation of probability. While to many people probability is more intuitive than measure theory, the technical definition of "probability one" is formally the same as that of "Lebesgue almost everywhere", but even more abstract, as it refers to the complement of a set of measure zero in an abstract probability space, rather than in Euclidean space.

## 4 Stochastic models

The probabilistic assumptions governing the spread of infection can be made precise in many ways. The analysis of the resulting model, however, becomes more difficult as the setting becomes more realistic. The following hypotheses are simplistic, but in the next section they lead to rigorous interpretations of the Lajmanovich-Yorke type equations.

In this section the total population $N$ is fixed. Time is discrete. At any time $k \in \mathbf{N}$, each member of the population is either uninfected and susceptible, or infected and not susceptible. The number of infecteds in $\mathcal{G}_i$ is a random variable $M_k^i \in \{0, \dots, N_i\}$, and the density of infecteds in $\mathcal{G}_i$ is the random variable

$$X_k^i = M_k^i/N_i \in \Big\{0, \frac{1}{N_i}, \dots, \frac{N_i - 1}{N_i}, 1\Big\}.$$

We also consider the random vectors $X_k = (X_k^1, \dots, X_k^n)$. The basic object of interest is the stochastic process $\{X_k\}$ taking values in $I^n$.

Fix $k \in \mathbf{N}^+$ and $y \in I^n$ and suppose $X_k = y$. At most one individual, called $\alpha_k$, changes her status (susceptible or infected) at time $k$. She is chosen at random from the uniform distribution on the whole population, so that the probability that $\alpha_k \in \mathcal{G}_i$ is $p^i = N_i/N$. Given that $\alpha_k \in \mathcal{G}_i$, the conditional probability that $\alpha_k$ is infected (respectively, susceptible) is therefore $y^i$ (respectively, $1 - y^i$).

Suppose $\alpha_k \in \mathcal{G}_i$. If infected, she is cured with probability $K_i(y^i)$, and at time $k+1$ she is susceptible. If $\alpha_k$ is susceptible, then either she meets exactly one other individual, called $\beta_k$, or she does not. Given a meeting, the probability that $\beta_k \in \mathcal{G}_j$ is $A_{ij}(y)$, assumed to depend only on $y^i$ and $y^j$. We suppose from now on that $\sum_j A_{ij} \le 1$. The probability that $\alpha_k$ doesn't meet anyone is $1 - \sum_j A_{ij}(y)$.

Suppose $\alpha_k$ is susceptible and meets an individual $\beta_k \in \mathcal{G}_j$. The probability that $\alpha_k$ becomes infected, conditioned on such a meeting, is the probability $y^j$ that

$\beta_k$ is infected, times the conditional probability $B_{ij} \in I$ that this contact infects $\alpha_k$. For simplicity we assume $B_{ij}$ is constant.

Set $\Delta M_k^i = M_{k+1}^i - M_k^i$; this random variable takes values in $\{-1, 0, 1\}$. The event $\Delta M_k^i = 1$ occurs if and only if $\alpha_k$ belongs to $\mathcal{G}_i$, is uninfected, meets an infected member $\beta_k$ of some $\mathcal{G}_j$, and thereupon becomes infected. The conditional probability of this, given the time $k$ and the state vector $X_k = y$, equals the probability that $\alpha_k \in \mathcal{G}_i$ (which is $p^i$), times the conditional probability that $\alpha_k$ is susceptible $(1 - y^i)$, times the sum over $j$ of the probability that $\beta_k$ is an infected member of $\mathcal{G}_j$ and that $\beta_k$ infects $\alpha_k$, which is $\sum_{j=1}^n A_{ij}(y)y^j B_{ij}$. Setting $C_{ij}(y) = A_{ij}(y)B_{ij}$, we thus have

$$\mathsf{P}_{X_k=y}\{\Delta M_k^i = 1\} = p^i(1 - y^i)\sum_j C_{ij}(y)y^j. \tag{4.1}$$

Similarly, $\Delta M_k^i = -1$ if and only if $\alpha_k$ belongs to $\mathcal{G}_i$, is infected, and becomes cured,

$$\mathsf{P}_{X_k=y}\{\Delta M_k^i = -1\} = p^i y^i K_i(y^i). \tag{4.2}$$

The conditional probability that $\Delta M_k^i = 0$ (the only other possibility) can be easily calculated, but we do not need it to obtain the conditional expected value of $\Delta M_k^i$,

$$\begin{aligned} \mathsf{E}_{X_k=y}\Delta M_k^i &= \mathsf{P}_{X_k=y}\{\Delta M_k^i = 1\} - \mathsf{P}_{X_k=y}\{\Delta M_k^i = -1\} \\ &= p^i\left[(1 - y^i)\sum_j C_{ij}(y)y^j - K_i(y^i)y^i\right]. \end{aligned}$$

Dividing by $N_i$, we obtain the conditional expected value of the changes in infection densities:

$$\begin{aligned} \mathsf{E}_{X_k=y}\left(X_{k+1}^i - X_k^i\right) &= \frac{p^i}{N_i}\left[(1 - y^i)\sum_j C_{ij}(y)y^j - K_i(y^i)y^i\right] \\ &= \frac{1}{N}\left[(1 - y^i)\sum_j C_{ij}(y)y^j - K_i(y^i)y^i\right] \\ &= \frac{1}{N}G_i(y). \end{aligned} \tag{4.3}$$

Notice that the right-hand side, which is independent of $k$, is the same as that of the generalized Lajmanovich-Yorke equations (3.4) with $Q_i = 0$, except for the factor $1/N$.

The sequence of vector random variables $\{X_k\}_{k\in\mathbf{N}}$ is a stationary Markov chain with values in the finite set

$$S_N = \prod_{i=1}^n \left\{0, \frac{1}{N_i}, \ldots, \frac{N_i - 1}{N_i}, 1\right\} \subset I^n.$$

It is convenient to take $I^n$ as the common state space, and to allow the initial state to be an arbitrary vector $X_0 \in S_N$ different from 0. The probability distribution of $X_{k+1}$ is determined by the value of $X_k \in S_N$.

Comparing (4.3) with Equation (3.4), we arrive at the following interpretation of the generalized Lajmanovich-Yorke vector field $G$ with $Q_i = 0$. Except for the factor $1/N$, it gives the conditional expected value of the change in state of the stochastic process $\{X_k\}$.

Now we adjust the stochastic model, for the following reason. With probability one, $X_k = 0$ for all sufficiently large $k$. In other words, the disease almost surely dies out! For regardless of the state, there is a positive probability, bounded away from 0, that all infecteds will be cured before there are any new infections, and from then on no one can be infected.

This does not invalidate the model, however, because for large populations this will take a very long time. Application of standard averaging theorems reveals that for the Lajmanovich-Yorke model there are long periods, exponential in $N$, during which infection densities are arbitrarily close to the unique stable equilibrium of $G$.

Rather than pursue this idea, we change the model slightly by adding the following assumption: If the selected individual $\alpha_k \in \mathcal{G}_i$ is uninfected and does not meet an infected individual, there is nevertheless a conditional probability $Q_i$, $0 < Q_i < K_i(1)$ that $\alpha_k$ becomes infected. We assume for simplicity that $Q_i$ is constant It is easy to think of scenarios for such infection, e.g., foreign travel, transmission by another species, dirty needles, contaminated blood banks, etc. We therefore replace Equation (4.1) by

$$\begin{aligned} \mathsf{P}_{X_k=y}\{\Delta M_k^i = 1\} &= p^i(1-y^i)\left[\sum_j C_{ij}(y)y^j + Q_i(1-\sum_j C_{ij}(y)y^j)\right] \\ &= p^i(1-y^i)\left[(1-Q_i)\sum_j C_{ij}(y)y^j + Q_i\right]. \end{aligned} \tag{4.4}$$

Equation (4.2) is unchanged. Equation (4.3) now becomes

$$\mathsf{E}_{X_k=y}(X_{k+1}^i - X_k^i) = \frac{1}{N}\left[(1-y^i)(Q_i + (1-Q_i)\sum_j C_{ij}(y)y^j) - K_i(y^i)y^i\right].$$

Setting $D_{ij}(y) = (1-Q_i)C_{ij}(y)$, we rewrite this as

$$\mathsf{E}_{X_k=y}(X_{k+1}^i - X_k^i) = \frac{1}{N}\left[Q_i + (1-y^i)\sum_j D_{ij}(y)y^j - K_i(y^i)y^i\right]. \tag{4.5}$$

Equation (4.5) shows that even if $X_k = 0$, the conditional expected value of $X_{k+1}$ is positive. Together with our other assumptions, this makes the Markov chain $\{X_k\}$ irreducible.

The mean field of the Markov process $\{X_k\}$ has its $i$th component given by the right-hand side of (4.5). Except for the scaling factor $1/N$, this is the same as the right-hand side of the generalized Lajmanovich-Yorke Equation (3.4). Thus the dynamics of the mean field is independent of $N$, and is given by Equation (3.4).

$$\begin{aligned} \frac{dy^i}{dt} &= Q^i + (1-y^i)\sum_j(1-Q_i)C_{ij}(y)y^j - K_i(y^i)y^i \\ &\equiv G_i(y), \;\; y \in I^n. \end{aligned} \tag{4.6}$$

We assume from now on that $C_{ij}, K_i$ and $Q_i$ satisfy hypotheses (i), (ii) and (iii) of Theorem 3.4, so that Theorems 3.4 and 3.5 apply to $G$. When $K_i$ and $C_{ij}$ are constants, the conclusion of the Lajmanovich-Yorke theorem applies, with the added fact that there is a unique equilibrium $e \in \mathrm{Int} I^n$ that attracts all solutions.

It is easy to see that this equilibrium depends continuously on the parameter $Q = (Q_1, \dots, Q_n)$ and tends to the Lajmanovich-Yorke equilibrium as $Q \to 0$.

Manipulating Equation (4.3), we obtain the recursion formula

$$X_{k+1} = X_k + \frac{1}{N}\big[G(X_k) + U_{k+1}\big], \tag{4.7}$$

where the $U_{k+1}$ are random variables in $\mathbf{R}^n$ satisfying

$$\mathsf{E}(U_{k+1} \mid X_k) = 0. \tag{4.8}$$

Equations (4.7), (4.8) define $U_{k+1}$, and it is not hard to verify that

$$U_{k+1} = N[X_{k+1} - \mathsf{E}(X_{k+1} \mid X_k)].$$

Equation (4.7) bears a striking resemblance to a Cauchy-Euler numerical integration scheme for the vector field $G$ with step size $1/N$; the only difference is the presence of the noise $\frac{1}{N}U_{k+1}$. Therefore we expect that with high probability, sample paths $\{X_k\}$ closely track integral curves of $G$. In the next section we make this intuition precise.

## 5 Invariant measures

For each $N \in \mathbf{N}^+$, Equations (4.4) and (4.2) and the resulting equation for $\mathsf{P}_{X_k=y}\{\Delta M_k^i = 0\}$ determine a stationary Markov chain $\mathcal{C}^N$ in $I^n$. The $k$th state vector of $\mathcal{C}^N$ is a random vector-valued variable $X_k^N \in I^n$ with components $X_k^{N,i} \in I$.

We take this family $\{\mathcal{C}^N\}_{N\in\mathbf{N}^+}$ of stochastic processes, parametrized by $N$, as the primary epidemic model. Heuristically, the differential equation (4.6) is some kind of limit of the family as $N \to \infty$— a model of a model, as it were. The rigorous link between the stochastics and the dynamics is provided by the theory of invariant measures.

Each Markov process $\mathcal{C}^N$ is irreducible. There is nonzero probability of eventually going from any state to any other state. Consequently it has a unique invariant Borel probability measure $\mu^N$ on $I^n$. Notice that the support of $\mu^N$ is the finite set $S^N$.

The set $\mathcal{M}(I^n)$ of Borel probability measures on $I^n$ is a compact set in the weak-* topology on the dual space to the Banach space $C(I^n)$ of continuous real-valued functions on $I^n$. This means a sequence $\{\lambda_N\}$ of measures converges to a measure $\lambda$ if and only if $\int f d\lambda_m \to \int f d\lambda$ for every $f \in C(I^n)$.

**Theorem 5.1** *Assume the hypotheses of Theorem 3.4 or 3.5 hold. Let $\rho \in \mathcal{M}(I^n)$ denote a measure that is a weak-* limit point of the sequence $\{\mu^N\}$. Then $\rho$ is invariant for $G$, and has support in $\mathcal{E}_{\text{stab}}(G)$.*

**Proof** This is a consequence of Theorem 3.2 of [**4**] and results in [**7**]. □

In view of Proposition 3.3, we immediately conclude:

**Corollary 5.2** *Let $\rho$ be as in Theorem 5.1. Then*

(i) *The support of $\rho$ lies in $\mathcal{E}_{\text{stab}}(G)$.*
(ii) *Every connected component of the support of $\rho$ is either a point or a totally ordered arc.*
(iii) *Suppose $\mathcal{E}_{\text{stab}}(G)$ is a finite set $\{p_1, \dots, p_r\}$. Then $\rho$ is a convex combination of the Dirac measures supported at the $p_i$.*

Set $J_m = \{1, \ldots, m\}$; let $\#$ denote cardinality. For each $m, N \in \mathbf{N}^+$ and each Borel set $U \subset I^n$, define random variables $T_m^N(U) \in \mathbf{N}$ and $\tau_m^N(U) \in [0,1]$.

$$T_m^N(U) = \#\{k \in J_m : X_k^N \in U\},$$

$$\tau_m^N(U) = \frac{1}{m} T_m^N(U).$$

For each $N, m, U$ the random measure $\tau_m^N(U)$ gives the empirical frequency with which the first $m$ states of $\mathcal{C}^N$ enter $U$.

The following consequence of Theorem 5.1 shows that as $N, m \to \infty$ the supports of the empirical occupation measures cluster at stable equilibria of the vector field $G$:

**Corollary 5.3** *Let $U \subset I^n$ be any neighborhood of $\mathcal{E}_{\text{stab}}(G)$. Then almost surely*

$$\lim_{N\to\infty} \lim_{m\to\infty} \tau_m^N(U) = 1.$$

**Proof** This follows easily from Corollary 5.2(i); details can be found in [**7**]. □

When there is only one stable equilibrium, Corollary 5.3 becomes:

**Corollary 5.4** *Suppose $G$ has a unique stable equilibrium $e$. Then with probability 1, for every $\epsilon > 0$ there exists $N_* \geq 1$ having the following property. For every $N \geq N_*$, there exists $m_* \geq 1$ such that if $m \geq m_*$, then*

$$\frac{\#\{k \in J_m : ||X_k^n - e|| > \epsilon\}}{m} < \epsilon.$$

## 6 Discussion

It is interesting to compare the biological interpretations of our differential equations models of epidemics with their stochastic counterparts. From Equation (4.6) with constant $K_i$ and $C_{ij}$, we infer:

*Regardless of the initial infection density, the state of the epidemic converges to the unique equilibrium $e$.*

But for the full probabilistic model, (the family $\{X_k^N\}_{N\in\mathbf{N}^+}$ of Markov processs parameterized by total population $N$) we have the following more complicated conclusion given by Corollary 5.4.

*It is highly likely that for very large populations, in the long run the state of the infection is very close to $e$ most of the time.*

This is perhaps a less attractive conclusion, but it is expressed in terms of finite populations, and it is more realistic than the impossibly smooth convergence of solutions to the differential equations. Most importantly, in our view, is that the stochastic model has better foundations than the differential equations model.

## References

[1] Arthur, B. [1989] *Competing technologies, increasing returns, and lock-in by historical events*, Economic Journal **99**, 116–131.

[2] Arthur, B., Ermol'ev, Y. & Kaniovskii, Y. [1987] *Adaptive growth processes modeled by urn schemes*, Cybernetics **23**, 779–789.

[3] Benaïm, M. [1996] *A dynamical system approach to stochastic approximations*, SIAM Journal of Optimisation and Control **34**, 437–472.

[4] Benaïm, M. [1998] *Recursive algorithms, urn processes and chaining number of chain recurrent sets*, Ergodic Theory and Dynamical Systems **18**, 53-87.

[5] Benaïm, M. and Hirsch, M. [1994] *Asymptotic pseudotrajectories and chain recurrent flows, with applications*, J. Dynamics Differential Equations, **8**, 141–176.

[6] Benaïm, M. and Hirsch, M. [1995] *Dynamics of Morse-Smale urn processes*, Ergodic Theory and Dynamical Systems, **15**, 1005–1030.

[7] Benaïm, M. and Hirsch, M. [1996] *Stochastic approximation algorithms with constant step size whose average is cooperative*, preprint.

[8] Benaïm, M. and Hirsch, M. [1998] *Dynamics of learning processes arising from repeated games*, preprint.

[9] Benveniste, A., Métivier, M. and Priouret, P. [1990] *Stochastic Approximations and Adaptive Algorithms*, Springer-Verlag, New York.

[10] Dancer, E. and Hess. P. [1991] *Stability of fixed points for order-preserving discrete-time dynamical systems*, J. reine angew. Math. **419**, 125-139.

[11] Duflo, M. [1996] *Algorithmes Stochastiques*, Springer-Verlag, Paris.

[12] Fomenko, I. [1993] *Stability for some equations of mathematical biology and monotone flows*, Bull. Austral. Math. Soc. **48**, 201–207.

[13] Fudenberg, D. and Kreps, K. [1993] *Learning mixed equilibria*, Games and Economic Behavior **5**, 320–367.

[14] Hethcote, H. and Yorke, J. [1984] *Gonorrhea transmission dynamics and control*, Lecture Notes in Biomathematics, vol. 56, Springer-Verlag, New York.

[15] Hirsch, M. [1984] *Systems of differential equations which are competitive or cooperative, II: Convergence almost everywhere.* SIAM J. Math. Anal. **16**, 423–439.

[16] Hirsch, M. [1985] *Attractors for discrete-time monotone dynamical systems in strongly ordered spaces*, Geometry and Topology, Lecture Notes in Mathematics, vol. 1167, (J. Alexander and J. Harer, eds.), 141–153. Springer-Verlag, New York.

[17] Hirsch, M. [1988] *Stability and convergence in strongly monotone dynamical systems.* J. Reine Angew. Math. **383**, 1–58.

[18] Hirsch, M. [1994] *Positive equilibria and convergence in subhomogeneous monotone dynamics*, in Comparison Methods and Stability Theory, Proceedings of the Conference on Differential Equations and Dynamical Systems, Waterloo, June 3-6, 1993, (Xinzhi Liu & David Siegel, eds.), 169–187. Marcel Dekker, New York.

[19] Hirsch, M. [1998] *Chain transitive sets in smooth strongly monotone dynamical systems*, Proceedings of Conference on Differential Equations and Dynamical Systems, Waterloo, August 1–4, 1997, Dynamics of Discrete, Continuous and Impulsive Systems, in press.

[20] Hirsch, M. [1998a] *Attractors and periodic orbits for strongly order-preserving maps*, preprint.

[21] Jiang, J.-F. [1991] *Attractors for strongly monotone flows*, J. Math. Anal. Appl. **162**, 210–222.

[22] Kunze, H. and Siegel, D. [1994] *A graph theoretic approach to monotonicity with respect to initial conditions*, in Comparison Methods and Stability Theory, Proceedings of the Conference on Differential Equations and Dynamical Systems, Waterloo, June 3-6, 1993, (Xinzhi Liu & David Siegel, eds.), 169–187. Marcel Dekker, New York.

[23] Kushner, H. and Clark, D. [1978] *Stochastic Approximation Methods for Constrained and Unconstrained Systems.* Springer-Verlag, New York.

[24] Lajmanovich, A. and Yorke, J. [1976] *A deterministic model for gonorrhea in a nonhomogeneous population*, Math. Biosciences **28**, 221–236.

[25] Ljung, L. [1977] *Analysis of recursive stochastic algorithms*, IEEE Trans. Automatic Control **AC22**, 698–712.

[26] Robbins, H. and Monro, S. [1951] *A stochastic approximation method*, Ann. Math. Statistics **22**, 400–407.

[27] Selgrade, J. [1980] *Asymptotic behavior of solutions to single loop positive feedback systems*, J. Diff. Eqns. **38**, 80–103.

[28] Smith, H. [1995] *Monotone Dynamical Systems*, Math. Surveys and Monographs, vol. 41, Amer. Math. Soc., Providence, Rhode Island.

[29] Smith, H. and Thieme, H. [1991] *Convergence for strongly order preserving semiflows*, SIAM J. Math. Anal. **22**, 1081-1101.

[30] Takáč, P. [1990] *Asymptotic behavior of discrete-time semigroups of sublinear, strongly increasing mappings with applications to biology*, Nonlinear Analysis **14**, 35–42.

[31] Walter, W. [1997] *On strongly monotone flows*, Annales Polonici Mathematici **LXVI**, 269-274.

Fields Institute Communications
Volume **21**, 1999

# General Recruitment Models for Sexually Transmitted Diseases

**Fred Brauer**
Current address: Department of Mathematics
University of Wisconsin
Madison, Wisconsin 53706, USA
brauer@math.wisc.edu
Department of Mathematics
University of British Columbia
Vancouver, B. C., Canada V6T 1Z2
brauer@math.ubc.ca

**Abstract.** We consider a variety of general models for sexually transmitted diseases in a small active core population with recruitment from a large inactive metapopulation. We include diseases with recovery without immunity and with removal through immunity, allowing disease-related deaths, and also universally fatal diseases. We are particularly interested in the possibility of unstable endemic equilibria, with sustained oscillations.

## 1 Introduction

In the classical basic models for the spread of communicable diseases [**6**], [**7**], and [**8**] it is assumed that the mixing of members of the population being studied is homogeneous. In practice this frequently is not so, and it is reasonable to incorporate additional structure in a model. This is especially significant in the study of sexually transmitted diseases, where the recognition of differences in the levels of sexual activity of different members is an essential ingredient in modelling. One approach is to divide the population into subgroups, each with a different level of sexual activity. This was the approach taken by Hethcote and Yorke in modelling gonorrhea [**5**], and their deduction that a small very active "core" group could be responsible for most of the infections was used to good effect in combatting the spread of the disease by concentrating efforts on treatment and education of core members.

We shall study models which allow different levels of sexual activity but are as simple as possible. We postulate that there is a "core" group, presumably small, with a very high level of sexual activity, and a "non-core" group, presumably large, with a much lower level of sexual activity. We shall confine our attention to models

1991 *Mathematics Subject Classification.* 92D30.

in which the non-core is completely inactive. A full model would allow sexual activity by non-core members and would require assumptions not only on the activity levels of members of the two groups but also on the pattern of mixing between the groups. Different mixing patterns would lead to different models, possibly exhibiting different qualitative behaviours. Two extreme patterns are where members mix only with members of the same group and where members mix only with members of the other group. Our models, with inactive non-core members, will be limiting cases of full models as the non-core activity level approaches zero. It has been shown that the qualitative behaviour of a limiting model approximates the qualitative behaviour of every full model, independent of the mixing pattern, provided the non-core activity level is sufficiently small [**2**].

To this structure we shall add a recruitment between classes in which the rate of recruitment of non-core members into the core is a decreasing function of infection prevalence. The rationale for such an assumption is that individuals are less likely to become highly sexually active if they perceive that such behaviour is dangerous. Studies of HIV/AIDS have frequently observed such behaviour changes [**3**], [**9**], and [**10**]. We assume also that there is a rate of departure from core to non-core proportional to core population size; this has the effect of setting a mean period of high sexual activity.

In the formulation of our models we will need epidemiological assumptions (concerning the rate of acquisition of infection, the rates of recovery and death caused by the disease, and whether recovered infectives are immune against re-infection), internal demographic assumptions (concerning the rates of recruitment from non-core to core and from core to non-core), and external demographic assumptions (concerning maturation into and departure from the population of potentially sexually active members).

Our goals in analyzing the various models are to identify the basic reproductive number $R_0$ and show that if $R_0 < 1$ there is a globally asymptotically stable equilibrium while if $R_0 > 1$ there is an endemic equilibrium which may be (locally) asymptotically stable but which may in some cases be unstable, leading to a Hopf bifurcation and oscillatory solutions.

There are obvious possibilities for generalization which we shall not consider here. These include an exposed period between acquiring infection and becoming infective, recovery with partial or temporary immunity, and two-sex models with heterosexual disease transmission.

## 2 Diseases with no immunity against re-infection (S-I-S models)

To model diseases with no immunity against reinfection, we consider a population of (possibly variable) size $Q$ divided into a core population of size $N$ consisting of $S$ susceptibles and $I$ infectives ($N = S + I$) and a non-core population of size $P = Q - N$ consisting of $T$ susceptibles and $J$ infectives. However, as we shall consider only limit models with no sexual activity or infection in the non-core, we have $J = 0$ and $P = T$. We make the following assumptions:

(external demographics): New members are added to the population (reach sexual maturity) at a rate $g(Q)$ and members leave the population for reasons not related to disease (discontinue sexual activity) at a rate $\alpha Q$ divided proportionally among classes. There is a carrying capacity $K$, with $g(K) = \alpha K$, $g'(K) < \alpha$, $g(Q) > \alpha Q$ for $0 < Q < K$.

(internal demographics): Susceptibles are recruited from non-core to core at a rate $(Q-N)r\left(\frac{I}{N}\right)$, where $r(\eta)$ is a non-negative, non-increasing function on $0 \le \eta \le 1$. Core susceptibles move to the non-core at a rate $\mu S$, so that $\frac{1}{\mu+\alpha}$ is the mean stay in the core.

(epidemiology): Core members make sexual contacts sufficient to transmit infection in unit time, while non-core members are sexually inactive. A fraction $\sigma$ of infectives die of infection and a fraction $\gamma$ of infectives recover and return to the core susceptible class in unit time (from which they may move to the non-core).

These assumptions lead to the model

$$\begin{aligned} S' &= (Q-N)r\left(\frac{I}{N}\right) - \mu S - a\frac{SI}{N} + \gamma I - \alpha S \\ I' &= a\frac{SI}{N} - (\gamma+\sigma)I - \alpha I \\ T' &= -(Q-N)r\left(\frac{I}{N}\right) + g(Q) - \alpha T + \mu S. \end{aligned}$$

It is convenient to express the model in terms of the variables $I$, $N = S + I$, and $Q = T + N$,

$$\begin{aligned} I' &= a\frac{I(N-I)}{N} - (\gamma+\sigma+\alpha)I \\ N' &= (Q-N)r\left(\frac{I}{N}\right) - \mu(N-I) - \sigma I - \alpha N \\ Q' &= g(Q) - \alpha Q - \sigma I. \end{aligned} \tag{2.1}$$

The model (2.1) goes beyond recruitment models studied earlier [**2**] by including disease-related deaths and external demographics.

Equilibria of (2.1) are given by $I = 0$ (disease-free) or $a(N-I) = (\gamma+\sigma+\alpha)N$ (endemic). If $I = 0$ at equilibrium, then $g(Q) = \alpha Q$, which gives $Q = K$, the carrying capacity of the population. In addition, from the second equation in (2.1), we have $(K-N)r(0) = (\mu+\alpha)N$, giving

$$I = 0, N = \frac{Kr(0)}{\mu+\alpha+r(0)}, Q - N = \frac{K(\mu+\alpha)}{\mu+\alpha+r(0)}. \tag{2.2}$$

By linearizing (2.1) about the disease-free equilibrium (2.2) and examining the eigenvalues of the coefficient matrix, it is easy to show that this equilibrium is (locally) asymptotically stable if and only if $a < \gamma+\sigma+\alpha$. Further, if $a < \gamma+\sigma+\alpha$, the first equation in (2.1) gives

$$\frac{I'}{I} = \{a - (\gamma+\sigma+\alpha)\} - \frac{aI}{N} \le \{a - (\gamma+\sigma+\alpha)\}$$

and the solution of this differential inequality shows that $I(t) \to 0$ exponentially, giving global asymptotic stability of the disease-free equilibrium.

The basic reproductive number $R_0$ is given by

$$R_0 = \frac{a}{\gamma+\sigma+\alpha}.$$

If $R_0 > 1$, so that the disease-free equilibrium is unstable, there is also an endemic equilibrium given by

$$a(N-I) = (\gamma+\sigma+\alpha)N.$$

We let $\eta = I/N$, so that $\eta$ is determined by

$$a(1-\eta) = \gamma + \sigma + \alpha.$$

Then

$$\begin{aligned} Qr(\eta) &= N[\mu(1-\eta) + r(\eta) + \eta\sigma + \alpha] \\ g(Q) &= \alpha Q + \sigma I = \alpha Q + \eta\sigma N \end{aligned}$$

determine $N$ and $Q$, with

$$\frac{Q-N}{N} = \frac{\mu(1-\eta) + \eta\sigma + \alpha}{r(\eta)}.$$

Again, linearization and examination of the eigenvalues of the coefficient matrix gives a necessary and sufficient condition

$$-\frac{r'(\eta)}{r(\eta)} < \frac{\alpha + \gamma + \sigma + r(\eta)}{\eta[\alpha + \mu(1-\eta)]} + \frac{1}{\eta(1-\eta)} \tag{2.3}$$

for asymptotic stability of the endemic equilibrium. We omit the details of the calculation as they are straightforward and similar to the calculations in [**2**]. The condition (2.3) is satisfied for $\eta$ sufficiently close to 0 or 1 but may be violated if $-\frac{r'(\eta)}{r(\eta)}$ is large for intermediate values of $\eta$. The quantity $-\frac{r'(\eta)}{r(\eta)}$ describes the response of recruitment to the prevalence of infection. Instability is possible if this response is strong, and numerical simulations indicate that when this occurs there is a periodic orbit whose amplitude may be large and whose period may be several times the mean stay $1/(\mu + \alpha)$ in the core.

For example, with the recruitment function $r(\eta) = 0.5e^{-15\eta}$, the parameter values $a = 6$, $\alpha = 0.05$, $\gamma = 2$, $\mu = 1$, $\sigma = 1$ and the external recruitment function $g(Q) = 5$, the equilibrium population size is $Q = 100$. We observe orbits with period approximately 10.5 (compared to a mean stay of 0.95 in the core and a mean lifetime in the population of 20). In these oscillations, the core population size varies from almost zero to 27 and the number of infectives in the core varies from almost zero to 5.5. The basic reproductive number is 1.97 and the equilibrium value of the core population size is approximately 0.005.

A periodic oscillation of large period and amplitude means that experimental data is meaningless unless extended over a long time, suggesting that instability of an endemic equilibrium is a situation to be avoided if possible. Our model suggests that oscillations occur only if $-r'(\eta)/r(\eta)$ is large, that is, if prevalence of infection in the core is a very strong influence in discouraging recruitment into the core. A counter-intuitive conclusion is that a strong campaign to discourage a high level of sexual activity may make the disease harder to control.

If there are no deaths caused by the disease, $\sigma = 0$, the equation for $Q$ in the model (2.1) is $Q' = g(Q) - \alpha Q$, uncoupled from the other two equations. In this case, $Q(t)$ always approaches $K$, the system is asymptotically autonomous, and may be reduced to the two-dimensional model

$$\begin{aligned} I' &= a\frac{I(N-I)}{N} - (\gamma + \alpha)I \\ N' &= (K-N)r\left(\frac{I}{N}\right) - \mu(N-I) - \alpha N. \end{aligned} \tag{2.4}$$

Another special case of possible interest is the case $\sigma = 0$, $\gamma = 0$ with no deaths caused by disease and no recovery. This S-I model is

$$\begin{aligned} I' &= a\frac{I(N-I)}{N} - \alpha I \\ N' &= (K-N)r\left(\frac{I}{N}\right) - \mu(N-I) - \alpha N. \end{aligned} \tag{2.5}$$

As the models (2.4) and (2.5) are special cases of (2.1), both may exhibit sustained oscillations if $R_0 > 1$.

## 3 Diseases with removal (S-I-R models)

There are two types of S-I-R models, depending on whether the removed class $R$ represents individuals who have been removed by death due to the disease or individuals who have been removed from epidemiological interactions through recovery with immunity against re-infection. If the disease is universally fatal, the structure constructed in the previous section is appropriate. The model is just the special case $\gamma = 0$ of (2.1), and we again obtain the possibility of oscillation about an unstable endemic equilibrium.

If some individuals recover from infection, we must incorporate a removed class $R$ in the core and a removed class $G$ in the non-core coming from back-recruitment of recovered core members. The resulting model, using the variables $I$, $R$, $N$, $G$, $Q$, with $S = N - I - R$, $T = P - G = Q - N - G$, is

$$\begin{aligned} I' &= a\frac{I(N-I)}{N} - (\gamma+\sigma+\alpha)I \\ R' &= \gamma I - (\alpha+\mu)R \\ N' &= (Q-N-G)r\left(\frac{I}{N}\right) - \mu(N-I) - \alpha N - \sigma I \\ G' &= \mu R - \alpha G \\ Q' &= g(Q) - \alpha Q - \sigma I. \end{aligned} \tag{3.1}$$

For the model (3.1), it is easy to see that the disease-free equilibrium $I = 0$, $R = 0$,

$$N = \frac{Kr(0)}{\mu+\alpha+r(0)}, G = 0, Q = N = \frac{K(\mu+\alpha)}{\mu+\alpha+r(0)}$$

is asymptotically stable if and only if

$$R_0 = \frac{a}{\gamma+\alpha+\sigma} < 1.$$

If $R_0 > 1$ there is an endemic equilibrium whose asymptotic stability is judged by determining the location of the roots of a fifth degree polynomial equation. We have not been successful in carrying out this analysis in general, but we conjecture that instability of the endemic equilibrium is possible if the response of recruitment to prevalence of infection in the core is strong, just as for S-I-S and fatal S-I-R models.

## 4 Conclusions

We have seen that for core-noncore recruitment models of sexually transmitted diseases, it is possible for the endemic equilibrium to be unstable if recruitment is affected sufficiently strongly by the fraction of the core population which is infective. This instability is manifested in oscillations of large amplitude and a long period, which would both make predictions of disease levels very difficult and strain the resources available for treatment and education. Thus such instability is to be avoided if possible, suggesting that campaigns to discourage risky behaviour could be dangerous if they are too successful.

If we had formulated our models with recruitment depending on the number of infectives in the core rather than on the fraction of the core which is infected, we would have found that the endemic equilibrium is always asymptotically stable. This would indicate that educational campaigns should concentrate on numbers of infectives and not on the fraction which is infected.

However, there is another side to this question, suggesting that the proper policy is not obvious. It is reasonable to assume that the response of recruitment to infection is not instantaneous, but that there is a time-lag. It has been seen ([**1**, **4**]) that a time-lag in recruitment tends to stabilize an equilibrium if recruitment depends on the infective fraction in the core and to destabilize an equilibrium if recruitment depends on the number of infectives in the core. Thus it is not at all clear whether a campaign to warn members of the dangers of high levels of sexual activity should focus more on the number or the fraction of infectives.

## References

[1] Brauer, F. [1997], *Time lag in disease models with recruitment*, Mathematical and Computer Modelling.

[2] Brauer, F., Castillo-Chavez, C., Velasco-Hernandez, J. [1997], *Recruitment into a core group and its effect on the spread of a sexually transmitted disease*, Proceedings of Fourth International Conference on Mathematical Population Dynamics, (Houston, 1995).

[3] Centers for Disease Control [1985], *Self-reported behavioral change among gay and bisexual men*, San Francisco MMWR **34**, 613–615.

[4] Heiderich, K. R., Huang, W., Castello-Chavez, C. [1994], *Nonlocal response in a simple epidemiological model*, Cornell University Mathematical Sciences Institute **TR** 94–29.

[5] Hethcote, H. W., Yorke, J. A. [1984], *Gonorrhea: transmission dynamics and control*, Springer-Verlag, Lecture Notes in Biomathematics **56**.

[6] Kermack, W. O., McKendrick, A. G. [1927], *A contribution to the mathematical theory of epidemics*, Proc. Royal Soc. London **A 115**, 200-721.

[7] Kermack, W. O., McKendrick, A. G. [1932], *Contributions to the mathematical theory of epidemics*, Part II, Proc. Royal Soc. London **A 138**, 55–83.

[8] Kermack, W. O., McKendrick, A. G. [1933], *Contributions to the mathematical theory of epidemics*, Part III, Proc. Royal Soc. London **A 141**, 94–122.

[9] McKusick, L., Horstman, W., Coates, T. J. [1987], *The impact of AIDS in gay male sexual behavior patterns in New York City*, Am. J. Public Health **77**, 578–581.

[10] Wilkenstein, J. A. Jr. et al. [1988], *The San Francisco Men's Health Study, continued decline in HIV seroconversion rates among homosexual/bisexual men*, Am. J. Public Health **78**, 1472–1474.

Fields Institute Communications
Volume **21**, 1999

# Asymptotic Behavior of Two Interacting Pioneer/Climax Species

**J. Robert Buchanan**
Mathematics Department
Millersville University
P.O. Box 1002, Millersville, PA 17551–0302, USA
jbuchana@marauder.millersv.edu

**Abstract.** The long-term behaviors of Kolmogorov systems of two interacting species with pioneer or climax fitnesses are studied. Simple inequalities involving the parameters of the system define nine equivalence classes on the nullclines of the system. These inequalities completely determine the asymptotic behavior of orbits in seven of the equivalence classes. All $\omega$-limit sets in these classes are isolated equilibria. Hopf bifurcations are possible in the remaining two equivalence classes. Conditions under which Hopf bifurcations can take place are described. In those cases, closed nontrivial periodic orbits and heteroclinic loops are possible.

## 1 Introduction

Explicit in many ecological models is the assumption that the growth rate of a population is proportional to the size of the population. The proportionality factor is called the *per capita growth rate* or *fitness function*. In this article we will discuss the asymptotic behavior of two-species interactions where the fitness functions are described as *pioneer* or *climax* fitnesses. A pioneer fitness function is a monotone decreasing function of population density and models the detrimental effects of crowding on reproduction and survival. Ricker in [**11**] concludes that certain fish populations have exponential pioneer fitnesses, and Hassell and Comins in [**9**] study pioneer fitnesses which are rational functions. We will assume a pioneer fitness function is a $\mathcal{C}^1$ function $f_1(z)$, which is monotonically decreasing and for which $f_1(0) > 0$ and there exists a (unique) $z_1 > 0$ such that $f_1(z_1) = 0$ and $f_1'(z_1) < 0$.

1991 *Mathematics Subject Classification.* Primary 92D25; Secondary 34C23.
Support for attending and presenting at this conference was provided by the Millersville University Faculty Grants Committee and Dr. Albert Hoffman, Dean of the School of Science and Mathematics and Dr. Charles Denlinger, Chair of the Department of Mathematics and Statistics.

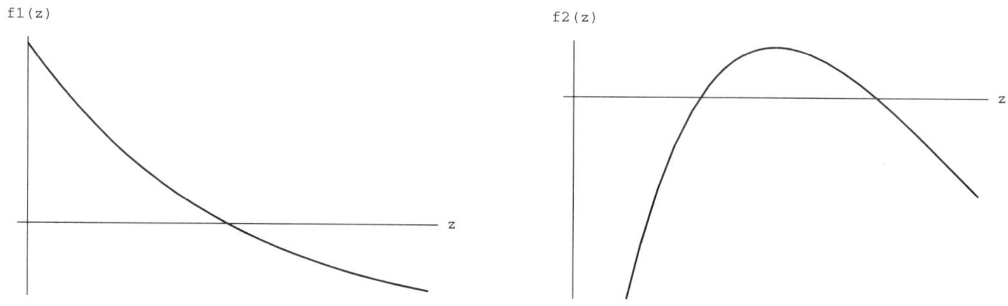


(a) pioneer (b) climax

**Figure 1** Plots of typical pioneer and climax fitness functions.

By contrast reproduction and survival rates of some species have been observed to increase in an interval of population densities. This may be due to the benefits of group defense or environmental conditioning resulting from higher densities. Extreme overcrowding remains detrimental to their reproductive rates, therefore these species exhibit a fitness function possessing a unique maximum. Per capita reproduction or survival rates with this feature are called climax fitnesses. ([**3**], [**12**], [**14**], [**16**], and [**13**]) discuss examples of climax fitness functions. A climax fitness function is a $\mathcal{C}^1$ function $f_2(z)$, such that $f_2(0) < 0$. The function has two positive roots $z_{2+} < z_{2-}$ with $f_2'(z_{2-}) < 0 < f_2'(z_{2+})$. For densities in the interval $(z_{2+}, z_{2-})$ the climax fitness function is positive, otherwise it is negative. Typical examples of pioneer and climax fitness functions are shown in Figure 1.

Consider the interaction between a pioneer species, $x_1$, and a climax species, $x_2$. A system describing this interaction can be written as

$$\begin{aligned} \frac{dx_1}{dt} &= x_1 f_1(c_{11}x_1 + c_{12}x_2) \\ \frac{dx_2}{dt} &= x_2 f_2(c_{21}x_1 + c_{22}x_2), \end{aligned} \tag{1.1}$$

where $c_{ij} \geq 0$. The matrix $C = (c_{ij})_{i,j=1}^2$ is called the *interaction matrix*. The parameter $c_{ij}$ specifies the intensity of the interaction of species $j$ on species $i$. The arguments to the fitness functions are chosen to be weighted densities of the two populations. This choice separates a population's response to density from the individual effects of each population simplifying the qualitative analysis of the behavior of solutions. This same assumption has been used in [**7, 8, 19, 18**] to study Hopf bifurcations arising in pioneer/climax systems and in [**10**] in the context of discrete systems. If $c_{12}$ and $c_{21}$ are positive then we may introduce two scaled variables

$$\tilde{x}_1 = c_{21}x_1 \text{ and } \tilde{x}_2 = c_{12}x_2.$$

Substituting into (1.1) yields

$$\begin{aligned} \frac{d\tilde{x}_1}{dt} &= \tilde{x}_1 f_1(\tilde{c}_{11}\tilde{x}_1 + \tilde{x}_2) \\ \frac{d\tilde{x}_2}{dt} &= \tilde{x}_2 f_2(\tilde{x}_1 + \tilde{c}_{22}\tilde{x}_2), \end{aligned} \tag{1.2}$$

where $\tilde{c}_{11} = c_{11}/c_{21}$ and $\tilde{c}_{22} = c_{22}/c_{12}$. We prefer to work with the scaled form of the equations and to keep the notation simple, the tildes over the variables and parameters will be dropped.

## 2 Equivalence classes

In this article we will describe the asymptotic behavior of two-dimensional systems of pioneer and climax species interactions. The approach taken and terminology used to achieve this description are similar to those in [**17**]. Asymptotic behavior in each of several equivalence classes is described in terms of the geometry of the nullclines of (1.2). The equivalence classes are characterized by algebraically invariant quantities of the system.

For the remainder of this article $\mathbb{R}^2_+$ will denote the quadrant of $\mathbb{R}^2$ where both the coordinates of a point are nonnegative. Let $PC$ denote the set of $\mathcal{C}^1$ vector fields on $\mathbb{R}^2_+$ such that if $F \in PC$ then $F_1(x) = x_1 f_1(c_{11}x_1 + x_2)$ and $f_1(z)$ is a pioneer fitness function, while $F_2(x) = x_2 f_2(x_1 + c_{22}x_2)$ and $f_2(z)$ is a climax fitness function as described in (1.2). If $F, G \in PC$ then $F$ and $G$ are said to be *topologically equivalent in* $PC$ if there is a homeomorphism of $\mathbb{R}^2_+$ which maps the orbits of $F$ onto the orbits of $G$ while preserving orientation. This defines an equivalence relation on $PC$. The distance $\|F - G\|$, between two vector fields is defined to be

$$\|F - G\| = \sup_{x \in \mathbb{R}^2_+} \{|F - G|, |DF - DG|\}.$$

An $\epsilon$-neighborhood of $F \in PC$ is

$$B_\epsilon(F) = \{G \in PC \,|\, \|F - G\| < \epsilon\}.$$

These open sets form a basis for a topology on $PC$. A vector field $F$, is *structurally stable in* $PC$ provided it has an open neighborhood of topological equivalents [**4**, p. 143].

By the form of (1.2), $\mathbb{R}^2_+$ is invariant. Furthermore each of the interior and the boundary of this set is invariant. For the remainder of this article we will assume that $x_i \geq 0$ for $i = 1, 2$. The study of the dynamical behavior of the orbits of $F \in PC$ is facilitated by an analysis of the nullclines of $F$. The pioneer species has a single nullcline in $\mathbb{R}^2_+$ while the climax species possesses two parallel nullclines. All the non-axial nullclines are lines of the form

$$c_{ii}x_i + x_j = z_i, \quad i \neq j.$$

The $x_i$ axial intercept of this nullcline has $x_i$ coordinate $z_i/c_{ii} > 0$ and is an equilibrium point of (1.2). The $x_j$ axial intercept of this nullcline has $x_j$ coordinate $z_i > 0$ and in general is not an equilibrium point. Since there are a total of three nullclines in $\mathbb{R}^2_+$, there are three intercepts of the nullclines with each coordinate axis. If $F \in PC$ the nullcline configuration of $F$ is determined by the following four

**Table 1** The signs of the algebraic expressions characterizing the nine stable nullcline classes are listed in columns one through four. These four quantities are in some cases sufficient to determine the sign of the determinant of the interaction matrix, $C$. The last column contains the number of a figure representing a typical member of each nullcline class. The order of presentation of the figures reflects increasing numbers of interior equilibria.

| $\frac{z_1}{c_{11}} - z_{2+}$ | $\frac{z_1}{c_{11}} - z_{2-}$ | $\frac{z_{2+}}{c_{22}} - z_1$ | $\frac{z_{2-}}{c_{22}} - z_1$ | $\det C$ | Figure No. |
|---|---|---|---|---|---|
| $-$ | $-$ | $+$ | $+$ | $\pm$ | 2 |
| $+$ | $-$ | $-$ | $+$ | $\pm$ | 3 |
| $+$ | $+$ | $-$ | $-$ | $\pm$ | 4 |
| $+$ | $+$ | $-$ | $+$ | $-$ | 5 |
| $+$ | $-$ | $-$ | $-$ | $+$ | 6 |
| $+$ | $-$ | $+$ | $+$ | $-$ | 7 |
| $-$ | $-$ | $-$ | $+$ | $+$ | 8 |
| $+$ | $+$ | $+$ | $+$ | $-$ | 9 |
| $-$ | $-$ | $-$ | $-$ | $+$ | 10 |

quantities:

$$\operatorname{sgn}\left(\frac{z_1}{c_{11}} - z_{2+}\right), \quad \operatorname{sgn}\left(\frac{z_1}{c_{11}} - z_{2-}\right), \quad \operatorname{sgn}\left(\frac{z_{2+}}{c_{22}} - z_1\right), \quad \operatorname{sgn}\left(\frac{z_{2-}}{c_{22}} - z_1\right).$$

If $F, G \in PC$ then $F$ and $G$ are *nullcline equivalent* if and only if they have the same nullcline configuration. This defines another equivalence relation on $PC$. $F \in PC$ is *nullcline stable* provided it has an open neighborhood of nullcline equivalents in $PC$. It is evident that $F$ is nullcline stable if and only if all of the following inequalities are true:

$$\operatorname{sgn}\left(\frac{z_1}{c_{11}} - z_{2+}\right) \neq 0, \operatorname{sgn}\left(\frac{z_1}{c_{11}} - z_{2-}\right) \neq 0, \operatorname{sgn}\left(\frac{z_{2+}}{c_{22}} - z_1\right) \neq 0, \operatorname{sgn}\left(\frac{z_{2-}}{c_{22}} - z_1\right) \neq 0.$$

**Lemma 2.1** *There are nine stable nullcline classes of $PC$ and they have open dense union in $PC$.*

**Proof** According to the assumptions on the zeroes of the climax fitness function $z_{2+} < z_{2-}$ and thus there are only three allowable permutations of the distinct nullcline axial intercepts on each coordinate axis. Hence there are $3^2$ stable nullcline classes. Each stable nullcline class (and hence their union) is open by its definition. That their union is dense in $PC$ also follows from the definition of nullcline stability. □

The stable nullcline classes are summarized in Table 1. We will show below that seven of the nine stable nullcline classes are structurally stable topological classes. The remaining two classes where the dynamics are not determined by the nullcline configuration are the ones in which a Hopf bifurcation can take place.

The linearization of (1.2) about a point $(x_1, x_2)$ is

$$DF(x_1, x_2) = \begin{pmatrix} f_1(y_1) & 0 \\ 0 & f_2(y_2) \end{pmatrix} + \begin{pmatrix} x_1 f_1'(y_1) & 0 \\ 0 & x_2 f_2'(y_2) \end{pmatrix} C. \tag{2.1}$$

Note that $y_1 = c_{11}x_1 + x_2$ and $y_2 = x_1 + c_{22}x_2$. If the Euclidean norm of $(x_1, x_2)$ is large enough that $y_1 > z_1$ and $y_2 > z_{2-}$ then $DF(x_1, x_2)_{ij} < 0$ for $i \neq j$ and thus the system is competitive in this region. By Kamke's comparison principle [**1**] flow in this unbounded region is monotone. Furthermore, in that region $dx_i/dt < 0$ for $i = 1, 2$ and hence the system is dissipative.

Since the coordinate axes are invariant for (1.2) the stable and unstable manifolds of axial equilibria lie along the corresponding axes. We will assume all the equilibria on the coordinates axes are nondegenerate and thus we can use (2.1) to determine their local stability properties and the tangent vectors to their stable and unstable manifolds. The equilibria and their associated eigenvalues $\lambda_i$, and eigenvectors $v_i$, are listed below.

$$(0,0) \quad : \quad \begin{cases} \lambda_1 = f_1(0), \quad v_1 = (1,0)^t \\ \lambda_2 = f_2(0), \quad v_2 = (0,1)^t, \end{cases} \tag{2.2}$$

$$\left(\frac{z_1}{c_{11}},0\right) \quad : \quad \begin{cases} \lambda_1 = f_2\left(\frac{z_1}{c_{11}}\right), \quad v_1 = \left(\frac{z_1 f_1'(z_1)}{c_{11}(f_2(\frac{z_1}{c_{11}}) - z_1 f_1'(z_1))}, 1\right)^t \\ \lambda_2 = z_1 f_1'(z_1), \qquad v_2 = (1,0)^t, \end{cases} \tag{2.3}$$

$$\left(0,\frac{z_{2\pm}}{c_{22}}\right) \quad : \quad \begin{cases} \lambda_1 = f_1\left(\frac{z_{2\pm}}{c_{22}}\right), \quad v_1 = \left(\frac{c_{22}(f_1(\frac{z_{2\pm}}{c_{22}}) - z_{2\pm} f_2'(z_{2\pm}))}{z_{2\pm} f_2'(z_{2\pm})}, 1\right)^t \\ \lambda_2 = z_{2\pm} f_2'(z_{2\pm}), \quad v_2 = (0,1)^t. \end{cases} \tag{2.4}$$

Thus by (2.2) the origin is always a saddle point with stable manifold along the $x_2$ axis and unstable manifold along the $x_1$ axis. Consider a small positive number $\epsilon$ such that

$$0 < \epsilon < \min\{z_1, z_1/c_{11}, z_{2+}, z_{2+}/c_{22}\}.$$

Call $Q_\epsilon$ the region where $y_1 \leq z_1 - \epsilon$ and $y_2 \leq z_{2+} - \epsilon$. $Q_\epsilon$ contains exactly one fixed point, the origin, which is a saddle point. The Stable Manifold Theorem [**6**] implies that only solutions on the $x_2$ axis approach the origin. Thus by the Poincaré-Bendixson theorem, no orbit passing through the interior of $Q_\epsilon$ remains in that region for all future time.

According to (2.3) the equilibrium located at $(z_1/c_{11}, 0)$ attracts along the $x_1$ axis. The complete determination of this equilibrium's stability properties depends on the size of $z_1/c_{11}$. The eigenvalue, $\lambda_1$ which equals $f_2(z_1/c_{11})$ will be negative whenever $z_1/c_{11} < z_{2+}$ or $z_1/c_{11} > z_{2-}$. Thus $(z_1/c_{11}, 0)$ is asymptotically stable if

$$\operatorname{sgn}\left(\frac{z_1}{c_{11}} - z_{2+}\right) = -1 \text{ or } \operatorname{sgn}\left(\frac{z_1}{c_{11}} - z_{2-}\right) = +1$$

(see Figures 2 and 4) and is a saddle point otherwise (see Figure 3).

There are two equilibria located on the positive part of the $x_2$ axis. The smaller, $(0, z_{2+}/c_{22})$, repels along the $x_2$ axis while the larger, $(0, z_{2-}/c_{22})$, attracts along the $x_2$ axis. As in the previous case, a complete determination of the stability properties of these equilibria depends on the sizes the quantities $z_{2\pm}/c_{22}$ relative to $z_1$. If $z_{2+}/c_{22} < z_1$ then the eigenvalue $\lambda_1$ associated with $(0, z_{2+}/c_{22})$ is positive while if $z_{2+}/c_{22} > z_1$ the eigenvalue $\lambda_1$ is negative. Thus $(0, z_{2+}/c_{22})$ is an unstable source if $\operatorname{sgn}(z_{2+}/c_{22} - z_1) = -1$ (see Figures 3 and 4) and otherwise is a saddle point (see Figure 2). The stability properties of the two equilibria on the $x_2$ axis are related through the inequality $z_{2-} > z_{2+}$. If $(0, z_{2+}/c_{22})$ is a saddle point, $(0, z_{2-}/c_{22})$ is asymptotically stable. If $z_{2-}/c_{22} < z_1$ the equilibrium $(0, z_{2-}/c_{22})$ is a saddle point and thus $(0, z_{2+}/c_{22})$ must be repelling (see Figure 4).

The next lemma summarizes the conditions under which non-axial, isolated equilibria are present.

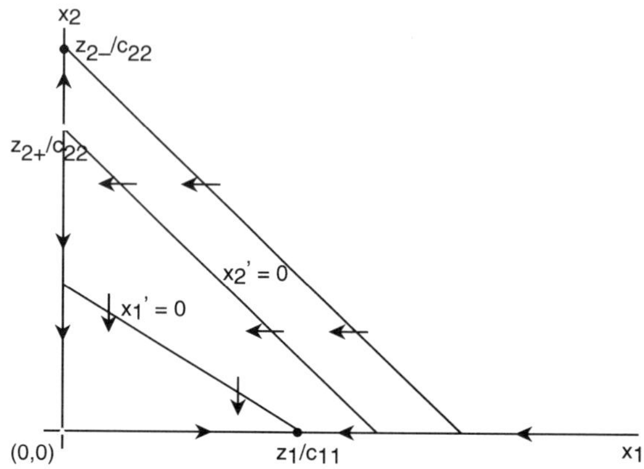


**Figure 2** $\frac{z_1}{c_{11}} < z_{2+}$, $\frac{z_1}{c_{11}} < z_{2-}$, $\frac{z_{2+}}{c_{22}} > z_1$, $\frac{z_{2-}}{c_{22}} > z_1$

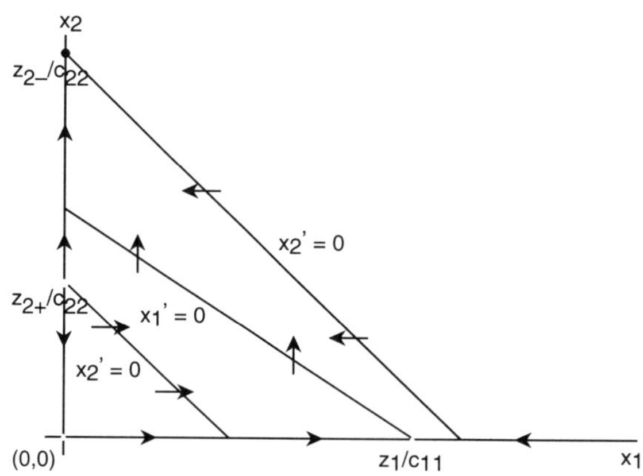


**Figure 3** $\frac{z_1}{c_{11}} > z_{2+}$, $\frac{z_1}{c_{11}} < z_{2-}$, $\frac{z_{2+}}{c_{22}} < z_1$, $\frac{z_{2-}}{c_{22}} > z_1$

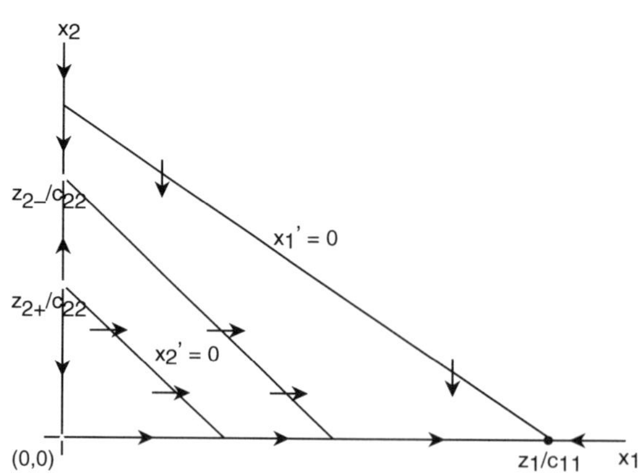


**Figure 4** $\frac{z_1}{c_{11}} > z_{2+}$, $\frac{z_1}{c_{11}} > z_{2-}$, $\frac{z_{2+}}{c_{22}} < z_1$, $\frac{z_{2-}}{c_{22}} < z_1$

**Lemma 2.2** *If* $\det C \neq 0$ *and if*

$$\operatorname{sgn}\left(\frac{z_1}{c_{11}} - z_{2\pm}\right) = \operatorname{sgn}\left(\frac{z_{2\pm}}{c_{22}} - z_1\right) \neq \operatorname{sgn}(\det C), \tag{2.5}$$

*then systems of the type represented by equation (1.2) have isolated equilibria in the interior of* $\mathbb{R}^2_+$.

**Proof** If $\det C \neq 0$ then the non-axial equilibria are isolated. Cramer's rule gives the coordinates of the equilibria as

$$(e_1, e_2) = \left(\frac{c_{22}z_1 - z_{2\pm}}{c_{11}c_{22} - 1}, \frac{c_{11}z_{2\pm} - z_1}{c_{11}c_{22} - 1}\right) = \left(\frac{-c_{22}\left(\frac{z_{2\pm}}{c_{22}} - z_1\right)}{\det C}, \frac{-c_{11}\left(\frac{z_1}{c_{11}} - z_{2\pm}\right)}{\det C}\right). \tag{2.6}$$

Since $c_{11}$ and $c_{22}$ are both assumed positive then $(e_1, e_2) \in \mathbb{R}^2_+$ if the $\operatorname{sgn}(\frac{z_i}{c_{ii}} - z_j)$'s agree for $i, j = 1, 2$ with $i \neq j$ and differ from $\det C$. □

Table 1 includes a column for the $\operatorname{sgn}(\det C)$ for those nullcline classes permitting interior fixed points. As will be shown below, the $\operatorname{sgn}(\det C)$ determines which nullcline classes can permit a Hopf bifurcation.

According to (2.5) there can only fail to exist a non-axial fixed point if $\operatorname{sgn}(z_i/c_{ii} - z_j) \neq \operatorname{sgn}(z_j/c_{jj} - z_i)$ for $i \neq j$. This implies that at least one of the equilibria, $(z_1/c_{11}, 0)$ or $(0, z_{2-}/c_{22})$, is always a sink in the absence of a fixed point in $\mathbb{R}^2_+$ (see Figures 2, 3, 4). If $(z_1/c_{11}, 0)$ is a saddle point then necessarily $(0, z_{2-}/c_{22})$ is a sink (see Figure 3). Furthermore if $(z_1/c_{11}, 0)$ is a saddle point then $\operatorname{sgn}(z_1/c_{11} - z_{2+}) = 1$ and the assumption of no interior fixed points implies that $\operatorname{sgn}(z_{2+}/c_{22} - z_1) = -1$, *i.e.* $(0, z_{2+}/c_{22})$ is a source. Thus for all $x \in \mathbb{R}^2_+$, $\omega(x) = (0, z_{2-}/c_{22})$. If $(0, z_{2-}/c_{22})$ is a saddle point then necessarily $(z_1/c_{11}, 0)$ is a sink and also $(0, z_{2+}/c_{22})$ is a source (see Figure 4). Hence for all $x \in \mathbb{R}^2_+$, $\omega(x) = (z_1/c_{11}, 0)$. If both $(z_1/c_{11}, 0)$ and $(0, z_{2-}/c_{22})$ are sinks then $\operatorname{sgn}(z_{2-}/c_{22} - z_1) = 1$ and $\operatorname{sgn}(z_1/c_{11} - z_{2+}) = -1$ (see Figure 2). The assumption of no interior equilibria implies that $\operatorname{sgn}(z_{2+}/c_{22} - z_1) = 1$ and hence $(0, z_{2+}/c_{22})$ is a saddle point. The stable manifold of $(0, z_{2+}/c_{22})$ separates the basins of attraction of the two sinks.

Table 2 summarizes the stability properties of the fixed points of systems of the form shown in (1.2).

**Table 2** The origin is a saddle point in every case and thus not included in the table. In the vector field diagrams the stability properties of a fixed point are indicated by the symbols: • sink, ○ source, △ saddle.

| $\left(\frac{z_1}{c_{11}}, 0\right)$ | $\left(0, \frac{z_{2+}}{c_{22}}\right)$ | $\left(0, \frac{z_{2-}}{c_{22}}\right)$ | $(e_{1+}, e_{2+})$ | $(e_{1-}, e_{2-})$ | Figure No. |
|---|---|---|---|---|---|
| sink | saddle | sink | | | 2 |
| saddle | source | sink | | | 3 |
| sink | source | saddle | | | 4 |
| sink | source | sink | | saddle | 5 |
| saddle | source | saddle | | sink | 6 |
| saddle | saddle | sink | Hopf | | 7 |
| sink | source | sink | saddle | | 8 |
| sink | saddle | sink | Hopf | saddle | 9 |
| sink | source | saddle | saddle | sink | 10 |

## 3 Hopf bifurcations

Since the coordinate axes are invariant, a nontrivial periodic solution must lie in the interior of $\mathbb{R}^2_+$ and will contain a non-axial fixed point. Now we determine some conditions necessary for a Hopf bifurcation. This requires examining the real parts of the eigenvalues of the linearized system at $(e_1, e_2)$. For an interior equilibrium the characteristic polynomial of $DF(e_1, e_2)$ is

$$\lambda^2 - (\operatorname{tr} DF(e_1, e_2))\lambda + \det DF(e_1, e_2).$$

Since a quadratic polynomial with a complex conjugate pair of roots can be factored as

$$(\lambda - (\alpha + i\beta))(\lambda - (\alpha - i\beta)) = \lambda^2 - 2\alpha\lambda + \alpha^2 + \beta^2,$$

where $\alpha$ and $\beta$ are real numbers. Then by equating powers of $\lambda$ we see that

$$2\alpha = \operatorname{tr} DF(e_1, e_2) = \frac{-c_{11}c_{22}}{\det C}\Big(\big(\frac{z_2}{c_{22}} - z_1\big)f_1'(z_1) + \big(\frac{z_1}{c_{11}} - z_2\big)f_2'(z_2)\Big) \tag{3.1}$$

$$\alpha^2 + \beta^2 = \det DF(e_1, e_2) = \frac{c_{11}c_{22}}{\det C}\big(\frac{z_1}{c_{11}} - z_2\big)\big(\frac{z_2}{c_{22}} - z_1\big)f_1'(z_1)f_2'(z_2). \tag{3.2}$$

A necessary condition for a Hopf bifurcation to occur is that as a parameter in the system is varied, the real parts of the complex conjugate pair of eigenvalues must pass through the imaginary axis while the imaginary parts are nonzero, i.e., we must have $\operatorname{tr} DF(e_1, e_2) = 0$ while $\det DF(e_1, e_2) > 0$. These constraints allow us to prove the following two lemmas (similar results are found in [**7, 18**]).

**Lemma 3.1** *A necessary condition for a Hopf bifurcation to occur about an interior equilibrium $(e_1, e_2)$ is that $(e_1, e_2)$ lie at the intersection of nullclines where $f_2'(z_2) > 0$.*

**Proof** If $(e_1, e_2)$ lies at the intersection of two nullclines where $f_2'(z_2) < 0$ then by (2.5) and (3.1), $\operatorname{tr} DF(e_1, e_2) \neq 0$ for all parameter values. Hence no Hopf bifurcation can take place. □

**Lemma 3.2** *If $\det C > 0$ then no Hopf bifurcation can occur around any interior equilibrium.*

**Proof** By Lemma 3.1 we know Hopf bifurcations cannot occur in vector fields in $PC$ around equilibria which lie at the intersections of two nullclines where $f_2'(z_2) < 0$. Hence if $f_2'(z_2) < 0$ and $(e_1, e_2) \in \mathbb{R}^2_+$, then by (2.5) and (3.2), $\det DF(e_1, e_2) < 0$. Therefore no Hopf bifurcation can take place. □

Thus if $\det C < 0$ and $f_2'(z_2) > 0$ then a Hopf bifurcation may take place. Upon setting the right–hand side of (3.1) equal to zero we may formally solve for either $c_{11}$ or $c_{22}$. The solutions are

$$\hat{c}_{ii} = \frac{z_i f_j'(z_j)}{z_j f_j'(z_j) - (\frac{z_j}{c_{jj}} - z_i)f_i'(z_i)} \tag{3.3}$$

for $i = 1$ or 2 and $i \neq j$. Assuming $\det C < 0$ and $f_2'(z_2) > 0$ then by (2.5) $\hat{c}_{ii}$ is positive. The eigenvalues of $DF(e_1, e_2)$ cross the imaginary axis as $c_{ii}$ is varied with speed given by

$$\left.\frac{d\alpha}{dc_{ii}}\right|_{\hat{c}_{ii}} = \frac{c_{jj}\Big(z_j f_j'(z_j) - (\frac{z_j}{c_{jj}} - z_i)f_i'(z_i)\Big)^2}{2(\frac{z_j}{c_{jj}} - z_i)\big(f_i'(z_i) - c_{jj}f_j'(z_j)\big)}. \tag{3.4}$$

The numerator of (3.4) is nonzero by (2.5) and the assumptions that $\det C < 0$ and $f_2'(z_2) > 0$. The direction the eigenvalues of $DF(e_1, e_2)$ cross the imaginary axis is determined by equation (3.4)

$$\operatorname{sgn}\left(\frac{d\alpha}{dc_{ii}}\Big|_{\hat{c}_{ii}}\right) = \operatorname{sgn}\left(f_i'(z_i) - c_{jj} f_j'(z_j)\right). \tag{3.5}$$

Thus if $f_1'(z_1) < 0$ and $f_2'(z_2) > 0$ the fixed point destabilizes as $c_{11}$ decreases or $c_{22}$ increases. The stability properties of the resulting periodic orbit are given by a stability coefficient which can be found in [**7**]. The derivation of the formula for this coefficient can be found in [**5**] or [**2**].

## 4 Omega-limit sets

If there exists a unique interior equilibrium at $(e_{1-}, e_{2-})$, this equilibrium behaves like the interior fixed point of a competitive system [**7**]. Along the isocline of species $x_i$, $c_{ii}x_i + x_j = z_i$ where $i \neq j$ (we have assumed in this case that if $i = 2$ then $z_2 = z_{2-}$). Thus on the isocline, the $\operatorname{sgn}(\frac{dx_j}{dt}) = \operatorname{sgn}(c_{jj}x_j + x_i - z_j)$. Thus the region between the nullclines is positively invariant (see Figures 5 and 6). The fixed point located at $(0, z_{2+}/c_{22})$ is a source. The asymptotic behavior of orbits in $\mathbb{R}_+^2$ is determined by the $\det C$. If $\det C < 0$ then by (2.5)

$$\operatorname{sgn}\left(\frac{z_1}{c_{11}} - z_{2-}\right) = \operatorname{sgn}\left(\frac{z_{2-}}{c_{22}} - z_1\right) = 1$$

which implies the axial equilibria $(z_1/c_{11}, 0)$ and $(0, z_{2-}/c_{22})$ are both sinks (see Figure 5). Hence $(e_{1-}, e_{2-})$ is a saddle point whose stable manifold separates the basins of attraction of the two sinks. If $\det C > 0$ then by (2.5)

$$\operatorname{sgn}\left(\frac{z_1}{c_{11}} - z_{2-}\right) = \operatorname{sgn}\left(\frac{z_{2-}}{c_{22}} - z_1\right) = -1$$

and thus $(z_1/c_{11}, 0)$ and $(0, z_{2-}/c_{22})$ are both saddle points (see Figure 6). Therefore $(e_{1-}, e_{2-})$ is a sink. In this case, if $x \in \mathbb{R}_+^2$, $\omega(x) = (e_{1-}, e_{2-})$. If there exists

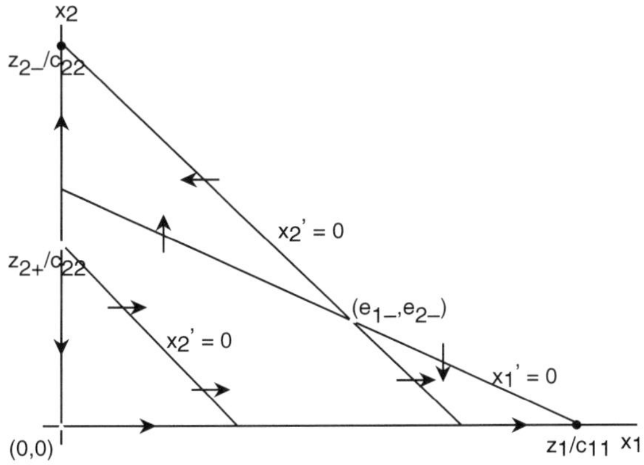


**Figure 5** $\frac{z_1}{c_{11}} > z_{2+}$, $\frac{z_1}{c_{11}} > z_{2-}$, $\frac{z_{2+}}{c_{22}} < z_1$, $\frac{z_{2-}}{c_{22}} > z_1$, $\det C < 0$

a unique interior equilibrium at $(e_{1+}, e_{2+})$ then the asymptotic behavior of the system depends on $\det C$. According to Lemma 3.2 and equation (3.3) if $\det C < 0$ a Hopf bifurcation can occur. When the interior equilibrium is hyperbolic it is either

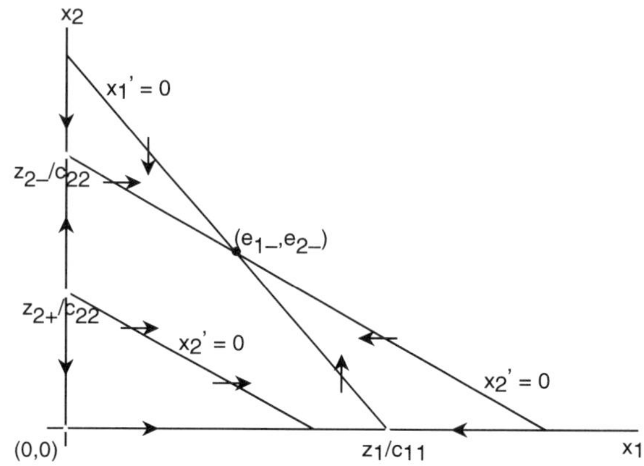


**Figure 6** $\frac{z_1}{c_{11}} > z_{2+}$, $\frac{z_1}{c_{11}} < z_{2-}$, $\frac{z_{2+}}{c_{22}} < z_1$, $\frac{z_{2-}}{c_{22}} < z_1$, $\det C > 0$

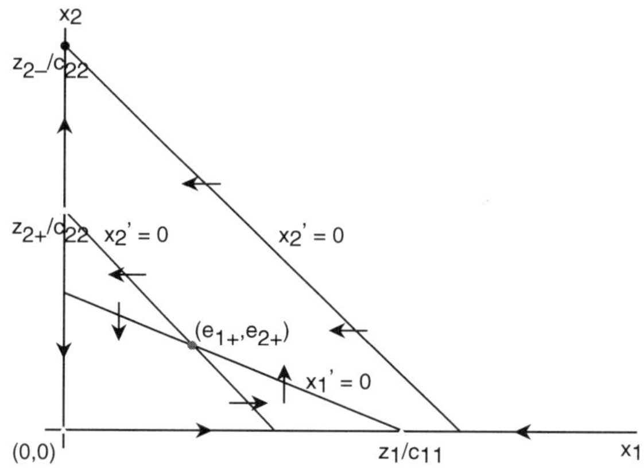


**Figure 7** $\frac{z_1}{c_{11}} > z_{2+}$, $\frac{z_1}{c_{11}} < z_{2-}$, $\frac{z_{2+}}{c_{22}} > z_1$, $\frac{z_{2-}}{c_{22}} > z_1$, $\det C < 0$

a spiral sink or source (see Figure 7). This behavior is exhibited by the system below:

$$\begin{aligned} \frac{dx_1}{dt} &= x_1(1 - \tfrac{9}{10}x_1 + x_2) \\ \frac{dx_2}{dt} &= -x_2(\tfrac{3}{4} - x_1 + c_{22}x_2)(\tfrac{5}{4} - x_1 + c_{22}x_2). \end{aligned} \tag{4.1}$$

Here $z_1 = 1$, $z_{2+} = \frac{3}{4}$, $z_{2-} = \frac{5}{4}$, and $c_{11} = \frac{9}{10}$. For a range of values of $c_{22}$, the system has a unique interior equilibrium point $(e_{1+}, e_{2+}) \in \mathbb{R}^2_+$. As $c_{22}$ increases through the value 54/85, the equilibrium point undergoes a Hopf bifurcation from spiral sink to spiral source.

If $\det C > 0$ then the interior fixed point is a saddle by (3.2) and implies $(z_1/c_{11}, 0)$ and $(0, z_{2-}/c_{22})$ are sinks while $(0, z_{2+}/c_{22})$ is a source. Thus the stable manifold of $(e_{1+}, e_{2+})$ separates the basins of attraction of $(z_1/c_{11}, 0)$ and $(0, z_{2-}/c_{22})$ (see Figure 8).

If two fixed points lie in the interior of $\mathbb{R}^2_+$ then the $\det C$ determines the asymptotic behavior of the system. If $\det C < 0$ then a Hopf bifurcation can occur. When $(e_{1+}, e_{2+})$ is hyperbolic it is either a spiral sink or source (see Figure 9). A numerical example of this behavior is given in [**15**, p. 26]. The fixed point $(e_{1-}, e_{2-})$ is a saddle point. If $\det C > 0$ then $(e_{1+}, e_{2+})$ is a saddle point while $(e_{1-}, e_{2-})$ is a sink (see Figure 10).

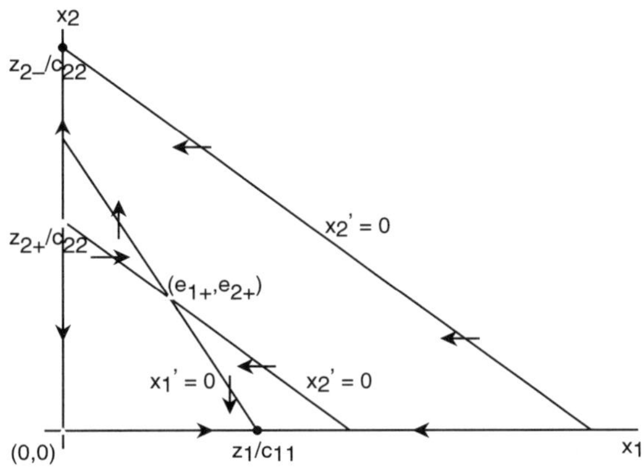


**Figure 8** $\frac{z_1}{c_{11}} < z_{2+}$, $\frac{z_1}{c_{11}} < z_{2-}$, $\frac{z_{2+}}{c_{22}} < z_1$, $\frac{z_{2-}}{c_{22}} > z_1$, $\det C > 0$

Thus in seven of the nine stable nullcline classes, the $\omega$-limit sets are single equilibria. For these seven equivalence classes (represented by Figures 2-6, 8, and 10) the nullclines and topological classes coincide. In the remaining two nullcline classes, Hopf bifurcations can take place and thus these classes do not correspond with the topological equivalence classes. The topological classes represented by Figures 7 and 9 contain vector fields which are not structurally stable. The following theorem summarizes the results established above.

**Theorem 4.1** *For systems in PC of two interacting species all $\omega$-limit sets are single equilibria, except when*

(i) $\det C < 0$,

*and*

(ii) *there exists an equilibrium at the transverse intersection of two nullclines where $f_2'(z_2) > 0$.*

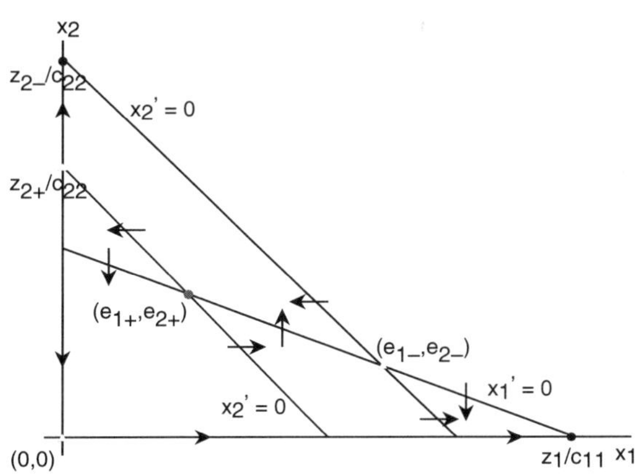


**Figure 9** $\frac{z_1}{c_{11}} > z_{2+}$, $\frac{z_1}{c_{11}} > z_{2-}$, $\frac{z_{2+}}{c_{22}} > z_1$, $\frac{z_{2-}}{c_{22}} > z_1$, $\det C < 0$

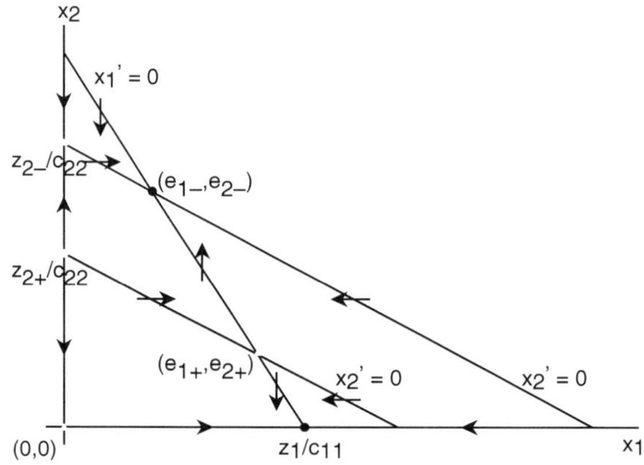


**Figure 10** $\frac{z_1}{c_{11}} < z_{2+}$, $\frac{z_1}{c_{11}} < z_{2-}$, $\frac{z_{2+}}{c_{22}} < z_1$, $\frac{z_{2-}}{c_{22}} < z_1$, $\det C > 0$

## 5 Conclusion

An examination of the positions of the nullclines of pioneer/climax interactions reveals a set of algebraically invariant quantities which can be used to classify the nullcline configurations. These same algebraic invariants in many cases determine the number of equilibrium points of the system and their stability properties. The same approach used here for pioneer/climax interactions can be used to classify the asymptotic dynamics of interactions between two climax species. There are eighteen stable nullcline classes (counting symmetry) to be examined in that case.

## References

[1] Kamke, E. [1932], *Zur Theorie der Systeme Gewöhnlicher Differential-Gleichungen, II*, Acta Mathematica **58**, 57-85.

[2] Guckenheimer, J. and Holmes, P. [1983], *Nonlinear Oscillations, Dynamical Systems, and Bifurcations of Vector Fields*, Applied Mathematical Sciences **42**, Springer–Verlag, New York.

[3] Allee, W. C. [1931], *Animal Aggregations*, University of Chicago Press, Chicago.

[4] Hale, J. K. [1980], *Ordinary Differential Equations*, Robert E. Krieger Publishing Company, 2nd edition.

[5] Hassard, B. D., Kazarinoff, N. D. and Wan, Y-H. [1981], *Theory and Applications of Hopf Bifurcation*, Cambridge University Press, New York.

[6] Miller, R. K. and Michel, A. N. [1982], *Ordinary Differential Equations*, Academic Press, New York,

[7] Selgrade, J. F. and Namkoong, G. [1990], *Stable periodic behavior in a pioneer–climax model*, Natural Resource Modeling **4**, no. 2, 215–227.

[8] Selgrade, J. F. [1994], *Planting and harvesting for pioneer–climax models*, Rocky Mountain Journal of Mathematics **24**, no. 1, 293–310.

[9] Hassell, M. P. Comins, H. N. [1976], *Discrete time models for two–species competition*, Theoretical Population Biology **9**, 202–221.

[10] Franke, J. E. and Yakubu, A.-A. [1991], *Mutual exclusion verses coexistence for discrete competitive systems*, Journal of Mathematical Biology **30**, 161–168.

[11] Ricker, W. E. [1954], *Stock and Recruitment*, Journal. Fisheries Research Board of Canada **11**, 559–623.

[12] Freedman, H. E. and Wolkowicz, G. [1986], *Predator-prey systems with group defense: the paradox of enrichment revisited*, Bulletin of Mathematical Biology **48**, 493–508.

[13] Wolkowicz, G. S. K. [1988], *Bifurcation analysis of a predator-prey system involving group defense*, SIAM Journal of Applied Mathematics **48**, 592–606,

[14] Tonkyn, D. W. [1986], *Predator-mediated mutualism: Theory and tests in the homoptera*, Journal of Theoretical Biology **118**, 15-31.

[15] Buchanan, J. R. [1993], *Asymptotic Behavior of n–dimensional Systems of Ordinary Differential Equations of Kolmogorov–type*, Ph. D. Thesis, North Carolina State University, Raleigh, NC. (March)

[16] Cushing, J. M. [1988], *The Allee effect in age-structured population dynamics*, in Proceedings of the Autumn Course Research Seminars: Mathematical Ecology (T. G. Hallam, L. J. Gross and S. A. Levin, eds.), World Scientific Publishing Company, Singapore, 479–505.

[17] Zeeman, M. L. [1993], *Hopf bifurcations in competitive three-dimensional Lotka–Volterra systems*, Dynamics and Stability of Systems **8**, no. 3, 189–217.

[18] Sumner, S. [1996] *Hopf bifurcation in pioneer-climax competing species models*, Mathematical Biosciences **137**, 1–24.

[19] Selgrade, J. F. and Roberds, J. H. [1996], *Lumped-density population models of pioneer-climax type and stability analysis of Hopf bifurcations*, Mathematical Biosciences **135**, 1–21.

Fields Institute Communications
Volume **21**, 1999

# Stability and Bifurcation of a Simple Neural Network with Multiple Time Delays

**Sue Ann Campbell**
Department of Applied Mathematics
University of Waterloo
Waterloo, Ontario, Canada N2L 3G1
and
Centre for Nonlinear Dynamics in Physiology and Medicine
McGill University
Montréal, Quebec
sacampbe@duchatelet.math.uwaterloo.ca

**Abstract.** A system of delay differential equations representing a simple model for a ring of neurons with time delayed connections between the neurons is studied. Conditions for the linear stability of fixed points of this system are represented in a parameter space consisting of the sum of the time delays between the elements and the product of the strengths of the connections between the elements. It is shown that both Hopf and steady state bifurcations may occur when a fixed point loses stability. Codimension two bifurcations are shown to exist and numerical simulations reveal the possibility of quasiperiodicity and multistability near such points.

## 1 Introduction

Hopfield in [**6**] considered a simplified neural network model in which each neuron is represented by a linear circuit consisting of a resistor and capacitor, and is connected to the other neurons via nonlinear sigmoidal activation functions. Assuming instantaneous updating of each neuron and communication between the neurons, Hopfield arrived at a system of first order ordinary differential equations. Not long afterward, Marcus and Westervelt in [**8**] considered the effect of including discrete time delays in the connection terms to represent the propagation time between neurons and/or processing time at a given neuron. Due to the complexity of the analysis, Marcus and Westervelt in [**8**] and most authors of subsequent work, for example [**4**], [**12**], [**2**] (see also references therein), have focussed on the situation where all connection terms in the network have the same time delay. In the work which has been done on Hopfield neural networks with multiple time delays the

1991 *Mathematics Subject Classification.* Primary 92B20, 34K20; Secondary 34K15.

This work was supported by a grant from the Natural Sciences and Engineering Research Council of Canada.

analysis is usually simplified by either restricting the size of the network (e.g., [**9**]), or considering networks with simple architectures (e.g., [**1**]). A notable exception is the paper by Ye, Michel and Wang [**13**] who have considered the global stability of fixed points of an arbitrary sized network with different time delays in each connection term.

Here we are interested in studying not only the stability of fixed points of the network but also the bifurcation of new solutions when stability is lost. We thus consider a Hopfield network of arbitrary size with multiple time delays but with a simple architecture. Our network consists of a ring of neurons where the $j$th element receives two time delayed inputs: one from itself with delay $\tau_s$, one from the previous element with delay $\tau_{j-1}$. The architecture of this system is illustrated in Figure 1.

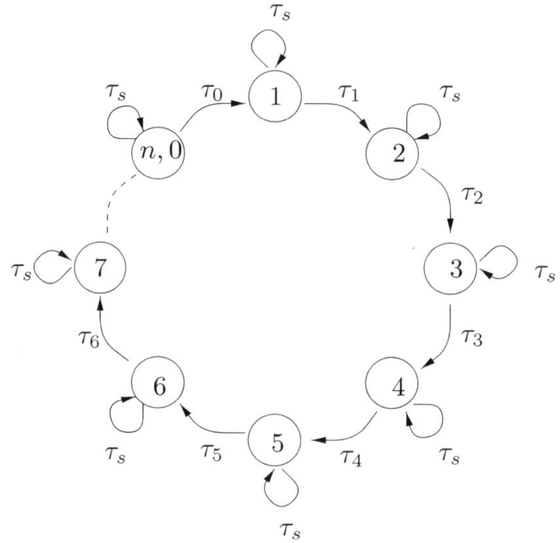


**Figure 1** Architecture of the neural network.

The equations for this system are

$$C_j \dot{u}_j(t) = -\frac{1}{R_j} u_j(t) + F_j(u_j(t-\tau_s)) + G_j(u_{j-1}(t-\tau_{j-1})), \; j = 1, \dots, n, \quad (1.1)$$

where $C_j > 0$, $R_j > 0$ represent the capacitances and resistances of the individual neurons, and $F_j$, $G_j$ are nonlinear functions representing, respectively, the feedback from neuron $j$ to itself, and the connection from neuron $j-1$ to neuron $j$; and the index 0 is taken equal to $n$. Normalizing this equation leads to

$$\dot{u}_j(t) = -d_j u_j(t) + f_j(u_j(t-\tau_s)) + g_j(u_{j-1}(t-\tau_{j-1})), \; j = 1, \dots, n, \quad (1.2)$$

where $d_j > 0$.

The plan for the article is as follows. In Section 2 we consider the linear stability analysis of Equation (1.2) and present some theorems about the region of stability of the fixed points as a function of the physical parameters in the model. In Section 3 we discuss the codimension one and two bifurcations which can occur when stability is lost and illustrate these with numerical simulations of a particular system. In the final section we will discuss the implications of our results and put them in the context of the related work mentioned above.

## 2 Linear stability analysis

Fixed points of (1.2) are solutions $\mathbf{u}(t) = \mathbf{u}^*$, $\forall t$, where

$$\mathbf{u}^* = [u_1^*, u_2^*, \dots, u_n^*]^T$$
$$d_j u_j^* - f_j(u_j^*) = g_j(u_{j-1}^*), \qquad j = 1, \dots, n.$$

The existence of such solutions depends, of course, on the particular functions $f_j$ and $g_j$ used in the model. Assuming that such a fixed point exists one can translate it to the origin via the transformation $\mathbf{u}(t) = \mathbf{u}^* + \mathbf{x}(t)$. If $f_j$ and $g_j$ are sufficiently smooth, one can expand these functions in Taylor series about $u(t) = u^*$. For $x(t)$ sufficiently small (solutions close to the fixed point) one can truncate this Taylor series to obtain the linearization of (1.2) about the fixed point:

$$\dot{x}_j(t) = -d_j x_j(t) + a_j x_j(t - \tau_s) + b_j x_{j-1}(t - \tau_{j-1}), \; j = 1, \dots, n, \tag{2.1}$$

where $a_j = f'(u_j^*)$, $b_j = g'(u_{j-1}^*)$. Physically, the $a_j$ and $b_j$ can be thought of as gains or strengths of the connections between neurons. To study the linearized stability of the fixed point $\mathbf{u}^*$ of (1.2), we consider solutions of (2.1) of the form $\mathbf{x}(t) = e^{\lambda t}\mathbf{c}$, where $\mathbf{c} = [c_1, c_2, \dots, c_n]^T$, $c_j$ constants. Substituting this expression into (2.1) yields the matrix equation

$$\left[\lambda I + D - Ae^{-\lambda\tau_s} - B\right] \mathbf{c} = \mathbf{0}. \tag{2.2}$$

Here $I$ is the $n \times n$ identity matrix, $D = diag(d_1, d_2, \dots, d_n)$, $A = diag(a_1, a_2, \dots, a_n)$ and $B = B_{jk}$ where $B_{jk} = b_j e^{-\lambda\tau_{j-1}}$, if $k = j-1$, $B_{jk} = 0$ otherwise. For nontrivial solutions of (2.2) we require that the determinant of the coefficient matrix be zero,

$$det\left[\lambda I + D - Ae^{-\lambda\tau_s} - B\right] = 0, \tag{2.3}$$

which leads to the characteristic equation associated with the delay differential equation (2.1):

$$\Delta(\lambda) = \prod_{j=1}^{n}(\lambda + d_j - a_j e^{-\lambda\tau_s}) - \prod_{j=1}^{n}(b_j e^{-\lambda\tau_{j-1}}) = 0. \tag{2.4}$$

We note that the latter product in this equation may be expanded giving a simplified equation

$$\prod_{j=1}^{n}(\lambda + d_j - a_j e^{-\lambda\tau_s}) = \beta e^{-\lambda\tau}, \tag{2.5}$$

where

$$\beta \stackrel{def}{=} \prod_{j=1}^{n} b_j \quad \text{and} \quad \tau \stackrel{def}{=} \sum_{j=1}^{n} \tau_j. \tag{2.6}$$

Thus the connections between the neurons act (so far as the linear stability is concerned) as a *single* feedback loop with gain $\beta$ and delay $\tau$. We shall see below that these are convenient and natural parameters to use in our stability and bifurcation analysis.

It is well known (see, e.g., [**7**] or [**11**]) that the trivial solution of Equation (2.1) will be asymptotically stable if all the roots of the characteristic equation (2.5) have negative real parts. This further implies [**5**] that the fixed point $\mathbf{u} = \mathbf{u}^*$ of nonlinear system (1.2) is locally stable. For brevity, we will say that the network modelled by (2.1) (by (1.2)) is *stable* (*locally stable*) in this case.

Another standard result [**7**] tells us that the trivial solution of (2.1) can only lose stability as parameters are varied by having a root of the characteristic equation pass through the imaginary axis. Thus changes of stability will occur at points in parameter space where (2.5) has roots with zero real parts. Consider first the zero roots, $\lambda = 0$. These will occur where

$$\beta = \prod_{j=1}^{n}(d_j - a_j) \stackrel{def}{=} \alpha. \tag{2.7}$$

Locating the pure imaginary roots is slightly more complicated. Substituting $\lambda = i\omega$ into (2.5) and separating real and imaginary parts yields

$$\begin{aligned} F_R &= \beta \cos \omega\tau \\ F_I &= -\beta \sin \omega\tau, \end{aligned} \tag{2.8}$$

where $F_R$ and $F_I$ are, respectively, the real and imaginary parts of $\prod_{j=1}^{n}(i\omega + d_j - a_j e^{-i\omega\tau_s})$. These are defined by the following

$$\begin{aligned} F_R \stackrel{def}{=} & \prod_{j=1}^{n}(d_j - a_j \cos\omega\tau_s)\left[1 + \sum_{l=1}^{k}(-1)^k \left\{\sum_{l1=1}^{n}\sum_{l2=1}^{n}\cdots\sum_{l2k=1}^{n}\right.\right. \\ & \left.\left.\frac{\omega + a_{l1}\sin\omega\tau_s}{d_{l1} - a_{l1}\cos\omega\tau_s}\,\frac{\omega + a_{l2}\sin\omega\tau_s}{d_{l2} - a_{l2}\cos\omega\tau_s}\cdots\frac{\omega + a_{l2k}\sin\omega\tau_s}{d_{l2k} - a_{l2k}\cos\omega\tau_s}\right\}\right], \\ & l2 \neq l1; \quad l3 \neq l1, l2; \quad \cdots \quad l2k \neq l1, l2, \ldots, l(2k-1). \end{aligned} \tag{2.9}$$

$$\begin{aligned} F_I \stackrel{def}{=} & \prod_{j=1}^{n}(d_j - a_j \cos\omega\tau_s)\sum_{l=1}^{m}(-1)^{m-1} \left\{\sum_{l1=1}^{n}\sum_{l2=1}^{n}\cdots\sum_{l(2m-1)=1}^{n}\right. \\ & \left.\frac{\omega + a_{l1}\sin\omega\tau_s}{d_{l1} - a_{l1}\cos\omega\tau_s}\,\frac{\omega + a_{l2}\sin\omega\tau_s}{d_{l2} - a_{l2}\cos\omega\tau_s}\cdots\frac{\omega + a_{l(2m-1)}\sin\omega\tau_s}{d_{l(2m-1)} - a_{l(2m-1)}\cos\omega\tau_s}\right\}, \\ & l2 \neq l1; \quad l3 \neq l1, l2; \quad \cdots \quad l(2m-1) \neq l1, l2, \ldots, l(2m-2). \end{aligned} \tag{2.10}$$

Here $k = \frac{n}{2} = m$ if $n$ is even and $k = \frac{n-1}{2}, m = \frac{n+1}{2}$ if $n$ is odd.

Equation (2.8) may be easily solved for $\tau$ and $\beta$ to yield

$$\beta = \beta^+ \stackrel{def}{=} \sqrt{F_R^2 + F_I^2}, \tag{2.11}$$

$$\tau = \tau^+ \stackrel{def}{=} \begin{cases} \dfrac{1}{\omega}\left[\operatorname{Arctan}\left(\dfrac{-F_I}{F_R}\right) + 2l\pi\right], & F_R > 0 \\ \dfrac{1}{\omega}\left[\operatorname{Arctan}\left(\dfrac{-F_I}{F_R}\right) + (2l+1)\pi\right], & F_R < 0 \end{cases} \tag{2.12}$$

and

$$\beta = \beta^- \stackrel{def}{=} -\sqrt{F_R^2 + F_I^2}, \tag{2.13}$$

$$\tau = \tau^- \stackrel{def}{=} \begin{cases} \dfrac{1}{\omega}\left[\text{Arctan}\left(\dfrac{F_I}{-F_R}\right) + 2l\pi\right], & F_R < 0 \\ \dfrac{1}{\omega}\left[\text{Arctan}\left(\dfrac{F_I}{-F_R}\right) + (2l+1)\pi\right], & F_R > 0. \end{cases} \tag{2.14}$$

Here $l = 0, \ldots, 1$ and Arctan denotes the inverse tangent function which has range $(-\frac{\pi}{2}, \frac{\pi}{2})$. For fixed $a_j, d_j$ and $\tau_s$ eqs. (2.11)–(2.12) and (2.13)–(2.14) describe curves which lie in the right and left half of the $\beta, \tau$ plane, respectively, and which are parameterized by $\omega$. If $n$ is large, $F_R$ and $F_I$ may be quite complicated functions of $\omega$ and the physical parameters $a_j, d_j, \tau_s$. Nevertheless, we can still make some assertions about the parameter values for which the trivial solution of (2.1) is asymptotically stable. We begin with some preliminary results about limiting cases.

**Lemma 2.1 (Single Element)** *Suppose there is just one element in the network. If*

$$-d_1 < a_1 < d_1 \quad \text{and} \quad \tau_s \geq 0$$

or

$$a_1 < -d_1 \quad \text{and} \quad \tau_s < \left\{\frac{1}{\sqrt{a_1^2 - d_1^2}}\left[\text{Arccos}\left(\frac{d_1}{a_1}\right)\right]\right\},$$

*then the network is stable.*

**Proof** If $n = 1$ then the characteristic equation reduces to

$$\lambda + d_1 - a_1 e^{-\lambda\tau_s} = 0. \tag{2.15}$$

This equation has been studied by many authors, most recently in the books by Kolmankovskii and Nosov [**7**] and Stépán [**11**]. For clarity we repeat the results here. When $a_1 = 0$ the equation has one negative real root $\lambda = -d_1$ and thus the trivial solution is stable. The trivial solution can only lose stability for $a_1 \neq 0$ by having a root which passes through the imaginary axis. Equation (2.15) has a zero root when $a_1 = d_1$ and pure imaginary roots, $\pm i\omega = \pm i\sqrt{a_1^2 - d_1^2}$, when

$$\tau_s = \tau_s^+ \stackrel{def}{=} \frac{1}{\sqrt{a_1^2 - d_1^2}}\left[(2l+1)\pi - \text{Arccos}\left(\frac{d_1}{a_1}\right)\right], \quad 0 < d_1 < a_1$$

or

$$\tau_s = \tau_s^- \stackrel{def}{=} \frac{1}{\sqrt{a_1^2 - d_1^2}}\left[2l\pi + \text{Arccos}\left(\frac{d_1}{a_1}\right)\right], \quad a_1 < -d_1 < 0.$$

Here $l = 0, 1, 2, \ldots$ and Arccos denotes the inverse cosine function which has range $[0, \pi]$. For fixed $d_1$ the curves $\tau_s^+$ are monotone decreasing in $a_1$ with $\lim_{a_1 \to d_1^+} \tau_s^+ = \infty$ and the $\tau_s^-$ are monotone increasing in $a_1$ with $\lim_{a_1 \to d_1^-} \tau_s^- = \infty$. Thus the trivial solution will lose stability for $a_1$ positive when $a_1 = d_1$ and for $a_1$ negative when $a_1$ crosses the $\tau_s^-$ curve closest to $a_1 = -d_1$. The result follows. □

**Corollary 2.2 (Chain of Neurons)** *Suppose the network is a chain, i.e., at least one of the $b_j = 0$. If for each $j$*

$$-d_j < a_j < d_j \quad \text{and} \quad \tau_s \geq 0$$

or

$$a_j < -d_j \quad \text{and} \quad \tau_s < \left\{\frac{1}{\sqrt{a_j^2 - d_j^2}}\left[\text{Arccos}\left(\frac{d_j}{a_j}\right)\right]\right\},$$

*then the network is stable.*

**Proof** If one of the $b_j = 0$ then $\beta = 0$ and the characteristic Equation (2.5) for the network becomes

$$\prod_{j=1}^{n}(\lambda + d_j - a_j e^{-\lambda\tau_s}) = 0.$$

All the roots of this equation will have negative real parts if and only if, for each $j$, all the roots of

$$\lambda + d_j - a_j e^{-\lambda\tau_s} = 0$$

have negative real parts. This is exactly the equation studied in Theorem 2.1 with $j = 1$. The result follows. □

Thus a chain of neurons with unidirectional connections will be stable if each element in the chain is stable when it is isolated. Note that under the conditions of this theorem, $\alpha$ as defined by (2.7) is always positive.

**Lemma 2.3** *For fixed $a_j, d_j$ and $\tau_s$, if $\beta^{\pm}$ as defined by* (2.11) *and* (2.13) *are monotone in $\omega$ then there are no intersection points of the curves defined by* (2.11)–(2.12) *or of the curves defined by* (2.13)–(2.14)*. If $\beta^{+}$ is increasing in $\omega$ then there are no intersection points of the line* (2.7) *with the curves defined by* (2.11)–(2.12).

**Proof** Consider first intersections of the curves defined by (2.11)–(2.12) or between the curves defined by (2.13)–(2.14). Since these curves are defined parametrically in terms of $\omega$ intersection points of two different curves will occur when the *same* values of $\beta$ and $\tau$ occur for two *different* values of $\omega$. If $\beta^{\pm}$ are monotone in $\omega$ then this is impossible.

Now consider the curves defined by (2.11)–(2.12). Consideration of (2.9)–(2.10) and (2.11) shows that $\lim_{\omega\to 0^+}\beta^{+}(\omega) = |\alpha|_{+}$, thus if $\beta^{+}$ is increasing with $\omega$ then there can be no intersection points of the line (2.7) with these curves. □

**Theorem 2.4** *Let $a_j, d_j, \tau_s$, be fixed and let the conditions of Corollary 2.2 be satisfied. If $\beta^{+}$ is monotone increasing with $\omega$ and $\beta^{-}$ is monotone decreasing with $\omega$ then the stability region of the network modelled by* (2.1) *is given by the following:*

$$-\alpha \le \beta < \alpha \quad \textit{and} \quad 0 < \tau,$$

$$\beta < -\alpha \quad \textit{and} \quad 0 < \tau < \begin{cases} \dfrac{1}{\omega}\left[\operatorname{Arctan}\left(\dfrac{F_I}{-F_R}\right)\right] & \textit{if } F_R < 0, \\[2ex] \dfrac{1}{\omega}\left[\operatorname{Arctan}\left(\dfrac{F_I}{-F_R}\right) + \pi\right] & \textit{if } F_R > 0, \end{cases}$$

*where $\alpha > 0$ is defined as in* (2.7).

**Proof** From Corollary 2.2 the trivial solution of (2.1) will be asymptotically stable for $\beta = 0$. For $\beta \neq 0$ it can lose stability only when one of the roots of the characteristic equation (2.5) passes through the imaginary axis. Recall that these roots lie along the line (2.7) and the curves defined by (2.11)–(2.12) and (2.13)–(2.14). From Lemma 2.3 there can be no intersection points of the curves defined by (2.11)–(2.12) and (2.13)–(2.14) or of the line (2.7) and the curves defined by (2.11)–(2.12). From (2.9)–(2.10), $\lim_{\omega\to 0^+} F_R = \alpha = \lim_{\omega\to 0^+} F_I$; thus $\lim_{\omega\to 0^+}\beta^{\pm}(\omega) = \pm\alpha^{\pm}$ and $\lim_{\omega\to 0^+}\tau^{\pm} = \infty$. It follows that the curves will be nested and the boundary of the stability region will be defined by the line $\beta = \alpha$ in the right half plane and the curve defined by (2.13)–(2.14) closest to $\beta = -\alpha$ in the left half plane. The result follows. □

In general, $\beta^{\pm}$ may be quite complicated functions of $\omega$, however, we can establish the following theorem about their monotonicity.

**Theorem 2.5** *For fixed $a_j$ and $d_j$ if*

$$\tau_s < \min_j \frac{1}{d_j}\left(\sqrt{1+\frac{d_j}{|a_j|}}-1\right),$$

*then $\beta^+$ is monotone increasing in $\omega$ and $\beta^-$ is monotone decreasing in $\omega$.*

**Proof** From (2.11) and (2.13) we have

$$\begin{aligned}
\beta^{\pm 2} &= F_R^2 + F_I^2 \\
&= \left[\prod_{j=1}^{n}(d_j - a_j\cos\omega\tau_s) + i(\omega + a_j\sin\omega\tau_s)\right] \\
&\quad \times \left[\prod_{j=1}^{n}(d_j - a_j\cos\omega\tau_s) - i(\omega + a_j\sin\omega\tau_s)\right] \\
&= \prod_{j=1}^{n}\left[(d_j - a_j\cos\omega\tau_s)^2 + (\omega + a_j\sin\omega\tau_s)^2\right].
\end{aligned}$$

Taking the derivative with respect to $\omega$ yields

$$\begin{aligned}
\beta^{\pm}\tfrac{d\beta^{\pm}}{d\omega} &= \left[\sum_{j=1}^{n}(d_j - a_j\cos\omega\tau_s)(a_j\tau_s\sin\omega\tau_s) + (\omega + a_j\sin\omega\tau_s)(1 + a_j\tau_s\cos\omega\tau_s)\right] \\
&\quad \times \left[\prod_{k=1,k\neq j}^{n}(d_k - a_k\cos\omega\tau_s)^2 + (\omega + a_k\sin\omega\tau_s)^2\right] \\
&= \beta^2\sum_{j=1}^{n}\frac{h_j(\omega)}{(d_j - a_j\cos\omega\tau_s)^2 + (\omega + a_j\sin\omega\tau_s)^2},
\end{aligned}$$

where

$$h_j(\omega) \quad = \omega + a_j d_j\tau_s\sin\omega\tau_s + a_j\omega_j\tau_s\cos\omega\tau_s + a_j\sin\omega\tau_s.$$

Note from Equations (2.11) and (2.13) that $\beta^+$ is strictly positive and $\beta^-$ strictly negative. Thus if each of the $h_j(\omega)$ is of the same sign for all $\omega$ then $\beta^{\pm}$ will be monotone in $\omega$. Now,

$$\begin{aligned} h_j(\omega) &= \omega + a_j(d_j\tau_s + 1)\sin\omega\tau_s + a_j\omega_j\tau_s\cos\omega\tau_s \\ &\geq \omega - |a_j|(d_j\tau_s + 1)\omega\tau_s - |a_j|\omega\tau_s \\ &= \omega(1 - 2|a_j|\tau_s - |a_j|d_j\tau_s^2) \\ &> \omega\left[1 - 2|a_j|\tfrac{1}{d_j}\left(\sqrt{1+\tfrac{d_j}{|a_j|}} - 1\right) - |a_j|\tfrac{1}{d_j}\left(2 + \tfrac{d_j}{|a_j|} - 2\sqrt{1+\tfrac{d_j}{|a_j|}}\right)\right] \\ &= 0. \end{aligned}$$

Since $h_j(\omega) > 0$ for each $j$, $\beta^+$ is monotone increasing in $\omega$ and $\beta^-$ is monotone decreasing in $\omega$. □

This result is illustrated in the following figures which display the region of stability of the trivial solution of (2.1) in the $\beta$, $\tau$ plane for various values of the parameters $a_j, d_j$ and $\tau_s$. Figures 2–4 show the results for a two element ring ($n = 2$) with $d_j = 1,\ j = 1, 2$ and $a_1 = -0.1,\ a_2 = -0.2$ (Figure 2), $a_1 = 0.1,\ a_2 = -0.2$ (Figure 3), and $a_1 = 0.1,\ a_2 = 0.2$ (Figure 4). The stability plots for two values of $\tau_s$ are shown, illustrating the difference when the $\beta^\pm$ are monotone in $\omega$ from when they are not. The critical value of $\tau_s$ predicted by Theorem 2.5 is $\tau_s \approx 1.45$ for all cases.

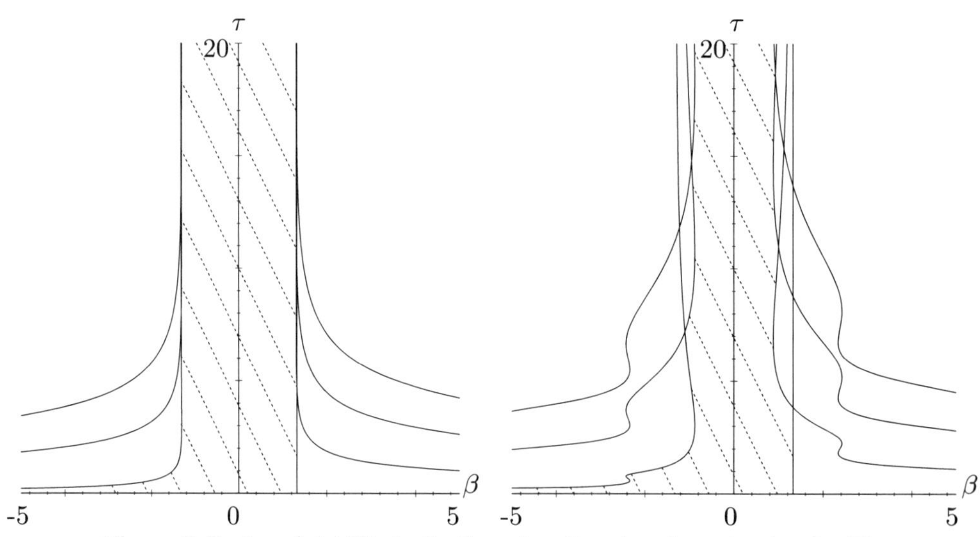


**Figure 2** Region of stability in the $\beta$, $\tau$ plane for a two element network with negative self connections.(a) $\tau_s = 2$ (b) $\tau_s = 6$. Other parameter values are give in the text. Dashed lines indicate where the fixed point is stable.

Figures 5 and 6 show the results for a three element ring ($n = 3$) with $d_j = 1,\ j = 1, 2, 3$ and $a_1 = -0.1,\ a_2 = -0.2,\ a_3 = -0.3$ (Figure 5) and $a_1 = 0.1,\ a_2 = 0.2,\ a_3 = 0.3$ (Figure 6). The stability plots for two values of $\tau_s$ are shown, illustrating the difference when the $\beta^\pm$ are monotone in $\omega$ from when they are not. The critical value of $\tau_s$ predicted by Theorem 2.5 is $\tau_s \approx 1.08$ for both cases.

It is clear from Figures 2(b)–6(b) that when $\beta^\pm$ are not monotone in $\omega$ the boundary of the stability region can be quite complicated as it is made up of pieces

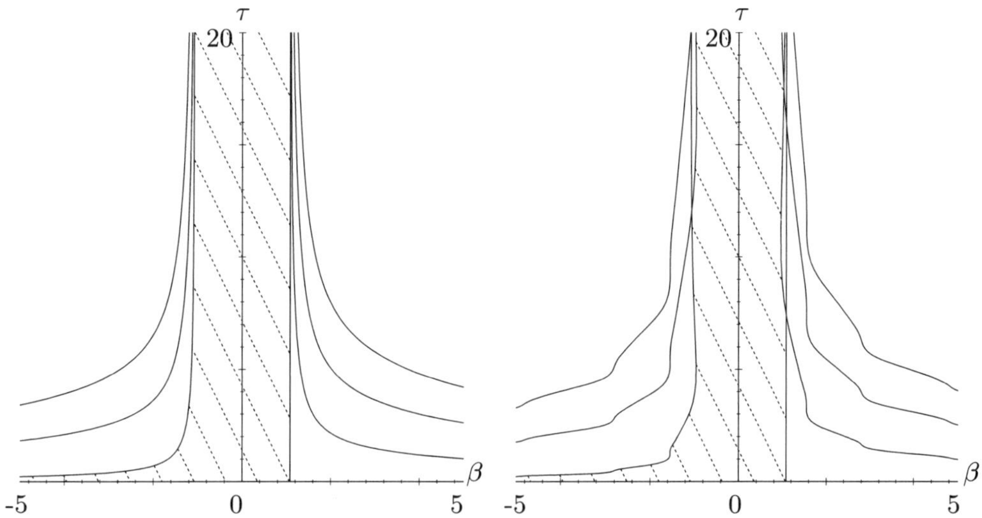


**Figure 3** Region of stability in the $\beta$, $\tau$ plane for a two element network with one positive and one negative self connection.(a) $\tau_s = 5$ (b) $\tau_s = 9$. Dashed lines indicate where the fixed point is stable.

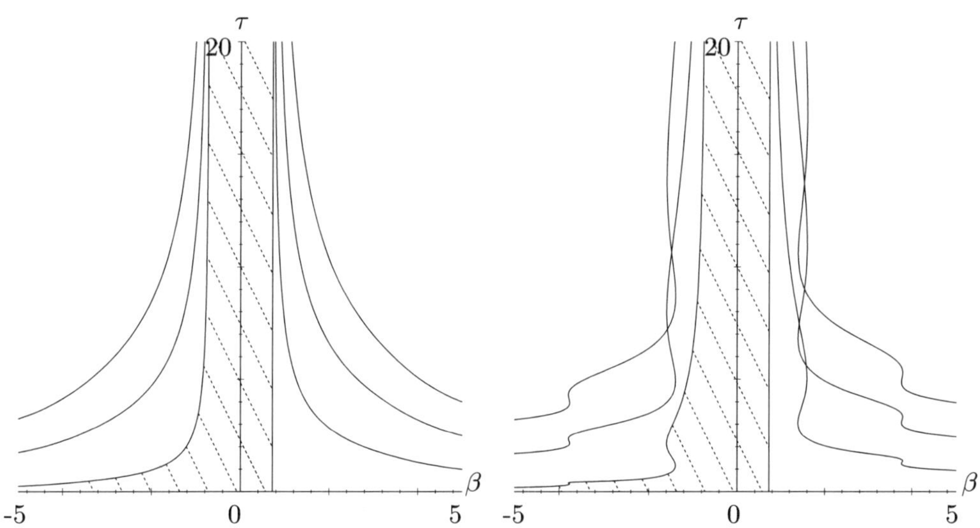


**Figure 4** Region of stability in the $\beta$, $\tau$ plane for a two element network with positive self connections.(a) $\tau_s = 2$ (b) $\tau_s = 6$. Other parameter values are give in the text. Dashed lines indicate where the fixed point is stable.

of the line (2.7) and arcs of the the curves defined by (2.11)–(2.12) and (2.13)–(2.14). It would be quite difficult to define this boundary analytically, however, we can state the following theorem about the region where the stability is independent of the delays $\tau_s$ and $\tau$.

**Theorem 2.6** *Let $a_j, d_j$, be fixed. If $|a_j| < d_j$, $j = 1, \ldots n$, then the trivial solution of (2.1) is stable for $-\beta_{min} < \beta < \beta_{min}$ and all $\tau_s > 0, \tau > 0$, where $\beta_{min} \overset{def}{=} \sqrt{\prod_{j=1}^{n}(d_j - |a_j|)^2}$. If $|a_j| > d_j$ for at least one $j$ then there is no such region of delay independent stability.*

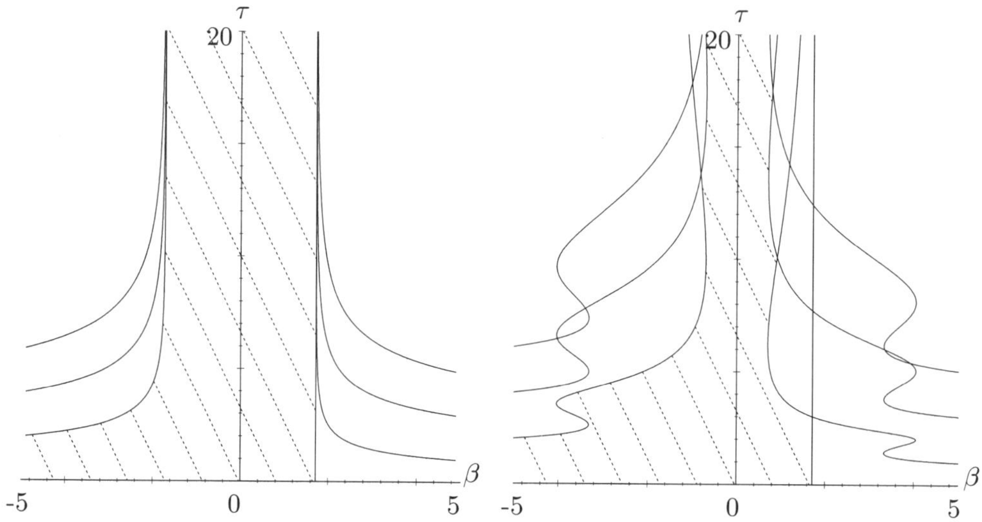


**Figure 5** Region of stability in the $\beta$, $\tau$ plane for a three element network with negative self connections. (a) $\tau_s = 1.5$ (b) $\tau_s = 6$. Other parameter values are given in text. Dashed lines indicate where the fixed point is stable.

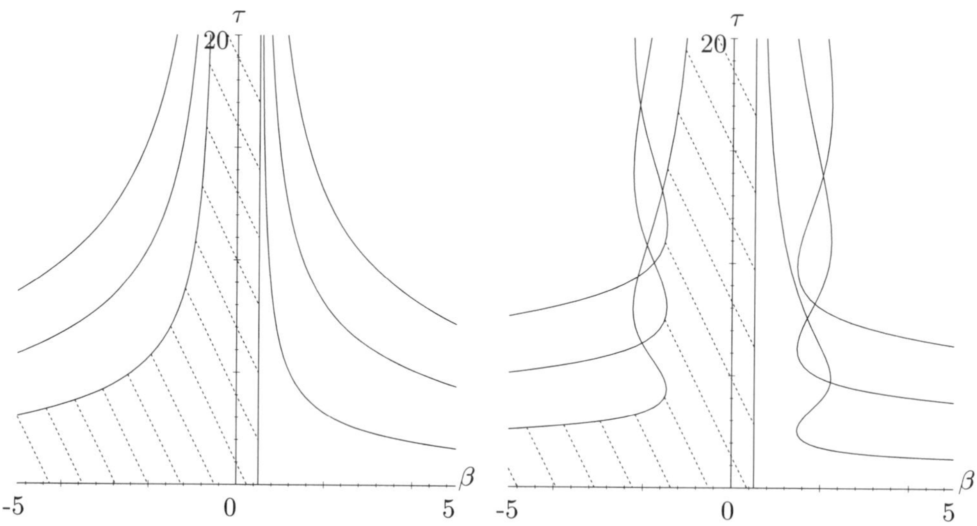


**Figure 6** Region of stability in the $\beta$, $\tau$ plane for a three element network with positive self connections. (a) $\tau_s = 2$ (b) $\tau_s = 6$. Other parameter values are given in text. Dashed lines indicate where the fixed point is stable.

**Proof** Recall from Theorem 2.5 that

$$\beta^{\pm 2} = \prod_{j=1}^{n} \left[ (d_j - a_j \cos \omega\tau_s)^2 + (\omega + a_j \sin \omega\tau_s)^2 \right].$$

Now if $|a_j| < d_j,\ j = 1, \ldots n$, then

$$(d_j - a_j \cos \omega\tau_s)^2 + (\omega + a_j \sin \omega\tau_s)^2 \geq (d_j - |a_j|)^2, \qquad j = 1, \ldots n,$$

and

$$\beta^{\pm 2} \geq \prod_{j=1}^{n} (d_j - |a_j|)^2.$$

Using a similar argument to that of Theorem 2.4 we see that the trivial solution of (2.1) is stable for $-\beta_{min} < \beta < \beta_{min}$ and all $\tau_s > 0, \tau > 0$.

However, if $|a_j| > d_j$ for some $j$, then $\beta^{\pm 2} = 0$ when $\omega = \sqrt{a_j^2 - d_j^2}$, and

$$\tau_s = \begin{cases} \dfrac{1}{\sqrt{a_j^2 - d_j^2}} \left[ (2l+1)\pi - \mathrm{Arccos}\left(\dfrac{d_j}{a_j}\right)\right], & 0 < d_j < a_j \\ \dfrac{1}{\sqrt{a_j^2 - d_j^2}} \left[ 2l\pi + \mathrm{Arccos}\left(\dfrac{d_j}{a_j}\right)\right], & a_j < -d_j < 0 \end{cases}$$

for $l = 0, 1, 2, \ldots$. Hence there is no region for which the trivial solution is stable for all positive values of $\tau_s$ and $\tau$. □

**Remark 2.7** Suppose that $|a_j| < d_j$, $j = 1, \ldots n$. Note that if $a_j > 0$, $j = 1, \ldots, n$ then $\beta_{min} = \alpha$ and this line will be the boundary of the region of stability for $\beta > 0$ for all values of $\tau_s$. This situation is shown in Figures 4(b) and 6(b). Alternatively, if at least one $a_j < 0$ then $d_j - |a_j| < d_j - a_j$ for this $j$ and $\beta_{min} < \alpha$. In this case the boundary of the region of stability for $\beta > 0$ will be more complicated. This situation is shown in the other figures.

## 3 Bifurcations

One can verify that the roots of the characteristic equation as described above are simple and that they cross the imaginary axis with nonzero speed. To do this we consider the following derivatives of the characteristic polynomial $\Delta(\lambda)$ as defined by (2.4):

$$\frac{d\Delta}{d\lambda} = \beta e^{-\lambda\tau} \left[ \tau + \sum_{j=1}^{n} \frac{1 + \tau_s a_j e^{-\lambda\tau_s}}{\lambda + d_j - a_j e^{-\lambda\tau_s}} \right] \quad \text{and} \quad \frac{d\Delta}{d\beta} = -e^{-\lambda\tau}, \tag{3.1}$$

where $\beta$ and $\tau$ are as defined in (2.6). Consideration of these derivatives where the characteristic equation has a zero root, i.e., along the line $\beta = \alpha$ yields:

$$\left.\frac{d\Delta}{d\lambda}\right|_{\lambda=0} = \alpha \left[ \tau + \sum_{j=1}^{n} \frac{1 + \tau_s a_j}{d_j - a_j} \right] \tag{3.2}$$

and

$$\left.\frac{d\lambda}{d\beta}\right|_{\lambda=0} = -\left[ \left.\frac{d\Delta}{d\lambda}\right|_{\lambda=0} \right]^{-1}. \tag{3.3}$$

Thus for fixed $a_j, b_j, \tau_s$ the roots are simple and pass through zero with nonzero speed everywhere on the line $\beta = \alpha$ except possibly for one isolated point given by

$$\tau = -\sum_{j=1}^{n} \frac{1 + \tau_s a_j}{d_j - a_j}. \tag{3.4}$$

Consideration of the derivatives (3.1) when the characteristic equation has a pure imaginary root, i.e., along the curves defined by (2.11)–(2.12) and (2.13)–(2.14) yields:

$$Re\left[\left.\frac{d\lambda}{d\beta}\right|_{\lambda=i\omega}\right] = -\frac{1}{\beta S^2}\left[\tau + \sum_{j=1}^{n}\frac{d_j - \tau_s a_j^2 + (d_j\tau_s - 1)a_j\cos\omega\tau_s - \omega\tau_s a_j \sin\omega\tau_s}{K_j^2}\right], \tag{3.5}$$

where

$$S^2 = \left[\sum_{j=1}^{n}\frac{d_j - \tau_s a_j^2 + (d_j\tau_s - 1)a_j\cos\omega\tau_s - \omega\tau_s a_j\sin\omega\tau_s}{K_j^2} + \tau\right]^2 + \left[\sum_{j=1}^{n}\frac{\omega + (d_j\tau_s + 1)a_j\sin\omega\tau_s + \omega\tau_s a_j\cos\omega\tau_s}{K_j^2}\right]^2,$$

$$K_j^2 = (d_j - a_j\cos\omega\tau_s)^2 + (\omega + a_j\sin\omega\tau_s)^2.$$

Thus the pure imaginary roots will pass through the imaginary axis with nonzero speed everywhere except at isolated points $(\beta^{\pm}(\omega_s), \tau^{\pm}(\omega_s))$ where $\omega_s$ is found by solving

$$\tau^{\pm}(\omega) = -\sum_{j=1}^{n}\frac{d_j - \tau_s a_j^2 + (d_j\tau_s - 1)a_j\cos\omega\tau_s - \omega\tau_s a_j\sin\omega\tau_s}{K_j^2},$$

$\beta^{\pm}$ are given by (2.11) and (2.13), $\tau^{\pm}$ are given by (2.12) and (2.14). From (3.1) if $Re\left[\left.\frac{d\lambda}{d\beta}\right|_{\lambda=i\omega}\right] \neq 0$ then $\left.\frac{d\Delta}{d\lambda}\right|_{\lambda=i\omega} \neq 0$ also, thus the pure imaginary roots will certainly be simple if they cross the imaginary axis with nonzero speed.

We conclude that if the nonlinearities of Equation (1.2) satisfy appropriate non-degeneracy conditions, the curves (2.11)–(2.12) and (2.13)–(2.14) define Hopf bifurcation curves in the $\beta, \tau$ plane, and the vertical line $\beta = \alpha$ defines a steady state bifurcation. The exact nature of these bifurcations, i.e., whether the Hopf bifurcations are supercritical or subcritical, and what type of steady state bifurcation occurs, would depend on the nonlinearities $f_j, g_j$. As an example, consider the case when $n = 2$ and the nonlinearities are given by $f_j(u) = a_j \tanh(u)$, $g_j(u) = b_j \tanh(u)$. Then $\mathbf{u}^* = \mathbf{0}$ and it can be shown [**10**] that the steady state bifurcation is a supercritical pitchfork. One way to determine the nature of the Hopf bifurcations is to analyze the center manifold for the equations. In [**3**] we describe a Maple program which will carry out such an analysis. Applying our program to this example shows that for fixed $d_j, a_j$ and $\tau_s$ the criticality of the Hopf bifurcation may vary with $\omega$ along the curves shown in Figures 2 – 6. More significantly, even if $\tau$ and $\beta$ are fixed, corresponding to a particular point on one of these curves, the criticality may still change if $b_j$ or $\tau_j$ change.

Codimension two bifurcation points occur when two of these different bifurcations happen simultaneously. Points of Hopf-Hopf interaction exist when the characteristic equation has two pairs of pure imaginary roots $\pm i\omega_1$, $\pm i\omega_2$. For the system we are considering these points can occur for any set of values of the parameters $a_j, d_j, \tau_s$ such that $\beta^{\pm}(\omega_1) = \beta^{\pm}(\omega_2)$ and $\tau_j^{\pm}(\omega_1) = \tau_k^{\pm}(\omega_1)$, for some $j, k \in \mathbb{Z}$. These points cannot, in general, be solved for in closed form. They can, however, be computed numerically and are easily seen in Figures 2(b)–6(b), where they appear as intersection points of two of the Hopf bifurcation curves. Hopf-steady state interactions exist when the characteristic equations has both a zero root and a pure imaginary pair. For our system these points may be found by solving $\beta(\omega) = \alpha$ for $\omega$ and substituting in $\tau^+$ as given by (2.12). Again these are readily visible in Figures 2(b)–6(b) where they appear as intersection points of the Hopf bifurcation curves with the line $\beta = \alpha$. From Theorem 2.5 it is clear that neither type of codimension two point can occur in our system if

$$\tau_s < \min_j \frac{1}{d_j}\left(\sqrt{1+\frac{d_j}{|a_j|}} - 1\right).$$

Codimension two points can be the source of more complicated dynamics such as multistability and quasiperiodicity. To illustrate what may occur in the system we are studying, we performed numerical simulations near one Hopf-Hopf and the Hopf-steady state interaction points which occur next to the stability region in Figure 2(b). We chose our nonlinearities as in the example discussed above. Plots in the $u_1$, $u_2$ plane of two of these simulations are shown in Figure 3. Figure 3(a) shows the existence of a stable 2-torus near the Hopf-Hopf interaction at $\tau \approx 7$, $\beta \approx -1.2$. Figure 3(b) shows the coexistence of a stable limit cycle with two stable fixed points near the Hopf-pitchfork interaction at $\tau \approx 4$, $\beta = 1.32$.

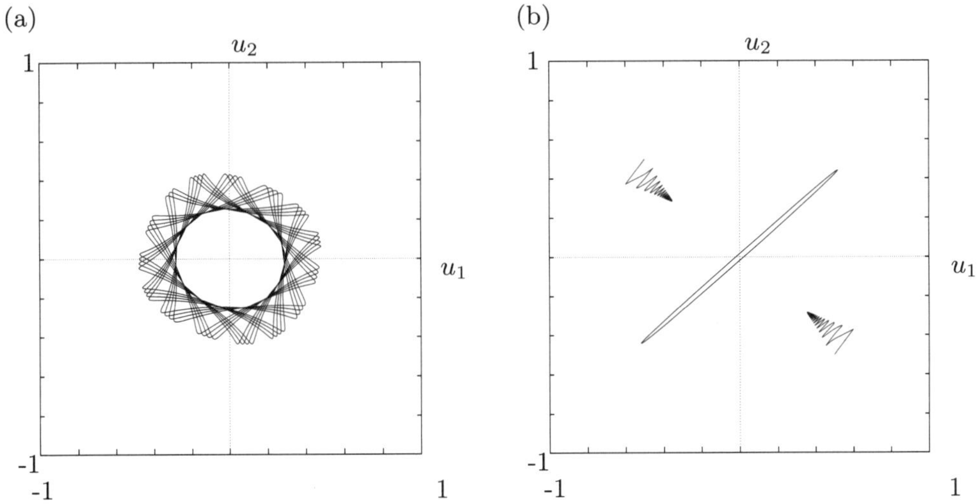


**Figure 7** Numerical simulations of a two element network given by (1.2) with $d_j = 1$, $\tau_s = 6$, $f_1(x) = -0.1\tanh(x)$, $f_2(x) = -0.2\tanh(x)$ and $g_j(x) = b_j\tanh(x)$. Other parameter values are (a) $b_1 = 1.2$, $b_2 = -1$, (giving $\beta = -1.2$) and $\tau = 6$ (b) $b_1 = -1.4$, $b_2 = -1$ (giving $\beta = 1.4$) and $\tau = 4$.

## 4 Conclusions

We have studied a system of delay differential equations representing a simple model for a ring of neurons with time delayed connections between each neuron and its immediate predecessor in the ring and time delayed feedback from each neuron to itself. We showed how conditions for the linear stability of fixed points of this system may be easily represented in a parameter space consisting of the sum of the time delays between the elements and the product of the strengths of the connections between the elements. A connection was made between the stability of the fixed point in the ring and the stability of fixed points in the individual neurons which comprise it. It was shown that both Hopf and steady state bifurcations may take place when a fixed point loses stability. Conditions under which interactions between these bifurcations may take place were given and numerical simulations revealed the possibility of quasiperiodicity and multistability near such points.

Our work is complementary to that of Ye, Michel and Wang in [**13**] who analyzed the global stability of a general network of $n$ neurons with different time delays in each connection. One should expect their regions of global stability to lie within our region of local stability. This has been checked (and found to be true) by Shayer in [**10**] in the case $n = 2$ with nonlinearity $f_j(u) = a_j \tanh(u)$, $g_j(u) = b_j \tanh(u)$. Olien and Bélair in [**9**] have made a detailed study of the stability and bifurcations of a network consisting of two neurons with two time delays. However, as they choose to identify a particular time delay with a given neuron (representing the processing time at that neuron) their work is not directly comparable to ours, even in the case $n = 2$. Baldi and Atiya in [**1**] have considered a ring of $n$ neurons with $n$ time delays. The main difference between their work and our is that we include delayed self connection terms (represented by $f_j(u_j(t - \tau_s))$ in (1.2)). As far as the linear stability analysis is concerned, their system effectively has one time delay: $\tau$, the product of the delays in the connections between neurons, whereas ours effectively has two: $\tau$ and $\tau_s$. Our analysis can be applied to their system by taking $a_j = 0$ for all $j$. Consideration of Theorems 2.4 and 2.5 shows that the stability diagrams in this case will always be qualitatively like Figure 2(a). That is, the stability region will be bounded on one side by the steady state bifurcation curve and on the other by one curve of Hopf bifurcation. As the codimension two points cannot exist one would not expect more complex dynamics than periodic solutions and fixed points. This is consistent with the results given in [**1**].

While ring networks are of limited biological relevance, they may be regarded as building blocks for networks with more realistic connection topologies. One approach which seems promising is to decompose larger networks into sets of connected rings and use studies such as ours to gain insight into the dynamics of the networks. For example, many of our results have shown how stability of the ring depends on the stability of the individual neurons. It is possible this may be extended to show how the stability of a network depends on the stability of the individual rings. See [**1**] for further discussion and example applications of similar ideas.

## Acknowledgements

This work has benefitted from the support of the Natural Sciences and Engineering Research Council of Canada and from useful discussions with L. Shayer and J. Wainwright. The numerical simulations were performed with the package XPP written by Bard Ermentrout.

## References

[1] Baldi, P. and Atiya, A. [1994], *How delays affect neural dynamics and learning*, IEEE Trans. Neural Networks, **5**(4), 612–621.

[2] Bélair, J., Campbell, S., and van den Driessche, P. [1996], *Frustration, stability and delay-induced oscillations in a neural network model*, SIAM J. Appl. Math., **56**(1), 245–255.

[3] Campbell, S. and Bélair, J. [1995], *Analytical and symbolically-assisted investigation of Hopf bifurcations in delay-differential equations*, Can. Appl. Math. Quart., **3**(2), 137–154.

[4] Gopalsamy, K. and Leung, I. [1996], *Delay induced periodicity in a neural netlet of excitation and inhibition*, Physica D, **89**, 395–426.

[5] Hale, J. and Lunel, S. V. [1993], *Introduction to Functional Differential Equations*. Springer Verlag, New York.

[6] Hopfield, J. [1984], *Neurons with graded response have collective computational properties like those of two-state neurons*, Proc. Nat. Acad. Sci. USA, **81**, 3088–3092.

[7] Kolmanovskii, V. and Nosov, V. [1986], *Stability of Functional Differential Equations*, Academic Press, London.

[8] Marcus, C. and Westervelt, R. [1989], *Stability of analog neural networks with delay*, Phys. Rev. A, **39**(1), 347–359.

[9] Olien, L. and Bélair, J. [1997], *Bifurcations, stability, and monotonicity properties of a delayed neural network model*, Physica D, **102**(3-4), 349–363.

[10] Shayer, L. [1998], *Stability analysis of a two dimensional neural network with two time delays*, Master's thesis, University of Waterloo, (in preparation).

[11] Stépán, G. [1989], *Retarded Dynamical Systems*, Pitman Research Notes in Mathematics, vol. 210, Longman Group, Essex.

[12] Ye, H., Michel, A. N., and Wang, K. [1994], *Global stability and local stability of Hopfield neural networks with delays*, Phys. Rev. E, **50**(5), 4206–4213.

[13] Ye, H., Michel, A. N., and Wang, K. [1995], *Qualitative analysis of Cohen-Grossberg neural networks with multiple delays*, Phys. Rev. E, **51**(3), 2611–2618.

Fields Institute Communications
Volume **21**, 1999

# Herding Behavior: The Emergence of Large-Scale Phenomena from Local Interactions

**Dennis Chao**
EPSON Research and Development, Inc.
3145 Porter Dr., Ste. 104
Palo Alto, CA 94304-1224, USA
dlchao@alumni.princeton.edu

**Simon A. Levin**
Department of Ecology and Evolutionary Biology
Princeton University
Princeton, NJ 08544-1003, USA
slevin@eno.princeton.edu

**Abstract.** Herding behavior arises as the collective action by individual agents that base their movement only on information about their neighbors; global properties arise from strictly local interactions. In this paper, we use simulation models to explore the influence of individual behavioral rules on group pattern, and to investigate the mechanisms underlying such patterns.

## 1 Introduction

One of the fundamental challenges in mathematical biology is to relate phenomena across scales, e.g., from individuals to populations to ecosystems (e.g., [**6**]). A prototypical example is provided by animal grouping patterns where behaviors are specified at the level of an individual, whereas recognizable and ecologically important patterns emerge at the level of the collective dynamics of large numbers of individuals [**1**].

The phenomenon of herding is a common one among large land mammals. Herd members can realize a variety of benefits by joining large groups, many of which relate to protection from predation and reduction of search time for patchy resources [**7**]. Although numerous theories exist to explain the advantages of grouping, the proximate mechanisms that form and maintain them are not obvious. The structure of herds emerges from decisions taken by individuals, and a fundamental problem is to explain how grouping patterns relate to individual behaviors.

It is nearly impossible, and unnecessary for the maintenance of groups, for an individual in a large herd to assess the group's global structure. Herding animals

---

1991 *Mathematics Subject Classification.* Primary 92A17; Secondary 92A18.
All correspondence should be addressed to SAL.

typically react to a small subset of their neighbors to reduce the amount of information that they must process. It is the degree of information sharing and the size of a herd member's perceptual neighborhood that concern us.

Many models have been constructed to demonstrate that stable herd-like behavior in large groups can emerge even when individuals only interact with their nearest neighbors ([**8**], [**5**], [**3**], [**2**]). Similar models also have found wide application to fish and invertebrate swarms and schools in the oceans ([**2**], [**1**]). In this paper, individual-based models are applied specifically to data from the observation of wildebeest herds, and used to investigate how changing local rules changes observed patterns. Properties such as spatial distribution and stability are examined. Wildebeest are ideal study organisms for such work because of their extremely large herd sizes during the migratory season.

## 2 The model

A variety of patterns characterize observations on wildebeest herds ([**10**] and [**9**]). Wave-like fronts, undulating trails, and patchy distributions all have been observed. How can these be understood as emerging from the aggregate behaviors of individuals? How do changes in individual behaviors affect global patterns? To approach this question, we begin from a collection of agents that base their movements on the states of their surrounding neighbors. This approach follows the general development of [**2**], [**3**], and [**5**]. All agents follow the same set of rules, with no special kin relationships or other such affinities. This assumption seems reasonable given Sinclair's observation in [**10**] that:

> "The main migratory population of three-quarters of a million animals acts like one enormous herd, with individuals showing no particular attachment to any subgroup. In fact, the evidence suggests that there are no well-defined subgroups, (except for the mother, her calf, and for a brief period her penultimate calf). At times in the wet season they may all combine into one herd." (p.152)

For our simulation, agents have information relating to their own location on a cartesian plane, speed, and a directional bias. The bias is the direction the individual would travel in the absence of neighbors. This direction of the bias shifts in time, reflecting the imprecision of its directional sense. It is randomly set between -theta and +theta. During each time step, each individual calculates its desired speed and accelerates to match it. An agent's intrinsic walking speed, its speed when it has no neighbors, has a magnitude equal to its normal walking speed in the direction of its "bias." If there are neighbors nearby, their states may affect the individual's desired speed.

The interaction paradigm used in this study is similar to [**5**] and [**3**]. Each individual is surrounded by three concentric spatial "zones" (Figure 1). The three zones are designated the stress zone, the neutral zone, and the attraction zone. These represent the three classes of reactions an individual will have to a neighbor, including attraction and repulsion. Attraction holds the herd together, while repulsion keeps animals from colliding. The desired behavior in the neutral zone is less obvious, and different strategies were implemented, which are discussed later. An individual will only react to neighbors in its nearest occupied zone. For example, an animal will only react to a neighbor in its attraction zone if it sees no one in its stress zone (Figure 2). We take zones to be rectangular rather than circular to allow faster computation, and with the individual at the center. Wildebeest have essentially 360 degree vision.

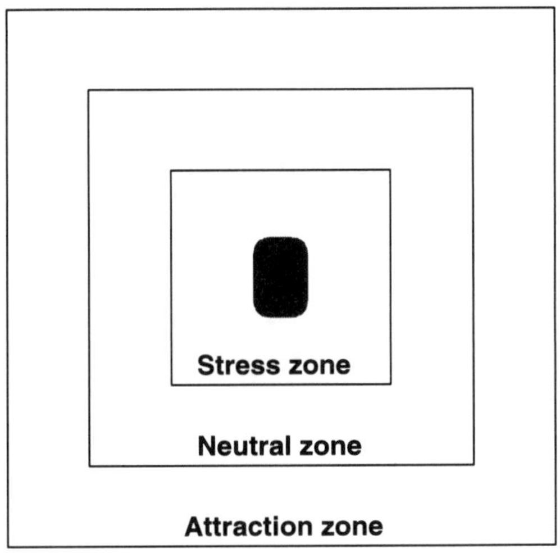


**Figure 1** The three concentric zones. The agent's reaction to a neighbor will depend on its zone.

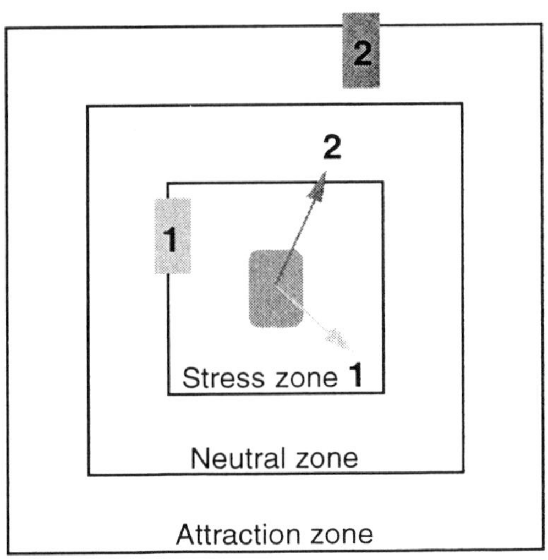


**Figure 2** A sample interaction. The agent will want to avoid neighbor 1 and move toward neighbor 2. Since the agent only reacts to neighbors in its nearest occupied zone, it will ignore 2.

The "stress zone" immediately surrounds the individual. This zone represents an animal's need for space, and the size of the zone is the minimum distance an animal will tolerate a neighbor to approach. If an individual finds a neighbor in its stress zone, it tries to move diametrically away from the intruder. The exact quantitative rules are not known, and in this paper our purpose is simply to illustrate the consequences of avoidance behavior. Hence, we assume that if there are one or more neighbors in its stress zone, the individual calculates its "desired" direction in the following manner: it determines the vectors from itself to the intruders, takes the sum of these vectors divided by their magnitudes cubed, then adjusts the magnitude to 1.5 times its normal walking speed. The individual then reverses its direction at this speed. We adjust the speed to a level higher than normal walking speed, so that an agent can walk fast enough to evade an intruder entering its stress zone from the rear. The influences of the neighbors are weighted by the inverse of their distances cubed so that the closest have the strongest effect by an inverse-square law. Obviously, the choice of parameters as well as the cubic dependence are arbitrary; they emphasize the need for data as to how individuals respond to their neighbors. The simulations should be understood as an initial and qualitative inquiry, and sensitivity analyses are essential to determine how observed behavior depends on the assumptions. Our objective is simply to demonstrate the wealth of patterns that can result from simple local rules.

If there are no neighbors in the individual's stress zone, it will search its neutral zone. The three strategies of neutrality, velocity matching, and centering for neutral zone interactions were tested in this study. In the "neutral" model, agents do not react to neutral zone neighbors. In the "velocity-matching" model, the agent tries to match its velocity to the average of those in the neutral zone. This tendency should improve group cohesiveness and directionality [**5**]. The "centering" model causes an agent to move towards the "center of mass" of its neighbors in the neutral zone, in effect "centering" itself within its neutral zone neighbors. This tendency will have little effect on individuals in the center of a group but will have a great effect on those on the edge of a group [**8**]. Those on the edge will tend to move towards the center of the group. This cohesion reflects an animal's attempt to avoid group boundaries, where predation may be highest [**4**].

In each of these models, an agent's desired velocity is set to the weighted sum of the agent's intrinsic walking velocity, designated to be the "neutral" tendency, and a vector derived from one of the above strategies, with weights summing to 1.0. If the weight of the neutral tendency is set to 1.0, the agent exhibits the "neutral" model behavior. Neutral tendency is never set to 0.0, since the agent would lose its intrinsic directional sense and the herd would not move. The influences of neighbors are not weighted by the distance from the agent because it is assumed that all neighbors in the neutral zone have the same importance.

The next area that an individual will search is its attraction zone. As with the stress zone, the individual calculates a vector by summing the distances of the neighbors in the zone weighted by the inverses of their distances cubed, and again adjusts its speed to 1.5 times its normal walking speed. Hence, the agent will accelerate towards its attraction zone neighbors. This tendency helps solitary individuals form herds or join currently-existing ones.

The herd is initialized by placing a number of agents randomly in a square of a given size with all agents travelling at their intrinsic walking velocities. The square's size and the number of agents in the simulation determine the initial density of the herd, which remains constant throughout the simulation. The model is run in discrete time steps for a given length of time. At each step, all agents simultaneously determine their desired velocities and accelerate to match them. The time steps are sufficiently small that the agents will react to each other quickly enough to prevent collisions. A small time step size also helps stabilize the speed of the agents.

## 3 Methods

The sizes of the zones and velocities of the agents in the model were based on analysis of tapes of wildebeest herds. Measurements of velocity and distance were taken in units of adult body lengths, roughly 2 meters. Wildebeest were found to travel at 0.6 body lengths per second. The boundaries of the zones were derived by noting where changes in inter-animal interaction seemed to occur.

The size of the stress zone determines the closest approach a wildebeest will tolerate. This figure can be estimated by observing the spacing of animals in a dense configuration. This appeared to be between 0.5 and 2.0 body lengths for walking wildebeest. In the model, we use the larger figure.

The neutral zone is the region in which neighbors will neither speed up nor slow down in response to the proximity of each other. We determined this distance by measuring the distances between animals walking in a single file. If these inter-animal distances remained stable over a significant distance, we assumed that the animals were in each others' neutral zones. These distances tended to be between 3 and 5 body lengths, so we use 5 as the radius of the neutral zone.

To determine the boundaries of the attraction zone, we observed changes in an individual's behavior as the distance to its nearest neighbor changed. If a wildebeest makes a sharp change in direction or velocity towards another animal, the location of this change marks the outer boundary of the wildebeest's attraction zone. When the wildebeest then settles into a steady walk that matches the direction and velocity of its neighbor, we postulated that the neighbor crossed the boundary between the attraction and neutral zones. This distance proved harder to measure, since the moment of decision was difficult to determine. However, the size of the attraction zone is not very important in the simulation, since agents will almost always have neighbors in their neutral or stress zones in sufficiently dense herds. The attraction zone only needs to be large enough so that stragglers can detect nearby animals and move towards them. We believe that the outer boundary of the attraction zone is roughly 10 body lengths.

We ran the model under a variety of initial conditions. Most tests were performed with all agents having the same intrinsic speed and directional bias in order to determine the effects of varying the initial density, neutral zone size, and neutral tendency strength. The speed and bearing variations were later introduced to test the robustness of the behaviors observed. To quantify the scattering of the herds in some cases, the average of the number of neighbors within 10 units of each agent was calculated. This average is positively correlated to the density of the herd; but because a complex, two dimensional structure is reduced to such a simple metric, the results must be treated with some skepticism.

## 4 Results

The initial herd density had noticeable effects on early herd structure in all models. In the neutral model, these effects were persistent. After agents had adjusted their inter-agent distances so that their stress zones were empty, the overall herd patterns stabilized because the agents had perfect directionality and no variation in walking speed. Sparse starting herds tended to remain sparse; dense configurations expanded to satisfy the need for individual agents' space, then stabilized in relatively tight aggregations (Figure 3). In the centering model, herds fragmented earlier when they had sparser initial configurations (Figure 4). It does not appear

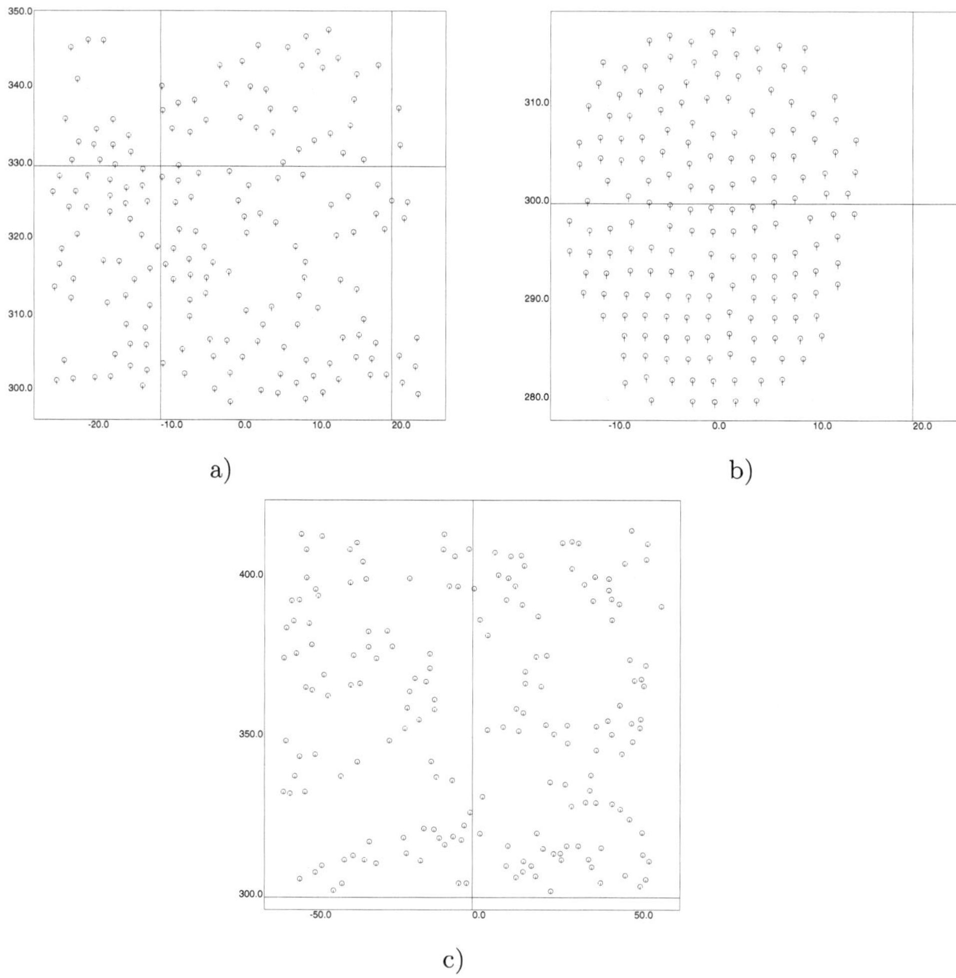


**Figure 3** Initial herd density effect on the neutral model at time 500. Sparser herds appear more fragmentary, but no subgroups form. If the initial density is too high, the agents will space themselves out, as in b). All herds contain 180 agents with the initial herd radii: a) 24.0, b) 8.0, and c) 60.0.

that steady-state subgroup sizes depended on initial density unless the initial configuration was so sparse that only small, widely-spaced aggregations could form. Unlike the neutral model, the centering model usually did not reach a steady state immediately after all agents had empty stress zones. Large aggregations tended to fragment. The agents in the velocity-matching model were given a $\pm 10\%$ speed and a $\pm 0.2$ radian directional bias variation because the model would be almost identical to the neutral model otherwise. The effects of initial density resembled those observed in the neutral model, but they were blurred by the variation (Figure 5). The herd with a high initial density had a larger central grouping, while that with the sparse initial density remained sparse.

The neutral and velocity-matching models did not generate herds with identifiable large-scale features. Neutral zone size in the neutral model did not seem to contribute to the spatial distribution of the agents (Figure 6). The velocity-matching model displayed a less-scattered herd when the weight of the matching

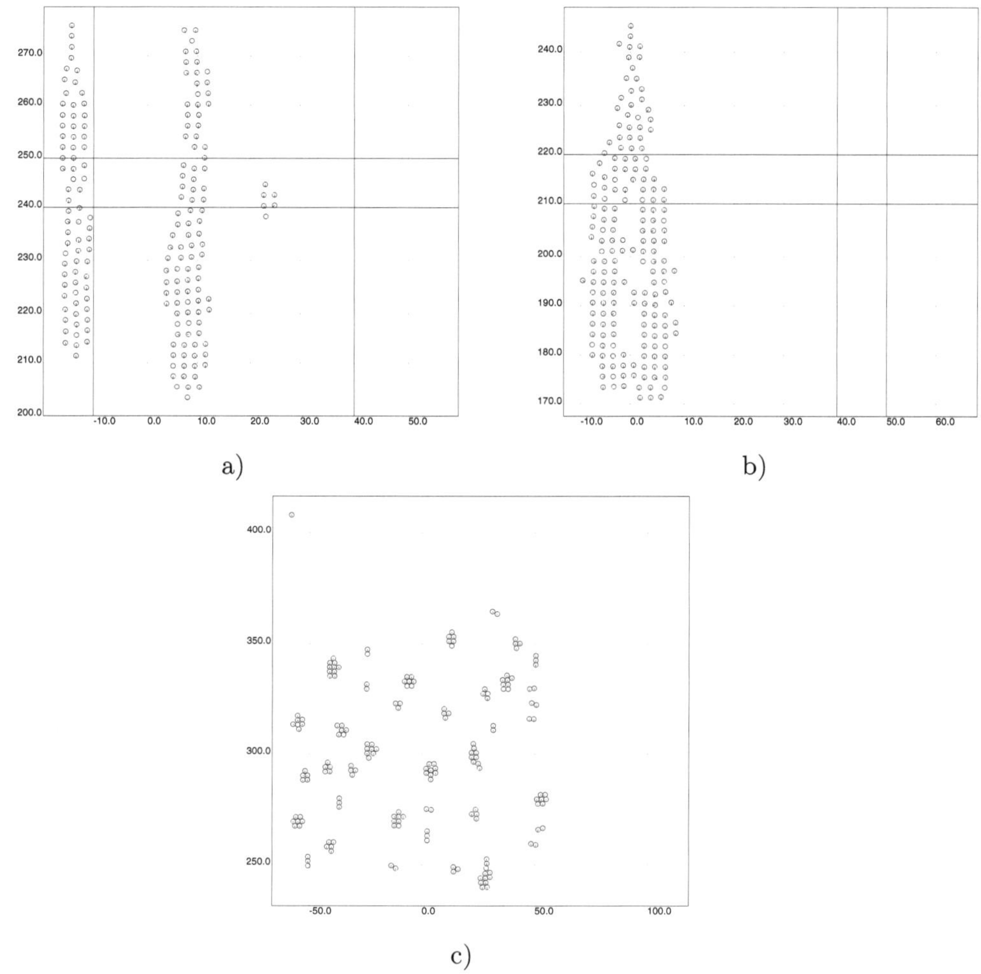


**Figure 4** The centering model is sensitive to initial herd density only in the beginning stages under most circumstances. Sparser herds fragment earlier. The herd in b) will converge to a shape more like a). Herd c) started too sparse and will remain sparse. The herds of 180 agents had initial radii: a) 24.0, b) 8.0, and c) 60.0.

tendency was increased from 30% to 70% (Figure 7). The neighborhood occupancy metric, averaged over 10 runs for each of these two conditions, appeared to detect this difference. With a velocity-matching tendency of 30%, the average number of neighbors within 10 units of each agent was 9.78, while at 70% the average was 11.89. Similar effects could be seen when the size of the neutral zone was decreased from 8.0 to 3.0, while other zone radii were kept constant (Figure 8). The 10-unit-radius neighborhood occupancy was 9.39 for the 8.0 unit neutral zone, and 9.92 for the 3.0.

The centering model produced identifiable large-scale phenomena. If the centering tendency was given a greater weight, the herd became more orderly and cohesive at the expense of speed (Figure 9). A larger neutral zone resulted in larger aggregations (Figure 10). The herds of size 45 and 720 featured aggregations of the same scale (Figure 11). The introduction of a ±0.2 radian variation in the agents'

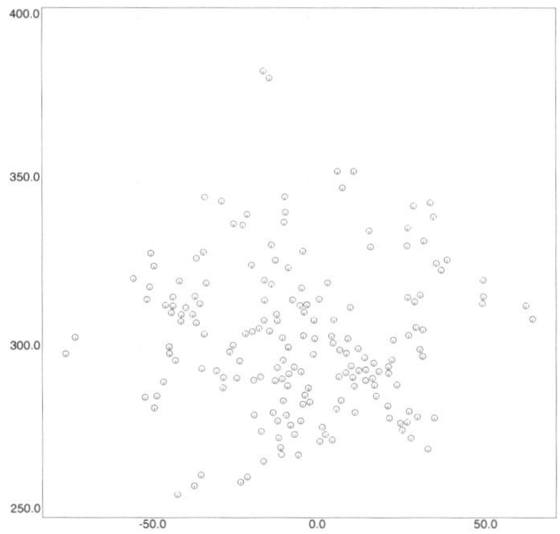


a)

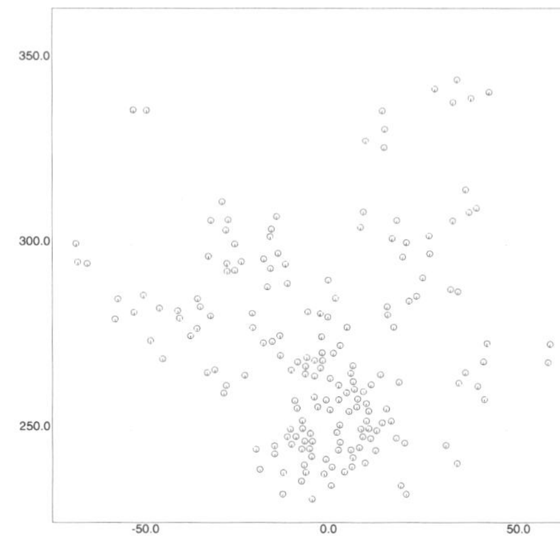


b)

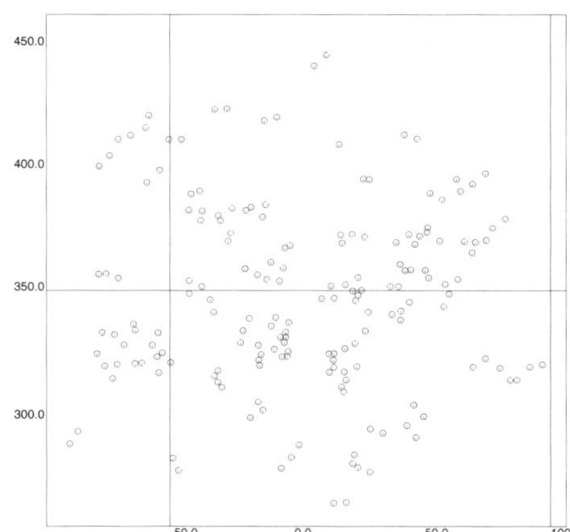


c)

**Figure 5** Initial herd density effects on the velocity-matching model. The effects of initial density are not as obvious as in the neutral model. The herds of 180 agents had the following initial radii: a) 24.0, b) 8.0, and c) 60.0.

directional biases effectively added noise to agent directionality. The neutral and velocity-matching models produced similar results: herds that were broader and shorter (Figure 12). The centering model displayed a noticeable scattering, but distinct subgroups still formed. The varying of intrinsic speed by ±10% elongated all herds along the direction they were travelling but did little to alter their shapes (Figure 13).

## 5 Discussion

These simulations are meant to tease, rather than to resolve the problem of grouping, demonstrating the range of patterns that can emerge from simple variations of local rules. The mathematical issues are explored in greater detail in [**1**].

As established by other investigations, local rules of interaction are sufficient to maintain a wide variety of global patterns of spatial organization. By only reacting

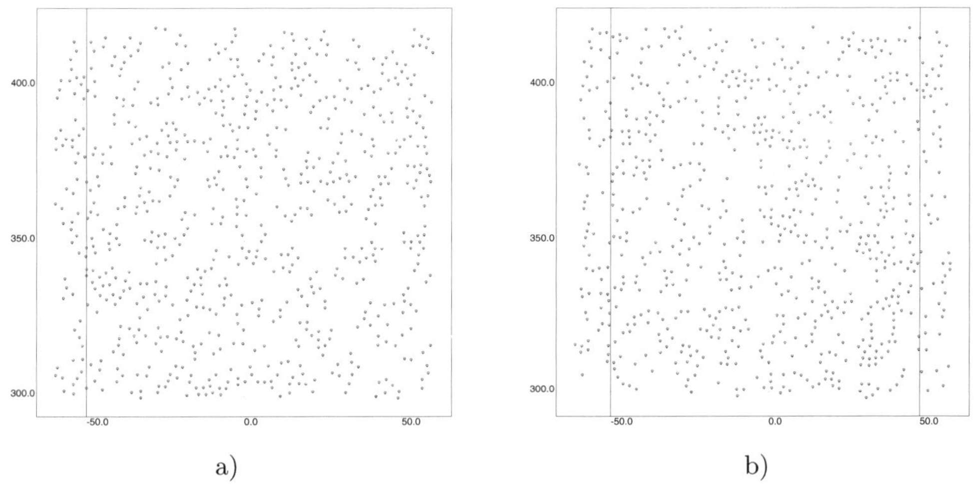


**Figure 6** Neutral zone size expansion has little effect on the neutral model. The neutral zone sizes are: a) 3.0 and b) 8.0.

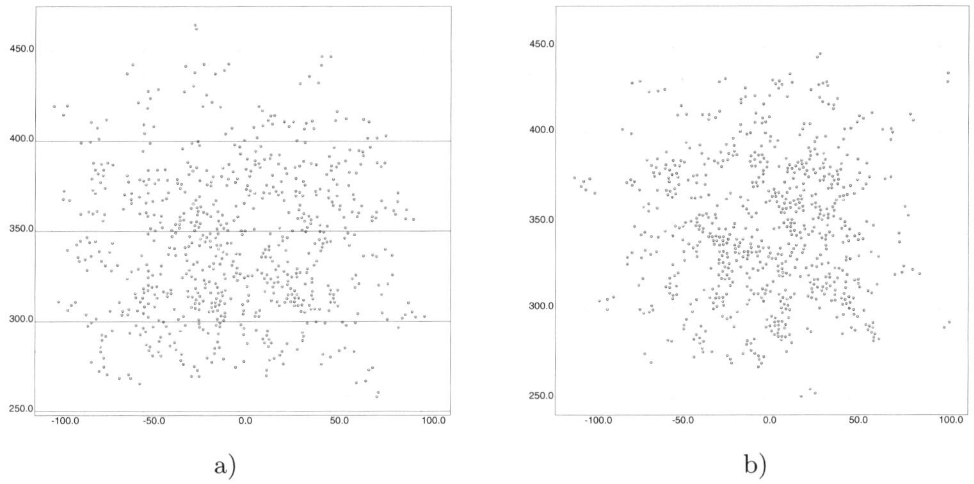


**Figure 7** Effects of velocity-matching tendency strength. A stronger velocity-matching strength may result in somewhat more clustered groups. Velocity-matching is 30% in a) and 70% in b).

to neighbors in a limited range, individual agents are able to maintain cohesion and order. Three zones, observed from animal behavior studies, seem necessary and sufficient to the maintenance of that order. A stress zone keeps inter-animal distances above a certain limit, reducing collisions and traffic jams; an attraction zone reduces the number of strays. A neutral zone also seems to play a crucial role in the overall spatial characteristics of the herd.

The herds generated by the neutral and velocity-matching models shared many characteristics. The stability of sparse formations allowed initial herd density to have a persistent effect. In the neutral model, herds that started sparse remained sparse. Velocity-matching is a unifying force that polarizes the velocity of agents in aggregations. However, since the simulation represents migrating animals, the

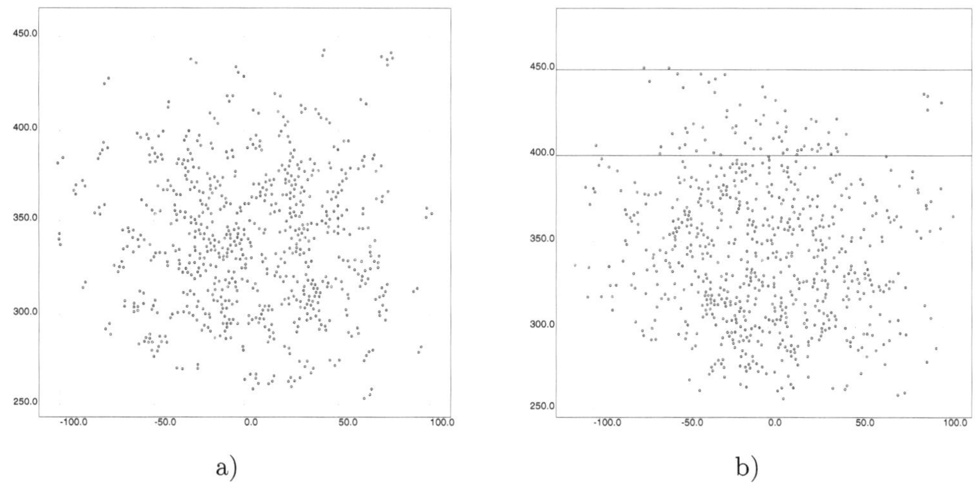


**Figure 8** The size of the neutral zone may be correlated with cohesion in the velocity-matching model, but the effects are not obvious. The neutral zone sizes are: a) 3.0 and b) 8.0.

agents shared roughly the same directional biases. A velocity-matching tendency amplified by either increasing the strength of the tendency or increasing the neutral zone size produced more coherent herds, though the effect was not strong.

The centering model produced large-scale formations independent of herd size and relatively insensitive to initial herd density up to a certain point. The size of the formations was positively correlated with the neutral zone size. The strength of the centering tendency also affected the size of the aggregations, with aggregation size increasing with the centering tendency strength. Although large aggregations could form in runs with small neutral zones or weak centering tendencies, these groups were not persistent. Both a large neutral zone and a strong centering tendency make large groups more stable. When either of these parameters was weakened, large aggregations tended to fragment.

Agents in all models used the position of neighbors implicitly by having spatial zones, but the centering model used neighbor positions explicitly. The centering strategy displayed the most interesting herd structures, and the use of the relative locations of neighbors appears to be necessary for the formation of regular spatial patterns. The neutral and velocity-matching models had a cohesive force that allowed aggregations to form, but they displayed no distinct spatial features. The velocity-matching model may be more suited to non-migrating animals, for which directionality is not as well-defined.

The actions and reactions of agents to each other can be thought of as information transfer. When an individual causes a neighbor to react, the neighbor's behavior may influence others, with the consequences of the initial action rippling through other agents. The agents are a medium through which messages from individuals travel. The stress and attraction zone sizes establish the density of the medium by regulating the spacing among individuals. The neutral zone size relative to the attraction zone size determines the reliability of signal transfer over a distance, with a large neutral zone leading to higher reliability and a small neutral zone leading to lower reliability. Reliability, and therefore neutral zone size,

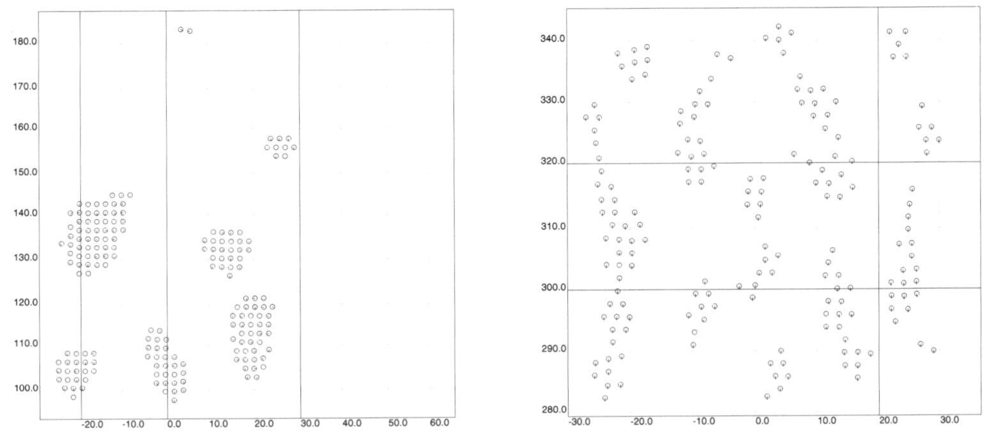


a) b)

**Figure 9** A stronger centering tendency leads to a more orderly and cohesive herd at the expense of speed. Note that the herd in a) did not move as far. The centering strengths are: a) 40% and b) 2.5%.

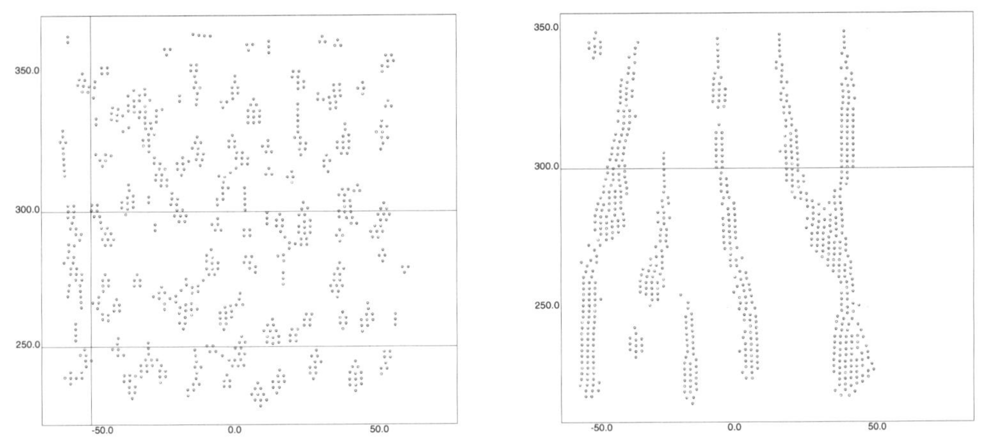


a) b)

**Figure 10** The subgroup size is correlated with the size of the neutral zone in the centering model. The neutral zone sizes are: a) 3.0 and b) 8.0.

corresponds to the distance a signal can travel. The strength of the "neutral" tendency is analogous to the decay rate of the signal because a strong neutral tendency diminishes the reaction an individual would have to a neighbor.

Stated more strongly, the agents are part of a self-organizing structure, which has properties that are encoded in each individual. The agents themselves are the mode of communication that gives the herd a coherent global behavior. Once a stable density is established by the stress and attraction zones, the nature of the "messages" transmitted through the herd depend entirely on neutral zone behavior and are manifest in the herd structures that emerge. In the neutral model, no information is passed through neutral zone interaction, so the herd will stabilize at a certain density but will have no characteristic form. The neutral model herd

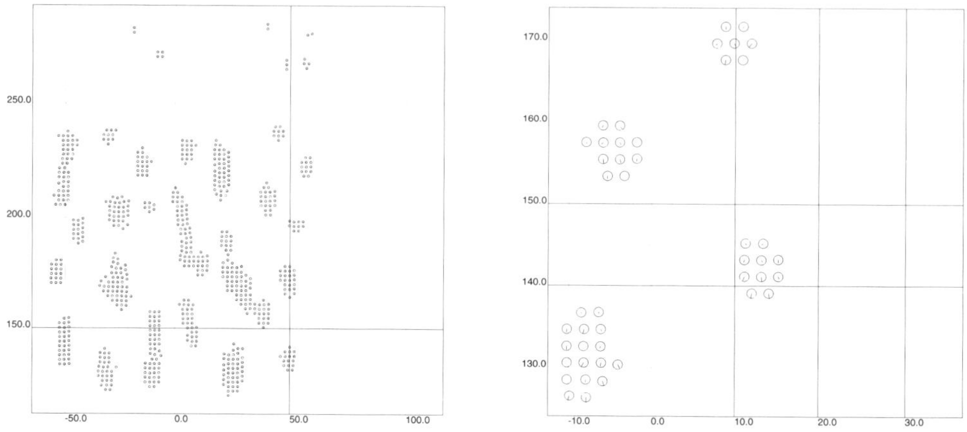


a) b)

**Figure 11** The features that emerge in the centering model are independent of initial herd size. Herd sizes are 45 in a) and 720 in b).

shown in Figure 3c does not reorganize — it basically retains its initial shape, a square. Velocity-matching behavior has the effect of polarizing the direction of the herd. If the neutral zone were infinitely large and the neutral tendency were set to zero, then every agent would be travelling at the same velocity, no matter how large the scatter in the intrinsic walking velocities of the individuals. The medium would have perfect transmission of messages that enable the herd to polarize direction perfectly. In the centering model, perfect transmission would cause the herd to converge to a perfect circle, similar to the sub-herds shown in Figure 9a.

The structure of the self-organizing herd is corrupted by the decay of signals from the imperfect transmission properties of the agents and from the sparsity of the herd. If the herd is small enough, even poor transmission may be sufficient to bring about a coherent global structure, as can be seen in the circular subherds forming in the centering model herd in Figure 11b. The strength of the centering tendency and the size of the neutral zone are sufficiently large to give the small herds circular shapes. This simple shape is unstable for a larger herd with the same parameters because the transmission properties are insufficient to carry signals great distances. To make a large circular herd, one must either increase the centering tendency (Figure 9a) or the neutral zone size (Figure 10b), both having the effect of increasing the interconnectedness of agents from an information transfer point of view. In a sparse herd, signals are not capable of travelling far. The self-organizing properties would therefore operate on smaller subgroups, rather than over the entire herd. This effect can be seen in the large herd in Figure 4c, which has a large number of unconnected sub-herds because the initial density was too low to establish a coherent transmission medium.

The concept of animal aggregations as information transfer mediums may be useful to ethologists (e.g., [**11**]). From the nature of the herd, one can deduce the nature of an individual's behavior and vice versa. If an aggregation has a strong spatial order, as in the centering model, each individual may be using the positions of its neighbors to adjust its behavior. If an aggregation has a strongly polarized direction that seems self-organized in nature, as in fish schools, individuals may be using the velocities of neighbors. In the case of wildebeest, it appears that

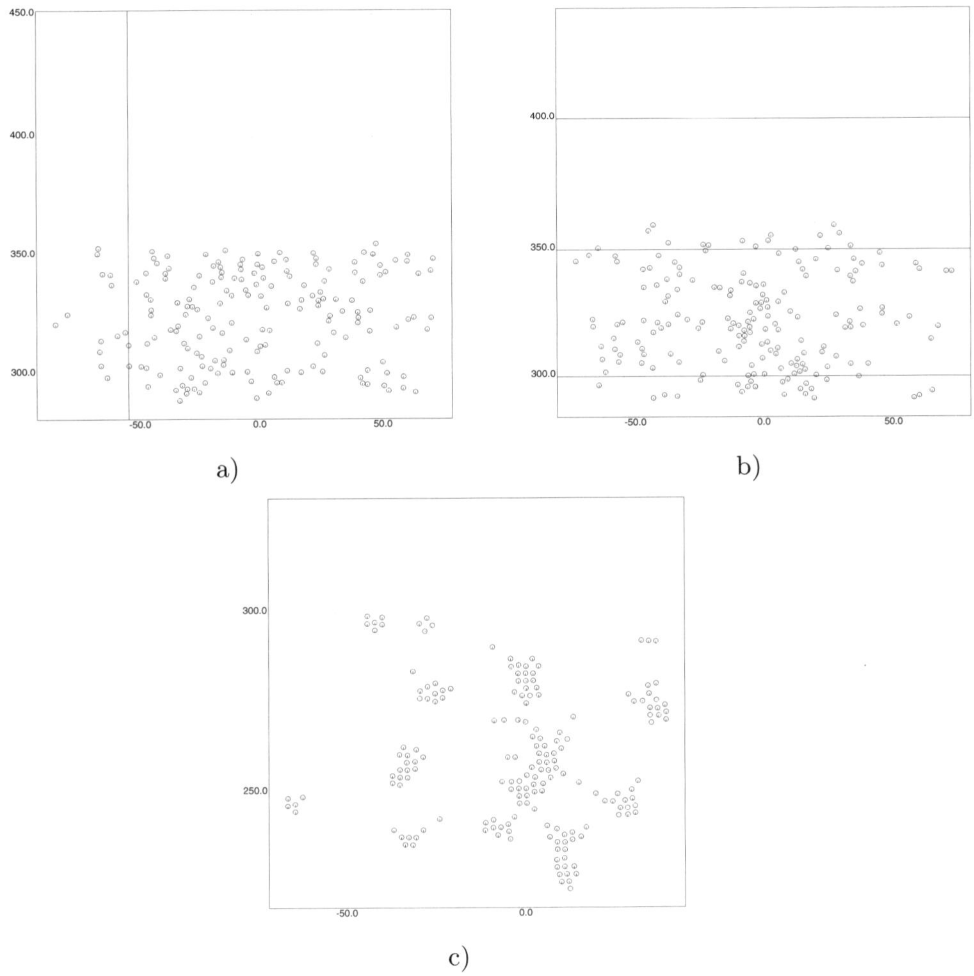


**Figure 12** The effects of a $\pm 0.2$ radian variation in intrinsic directionality of agents in a) neutral, b) velocity-matching, and c) centering models. All herds were broader and shorter, but otherwise closely resembled the herds without directionality variation.

individuals are using the position of neighbors when distinct herd fronts form. The "fingers" that can appear in these herds might be the manifestation of the limits of information transfer among the wildebeest, which may in fact be trying to minimize the herd boundary. The situation is naturally complicated by environmental noise and differences among the individuals, but we can learn much about the behavior of individuals from the observation of groups.

## Acknowledgements

We would like to thank Dan Rubenstein, Shay Gueron, Dan Grünbaum and Marc Mangel for their invaluable advice, and the Office of Naval Research through its University Research Initiative Program Award (ONR-URIP N00014-92-J-1527) to Woods Hole Oceanographic Institution. We also thank an anonymous referee for helpful comments.

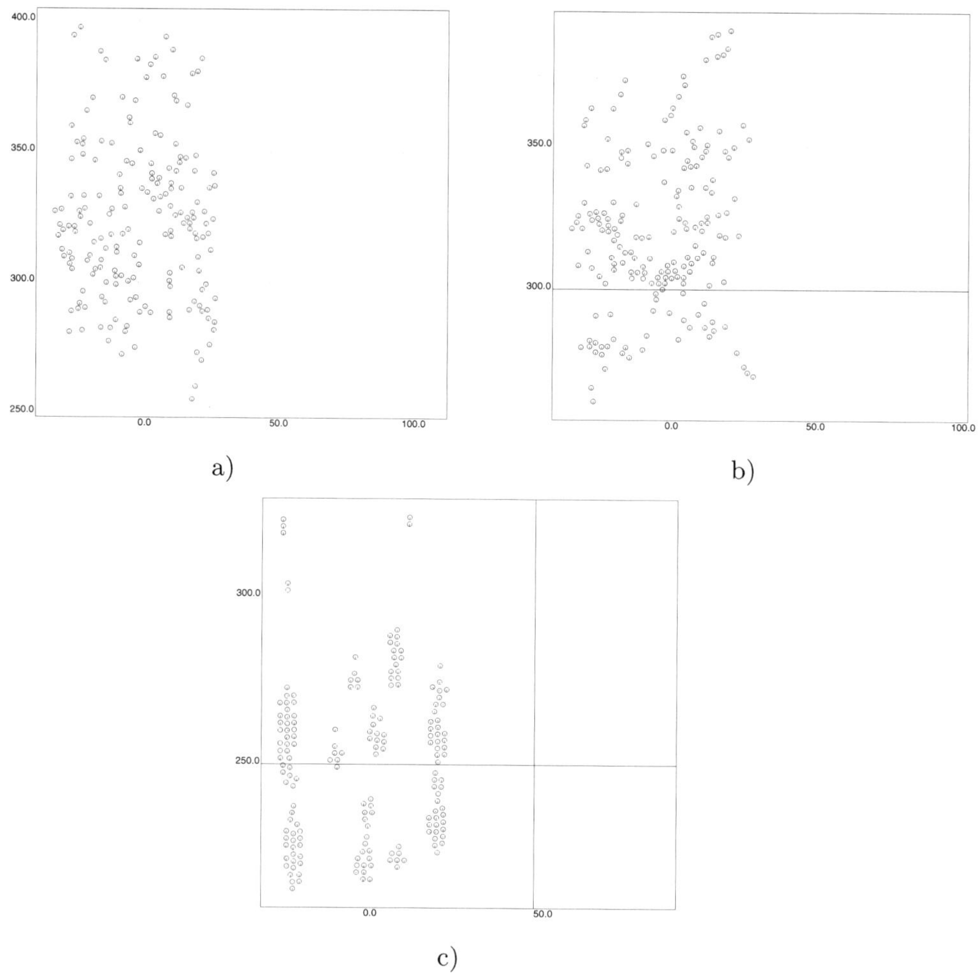


**Figure 13** The effects of a $\pm 10\%$ speed variation in a) neutral, b) velocity-matching, and c) centering models. The herds were elongated, but otherwise resembled herds without speed variation.

# References

[1] Flierl, G., Grünbaum, D., Levin, S. A. and Olson, D. [1998], *Individual-based perspectives on grouping*, J. theor. Biol., in press.

[2] Grünbaum, D. [1997], *Schooling as a strategy for taxis in a noisy environment*, in Animal Groups in Three Dimensions (J. K. Parrish and W. M. Hamner, eds.), Cambridge University Press, Cambridge, UK, 257–281.

[3] Gueron, S., Levin, S. A. and Rubenstein, D. [1996], *The dynamics of mammalian herds: From individuals to aggregations*, J. theor. Biol. **182**, 85–98.

[4] Hamilton, W. D. [1971], *Geometry for the selfish herd*, J. theor. Biol. **31**, 295–311.

[5] Huth, A. and Wissel, C. [1992], *The simulation of movement of fish schools*, J. theor. Biol. **156**, 365–385.

[6] Levin, S. A., Grenfell, B., Hastings, A. and Perelson, A. S. [1997], *Mathematical and computational challenges in population biology and ecosystem science*, Science **275**, 334–343.

[7] Pulliam, H. and Caraco, T. [1984], *Living in groups: Is there an optimal size?*, in Behavioral Ecology: an evolutionary approach, 2nd edition (J. R. Krebs and N. B. Davies, eds.), Blackwell Scientific, Oxford, 122–147.

[8] Reynolds, C. W. [1987], *Flocks, herds, and schools: A distributed behavior model*, Computer Graphics **21**, 25–34.

[9] Scott, J. [1988], *The Great Migration.* Rodale Press, Emmaus, PA.

[10] Sinclair, A. R. E. [1977] , *The African Buffalo: A Study of Resource Limitation of Populations*, The University of Chicago Press, Chicago, IL.

[11] Ward, P. and Zahavi, A. [1973], *The importance of certain assemblages of birds as 'information centres' for food finding*, Ibis **115**, 517–534.

Fields Institute Communications
Volume **21**, 1999

# On the Formation of Scroll Waves in an Anisotropic Ventricular Myocardium

**Clyde J. Clements**
Department of Physiology and Biophysics
Dalhousie University
Halifax, Nova Scotia, Canada B3H 3J5
Clyde.Clements@Dal.Ca

**John C. Clements**
Department of Mathematics, Statistics and Computing Science
Dalhousie University
Halifax, Nova Scotia, Canada B3H 3J5
John.Clements@Dal.Ca

**B. Milan Horáček**
Department of Physiology and Biophysics and Department of Medicine
Dalhousie University
Halifax, Nova Scotia, Canada B3H 3J5
Milan.Horacek@Dal.Ca

**Abstract.** Using a mathematical model based on anisotropic bidomain theory and a physiologically accurate transmembrane ionic current term, we investigated action-potential propagation in a three-dimensional domain representing the ventricular myocardium. The model, formulated under the assumption of an equal anisotropy ratio, consists of a single nonhomogeneous quasilinear parabolic partial differential equation which describes the flow of current in the cardiac tissue, coupled with a system of seven nonlinear ordinary differential equations which determine the time-varying ionic current at each point in the domain. The total ionic current is governed by Hodgkin-Huxley-type gating equations as defined in the Luo-Rudy model. Employing numerical methods based on the Douglas-Brian finite-difference scheme, we performed simulations to investigate the formation of vortex-like reentry in the form of scroll waves. We found that Luo-Rudy media with uniform anisotropy can support scroll-wave activity.

---

1991 *Mathematics Subject Classification.* Primary 35K60, 92C05; Secondary 35Q80, 92C30.
This work was supported in part by research grants from the Medical Research Council of Canada, the Heart and Stroke Foundation of Nova Scotia, and the Natural Sciences and Engineering Research Council of Canada.

## 1 Introduction

Mathematical models for simulating the electrical propagation phenomena in the heart can provide valuable insight into the normal and pathological process of cardiac depolarization and repolarization.[1] Of particular importance is the formation and termination of reentrant excitation. Reentry occurs when the propagating wave of electrical activity re-invades previously excited tissue and excites it again. If this happens repeatedly, it can lead to rapid, self-sustaining electrical activity in the heart which underlies the life-threatening cardiac arrhythmias such as ventricular tachycardia, and atrial and ventricular fibrillation. Spiral-wave activity has been proposed as a mechanism for this reentrant excitation in two-dimensional media [**13, 17, 20, 21**]. Recent experimental evidence [**7, 8, 9, 11, 16**] supports this claim. Scroll waves, the three-dimensional analogue of spiral waves, have similarly been proposed as a mechanism for reentry in three-dimensional tissue [**20, 21**], but experimental evidence for this is still scanty [**11**], mostly due to the difficulty of measuring electrical activity in three-dimensional tissue.

Reentrant rotating waves in mathematical models of cardiac electrical activity have been studied by many investigators [**6, 7, 9, 15, 16, 18**]. Much of this work is restricted to two-dimensional simulations using the simple FitzHugh-Nagumo (FHN) membrane model. Some exceptions do apply. Courtemance and Winfree in [**6**] showed reentrant patterns in a two-dimensional sheet of cells using the ionic model of [**1**], and [**18**] used FHN equations to demonstrate vortex-like reentry in three dimensions.

The primary goal of this study is to investigate the formation of scroll waves using the realistic action-potential model of [**14**], which is based on Hodgkin-Huxley-type formalism for describing the gating of ionic currents. Since our aim is only to investigate whether Luo-Rudy media can support scroll waves, we neglect the complex geometry of the ventricles in favour of a simplified slab model of the ventricular myocardium.

## 2 Methods

**Mathematical Model.** The spread of electrical activity in cardiac tissue can be described as the propagation of action potentials. For describing the space-clamped action potential, we use the mathematical model of ventricular cardiac action potential developed by [**14**]. It is an extension of a model developed by [**1**], and is based upon Hodgkin-Huxley formalism [**12**]. Briefly, in Hodgkin-Huxley formalism the transmembrane current per unit area $I_s$ generated by an ionic species $s$ as it passes through its respective channel is the product of the conductance of the channel $G_s$, and the difference between the transmembrane potential $V$, and the reversal potential $E_s$ for the ion species $s$. That is,

$$I_s = G_s \cdot (V - E_s).$$

The conductance of the channel varies with time and transmembrane potential, and is typically formulated as

$$G_s = \bar{G}_s \cdot g(u_1, u_2, \ldots),$$

where $\bar{G}_s$ is the maximum channel conductance and $u_1, u_2, \ldots$ represent gating parameters that control the opening and closing of the channel $s$. The rate of

[1] Hereafter, we will use the term excitation to refer summarily to the entire depolarization-repolarization cycle.

change of each gating parameter $u$ depends upon its steady-state value $u_\infty$ and its time constant $\tau_u$ as follows

$$\frac{du}{dt} = \frac{u_\infty - u}{\tau_u}. \tag{2.1}$$

In turn, the steady-state value and time constant are defined as

$$u_\infty = \frac{\alpha_u}{\alpha_u + \beta_u}, \quad \tau_u = \frac{1}{\alpha_u + \beta_u},$$

where $\alpha_u$ and $\beta_u$ are voltage-dependent rate parameters. The total ionic current per unit area $I_{\text{ion}}$, is then defined as the sum of individual ionic currents,

$$I_{\text{ion}} = \sum_s I_s.$$

To complete the action-potential model, the rate of change of the transmembrane potential is given by

$$\frac{dV}{dt} = -\frac{1}{C}(I_{\text{ion}} - I_{\text{app}}), \tag{2.2}$$

where $C$ is the membrane capacitance per unit area and $I_{\text{app}}$ is an applied stimulus current per unit area. In the Luo-Rudy (LR) model, there is a total of six ionic currents: $I_{\text{Na}}$, a fast sodium current; $I_{\text{si}}$, a slow inward current carried primarily by calcium ions; $I_{\text{K}}$, a time-dependent potassium current; $I_{\text{K1}}$, a time-independent potassium current; $I_{\text{Kp}}$, a plateau potassium current; and $I_{\text{b}}$, a background potassium current. The latter three currents are time-independent and so contribute no ODEs to the model. The first three currents are time-dependent and contribute a total of six ODEs of the form of Equation (2.1). As a further point, the reversal potential $E_{\text{si}}$ of the slow inward current is not a constant; it is modeled as a function of the intracellular calcium concentration, $[\text{Ca}]_i$, which itself is governed by the following ODE

$$\frac{d[\text{Ca}]_i}{dt} = -10^4 I_{\text{si}} + 0.07(10^4 - [\text{Ca}]_i). \tag{2.3}$$

In summary, simulating the action potential according to the LR model involves integrating a system of eight ODEs: six of the form of Equation (2.1) coupled with Equations (2.2) and (2.3).

With the space-clamped action potential described mathematically, we now turn our attention to modeling the *propagation* of these action potentials. Let $H$ represent the ventricular heart volume of interest, comprised of cylindrical myocardial fibers and defined as $H = \{\boldsymbol{x} = (x, y, z) | 0 \leq x \leq L_1, 0 \leq y \leq L_2, 0 \leq z \leq L_3\}$. We assume that $H$ is a bidomain region composed of interpenetrating media — the intracellular medium, denoted by (sub)superscript $i$, and the interstitial medium, denoted by (sub)superscript $e$ — everywhere connected by a continuous cardiac cell membrane. Electric potentials within the intracellular and interstitial media are denoted by $\Phi_i$ and $\Phi_e$, respectively. The transmembrane potential, $V$, is then defined as $V = \Phi_i - \Phi_e$. With the assumption of axial symmetry of the myocardial fibers, electrical conductivities in the local basis are the same in any direction perpendicular to the fiber axis (referred to as the transverse fiber direction), and are denoted by $\sigma_t^i$ and $\sigma_t^e$. However, due to the anisotropic nature of cardiac muscle, these conductivities are respectively less than $\sigma_l^i$ and $\sigma_l^e$, the conductivities measured along any direction parallel to the fiber axis (referred to as the longitudinal fiber direction). Because of the complex geometry of the myocardial fibers, such as rotation and curvature, electrical conductivities in the global coordinate space are given by conductivity tensors, $\boldsymbol{D}_i$ and $\boldsymbol{D}_e$, as follows

$$\begin{aligned} \boldsymbol{D}_i &= \sigma_t^i \boldsymbol{I} + (\sigma_l^i - \sigma_t^i)\boldsymbol{a}_l(\boldsymbol{x})\boldsymbol{a}_l^T(\boldsymbol{x}) & (2.4)\\ \boldsymbol{D}_e &= \sigma_t^e \boldsymbol{I} + (\sigma_l^e - \sigma_t^e)\boldsymbol{a}_l(\boldsymbol{x})\boldsymbol{a}_l^T(\boldsymbol{x}), & (2.5)\end{aligned}$$

where $\boldsymbol{I}$ is the identity matrix and $\boldsymbol{a}_l(\boldsymbol{x})$ is a unit vector corresponding to the longitudinal fiber direction at each point $\boldsymbol{x}$ in the bidomain region $H$. Application of bidomain theory then yields the following system of partial differential equations (PDEs) [**4, 5**]

$$\begin{aligned} \chi C\,\frac{\partial V}{\partial t} &= \nabla\cdot\boldsymbol{D}_i\nabla V + \nabla\cdot\boldsymbol{D}_i\nabla\Phi_e - \chi I_{\text{ion}}(\boldsymbol{x},t,V,\boldsymbol{U}) + \chi I_{\text{app}}(\boldsymbol{x},t) & (2.6)\\ -\nabla\cdot\boldsymbol{D}_i\nabla V &= \nabla\cdot(\boldsymbol{D}_i+\boldsymbol{D}_e)\nabla\Phi_e, & (2.7)\end{aligned}$$

where $\chi$ is the membrane surface area per unit volume of tissue; $C$, $I_{\text{ion}}$ and $I_{\text{app}}$ are as defined previously.

To make the problem more tractable, we make an assumption which reduces the above system of equations to a single PDE in the transmembrane potential $V$ [**3**]. We assume that the ratio of extracellular to intracellular conductivities is the same in any direction; that is, for some constant $r$,

$$\frac{\sigma_t^e}{\sigma_t^i} = \frac{\sigma_l^e}{\sigma_l^i} = r.$$

As a consequence of this restriction, we have $\boldsymbol{D}_e = r\boldsymbol{D}_i$. Substituting this into Equation (2.7) gives

$$\nabla\cdot\boldsymbol{D}_i\nabla\Phi_e = -\frac{1}{1+r}\nabla\cdot\boldsymbol{D}_i\nabla V,$$

and with this result Equation (2.6) can be written as

$$\chi C\frac{\partial V}{\partial t} = \frac{r}{1+r}\nabla\cdot\boldsymbol{D}_i\nabla V - \chi I_{\text{ion}} + \chi I_{\text{app}}. \tag{2.8}$$

This is a nonhomogeneous, quasi-linear, parabolic PDE in the transmembrane potential $V$ and it is used in our model to describe the spread of action potentials in $H$.

To complete the model, we impose standard initial and boundary conditions. All cells are initially assumed to be at resting potential; therefore,

$$V(\boldsymbol{x},0) = V_{\text{rest}}$$

for all $\boldsymbol{x}\in H$, where for simplicity we specify the initial time to be $t_0 = 0$. The boundary conditions reflect the fact that there is no current flow across the boundary of $H$ and are given by

$$\boldsymbol{n}\cdot\nabla V \equiv \frac{\partial V}{\partial \boldsymbol{n}} = 0,$$

where $\boldsymbol{n}$ is the unit outward normal vector on $\partial H$, the boundary of $H$. Thus, the full mathematical model used in this study to describe the propagation of electrical activity in the ventricular myocardium is

$$\begin{aligned}
\chi C_m \frac{\partial V}{\partial t} &= \nabla \cdot \boldsymbol{D}_r \nabla V - \chi I_{\text{ion}}(V, \boldsymbol{U}) + \chi I_{\text{app}}(\boldsymbol{x}, t) && (2.9)\\
\frac{d\boldsymbol{U}}{dt} &= \boldsymbol{F}(V, \boldsymbol{U}) && (2.10)\\
\frac{\partial V}{\partial \boldsymbol{n}} &= 0 && (2.11)\\
V(\boldsymbol{x}, 0) &= V_{\text{rest}} && (2.12)\\
\boldsymbol{U}(0) &= \boldsymbol{U}_0. && (2.13)
\end{aligned}$$

The vector $\boldsymbol{U} = [u_1, \ldots, u_7]^T$ represents the variables of the LR model, namely, $u_1$, $u_2$, ..., $u_6$ are gating parameters and $u_7 = [\text{Ca}]_i$, the intracellular calcium concentration.

**Numerical Methods.** Simulations of the mathematical model involve updating the seven ODEs of the LR model at each point in the spatial domain, as well as integrating the parabolic PDE in the transmembrane potential $V$. Six of the seven LR variables are gating parameters, where each variable $u$ is governed by the ODE

$$\frac{du}{dt} = \frac{u_\infty(V) - u}{\tau_u(V)}.$$

Following the method of [**19**], we assume that $V$ is constant over the time step necessary to update $u$. Then the above equation can be integrated explicitly to yield

$$u(t_1) = u_\infty - [u_\infty - u(t_0)]\, e^{-k/\tau_u}$$

and in general,

$$u(t_{n+1}) = u_\infty - [u_\infty - u(t_n)]\, e^{-k/\tau_u},$$

where $k$ represents the time step and $t_n = n \cdot k$. Updating the intracellular calcium concentration is done by a fourth-order Runge-Kutta method.

For the numerical solution of the PDE, we use the Douglas-Brian finite difference scheme [**2, 10**] with modifications to incorporate the mixed and first-order spatial derivatives of $V$ [**3**]. Equation (2.8) can be rewritten as

$$\begin{aligned}
\frac{\partial V}{\partial t} &= c^1 \frac{\partial^2 V}{\partial x^2} + c^2 \frac{\partial^2 V}{\partial y^2} + c^3 \frac{\partial^2 V}{\partial z^2} + c^4 \frac{\partial^2 V}{\partial x \partial y} + c^5 \frac{\partial^2 V}{\partial x \partial z} + c^6 \frac{\partial^2 V}{\partial y \partial z}\\
&\quad + c^7 \frac{\partial V}{\partial x} + c^8 \frac{\partial V}{\partial y} + c^9 \frac{\partial V}{\partial z} + f(x, y, z, t, V, \boldsymbol{U}), \qquad (2.14)
\end{aligned}$$

where the coefficients $c^1, c^2, \ldots, c^9$ are functions of the spatial variables (i.e., $c^l = c^l(x, y, z)$), and $f$ represents the current source term. Let $hx$, $hy$ and $hz$ represent the spatial step sizes, and again $k$ will denote the time step. Furthermore, let $\rho$ represent mesh ratios such that $\rho_{xx} = k/(hx)^2$, $\rho_x = k/hx$ and similarly for other parameters. We impose a grid on the heart volume $H$ such that $(x_i, y_j, z_m) = (i \cdot hx, j \cdot hy, m \cdot hz)$ and then the values of the transmembrane potential $V$ on this grid are approximated at time $t_{n+1}$ from the values of $V$ at time $t_n$ according to the following finite-difference scheme

$$\begin{aligned}
\left[1 - \frac{\rho_{xx}}{2} c^1 \delta_x^2\right] V_{n+1}^* &= \Big[1 + \frac{\rho_{xx}}{2} c^1 \delta_x^2 + \rho_{yy} c^2 \delta_y^2 + \rho_{zz} c^3 \delta_z^2 + \rho_{xy} c^4 \mu_x \delta_x \mu_y \delta_y \\
&\quad + \rho_{xz} c^5 \mu_x \delta_x \mu_z \delta_z + \rho_{yz} c^6 \mu_y \delta_y \mu_z \delta_z + \rho_x c^7 \mu_x \delta_x \\
&\quad + \rho_y c^8 \mu_y \delta_y + \rho_z c^9 \mu_z \delta_z\Big] V_n + k\hat{f}_n \qquad (2.15) \\
\left[1 - \frac{\rho_{yy}}{2} c^2 \delta_y^2\right] V_{n+1}^{**} &= -\frac{\rho_{yy}}{2} c^2 \delta_y^2 V_n + V_{n+1}^* \qquad (2.16) \\
\left[1 - \frac{\rho_{zz}}{2} c^3 \delta_z^2\right] V_{n+1} &= -\frac{\rho_{zz}}{2} c^3 \delta_z^2 V_n + V_{n+1}^{**}, \qquad (2.17)
\end{aligned}$$

where

$$\hat{f}_n = \hat{f}_{i,j,m,n} \equiv f(x_i, y_j, z_m, t_{n+1/2}, \hat{V}_{i,j,m,n+1/2}, \boldsymbol{U}_{i,j,m,n})$$

and $\mu$ and $\delta$ are, respectively, the average and central-difference operators. As well, in the above equations the $i$, $j$ and $m$ subscripts have been suppressed, so that $c^1 = c^1_{i,j,m} \equiv c^1(x_i, y_j, z_m)$ and $V_n = V_{i,j,m,n} \equiv V(x_i, y_j, z_m, t_n)$, and similarly for other values. This scheme gives rise to three tridiagonal systems of equations which are easy to implement and solve.

## 3 Results and discussion

A rough estimate of the wavelength, and hence the size of rotating waves for LR media is given by $s \times$ APD, where $s$ is the wave speed and APD is the action potential duration. For our propagation model, the wave speeds in the transverse and longitudinal directions are approximately 0.025 cm/ms and 0.035 cm/ms, respectively, and the APD of the LR model is 366 ms. An investigation of reentrant rotating waves would therefore require simulations of blocks of tissue on the order of 10 cm $\times$ 10 cm. To reduce this requirement, and hence the computational load, we modified the LR model to produce an action potential with the same essential shape but with a reduced duration. To reduce its action potential duration to approximately 200 ms, the time constants for calcium channel activation and inactivation were replaced by $\tau_{d,\text{new}} = 0.5\tau_d$ and $\tau_{f,\text{new}} = 0.5\tau_f$, respectively; the time constant for activation of the time-dependent potassium current was changed to $\tau_{X,\text{new}} = 0.4\tau_X$; and the maximum conductance for the plateau potassium current was increased to $\hat{G}_{\text{Kp,new}} = 0.0583\text{mS/cm}^2$. Figure 1 shows the action potential as calculated from both the modified and unmodified LR model.

*Tissue Block for Cross-field Stimulation.* We consider a 5 cm $\times$ 5 cm $\times$ 0.5 cm block that consists of homogeneous myocardial tissue with *uniform anisotropy*; that is, we neglect the rotation and curvature of the fibers, and furthermore, the fibers are assumed to run parallel with the $x$-axis. The $z$-axis is vertical, and the planes $z = 0$ and $z = 0.5$ represent the endo- and epicardial surfaces, respectively. Using a uniform spatial step of 0.05 cm, the block was discretized by 100 $\times$ 100 $\times$ 10 nodes. To initiate rotating waves, we used the cross-field stimulation method [**7**]. Excitation was begun with a basic stimulus applied to the entire left lateral face of the block of tissue. This initiated a planar wave that propagated from left to right. A second premature stimulus was then applied to the entire front lateral face. The propagation parameters are given in Table 1.

*Tissue Block for Pinwheel Stimulation.* Here we used the exact same setup as above, except the block size was increased to 5 cm $\times$ 5 cm $\times$ 1 cm and subsequently

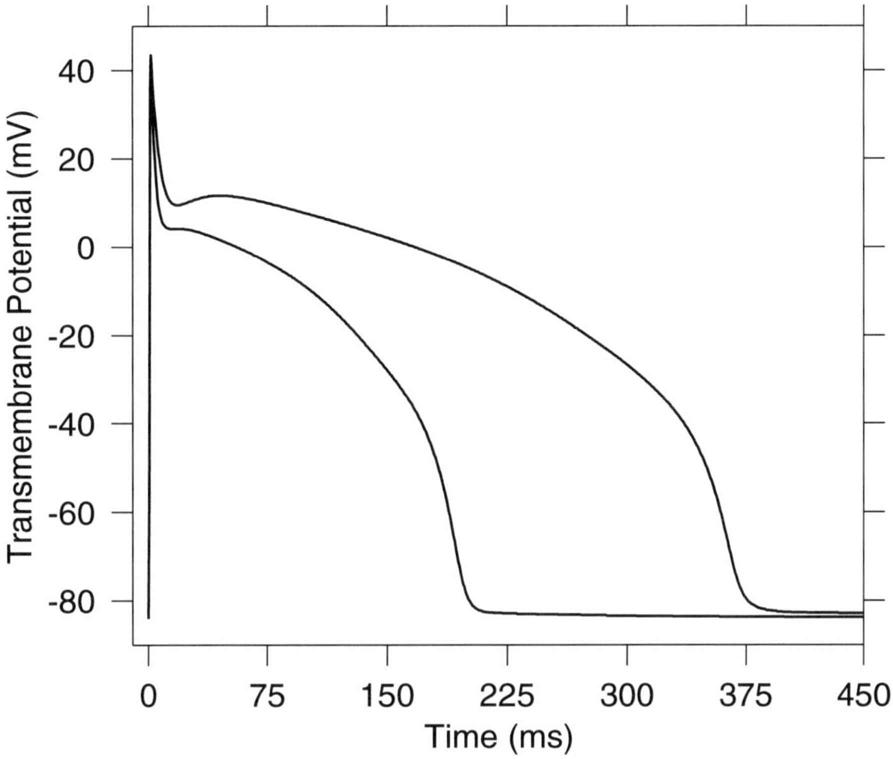


**Figure 1** Action potentials from the original and modified LR model.

**Table 1** Propagation parameters for reentrant rotating wave simulations.

| | |
|---|---|
| Equal anisotropy ratio $r$ | 1.0 |
| Intracellular transverse conductivity $\sigma_t^i$ | 0.5 mS/cm |
| Intracellular longitudinal conductivity $\sigma_l^i$ | 1.0 mS/cm |
| Membrane capacitance $C_m$ | 1.0 $\mu$F/cm$^2$ |
| Surface-to-volume ratio $\chi$ | 1000.0 cm$^{-1}$ |
| Spatial step $hx = hy = hz$ | 0.05 cm |
| Time step $k$ | 0.2 ms |
| Basic current stimulus $I_{\text{app}}$ | 30.0 $\mu$A/cm$^2$ |
| Duration of basic stimulus $t_{\text{off}}$ | 2.5 ms |
| Premature current stimulus $I_{\text{app}}$ | 12.0 $\mu$A/cm$^2$ |
| Duration of premature stimulus $t_{\text{off}}$ | 4.0 ms |

discretized by $100 \times 100 \times 20$ nodes. Rotating waves were initiated with the pinwheel stimulation protocol as follows. As before, excitation was begun with a basic stimulus applied to the entire left lateral face of the block, which initiated a planar wave that propagated from left to right. However, the second premature stimulus was applied to an area on the top face of the block just behind the propagating refractory tail of the first wave. The propagation parameters are given in Table 1.

All computer programs were written in Fortran 77. The simulations were performed on an IBM RS/6000 model 590 with version 3.1.2 of the IBM XL Fortran compiler.

Figure 2 shows the sequential time plots of the electrical activity resulting from cross-field stimulation on the top face or epicardial surface of a 5 cm × 5 cm × 0.5 cm block. The first panel shows the propagating refractory tail of the first wave that propagated from left to right just before onset of the second stimulus occurred. Panels 2 through 4 show the initiation of a second planar wave that propagated from front to back except at the right where it encountered the refractory tail of the first wave. In panel 5, we see the first formation of a wave break that propagated to the right in the now recovered zone. The remaining panels show that this wave break began to curl and led to the formation of a clockwise rotating scroll wave.

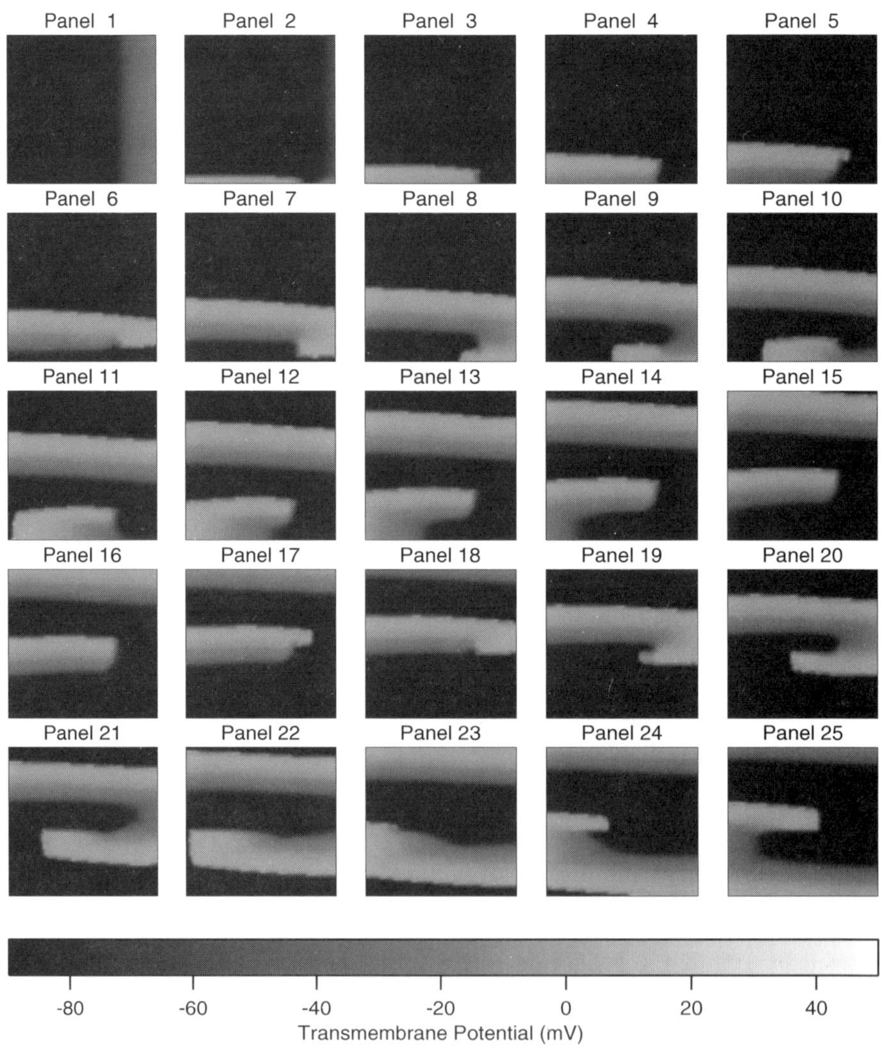


**Figure 2** Initiation of reentrant activity in the form of a scroll wave in a 5 cm × 5 cm × 0.5 cm homogeneous block of myocardial tissue with uniform anisotropy. Panels 1 through 25 show the time variation of electrical activity on the top face of the block only. Panel 1 is a snapshot at $t = 400$ ms and the time interval between subsequent panels is 50 ms. Transmembrane voltage is shaded according to the grey-scale shown in the bottom panel.

Figure 3 shows the sequential time plots of the electrical activity on the epicardial surface resulting from the pinwheel stimulation protocol. As before, panel 1 shows the propagating refractory tail of the first wave that propagated from left to right. Panel 2 shows the activated region resulting from the second stimulus. This premature stimulus was applied to a rectangular patch of cells ($30 \times 40$) on the top layer of the block of tissue. Excitation from this stimulus initially propagates in all directions, except towards the right. In panel 6 we see the formation of two counter-rotating waves that are subsequently terminated when they hit the left border before curling around to complete another rotation.

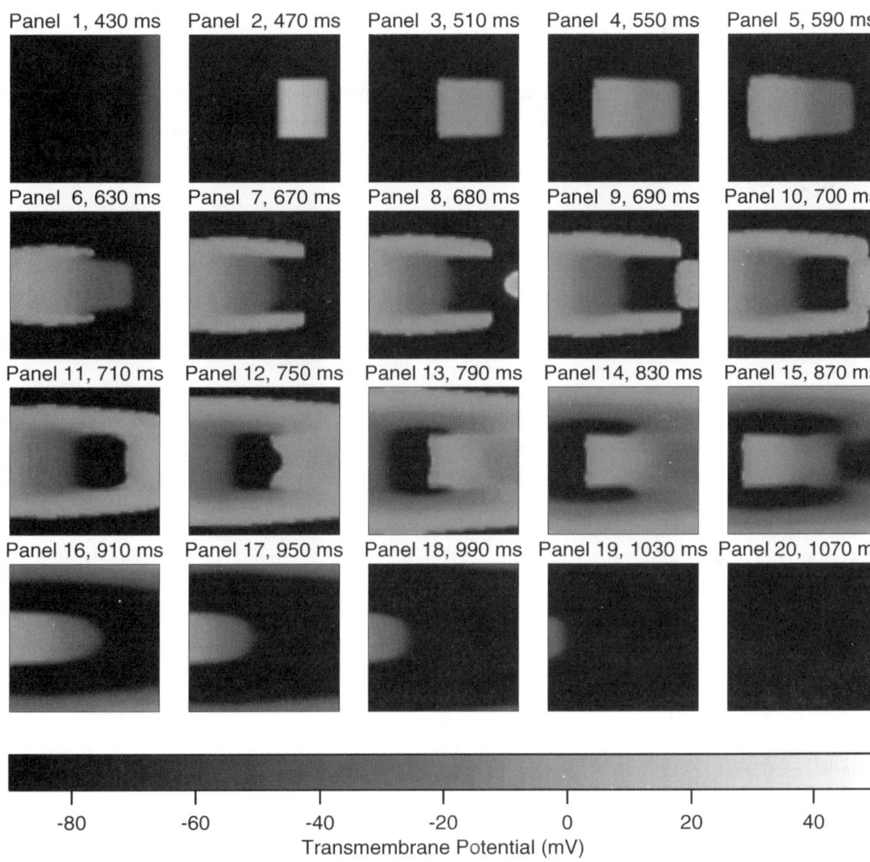


**Figure 3** Initiation of reentrant activity in the form of a counter-rotating waves in a 5 cm $\times$ 5 cm $\times$ 1.0 cm homogeneous block of myocardial tissue with uniform anisotropy. Other explanations as in Figure 2.

Despite the uniform anisotropy of the medium, the scroll waves were not stationary. In the case of the cross-field stimulation, the resulting scroll wave drifted from front to back, and then right to left and left to right several times. At approximately $t = 2900$ ms, the wave front struck the left border before it could curl around. This led to termination of the scroll wave. In addition, timing of the second, premature stimulus was critical in initiating the rotating waves. Applying the second stimulus at a later time shifts the wave break seen in panel 5 of Figure 2, for example, towards the right. If the stimulus was applied too late, this wave break struck the right border and was extinguished before it could curl around. In a similar fashion, applying the premature stimulus too soon shifted the wave break towards the left. In this case, the wavefront did curl around, but it subsequently struck the left border and was extinguished before it could curl around again to complete a full cycle.

We have investigated the propagation of electrical activation in the three-dimensional anisotropic ventricular myocardium using a mathematical model based on bidomain theory. A key feature of the model is a physiologically accurate formulation of the ionic currents describing the transmembrane action potential. This feature allows for simulation of the phenomena of action-potential propagation which fully incorporates the recovery phase of the process. Many propagation models to date often lack this feature, particularly in three-dimensional simulations. Finally, we have shown that Luo-Rudy media with uniform anisotropy can support reentrant activity in the form of scroll waves. Though we present only the results for the case of uniform anisotropy, the mathematical model and corresponding numerical methods allow incorporation of anatomically accurate properties of cardiac tissue such as fiber rotation and curvature.

## References


[1] Beeler, G. W. and Reuter, H. [1977], *Reconstruction of the action potential of ventricular myocardial fibres*, J. Physiol. **268**, 177–210.

[2] Brian, P. L. T. [1961], *A finite difference method of high-order accuracy for the solution of three-dimensional transient heat conduction problems*, AIChE J. **1**, 367.

[3] Clements, C. J. [1996], *Nonlinear wave propagation in an anisotropic medium*, Master's thesis, Department of Mathematics, Statistics, and Computing Science, Dalhousie University, Halifax, Nova Scotia, Canada, 1996.

[4] Colli Franzone, P. and Guerri, L. [1990], *Mathematical modeling of the excitation process in myocardial tissue: Influence of fiber rotation on wavefront propagation and potential field*, Math. Biosci. **101**, 155–235.

[5] Colli Franzone, P. [1993], *Spreading of excitation in 3-D models of the anisotropic cardiac tissue. I. Validation of the eikonal model*, Math. Biosci. **113**, 145–209.

[6] Courtemanche, M. and Winfree, A. T. [1991], *Re-entrant rotating waves in a Beeler-Reuter based model of two-dimensional cardiac electrical activity*, Int. J. Bifur. Chaos **1**, 431–444.

[7] Davidenko, J. M., Kent, P. F., Chialvo, D. R., Michaels, D. C. and Jalife, J. [1990], *Sustained vortex-like waves of excitation in normal isolated ventricular muscle*, Proc. Natl. Acad. Sci. USA **87**, 8785–8789.

[8] Davidenko, J. M., Pertsov, A. M., Salomonsz, R., Baxter, W. P. and Jalife, J. [1992], *Spatiotemporal irregularities of spiral wave activity in isolated ventricular muscle*, J. Electrocardiol. **24 (Suppl.)**, 113–122.

[9] Davidenko, J. M., Pertsov, A. V., Salomonsz, R., Baxter, W. and Jalife, J. [1992], *Stationary and drifting spiral waves of excitation in isolated cardiac muscle*, Nature **355**, 349–351.

[10] Douglas, Jr., J. [1962], *Alternating direction methods for three space variables*, Numer. Math. **4**, 41–63.

[11] Frazier, D. W., Wolf, P. D., Wharton, J. M., Tabg, A. S. L., Smith, W. M. and Ideker, R. E. [1989], *Stimulus-induced critical point: Mechanism for the electrical initiation of reentry in normal canine myocardium*, J. Clin. Invest. **83**, 1039–1052.

[12] Hodgkin, A. L. and Huxley, A. F. [1952], *A quantitative description of membrane current and its application to conduction and excitation in nerve*, J. Physiol. **117**, 500–544.

[13] Krinsky, V. I. [1978], *Mathematical models of cardiac arrhythmias (spiral waves)*, J. Pharmacol. Ther. B **3**, 539–555.

[14] Luo, C.-H. and Rudy, Y. [1991], *A model of the ventricular cardiac action potential: Depolarization, repolarization, and their interaction*, Circ. Res. **68**, 1501–1526.

[15] Panfilov, A. V. and Holden, A. V. [1991],*Spatiotemporal irregularity in a two-dimensional model of cardiac tissue*, Int. J. Bifur. Chaos **1**, 219–225.

[16] Pertsov, A. M., Davidenko, J. M., Salomonsz, R., Baxter, W. T. and Jalife, J. [1993], *Spiral waves of excitation underlie reentrant activity in isolated cardiac muscle*, Circ. Res. **72**, 631–650.

[17] Pertsov, A. M., Emarkova, E. A. and Panfilov, A. V. [1984], *Rotating spiral waves in modified FitzHugh-Nagumo model*, Physica D **14**, 117–124.

[18] Pertsov, A. M. and Jalife, J. [1995], *Three-dimensional vortex-like reentry*, Cardiac Electrophysiology: From Cell to Bedside (D. P. Zipes and J. Jalife, eds.), W. B. Saunders Co., Philadelphia, 2nd ed., pp. 403–410.

[19] Rush, S. and Larsen, H. [1978], *A practical algorithm for solving dynamic membrane equations*, IEEE Trans. Biomed. Eng **BME-25**, 389–392.

[20] Winfree, A. T. [1987], *When time breaks down*, Princeton University Press, Princeton, New Jersey.

[21] Winfree, A. T. [1989], *Electrical instability in cardiac muscle: Phase singularities and rotors*, J. Theor. Biol. **138**, 353–405.

Fields Institute Communications
Volume **21**, 1999

# An Optimal Control Theory Approach to the Inverse Problem of Electrocardiography

**John C. Clements**
Department of Mathematics, Statistics and Computing Science
Dalhousie University
Halifax, Nova Scotia, Canada B3H 3J5
clements@mscs.dal.ca

**Abstract.** Many ill-posed problems in the biomedical and related sciences reduce to solving systems of the form $A\boldsymbol{x} = \boldsymbol{b}, A : E^n \to E^m$ where $A$ is ill-conditioned with singular values decaying to zero in such a way that standard methods cannot yield solutions which are stable with respect to small perturbations in the components of $\boldsymbol{b}$. One approach to resolving this problem is Tikhonov regularization in which $A\boldsymbol{x} = \boldsymbol{b}$ is replaced by the perturbed least squares problem $\min_{\boldsymbol{x}\in E^n}\{\|A\boldsymbol{x}-\boldsymbol{b}\|^2 + \lambda^2\|B\boldsymbol{x}\|^2\}$ where $\|.\|$ denotes the usual Euclidean norm, $\lambda$ is the regularization or smoothing parameter, and $B$ is a $p \times n$ regularizing operator. The key to this approach is the determination of a value for $\lambda$ which strikes a good compromise between minimizing the residual and regularizing the solution. We give here a brief survey of the most effective methods currently available for estimating the optimal value $\lambda^*$. These methods are applied to the inverse problem of electrocardiography based on formulas generated for the components $\|A\boldsymbol{x} - \boldsymbol{b}\|$ and $\|B\boldsymbol{x}\|$ of the objective function using the generalized singular value decomposition (GSVD) of $A = UDRQ^T$ and the regularizing operator $B = VZRQ^T$. The problem is then reformulated as a well-posed optimal control problem and solved using the regularized solutions $\boldsymbol{x}(\lambda)$ as starting values for the control problem. Preliminary results indicate that this approach can lead to a solution which is better than the best possible Tikhonov solution $\boldsymbol{x}(\lambda^*)$.

## 1 Introduction

An important problem in the biomedical sciences is the inverse problem of electrocardiography, the objective of which is the non-invasive recovery of epicardial potentials from body surface potentials. In Section 2, this ill-posed problem is reduced to the solution of a discrete system $A\boldsymbol{x} = \boldsymbol{b}, A : E^n \to E^m$ where $A$ is

1991 *Mathematics Subject Classification.* Primary 65F20, 65F30; Secondary 49A22.

This work was partially supported by a research grant from the Natural Sciences and Engineering Council of Canada.

ill-conditioned with singular values decaying to zero in such a way that standard methods cannot yield solutions which are stable with respect to small perturbations in the components of $\boldsymbol{b}$. The most successful approach to resolving this instability with respect to perturbations has been the application of Tikhonov regularization [**2, 6, 3, 7, 14, 15**]. The $L$-curve, CRESO and Zero-crossing methods for estimating the optimal regularization parameter when Tikhonov regularization is used to solve the system $A\boldsymbol{x} = \boldsymbol{b}$ are described in Section 3. These parameter estimates are denoted by $\lambda_{LC}$, $\lambda_{CR}$ and $\lambda_{ZC}$, respectively and are computed using the formulas derived in Section 3. In Section 4, the inverse problem is reformulated as a well-posed optimal control problem and a method of solution is introduced which requires a regularized solution $\boldsymbol{x}(\lambda)$ as a starting value. A specific numerical example is considered in Section 5 and the preliminary results indicate that this approach can lead to a control solution which is better than the best possible Tikhonov solution $\boldsymbol{x}(\lambda^*)$.

## 2 The inverse problem of electrocardiography

The wave of electrical activation which precedes each human heart beat can be recorded as a sequence of body-surface electrocardiograms (ECGs) or body surface potential maps (BSPMs). Each BSPM can be represented as a potential distribution $\boldsymbol{\Phi}_B = (\phi_B^1, \ldots, \phi_B^m)$ at $m$ points (possibly ECG electrode locations) $\boldsymbol{Q}_j, j = 1, \ldots, m$ on the body surface. The inverse problem of electrocardiography constitutes the recovery of the corresponding epicardial potential distribution $\boldsymbol{\Phi}_H = (\phi_H^1, \ldots, \phi_H^n)$ at $n$ designated points $\boldsymbol{P}_i, i = 1, \ldots, n$ on the epicardial surface. Epicardial potential mappings have become a useful clinical tool for the analysis of various cardiovascular abnormalities such as the effects of myocardial ischemia and the mechanisms of ventricular arrhythmias [**6, 10, 12, 13**].

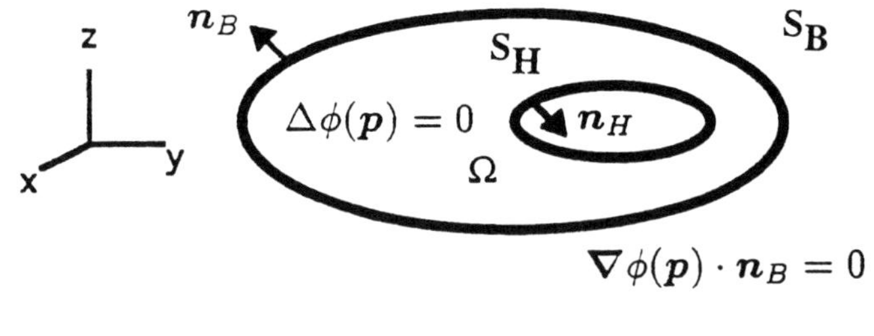


**Figure 1** The volume conductor $\Omega$ bounded by closed surfaces $S_H$ and $S_B$.

In the standard quasi-static potential theoretic model for the propagation of the excitation [**2, 6, 15**], the human torso is represented by a doubly connected, homogeneous, isotropic volume conductor $\Omega$ bounded by smooth closed surfaces $S_H$ (the epicardial surface) and $S_B$ (the torso surface) with unit outer normals $\boldsymbol{n}_H$ and $\boldsymbol{n}_B$, respectively as in Figure 1. Let $\phi(\boldsymbol{p})$ be the electric potential at a field point $\boldsymbol{p} = (x, y, z)$ in $\Omega \cup S_H \cup S_B$. Then $\phi$ satisfies the boundary-value problem

$$\Delta\phi(\boldsymbol{p}) = 0, \qquad \boldsymbol{p} \in \Omega; \qquad \boldsymbol{\nabla}\phi(\boldsymbol{p}) \cdot \boldsymbol{n}_B = 0, \qquad \boldsymbol{p} \in S_B, \tag{2.1}$$

where $\Delta$ and $\boldsymbol{\nabla}$ denote the three-dimensional Laplace and gradient operators, respectively.

The problem of solving for epicardial potentials $\phi_H(\boldsymbol{p})$ given the torso-surface potentials $\phi_B(\boldsymbol{p})$ is mathematically ill-posed. The equivalent boundary-integral equation for the potential $\phi$ at any field point $\boldsymbol{p}$ on $S_H \cup S_B$ is

$$2\pi\phi(\boldsymbol{p}) = \int_{S_B^-} \phi_B \boldsymbol{\nabla} r^{-1} \cdot \boldsymbol{n}_B dS - \int_{S_H^-} \phi_H \boldsymbol{\nabla} r^{-1} \cdot \boldsymbol{n}_H dS - \int_{S_H^-} r^{-1} \boldsymbol{\nabla}\phi_H \cdot \boldsymbol{n}_H dS, \tag{2.2}$$

where $S_B^-$ and $S_H^-$ denote integration over the surfaces $S_B$ and $S_H$, with the singularity removed; $dS = dS(\boldsymbol{q})$ is the differential of the integration surface; $\boldsymbol{r} = \boldsymbol{q} - \boldsymbol{p}$ is the directed distance from the field point $\boldsymbol{p}$ to the source point $\boldsymbol{q}$ on $S_H \cup S_B$ and $r = |\boldsymbol{r}|$. Discretization of $S_H$ and $S_B$ as in [**6**] gives for each $i = 1, \dots, n$ on $S_H$ and $j = 1, \dots, m$ on $S_B$

$$\sum_{k=1}^{m} p_{HB}^{ik}\phi_B^k + \sum_{k=1}^{n} p_{HH}^{ik}\phi_H^k + \sum_{k=1}^{n} g_{HH}^{ik}\Gamma_H^k = 0 \tag{2.3}$$

$$\sum_{k=1}^{m} p_{BB}^{jk}\phi_B^k + \sum_{k=1}^{n} p_{BH}^{jk}\phi_H^k + \sum_{k=1}^{n} g_{BH}^{jk}\Gamma_H^k = 0, \tag{2.4}$$

where each coefficient $p_{Hs}^{ik}$, $p_{Bs}^{jk}$, $g_{HH}^{ik}$ and $g_{BH}^{jk}$ depends only on the geometry of $S_H$ and $S_B$ and $\Gamma_H^k = \boldsymbol{\nabla}\phi_H^k \cdot \boldsymbol{n}_H$, $k = 1, \dots, n$.

This $m + n$ system of equations can be expressed in the form

$$P_{HB(n\times m)}\boldsymbol{\Phi}_B + P_{HH(n\times n)}\boldsymbol{\Phi}_H + G_{HH(n\times n)}\boldsymbol{\Gamma}_H = 0 \tag{2.5}$$

$$P_{BB(m\times m)}\boldsymbol{\Phi}_B + P_{BH(m\times n)}\boldsymbol{\Phi}_H + G_{BH(m\times n)}\boldsymbol{\Gamma}_H = 0 \tag{2.6}$$

for matrices $P_{rs}$ and $G_{rH}$, where $\boldsymbol{\Phi}_H = (\phi_H^1, \dots, \phi_H^n)$, $\boldsymbol{\Phi}_B = (\phi_B^1, \dots, \phi_B^m)$ and $\boldsymbol{\Gamma}_H = (\Gamma_H^1, \dots, \Gamma_H^n)$.

In order to eliminate the unknown gradients $\boldsymbol{\Gamma}_H$ on the epicardial surface $S_H$, we solve (2.5) for $\boldsymbol{\Gamma}_H = -G_{HH}^{-1}(P_{HB}\boldsymbol{\Phi}_B + P_{HH}\boldsymbol{\Phi}_H)$ and then substitute the result into (2.6) to obtain the $(m \times n)$ system

$$A\boldsymbol{\Phi}_H = (P_{BB} - G_{BH}G_{HH}^{-1}P_{HB})^{-1}(G_{BH}G_{HH}^{-1}P_{HH} - P_{HB})\boldsymbol{\Phi}_H = \boldsymbol{\Phi}_B. \tag{2.7}$$

Thus, the inverse problem in this particular formulation becomes one of solving (2.7) for $\boldsymbol{\Phi}_H$ given $\boldsymbol{\Phi}_B = \boldsymbol{b}$. This problem is ill-posed and numerically ill-conditioned [**6, 10**].

## 3 Parameter estimation methods

Let $A \in E^{m\times n}$ be any $m \times n$ $(m \geq n)$ real matrix, let $B \in E^{p\times n}$ be a $p \times n$ $(p \leq n)$ real matrix of rank $q \leq p$ and let $k + q$ be the numerical effective rank of $\begin{pmatrix} A \\ B \end{pmatrix}$ with $k + q = n$.

In Tikhonov regularization, the problem $A\boldsymbol{x} = \boldsymbol{b}$ is replaced by the perturbed least squares problem

$$\min_{\boldsymbol{x}\in E^n} \{\|A\boldsymbol{x} - \boldsymbol{b}\|^2 + \lambda^2\|B\boldsymbol{x}\|^2\} \tag{3.1}$$

where $\|.\|$ denotes the usual Euclidean norm, $\lambda$ is the regularization (smoothing) parameter and $B$ is the $p \times n$ $(p \leq n)$ regularizing operator. Thus, for any $A \in E^{m\times n}$ and each $\lambda > 0$, a solution $\boldsymbol{x}(\lambda)$ of the perturbed problem (3.1) satisfies the generalized normal equations

$$(A^TA + \lambda^2 B^TB)\boldsymbol{x}(\lambda) = A^T\boldsymbol{b}. \tag{3.2}$$

Zero, first and second-order Tikhonov regularization are the terms normally associated with the choice of smoothing operators $B = I_N$, the identity, and $B = \boldsymbol{\nabla}$ and $B = \Delta$, the discretized gradient and Laplace differential operators, respectively. The key step in this approach is the assignment of a value for $\lambda$ which strikes a good compromise between minimizing the residual and regularizing the solution.

The generalized singular-value decomposition (GSVD) of $A$ and $B$ is given by

$$A = UDRQ^T, \qquad B = VZRQ^T, \tag{3.3}$$

where

$$D = \begin{pmatrix} I_k & O_{k\times q} \\ O_{q\times k} & C_{q\times q} \\ O_{(m-n)\times k} & O_{(m-n)\times q} \end{pmatrix}, \qquad Z = \begin{pmatrix} O_{q\times k} & S_{q\times q} \\ O_{(p-q)\times k} & O_{(p-q)\times q} \end{pmatrix},$$

$$R = \begin{pmatrix} R1_{k\times k} & R2_{k\times q} \\ O_{q\times k} & R3_{q\times q} \end{pmatrix}.$$

Here $C_{q\times q} = \operatorname{diag}(\alpha_{(k+1)}, \cdots, \alpha_n)$ and $S_{q\times q} = \operatorname{diag}(\beta_{(k+1)}, \cdots, \beta_n)$ are diagonal matrices, $O_{q\times k}$ denotes the $q \times k$ additive identity, $U \in E^{m\times m}$, $V \in E^{p\times p}$, and $Q \in E^{n\times n}$ are orthogonal and $R \in E^{n\times n}$ is a nonsingular upper tridiagonal matrix [**1**]. The generalized singular values $\mu_i = \dfrac{\alpha_i}{\beta_i}, i = k+1, \dots, n$ satisfy $0 \leq \alpha_{k+1}, \cdots, \alpha_n \leq 1, 1 \geq \beta_{k+1}, \dots, \beta_n > 0$ and $\alpha_i^2 + \beta_i^2 = 1, i = k+1, ..., n$. As in [**6**], substitution of (3.3) into (3.2) gives the unique solution $\boldsymbol{x}(\lambda)$ of (3.1) for each $\lambda > 0$

$$\boldsymbol{x}(\lambda) = \sum_{i=1}^{k} (\boldsymbol{u}_i \cdot \boldsymbol{b})\boldsymbol{y}_i + \sum_{i=k+1}^{n} \frac{\mu_i}{\mu_i^2 + \lambda^2}\left(\frac{1}{\beta_i}\right)(\boldsymbol{u}_i \cdot \boldsymbol{b})\boldsymbol{y}_i \tag{3.4}$$

and it follows that the residual $\|A\boldsymbol{x}(\lambda) - \boldsymbol{b}\|$ and solution seminorm $\|B\boldsymbol{x}(\lambda)\|$ in (3.1) can be expressed in the form

$$\|A\boldsymbol{x}(\lambda) - \boldsymbol{b}\| = \left(\sum_{i=k+1}^{n} \left[\frac{\lambda^2}{\mu_i^2 + \lambda^2}(\boldsymbol{u}_i \cdot \boldsymbol{b})\right]^2 + \delta_o^2\right)^{1/2}, \tag{3.5}$$

$$\|B\boldsymbol{x}(\lambda)\| = \left(\sum_{i=k+1}^{n} \left[\frac{\mu_i}{\mu_i^2 + \lambda^2}(\boldsymbol{u}_i \cdot \boldsymbol{b})\right]^2\right)^{1/2}, \tag{3.6}$$

where the $\boldsymbol{u}_i$ are the columns of $U$, the $\boldsymbol{y}_i$ are the columns of $Y = (RQ^T)^{-1}$, $i = 1, \dots, n$ and $\delta_o^2 = \sum_{i=n+1}^{m}(\boldsymbol{u}_i \cdot \boldsymbol{b})^2$.

If $\boldsymbol{b} \in range(A)$, the system is termed *consistent* and $\delta_o^2 = 0$ in (3.5). We now give a brief description of the $L$-curve, CRESO and Zero-crossing methods for estimating the regularization parameter based on formulas (3.5) and (3.6).

**The $L$-curve method for computing $\lambda_{LC}$.** The best-known method for estimating a value for the regularization parameter is defined in terms of the $L$-curve $C := \{(\|A\boldsymbol{x}(\lambda) - \boldsymbol{b}\|, \|B\boldsymbol{x}(\lambda)\|)|, \lambda > 0\}$ for (3.1). The criterion proposed by Hansen and O'Leary [**4, 5**] is that value $\lambda_{LC}$ which corresponds to the "corner" point on the $L$-curve possessing maximum curvature $\kappa(\lambda)$. As in [**2**], an explicit expression for the curvature $\kappa(\lambda)$ of $C$ can be obtained using (3.5) and (3.6) and that $\lambda_{LC}$ corresponding to the "corner" point can be computed as the solution of a single-variable optimization problem using the NAG routine E04ABF [**11**]. A particular advantage associated with the $L$-curve method is that it has been shown to be more robust than generalized cross-validation in the presence of correlated errors [**5**].

**The CRESO method for computing $\lambda_{CR}$.** The CRESO (or composite residual and smoothing operator) technique for estimating the regularization parameter was developed by Colli-Franzone *et al.* in [**3**]. The method was presented as an empirical heuristic but has become widely accepted as the preferred method of parameter estimation in a large class of ill-posed bioelectric inverse problems (see e.g., [**2, 14, 15**]). The CRESO method chooses that parameter value $\lambda_{CR}$ which generates the first local maximum of the difference between the derivative of the constraint term and the derivative of the residual term

$$C(\lambda) = \frac{d}{d(\lambda^2)}(\lambda^2\|B\boldsymbol{x}(\lambda)\|^2) - \frac{d}{d(\lambda^2)}\|A\boldsymbol{x}(\lambda) - \boldsymbol{b}\|^2 \ , \ \lambda > 0. \tag{3.7}$$

It is straightforward to show using (3.5) and (3.6) that the CRESO function has the explicit form

$$C(\lambda) = \sum_{i=k+1}^{n} \frac{\mu_i^2(\mu_i^2 - 3\lambda^2)}{(\mu_i^2 + \lambda^2)^3}(\boldsymbol{u}_i \cdot \boldsymbol{b})^2. \tag{3.8}$$

The first local maximum of $C(\lambda)$, if it exists, can be readily computed from the appropriate derivative formulas generated using (3.8). A more detailed analysis of the CRESO method is given in [**2**].

**The Zero-crossing method for computing $\lambda_{ZC}$.** A simple but elegant technique for estimating the regularization parameter called the Zero-crossing method was recently introduced by Johnston and Gulrajani [**7**]. This method is an adaptation of the CRESO method and chooses as the regularization parameter the first zero $\lambda_{ZC}$ of the function

$$Z(\lambda) = \lambda^2\|B\boldsymbol{x}(\lambda)\|^2 - \|A\boldsymbol{x}(\lambda) - \boldsymbol{b}\|^2 \ , \ \lambda > 0. \tag{3.9}$$

In this approach equal weight is assigned to the residual and $\lambda^2$ times the solution seminorm and, provided $\delta_o^2 = \sum_{i=n+1}^{m}(\boldsymbol{u}_i \cdot \boldsymbol{b})^2 > 0$, this first zero $\lambda_{ZC}$ exists and can be computed using the NAG routine C05ADF [**11**] to solve the fixed point equation

$$Z(\lambda) = \sum_{i=k+1}^{n} \frac{\lambda^2(\mu_i^2 - \lambda^2)}{(\mu_i^2 + \lambda^2)^2}(\boldsymbol{u}_i \cdot \boldsymbol{b})^2 - \delta_o^2 = 0. \tag{3.10}$$

**The optimal regularization parameter $\lambda^*$.** If the exact solution $\boldsymbol{x}^*$ of $A\boldsymbol{x} = \boldsymbol{b}$ is known *a priori*, then the optimal regularization parameter $\lambda^*$ and the best-possible Tikhonov solution $\boldsymbol{x}(\lambda^*)$ to $A\boldsymbol{x} = \boldsymbol{b}$ for any given $\boldsymbol{b}$ and regularizing operator $B$ can be determined by using (3.4) to compute that $\lambda = \lambda^*$ for which the relative error $\epsilon(\lambda) = \|\boldsymbol{x}(\lambda) - \boldsymbol{x}^*\|/\|\boldsymbol{x}^*\|$ is a minimum. In the numerical example considered in Section 5, this approach is used to assess the effectiveness of the parameter estimation methods as well as the accuracy of the solution for the optimal control formulation of the inverse problem of electrocardiography.

## 4 Reformulation as an optimal control problem

The following optimal control theory formulation of the inverse problem of electrocardiography is based on work by Kabanikhin and Karchevsky [**8**] and is an extension of the quasi-inversion method first proposed by Lattes and Lions [**9**].

We first consider the well-posed linear boundary control problem. Given $\boldsymbol{b}$ on $S_B$, determine $\phi(\boldsymbol{p})$ on $\Omega \cup S_H \cup S_B$ and the control $\chi$ on $S_H$ which minimizes the functional

$$J(\chi) = \int_{S_H \cup S_B} [\phi(\boldsymbol{p}) - \boldsymbol{b}]^2 dS \tag{4.1}$$

subject to

$$\begin{aligned} \Delta\phi(\boldsymbol{p}) &= 0 & \boldsymbol{p} &\in \Omega \\ \boldsymbol{\nabla}\phi(\boldsymbol{p}) \cdot \boldsymbol{n}_B &= 0 & \boldsymbol{p} &\in S_B \\ \boldsymbol{\nabla}\phi(\boldsymbol{p}) \cdot \boldsymbol{n}_H + \phi(\boldsymbol{p}) &= \chi(\boldsymbol{p}) & \boldsymbol{p} &\in S_H. \end{aligned} \tag{4.2}$$

For a broad class of bounded domains and associated boundaries, this problem has a unique solution which depends continuously on the boundary (control) data $\chi(\boldsymbol{p})$.

As in Section 2, application of Green's integral identities and collocation at points of $S_H \cup S_B$ gives

$$P_{HB(n\times m)}\boldsymbol{\Phi}_B + P_{HH(n\times n)}\boldsymbol{\Phi}_H + G_{HH(n\times n)}(\boldsymbol{\chi} - \boldsymbol{\Phi}_H) = 0 \tag{4.3}$$

$$P_{BB(m\times m)}\boldsymbol{\Phi}_B + P_{BH(m\times n)}\boldsymbol{\Phi}_H + G_{BH(m\times n)}(\boldsymbol{\chi} - \boldsymbol{\Phi}_H) = 0 \tag{4.4}$$

for (4.2) where $\boldsymbol{\chi} = (\chi^1, \ldots, \chi^n)$ on $S_H$.

Solving (4.4) for $\Phi_B$ gives

$$\Phi_B = Z_{(m\times 2n)}\begin{pmatrix} \Phi_H \\ \boldsymbol{\chi} \end{pmatrix} = \begin{pmatrix} P_{BB}^{-1}(G_{BH} - P_{BH}) & -P_{BB}^{-1}G_{BH} \end{pmatrix}\begin{pmatrix} \Phi_H \\ \boldsymbol{\chi} \end{pmatrix} \tag{4.5}$$

and substituting (4.5) into (4.3) gives the system of $n$ linear constraints

$$C_{(n\times 2n)}\begin{pmatrix} \Phi_H \\ \boldsymbol{\chi} \end{pmatrix} = 0 \tag{4.6}$$

where

$$C_{(n\times 2n)} = \begin{pmatrix} P_{HH} - G_{HH} + P_{HB}P_{BB}^{-1}(G_{BH} - P_{BH}) & G_{HH} - P_{HB}P_{BB}^{-1}G_{BH} \end{pmatrix}.$$

Thus, in this formulation the inverse problem reduces to a linear constrained optimization problem

$$\min_{\begin{pmatrix} \Phi_H \\ \boldsymbol{\chi} \end{pmatrix} \in E^{2n}} \left\|Z_{(m\times 2n)}\begin{pmatrix} \Phi_H \\ \boldsymbol{\chi} \end{pmatrix} - \boldsymbol{b}\right\|^2 \qquad s.t. \qquad C_{(n\times 2n)}\begin{pmatrix} \Phi_H \\ \boldsymbol{\chi} \end{pmatrix} = 0. \tag{4.7}$$

The solution $\begin{pmatrix} \Phi^c_H \\ \chi^c \end{pmatrix}$ to (4.7) is computed using the NAG routine E04NCF [11] where the starting values $\Phi^c_{Ho}$ and $\chi^c_o = -G^{-1}_{HH}(P_{HB}\boldsymbol{b} + P_{HH}\Phi^c_{Ho}) + \Phi^c_{Ho}$ for the routine are chosen from the candidates $\Phi^c_{Ho} = \boldsymbol{x}(\lambda)$ computed using the regularization methods described in Section 3.

## 5 A numerical example

As described in [6], $S_B$ and $S_H$ for a realistic human thorax and heart geometry are subdivided into triangular elements defined by $n = 202$ and $m = 352$ node locations, respectively and the problem reduced as in Section 2 to the discrete ill-conditioned system (2.7) as illustrated in Figure 2.

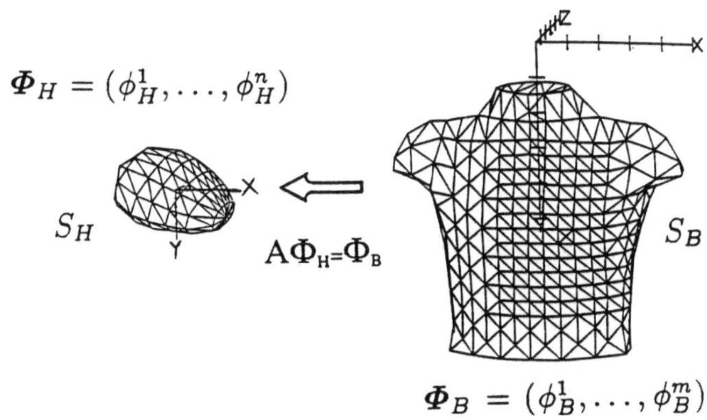


**Figure 2** Boundary element representation of the epicardial and torso surfaces.

The precise calculation of the coefficients of the transfer matrix $A$ and the discretized Laplace regularizing operator $B = \Delta_{202}$ are given in Appendix II of [6]. Using a representative single dipole source located near the centre of gravity of the region bounded by $S_H$, an (*a priori*) potential distribution $\boldsymbol{x}^* = \boldsymbol{\Phi}^*_H$ was generated on $S_H$ and the corresponding distribution $\boldsymbol{\Phi}_B = \boldsymbol{b}$ on $S_B$ computed. The problem is then to solve (2.7) given $\boldsymbol{\Phi}_B = \boldsymbol{b}$.

For the particular dipole configuration selected here, the $L$-curve, CRESO and Zero-crossing methods were used to compute regularization parameters $\lambda$ and the corresponding solutions $\boldsymbol{x}(\lambda) = \boldsymbol{\Phi}_H(\lambda)$. From the values obtained for $\lambda_{LC}$, $\lambda_{CR}$ and $\lambda_{ZC}$, $\lambda_s = 8.5$ was chosen as the average and $\boldsymbol{x}(\lambda_s)$ given by (3.4) employed as the starting solution for the optimal control formulation of the problem. The solution $\boldsymbol{x}^c = \boldsymbol{\Phi}^c_H$ (and $\chi^c$) of (4.7) was computed using E04NCF. The optimal regularization parameter $\lambda^*$ and the best possible Tikhonov solution $\boldsymbol{x}(\lambda^*) = \boldsymbol{\Phi}_H(\lambda^*)$ were also computed by minimizing the relative error function $\epsilon(\lambda) = \|\boldsymbol{x}(\lambda) - \boldsymbol{x}^*\|/\|\boldsymbol{x}^*\|$ with $\boldsymbol{x}(\lambda) = \boldsymbol{\Phi}_H(\lambda)$ given by (3.4). The results obtained are displayed in Table 1 in terms of the relative error $RE = \|\boldsymbol{x} - \boldsymbol{x}^*\|/\|\boldsymbol{x}^*\|$ and correlation coefficient $CC = \frac{\boldsymbol{x}\cdot\boldsymbol{x}^*}{\|\boldsymbol{x}\| \ \|\boldsymbol{x}^*\|}$ for $\boldsymbol{x} = \boldsymbol{\Phi}_H(\lambda^*)$, $\boldsymbol{x} = \boldsymbol{\Phi}_H(\lambda_{LC})$, $\boldsymbol{x} = \boldsymbol{\Phi}_H(\lambda_{CR})$, $\boldsymbol{x} = \boldsymbol{\Phi}_H(\lambda_{ZC})$ and $\boldsymbol{x} = \boldsymbol{\Phi}^c_H$ respectively. These results show that the three parameter estimation methods are essentially equivalent for this example and yield solutions reasonably close to the best possible Tikhonov solution $\boldsymbol{x}(\lambda^*) = \boldsymbol{\Phi}_H(\lambda^*)$. The optimal control

**Table 1. Solution Results for Example Application**

| Example | Regularizing Operator | Parameter $\lambda$ | Solution: $RE$, % | $\boldsymbol{x} = \boldsymbol{\Phi}_H$ $CC$, % |
|---|---|---|---|---|
| 1 | $B = \Delta_{202}$ | $\lambda^* = 13.77$ | 13.1 | 0.991 |
| | $B = \Delta_{202}$ | $\lambda_{LC} = 1.445$ | 20.4 | 0.979 |
| | $B = \Delta_{202}$ | $\lambda_{CR} = 13.85$ | 18.2 | 0.983 |
| | $B = \Delta_{202}$ | $\lambda_{ZC} = 10.23$ | 17.0 | 0.986 |
| | | Control Formulation | 12.2 | 0.996 |

$$RE = \|\boldsymbol{x} - \boldsymbol{x}^*\|/\|\boldsymbol{x}^*\| \quad \text{and} \quad CC = \frac{\boldsymbol{x} \cdot \boldsymbol{x}^*}{\|\boldsymbol{x}\| \quad \|\boldsymbol{x}^*\|}$$

where $\boldsymbol{x}^*$ is the known *a priori* solution.

formulation method improved the starting solution $\boldsymbol{x}(\lambda_s)$ but only marginally over $\boldsymbol{\Phi}_H(\lambda^*)$.

## 6 Comments and conclusions

In the actual clinical applications of the methods described here for solving the inverse problem of electrocardiography [**10, 12, 13**], neither *a priori* solutions $x^* = \Phi_H^*$ nor best Tikhonov solutions $x\lambda^* = \Phi_H(\lambda^*)$ of $Ax = b$ are available and regularizaton methods must be used to obtain approximate solutions which are accurate and stable with respect ot the measurement noise and error which is always present in the recorded body potential data $b$. From Table 1 for the specific numerical example considered in Section 5, the $L$-curve, CRESO and Zero-crossing methods yielded similar estimates for the optimal regularization parameter and produced approximate solutions with a similar percentage range (17-21%) for the relative error $RE$. For some objectives such as isolating pre-excitation cites, this is within the acceptable range for clinical applications [**12**]. Choosing an average of these values $\lambda_s = 8.5$ and the starting solution $x(\lambda_s)$ given by (3.4), the example problem was solved again using the control theory formulation given in Section 4. While the optimal control solution $x = \Phi_H^c$ exhibited an improvement over the starting value $x([\lambda_s)$, only a modest improvement over the best possible Tikhonov solution $x(\lambda^*)$ was achieved. Despite this, the optimal control theory approach appears to have the potential to yield significantly improved results because it also allows for the adjustment of the solution values $\phi_B$ on the torso surface to account for possible errors or inaccuracies in the measured data $b$. This makes it unique among the other constrained optimization approaches that have been considered (cf. [**6**]) and consequently merits some further study. Certainly one avenue which should be explored is the elimination of $\chi$ using (4.3) rather than $\Phi_B$ using (4.4) as was done in Section 4. This would result in a constrained least square problem involving $\begin{pmatrix} \Phi_H \\ \Phi_B \end{pmatrix} \in E^{n+m}$ directly and might allow for greater control over the amount of change which occurs in the solution values $\Phi_B$ at the torso surface $S_B$.

## References

[1] Anderson, E., Bai, Z., Bischof, C., Demmel, J. W., Dongarra, J. J., Du Croz, J., Greenbaum, A., Hammarling, S., McKenney, A., Ostrouchov, S. and Sorensen, D. [1992], *LAPACK User's Guide*, Society for Industrial and Applied Mathematics, Philadelphia.

[2] Clements, J. C., Carroll, R. and Horáček, B. M. [1993], *On computing regularization parameters for inverse problems in electrocardiography*, Biomedical and Life Physics, in Proceedings of the Second Gauss Symposium (D. N. Ghista, ed.), Munich, Vieweg Verlag, Wiesbaden, 235-246.

[3] Colli-Franzone, P., Guerri, L., Taccardi, B. and Viganotti, C. [1985], *Finite element approximation of regularized solution of the inverse potential problem of electrocardiography and applications to experimental data*, Calcolo, XXII(I), 91-186.

[4] Hansen, P. C. [1992] *Analysis of discrete ill-posed problems by means of the L-curve*, SIAM Review **34**, 561-580.

[5] Hansen, P. C. and O'Leary, D. P. [1993], *The use of the L-curve in the regularization of discrete ill-posed problems*, SIAM J. Sci. Comput. **14**, No. 6, 1487-1503.

[6] Horáček, B. M. and Clements, J. C. [1997], *The inverse problem of electrocardiography: A solution in terms of single-and double-layer sources on the epicardial surface*, Mathematical Biosciences **144**, 19-154.

[7] Johnston, P. R. and Gulrajani, R. M. [1997], *A new method for regularization parameter determination in the inverse problem of electrocardiography*, IEEE Trans. Biomed. Eng. **44**, 19-39.

[8] Kabanikhin, S. I. and Karchevsky, A. L. [1995], *Optimizational method for solving the Cauchy problem for an elliptic equation*, J. Inv. Ill-Posed Problems **3**, No. 1, 21-46.

[9] Lattes, R. and Lions, J.-L. [1967], *Methodé de Quasi Reversibilité et Applications*, Dunod, Paris.

[10] MacLeod, R. S., Gardner, M. J., Miller, R. M. and Horáček, B. M. [1995], *Application of an electrocardiographic inverse solution to localize ischemia during coronary angioplasty*, J. Cardiovasc. Electrophysiol. **6**, No. 1, 2-18.

[11] *The NAG Fortran Library Manual – Mark 15* [1988], The Numerical Algorithms Group Ltd., Oxford, UK.

[12] Penney, C. J., Clements, J. C., Gardner, M. J. Sterns, L. and Horáček, B. M. [1995], *The inverse problem of electrocardiography: application to localization of Wolff-Parkinson-White pre-excitation cites*, Proceedings IEEE EMBS, Montreal, Quebec, 1-3.

[13] Penney, C. J., Title, L. M., Iles, S. E., Clements, J. C., Gardner, M. J. and Horáček, B. M. [1995], *Comparison of 'electrophysiological imaging' and radionuclide imaging in quantification of controlled cardiac ischemia*, Proceedings of Computers in Cardiology, Vienna, 79-82.

[14] Rudy, Y. and Oster H. S. [1993], *The electrocardiographic inverse problem*, in High-Performance Computing in Biomedical Research (T. C. Pilkington, B. Loftis, J. F. Thompson, S. L-Y Woo, T. C. Palmer, and T. F. Budinger, eds.), CRC Press, Boca Raton, 135-155.

[15] Rudy, Y. and Messinger-Rapport, B. J. [1988], *The inverse solution in electrocardiography: Solutions in terms of epicardial potentials*, CRC Crit. Rev. Biomed. Eng. **16**, 215-268.

Fields Institute Communications
Volume **21**, 1999

# Natural Selection as an Extensive Form Game

**R. Cressman**
Department of Mathematics
Wilfrid Laurier University
Waterloo, Ontario, Canada N2L 3C5
rcressma@mach1.wlu.ca

**Abstract.** The paper develops dynamic extensive form evolutionary games as models of multi-locus natural selection. The correspondence between the subgame structure of these games and the structure of the loci is exploited to gain a better understanding of both the theory of natural selection and dynamic evolutionay games.

## 1 Introduction

It is well known that natural selection at a single locus can be modelled as a dynamic evolutionay game where payoffs are equated to genotypic fitnesses [**9, 3**]. Typically, these games have been analyzed through their normal form payoff matrix. However, this paper emphasizes the extensive form analogy and then generalizes the connection to multi-locus systems.

Section 2 briefly summarizes the single locus models for both discrete and overlapping generations. Standard results concerning the convergence of natural selection are reinterpreted in terms of the Nash equilibria and of the evolutionarily stable strategies and sets from evolutionary game theory (Theorems 2.3 and 2.4). Section 3 analyzes multi-locus systems, with no recombination and additive fitness, through a heuristic correspondence between each locus and a subgame of the extensive form (Figure 2). The intuition gained from this perspective leads to the conclusion (Theorem 3.2) that these systems are better understood by separating the loci rather that the traditional way that equates this system to a single locus with many alleles. In particular, natural selection leads to a population where the allelic frequencies at each locus are usually at an evolutionarily stable strategy.

Sections 4 and 5 analyze the effects of recombination on natural selection. The recombination process can also be viewed as a dynamic on the extensive form that restricts naturally to dynamics on subgames. In the long run, the population evolves to equilibria where strategies for different subgames are independently distributed (Theorem 4.2). That is, recombination leads to the linkage-equilibrium manifold or Wright manifold (Definition 4.1) that can be generalized to all extensive form games [**4**]. Recombination together with additive fitness leads generically to points

1991 *Mathematics Subject Classification.* Primary 92D15; Secondary 90D40.

on the Wright manifold where allelic frequencies are evolutionarily stable. Section 6 briefly considers an extensive form representation when fitness is not additive.

## 2 Single locus model

The standard discrete-generation viability selection model at a single autosomal diploid locus assumes there are $m$ possible alleles $A_1, \ldots, A_m$ at this locus. With random mating, no fertility differences and offspring produced through Mendelian segregation in equal sex ratio, the initial frequencies $p_{ij}$ of ordered genotypic zygotes $A_iA_j$ in the next generation will be in Hardy-Weinberg proportions (i.e., $p_{ij} = p_i p_j$ where $p_i$ is the frequency of allele $A_i$). If $w_{ij}$ is the viability (i.e., the probability of survival to reproductive age) of an $A_iA_j$ individual then

$$p_i(t+1) = p_i(t)\frac{w_i}{\bar{w}} \tag{2.1}$$

where $w_i = \sum\limits_j w_{ij}p_j(t)$, $\bar{w} = \sum\limits_i w_i p_i(t) = \sum\limits_{ij} w_{ij}p_i(t)p_j(t)$ and time $t$ is measured in generations.[1] We assume $w_{ij} = w_{ji}$ for all $1 \leq i, j \leq m$.

Throughout the paper, we often deal with the continuous-time approximation of (2.1) when the time between generations is small and/or selection is weak (i.e., all viabilities are close to the same (positive) constant). Then $\dot{p}_i \cong p_i(t+1) - p_i(t) = p_i\left(\frac{w_i - \bar{w}}{\bar{w}}\right)$ where $\dot{p}_i$ is the time derivative of $p_i$ and the explicit time dependence of $p_i$ on $t$ is usually suppressed. With the assumption that all viabilities are positive, the trajectories of the continuous-time model are the same as those of the dynamic system $\dot{p}_i = p_i(w_i - \bar{w})$ which we will rewrite in the matrix form

$$\dot{p}_i = p_i(e_i \cdot Mp - p \cdot Mp). \tag{2.2}$$

Here $M$ is the $m \times m$ viability matrix with entries $m_{ij} = w_{ij}$; $p \in \Delta^m = \{(p_1, \ldots p_m) \mid \sum\limits_i p_i = 1, p_i \geq 0\}$ is the state of the system (at time $t$); $e_i = (0, \ldots, 0, 1, 0, \ldots, 0)$ is the unit coordinate vector in $\Delta^m$ with 1 in the $i^{th}$ component and 0 everywhere else; and the dot product on the right-hand side is the standard inner product $u \cdot v = \sum\limits_{i=1} u_i v_i$ for vectors $u, v \in \mathbf{R}^m$.

The continuous selection Equation (2.2) can also be developed from first principles as a model of selection with overlapping generations [**7, 1, 9**]. With the assumption that the population is always in Hardy-Weinberg proportions[2], the entries in $M$ are the Malthusian fitness parameters that specify the difference between the birth rate and death rate of $A_iA_j$ individuals. These entries have traditionally been denoted by $m_{ij}$ (as opposed to $w_{ij}$) and can be positive or negative.

To formulate the single locus model as an extensive form game, consider a two-player game where each player has a single decision with $m$ possible choices indexed by $A_1, \ldots, A_m$. The symmetric extensive form of Figure 1 represents this

---

[1] The discrete dynamic (2.1) is also valid if the fertility (i.e., expected number of offspring) of the mating pair $A_iA_j$ with $A_kA_\ell$ is multiplicative and the viability parameter is adjusted by these zygotic "birth" rates [**11**].

[2] Although this assumption cannot be fully justified in all overlapping generations models [**1, 11**] unless the death rates are independent of genotype, we will take (2.2) as our standard model of continuous-time selection since the connection with evolutionary game theory is most clear through this dynamic.

as a single-stage simultaneity game[3] where the payoff to player one (player two respectively) is $a_{ij}$ ($b_{ji}$ respectively) if player one chooses $A_i$ and player two chooses $A_j$. If $a_{ij} = b_{ij}$ for all $1 \leq i,j \leq m$, the game is symmetric. In our case, we have $m_{ij} = a_{ij} = a_{ji} = b_{ij} = b_{ji}$ since the path $A_i$ followed by $A_j$ in Figure 1 represents the zygote $A_iA_j$ whose payoff to both players is the fitness $m_{ij}$. The game tree of Figure 1 can be interpreted as player one making the first decision followed by a choice of player two. However, since player two's only information set includes all its decision points (indicated by the dashed curve in Figure 1), player two's choice cannot depend on that of player one and so the usual interpretation is that both players make their choice simultaneously.

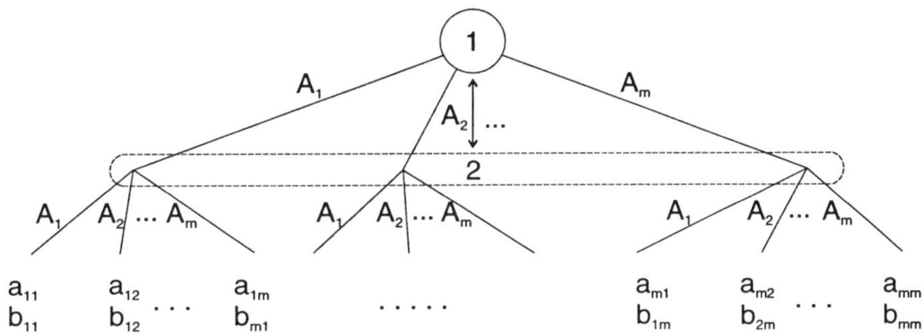


**Figure 1** The symmetric extensive form of a single-locus selection model.

The symmetric normal form of Figure 1 is the $m \times m$ viability matrix $M$ of (2.2) that is now called a payoff matrix. In fact, since $M$ is a symmetric matrix, the normal form is called a doubly symmetric game by [**14**] (or a partnership game by [**9**]). The first connection between evolutionary game theory and natural selection is that the continuous (respectively, discrete) replicator dynamic [**9, 14**] of the symmetric extensive form game of Figure 1 is none other than the dynamical system (2.2) (respectively, (2.1)). The replicator dynamic is the standard model for the evolution of strategy frequencies in a large, effectively infinite, population where individuals compete at random in pairs and strategies (i.e., choices) with higher expected payoff increase in relative frequency. That is, as an evolutionary game, selection at a single locus is the same as pairwise interactions between alleles whose payoff is the fitness of the corresponding genotypic zygote.

Much is known about the behavior of trajectories of (2.2) and/or (2.1) either from a biological perspective [**10, 11**] or as an evolutionary game [**9, 3, 14**]. The remainder of this section summarizes results that emphasize connections between the two perspectives. The proofs, which are omitted, rely primarily on the fact the mean fitness, $p \cdot Mp$, is strictly increasing as a function of $t$ for both (2.2) and (2.1) unless the population is in equilibrium. That is, the proofs rely on the Fundamental Theorem of Natural Selection.

**Definition 2.1** *A state $p^* \in \Delta^m$ is a Nash equilibrium (NE) of the payoff matrix $M$ if $p \cdot Mp^* \leq p^* \cdot Mp^*$ for all $p \in \Delta^m$. A set $E$ is a NE component if it is a connected set of Nash equilibria that is not contained in a larger connected set of NE. A subset $E$ of $\Delta^m$ is an evolutionary stable set (ES set) if all elements in*

[3]The terminology and notation for symmetric extensive form games used in this paper follows the formalism of [**12**] most closely (see also [**13**]). Less technical approaches that suffice for our purposes are found in [**3, 4, 5**].

*$E$ are NE and $p^* \cdot Mp > p \cdot Mp$ whenever $p^* \in E$, $p \notin E$ and $p \cdot Mp^* = p^* \cdot Mp^*$. The state $p^* \in \Delta^m$ is an evolutionarily stable strategy (ESS) if $\{p^*\}$ is an ES set.*

**Definition 2.2** *A rest point (or dynamic equilibrium) for (2.2) (respectively, (2.1)) is a state $p^* \in \Delta^m$ such that $\dot{p}_i = 0$ for all $i$ when $p = p^*$ (respectively, $p_i(t+1) = p_i(t)$ for all $i$ when $p = p^*$).*

*A subset $E$ of $\Delta^m$ consisting of rest points is (Lyapunov) stable if, for every neighborhood $U$ of $E$, there exists another neighborhood $O$ of $E$ such that, $p(t) \in U$ for all positive $t$ if $p(0) \in O \cap \Delta^m$. A state $p^* \in \Delta^m$ is (Lyapunov) stable if $\{p^*\}$ is stable.*

*A subset $E$ of $\Delta^m$ consisting of rest points is asymptotically stable if $E$ is stable and there is some neighborhood $U$ of $E$ such that all $\omega$-limits of trajectories with $p(0) \in U \cap \Delta^m$ are elements of $E$. A state $p^* \in \Delta^m$ is asymptotically stable if $\{p^*\}$ is asymptotically stable.*

**Theorem 2.3** *The mean fitness function is constant on each NE component. Every ES set is a finite union of disjoint connected ES sets, each of which is a NE component. For a generic set of payoff assignments in Figure 1 (e.g., if all principal minors of the payoff matrix $M$ are nonzero), every ES set is a finite union of ESS's.*

**Theorem 2.4** *Every initial polymorphism[4] (i.e., $p_i(0) > 0$ for all $i$) evolves to a NE. If a state $p^*$ in the interior of $\Delta^m$ is in an ES set $E$, then every initial polymorphism evolves to $E$. A state $p^* \in \Delta^m$ is stable if and only if (iff) $p^*$ is in an ES set iff $p^*$ is a local maximum of the mean fitness function, $p \cdot Mp$, defined on $\Delta^m$. A set of rest points $E$ is stable iff each point in $E$ is stable.*

*A state $p^* \in \Delta^m$ is asymptotically stable iff $p^*$ is an ESS iff $p^*$ is a strict local maximum of $p \cdot Mp$. A set of rest points $E$ is asymptotically stable iff $E$ is an ES set.*

The above results translate the heuristic understanding of natural selection as an adaptive landscape whereby evolution works to locally optimize the growth rate of the population (i.e., the mean fitness). For almost all initial polymorphisms, the population evolves to an ES set. Thus, generically by Theorem 2.3, the effect of natural selection can be predicted by the evaluation of the set of ESS's for the payoff matrix $M-$ a purely algebraic exercise that is trivial compared to solving the dynamical system either analytically or numerically.

## 3 A symmetric extensive form for a multi-locus system

Suppose there are $n$ loci ($n \geq 2$) and locus $k$ has $m_k$ possible alleles $A_1^{(k)}, \dots A_{m_k}^{(k)}$. An extensive form game $\Gamma$ that contains this information is a game whose initial decision point is a move by nature (see player 0 in Figure 2) that takes each of $n$ possible choices with probability $\frac{1}{n}$. The $k^{th}$ action chosen by nature leads to a symmetric subgame of the type considered in Figure 1 that now has a corresponding $m_k \times m_k$ payoff matrix $M^{(k)}$. The pure strategies $e_i$ of player one specify an action $A_{i_k}^{(k)}$ at each of the $n$ information sets of player one. More specifically, $e_i$ is specified by the multi-index $i = (i_1, \dots, i_n)$ where $1 \leq i_k \leq m_k$ for all $1 \leq k \leq n$.

[4]If $p(0)$ is not polymorphic, then $p(t)$ evolves to a NE of the game restricted to strategies $A_i$ for which $p_i(0) > 0$ (i.e., strategies in the support of $p(0)$). The selection dynamic leaves the face that contains the support of $p(0)$ forward invariant.

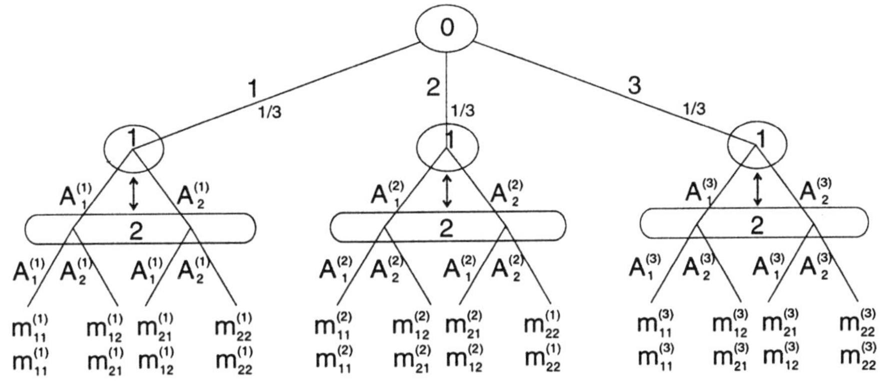


**Figure 2** A symmetric extensive form of a three-locus two-allele system.

The normal form of this game is an $N \times N$ payoff matrix $M$ where $N = \prod_{k=1}^{n} m_k$ and

$$m_{ij} = \overset{(1)}{m}_{i_1 j_1} + \ldots + \overset{(n)}{m}_{i_n j_n} \tag{3.1}$$

when $i = (i_1, \ldots, i_n)$ and $j = (j_1, \ldots, j_n)$. [Strictly speaking, $m_{ij} = \frac{1}{n}(\overset{(1)}{m}_{i_1 j_1} + \ldots + \overset{(n)}{m}_{i_n j_n})$. This rescaling of payoffs makes no difference to the analysis and so is ignored here.] That is, $M$ models the situation where fitness is additive between loci. Furthermore, the replicator dynamic with respect to $M$ is then natural selection with additive fitness and no recombination - a case that is usually dismissed as uninteresting since such a system is equivalent to single-locus natural selection with $N$ alleles.

However, considered in this traditional way as a single-locus model, the system is highly nongeneric compared to the typical system in Section 2. For instance, in Figure 2, if each of the three loci separately have an interior ESS, then the $8 \times 8$ payoff matrix has a single four-dimensional ES set and no ESS. Specifically, if $M^{(k)} = \begin{bmatrix} 1 & 2 \\ 2 & 1 \end{bmatrix}$ for $k \in \{1, 2, 3\}$, each $M^{(k)}$ has ESS $(\frac{1}{2}, \frac{1}{2})$ whereas the only ES set of $M$ is

$$E = \left\{ p \in \Delta^8 \mid \frac{1}{2} = \sum_{jk} p_{1jk} = \sum_{ik} p_{i1k} = \sum_{ij} p_{ij1} \right\},$$

where $p_{ijk}$ is the frequency of the strategy with multi-index $(i, j, k)$. Thus, to predict the effect of natural selection, we must determine the NE structure and, in particular, the ES structure of these higher dimensional payoff matrices.

These structures can best be described (Theorem 3.1 below) by treating the loci separately. To this end, for each $p \in \Delta^N$ of our n-locus system, the frequency of action $A^{(k)}_{i_k}$ at locus $k$ is given by

$$p^{(k)}_{i_k} = \sum_j \{p_j \mid j_k = i_k\}. \tag{3.2}$$

Then $p^{(k)} = (p_1^{(k)}, \ldots, p_{m_k}^{(k)}) \in \Delta^{m_k}$ is the state vector for the $k^{th}$ subgame of $\Gamma$ and the map $p \mapsto (p^{(1)}, ..., p^{(n)})$ from $\Delta^N$ to $\Delta^{m_1} \times ... \times \Delta^{m_n}$ is surjective. The following theorem extends a result of [**6**] who proved it in the special case $n = 2$.[5]

**Theorem 3.1** *(a) $p^* \in \Delta^N$ is a NE of $M$ iff $p^{*(k)}$ is a NE of $M^{(k)}$ for all $1 \leq k \leq n$.*

*(b) $E$ is a connected ES set with respect to $M$ iff $E^{(k)} = \{p^{(k)} \mid p \in E\}$ is a connected ES set of $M^{(k)}$ for all $1 \leq k \leq n$ and $E = \{p \mid p^{(k)} \in E^{(k)}$ for all $1 \leq k \leq n\}$.*

*(c) $p^* \in \Delta^N$ is an ESS of $M$ iff $p^{*(k)}$ is an ESS for all $1 \leq k \leq n$ and there is at most one $p^{*(k)}$ which is not a pure strategy in the $k^{th}$ subgame.*

**Proof** (a) From (3.1) and (3.2), for every $p, q \in \Delta^N$,

$$p \cdot Mq = \sum_{k=1}^{n} p^{(k)} \cdot M^{(k)} q^{(k)}.$$

Suppose $p^*$ is a NE of $M$. Choose $p \in \Delta^N$ such that $p^{(1)}$ is arbitary and $p^{(k)} = p^{*(k)}$ for $k = 2, \ldots, n$. Since $p \cdot Mp^* - p^* \cdot Mp^* = p^{(1)} \cdot M^{(1)} p^{*(1)} - p^{*(1)} \cdot M^{(1)} p^{*(1)}$, $p^{(1)} M^{(1)} p^{*(1)} \leq p^{*(1)} \cdot M^{(1)} p^{*(1)}$ for all $p^{(1)} \in \Delta^{m_1}$. Thus $p^{*(1)}$ is a NE of $M^{(1)}$ and, by a similar argument, $p^{*(k)}$ is a NE of $M^{(k)}$. Conversely, if $p^{*(k)}$ is a NE of $M^{(k)}$ for all $1 \leq k \leq n$, then $p \cdot Mp^* = \sum p^{(k)} \cdot M^{(k)} p^{*(k)} \leq \sum p^{*(k)}, M^{(k)} p^{*(k)} = p^* \cdot Mp^*$ for all $p \in \Delta^N$. That is, $p^*$ is a NE of $M$.

(b) Suppose E is an ES set for $M$. From part(a), $E^{(k)}$ is a set of NE of $M^{(k)}$. Furthermore, if $p^{*(k)} \in E^{(k)}, p^{(k)} \notin E^{(k)}$ and $p^{(k)} \cdot M^{(k)} p^{*(k)} = p^{*(k)} \cdot M^{(k)} p^{*(k)}$, then choose $p \in \Delta^N$ such that $p^{(k)}$ is given above and $p^{(\ell)} = p^{*(\ell)}$ if $\ell \neq k$. Then $p \cdot Mp^* = p^* \cdot Mp^*$ and $p \notin E$ and so $p^* \cdot Mp > p \cdot Mp$. Thus $p^{*(k)} \cdot M^{(k)} p^{(k)} > p^{(k)} \cdot M^{(k)} p^{(k)}$ for all such $p^{(k)}$ and so $E^{(k)}$ is an ES set. If $E$ is connected, clearly $E^{(k)}$ is also connected.

Conversely, if $E^{(k)}$ is a connected ES set for all $k$, then $E = \{p \mid p^{(k)} \in E^{(k)}\}$ is also connected. Furthermore, if $p \cdot Mp^* = p^* \cdot Mp^*$ with $p \notin E$ and $p^* \in E$, then $p^{(k)} \cdot M^{(k)} p^{*(k)} = p^{*(k)} \cdot M^{(k)} p^{*(k)}$ for all $k$ and $p^{(k)} \notin E^{(k)}$ for at least one $k$. Thus, if $p^{(k)} \notin E^{(k)}$, then $p^{*(k)} \cdot M^{(k)} p^{(k)} > p^{(k)} \cdot M^{(k)} p^{(k)}$. On the other hand, if $p^{(k)} \in E^{(k)}$, then $p^{*(k)} \cdot M^{(k)} p^{(k)} = p^{(k)} \cdot M^{(k)} p^{(k)}$ since the mean fitness at locus $k$ is constant on each connected $E^{(k)}$. Thus $p^* \cdot Mp > p \cdot Mp$ and so $E$ is an ES set.

(c) If $p^* \in \Delta^N$ is an ESS, then $p^{*(k)}$ is an ESS for each $k$ by part (b). Furthermore, $E = \{p \mid p^{(k)} = p^{*(k)}\}$ is a singleton set iff $p^{*(k)}$ is a pure strategy for all $k$ except possibly one value. The remainder of the proof is a straightforward application of these facts combined with the result in part (b). □

By Theorem 3.1, the NE and ES set structure of a multi-locus system can be quite complicated in general (especially when some subgames of the extensive form have multiple NE). However, an interesting special case occurs when each of the $n$ subgames $\Gamma_k$ has an ESS $p^{*(k)}$ in the interior of $\Delta^{m_k}$.[6] Then $p^{*(k)}$ is also the unique NE of $M^{(k)}$. The only NE component of $M$ (and hence the only ES set) is the set $E = \{p \in \Delta^N \mid p^{(k)} = p^{*(k)}$ for all $k\}$ that is the intersection of $\Delta^N$ with a

[5]The result was proven without the assumption that the payoff matrices, $M^{(k)}$, of the subgames are symmetric. This assumption is not needed in our statements either.

[6]For multi-alleles at a single locus $k$, polymorphic ESS's are unexpected if $m_k \geq 3$ and payoff entries are random [**8**]. For $m_k = 2$, an interior ESS exists iff the heterozygote at locus $k$ is overdominant.

hyperplane in $\mathbf{R}^N$. By Theorem 2.4, every initial polymorphism evolves to a point in $E$ where each locus attains its maximum mean fitness. For the continuous-time replicator dynamic, (2.2), we will see in Section 5 below that the limit point in $E$ is easily specified by $p(0)$.

In general, Theorems 2.3, 2.4 and 3.1 combine to prove the following result.

**Theorem 3.2** *Every initial polymorphism of a multi-locus system with additive fitness and no recombination evolves under (2.1) or (2.2) to a point where the allelic frequencies at each locus are at a Nash equilibrium. Generically, these frequencies will be an ESS of the subgame at the corresponding locus.*

## 4 Recombination

In this section, we consider recombination in the absence of selection for a continuous-time model. That is, the evolution of the frequency $p_i$ in Equation (4.1) below of pure strategy $e_i$ for multi-index $i = (i_1, ..., i_n)$ is given solely through the reassociation of the genes at the different loci when a diploid individual is formed from its two parental gametes. The multi-locus notation adopted here is a mixture of that in [**10**], [**11**], [**2**] and [**4**].

Let $L = \{1, 2, \ldots, n\}$ denote the set of $n$ loci, I be a proper subset of $L$ that has $\{1\}$ as a subset and $L \backslash I = \{k \in L \mid k \notin I\}$. Let $r_I$ denote the probability of reassociation of the genes at the loci in I, inherited from one parent, with the genes at the loci in $L \backslash I$ inherited from the other parent. For example, a zygote formed from gametes $A_{i_1}^{(1)} A_{i_2}^{(2)} \cdots A_{i_n}^{(n)}$ and $A_{j_1}^{(1)} A_{j_2}^{(2)} \cdots A_{j_n}^{(n)}$ will have unordered genotype $A_{i_1}^{(1)} A_{i_2}^{(2)} \cdots A_{i_n}^{(n)} \backslash A_{j_1}^{(1)} A_{j_2}^{(2)} \cdots A_{j_n}^{(n)}$ with probability $1 - \sum_I r_I$ and genotype $A_{i_1}^{(1)} A_{j_2}^{(2)} \cdots A_{j_n}^{(n)} \backslash A_{j_1}^{(1)} A_{i_1}^{(2)} \cdots A_{i_n}^{(n)}$ with probability $r_{\{1\}}$ etc. (This latter genotype is also denoted as $A_{i \backslash j}^{I} \backslash A_{j \backslash i}^{I}$ where $I = \{1\}$.) If all $r_I = 0$, there is no recombination (i.e. there is tight linkage between all the loci) and we are in the situation of Section 3. For simplicity, we will assume all $r_I > 0$ in this section.

The continuous-time recombination equation[7] for $n$ loci generalizes the two-locus equation of [**9**] to

$$\dot{p}_i = -\sum_I r_I \sum_j \left( \overset{I}{p}_{i \backslash i} \overset{I}{p}_{i \backslash j} - \overset{I}{p}_{i \backslash j} \overset{I}{p}_{j \backslash i} \right). \tag{4.1}$$

where $\overset{I}{p}_{i \backslash j}$ is the frequency of the strategy that takes action $A_{i_k}^{(k)}$ for all $k \in I$ and action $A_{j_k}^{(k)}$ for all $k \in L \backslash I$. For every subset $J$ of $L$, let

$$p_i^J = \sum_j \{p_j | j_k = i_k \text{ for all } k \in J\} = \sum_j \{\overset{J}{p}_{i \backslash j} \, | j_k = i_k \text{ for all } k \in J\}. \tag{4.2}$$

Notice that $\overset{J}{p}_{i \backslash i} = p_i$ but $\overset{J}{p}_{i \backslash i} \neq p_i^J$ in general. Also notice that (3.2) is a special case of (4.2) in that $p_{i_k}^{(k)} = p_i^{\{k\}}$.

From (4.1), it is clear that every point $p \in \Delta^N$ that satisfies $\overset{I}{p}_{i \backslash i} \overset{I}{p}_{j \backslash j} = \overset{I}{p}_{i \backslash j} \overset{I}{p}_{j \backslash i}$ for every $i, j$ and $I$ is an equilibrium of the recombination equation. Our main result (Theorem 4.2 below) shows the converse is also true and this motivates the following definition.

---

[7]Since there is no selection, the assumption of Hardy-Weinberg proportions for zygotic frequencies is innocuous in this section.

**Definition 4.1** *The Wright manifold (also called the linkage-equilibrium manifold) is* $W = \{p \in \Delta^N \,|\, \overset{I}{p}_{i\backslash i}\overset{I}{p}_{j\backslash j} = \overset{I}{p}_{i\backslash j}\overset{I}{p}_{j\backslash i}$ *for all subsets* $I$ *of* $L$ *and all pure strategies* $e_i$ *and* $e_j\}$.

Lyubich in [**10**] (see also [**4**] who generalized this definition to all extensive form games) showed that $W = \{p \in \Delta^N | p_i = p_i^I p_i^{L\backslash I}$ for all $I$ and $i\} = \{p \in \Delta^N | p_i = p_{i_1}^{(1)} p_{i_2}^{(2)} \cdot \ldots \cdot p_{i_k}^{(k)}$ for all $i\}$. The proof of the main result relies on this as well as the fact that (4.1) induces a recombination equation on $p_i^J$ for each subset $J$ of $L$. Specifically if $1 \in J$, then [**1**]

$$\dot{p}_i^J = -\sum_{I \subseteq J} r_I \sum_j \left( \overset{JI}{p}_{i\backslash i}\overset{JI}{p}_{j\backslash j} - \overset{JI}{p}_{i\backslash j}\overset{JI}{p}_{j\backslash i} \right), \tag{4.3}$$

where $\overset{JI}{p}_{i\backslash j} = \sum_\ell \{p_\ell | \ell_k = i_k \text{ if } k \in I \text{ and } \ell_k = j_k \text{ if } k \in J\backslash I\}$. To illustrate this result, first consider the case $J = \{1\}$. Then $\dot{p}_i^{\{1\}} = \dot{p}_{i_1}^{(1)} = 0$ since there are no proper subsets of $J$. This is also clear from (4.1) and (3.2).

As a further illustration, consider the two-locus case. Then (4.1) becomes $\dot{p}_i = -r \sum_{(j_1,j_2)} (p_{(i_1,i_2)}p_{(j_1,j_2)} - p_{(i_1,j_2)}p_{(j_1,i_2)})$ where $I = \{1\}$ and $r = r_I$ is the crossover rate between the two loci. Since $\sum_{(j_1,j_2)} p_{(i_1,j_2)}p_{(j_1,i_2)} = p_{i_1}^{(1)} p_{i_2}^{(2)}$, we can rewrite this as $\dot{p}_i = -D_i$ where $D_i = -r(p_i - p_{i_1}^{(1)} p_{i_2}^{(2)})$ measures the linkage disequilibrium in gamete $i$ [**11**]. Since $\dot{p}_{i_k}^{(k)} = 0$, we have $\frac{d}{dt}(p_i - p_{i_1}^{(1)} p_{i_2}^{(2)}) = -r\ (p_i - p_{i_1}^{(1)} p_{i_2}^{(2)})$ and so $p_i$ converges to $p_{i_1}^{(1)} p_{i_2}^{(2)}$ for all strategies $e_i$. That is, $p$ evolves to the unique point of $W$ whose gene frequency at each locus equals the initial gene frequency. It is also important to note here that the convergence is exponential in that $|p_i(t) - p_{i_1}^{(1)} p_{i_2}^{(2)}| \leq Ae^{-ct}$ for some positive constants $A$ and $c$.

**Theorem 4.2** *For an* $n$*-locus continuous-time*[8] *recombination equation with all reassociation rates positive, every trajectory of (4.1) converges to the unique point of* $W$ *whose gene frequency at each locus equals the initial gene frequency at that locus. Moreover, convergence to this point is exponential.*

**Proof** The proof is by induction on the number of loci. By the paragraph immediately preceeding the theorem, the statement is true for $n = 2$. Assume the statement is true for any recombination equation with at most $n - 1$ loci. From (4.1), $\dot{p}_i = -(\sum_I r_I)p_i + \sum_I r_I p_i^I p_i^{L\backslash I}$.

From (4.3) and our induction assumption, $p_i^I$ and $p_i^{L\backslash I}$ both converge exponentially and their product in the limit is $p_{i_1}^{(1)} p_{i_2}^{(2)} \cdot \ldots \cdot p_{i_n}^{(n)} \equiv p_i^*$ Thus

$$\frac{d}{dt}(p_i - p_i^*) \;=\; -\left(\sum_I r_I\right)(p_i - p_i^*) + \sum_I r_I \left(p_i^I p_i^{L\backslash I} - p_i^*\right).$$

The nonhomogeneous term (i.e., the final summation) in this ordinary differential equation is bounded by an exponentially decreasing function and the homogeneous

[8] Lyubich in [**10**] proves the same result for the discrete-time recombination equation.

solution is also decreasing exponentially. By the variation of parameters technique, $p_i - p_i^*$ converges to zero exponentially. $\square$

## 5 Recombination and additive fitness for the multi-locus continuous selection model

Although the Wright manifold was not introduced until Section 4, in retrospect, it also plays an important role when there is no recombination and fitness is additive as in Section 3. Specifically, the continuous replicator dynamic (2.2) for $M$ given by (3.1) is invariant on $W$ [**4**].[9] In fact, on $W$, (2.2) induces the replicator dynamic (with respect to $M^{(k)}$) for $p^{(k)}$ at each locus $k$. That is, on $W$, continuous-time natural selection reduces to analyzing natural selection acting on the gene frequency at each locus seperately.

What is important for this section is that, since $W$ consists of equilibrium points of the recombination equation, the results of the preceding paragraph are also true for the recombination-selection equation under additive fitness. The main result, Theorem 5.1 below, shows this intuition goes a long way to understanding trajectories that are not on $W$ as well.

Before proceeding, it should be pointed out that generalized Wright manifolds $W_K$ exist on which (2.2) is invariant for (3.1) [**6**]. Here $K$ consists of a set of positive constants $K_{ij}^I$ for which $W_K = \{p \in \Delta^N \mid p_{i\backslash i}^{I} p_{j\backslash j}^{I} = K_{ij}^I p_{i\backslash j}^I p_{j\backslash i}^I$ for all $i, j, I\}$ is nonempty (the Wright manifold is then $W_K$ where $K_{ij}^I = 1$ for all $i, j, I$). Although each such $W_K$ is diffeomorphic to $W$, (2.2) does not induce the replicator at the separate loci and this complicates the dynamic analysis.

As derived by [**1**] (see also [**9**]) under the assumption of Hardy-Weinberg proportions, the continuous-time recombination-selection equation becomes

$$\dot{p}_i = p_i(e_i - p) \cdot Mp - \sum_I r_I \sum_j (b_{i\backslash i,j\backslash j}^I p_{i\backslash i}^I p_{j\backslash j}^I - b_{i\backslash j,j\backslash i}^I p_{i\backslash j}^I p_{j\backslash i}^I). \tag{5.1}$$

This combines the selection equation (2.2) with the recombination equation (4.1) altered by the factors $b_{i\backslash j,j\backslash i}^I$ that are the birth rates of individuals with genotype $A_{i\backslash j}^I \backslash A_{j\backslash i}^I$. We will assume that there are no position effects in these factors and so $b_{i\backslash j,j\backslash i}^I = b_{i\backslash i,j\backslash j}^I$ for all $i, j$, and $I$. Suppose $p(t)$ is a trajectory of (5.1) in $\Delta^N$. Let $\Omega$ be the set of limit points of this trajectory (i.e., $\Omega = \{p^* \in \Delta^N \mid p(t_n) \longrightarrow p^*$ for some sequence $t_n \longrightarrow \infty\}$). We want to show that $\Omega$ consists of a single point $p^*$ and that $p^* \in W$ is an equilibrium of the selection equation with no recombination.[10]

There are several key steps to show Theorem 5.1. The first is to notice that, since fitness is additive between loci, recombination has no effect on the population mean fitness. Thus, $p \cdot Mp$ is increasing through the selection component of (5.1) and so all points in $\Omega$ have the same mean fitness. Secondly, for any continuous-time dynamical system on a compact space, $\Omega$ is a closed connected subset of $\Delta^N$ that is invariant under (5.1). Furthermore, each $p \in \Omega$ is a rest point of (2.2) since fitness would increase with initial point $p$ otherwise. That is, $\Omega$ is a connected set

[9]The invariance of $W$ is no longer true for the discrete replicator dynamic (2.1) under additive fitness. In the discrete case, what is needed is multiplicative fitnesses (e.g., for two-loci, $m_{ij} = m_{i_1 j_1}^{(1)} m_{i_2 j_2}^{(2)}$).

[10]Lyubich in [**10**] proved this for the discrete-time recombination-selection with additive fitness. Although our method for (5.1) assumes $M$ is generic in that all equilibria of $M^{(k)}$ at each separate locus are isolated, it can be extended to give the general proof of Theorem 5.1.

of equilibria with respect to $M$ on which mean fitness is constant. Given $p \in \Omega$, the trajectory of (5.1) with initial point $p$ lies entirely in $\Omega$. Since the dynamic reduces to the recombination term on $\Omega$, there is a $p^* \in \Omega$ such that $p_i^* \neq 0$ whenever $p_{i_k}^{*(k)} \neq 0$ for all $k$. By a straight forward generalization of part (a) of Theorem 3.1, $p^*$ is a rest point of $M$ if and only if $p^{*(k)}$ is a rest point of $M^{(k)}$ for every $1 \leq k \leq n$. Thus, in the generic case where each $M^{(k)}$ has only isolated equilibria, every $p$ in the connected set $\Omega$ is in $\{p \in \Delta^N \mid p^{(k)} = p_0^{*(k)} \text{ for all } k\}$ for some $p_0^* \in W$. Finally, it can be shown by a generalization of the method of proof for the discrete recombination-selection equation [**10**] that the "entropy" function $H(p) = -\sum p_i \ln p_i$ is strictly increasing at every $p \in \Omega$ except at this $p_0^*$. This forces $\Omega$ to be the singleton set $\{p_0^*\}$ and so the original trajectory converges to $p_0^*$. That is, we have the following result that combines Theorems 3.2 and 4.2.

**Theorem 5.1** *For an $n$-locus continuous-time recombination-selection equation with additive fitness and with all reassociation rates positive, every trajectory of (5.1) converges to a unique point $p^*$ on the Wright manifold $W$. Moreover, $p^*$ is a rest point of the corresponding selection equation with no recombination that generically yields ESS allelic frequencies at each locus.*

## 6 A symmetric extensive form for a two-locus, two-allele system with additive epistasis or weak selection

Epistasis refers to interactions between loci. These interactions may model various genetic phenomena such as the presence of a modifier locus where a specific allele at one locus affects the fitness of genes at another locus. In any case, additive epistasis refers to the situation where fitness is not additive between loci. That is, Sections 3 and 5 considered selection in the absence of additive epistasis. Weak selection refers to the situation where all fitness parameters are within $\varepsilon$ of the same (positive) constant for sufficiently small $\varepsilon$. Section 4 considered the case when $\varepsilon = 0$.

It is well-known [**11**] that trajectories of (5.1) need not converge for two-locus, two-allele systems with additive epistasis; that mean fitness need not increase; and that the Wright manifold is not necessarily invariant. Thus, the extensive form of this system cannot consist of two subgames corresponding to independent loci. The dependence of payoffs at one locus on the alleles present at the other locus can be modelled in extensive form by Figure 3.

As a two-player extensive form game, Figure 3 models the situation where each player has two sequential decisions to make (the first choice is between $A_1$ and $A_2$, the second between $B_1$ and $B_2$) on each play without knowing what the opponent chose. Each player has three information sets (e.g., $u_1, u_2, u_3$ for player one) and eight pure strategies indexed by $[A_i; B_j, B_k]$ where $i, j, k$ refers to the choice at $1, 2, 3$ respectively. However, the choice at $u_1$ effects what choices are relevant at player one's second decision. For example, if $A_1$ is chosen at $u_1$, then information set $u_3$ is not reached and so the decision $B_k$ by player one at $u_3$ is irrelevant. This leads to the following semi-reduced normal form [**13**] of Figure 3 where $m_{ij,k\ell}$ is the fitness of genotype $A_iB_j/A_kB_\ell$.

$$
\begin{array}{c}
[A_1; B_1, \cdot] \\
[A_1; B_2, \cdot] \\
[A_2; \cdot, B_1] \\
[A_2; \cdot, B_2]
\end{array}
\quad
\begin{bmatrix}
m_{11,11} & m_{11,12} & m_{11,21} & m_{11,22} \\
m_{12,11} & m_{12,12} & m_{12,21} & m_{12,22} \\
m_{21,11} & m_{21,12} & m_{21,21} & m_{21,22} \\
m_{22,11} & m_{22,12} & m_{22,21} & m_{22,22}
\end{bmatrix}
$$

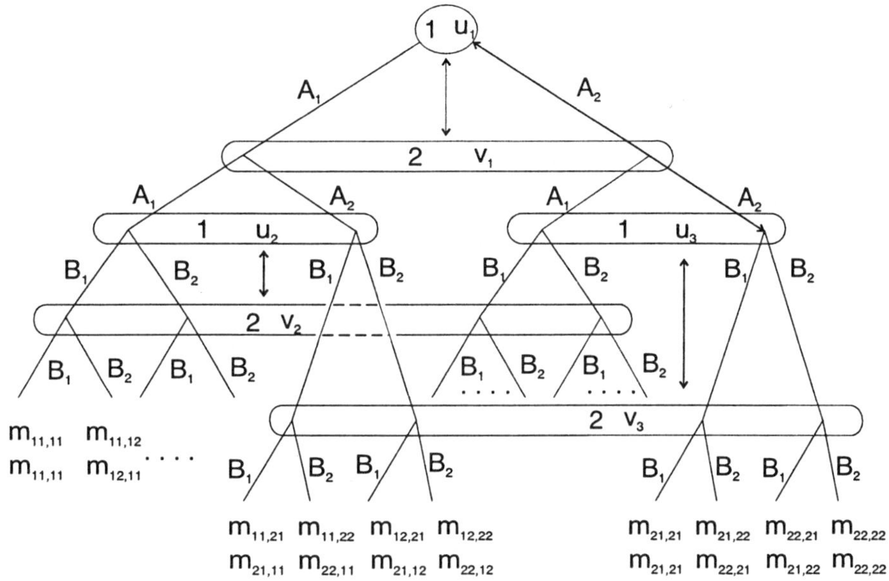


**Figure 3** A two-locus, two allele system.

This is a symmetric normal form game with symmetric payoff matrix $M$ and so, without recombination, all the results of Section 2 are valid. With recombination, it is usually assumed that there are no position effects for the fitness parameters (i.e., $m_{ij,k\ell} = m_{i\ell,kj}$).[11] Little is known then about the trajectories of (5.1) for the general system. However, if there is weak epistasis (i.e., the $m_{ij}$ satisfy (3.1) approximately with errors of order $\varepsilon$ sufficiently small), then generically every trajectory converges [**2**]. The same is true when there is weak selection in which case there is also a quasi-linkage-equilibrium manifold close to the Wright manifold that is invariant and contains all the limit points of these trajectories.

## References

[1] Akin, E. [1979], *The Geometry of Population Genetics*, Vol. 31, Lecture Notes in Biomathematics, Springer-Verlag, Berlin.

[2] Brunovský, P., Hofbauer, J. and Nagylaki, T. [1997], *Convergence of multilocus systems under weak epistasis or weak selection*, Mimeo, University of Vienna.

[3] Cressman, R. [1992], *The Stability Concept of Evolutionary Game Theory (A Dynamic Approach)*, Lecture Notes in Biomathematics, Vol. 94, Springer-Verlag, Berlin.

[4] Cressman, R. [1996], *Subgame monotonicity in two-person extensive form evolutionary games*, Mimeo, Wilfrid Laurier University.

[5] Cressman, R. [1997], *Dynamic stability in symmetric extensive form games*, Int. J. of Game Theory **26**, 525-547.

[6] Cressman, R., Gaunersdorfer, A. and Wen, J. F. [1997], *Evolutionary and dynamic stability in symmetric evolutionary games with two independent decisions*, Mimeo, Wilfrid Laurier University.

[7] Crow J. F. and Kimura, M. [1970], *An Introduction to Population Genetics Theory*, Harper and Row, New York.

[8] Haigh, J. [1989], *How large is the support of an ESS?* J. App. Prob. **26**, 164-170.

---

[11] These conditions are automatically satisfied if fitness is additive since $m_{ij,k\ell} = \overset{(1)}{m}_{ik} + \overset{(2)}{m}_{j\ell}$. For multi-locus models, fitness has no position effects if $m_{ij} = \overset{I}{m}_{i\backslash jj\backslash i}$ for all subsets I of the set of loci.

[9] Hofbauer, J. and Sigmund, K. [1988], *The Theory of Evolution and Dynamical Systems*, Cambridge University Press, Cambridge.
[10] Lyubich, Y. I. [1992], *Mathematical Structures in Population Genetics*, Biomathematics Texts, Vol. 22, Springer-Verlag, Berlin.
[11] Nagylaki, N. [1992], *Introduction to Theoretical Population Genetics*, Biomathematics Texts, Vol. 21, Springer-Verlag, Berlin.
[12] Selten, R. [1983], *Evolutionary stability in extensive two-person games*, Math. Social Sciences **5**, 269-363.
[13] van Damme, E. [1991], *Stability and Perfection of Nash Equilibria*, Second Edition. Springer-Verlag, Berlin.
[14] Weibull, J. [1995] *Evolutionary Game Theory*, MIT Press, Cambridge, MA.

Fields Institute Communications
Volume **21**, 1999

# Growth in the Unstirred Chemostat with Different Diffusion Rates

**Le Dung**
School of Mathematics
Georgia Institute of Technology
Atlanta, GA 30332, USA
ledung@math.gatech.edu

**H. L. Smith**
Department of Mathematics
Arizona State University
Tempe, AZ 85287–1804, USA
halsmith@math.la.asu.edu

**P. Waltman**
Department of Mathematics and Computer Science
Emory University
Atlanta, GA 30322 USA
waltman@mathcs.emory.edu

*Dedicated to Herb Freedman and Lynn Erbe on the occasion of their retirement*

**Abstract.** The paper considers growth in the unstirred chemostat where the nutrient and organism have different diffusion coefficients. The usual reduction of such a chemostat to one equation does not apply in this case. The problem is treated by a recently established perturbation theorem that preserves the global attractivity of a rest point. The application of this theorem is complicated by the presence of a rest point of the boundary.

## 1 Introduction

The unstirred chemostat has been investigated in a series of papers, (see [**2, 4, 5, 13**]) and has been a model for generalizations to an abstract setting [**6**] of competition problems. One views the chemostat as being a thin tube along the x-axis with the input of constant nutrient at $x = 0$ and the washout at $x = 1$. The vessel is not mixed and the nutrient and organism diffuse through the vessel. The

1991 *Mathematics Subject Classification.* Primary 35K57, 92D25; Secondary 35Q80.
HLS was supported by NSF Grant DMS 9700910.
PW was supported by NSF Grant DMS 9424592 and an award from the University Research Council of Emory University.

constant flow rate $F$ keeps the volume constant. The model takes the form for a single population u(x,t) consuming a nutrient S(x,t)

$$\begin{aligned} S_t &= dS_{xx} - uf(S) \\ u_t &= du_{xx} + uf(S) \end{aligned} \tag{1.1}$$

with boundary conditions

$$\begin{aligned} S_x(0,t) &= -S^{(0)} \\ u_x(0,t) &= 0 \\ S_x(1,t) + rS(1,t) &= 0 \\ u_x(1,t) + ru(1,t) &= 0 \end{aligned}$$

and initial conditions

$$S(x,0) = S_0(x), u(x,0) = u_0(x).$$

The quantity $S^{(0)}$ is the input flux of the nutrient, d is a diffusion coefficient, and r is a function of the flow rate, $F/d$. A more complete description can be found in [**11**].

The growth function $f(S)$ will be assumed to satisfy

(i) $f : R_+ \to R_+$ and $f(0) = 0$;
(ii) $f \in C^2$ with $f'(S) > 0$;
(iii) $f$ is a bounded function.

Adding the two equations yields a simple heat equation in the sum of the variables which in turn can be used to reduce the problem to a single parabolic equation with nice monotonicity properties. This device depends on the fact that each dependent variable has the same diffusion coefficient. Except for [**2**], all of the above cited work on competition in the unstirred chemostat assumes equal diffusion coefficients. This assumption, however, is not biologically realistic.

In this work we deal with a single population growing on a nutrient and we weaken the assumption that the diffusion coefficients are the same to the assumption that they are close. More directly we treat the problem as a perturbation problem in the diffusion coefficient. It turns out also not to be any more difficult to also introduce a small death rate for the population u(x,t) as well.

The vehicle for this perturbation is a theorem established in [**12**] for abstract dynamical systems. We state that theorem in order to discuss the hypotheses. (See [**9**] for an earlier theorem in this direction.)

**Theorem 1.1** *Let $T$ be a family of semi-dynamical systems parameterized by $\Lambda$. Let $(x_0, \lambda_0) \in U \times \Lambda$, $B_X(x_0, \delta) \subset U$ for some $\delta > 0$ and assume that $D_x T(x,t,\lambda)$ exists for $(x,t,\lambda) \in B_X(x_0,\delta) \times [0,\infty) \times \Lambda$ and for each fixed $t \geq 0$, $D_x T(x,t,\lambda)$ is continuous on $B_X(x_0,\delta) \times \Lambda$. Suppose that $T(x_0,t,\lambda_0) \equiv T^t_{\lambda_0} x_0 = x_0$ for all $t \geq 0$, $U(t) \equiv D_x T^t_{\lambda_0} x_0$ defines a strongly continuous semi-group with negative growth bound ($r(U(t)) = \exp(-\omega t)$ with $\omega > 0$), and $T^t_{\lambda_0}(x) \to x_0$ as $t \to \infty$ for each $x \in U$. In addition, suppose that:*

(H1) *There is a subset $B$ of $U$ such that for each $x \in U$ and $\lambda \in \Lambda$, $T^t_\lambda(x) \in B$ for all large $t$.*
(H2) *$C \equiv \overline{\cup_{\lambda \in \Lambda} T^s_\lambda(B)}$ is compact in $U$ for some $s > 0$.*

*Then there exists* $\epsilon_0 > 0$ *and a continuous map* $\hat{x} : B_\Lambda(\lambda_0, \epsilon_0) \to U$ *such that* $\hat{x}(\lambda_0) = x_0$, $T^t_\lambda \hat{x}(\lambda) = \hat{x}(\lambda)$ *for* $t \geq 0$, *and*

$$\lim_{t\to\infty} T^t_\lambda x = \hat{x}(\lambda), \quad x \in U, \ \lambda \in B_\Lambda(\lambda_0, \epsilon_0). \tag{1.2}$$

The idea is to fix a diffusion coefficient $d$ and use the fact that (with some conditions) all solutions of (1.1) tend to a rest point. Then we seek to apply the above theorem to the perturbed system

$$\begin{aligned} S_t &= d_0 S_{xx} - uf(S) \\ u_t &= du_{xx} + u(f(S) - k) \end{aligned} \tag{1.3}$$

with the perturbed boundary conditions

$$\begin{aligned} S_x(0,t) = -S^{(0)}, u_x(0,t) &= 0 \\ S_x(1,t) + r_0 S(1,t) &= 0 \\ u_x(1,t) + ru(1,t) &= 0 \end{aligned} \tag{1.4}$$

and initial conditions

$$\begin{aligned} S(x,0) &= S_0(x), \\ u(x,0) &= u_0(x). \end{aligned} \tag{1.5}$$

where $d$ is close to $d_0$ and k is close to zero. (It follows that $r$ will be close to $r_0$.)

The difficulty is that in this problem the dynamical system is defined on a (relatively) open set plus the boundary which contains a rest point. The above compactness condition then is difficult to verify. We do this by showing that the system (1.3) has a global attractor in the interior of the positive cone in the product of two copies of the space of continuous functions. This result essentially appeared in [**2**] but there is a question of the dependence of the location of the global attractor on the parameters.

The basic theorem and some necessary propositions are stated in Section 2 which also contains the proof of the main theorem using Theorem 1.1. The proofs of the necessary technical propositions are in Section 3.

## 2 Statement of results

We begin with some basic notation. Fix the parameter $d_0 > 0$. This in turn determines the parameter $r_0$ since $r_0 = F/d_0$ where as noted above $F$ is the flow rate. We a priori restrict the parameters. Define the set $\Lambda = \{(d,k,r) | \frac{d_0}{2} \leq d \leq 2d_0, \frac{r_0}{2} \leq r \leq 2r_0, 0 \leq k \leq k_0\}$. The perturbation parameter will be $\lambda = (d,k,r) \in \Lambda$ and the reference parameter $\lambda_0 = (d_0, 0, r_0)$. The application will restrict $\lambda$ to be close to $\lambda_0$.

We begin first with a proposition that gives the known properties for the unperturbed equation. (See ([**5**], Proposition 4.1) for the proof).

**Proposition 2.1** *Let* $d = d_0, k = 0, r_0 = r$. *Let* $\mu_0$ *be the principal eigenvalue of the regular Sturm-Liouville problem*

$$\begin{aligned} d_0 w'' + f(\phi)w &= \lambda w \\ w'(0) = w'(1) + r_0 w(1) &= 0, \end{aligned}$$

*where* $\phi(x)$ *is given (see* [**11**]*) by*

$$\phi(x) = S^{(0)} \left( \frac{1 + r_0}{r_0} - x \right), \quad 0 < x < 1. \tag{2.1}$$

*Then $\mu_0 > 0$ implies that all solutions of (1.1) with $u_0(x) \geq 0$, $u_0(x) \not\equiv 0, S_0(x) \geq 0$ converge uniformly to the unique, strictly positive steady state $(\hat{S}, \hat{u})$ in $X_+$.*

The equilibrium $(\phi, 0)$ of (1.3)-(1.4) is called the "washout equilibrium". It attracts all solutions with $u_0(x) \equiv 0$. The condition $\mu_0 > 0$ reflects the assumption that the washout equilibrium is unstable.

**Proposition 2.2** *Let $d, k, r$ be as in Proposition 2.1 and let $\mu_0$ be positive. Then the linearization of (1.3) - (1.5) around the steady state $(\hat{S}, \hat{u})$ has a negative principal eigenvalue.*

For each $\lambda \in \Lambda$, denote by $\{T_\lambda(t)\}_{t\geq 0}$ the semi-flow on

$$X_+ = \{(S, u) \in C_+[0,1] \times C_+[0,1]\}$$

defined by solutions of (1.3), (1.4) and (1.5). (We intend "mild" solutions, of course.) $C_+[0,1]$ denotes the non-negative continuous functions on $[0,1]$. The smoothness properties needed for Theorem 1.1 follow from basic properties of parabolic equations. Our main result is

**Theorem 2.3** *Let $\mu_0 > 0$ be defined as above. There exists a $\delta > 0$ such that for each $\lambda \in \Lambda$ with $|\lambda - \lambda_0| < \delta$ there is a unique steady state solution $(\hat{S}(x;\lambda), \hat{u}(x;\lambda))$ with $\hat{u}(x;\lambda) > 0$, $(\hat{S}(x;\lambda_0), \hat{u}(x;\lambda_0)) = (\hat{S}(x), \hat{u}(x))$, and $\lambda \to (\hat{S}(\cdot;\lambda), \hat{u}(\cdot;\lambda))$ is continuous.*

*Furthermore*

$$T_\lambda(t)(S_0, u_0) \to (\hat{S}(\cdot;\lambda), \hat{u}(\cdot;\lambda))$$

*as $t \to \infty$ for every $(S_0, u_0) \in X_+$ with $u_0 \not\equiv 0$, $\lambda \in \Lambda$ with $|\lambda - \lambda_0| < \delta$.*

To prove the theorem two additional propositions will be needed.

**Proposition 2.4** *Let $T_\lambda(t)$ be defined as above.*

(i) *There exists a bounded set $B \subset X_+$ such that for all $\lambda \in \Lambda$ and all initial conditions $(S_0, u_0) \in X_+$,*

$$T_\lambda(t)(S_0, u_0) \in B$$

*for all large $t$.*

(ii) *$\cup_{\lambda\in\Lambda}\{T_\lambda(t)C | t > 1\}$ is bounded for any bounded set $C$.*

(iii) *$\overline{\cup_{\lambda\in\Lambda}T_\lambda(t)C}$ is compact for any bounded set $C$ and any $t > 1$.*

**Proposition 2.5** *There exists a $\delta > 0$ and an $\epsilon > 0$ such that*

$$\liminf_{t\to\infty} ||u(\cdot, t, \lambda)||_\infty \geq \epsilon$$

*for all solutions of (1.3), $(S(x,t;\lambda), u(x,t;\lambda))$, such that $(S(\cdot,t;\lambda), u(\cdot,t;\lambda)) \in X_+$, $u(\cdot, 0;\lambda) \not\equiv 0$, and $|\lambda_0 - \lambda| \leq \delta$.*

**Proof of Theorem 2.3**

We seek to apply Theorem 1.1 with $\delta$ as in Proposition 2.5. Take the region $U$ to be the non-negative cone in $X_+$ with $||u_0|| \not\equiv 0$. The existence of a semi-group follows from the uniformly parabolic nature of the problem, and the continuity of the derivative from the fact that $f \in C^2$. Proposition 2.1 establishes the uniqueness of a steady state for the unperturbed problem $(\lambda = \lambda_0)$ and the convergence of all trajectories in $U$. Proposition 2.2 yields the negative growth bound on the linearization.

In hypothesis (H1) let

$$\tilde{B} = B \cap \left\{ (S, u) \in X_+ | : \| u \|_\infty > \frac{\epsilon}{2} \right\},$$

where $B$ is given by Proposition 2.4 and $\epsilon$ is given in Proposition 2.5. Thus (H1) holds.

For (H2), let

$$C \equiv \overline{\cup_{\lambda \in \Lambda,\ |\lambda - \lambda_0| \leq \delta}\ T_\lambda(s)\tilde{B}}.$$

$C$ is compact in $U$ by Proposition 2.4. Theorem 1.1 applies and the theorem is established.

There remains the task of establishing the technical propositions.

## 3 Proofs

**Proof of Proposition 2.2**

The existence, uniqueness and global stability of the interior rest point $(\hat{S}, \hat{u}) \equiv (\hat{S}(x), \hat{u}(x))$ was established in Proposition 2.1. Theorem 1.1 requires linearized stability with negative eigenvalues. We investigate the spectrum of the linearization of (1.1) about $(\hat{S}, \hat{u})$.

First, note that since $\hat{u} > 0$ satisfies

$$d_0 u'' + f(\hat{S}(x))u = 0$$

and the boundary conditions, then $\lambda = 0$ is the principal eigenvalue of

$$d_0 u'' + f(\hat{S}(x))u = \lambda u. \tag{3.1}$$

Make the change of variables $T = \phi - S - u$ in (1.1) to obtain

$$\begin{aligned} T_t &= d_0 T_{xx} \\ u_t &= d_0 u_{xx} + f(\phi - T - u)u, \end{aligned}$$

where $T$ satisfies homogeneous Neumann conditions at $x = 0$.

Linearize this systems about the rest point $(\hat{S}, \hat{u})$ to obtain

$$\begin{aligned} \lambda v_1 &= d_0(v_1)_{xx} \\ \lambda v_2 &= d_0(v_2)_{xx} - \hat{u} f'(\phi - \hat{T} - \hat{u})v_1 \\ &\quad + [f(\phi - \hat{T} - \hat{u}) - \hat{u} f'(\phi - \hat{T} - \hat{u})]v_2. \end{aligned}$$

If $v_1 \neq 0$ then $\lambda$ is in the spectrum of the operator $d_0 \frac{d^2}{dx^2}$ and the spectrum of this operator is negative by comparison with (3.1).

If $v_1 \equiv 0$, then $v_2$ satisfies

$$\begin{aligned} \lambda v_2 &= d_0(v_2)_{xx} + [f(\phi - \hat{T} - \hat{u}) - \hat{u} f'(\phi - \hat{T} - \hat{u})]v_2 \\ &= d_0(v_2)_{xx} + q(x)v_2. \end{aligned} \tag{3.2}$$

Comparing the eigenvalues of (3.2) with the eigenvalue problem (3.1), one concludes that the principal eigenvalue of (3.2) is negative since $q(x) < f(\phi(x) - \hat{T}(x) - \hat{u}(x)) = f(\hat{S}(x))$ and the principal eigenvalue of (3.1) is zero.

**Proof of Proposition 2.4** We begin the proof of Proposition 2.4 by establishing the necessary dissipativeness condition. We note first, however, that both $S(x,t)$ and $u(x,t)$ are nonnegative for $t > 0$. (If $u(x,t) \not\equiv 0$, then they are strictly positive.) Note that $S(x,t)$ satisfies the differential inequality

$$S_t(x,t) \leq d_0 S_{xx}(x,t).$$

Solutions of the inequality may be compared to solutions of the corresponding equality (with the same boundary and initial conditions) which converge to the function $\phi$ defined in Section 2. Hence, $S(x,t)$ is uniformly bounded for large t, and one needs only to bound $u(x,t)$. This is done in the three following lemmas which give bounds on $u(x,t)$ in $L^1, L^2$ and $W_{1,2}$ respectively. The estimates hold uniformly for $\lambda \in \Lambda$ and $|\lambda - \lambda_0|$ sufficiently small. As previously noted, this sequence of estimates follows the work in [**2**]. We include the details because the estimates are simpler (we are in one space dimension) and because we need to verify that the estimates are uniform in the parameters

**Lemma 3.1** *There exist positive constants $C_1, \alpha$, independent of initial data, and $C_2$ dependent on the $L_1$ norm of the initial data, both independent of $\lambda \in \Lambda$, such that*

$$||u(\cdot,t)||_1 \equiv \int_0^1 u(x,t)\,dx \le C_1 + C_2 e^{-\alpha t}.$$

**Proof** Let $\mu$ be the unique positive root of

$$\frac{r_0}{2} cos(\mu) - \mu sin(\mu) = 0.$$

(Alternatively, this is $tan(\mu) = \frac{r_0}{2\mu}$.) Note that $0 < \mu < \frac{\pi}{2}$. Multiply both equations of (1.3) by $cos(\mu x)$ and integrate over $[0,1]$ to obtain (using the boundary conditions, integration by parts, and the definition of $\mu$)

$$\begin{aligned}
\frac{d}{dt}\int_0^1 S(x,t)cos(\mu x)\,dx &\le d_0 S^{(0)} - d_0\mu^2 \int_0^1 S(x,t)cos(\mu x)\,dx \\
&\quad - \int_0^1 u(x,t) f(S(x,t))\, cos(\mu x)\,dx \\
\frac{d}{dt}\int_0^1 u(x,t)cos(\mu x)\,dx &\le -d\mu^2 \int_0^1 u(x,t)cos(\mu x)\,dx \\
&\quad + \int_0^1 u(x,t)[f(S(x,t)) - k]cos(\mu x)\,dx.
\end{aligned}$$

Add the two equations to obtain

$$\begin{aligned}
\frac{d}{dt}\int_0^1 [S(x,t)+u(x,t)]cos(\mu x)\,dx &\le d_0 S^{(0)} - d_0\mu^2 \int_0^1 S(x,t)cos(\mu x)\,dx \\
&\quad -d\mu^2 \int_0^1 u(x,t)cos(\mu x)\,dx \\
&\quad -\int_0^1 k u(x,t)\, cos(\mu x)\,dx.
\end{aligned}$$

Letting $D = \frac{d_0}{2}\mu^2$, using $d \ge \frac{d_0}{2}$, and dropping the last term yields the inequality

$$\frac{d}{dt}\int_0^1 [S(x,t)+u(x,t)]cos(\mu x)\,dx \le d_0 S^{(0)} - D\int_0^1 [S(x,t)+u(x,t)]cos(\mu x)\,dx.$$

This is just an ordinary differential inequality of the form

$$y' + Dy \le d_0 S^{(0)},$$

so the comparison theorem yields that

$$\int_0^1 [S(x,t)+u(x,t)]cos(\mu x)\,dx \le \frac{d_0 S^{(0)}}{D}(1-e^{-Dt}) + e^{-Dt}\int_0^1 [S(x,0)+u(x,0)]cos(\mu x)\,dx.$$

Since $S(x,t), u(x,t)$, and $cos(\mu x)$ are positive for $x \in [0,1]$ and all $t$, this yields the $L^1$ bound on $u(x,t)$with $\alpha = D$. The constants are independent of $\lambda$ for $\lambda \in \Lambda$ and $C_2$ depends on the initial data.

Note that

$$\limsup_{t\to\infty} \int_0^1 u(x,t)\,dx \le C_1$$

where $C_1$ is independent of the initial conditions and the parameter $\lambda$.

In the next lemma it is convenient to use $\hat{f}(S) = f(S) - k$.

**Lemma 3.2** *Let $(S,u)$ be a solution of (1.3) with $\lambda \in \Lambda$ and with initial conditions $(S_0, u_0)$. There exist constants $K_1$, $K_2$, and $\beta > 0$, $K_2$ depending on the $L_1$ and $L_2$ norms of the initial data, such that*

$$\int_0^1 u^2(x,t)\,dx \le K_1 + K_2 e^{-\beta t}, \tag{3.3}$$

$$\limsup_{t\to\infty} \int_0^1 u^2(x,t)\,dx \le K_1. \tag{3.4}$$

*Moreover, there exist constants $K_3$ and $K_4$, $K_4$ depending on the initial data, such that*

$$\int_t^{t+1}\int_0^1 u_x^2(x,t)\,dx \le K_3 + K_4 e^{-\beta t}, \tag{3.5}$$

$$\limsup_{t\to\infty} \int_t^{t+1}\int_0^1 u_x^2(x,t)\,dx \le K_3. \tag{3.6}$$

**Proof** Multiply the second equation in (1.3) by $u$ and integrate from zero to one producing

$$\begin{aligned}\frac{1}{2}\frac{d}{dt}\int_0^1 u^2(x,t)\,dx + d\int_0^1 u_x^2(x,t)\,dx &\\ &= du_x(1,t)u(1,t) + \int_0^1 u^2(x,t)\hat{f}(S(x,t))\,dx \\ &= -rdu^2(1,t) + \int_0^1 u^2(x,t)\hat{f}(S(x,t))\,dx \\ &\le C_3\int_0^1 u^2(x,t)\,dx,\end{aligned} \tag{3.7}$$

where we have used the bound on $f$ and the fact that $u$ is non-negative. We now use the Gagliardo-Nirenberg inequality ([**16**], pp.1034), with $p=2, \theta=\frac{1}{3}, q=1, r=2, m=1$ to obtain

$$||u(\cdot,t)||_2 \le C_4||u||_{W_{1,2}}^{\frac{1}{3}}||u(\cdot,t)||_1^{\frac{2}{3}}.$$

Using as norm for $W_{1,2}$ ([**16**], pp.1033),

$$||z||^2_{W_{1,2}} = \int_0^1 z_x^2(x)\,dx + \left(\int_0^1 z(x)\,dx\right)^2,$$

this becomes

$$\int_0^1 u^2(x,t)\,dx \le C_4 \left[ \int_0^1 u_x^2(x,t))\,dx) + \left( \int_0^1 u(x,t)\,dx \right)^2 \right]^{\frac{1}{3}} \|u(\cdot,t)\|_1^{\frac{4}{3}}.$$

We now use Young's inequality ([**16**], pp. 1020) with $p = 3$, on the two factors on the right (and introducing $(p\epsilon)$ and $(p\epsilon)^{-1}$ for $\epsilon > 0$ into the first and second factors respectively) to obtain

$$\int_0^1 u^2(x,t)\,dx \le C_4\epsilon \left[ \int_0^1 u_x^2(x,t)\,dx + \left( \int_0^1 u(x,t)\,dx \right)^2 \right] + C(\epsilon)\|u(\cdot,t)\|_1^2.$$

The function $C(\epsilon)$ contains inverse powers of $\epsilon$ and so is bounded when $\epsilon$ is bounded away from zero. We choose $\epsilon = \frac{d}{2C_3C_4}$ which fixes $C(\epsilon)$. Insert this estimate into (3.7) to obtain the inequality (combining constants and using Lemma 3.1)

$$\frac{1}{2}\frac{d}{dt}\int_0^1 u^2(x,t)\,dx + d\int_0^1 u_x^2(x,t)\,dx \le \frac{d}{2}\int u_x^2(x,t)\,dx + \hat{C}_1 + \hat{C}_2 e^{-\alpha t}$$

or

$$\frac{1}{2}\frac{d}{dt}\int_0^1 u^2(x,t)\,dx + \frac{d}{2}\int_0^1 u_x^2(x,t)\,dx \le \hat{C}_1 + \hat{C}_2 e^{-\alpha t}.$$

Both constants are independent of $\lambda$ for $\lambda \in \Lambda$ and $C_2$ depends on the $L_1$ norm of the initial data.

The fact that the two norms for $W_{1,2}$ are equivalent means that there exists a $\gamma > 0$ such that

$$\int_0^1 z_x^2(x)\,dx + \left( \int_0^1 z(x)\,dx \right)^2 \ge \gamma \left[ \int_0^1 z^2(x)\,dx + \int_0^1 z_x^2(x)\,dx \right],$$

or, for our case, that we have

$$\int_0^1 u_x^2(x,t)\,dx \ge \gamma \int_0^1 u^2(x,t)\,dx - \left( \int_0^1 u(x,t)\,dx \right)^2.$$

This leads to

$$\frac{d}{dt}\int_0^1 u^2(x,t)\,dx + d\gamma \int_0^1 u^2(x,t)\,dx \le \bar{C}_1 + \bar{C}_2 e^{-\alpha t}$$

where $\bar{C}_2$ depends on the $L_1$ norm of the initial data. This is just an ordinary differential inequality of the form (bounding $d$ below)

$$y'(t) + \frac{d_0\gamma}{2} y(t) \le \bar{C}_1 + \bar{C}_2 e^{-\alpha t}.$$

The comparison theorem yields

$$y(t) \le K_1 + K_2 e^{-\beta t}$$

where none of the constants depend on $\lambda \in \Lambda$, $K_2$ depends on the $L_1$ and $L_2$ norms of the initial data, and $\beta > 0$ is the smaller of $\alpha$ and $\frac{d_0\gamma}{2}$ . The first conclusion (3.3) follows and (3.4) follows by taking lim sup and noting that the dependence on initial conditions vanishes.

Integrating (3.7) from $t$ to $t+1$ yields

$$\int_0^1 u^2(x,t+1)\,dx \quad + \quad 2d\int_t^{t+1}\int_0^1 u_x^2(x,t)\,dxdt$$
$$\leq \int_0^1 u^2(x,t)\,dx + 2C_3\int_t^{t+1}\int_0^1 u^2(x,t)\,dx\,dt.$$

(3.5) now follows directly using (3.3), and (3.6) follows since the lim sup of $\int_t^{t+1}\int_0^1 u^2(x,t)\,dxdt$ is finite by (3.4) and $d \geq \frac{d_0}{2}$.

**Lemma 3.3** *There exist constants $D_1$ and $D_2$, where $D_2$ depends on the initial data and both are independent of $\lambda \in \Lambda$, such that*

$$\int_0^1 u_x^2(x,t)\,dx \leq D_1 + D_2 e^{-\beta(t-1)}, \quad t > 1 \tag{3.8}$$

*and*

$$\limsup_{t\to\infty} \int_0^1 u_x^2(x,t)\,dx \leq D_1. \tag{3.9}$$

**Proof** We begin with a change of variables [**8**]

$$w(x,t) = e^{\frac{1}{2}rx^2}u(x,t)$$

in the equation and boundary conditions for $u$ in (1.3). A straightforward calculation shows that the function $w(x,t)$ satisfies

$$w_t(x,t) = dw_{xx}(x,t) - 2drxw_x(x,t) + w(x,t)(d(rx)^2 - dr + f(S) - k) \tag{3.10}$$

with homogeneous Neumann boundary conditions

$$\begin{aligned} w_x(0,t) &= 0 \\ w_x(1,t) &= 0. \end{aligned} \tag{3.11}$$

Since

$$w_x(x,t) = e^{\frac{1}{2}rx^2}[u_x(x,t) + rxu(x,t)],$$

the conclusions of Lemmas 3.1 and 3.2 apply to $w(x,t)$; specifically the quantities $\int_0^1 w(x,t)\,dx$, $\int_0^1 w^2(x,t)\,dx$ and $\limsup_{t\to\infty}\int_x^{x+1}\int_0^1 w_x^2(x,t)\,dx$ are all bounded, and we will use the same notation for the constants when we apply the Lemmas to $w(x,t)$. To prove the lemma one needs only to establish it for $\int_0^1 w_x^2(x,t)\,dx$.

Let $T > 1$ and $\eta : R_+ \to R_+$ be a smooth increasing function such that $\eta(s) \equiv 0$ if $s \leq T-1$, $\eta(s) \equiv 1$ if $s \geq T$, and $|\eta'(s)| \leq 2$ for all $s \in R_+$. Multiply the equation for $w(x,t)$ by $w_{xx}(x,t)\eta(t)$ and integrate over the domain to obtain

$$\frac{1}{2}\frac{d}{dt}\int_0^1 w_x^2(x,t)\eta(t)\,dx + d\int_0^1 w_{xx}^2(x,t)\eta(t)\,dx$$
$$\leq C\left[\int_0^1 |w_x(x,t)||w_{xx}(x,t)|\eta(t)\,dx + \int_0^1 |w(x,t)||w_{xx}(x,t)|\eta(t)\,dx\right]$$
$$+\frac{1}{2}\int_0^1 w_x^2(x,t)\eta'(t)\,dx, \tag{3.12}$$

where we have used an integration by parts and the Neumann conditions (3.11). C depends on $r_0, k_0$ and the bound on $f$ but not on $\lambda$. The integrands on the right-hand side will be estimated with Young's inequality using the same "epsilon" trick as in Lemma 3.2 to obtain

$$|w_x(x,t)||w_{xx}(x,t)| \le \epsilon w_{xx}^2(x,t) + C(\epsilon)w_x^2(x,t),$$
$$|w(x,t)||w_{xx}(x,t)| \le \epsilon w_{xx}^2(x,t) + C(\epsilon)w^2(x,t).$$

The choice $\epsilon = \frac{d_0}{4C}$ allows the terms in $w_{xx}(x,t)$ on both sides of (3.12) to be combined and then dropped. Using also the fact that $\eta, \eta'$ are bounded we obtain the basic inequality

$$\begin{aligned}\tfrac{d}{dt}\textstyle\int_0^1 w_x^2(x,t)\eta(t)\,dx &\le D_3 \textstyle\int_0^1 w_x^2(x,t)dx + D_4 \int_0^1 w^2(x,t)dx\\ &\le D_3 \textstyle\int_0^1 w_x^2(x,t)dx + D_5 + D_6 e^{-\beta t},\end{aligned}$$

where only $D_6$ depends on the initial data. Integrating this inequality from $T-1$ to $T$, because $\eta(T-1)=0$, we deduce

$$\int_0^1 w_x^2(x,T)\,dx \le D_3 \int_{T-1}^T \int_0^1 w_x^2(x,t)\,dxdt + D_5 + \frac{D_6}{\beta}\left(e^{-\beta(T-1)} - e^{-\beta T}\right).$$

Combining this with (3.5) gives (3.8). Letting $T$ tend to infinity, we obtain (3.9). Note that this bound is independent of the initial conditions.

**Proof of Proposition 2.4 (Continued)** Lemmas 3.2 and 3.3 along with the bound on $S(x,t)$ show that for any initial condition $(S_0, u_0)$, the corresponding solution eventually lies in a bounded set in $W_{1,2}$. The Sobolev Embedding Theorem ([**7**], p.208) allows one to conclude that

$$\limsup_{t\to\infty}[||u(\cdot,t)||_\infty + ||S(\cdot,t)||_\infty] \le K$$

since we are in one space dimension. The estimate is independent of $\lambda$ for $\lambda \in \Lambda$ and independent of the initial condition $(S_0(x), u_0(x))$. (i) of Proposition 2.4 is established.

Given a bounded set $C$ and $t \ge 1$, the solutions with initial data in $C$ satisfy (3.3) and (3.8) where the constants that depend on initial conditions are fixed. Since $t > 1$, (ii) follows.

To obtain (iii) we note that the embedding theorem actually gives a stronger conclusion than that used above, namely that $W_{1,2}$ is compactly embedded in $C_\alpha$ for $0 < \alpha < \frac{1}{2}$ so bounded sets $W_{1,2}$ are pre-compact in $C[0,1]$. For a given bounded set $C$, and for $t > 1$, Lemmas 3.2 and 3.3 yield a bounded set $T(t)C$ in $W_{1,2}$ which as noted is pre-compact in $C[0,1]$. The bound depends only on the initial data and $t$, so it holds for all $\lambda \in \Lambda$. The closure of the union lies in a compact set, so it is compact.

**Remark 3.1** Parts (ii) and (iii) are proved for $t > 1$. By changing the integration in Lemma 3.3 to $t-r$ to $t$, $r > 0$, one can make the same conclusions for any $t > r$. The stronger conclusions (i.e., for any $t > 0$) are not needed since it is known [**2**] that solutions extend for all time.

**Proof of Proposition 2.5** Choose $\delta > 0$ so that the principal eigenvalue, $\beta$, of

$$\begin{aligned} dw'' + [f(\phi-\delta)-\delta]w &= \beta w\\ w'(0) &= 0\\ w'(1)+rw(1) &= 0\end{aligned} \tag{3.13}$$

is positive for all $r, d$ with $|r-r_0| < \delta$, $|d-d_0| < \delta$, and the principal eigenfunction, $w$, satisfies $w > 0$ . Here we use Proposition 2.1 and the continuity of the principal eigenvalue under perturbation of the potential and the boundary conditions. Then choose $\eta$ so that the solution, $z(x)$ of the boundary value problem

$$\begin{aligned} d_0 z'' - \eta f(\phi(x) + \eta) &= 0 \\ z'(0) &= -S^{(0)} \\ z'(1) + r_0 z(1) &= 0 \end{aligned}$$

satisfies $z \geq \phi - \frac{\delta}{2}$ which can be done either by direct computation or by continuity in the parameters.

Suppose there existed a solution of (1.3) with $u(x,t)$ such that

$$\limsup_{t\to\infty} \|u(x,t)\|_\infty \leq \frac{\eta}{2}.$$

Then $||u(x,t)||_\infty \leq \eta$ for t sufficiently large, say for $t > T$. $S(x,t)$ satisfies the differential inequality

$$\begin{aligned} S_t &\leq d_0 S_{xx} \\ S_x(0) &= -S^0 \\ S_x(1) + r_0 S(1) &= 0 \end{aligned}$$

so that a comparison with solutions of the corresponding equality holds. However, all solutions of the system with equality converge to $\phi$, the $S$ component of the washout equilibrium. Thus, for large enough t (which we continue to take as T), $0 \leq S(x,t) \leq \phi(x) + \eta$. Estimating from below, one has

$$\begin{aligned} S_t &\geq d_0 S_{xx} - \eta f(\phi(x) + \eta), \quad t > T. \\ S_x(0) &= -S^0 \\ S_x(1) + r_0 S(1) &= 0. \end{aligned}$$

Let $\bar{S}$ denote the solution of the corresponding equation (with the same initial conditions) and one has $S(x,t) \geq \bar{S}(x,t)$; however, $\lim_{t\to\infty}\bar{S}(x,t) = z(x) \geq \phi - \frac{\delta}{2}$. Then one has that (for large t)

$$\begin{aligned} u_t &= du_{xx} + u[f(S) - k], \\ &\geq du_{xx} + u[f(\phi - \delta) - \delta], \quad t > T. \end{aligned}$$

The corresponding linear equation can be solved by separation of variables and has a positive solution

$$ce^{\beta(t-T)}w(x)$$

where $\beta$ and $w$ are the principal eigenvalue and eigenfunction of (3.13). There exists a $c > 0$ such that $u(x,T) > cw(x)$. Thus

$$u(x,t) \geq ce^{\beta(t-T)}w(x), \ t > T.$$

This shows that $||u(\cdot,t)||_\infty \to \infty$ which is a contradiction. This establishes the proposition with lim sup instead of lim inf.

To complete the proof one adopts one of the proofs that weak repulsion implies strong repulsion (weak persistence implies strong persistence). There is literature in this direction, for example, [**1, 3, 15**]. Proposition 1.2 of [**15**] can be adapted for this purpose to complete the proof.

Let $X_2 = \{(S_0, u_0) | u_0 = 0\}$ and $X_1 = X_+ \backslash X_2$. We proved above that there are positive constants $\epsilon, \delta$ such that if $|\lambda - \lambda_0| < \delta$, then

$$\limsup_{t\to\infty} dist(T_\lambda(t)(S_0, u_0), X_2) \geq \epsilon$$

for all $(S_0, u_0) \in X_1$. Let $\Lambda' = \{\lambda \in \Lambda | \ |\lambda - \lambda_0| \leq \delta\}$ and $\hat{X} = X_+ \times \Lambda'$. Let $w_0$ denote $(S_0, u_0) \in X_+$. Define $\Phi(t) : \hat{X} \to \hat{X}$ by $\Phi(t)(w_0, \lambda) \equiv (T_\lambda(t)\,(w_0, \lambda))$, $t > 0$.

The continuity of T implies that $\{\Phi(t)\}$ is a semiflow on $\hat{X}$. Put $\hat{X}_i = \{(w_0, \lambda) | w_0 \in X_i\}, i = 1, 2$. Then

$$\limsup_{t \to \infty} \; dist(\Phi(t)(w_0, \lambda), \hat{X}_2) \geq \epsilon$$

for all $(w_0, \lambda) \in \hat{X}_1$. Proposition 1.2 of [**15**] does not have a dissipative condition and so the required compactness assumption $C_{1.1}$ is stronger than what we have. However, we begin with a pre-compact, globally attracting set. The only question is whether there are points in the closure of this set whose orbits fail the conclusion of Proposition 2.5. With initial points so restricted, the proof of Proposition (1.2) of [**15**] finishes the proof of our proposition. In particular, the proof implies the existence of $\epsilon' > 0$ such that

$$\liminf_{t \to \infty} \; dist(\Phi(t)(w_0, \lambda), \hat{X}_2) \geq \epsilon'$$

for all $(w_0, \lambda) \in \hat{X}_1$ as asserted.

## References


[1] Butler, G., Freedman, H. and Waltman, P. [1986], *Uniformly persistent systems*, Proc. Amer. Math. Soc. **96**, 425-430.

[2] Dung, L. and Smith, H. L. [1996], *A parabolic system modeling microbial competition in an unmixed bio-reactor*, JDE **160**, 59-91.

[3] Hale, J. and Waltman, P. [1989], *Persistence in infinite dimensional systems*, SIAM J. Math. Anal. **20**, 388-395.

[4] Hsu, S. B. and Waltman, P. [1993], On a system of reaction-diffusion equations arising from competition in an unstirred chemostat, SIAM J. Appl. Math. **53**, 1026-1044.

[5] Hsu, S. B., Smith, H. L. and Waltman, P. [1994], *Dynamics of Competition in the unstirred chemostat*, Can. Appl. Math. Quart. **2**, 461-483.

[6] Hsu, S. B., Smith, H. L. and Waltman, P. [1996], *Competitive exclusion and coexistence for competitive systems on ordered Banach spaces*, Trans. AMS **348**, 4083-4094.

[7] Pazy, A. [1983], *Semigroups of Linear Operators and Applications to Partial Differential Equations*, Springer-Verlag, New York.

[8] Pilyugin, S. [1997], *Two models of a chemostat*, Ph.D. Thesis, Emory University.

[9] Smith, H. L. [1982], *On the basin of attraction of a perturbed attractor*, Nonlinear Analysis **6**, 911-917.

[10] Smith, H. L. [1995], *Monotone Dynamical Systems: An introduction to the Theory of Competitive and Cooperative Systems*, AMS Math. Surv.& Monographs **41**, Providence, R.I.

[11] Smith, H. L. [1995] and Waltman, P., *The Theory of the Chemostat*, Cambridge Univ. Press.

[12] Smith, H. L. and Waltman, P., *Perturbation of a globally stable steady state*, Proc. AMS, to appear.

[13] So, J. W-H. and Waltman, P. [1989], *A nonlinear boundary value problem arising from competition in the chemostat*, App. Math. and Comput. **32**, 169-183.

[14] Temam, R. [1998], *Infinite Dimensional Dynamical Systems in Mechanics and Physics*, Springer-Verlag, New York.

[15] Thieme, H. R. [1993], *Persistence under relaxed point-dissipativity (with application to an epidemic model)*, SIAM J. Math. Anal. **24**, 407-435.

[16] Zeidler, E. [1990], *Nonlinear Functional Analysis and Its Applications*, Vol. IIB, Springer-Verlag, New York.

Fields Institute Communications
Volume **21**, 1999

# Bifurcation Aspects for some Delayed Population Models with Diffusion

**Teresa Faria**
Departamento de Matemática
Faculdade de Ciências / CMAF
Universidade de Lisboa
R. Ernesto Vasconcelos
1700 Lisboa, Portugal
tfaria@lmc.fc.ul.pt

**Abstract.** In this paper, a previous theory of normal forms for Functional Differential Equations with diffusion near equilibria is applied to the study of some bifurcation problems. This theory is also improved, in order to include some situations where the operator giving the linear part of the equation mixes the modes of eigenvalues of the Laplacian.

## 1 Introduction

In a former paper [**4**], a theory of normal forms on center manifolds (or other invariant manifolds) for Partial Functional Differential Equations (PFEDs) near equilibrium points was addressed and used to study the qualitative behavior of solutions on those manifolds. This theory was based on a previous normal form procedure for retarded Functional Differential Equations (FDEs) in [**5, 6**] and on a center manifold theory for FDEs with diffusion. In [**15**], center manifold results were applied to a particular PFDE. Taking the same equation as a model, [**13**] considered stable and unstable manifolds near a hyperbolic equilibrium. In [**11**], a center manifold theory for reaction–diffusion equations with delays was developed and a coupled system of scalar ordinary differential equations as the equation on the center manifold was obtained. However, this system is given implicitly in terms of the considered PFDE. With the approach of normal forms in [**4**], *explicit* normal forms (in the usual sense of Ordinary Differential Equations (ODEs)) for the equation giving the flow on the center manifold are obtained, without having to compute the manifold beforehand. Also, that normal form coincides (up to a certain order of terms) with the normal form for an FDE, which is *associated* (in a precise and natural way) with the given PFDE. The main results of this normal form theory are exposed in Section 2, with some details, since we shall need it in Section 4.

1991 *Mathematics Subject Classification.* Primary 35B32, 34K15.

Work partially supported by F.C.T., PRAXIS XXI, FEDER, under projects PRAXIS/PCEX/P/MAT/36/96 and PRAXIS/2/2.1/MAT/125 /94.

In Section 3, we illustrate the theory by a scalar FDE with diffusion where a Bogdanov-Takens bifurcation occurs.

For developing a center manifold theory for FDEs with diffusion, Lin et al. in [**11**] had to make several hypotheses, also assumed in [**4**]. One of the restrictions was that the operator giving the linear terms does not mix the modes of eigenvalues of the Laplacian. A weaker hypothesis is considered in Section 4, improving the application of the theory of normal forms for PFDEs.

In Section 5, we analize a 2-dimensional Lotka-Volterra equation with diffusion that has mixed modes. It turns out that a Hopf bifurcation occurs and results are easily deduced from the situation without diffusion.

## 2 Preliminaries

**2.1 Notation and adjoint theory for FDEs with diffusion.** In the following, $\Omega \subset \mathbb{R}^n$ is open, $X$ is a Hilbert space of functions from $\overline{\Omega}$ to $\mathbb{R}^m$ with inner product $< \cdot, \cdot >$ and $\mathcal{C} = C([-r, 0]; X)$ $(r > 0)$ is the Banach space of continuous maps from $[-r, 0]$ to $X$ with the sup norm. Nevertheless, all the theory can be directly applied to the case where the functions in $X$ have values in $\mathbb{C}^m$. We write $u_t \in \mathcal{C}$ for $u_t(\theta) = u(t+\theta), -r \leq \theta \leq 0$ (throughout this paper, see [**9**] for standard notations and results about FDEs). We consider PFDEs with an equilibrium point at the origin, given in abstract form (i.e., in the phase space $\mathcal{C}$) as

$$\frac{d}{dt}u(t) = d\triangle u(t) + L(u_t) + F(u_t) \qquad (t > 0), \tag{2.1}$$

where $d > 0, \text{dom}(\triangle) \subset X, L \in \mathcal{L}(\mathcal{C}; X)$, i.e., $L : \mathcal{C} \to X$ is a bounded linear operator, and $F : \mathcal{C} \to X$ is a $C^k$ function $(k \geq 2)$ such that $F(0) = 0, DF(0) = 0$.

For the linearized equation about the equilibrium zero

$$\frac{d}{dt}u(t) = d\triangle u(t) + L(u_t), \tag{2.2}$$

we assume hypotheses (H1)-(H4) of [**11**] (see also [**13**]):

(H1) $d\triangle$ generates a $C_0$ semigroup $\{T(t)\}_{t\geq 0}$ on $X$ with $|T(t)| \leq e^{\omega t}$ for all $t \geq 0$, and $T(t)$ is a compact operator for $t > 0$;

(H2) the eigenfunctions $\{\beta_k\}_{k=1}^{\infty}$ of $d\triangle$, with corresponding eigenvalues $\{\mu_k\}_{k=1}^{\infty}$, form an orthonormal basis for $X$ and $\mu_k \to -\infty$ as $k \to \infty$;

(H3) the subspaces $\mathcal{B}_k$ of $\mathcal{C}$, $\mathcal{B}_k := \{< v(\cdot), \beta_k > \beta_k \,|\, v \in \mathcal{C}\}$ satisfy $L(\mathcal{B}_k) \subset \text{span}\{\beta_k\}$;

(H4) $L$ can be extended to a bounded linear operator from $B\mathcal{C}$ to $X$, where $B\mathcal{C} = \{\psi : [-r, 0] \longrightarrow X \,|\psi$ is continuous on $[-r, 0), \exists \lim_{\theta \to 0^-} \psi(\theta) \in X\}$, with the sup norm.

Let $A$ be the infinitesimal generator associated with the semiflow of the linearized equation (2.2). It is known (e.g., [**14**]) that $A$ is given by

$$(A\phi)(\theta) = \dot{\phi}(\theta),\ \text{dom}(A) = \{\phi \in \mathcal{C} : \dot{\phi} \in \mathcal{C}, \phi(0) \in \text{dom}(\triangle), \dot{\phi}(0) = d\triangle\phi(0) + L\phi\},$$

that the spectrum $\sigma(A)$ of $A$ coincides with its point spectrum $\sigma_P(A)$ and that $\lambda \in \mathbb{C}$ is in $\sigma_P(A)$ if and only if $\lambda$ satisfies the characteristic equation

$$\boldsymbol{\Delta}(\lambda)y := \lambda y - d\triangle y - L(e^{\lambda \cdot}y) = 0,\ \text{for some } y \in \text{dom}(\triangle) \setminus \{0\}. \tag{2.3}$$

Furthermore, for any $a \in \mathbb{R}$, the number of solutions of (2.3) such that $\mathrm{Re}\,\lambda \geq a$ is finite. Since (H3) states that $L$ does not mix the modes of eigenvalues of $\triangle$, using the decomposition of $X$ by $\{\beta_k\}_{k=1}^{\infty}$ we deduce that equation $\triangle(\lambda)y = 0$ is equivalent to the sequence of "characteristic" equations

$$\lambda - \mu_k - L_k(e^{\lambda \cdot}) = 0 \quad (k \in \mathbb{N}), \tag{2.4$_k$}$$

where $L_k : C \longrightarrow \mathbb{R}$, $C := C([-r, 0]; \mathbb{R})$, are defined by $L_k(\psi)\beta_k = L(\psi\beta_k)$, for $k \in \mathbb{N}$.

If $\lambda$ is a solution of (2.4$_k$) and $\mathrm{Re}\lambda \geq 0$, then $0 \leq \mu_k + \mathrm{Re}\,(L_k(e^{\lambda \cdot})) \leq \mu_k + \|L\|$. As $\mu_k \to -\infty$, then there exists an $N$ such that all the solutions of (2.4$_k$) satisfy $\mathrm{Re}\lambda < 0$ for $k > N$. In [**11**], the existence of a local center manifold was proved under Hypotheses (H1)-(H4). Its dimension is equal to the number of $\lambda \in \sigma(A)$ with real part zero, counting multiplicities.

Let $\Lambda$ be the finite set $\Lambda = \{\lambda \in \sigma(A) \ : \mathrm{Re}\,\lambda = 0\}$, or another finite subset of $\sigma(A)$ with which an invariant manifold is associated (e.g., in [**13**] results on the unstable manifold are stated, corresponding to the choice $\Lambda = \{\lambda \in \sigma(A) : \mathrm{Re}\lambda > 0\}$), and suppose $\Lambda \neq \emptyset$. Throughout this paper, unless otherwise stated, we assume that $\Lambda = \{\lambda \in \sigma(A) : \mathrm{Re}\,\lambda = 0\}$ and that the basis $\{\beta_k\}_{k=1}^{\infty}$ is ordered in such a way that $\Lambda$ contains exactly the solutions on the imaginary axis of the first $N$ equations (2.4$_k$): $\Lambda = \{\lambda \in \mathbb{C} : \lambda$ is a solution of (2.4$_k$) with $\mathrm{Re}\,\lambda = 0$, for some $k \in \{1, \dots, N\}\}$.

Now, we describe an adjoint theory for FDEs with diffusion, assuming the former hypotheses (cf. [**11**], [**13**], [**15**], [**14**]). The main idea is to relate the eigenvalues of the infinitesimal generator $A$ (in this case, only the elements of $\Lambda$) with the eigenvalues of certain FDEs without diffusion. On $\mathcal{B}_k$, the linear equation $\frac{d}{dt}u(t) = d\triangle u(t) + L(u_t)$ is equivalent to the FDE on $\mathbb{R}$

$$\dot{z}(t) = \mu_k z(t) + L_k z_t, \tag{2.5$_k$}$$

with characteristic equation given by (2.4$_k$). By this identification, the standard adjoint theory for retarded FDEs is used to decompose $\mathcal{C}$, in the following way. For $1 \leq k \leq N$, define $(\cdot,\cdot)_k$ as the adjoint bilinear form on $C^* \times C, C^* := C([0, r]; \mathbb{R})$, and decompose $C$ by $\Lambda_k := \{\lambda \in \mathbb{C} : \lambda$ satisfies (2.4$_k$) and $\mathrm{Re}\,\lambda = 0\}$ as in [**9**]:

$$C = P_k \oplus Q_k, \quad P_k = \mathrm{span}\Phi_k, \quad P_k^* = \mathrm{span}\Psi_k,$$
$$(\Psi_k, \Phi_k)_k = I, \quad \dim P_k = \dim P_k^* := m_k, \quad \dot{\Phi}_k = \Phi_k B_k \ ,$$

where $P_k$ is the generalized eigenspace for Equation (2.5$_k$) associated with $\Lambda_k$ and $B_k$ is an $m_k \times m_k$ constant matrix. We use the above decompositions to decompose $\mathcal{C}$ by $\Lambda$:

$$\mathcal{C} = \mathcal{P} \oplus \mathcal{Q}, \quad \mathcal{P} = \mathrm{Im}\,\pi, \quad \mathcal{Q} = \mathrm{Ker}\,\pi,$$

where $\dim \mathcal{P} = \sum_{k=1}^{N} m_k := M$ and $\pi : \mathcal{C} \longrightarrow \mathcal{P}$ is the projection defined by

$$\pi\phi = \sum_{k=1}^{N} \Phi_k(\Psi_k, < \phi(\cdot), \beta_k >)_k \beta_k. \tag{2.6}$$

**2.2 Normal form theory and the associated FDE.** In [**4**], a normal form theory for FDEs with diffusion was developed following closely the work done in [**5, 6**] for autonomous retarded FDEs. First, the phase space $\mathcal{C}$ was enlarged in such a way that Equation (2.1) can be written as an abstract ODE in a Banach space. Actually, rewriting Equation (2.1) with $v = u_t \in \mathcal{C}$, we obtain the system

$$\frac{dv}{dt}(0) = d\triangle v(0) + L(v) + F(v), \; \frac{dv}{dt}(\theta) = \dot{v}(\theta), \; \theta \in [-r, 0).$$

So, it is natural to consider for phase space the Banach space $BC$ introduced in (H4). In terms of the function $X_0$ defined by $X_0(0) = I, X_0(\theta) = 0$ , $-r \leq \theta < 0$, the elements of $BC$ have the form $\psi = \phi + X_0\alpha$, with $\phi \in \mathcal{C}, \alpha \in X$, so that $BC \equiv \mathcal{C} \times X$; its norm is equivalent to the norm $|\phi + X_0\alpha| = |\phi|_\mathcal{C} + |\alpha|_X$. Then, the abstract ODE in $BC$ equivalent to Equation (2.1) is

$$\frac{d}{dt}v = \tilde{A}v + X_0F(v), \tag{2.7}$$

where $\tilde{A}$ is an extension of the infinitesimal generator $A$,

$$\tilde{A} : \mathcal{C}_0^1 \subset BC \longrightarrow BC, \quad \tilde{A}\phi = \dot{\phi} + X_0[L(\phi) + d\triangle\phi(0) - \dot{\phi}(0)],$$

defined on $\mathcal{C}_0^1 := \{\phi \in \mathcal{C} : \dot{\phi} \in \mathcal{C}, \phi(0) \in \text{dom}(\triangle)\}$. On the other hand, it is easy to see that $\pi$, as defined in (2.6), is extended to a continuous projection (which we still denote by $\pi$), $\pi : BC \longrightarrow \mathcal{P}$, by defining

$$\pi(X_0\alpha) = \sum_{k=1}^{N} \Phi_k\Psi_k(0) < \alpha, \beta_k > \beta_k, \quad \alpha \in X. \tag{2.8}$$

The projection $\pi$ leads to the topological decomposition

$$BC = \mathcal{P} \oplus \text{Ker}\,\pi, \tag{2.9}$$

with $\mathcal{Q} \subsetneq \text{Ker}\,\pi), \mathcal{P} \subset \mathcal{C}_0^1$ and $\pi$ commuting with $\tilde{A}$ in $\mathcal{C}_0^1$. This allows us to decompose Equation (2.7) as a system of abstract ODEs on $\mathbb{R}^M \times \text{Ker}\,\pi$, with linear and nonlinear parts separated and with finite and infinite dimensional variables also separated in the linear term. More precisely, decomposing $v \in \mathcal{C}_0^1$ according to (2.9) as $v(t) = \sum_{k=1}^N \Phi_k z_k(t)\beta_k + y(t)$, where $z_k(t) = (\Psi_k, < v(t)(\cdot), \beta_k >)_k \in \mathbb{R}^{m_k}, 1 \leq k \leq N, y(t) \in \mathcal{C}_0^1 \cap \text{Ker}\,\pi = \mathcal{C}_0^1 \cap \mathcal{Q} := \mathcal{Q}^1$, in $BC \equiv \mathbb{R}^M \times \text{Ker}\,\pi$, Equation (2.7) is equivalent to the system

$$\begin{aligned}
\dot{z} &= Bz + \Psi(0)\Big( < F\left(\sum_{k=1}^{N} \Phi_k z_k \beta_k + y\right), \beta_p > \Big)_{p=1}^{N} \\
\frac{d}{dt}y &= A_1 y + (I - \pi)X_0 F\left(\sum_{k=1}^{N} \Phi_k z_k \beta_k + y\right), \\
z &= (z_1, \ldots, z_N) \in \mathbb{R}^M, y \in \mathcal{Q}^1,
\end{aligned} \tag{2.10}$$

where: $B$ is the $M \times M$ constant matrix $B = \text{diag}\,(B_1, \ldots, B_N)$,$\Phi$ is the $N \times M$ matrix $\Phi = \text{diag}(\Phi_1, \ldots, \Phi_N)$, $\Psi$ is the $M \times N$ matrix $\Psi$ =diag $(\Psi_1, \ldots, \Psi_N)$ and $A_1$ is defined by $A_1 : \mathcal{Q}^1 \subset \text{Ker}\,\pi \longrightarrow \text{Ker}\,\pi, A_1\phi = \tilde{A}\phi$, for $\phi \in \mathcal{Q}^1$.

We consider now the formal Taylor expansion

$$F(v) = \sum_{j \geq 2} \frac{1}{j!} F_j(v), \quad v \in \mathcal{C},$$

where $F_j$ is the $j$th Fréchet derivative of $F$. Then, Equation (2.10) (i.e., Equation (2.1) in $BC \equiv \mathbb{R}^M \times \operatorname{Ker} \pi$) is written as

$$\begin{aligned} \dot{z} &= Bz + \sum_{j\geq 2} \frac{1}{j!} f_j^1(z,y) \\ \frac{d}{dt} y &= A_1 y + \sum_{j\geq 2} \frac{1}{j!} f_j^2(z,y), \end{aligned} \tag{2.11}$$

where $z = (z_1, \ldots, z_N) \in \mathbb{R}^M, y \in \mathcal{Q}^1$ and $f_j = (f_j^1, f_j^2), j \geq 2$, are defined by

$$f_j^1(z,y) = \Psi(0)\Big( < F_j \left( \sum_{k=1}^{N} \Phi_k z_k \beta_k + y \right), \beta_p > \Big)_{p=1}^{N}$$

$$f_j^2(z,y) = (I - \pi) X_0 F_j \left( \sum_{k=1}^{N} \Phi_k z_k \beta_k + y \right).$$

Normal forms for Equation (2.10) were obtained in [**4**] by a recursive process of change of variables that simplify the equation, computing at each step the terms of order $j \geq 2$ from the terms of the same order and the terms of lower orders already computed in previous steps, in such a way that the *nonresonant terms* of order $j$ are eliminated. The result of this procedure is resumed in the following theorem:

**Theorem 2.1** *Suppose that (H1)-(H4) hold and* $\Lambda = \{\lambda \in \sigma(A) : \operatorname{Re} \lambda = 0\} \neq \emptyset$. *Then, there exists a formal change of variables* $(z,y) = (\overline{z}, \overline{y}) + O(|\overline{z}|^2)$ *such that: (i) Equation (2.11) is transformed into*

$$\begin{aligned} \dot{\overline{z}} &= B\overline{z} + \sum_{j\geq 2} \frac{1}{j!} g_j^1(\overline{z}, \overline{y}) \\ \frac{d}{dt} \overline{y} &= A_1 \overline{y} + \sum_{j\geq 2} \frac{1}{j!} g_j^2(\overline{z}, \overline{y}), \end{aligned} \tag{2.12}$$

*where the new terms of order* $j$, $g_j = (g_j^1, g_j^2), j \geq 2$, *satisfy* $g_j^2(\overline{z}, 0) \equiv 0, j \geq 2$;

*(ii) a locally center manifold for Equation (2.1) at zero satisfies* $\overline{y} = 0$ *and the flow on it is given by the* $M$*-dimensional ODE*

$$\dot{\overline{z}} = B\overline{z} + \sum_{j\geq 2} \frac{1}{j!} g_j^1(\overline{z}, 0), \tag{2.13}$$

*which is in normal form (in the usual sense of ODEs).*

Equation (2.12) is called a **normal form relative to** $\Lambda$ for Equation (2.1).

**Remark 2.2** The operators appearing in the normal form process, and that are associated with the sequence of changes of variables referred to, are extensions of the operators constructed for autonomous retarded FDEs without diffusion in [**5, 6**]. In particular, the operators $f_j^1 \longmapsto g_j^1$ are precisely those operators appearing in the computation of normal forms for ODEs in $\mathbb{R}^M$ (cf. [**1**], [**3**], [**8**]).

**Remark 2.3** For the sake of simplicity and because of the applications, we have considered $\Lambda = \{\lambda \in \sigma(A) : \operatorname{Re} \lambda = 0\}$. In this case, the existence of a local center manifold for Equation (2.1), tangent to $\mathcal{P}$ at zero, is guaranteed by (H1)-(H4). Now, let $\Lambda$ be another nonempty finite subset of $\sigma(A)$ and suppose, as before, that $\mathcal{P}$ is the invariant space of the linearized equation $\frac{d}{dt}u(t) = d\Delta u(t) + L(u_t)$

associated with $\Lambda$. For this case, we also obtain the above decomposition $\mathcal{C} = \mathcal{P} \oplus \mathcal{Q}$, as well as the procedure leading to Equation (2.12). If there exists a locally invariant manifold $\mathcal{M}_{\Lambda,F}$ for Equation (2.1) tangent to $\mathcal{P}$ at zero, for that manifold we achieve a similiar result to the one stated in Theorem 2.1 for the center manifold, if we guarantee that the following *nonresonance conditions relative to* $\Lambda$ are satisfied:

$$(q, \overline{\lambda}) \neq \mu \quad , \text{ for all } \mu \in \sigma(A) \setminus \Lambda, \ q \in \mathbb{N}_0^M, |q| \geq 2, \tag{2.14}$$

where $\overline{\lambda} = (\lambda_1, \dots, \lambda_M), \lambda_1, \dots, \lambda_M$ are the elements of $\Lambda$, each of them appearing as many times as its multiplicity as a root of the associated characteristic equation, $(q, \overline{\lambda}) = q_1\lambda_1 + \cdots + q_M\lambda_M$, $|q| = q_1 + \cdots + q_M$, for $q = (q_1, \dots, q_M) \in \mathbb{N}_0^M$. For instance, if $\Lambda = \{\lambda \in \sigma(A) : \operatorname{Re}\lambda > 0\} \neq \emptyset$, then Theorem 2.1 is valid with "unstable manifold" instead of "center manifold". (Note that the existence of this manifold under (H1)-(H4) follows from [**13**].)

**Remark 2.4** Consider equations with parameters of the type

$$\frac{d}{dt}u(t) = d\Delta u(t) + L(\alpha)(u_t) + F(u_t, \alpha), \tag{2.15}$$

where $\alpha \in \mathbb{R}^p, L : \mathbb{R}^p \longrightarrow \mathcal{L}(\mathcal{C}; X), F : \mathcal{C} \times \mathbb{R}^p \longrightarrow X$ are $C^k$ functions, $k \geq 2$, with $F(0, \alpha) = 0, D_1F(0, \alpha) = 0$ for all $\alpha \in \mathbb{R}^p$. For these equations, the normal form procedure is reduced to that described above by introducing the parameter $\alpha$ as a variable with $\dot{\alpha}(t) = 0$.

We shall now define an FDE with an equilibrium at zero, in such a way that, on one hand, the eigenvalues on the imaginary axis for its linearized equation at zero are precisely the elements of $\Lambda$; and, on the other hand, and eventually under appropriate additional hypotheses, the normal form on the center manifold for that FDE can be chosen so that it coincides with the normal form (2.13) on the center manifold of the original PFDE (2.1), at least up to the same finite order. Here, we shall state what happens up to second and third order only, because we shall apply the results to equations for which a Bogdanov-Takens or a Hopf bifurcation occur — so, the singularities are generically determined to second or to third orders, respectively.

It is natural to associate an FDE with the original PFDE (2.1). Formally, if one dropped the eigenfunctions $\beta_k$ and the inner product in Equation (2.10), we see that the system would become the system appearing when considering normal forms for the following FDE without diffusion in $C_N := C([-r, 0]; \mathbb{R}^N)$:

$$\dot{x}(t) = R(x_t) + \Big( < F(\sum_{k=1}^{N} x_{t,k}\beta_k), \beta_p > \Big)_{p=1}^{N}, \tag{2.16}$$

where: $x(t) = \big(x_k(t)\big)_{k=1}^N, x_t = \big(x_{t,k}\big)_{k=1}^N$, $R \in \mathcal{L}(C_N; \mathbb{R}^N)$ is the linear operator defined by

$$R(\phi) = \big(\mu_k\phi_k(0) + L_k(\phi_k)\big)_{k=1}^N,$$

for $\phi = (\phi_1, \dots, \phi_N) \in C_N$. Note that the characteristic equation for the linearized equation $\dot{x}(t) = R(x_t)$ is det $[\lambda I - R(e^{\lambda \cdot}I)] = 0$, which is equivalent to $\lambda - \mu_k - L_k(e^{\lambda \cdot}) = 0$, for some $1 \leq k \leq N$. Then, $\Lambda = \{\lambda \in \sigma(A) : \operatorname{Re}\lambda = 0\}$ is also the set of characteristic eigenvalues of $R$ with real part zero. Equation (2.16) in $C_N$ is called the **FDE associated with** Equation (2.1) **by** $\Lambda$ at zero. The next result establishes a relationship bweteen (2.1) and (2.16).

**Theorem 2.5** *Suppose (H1)-(H4) hold and let* $\Lambda = \{\lambda \in \sigma(A) : \operatorname{Re}\lambda = 0\} \neq \emptyset$. *Consider Equation (2.1) with* $F$ *of class* $C^2$. *Then, for a suitable change of variables, the normal forms on the center manifold for both Equation (2.1) and Equation (2.16) are the same, up to second order terms. If* $F$ *is of class* $C^3$, $F(v) = \frac{1}{2!}F_2(v) + \frac{1}{3!}F_3(v) + o(|v|^3)$, *and assuming also (with the notations above)*

$$< DF_2(u)(\phi\beta_j), \beta_p >= 0 \;, \text{ for } 1 \leq p \leq N, j > N \text{ and all } u \in \mathcal{P},\ \phi \in C, \qquad \text{(H5)}$$

*then, for a suitable change of variables, the normal forms on the center manifold for both Equation (2.1) and Equation (2.16) are the same, up to third order terms.*

The proof of this theorem in [**4**] is based on the possibility of identification of the operators referred in Remark 2.2 with the corresponding operators appearing in the computation of normal forms relative to $\Lambda$ for Equation (2.16). We point out that even when (H5) fails, the associated FDE by $\Lambda$ still carries much information that can be transfered to the computation of normal forms for the original PFDE (see [**4**] for details).

**Remark 2.6** As in Remark 2.3, instead of center manifolds we can consider other invariant manifolds associated with $\Lambda$, provided their existence and that the nonresonance conditions (2.14) are satisfied (which is the case for unstable or center-unstable manifolds).

Here, we shall first apply the theory of normal forms to a scalar PFDE of type (2.15) with a generic Bogdanov-Takens bifurcation (whose dynamics are then determined by the 2-jet). A second example will be presented, that is more complicated, because it is not scalar (it is a PFDE with two spatial dimensions) and hypothesis (H3) fails, which will oblige us to improve the theory.

## 3 An example of a scalar population model with diffusion

Consider the following example of a population equation

$$\begin{aligned} &\frac{\partial u(t,x)}{\partial t} = d\frac{\partial^2 u(t,x)}{\partial x^2} + u(t,x) - au(t-1,x)[1+u(t,x)],\ t \geq 0, x \in (0,\pi) \\ &\frac{\partial u(t,x)}{\partial x} = 0,\ x = 0, \pi, \end{aligned} \tag{3.1}$$

where $d > 0, a > 0$ (see e.g., [**14**], [**7**]). This equation can be written in abstract form in $\mathcal{C} = C([-1,0]; X)$ as

$$\frac{d}{dt}u(t) = d\Delta u(t) + L(a)(u_t) + f(u_t, a) \tag{3.2}$$

where $X = \{v \in W^{2,2}(0,\pi) : \frac{dv}{dx} = 0 \text{ at } x = 0, \pi\}$, $L(a)(v) = v(0) - av(-1), f(v,a) = -av(0)v(-1)$. However, in the following, we shall begin by considering the general case of any $f : \mathcal{C} \times \mathbb{R} \longrightarrow X$ such that $f \in C^2, f(0,a) = 0, D_1 f(0,a) = 0$, for $a > 0$.

The functions $\beta_{k+1}(x) = \frac{\cos(kx)}{\|\cos(kx)\|_{2,2}}$ are normalized eigenfunctions of $d\Delta = d\frac{\partial^2}{\partial x^2}$ on $X$, with corresponding eigenvalues $\mu_{k+1} = -dk^2, k \geq 0$. For the linearized equation about the equilibrium $u = 0$, $\frac{d}{dt}u(t) = d\Delta u(t) + L(a)(u_t)$, consider the characteristic equations $(2.4_{k+1})$:

$$\boldsymbol{\Delta}_k(\lambda, a) := \lambda + ae^{-\lambda} + dk^2 - 1 = 0 \ , \ (k = 0, 1, \dots). \tag{$3.3_k$}$$

Let $u, v \geq 0$ and suppose that $\lambda = u + iv$ is a solution of $(3.3_k)$. Then, $g(u, v, a) := v - ae^{-u} \sin v = 0$. Since $g'_v(u, v, a) > 0$ if $u > 0, a \leq 1$ and $g(u, 0, a) = 0$, we deduce that $v = 0$ if $a \leq 1$; in other words, for $a \leq 1$, all the roots of $(3.3_k)$ with positive real parts are necessarily real. Suppose now $a \leq 1, \lambda \in \mathbb{R}, \lambda \geq 0$. We have $\boldsymbol{\Delta}_k(0, a) = 0$ if and only if $a = a_k := 1 - dk^2$, with $dk^2 < 1$, $\frac{d}{d\lambda}\boldsymbol{\Delta}_k(\lambda, a) > 0$ if $\lambda > 0$, $\frac{d}{d\lambda}\boldsymbol{\Delta}_k(\lambda, a) = 0$ if and only if $\lambda = 0, a = 1$. Thus, $\boldsymbol{\Delta}_0(0, a) < 0, \dots, \boldsymbol{\Delta}_k(0, a) < 0$, for $0 < a < a_k (k \in \mathbb{N}_0)$. We conclude that $\lambda = 0$ is a double eigenvalue of $(3.3_k)$ if and only if $a = 1$ and $k = 0$, in which case all the other roots of all equations $(3.3_k)$ have negative real parts. The points $a_k = 1 - dk^2$ such that $dk^2 < 1$ are bifurcation points. Futhermore, if $k_0 \in \mathbb{N}, dk_0^2 < 1$ and $a \in (a_{k_0}, a_{k_0 - 1})$, there are exactly $k_0$ eigenvalues that are real and positive and all the other eigenvalues have negative real parts; for $a = a_{k_0}$, $\lambda = 0$ is a simple eigenvalue and there are exactly $k_0$ (real) eigenvalues with positive real parts. Therefore, the smaller $d$ is, i.e., the smaller the diffusion term is, the more bifurcation points $a_k$ appear and the more unstable the equilibrium zero is.

To study the qualitative behavior near the critical point $a_0 = 1$, let $a = 1 + \alpha$. With the notations of Section 2, let $\Lambda = \{0\}$. Defining $L := L(1)$ and $F(v, \alpha) = -\alpha v(-1) + f(v, 1 + \alpha)$, Equation (3.2) is written as

$$\frac{d}{dt}u(t) = d\triangle u(t) + L(u_t) + F(u_t, \alpha). \tag{3.4}$$

Hypotheses (H3)-(H4) hold for $L := L(1)$. Since $L(\psi\beta_1) = (\psi(0) - \psi(-1))\beta_1$, the operator $L_1 : C([-1, 0]; \mathbb{R}) \longrightarrow \mathbb{R}$ corresponding to the eigenvalue $\mu_1 = 0$ and defined in $(2.4_1)$ is $L_1(\psi) = \psi(0) - \psi(-1)$. Following Section 2, we obtain $P = P_1$, $\dim \mathcal{P} = \dim P_1 = 2$, $P_1 = \operatorname{span} \Phi$, where

$$\Phi(\theta) = (\phi_1(\theta), \phi_2(\theta)) = (1, \theta), \quad B = \begin{pmatrix} 0 & 1 \\ 0 & 0 \end{pmatrix},$$
$$\Psi(0) = \begin{pmatrix} \psi_1(0) \\ \psi_2(0) \end{pmatrix} = \begin{pmatrix} 2/3 \\ 2 \end{pmatrix}.$$

In $BC = \mathcal{P} \oplus \operatorname{Ker} \pi$ decomposed by $\Lambda$, Equation (3.4) becomes (cf. (2.10))

$$\begin{aligned} \dot{z} &= Bz + \Psi(0)\langle F(\Phi z \beta_1 + y, \alpha), \beta_1 \rangle \\ \frac{d}{dt}y &= A_1 y + (I - \pi)X_0 F(\Phi z \beta_1 + y, \alpha) \ , \quad z \in \mathbb{R}^2, y \in \mathcal{Q}^1. \end{aligned} \tag{3.5}$$

Thus, for Equation (3.4) there exists a two–dimensional local center manifold tangent to $\mathcal{P}$ at $u = 0, \alpha = 0$, which is stable, since all the nonzero eigenvalues have negative real parts. On that manifold, a Bogdanov-Takens bifurcation occurs at $\alpha = 0, u = 0$.

On the other hand, the FDE associated with Equation (3.4) by $\Lambda$ at the equilibrium point $u = 0$ for $\alpha = 0$ is $\dot{x}(t) = L_1(x_t) + \langle F(x_t\beta_1, \alpha), \beta_1 \rangle$, i.e., the FDE in $C([-1, 0]; \mathbb{R})$

$$\dot{x}(t) = x(t) - x(t-1) + \langle F(x_t \beta_1, \alpha), \beta_1 \rangle,$$

or, since $\beta_1 = 1/\sqrt{\pi}$ and $\langle \beta_1, \beta_1 \rangle = 1$,

$$\dot{x}(t) = x(t) - (1 + \alpha)x(t-1) + \langle f(\frac{1}{\sqrt{\pi}}x_t, 1 + \alpha), \beta_1 \rangle. \tag{3.6}$$

From Theorem 2.5, up to the second order terms (which are sufficient to determine a generic Bogdanov-Takens bifurcation), the normal forms on the center manifolds for Equations (3.5) and (3.6) coincide. But, for Equation (3.6), then also for Equation (3.5), we are able to write that equation without further calculation since it was already computed in [**5**]. For instance, for $f(v,a) = h(v(0), v(-1), a)$, with $h(x_1, x_2, a) = \frac{1}{2}[h_{(2,0)}x_1^2 + h_{(1,1)}x_1x_2 + h_{(0,2)}x_2^2] + O((|a-1|+|x|)|x|^2)$, Equation (3.6) is

$$\begin{aligned}\dot{x}(t) = & x(t) - (1+\alpha)x(t-1) + \frac{1}{2\sqrt{\pi}}[h_{(2,0)}x^2(t) \\ & + h_{(1,1)}x(t)x(t-1) + h_{(0,2)}x^2(t-1)] + O((|\alpha| + |x|)|x|^2)\end{aligned}$$

and the flow on the center manifold is given by

$$\begin{aligned}\dot{x}_1 &= x_2 + O((|\alpha| + |x|)|x|^2) \\ \dot{x}_2 &= \lambda_1 x_1 + \lambda_2 x_2 + bx_1^2 + cx_1x_2 + O((|\alpha| + |x|)|x|^2),\end{aligned}$$

where

$$\begin{aligned}b &= \frac{1}{\sqrt{\pi}}[h_{(2,0)} + h_{(1,1)} + h_{(0,2)}], \ c = \frac{1}{3\sqrt{\pi}}[2h_{(2,0)} - h_{(1,1)} - 4h_{(0,2)}], \\ \lambda_1 &= -2\alpha, \ \lambda_2 = \frac{4}{3}\alpha.\end{aligned}$$

In the case of Equation (3.1), we have $h_{(2,0)} = h_{(0,2)} = 0, h_{(1,1)} = -1$, so $bc < 0$, which restricts the possible topological types of the phase portraits around the origin.

## 4 A more general framework

Consider again an equation of type (2.1) and its linearized equation at the equilibrium zero, Equation (2.2). In many situations, the operator $L$ does mix the modes of eigenvalues of the Laplacian, and the decomposition (2.9) based on the decompositions of $C$ by $\Lambda_k$ as described in Section 2 is not possible. For instance, consider the very simple case of the scalar PFDE

$$\frac{\partial u(t,x)}{\partial t} = \frac{\partial^2 u(t,x)}{\partial x^2} + b(x)u(t-1,x) + \dots, \quad t \geq 0, x \in (a_1, a_2)$$

with Neumann or Dirichlet conditions, where $b$ is a continuous function in $[a_1, a_2] \subset \mathbb{R}$ and the dots stand for nonlinear terms. Suppose that the Hilbert space $X$ is chosen to take into account the Neumann or Dirichlet conditions, and write the above equation in the form (2.1), where $L(v) = bv(-1), v \in \mathcal{C}$. The characteristic equation for the linearized equation is $\mathbf{\Delta}(\lambda)y = 0$, for some $y \in \mathrm{dom}(\triangle) \setminus \{0\}$, where

$$\mathbf{\Delta}(\lambda)y = \lambda y - \triangle y - be^{\lambda}y.$$

For a generic function $b$ it is not possible to relate the eigenvalues of the infinitesimal generator $A$ associated with the flow of the linearized equation, i.e., the solutions $\lambda$ of the above characteristic equation, with the eigenvalues of $\triangle$, because there is no general method to solve linear ODEs with variable coefficients. Therefore, the procedure in Section 2 is not applied. However, in certain cases (H3) fails and we are still able to relate the eigenvalues of the characteristic equation (2.3) with the eigenvalues of $\triangle$.

Although we cannot completely remove hypothesis (H3), we shall consider a weaker condition. For the sake of simplicity of exposition, we shall also restate our hypothesis (H2) in another form.

For Equation (2.2), assume (H1), (H4) and the following two conditions:

(H2′) Let $\{\mu_k^{i_k} : k \in \mathbb{N}, i_k = 1, \ldots, p_k\}$ be the eigenvalues of $d\triangle$ (distinct or not) and $\beta_k^{i_k}$ be eigenfunctions corresponding to $\mu_k^{i_k}$, such that $\{\beta_k^{i_k} : k \in \mathbb{N}, i_k = 1, \ldots, p_k\}$ form an orthonormal basis for $X$;

(H3′) the subspaces $\mathcal{B}_k$ of $\mathcal{C}$, $\mathcal{B}_k := \operatorname{span}\{< v(\cdot), \beta_k^{i_k} > \beta_k^{i_k} \,|\, v \in \mathcal{C}, i_k = 1, \ldots, p_k\}$ satisfy $L(\mathcal{B}_k) \subset \operatorname{span}\{\beta_k^1, \ldots, \beta_k^{p_k}\}$.

With these assumptions, for each $k \in \mathbb{N}$ define the linear operators $L_k : C_{p_k} \longrightarrow \mathbb{R}^{p_k}$, $C_{p_k} = C([-r, 0]; \mathbb{R}^{p_k})$, by

$$
\begin{aligned}
&L_k(\psi) = (L_k^1(\psi), \ldots, L_k^{p_k}(\psi)) \\
&L(\psi_1 \beta_k^1 + \cdots + \psi_{p_k} \beta_k^{p_k}) = \sum_{i_k=1}^{p_k} L_k^{i_k}(\psi) \beta_k^{i_k},
\end{aligned}
\tag{4.1}
$$

where $\psi = (\psi_1, \ldots, \psi_{p_k}) \in C_{p_k}$. Note that $L_k$ are well defined because of (H3'). Then, (H2') implies that the characteristic equation (2.3) is equivalent to the sequence of characteristic equations

$$
\det \mathbf{\Delta}_k(\lambda) = 0, \quad \text{where} \quad \mathbf{\Delta}_k(\lambda) := \lambda I - M_k - L_k(e^{\lambda \cdot} I) \quad (k \in \mathbb{N}), \tag{4.2$_k$}
$$

where $M_k = \operatorname{diag}(\mu_k^1, \ldots, \mu_k^{p_k})$. Since (4.2$_k$) is the characteristic equation for the retarded FDE in $C_{p_k}$

$$
\dot{z}(t) = M_k z(t) + L_k z_t, \tag{4.3$_k$}
$$

the procedure of Section 2 can now be followed with a few changes, described below.

Suppose once more that the set $\Lambda$ of the eigenvalues of $A$ on the imaginary axis is given by $\Lambda = \{\lambda \in \mathbb{C} : \lambda \text{ is a solution of } (4.2_k) \text{ with } \operatorname{Re}\lambda = 0, \text{ for some } k \in \{1, \ldots, N\}\}$, and that $\Lambda_k := \{\lambda \in \mathbb{C} : \lambda \text{ satisfies } (4.2_k) \text{ and } \operatorname{Re}\lambda = 0\}$. We adopt the same notations of Section 2 with some necessary modifications. Namely, we have now $C_{p_k}$ decomposed by $\Lambda_k$, $C_{p_k} = P_k \oplus Q_k$, the adjoint bilinear form is $(\cdot, \cdot)_k : C_{p_k}^* \times C_{p_k} \longrightarrow \mathbb{R}$, $\Phi_k$ is a $p_k \times m_k$ matrix and $\Psi_k$ is an $m_k \times p_k$ matrix (instead of $1 \times m_k$ and $m_k \times 1$, respectively). The decomposition $\mathcal{C} = \mathcal{P} \oplus \mathcal{Q}$, $\mathcal{P} = \operatorname{Im}\pi$, $\mathcal{Q} = \operatorname{Ker}\pi$, is now achieved by a projection $\pi : \mathcal{C} \longrightarrow \mathcal{P}$ with a more complicated expression to take into account the coefficients of $\beta_k^1, \ldots, \beta_k^{p_k}$ all together (cf. (2.6)):

$$
\begin{aligned}
\pi\phi &= \sum_{k=1}^{N} \sum_{i_k=1}^{p_k} c_k^{i_k}(\phi) \beta_k^{i_k} \text{ with} \\
\left(c_k^{i_k}(\phi)\right)_{i_k=1}^{p_k} &= \Phi_k \Big( \Psi_k, \big( < \phi(\cdot), \beta_k^{i_k} > \big)_{i_k=1}^{p_k} \Big)_k.
\end{aligned}
$$

Similarly to what was done in (2.8), one can define $\pi(X_0\alpha)$ for $\alpha \in X$ as above, where

$$
\left(c_k^{i_k}(X_0\alpha)\right)_{i_k=1}^{p_k} = \Phi_k \Psi_k(0) \big( < \alpha, \beta_k^{i_k} > \big)_{i_k=1}^{p_k}.
$$

The projection $\pi : BC \longrightarrow \mathcal{P}$ defined above can be written as

$$\pi(\phi) = \sum_{k=1}^{N} \left[\Phi_k\Big(\Psi_k, \big( < \phi(\cdot), \beta_k^{i_k} > \big)_{i_k=1}^{p_k}\Big)_k\right]^T \begin{pmatrix} \beta_k^1 \\ \vdots \\ \beta_k^{p_k} \end{pmatrix}, \quad \phi \in \mathcal{C},$$
$$\pi(X_0\alpha) = \sum_{k=1}^{N} \left[\Phi_k \Psi_k(0)\big( < \alpha, \beta_k^{i_k} > \big)_{i_k=1}^{p_k}\right]^T \begin{pmatrix} \beta_k^1 \\ \vdots \\ \beta_k^{p_k} \end{pmatrix}, \quad \alpha \in X. \tag{4.4}$$

It is easy to check that decomposing $v \in \mathcal{C}_0^1$ according to (2.9) as

$$v(t) = \sum_{k=1}^{N} \Big[\Phi_k z_k(t)\Big]^T, \begin{pmatrix} \beta_k^1 \\ \vdots \\ \beta_k^{p_k} \end{pmatrix} + y(t), \tag{4.5}$$

where $z_k(t) = \Big(\Psi_k, \big( < v(t)(\cdot), \beta_k^{i_k} > \big)_{i_k=1}^{p_k}\Big)_k \in \mathbb{R}^{m_k}, 1 \leq k \leq N, y(t) \in \mathcal{Q}^1$, on $BC \equiv \mathbb{R}^M \times \operatorname{Ker} \pi$, Equation (2.7) is equivalent to the system

$$\dot{z} = Bz + \Psi(0)\Bigg(\Bigg( < F\Big(\sum_{k=1}^{N} \Big[\Phi_k z_k\Big]^T \begin{pmatrix} \beta_k^1 \\ \vdots \\ \beta_k^{p_k} \end{pmatrix} + y\Big), \beta_n^{i_n} > \Bigg)_{i_n=1}^{p_n}\Bigg)_{n=1}^{N}$$
$$\frac{d}{dt}y = A_1 y + (I - \pi)X_0 F\Big(\sum_{k=1}^{N} \Big[\Phi_k z_k\Big]^T \begin{pmatrix} \beta_k^1 \\ \vdots \\ \beta_k^{p_k} \end{pmatrix} + y\Big), \tag{4.6}$$

where $z = (z_1, \ldots, z_N) \in \mathbb{R}^M, y \in \mathcal{Q}^1$ (cf. Equation (2.10)). Normal form computations follow now along the lines previously described. In this case, the associated FDE by $\Lambda$ is

$$\dot{x}(t) = R(x_t) + G(x_t), \tag{4.7}$$

where: $x(t) = \big(x_k(t)\big)_{k=1}^{N}, x_k \in \mathbb{R}^{p_k}$, and $R : C_J \longrightarrow \mathbb{R}^J, G : C_J \longrightarrow \mathbb{R}^J$, for $J := \sum_{k=1}^{N} p_k$, are defined by

$$R(\phi) = \big(M_k\phi_k(0) + L_k(\phi_k)\big)_{k=1}^{N},$$
$$G(\phi) = \Bigg(\Bigg( < F\Big(\sum_{k=1}^{N} [\phi_k]^T \begin{pmatrix} \beta_k^1 \\ \vdots \\ \beta_k^{p_k} \end{pmatrix}\Big), \beta_n^{i_n} > \Bigg)_{i_n=1}^{p_n}\Bigg)_{n=1}^{N},$$

for $\phi = (\phi_1, \ldots, \phi_N) \in C_J, \phi_k \in C_{p_k}, k = 1, \ldots, N$. Clearly, hypothesis (H5) is the same, but reformulated according to (H2) rewritten as (H2'):

$$\begin{aligned} < DF_2(u)(\phi\beta_j^{i_j}), \beta_n^{i_n} >= 0 \text{ for } 1 \leq n \leq N, 1 \leq i_n \leq p_n, \\ j > N, 1 \leq i_j \leq p_j \text{ and all } u \in \mathcal{P}, \phi \in C. \end{aligned} \tag{H5'}$$

Within this context, we present the new version of Theorem 2.5.

**Theorem 4.1** *Suppose (H1), (H2'), (H3'), (H4) hold and let* $\Lambda = \{\lambda \in \sigma(A) : \operatorname{Re}\lambda = 0\} \neq \emptyset$. *Consider Equation (2.1) with* $F$ *of class* $C^2$. *Then, for a suitable change of variables, the normal forms on the center manifold for both Equation (2.1) and Equation (4.7) are the same, up to second order terms. Futhermore, if* $F$

*is of class* $C^3$, $F(v) = \frac{1}{2!}F_2(v) + \frac{1}{3!}F_3(v) + o(|v|^3)$ *and (H5′) holds, also the third order terms of those equations on the center manifold for both Equation (2.1) and Equation (4.7) are the same.*

**Proof** It follows from the identification of operators referred to in Remark 2.2 that the second order terms of the equations on the center manifold for both Equation (2.1) and Equation (4.7) are the same. It is straightforward to show the statement concerning the third order terms using the proof of Theorem 4.1 in [**4**], so we omit the argument. □

## 5 On a predator-prey model

Consider the following Lotka-Volterra predator-prey system

$$\begin{cases} \dot{u}(t) = u(t)[r_1 - a_1u(t) - a_2v(t)] \\ \dot{v}(t) = v(t)[-r_2 + bu(t-\tau)], \end{cases} \tag{5.1}$$

and the corresponding reaction-diffusion equation with Neumann conditions in $[0,\pi]$

$$\begin{cases} \dfrac{\partial u(t,x)}{\partial t} = \dfrac{\partial^2 u(t,x)}{\partial x^2} + u(t,x)[r_1 - a_1u(t,x) - a_2v(t,x)] \\[2ex] \dfrac{\partial v(t,x)}{\partial t} = \dfrac{\partial^2 v(t,x)}{\partial x^2} + v(t,x)[-r_2 + bu(t-\tau,x)],\ t>0, x\in(0,\pi) \\[2ex] \dfrac{\partial u(t,x)}{\partial x} = \dfrac{\partial v(t,x)}{\partial x} = 0,\ x = 0,\pi, \end{cases} \tag{5.2}$$

where $r_1, r_2, a_1, a_2, b, \tau$ are positive constants. System (5.1) and more generally systems of type (5.1) with distributed delays have been widely studied (e.g., [**2**], [**10**], [**12**], and references therein). In a forthcoming paper, we shall consider the general cases (5.1) and (5.2), but here only the particular example of $r_1 = 2, r_2 = a_1 = a_2 = b = 1$ will be pursued, and the delay $\tau > 0$ will be taken as a parameter.

Assuming $r_1 = 2, r_2 = a_1 = a_2 = b = 1$, system (5.1) has a unique positive equilibrium $E^* = (1,1)$, with corresponding stationary solution $E^* = (1,1)$ of (5.2). The purpose here is to relate the dynamics of (5.1) and (5.2) in the neighborhood of $E^*$, at a bifurcation point of the parameter $\tau$. To avoid considering neutral FDEs, one can normalize the delay $\tau$ by the time-scaling $t \to t/\tau$. After this scaling, Equation (5.1) is transformed into the FDE in $C_2 = C([-1,0];\mathbb{R}^2)$

$$\begin{cases} \dot{u}(t) = \tau u(t)[2 - u(t) - v(t)] \\ \dot{v}(t) = \tau v(t)[-1 + u(t-1)] \end{cases} \tag{5.3}$$

and Equation (5.2) is written in abstract form in $\mathcal{C} = C([-1,0];X)$ as

$$\begin{cases} \dfrac{d}{dt}u(t) = \triangle u(t) + \tau u(t)[2 - u(t) - v(t)] \\[2ex] \dfrac{d}{dt}v(t) = \triangle v(t) + \tau v(t)[-1 + u(t-1)], \end{cases} \tag{5.4}$$

where $X$ is the Hilbert space $X = \{(u,v)\ :\ u,v \in W^{2,2}(0,\pi), \frac{du}{dx} = \frac{dv}{dx} = 0 \text{ at } x = 0,\pi\}$, with the inner product $< \cdot,\cdot >$ induced by the inner product of the Sobolev space $W^{2,2}(0,\pi)$.

Let us set $z(t) = (u(t), v(t)) - E^* \in \mathbb{R}^2$. Then, Equation (5.3) becomes

$$\dot{z}(t) = N(\tau)(z_t) + f_0(z_t, \tau), \tag{5.5}$$

where $N(\tau) : C_2 \longrightarrow \mathbb{R}^2, f_0 : C_2 \times \mathbb{R}^+ \longrightarrow \mathbb{R}^2$ are given by

$$N(\tau)(\varphi) = \tau \begin{pmatrix} -\varphi_1(0) - \varphi_2(0) \\ \varphi_1(-1) \end{pmatrix}, \quad f_0(\varphi, \tau) = \tau \begin{pmatrix} -\varphi_1^2(0) - \varphi_1(0)\varphi_2(0) \\ \varphi_1(-1)\varphi_2(0) \end{pmatrix}$$

for $\varphi = (\varphi_1, \varphi_2) \in C_2$. The characteristic equation for the linear equation $\dot{z}(t) = L(\tau)(z_t)$ is $\mathbf{\Delta}(\lambda, \tau) = 0$, where $\mathbf{\Delta}(\lambda, \tau) = \lambda(\lambda + \tau) + \tau^2 e^{-\lambda}$.

**Theorem 5.1** *Let $v_0 \in (0, \pi/2)$, $\tau_0$ be defined by*

$$v_0 = \arcsin\left(\sqrt{\frac{-1+\sqrt{5}}{2}}\right), \quad \tau_0 = \sqrt{\frac{2}{-1+\sqrt{5}}}v_0.$$

*Then (i) for $\tau \in (0, \tau_0)$, all roots of $\mathbf{\Delta}(\lambda, \tau) = 0$ have negative real parts; (ii) for $\tau = \tau_0$, $\pm iv_0$ are simple roots of the characteristic equation $\mathbf{\Delta}(\lambda, \tau_0) = 0$ and the remaining roots have negative real parts. Furthermore, for Equation (5.5) there is a supercritical Hopf bifurcation at $\tau = \tau_0$, with the periodic orbits being stable.*

**Proof** If $v > 0$ and $\mathbf{\Delta}(iv, \tau) = 0$, then $\cos v = (v/\tau)^2, \sin v = v/\tau$ and $v/\tau$ is the positive solution of $x^4 + x^2 = 1$:

$$\frac{v}{\tau} = \sqrt{\frac{-1+\sqrt{5}}{2}}.$$

Moreover, defining

$$v_k = \arcsin\left(\sqrt{\frac{-1+\sqrt{5}}{2}}\right) + 2k\pi, \tau_k = \sqrt{\frac{2}{-1+\sqrt{5}}}v_k, \; k \in \mathbb{N}_0,$$

we have $\mathbf{\Delta}(iv, \tau) = 0$ if and only if $v = \pm v_k, \tau = \tau_k$, for some $k \in \mathbb{N}_0$.

Now, let $H_1(u, v, \tau) = \operatorname{Re} \mathbf{\Delta}(u + iv, \tau), H_2(u, v, \tau) = \operatorname{Im} \mathbf{\Delta}(u + iv, \tau)$ and $H(u, v, \tau) = (H_1(u, v, \tau), H_2(u, v, \tau))$. Using the implicit function theorem, we deduce that $H(u, v, \tau) = 0$ defines $u, v$ as functions of $\tau$, $u = u(\tau), v = v(\tau)$, in a neighborhood of $\tau_k$, such that $u(\tau_k) = 0, v(\tau_k) = v_k, u'(\tau_k) > 0$. Therefore, the Hopf condition of transversality is satisfied and the eigenvalues $\lambda_\pm(\tau) = u(\tau) \pm iv(\tau)$ cross the imaginary axis to the right. That is, a supercritical Hopf bifurcation occurs at the points $\tau = \tau_k, k \in \mathbb{N}_0$.

For $u \geq 0, v > 0$, $H_2(u, v, \tau) \geq vg(u, \tau)$, where $g(u, \tau) := 2u + \tau - \tau^2 e^{-u}$, with $g'_u(u, \tau) > 0$ and $g(0, \tau) > 0$ if $\tau \in (0, 1)$, which implies $H_2(u, v, \tau) > 0$ for $u \geq 0, v > 0, \tau \in (0, 1)$. Since the eigenvalues depend continuously on the parameter $\tau$, we conclude that for $\tau \in (0, \tau_0)$ all solutions of $\mathbf{\Delta}(\lambda, \tau_0) = 0$ have negative real parts.

It remains to prove that the periodic solutions bifurcating from $\tau = \tau_0$ are stable. That can be done by evaluating the third order term of the normal form on the center manifold, using the theory in [**6**]. This will be proved in a forthcoming paper for the generic case (5.1) (with some restrictions on the parameters), so it is omitted. □

Now, let us consider the equation with diffusion. Translating $E^*$ to the origin by setting $U(t) = (u(t), v(t)) - E^* \in X$, Equation (5.4) is transformed into

$$\frac{d}{dt}U(t) = \triangle U(t) + L(\tau)(U_t) + f(U_t, \tau), \tag{5.6}$$

where now $L(\tau) : \mathcal{C} \longrightarrow \mathbb{R}^2, f : \mathcal{C} \times \mathbb{R}^+ \longrightarrow \mathbb{R}^2$ are given by

$$L(\tau)(\varphi) = \tau \begin{pmatrix} -\varphi_1(0) - \varphi_2(0) \\ \varphi_1(-1) \end{pmatrix}, \quad f(\varphi, \tau) = \tau \begin{pmatrix} -\varphi_1^2(0) - \varphi_1(0)\varphi_2(0) \\ \varphi_1(-1)\varphi_2(0) \end{pmatrix}$$

for $\varphi = (\varphi_1, \varphi_2) \in \mathcal{C}$.

The eigenvalues of $\triangle = \frac{\partial^2}{\partial x^2}$ on $X$ are $\mu_{k+1} = -k^2, k \geq 0$, with corresponding normalized eigenfunctions $\beta^1_{k+1}, \beta^2_{k+1}$, where

$$\beta^1_{k+1} = \begin{pmatrix} \gamma_{k+1} \\ 0 \end{pmatrix}, \; \beta^2_{k+1} = \begin{pmatrix} 0 \\ \gamma_{k+1} \end{pmatrix}, \; \gamma_{k+1}(x) = \frac{\cos(kx)}{\| \cos(kx) \|_{2,2}}, \; k \in \mathbb{N}_0.$$

With the notations of Section 4, $p_{k+1} = 2, \mu^1_{k+1} = \mu^2_{k+1} = \mu_{k+1}$ for all $k \in \mathbb{N}_0$; for $\varphi_1, \varphi_2 \in C = C([-1,0]; \mathbb{R})$

$$L(\tau)(\varphi_1\beta^1_{k+1} + \varphi_2\beta^2_{k+1}) = -\tau(\varphi_1(0) + \varphi_2(0))\beta^1_{k+1} + \tau\varphi_1(-1)\beta^2_{k+1}$$

and (H1),(H2'),(H3'),(H4) hold. In this case, the operators $L_{k+1}(\tau)$ defined by (4.1) coincide with $N(\tau)$, for all $k \in \mathbb{N}_0$. Since $M_{k+1} = \mu_{k+1}I$, the characteristic equations $(4.2_{k+1})$ are

$$\det\left( (\lambda + k^2)I - \tau \begin{pmatrix} -1 & -1 \\ e^{-\lambda} & 0 \end{pmatrix} \right) = 0,$$

or, equivalently,

$$\mathbf{\Delta}_k(\lambda, \tau) := (\lambda + k^2 + \tau)(\lambda + k^2) + \tau^2 e^{-\lambda} = 0, \quad k = 0, 1, 2, \ldots . \tag{$5.7_k$}$$

Note that for $k = 0$ the above equation is the characteristic equation for the linearized equation for (5.3) about the equilibrium $E^*$. Let us now consider $k \in \mathbb{N}$ and $\lambda = u + iv, u, v \geq 0$. If $v = 0$, then $\mathbf{\Delta}_k(\lambda, \tau) = \mathbf{\Delta}_k(u, \tau) > 0$. For $v > 0$, we have $\operatorname{Im} \mathbf{\Delta}_k(u + iv, \tau) \geq vg(u, \tau)$, for all $k \in \mathbb{N}$, where $g(u, \tau) := 2u + 2 + \tau - \tau^2 e^{-u}$. Since $g'_u(u, \tau) > 0, g(0, \tau) > 0$ for $\tau \in (0, 2)$, $g(0, \tau) < 0$, for $\tau > 2$, we deduce that $\operatorname{Im} \mathbf{\Delta}_k(u + iv, \tau) > 0$, for $\tau \in (0, 2), v > 0, u \geq 0$. In particular, since $\tau_0 = 1.15\ldots < 2$, Theorem 5.1 yields the following result:

**Lemma 5.2** *For $0 < \tau < \tau_0$, all the characteristic values of $\frac{d}{dt}U(t) = \triangle U(t) + L(\tau)(U_t)$ have negative real parts; for $\tau = \tau_0$, we have $\mathbf{\Delta}_0(\pm iv_0, \tau_0) = 0$ (then $\pm iv_0$ is a pair of characteristic roots on the imaginary axis), and there are no other roots of the characteristic equations $(5.7_k)$ with Re $\lambda \geq 0$.*

To study the qualitative behavior of Equation (5.4) near the stationary solution $E^*$ at the critical point $\tau = \tau_0$, let $\tau = \tau_0 + \alpha$ and consider $\Lambda = \{-iv_0, iv_0\}$. Defining

$$L := L(\tau_0), F(v, \alpha) := L(\alpha)(v) + f(v, \tau_0 + \alpha),$$

Equation (5.6) is written as

$$\frac{d}{dt}U(t) = \triangle U(t) + L(U_t) + F(U_t, \alpha). \tag{5.8}$$

With the notations of Section 4, we observe that $N = 1, \mu_1 = 0$ and the operator $L_1 := L_1(\tau_0)$ defined by (4.1) is $L_1 = N(\tau_0)$. In (4.7), we have $J = 2$, $R(\phi) = L_1(\phi)$,

$$G(\phi,\alpha) = \Big( < F(\phi^T \begin{pmatrix}\beta_1^1\\ \beta_1^2\end{pmatrix}, \alpha), \beta_1^i > \Big)_{i=1}^2 = \Big( < F(\phi_1\beta_1^1 + \phi_2\beta_1^2, \alpha), \beta_1^i > \Big)_{i=1}^2$$

for $\phi = (\phi_1, \phi_2) \in C_2$. Since $\beta_1^1 = (1/\sqrt{\pi}, 0), \beta_1^2 = (0, 1/\sqrt{\pi})$ and $< \beta_1^i, \beta_1^i >= 1$, $< \beta_1^i, \beta_1^j >= 0$, if $i \neq j, i, j = 1, 2$, it is $G(\phi, \alpha) = N(\alpha)(\phi) + \frac{1}{\sqrt{\pi}} f_0(\phi, \tau_0 + \alpha)$ and the FDE associated with (5.6) by $\Lambda$ at $U = 0, \alpha = 0$ is

$$\dot{x}(t) = N(\tau_0 + \alpha)(x_t) + \frac{1}{\sqrt{\pi}} f_0(x_t, \tau_0 + \alpha). \tag{5.9}$$

From the above definition of $F$, we have

$$F(v,\alpha) = \frac{1}{2}F_2(v,\alpha) + O(\alpha|v|^2), \quad v = (v_1, v_2) \in \mathcal{C}$$
$$\frac{1}{2}F_2(v,\alpha) = \alpha \begin{pmatrix} -v_1(0) - v_2(0) \\ v_1(-1) \end{pmatrix} + \tau_0 \begin{pmatrix} -v_1^2(0) - v_1(0)v_2(0) \\ v_1(-1)v_2(0) \end{pmatrix}.$$

Let $u \in \mathcal{P}$. Necessarily $u$ has the form $u = \frac{1}{\sqrt{\pi}}(\varphi_1, \varphi_2), \varphi_1, \varphi_2 \in C$, and for $k \geq 2, i = 1, 2, \psi_1, \psi_2 \in C$ we obtain

$$\begin{aligned}\frac{1}{2}D_1F_2(u,\alpha)(\psi_1\beta_k^1 + \psi_2\beta_k^2) &= \alpha(-\psi_1(0) - \psi_2(0))\beta_k^1 + \alpha\psi_1(-1)\beta_k^2 \\ &+ \frac{\tau_0}{\sqrt{\pi}}\Big( - 2\varphi_1(0)\psi_1(0) - \varphi_1(0)\psi_2(0) - \varphi_2(0)\psi_1(0)\Big)\beta_k^1 \\ &+ \frac{\tau_0}{\sqrt{\pi}}\Big(\varphi_1(-1)\psi_2(0) - \varphi_2(0)\psi_1(-1)\Big)\beta_k^2;\end{aligned}$$

since $< \beta_k^i, \beta_1^j >= 0$, then $< D_1F_2(u,\alpha)(\phi\beta_k^i), \beta_1^j >= 0$, for $k \geq 2, i = 1, 2, j = 1, 2, u \in \mathcal{P}$, and clearly (H5') is fulfilled. Thus, Theorem 4.1 yields that, up to third order terms, the equations on the center manifold at zero for both Equation (5.8) and the associated FDE by $\Lambda$ (5.9) are the same. But Equation (5.9) is Equation (5.5) after the change of parameter $\tau = \tau_0 + \alpha$ and the scaling $z \longmapsto \frac{1}{\sqrt{\pi}}z$. Since the third order terms are sufficient to determine the dynamics of a generic Hopf bifurcation, Theorem 5.1, Lemma 5.2 and the above results allow us to state the following theorem:

**Theorem 5.3** *Let $\tau_0$ be as in Theorem 5.1. For Equation (5.4) a 2-dimensional locally stable center manifold exists at the stationary point $E^* = (1,1)$ for $\tau = \tau_0$, where a generic supercritical Hopf bifurcation occurs with the associated periodic solutions being stable.*

## References

[1] Arnold, V. I. [1983], *Geometrical Methods in the Theory of Ordinary Differential Equations*, Springer-Verlag, New York.

[2] Beretta, E. and Kuang, Y. [1996], *Convergence results in a well-known delayed predator-prey system*, J. Math. Anal. Appl. **204**, 840–853.

[3] Chow, S.–N. and Hale, J. K. [1982], *Methods of Bifurcation Theory*, Springer-Verlag, New York.

[4] Faria, T. [1997], *Normal forms and Hopf bifurcation for partial differential equations with delays*, Transactions of the A.M.S., to appear.

[5] Faria, T. and Magalhães, L. T. [1995], *Normal forms for retarded functional differential equations and applications to Bogdanov–Takens singularity,* J. Differential Equations **122**, 201–224.

[6] Faria, T. and Magalhães, L. T. [1995], *Normal forms for retarded functional differential equations with parameters and applications to Hopf bifurcation*, J. Differential Equations **122**, 181–200.

[7] Grabosch, A. and Moustakas, U. [1986], in One-parameter Semigroups of Positive Operators (R. Nagel, ed.), Lectures Notes in Math. **1184**, Springer-Verlag, Berlin, Heidelberg, Tokyo.

[8] Guckenheimer, J. and Holmes, P. [1983], *Nonlinear Oscillations, Dynamical Systems and Bifurcations of Vector Fields*, Springer-Verlag, New York.

[9] Hale, J. K. [1977], *Theory of Functional Differential Equations*, Springer-Verlag, Berlin.

[10] Kuang, Y. [1993], *Delay Differential Equations with Applications in Population Dynamics*, Academic Press, Boston.

[11] Lin, X., So, J. W.-H., and Wu, J. [1992] , *Centre manifolds for partial differential equations with delays*, Proc. Roy. Soc. Edinburgh **122A**, 237–254.

[12] Ma, W. and Takeuchi, Y. [1997], *Stability Analysis on a predator-prey system with distributed delays*, preprint.

[13] Memory, M. C. [1991], *Stable and unstable manifolds for partial functional differential equations*, Nonlinear Anal. **16**, 131–142.

[14] Travis, C. C. and Webb, G. F. [1974], *Existence and stability for partial functional differential equations*, Transactions of the A.M.S. **200**, 395–418.

[15] Yoshida, K. [1982], *The Hopf bifurcation and its stability for semilinear diffusion equations with time delay arising in ecology*, Hiroshima Math. J. **12**, 321–348.

Fields Institute Communications
Volume **21**, 1999

# Extinction and Persistence in a Size-Structured, Discrete Competitive System with Dispersion

**John E. Franke**
Department of Mathematics
North Carolina State University
Raleigh, North Carolina 27695-8205, USA
franke@math.ncsu.edu

**Abdul-Aziz Yakubu**
Department of Mathematics
Howard University
Washington, DC 20059, USA
aziz@scs.howard.edu

**Abstract.** We consider a discrete multi-species size-structured spatially structured ecological model. The principal focus is on competitive interactions and how these affect species survival and extinction. In particular, we obtain conditions for the global convergence to the extinction state of a dominated species. The relationship between the persistence of species and their effective growth rates is also explored.

## 1 Introduction

The importance of spatial structure and the physiological traits of individual species, both in controlling the total population sizes and the local variations in densities, on real-world populations is well-established ([**3**], [**11**], [**20**], and [**22, 23, 24**]). Body size and chronological age of individual species are examples of such physiological traits. However, for most insect species, it is body size and not chronological age that causes the development through various stages of an instar. In a recent paper [**13**], we studied size structured discrete competitive systems without spatial structure. The model for this study is a 2 species discrete competition model with 2 size classes, small and large, and with spatial structure. In contrast to the study in [**13**], our model consists of 2 patches, patch $A$ and patch $B$, with patch $B$ being a refuge for the small size individuals of one of the 2 competing species. The patches are coupled by an exchange of a fraction of the difference of the population densities of the small size individuals per generation.

For each species $i \in \{1, 2\}$, in patch $A$, $x_i(n)$ and $y_i(n)$ respectively denote the population densities of its small and large individuals at generation $n$ with the positive continuous function $g_i : [0, \infty) \to (0, \infty)$ being its growth rate. In

1991 *Mathematics Subject Classification.* 34C35, 39A10, 39A12, 92D25, 92D40.

patch $B$, $x_3(n)$ and $y_3(n)$ respectively denote the population densities of small and large individuals of species 2 at generation $n$ with the positive continuous function $g_3 : [0,\infty) \to (0,\infty)$ being its growth function. To insure the self-regulation of species, we assume throughout that for each $i \in \{1,2,3\}$, $g_i : [0,\infty) \to (0,\infty)$ is a strictly decreasing positive continuous function with $\lim_{x_i\to\infty} g_i(x_i) = 0$.

The model, system (1), is the following system of nonlinear difference equations:

$$\left.\begin{aligned}
x_1(n+1) &= y_1(n)g_1\left(\sum_{j=1}^{2}\{\alpha_{1j}x_j(n)+\beta_{1j}y_j(n)\}\right)+a_1x_1(n)\\
x_2(n+1) &= y_2(n)g_2\left(\sum_{j=1}^{2}\{\alpha_{2j}x_j(n)+\beta_{2j}y_j(n)\}\right)+a_2x_2(n)\\
x_3(n+1) &= y_3(n)g_3(\alpha_{33}x_3(n)+\beta_{33}y_3(n))+a_3x_3(n)\\
y_1(n+1) &= b_1x_1(n)+s_1y_1(n)\\
y_2(n+1) &= b_2x_2(n)-d[b_2x_2(n)-b_3x_3(n)]+s_2y_2(n)\\
y_3(n+1) &= b_3x_3(n)+d[b_2x_2(n)-b_3x_3(n)]+s_3y_3(n)
\end{aligned}\right\} \quad (1)$$

where each $\alpha_{ij}$, $\beta_{ij}$, $a_i$, $b_i$, $s_i$ and $d$ is a positive constant with $\beta_{ii} = 1$, $0 < a_i + b_i < 1$, $0 < s_i < 1$ and the diffusion coefficient $d \in (0, \frac{1}{2}]$.

In system (1), for each $i \in \{1,2,3\}$, after one generation, the constants $a_i$ and $b_i$ respectively denote the fraction of a species' small size individuals that remain in that class size (and patch) and the fraction that move out of the small size class into the large one. Also, for each $i$, the fraction $s_i$ denotes the fraction of a species' large size individuals that are alive in the large size class after one generation. Since $0 < a_i + b_i < 1$ and $0 < s_i < 1$, system (1) allows the death of individuals in each of the two size classes.

Density effects are modeled, in system (1), by assuming that the argument of each growth function in a patch is a linear combination of the densities of the individual species in that patch. In system (1), the 2 competing species interact in patch $A$. Due to population pressure, as the small size individuals of species 2 become large, they diffuse between the 2 patches, patch $A$ and patch $B$. When the dispersion constant is $1/2$, the small size individuals of species 2 who become large completely balance between the 2 patches. To realistically model the diffusion, we assume throughout that the diffusion coefficient $d \in (0, \frac{1}{2}]$. Species 1 is confined to patch $A$, making patch $B$ a refuge from competition for species 2.

If there is no diffusion and patch $B$ is empty ($d = x_3(n) = y_3(n) = 0$), then system (1) reduces to the following system of equations:

$$\left.\begin{aligned}
x_i(n+1) &= y_i(n)g_i\left(\sum_{j=1}^{2}\{\alpha_{ij}x_j(n)+\beta_{ij}y_j(n)\}\right)+a_ix_i(n)\\
y_i(n+1) &= b_ix_i(n)+s_iy_i(n) \qquad\qquad (i=1,2).
\end{aligned}\right\} \quad (2)$$

Franke and Yakubu have obtained various biologically significant conditions for global convergence to the extinction state in an $n$-dimensional version of system (2) [**13**]. Notice that in system (1) there is only density dependence in the population of the small size individuals through the growth function of the large size individuals, $g_i$. If $a_i = s_i = 0$, then system (2) reduces to Ebenman's model with two age classes ([**3**], [**4**], [**5**], [**6**], [**7**], [**8**], [**9**], [**11**], [**12**], and [**13**]). In that biological situation, for each species $i$, the density dependent growth function, $g_i$, is the growth function of its adults with $b_i$ being the constant survival function of its juveniles.

In system (1), we study the effect of competitive interactions on species survival. In particular, we obtain that both species persist in system (1) whenever the competition is very weak and the effective growth rate of each of the species is large (Corollary 4.4). We also obtain conditions that lead to extinction of species in system (1) [Theorems 5.2 and 6.1]. These results are applied to system (1) with all growth functions being exponential.

## 2 Notation and preliminaries

To study the long term behavior of system (1) in terms of the $\omega$-limit set of points, we denote the vector of population densities $x(n) = (x_1(n), x_2(n), x_3(n))$ and $y(n) = (y_1(n), y_2(n), y_3(n))$ by $x = (x_1, x_2, x_3)$ and $y = (y_1, y_2, y_3)$, respectively. From system (1), define the map $F : \mathbf{R}^3_+ \times \mathbf{R}^3_+ \to \mathbf{R}^3_+ \times \mathbf{R}^3_+$ by $F(x(n), y(n)) = (x(n+1), y(n+1))$.

For each species $i \in \{1, 2\}$ being modeled by system (1), after one generation, in patch $A$, $F_i(x, y) = y_i g_i(\sum_{j=1}^{2}\{\alpha_{ij}x_j + \beta_{ij}y_j\}) + a_i x_i$ and $F_{3+i}(x, y)$ [$F_4(x, y) = b_1x_1 + s_1y_1$ and $F_5(x, y) = b_2x_2 - d(b_2x_2 - b_3x_3) + s_2y_2$] respectively denote its small and large size populations. In patch $B$, after one generation, $F_3(x, y) = y_3g_3(\alpha_{33}x_3 + y_3) + a_3x_3$ and $F_6(x, y) = b_3x_3 + d(b_2x_2 - b_3x_3) + s_3y_3$ respectively denote the population densities of the small and large size individuals of species 2. Therefore, the sequence of iterates of the map $F$ is equivalent to the density sequence generated by system (1). The map $F$ composed with itself $n$ times is denoted by $F^n$, and $F^n_i(x, y)$ is the i-th component of $F^n$ evaluated at the point $(x, y) \in \mathbf{R}^3_+ \times \mathbf{R}^3_+$. Consequently, $F^n$ describes the state of population densities in generation $n$.

For each $i \in \{1, 2, 3\}$, define the map $f_i : \mathbf{R}_+ \times \mathbf{R}_+ \to \mathbf{R}_+ \times \mathbf{R}_+$ by $f_i(x_i, y_i) = (y_ig_i(\alpha_{ii}x_i + y_i) + a_ix_i,\ b_ix_i + s_iy_i)$. If all species are missing except species $i$ in either patch $A$ or patch $B$ and there is no diffusion between patches ($d = 0$), then the iterates of the map $f_i$ is equivalent to the density sequence generated by system (1). Also, define the map $h_i : \mathbf{R}_+ \to \mathbf{R}_+$ by $h_i(y_i) = y_ig_i(y_i)$. Observe that $f_i(0, y_i) = (h_i(y_i),\ s_iy_i)$. Thus, if an initial population density consists solely of large size individuals of species $i$ in either patch $A$ or patch $B$, then the dynamics of the single species model $h_i$ is closely related to that of the map $f_i$. In particular, if $g_i(y_i) = \exp(r_i - y_i)$, then $h_i$ is the classical model of [**21**]. May has shown period-doubling bifurcation route to chaos in Ricker's model, $h_i(y_i) = y_i \exp(r_i - y_i)$ ([**18, 19**]).

## 3 Boundedness of orbits

In this section we show that every point in system (1) has a bounded orbit. To construct compact invariant sets that contain the $\omega$-limit set of points, we introduce the following notation: For each $i \in \{1, 2\}$, let

$$N(g_i,\ 1 - s_i) = \left\{(x, y) \in \mathbf{R}^3_+ \times \mathbf{R}^3_+ | g_i\left(\sum_{j=1}^{2}\{\alpha_{ij}x_j + \beta_{ij}y_j\}\right) \geq 1 - s_i\right\}.$$

Also, let

$$N(g_3,\ 1 - s_3) = \{(x, y) \in \mathbf{R}^3_+ \times \mathbf{R}^3_+ | g_3(\alpha_{33}x_3 + y_3) \geq 1 - s_3\}.$$

Since each $g_i$ decreases to zero and $0 < s_i < 1$, $N(g_i,\ 1 - s_i)$ is the set

$$\left\{(x, y) \in \mathbf{R}^3_+ \times \mathbf{R}^3_+ | \sum_{j=1}^{2}\{\alpha_{ij}x_j + \beta_{ij}y_j\} \leq g_i^{-1}(1 - s_i)\right\}$$

and $N(g_3,\,1-s_3)$ is the set

$$\{(x,y)\in \mathbf{R}^3_+\times\mathbf{R}^3_+|\alpha_{33}x_3+y_3\le g_3^{-1}(1-s_3)\}\,.$$

Now, let

$$S_1=(\mathbf{R}_+\times\{0\}\times\{0\}\times\mathbf{R}_+\times\{0\}\times\{0\})\cap N(g_1,\,1-s_1)$$

and let

$$S_2=(\{0\}\times\mathbf{R}^2_+\times\{0\}\times\mathbf{R}^2_+)\cap(N(g_2,\,1-s_2)\cup N(g_3,\,1-s_3))\,.$$

Notice that each $S_i$ is a compact set in $\mathbf{R}^3_+\times\mathbf{R}^3_+$. Let

$$T_1=S_1\cup\{(x_1,0,0,y_1,0,0)\in\mathbf{R}^3_+\times\mathbf{R}^3_+|0\le x_1+y_1\le 2(\max_j F_j(S_1))\}$$

and let

$$T_2=S_2\cup\{(0,x_2,x_3,0,y_2,y_3)\in\mathbf{R}^3_+\times\mathbf{R}^3_+|0\le x_2+x_3+y_2+y_3\le 4(\max_j F_j(S_2))\}\,.$$

**Lemma 3.1** *$T_1$ is a compact positively $F$ invariant set. Moreover, if $(x,y)=(x_1,0,0,y_1,0,0)\in\mathbf{R}^3_+\times\mathbf{R}^3_+\backslash T_1$ then $F_1(x,y)+F_4(x,y)<x_1+y_1$.*

**Proof** Consider $(x,y)=(x_1,0,0,y_1,0,0)\in T_1$. If $(x,y)\in S_1$, then $F(x,y)\in T_1$. If $(x,y)\in T_1\backslash S_1$, then $g_1(\alpha_{11}x_1+y_1)<1-s_1$ and $F_1(x,y)+F_4(x,y)=y_1g_1(\alpha_{11}x_1+y_1)+a_1x_1+b_1x_1+s_1y_1\le(a_1+b_1)x_1+y_1\le x_1+y_1$. Therefore, if $(x,y)=(x_1,0,0,y_1,0,0)\in T_1\backslash S_1$, then $F_1(x,y)+F_4(x,y)\le x_1+y_1\le 2(\max_j F_j(S_1))$ and $F(x,y)\in T_1$. Hence, $T_1$ is compact and positively $F$ invariant. In addition, if $(x,y)=(x_1,0,0,y_1,0,0)\in(\mathbf{R}^3_+\times\mathbf{R}^3_+)\backslash T_1$, then $x_1+y_1>0$. Hence, $F_1(x,y)+F_4(x,y)<x_1+y_1$. □

**Lemma 3.2** *$T_2$ is a compact positively $F$ invariant set. Moreover, if $(x,y)=(0,x_2,x_3,0,y_2,y_3)\in\mathbf{R}^3_+\times\mathbf{R}^3_+\backslash T_2$ then $F_2(x,y)+F_3(x,y)+F_5(x,y)+F_6(x,y)<x_2+x_3+y_2+y_3$.*

**Proof** As in the proof of Lemma 3.1, consider $(x,y)=(0,x_2,x_3,0,y_2,y_3)\in T_2$. If $(x,y)\in S_2$, then $F(x,y)\in T_2$. If $(x,y)\in T_2\backslash S_2$, then $g_2(\alpha_{22}x_2+y_2)<1-s_2$ and $g_3(\alpha_{33}x_3+y_3)<1-s_3$. Therefore, $F_2(x,y)+F_3(x,y)+F_5(x,y)+F_6(x,y)=y_2g_2(\alpha_{22}x_2+y_2)+a_2x_2+y_3g_3(\alpha_{33}x_3+y_3)+a_3x_3+b_2x_2+s_2y_2+b_3x_3+s_3y_3\le(a_2+b_2)x_2+(a_3+b_3)x_3+y_2+y_3\le x_2+x_3+y_2+y_3$. Therefore, if $(x,y)=(0,x_2,x_3,0,y_2,y_3)\in T_2\backslash S_2$, then

$$F_2(x,y)+F_3(x,y)+F_5(x,y)+F_6(x,y)\le x_2+x_3+y_2+y_3\le 4(\max_j F_j(S_2))$$

and $F(x,y)\in T_2$. Hence, $T_2$ is compact and positively $F$ invariant. If $(x,y)=(0,x_2,x_3,0,y_2,y_3)\in(\mathbf{R}^3_+\times\mathbf{R}^3_+)\backslash T_2$, then $x_2+x_3+y_2+y_3>0$. Hence, $F_2(x,y)+F_3(x,y)+F_5(x,y)+F_6(x,y)<x_2+x_3+y_2+y_3$. □

Let $T=\{(x,y)=(x_1,x_2,x_3,y_1,y_2,y_3)\in\mathbf{R}^3_+\times\mathbf{R}^3_+|(x_1,0,0,y_1,0,0)\in T_1$ and $(0,x_2,x_3,0,y_2,y_3)\in T_2\}$. Lemmas 3.1 and 3.2 give

$$F(x_1,0,0,y_1,0,0)=(y_1g_1(\alpha_{11}x_1+y_1)+a_1x_1,0,0,b_1x_1+s_1y_1,0,0)\in T_1$$

and

$$\begin{aligned}F(0,x_2,x_3,0,y_2,y_3)=(0,y_2g_2(\alpha_{22}x_2+y_2)+a_2x_2,\;y_3g_3(\alpha_{33}x_3+y_3)+a_3x_3,0,b_2x_2\\ -d[b_2x_2-b_3x_3]+s_2y_2,b_3x_3+d[b_2x_2-b_3x_3]+s_3y_3)\in T_2.\end{aligned}$$

Since the $g_i$'s are decreasing, $(y_1 g_1\Big(\sum_{j=1}^{2}\{\alpha_{1j}x_j+\beta_{1j}y_j\}\Big)+a_1x_1, 0, 0, b_1x_1+s_1y_1, 0, 0)$ $\in T_1$ and $(0, y_2 g_2\left(\sum_{j=1}^{2}\{\alpha_{2j}x_j+\beta_{2j}y_j\}\right)+a_2x_2,\ y_3g_3(\alpha_{33}x_3+y_3)+a_3x_3, 0, b_2x_2 - d[b_2x_2-b_3x_3]+s_2y_2,\ b_3x_3+d[b_2x_2-b_3x_3]+s_3y_3) \in T_2$. Hence $T$ is a compact $F$ invariant set (Corollary 3.3).

**Corollary 3.3** $F(T) \subset T$.

**Lemma 3.4** *Every point in system (1) has a bounded orbit.*

**Proof** The set $T \subset \mathbf{R}_+^3 \times \mathbf{R}_+^3$ is a compact $F$ invariant set (Corollary 3.3). As a result, if an iterate of a point gets in $T$, then it remains in $T$ under all $F$ iterations. Hence, such an orbit is bounded.

Now, we consider an arbitrary point $(x, y) = (x_1, x_2, x_3, y_1, y_2, y_3) \in (\mathbf{R}_+^3 \times \mathbf{R}_+^3)\backslash T$. That is, $(x_1, 0, 0, y_1, 0, 0) \notin T_1$ or $(0, x_2, x_3, 0, y_2, y_3) \notin T_2$. By Lemmas 3.1 and 3.2 and the decreasing nature of each $g_i$, we have that $F_1(x,y)+F_4(x,y) \leq \max\{x_1+y_1,\ 2(\max_j F_j(S_1))\}$ and

$$F_2(x,y)+F_3(x,y)+F_5(x,y)+F_6(x,y) \leq \max\{x_2+x_3+y_2+y_3, 4(\max_j F_j(S_2))\}.$$

Thus, $(x, y)$ eventually gets mapped into $T$ or one of the sequences $\{(x_1(n), 0, 0, y_1(n), 0, 0)\}$ or $\{(0, x_2(n), x_3(n), 0, y_2(n), y_3(n))\}$ remains outside of $T$. If $\{(x_1(n), 0, 0, y_1(n), 0, 0)\}$ remains outside of $T$, then, by Lemma 3.1, $x_1(n)+y_1(n)$ is a decreasing sequence bounded below by $2\left(\max_j F_j(S_1)\right)$. If it does not converge to this value, there is a subsequence $\{(x_1(n_k), 0, 0, y_1(n_k), 0, 0)\}$ which converges to a point $(p_1, 0, 0, q_1, 0, 0)$ where $p_1+q_1 > 2\left(\max_j F_j(S_1)\right)$. If $(p_1, 0, 0, q_1, 0, 0)$ $\in S_1$ then $F_1(p_1, 0, 0, q_1, 0, 0)+F_4(p_1, 0, 0, q_1, 0, 0) \leq 2\left(\max_j F_j(S_1)\right) < p_1+q_1$. If $(p_1, 0, 0, q_1, 0, 0) \notin T_1$, then, by Lemma 3.1, we again have $F_1(p_1, 0, 0, q_1, 0, 0) + F_4(p_1, 0, 0, q_1, 0, 0) < p_1+q_1$. Now the continuity of the system gives $x_1(n)+y_1(n)$ must decrease below $p_1+q_1$. This contradiction shows that if $\{(x_1(n),0,0, y_1(n),0,0)\}$ remains outside of $T_1$, it must limit on $T_1$. A similar argument shows that if $\{(0, x_2(n), x_3(n), 0, y_2(n), y_3(n))\}$ remains outside of $T_2$, then it limits on $T_2$. Thus, each point outside of $T$ eventually gets mapped to $T$ or it limits on it. Therefore, no point in system (1) has an unbounded orbit. □

The next result, Corollary 3.5, is a consequence of the proof of Lemma 3.4.

**Corollary 3.5** *In system (1), the $\omega$-limit set of every point in $\mathbf{R}_+^3 \times \mathbf{R}_+^3$ is a nonempty subset of the compact invariant set $T$.*

We will use the following auxiliary results, Lemma 3.6 and Corollary 3.7, to describe the $\omega$-limit set of points in system (1).

**Lemma 3.6** *Let $(p, q) = (p_1, p_2, p_3, q_1, q_2, q_3) \in \mathbf{R}_+^3 \times \mathbf{R}_+^3$.*

1. If $p_1 = q_1 = 0$, then $\omega((p, q)) \subset T_2$.
2. If $p_2 = p_3 = q_2 = q_3 = 0$, then $\omega((p, q)) \subset T_1$.

**Proof** Let $p_1 = q_1 = 0$. Then $F_1(p,q) = F_4(p,q) = 0$. Therefore, $\{F_1^n(p,q)\}_{n\geq 0}$ and $\{F_4^n(p,q)\}_{n\geq 0}$ are sequences of zero and hence converge to zero. By Corollary 3.5, $\omega((p, q)) \subset T_2$. The proof of (2) is similar and is omitted. □

The above proof and the fact that each growth rate, $g_i$, as well as each $a_i, b_i, s_i$ and $d$ is strictly positive give the following result.

**Corollary 3.7** *Let* $(x, y) = (x_1, x_2, x_3, y_1, y_2, y_3) \in \mathbf{R}^3_+ \times \mathbf{R}^3_+$.

1. *If either* $x_1 \neq 0$ *or* $y_1 \neq 0$, *then* $F_1(x, y) \neq 0$ *and* $F_4(x, y) \neq 0$.
2. *If* $x_2 \neq 0$ *or* $x_3 \neq 0$ *or* $y_2 \neq 0$ *or* $y_3 \neq 0$, *then there exists a positive integer* $m$ *such that for each* $i \in \{2, 3, 5, 6\}$, $F_i^m(x, y) \neq 0$ *for all* $m \geq 3$.

## 4 Persistence

In this section, we obtain persistence results for system (1). We make the following additional assumptions throughout this section.

1. $(X, d)$ is a locally compact metric space with metric $d$,
2. $f : X \to X$ is a continuous map,
3. $C$ and $T$ are subsets of the metric space $X$,
4. $T$ is compact and positively invariant ($f(T) \subset T$),
5. $C$ is closed with $C$ and $X \backslash C$ positively invariant,
6. $M = C \cap T$.

**Definition 1** The map $f : X \to X$ is *uniformly persistent* (with respect to $C$) if there exists $\eta > 0$ such that for all $x \in X \backslash C$, $\liminf_{n \to \infty} d(f^n(x), C) > \eta$ ([**1**], [**2**], [**10**], [**12**], [**14**], and [**16**]).

The following result from [**12**], a generalization of a result of Fonda ([**10**] and [**16**]), will be useful in the proof of persistence results for system (1).

**Theorem 4.1** *Let* $\omega(x) \subset T\ \forall x \in X$. *Then* $\forall x \in X \backslash C$, $\omega(x) \cap C = \emptyset$ *if there exists a continuous function* $p : X \to [0, \infty)$ *such that:*

1. $p(x) = 0\ \forall\ x \in M$, *and*
2. *there exists a closed neighborhood* $V$ *of* $M$ *such that* $\forall x \in V \backslash C$ *there exists* $n = n(x) > 0$ *satisfying* $p(f^n(x)) > p(x)$.

*Moreover,* $f$ *is uniformly persistent with respect to* $C$.

The following result, Theorem 4.2, states that, if, in both patch $A$ and patch $B$, the effective growth rate of species 2 is sufficiently large, then species 2 persists provided the competition from the competitor, species 1, is sufficiently weak.

**Theorem 4.2** *In system (1), let* $\beta_{22} = \beta_{33} = 1$, $b_2 g_2(0) > 1 - s_2^2$ *and* $b_3 g_3(0) > 1 - s_3^2$. *If* $\alpha_{21}$ *and* $\beta_{21}$ *are small enough, then* $F$ *is uniformly persistent with respect to* $\mathbf{R}_+ \times \{0\} \times \{0\} \times \mathbf{R}_+ \times \{0\} \times \{0\}$. *Hence, for each point* $(x, y)$ *in the interior of* $\mathbf{R}^3_+ \times \mathbf{R}^3_+$, $\omega((x, y)) \cap \mathbf{R}_+ \times \{0\} \times \{0\} \times \mathbf{R}_+ \times \{0\} \times \{0\} = \emptyset$, *and species 2 persists.*

**Proof** We will show that all the hypotheses of Theorem 4.1 are satisfied. Theorem 4.2 then follows by a direct application of Theorem 4.1. Recall that

$$T \equiv \{(x, y) = (x_1, x_2, x_3, y_1, y_1, y_3) \in \mathbf{R}^3_+ \times \mathbf{R}^3_+ | (x_1, 0, 0, y_1, 0, 0) \in T_1$$

and $(0, x_2, x_3, 0, y_2, y_3) \in T_2\}$. By Corollary 3.5, $T$ is a nonempty compact, positively invariant set and $\omega((x, y)) \subset T$ for all points $(x, y) = (x_1, x_2, x_3, y_1, y_2, y_3)$ in $\mathbf{R}^3_+ \times \mathbf{R}^3_+$.

Let $C = \mathbf{R}_+ \times \{0\} \times \{0\} \times \mathbf{R}_+ \times \{0\} \times \{0\}$. Notice that $C$ is a closed positively $F$ invariant subset of $\mathbf{R}^3_+ \times \mathbf{R}^3_+$ and $(\mathbf{R}^3_+ \times \mathbf{R}^3_+) \backslash C$ is also $F$ invariant. Now, let $M = C \cap T$. Notice that $M = T_1$. Also, let $u_2 = \alpha_{21} x_1 + \alpha_{22} x_2 + \beta_{21} y_1 + y_2$ and $u_3 = \alpha_{33} x_3 + y_3$. Notice $u_2$ (respectively, $u_3$) is a continuous function of $x_1, x_2, y_1, y_2, \alpha_{21}, \alpha_{22}$ and $\beta_{21}$ (respectively, $x_3, y_3$ and $\alpha_{33}$). If $\alpha_{21} = \beta_{21} = 0$, then

$u_2 = u_3 = 0$ on $M = T_1$. By continuity, given any $\epsilon > 0$ there exists a compact neighborhood $V$ of $M$ in $\mathbf{R}^3_+ \times \mathbf{R}^3_+$ and positive constants $\alpha$ and $\beta$ such that, $u_2 < \epsilon$ and $u_3 < \epsilon$ on $V \times [0, \alpha] \times [0, \beta]$ in $(x, y, \alpha_{21}, \beta_{21})$-space. We will see that if $\alpha_{21} \leq \alpha$ and $\beta_{21} \leq \beta$ for a specific $\epsilon$, then the theorem is true.

Each $g_i : [0, \infty) \to (0, \infty)$ is continuous and strictly decreasing with $b_2 g_2(0) > 1 - s_2^2$ and $b_3 g_3(0) > 1 - s_3^2$. Now, choose $\epsilon > 0$ so that $b_2 g_2(\epsilon) > 1 - s_2^2$ and $b_3 g_3(\epsilon) > 1 - s_3^2$. Any such $\epsilon$ can be used as the specific $\epsilon$ of the last paragraph. The monotonicity condition on each growth function and the fact that $u_2 < \epsilon$ and $u_3 < \epsilon$ imply that $b_2 g_2(u_2) > 1 - s_2^2$ and $b_3 g_3(u_3) > 1 - s_3^2$ on $V \times [0, \alpha] \times [0, \beta]$.

Define the continuous function $p : \mathbf{R}^3_+ \times \mathbf{R}^3_+ \to \mathbf{R}_+$ by $p(x, y) = y_2 + y_3$. Notice that, for all points $(x, y) = (x_1, x_2, x_3, y_1, y_2, y_3) \in M$, $p(x, y) = 0$. Moreover, for all points $(x, y) = (x_1, x_2, x_3, y_1, y_2, y_3) \in V \backslash C$,

$$\begin{aligned} p(F^2(x, y)) &= y_2[b_2 g_2(u_2) + s_2^2] + y_3[b_3 g_3(u_3) + s_3^2] + [a_2 b_2 + s_2 b_2(1 - d) + \\ & ds_3 b_2] x_2 + [a_3 b_3 + s_3 b_3 (1 - d) + ds_2 b_3] x_3 > p((x, y)) = y_2 + y_3. \end{aligned}$$

Therefore, all the conditions of Theorem 4.1 are satisfied, and hence, $F$ is uniformly persistent with respect to the repelling set $C$. This completes the proof. □

On the other hand, if the effective growth rate of the sedentary species 1 is sufficiently large, then it persists provided the competition from the competitor, species 2, is sufficiently weak (Theorem 4.3).

**Theorem 4.3** *In system (1), let* $\beta_{11} = 1$, $b_1 g_1(0) > 1 - s_1^2$. *If* $\alpha_{12}$ *and* $\beta_{12}$ *are small enough, then* $F$ *is uniformly persistent with respect to* $\{0\} \times \mathbf{R}^2_+ \times \{0\} \times \mathbf{R}^2_+$. *Hence, for each point* $(x, y)$ *in the interior of* $\mathbf{R}^3_+ \times \mathbf{R}^3_+$, $\omega((x, y)) \cap \{0\} \times \mathbf{R}^2_+ \times \{0\} \times \mathbf{R}^2_+ = \emptyset$, *and species 1 persists.*

**Proof** This proof is similar to that of Theorem 4.2. As in the proof of Theorem 4.2, we proceed to show that all the conditions of Theorem 4.1 are satisfied.

By Corollary 3.5, $T$ is a nonempty compact, positively $F$ invariant set and $\omega((x, y)) \subset T$ for all points $(x, y) = (x_1, x_2, x_3, y_1, y_2, y_3) \in \mathbf{R}^3_+ \times \mathbf{R}^3_+$. Now, let $C = \{0\} \times \mathbf{R}^2_+ \times \{0\} \times \mathbf{R}^2_+$. The set $C$ is a closed positively $F$ invariant subset of $\mathbf{R}^3_+ \times \mathbf{R}^3_+$, and $(\mathbf{R}^3_+ \times \mathbf{R}^3_+) \backslash C$ is also $F$ invariant. Recall that $M = T \cap C = T_2$. Let $u_1 = \alpha_{11} x_1 + \alpha_{12} x_2 + y_1 + \beta_{12} y_2$. Thus, $u_1 = u_1(x, y, \alpha_{11}, \alpha_{12}, \beta_{12})$ is a continuous function of $\alpha_{11}, \alpha_{12}, \beta_{12}, x_1, x_2, y_1$ and $y_2$. If $\alpha_{12} = \beta_{12} = 0$, then $u_1 = 0$ on $M$.

Consequently, given any $\epsilon > 0$ there exist a compact neighborhood $V$ of $M$ and positive constants $\alpha$ and $\beta$ such that $u_1 < \epsilon$ on $V \times [0, \alpha] \times [0, \beta]$ in $(x, y, \alpha_{12}, \beta_{12})$-space. Now, choose $\epsilon > 0$ so that $b_1 g_1(\epsilon) > 1 - s_1^2$. With our choice of $\epsilon > 0$, $u_1 < \epsilon$ implies that $b_1 g_1(u_1) > 1 - s_1^2$ on $V \times [0, \alpha] \times [0, \beta]$. Take $\alpha_{12} < \alpha$, and $\beta_{12} < \beta$.

Define the continuous function $p : \mathbf{R}^3_+ \times \mathbf{R}^3_+ \to \mathbf{R}_+$ by $p(x, y) = y_1$. Notice that, for all points $(x, y) = (x_1, x_2, x_3, y_1, y_2, y_3) \in M$, $p(x, y) = 0$. Moreover, for all points $(x, y) = (x_1, x_2, x_3, y_1, y_2, y_3) \in V \backslash C$, $p(F^2(x, y)) = (b_1 g_1(u_1) + s_1^2) y_1 + b_1(a_1 + s_1) x_1 > p(x, y) = y_1$. As in the proof of Theorem 4.2, all the hypotheses of Theorem 4.1 are satisfied and the proof is done. □

By Theorems 4.2 and 4.3, if the effective growth rate of each of the species in system (1) is sufficiently large, then both competing species persist whenever the competition is very weak (Corollary 4.4).

**Corollary 4.4** *In system (1), let* $\beta_{ii} = 1$, $b_i g_i(0) > 1 - s_i^2$ *for each* $i \in \{1, 2, 3\}$. *Then there exist interspecific competition coefficients* $\alpha_{ij}$ *and* $\beta_{ij}$ *such that both* *species persist. Hence, $F$ is uniformly persistent with respect to* $\mathbf{R}_+ \times \{0\} \times \{0\} \times \mathbf{R}_+ \times \{0\} \times \{0\}$ *and* $\{0\} \times \mathbf{R}_+^2 \times \{0\} \times \mathbf{R}_+^2$.

To prove Corollary 4.4, use Theorems 4.2 and 4.3 with sufficiently small $\alpha_{12}, \alpha_{21}, \beta_{12}$ and $\beta_{21}$.

The persistence results in Theorems 4.2 and 4.3 and Corollary 4.4 can be compared with persistence results in the classic Lotka-Volterra system of ordinary differential equations. As in Theorems 4.2 and 4.3 and Corollary 4.4, persistence in Lotka-Volterra competition requires that intraspecific competition be greater than interspecific competition.

## 5 Extinction of species 1

In system (1), species 1 is confined to patch $A$ while the small size individuals of its competitor, species 2, diffuse between patch $A$ and patch $B$ as they mature into the larger size class. If patch $B$ is ideal for the growth and reproduction of species 2, then species 2 could dominate the system and drive species 1 to extinction. Here, we focus on sufficient conditions that guarantee the extinction of species 1 in system (1).

**Definition 2** In system (1), species 2 dominates 1 if there exists a positive constant $k \in (0,1]$ such that $a_1^k < \min\{a_2, a_3\}$, $b_1^k < \min\{b_2, b_3\}$, $s_1^k < \min\{s_2, s_3\}$ and $[g_1(u_1(x,y))]^k < \min\{dg_2(u_2(x,y)) + (1-d)g_3(\alpha_{33}x_3 + y_3),\ (1-d)g_2(u_2(x,y)) + dg_3(\alpha_{33}x_3 + y_3)\}$ for all points $(x,y) = (x_1, x_2, x_3, y_1, y_2, y_3) \in T$, where $u_i(x,y) = \sum_{j=1}^{2} \{\alpha_{ij}x_j + \beta_{ij}y_j\}$ for each $i \in \{1,2\}$.

In Definition 2, $a_1^k < \min\{a_2, a_3\}$ (respectively, $s_1^k < \min\{s_2, s_3\}$) means the fraction of the small (respectively, large) size individuals of species 1 that remain in the same class after one generation, raised to the positive power $k$, is less than the fraction of the small (respectively, large) size individuals of species 2. Also in Definition 2, $b_1^k < \min\{b_2, b_3\}$ means the fraction of the small size individuals of species 1 that move into the large size class after one generation, raised to the power $k$, is less than the fraction of the small size individuals of species 2 that move into the large size class. If, in addition, the growth rate of species 1 raised to the power $k$ is always less than a linear combination of the growth rate of species 2 in patch $A$ and patch $B$, then species 2 drives species 1 to extinction (Theorem 5.2). Consequently, Definition 2 is a sufficient condition for the dominance of species 2 over 1.

We will use the following simple but important inequality from [**13**], to obtain a Lyapunov function for system (1).

**Lemma 5.1** *If $p_i$, $q_i \geq 0$ and each $q_i \neq 0$, then*

$$\frac{\sum_{i=1}^{n} p_i}{\sum_{i=1}^{n} q_i} \leq \max_{j \in \{1,2,\cdots,n\}} \left(\frac{p_j}{q_j}\right).$$

**Theorem 5.2** *In system (1), if species 2 dominates 1, then $p_2 > 0$ or $p_3 > 0$ or $q_2 > 0$ or $q_3 > 0$ at the point $(p,q) = (p_1, p_2, p_3, q_1, q_2, q_3)$ in $\mathbf{R}_+^3 \times \mathbf{R}_+^3$ implies that $\omega((p,q)) \subset \{(x,y) = (x_1, x_2, x_3, y_1, y_2, y_3) \in \mathbf{R}_+^3 \times \mathbf{R}_+^3 | x_1 = y_1 = 0\}$. Hence, species 2 drives 1 to extinction.*

**Proof** First, we perform a change of variables on system (1). Let $\overline{x}_i(n) = b_i x_i(n)$ and $\overline{y}_i(n) = y_i(n)$ for each $i \in \{1,2,3\}$. Then system (1) transforms to the following system:

$$\left.\begin{array}{l}\overline{x}_1(n+1) = b_1\overline{y}_1(n)g_1\Big(\sum_{j=1}^{2}\{\overline{\alpha}_{1j}\overline{x}_j(n) + \overline{\beta}_{1j}\overline{y}_j(n)\}\Big) + a_1\overline{x}_1(n)\\ \overline{x}_2(n+1) = b_2\overline{y}_2(n)g_2\Big(\sum_{j=1}^{2}\{\overline{\alpha}_{2j}\overline{x}_j(n) + \overline{\beta}_{2j}\overline{y}_j(n)\}\Big) + a_2\overline{x}_2(n)\\ \overline{x}_3(n+1) = b_3\overline{y}_3(n)g_3(\overline{\alpha}_{33}\overline{x}_3(n) + \overline{y}_3(n)) + a_3\overline{x}_3(n)\\ \overline{y}_1(n+1) = \overline{x}_1(n) + s_1\overline{y}_1(n)\\ \overline{y}_2(n+1) = \overline{x}_2(n) - d(\overline{x}_2(n) - \overline{x}_3(n)) + s_2\overline{y}_2(n)\\ \overline{y}_3(n+1) = \overline{x}_3(n) + d(\overline{x}_2(n) - \overline{x}_3(n)) + s_3\overline{y}_3(n)\end{array}\right\} \quad (3)$$

where $\overline{\alpha}_{ij} = \frac{\alpha_{ij}}{b_j}$ and $\overline{\beta}_{ij} = \beta_{ij}$.

Define $\overline{F} : \mathbf{R}^3_+ \times \mathbf{R}^3_+ \to \mathbf{R}^3_+ \times \mathbf{R}^3_+$ by $\overline{F}_i(x,y) = b_i y_i g_i(\sum_{j=1}^{2}\{\overline{\alpha}_{ij}x_j + \overline{\beta}_{ij}y_j\}) + a_i x_i$ for each $i \in \{1,2\}$, $\overline{F}_3(x,y) = b_3 y_3 g_3(\overline{\alpha}_{33}x_3 + y_3) + a_3 x_3$, $\overline{F}_4(x,y) = x_1 + s_1 y_1$, $\overline{F}_5(x,y) = x_2 - d(x_2 - x_3) + s_2 y_2$ and $\overline{F}_6(x,y) = x_3 + d(x_2 - x_3) + s_3 y_3$. The iterates of the map $\overline{F}$ are equivalent to the density sequence generated by system (3).

By Corollary 3.5, the $\omega$-limit set using $F$ of every point in $\mathbf{R}^3_+ \times \mathbf{R}^3_+$ is a nonempty subset of the compact invariant set $T$. The change of variables is a topological conjugacy and gives a corresponding set $\overline{T}$ for $\overline{F}$. Therefore, under $\overline{F}$ iterations, every point eventually enters and remains inside small neighborhoods of $\overline{T}$. Note that the change of variables preserves the axes.

If $(x,y) \in \overline{T}$, then $(\frac{x_1}{b_1}, \frac{x_2}{b_2}, \frac{x_3}{b_3}, y_1, y_2, y_3) \in T$. Thus, by 2 dominance,

$$\left[g_1\left(\sum_{j=1}^{2}\{\overline{\alpha}_{1j}x_j + \overline{\beta}_{1j}y_j\}\right)\right]^k = \left[g_1\left(\sum_{j=1}^{2}\alpha_{1j}\frac{x_j}{b_j} + \beta_{1j}y_j\}\right)\right]^k$$

$$< \min\left\{dg_2\left(\sum_{j=1}^{2}\{\alpha_{2j}\frac{x_j}{b_j} + \beta_{2j}y_j\}\right) + (1-d)g_3(\alpha_{33}\frac{x_3}{b_3} + y_3),\right.$$

$$\left.(1-d)g_2\left(\sum_{j=1}^{2}\{\alpha_{2j}\frac{x_j}{b_j} + \beta_{2j}y_j\}\right) + dg_3(\alpha_{33}\frac{x_3}{b_3} + y_3)\right\}$$

$$= \min\left\{dg_2\left(\sum_{j=1}^{2}\{\overline{\alpha}_{2j}x_j + \overline{\beta}_{2j}y_j\}) + (1-d)g_3(\overline{\alpha}_{33}x_3 + y_3\right),\right.$$

$$\left.(1-d)g_2\left(\sum_{j=1}^{2}\{\overline{\alpha}_{2j}x_j + \overline{\beta}_{2j}y_j\}\right) + dg_3(\overline{\alpha}_{33}x_3 + y_3)\right\}.$$

By Corollary 3.7, if $p_2 > 0$ or $p_3 > 0$ or $q_2 > 0$ or $q_3 > 0$ then for each $i \in \{2,3,5,6\}$, $\overline{F}_i^m(x,y) > 0$ for all $m \geq 3$. Let $(p,q) = (p_1,p_2,p_3,q_1,q_2,q_3) \in \overline{T}$ with $p_2, p_3, q_2$ and $q_3 > 0$. By Corollary 3.5, $\omega((p,q))$ using $\overline{F}$ is a nonempty subset of $\overline{T}$. Let $(v,w) = (v_1,v_2,v_3,w_1,w_2,w_3) \in \omega((p,q))$. Then there exists $n_i \to +\infty$

such that $\overline{F}^{n_i}(p,q) \to (v,w)$. To establish the result, we show that $\overline{F}_1^{n_i}(p,q) \to 0$ and $\overline{F}_4^{n_i}(p,q) \to 0$ as $n_i \to +\infty$.

Define the map $V : \overline{T} \to \mathbf{R}_+$ by

$$V(x,y) = \max\left\{\frac{x_1^k}{(1-d)x_2 + dx_3}, \frac{x_1^k}{dx_2 + (1-d)x_3}, \frac{y_1^k}{y_2}, \frac{y_1^k}{y_3}\right\},$$

where each $x_2, x_3, y_2$ and $y_3$ is positive and $k \in (0,1]$ is obtained from the fact that species 2 dominates 1. Now, we show that $V : \overline{T} \to \mathbf{R}_+$ is a Lyapunov function under $\overline{F}$ iteration, so it decreases monotonically on orbits. To reduce the notation, let $\overline{u}_1(x,y) = \sum_{j=1}^{2}\{\overline{\alpha}_{1j}x_j + \overline{\beta}_{1j}y_j\}, \overline{u}_2(x,y) = \sum_{j=1}^{2}\{\overline{\alpha}_{2j}x_j + \overline{\beta}_{2j}y_j\}$ and $\overline{u}_3(x,y) = \overline{\alpha}_{33}x_3 + y_3$.

Notice that

$$V(\overline{F}(x,y)) = \max\left\{\frac{[\overline{F}_1(x,y)]^k}{(1-d)\overline{F}_2(x,y) + d\overline{F}_3(x,y)}, \right.$$

$$\left.\frac{[\overline{F}_1(x,y)]^k}{d\overline{F}_2(x,y) + (1-d)\overline{F}_3(x,y)}, \frac{[\overline{F}_4(x,y)]^k}{\overline{F}_5(x,y)}, \frac{[\overline{F}_4(x,y)]^k}{\overline{F}_6(x,y)}\right\}.$$

Moreover,

$$\frac{[\overline{F}_1(x,y)]^k}{(1-d)\overline{F}_2(x,y) + d\overline{F}_3(x,y)}$$

$$= \frac{[b_1y_1g_1(\overline{u}_1(x,y)) + a_1x_1]^k}{(1-d)[b_2y_2g_2(\overline{u}_2(x,y)) + a_2x_2] + d[b_3y_3g_3(\overline{u}_3(x,y)) + a_3x_3]}$$

$$\leq \frac{b_1^k y_1^k [g_1(\overline{u}_1(x,y))]^k + a_1^k x_1^k}{(1-d)b_2y_2g_2(\overline{u}_2(x,y)) + (1-d)a_2x_2 + db_3y_3g_3(\overline{u}_3(x,y)) + da_3x_3}$$

$$\leq \max\left\{\frac{b_1^k y_1^k [g_1(\overline{u}_1(x,y))]^k}{(1-d)b_2y_2g_2(\overline{u}_2(x,y)) + db_3y_3g_3(\overline{u}_3(x,y))}, \frac{a_1^k x_1^k}{(1-d)a_2x_2 + da_3x_3}\right\}$$

by Lemma 5.1 and the fact that $k \in (0,1]$.

Since species 2 dominates 1, if $y_1 > 0$, then

$$\frac{b_1^k y_1^k [g_1(\overline{u}_1(x,y))]^k}{(1-d)b_2y_2g_2(\overline{u}_2(x,y)) + db_3y_3g_3(\overline{u}_3(x,y))}$$

$$< \min\left\{\frac{b_1^k y_1^k [dg_2(\overline{u}_2(x,y)) + (1-d)g_3(\overline{u}_3(x,y))]}{(1-d)b_2y_2g_2(\overline{u}_2(x,y)) + db_3y_3g_3(\overline{u}_3(x,y))}, \right.$$

$$\left.\frac{b_1^k y_1^k [(1-d)g_2(\overline{u}_2(x,y)) + dg_3(\overline{u}_3(x,y))]}{(1-d)b_2y_2g_2(\overline{u}_2(x,y)) + db_3y_3g_3(\overline{u}_3(x,y))}\right\}$$

$$< \min\left\{\frac{y_1^k [dg_2(\overline{u}_2(x,y)) + (1-d)g_3(\overline{u}(x,y))]}{(1-d)y_2g_2(\overline{u}_2(x,y)) + dy_3g_3(\overline{u}(x,y))}, \right.$$

$$\left.\frac{y_1^k [(1-d)g_2(\overline{u}_2(x,y)) + dg_3(\overline{u}_3(x,y))]}{(1-d)y_2g_2(\overline{u}_2(x,y)) + dy_3g_3(\overline{u}_3(x,y))}\right\}.$$

On using Lemma 5.1, we obtain that

$$\frac{b_1^k y_1^k [(g_1(\overline{u}_1(x,y))]^k}{(1-d)b_2 y_2 g_2(\overline{u}_2(x,y)) + db_3 y_3 g_3(\overline{u}_3(x,y))} < \max\left\{\frac{y_1^k}{y_2}, \frac{y_1^k}{y_3}\right\}.$$

By the dominance condition, if $x_1 > 0$, then

$$\frac{a_1^k x_1^k}{(1-d)a_2 x_2 + da_3 x_3} < \frac{x_1^k}{(1-d)x_2 + dx_3}.$$

Hence,

$$\frac{[\overline{F}_1(x,y)]^k}{(1-d)\overline{F}_2(x,y) + d\overline{F}_3(x,y)} < \max\left\{\frac{x_1^k}{(1-d)x_2+dx_3}, \frac{x_1^k}{dx_2+(1-d)x_3}, \frac{y_1^k}{y_2}, \frac{y_1^k}{y_3}\right\}.$$

Proceeding exactly as with

$$\frac{[\overline{F}_1(x,y)]^k}{(1-d)\overline{F}_2(x,y) + d\overline{F}_3(x,y)},$$

we obtain that

$$\frac{[\overline{F}_1(x,y)]^k}{d\overline{F}_2(x,y) + (1-d)\overline{F}_3(x,y)} < \max\left\{\frac{x_1^k}{(1-d)x_2+dx_3}, \frac{x_1^k}{dx_2+(1-d)x_3}, \frac{y_1^k}{y_2}, \frac{y_1^k}{y_3}\right\}.$$

Since $k \in (0,1]$, we obtain that

$$\frac{[\overline{F}_4(x,y)]^k}{\overline{F}_5(x,y)} = \frac{[x_1 + s_1 y_1]^k}{(1-d)x_2 + dx_3 + s_2 y_2} \leq \frac{x_1^k + s_1^k y_1^k}{(1-d)x_2 + dx_3 + s_2 y_2}.$$

On using the dominance condition and Lemma 5.1, we obtain that

$$\frac{[\overline{F}_4(x,y)]^k}{\overline{F}_5(x,y)} < \max\left\{\frac{x_1^k}{(1-d)x_2+dx_3}, \frac{x_1^k}{dx_2+(1-d)x_3}, \frac{y_1^k}{y_2}, \frac{y_1^k}{y_3}\right\}.$$

Proceeding exactly as with $\dfrac{[\overline{F}_4(x,y)]^k}{\overline{F}_5(x,y)}$, we also obtain that

$$\frac{[\overline{F}_4(x,y)]^k}{\overline{F}_6(x,y)} < \max\left\{\frac{x_1^k}{(1-d)x_2+dx_3}, \frac{x_1^k}{dx_2+(1-d)x_3}, \frac{y_1^k}{y_2}, \frac{y_1^k}{y_3}\right\}.$$

Thus, for all points $(x,y) \in \overline{T}$ satisfying $x_2, x_3, y_2, y_3 > 0$ and $x_1, y_1 > 0$, we have that $V(\overline{F}(x,y)) < V(x,y)$.

Recall that the point $(v,w) = (v_1, v_2, v_3, w_1, w_2, w_3) \in \omega((p,q))$. If $v_2, v_3, w_2, w_3 > 0$ and $v_1$, $w_1 > 0$, then $V(\overline{F}(v,w)) < V(v,w)$. However, this is impossible for an $\omega$-limit point. Therefore, $v_1 = w_1 = 0$ if $v_2, v_3, w_2, w_3 > 0$. By Corollary 3.7, if $v_2$ or $v_3$ or $w_2$ or $w_3 > 0$ then for each $i \in \{2,3,5,6\}$, $\overline{F}_i^3(v,w) > 0$. Hence, if $v_2$ or $v_3$ or $w_2$ or $w_3 > 0$ and $v_1$ or $w_1 > 0$, then $V(\overline{F}^4(v,w)) < V(\overline{F}^3(v,w))$, a contradiction. Hence, $v_1 = w_1 = 0$ if $v_2$ or $v_3$ or $w_2$ or $w_3 > 0$.

If $v_2 = v_3 = w_2 = w_3 = 0$ and $v_1$ or $w_1 > 0$, then the sequence $\{V(\overline{F}^{n_i}(p,q))\}$ is unbounded, a contradiction. Hence, $\omega((p,q)) \subset \{(x,y) = (x_1,x_2,x_3,y_1,y_2,y_3) \in \mathbf{R}_+^3 \times \mathbf{R}_+^3 | x_1 = y_1 = 0\}$ for all points in $\overline{T}$.

Now we consider points that do not enter the compact invariant set $\overline{T}$. Continuity of the $g_i$'s guarantees the existence of a neighborhood of $T$ in $\mathbf{R}_+^3 \times \mathbf{R}_+^3$ on which the inequalities given by species 2 dominates 1 hold true. By Corollary 3.5, under $F$ iteration every point eventually enters and remains in this neighborhood. Our change of coordinates gives a corresponding neighborhood of $\overline{T}$ using $\overline{F}$. The Lyapunov function $V$ is eventually decreasing on the $\overline{F}$ orbit of every point

$(p,q)=(p_1,p_2,p_3,q_1,q_2,q_3)\in \mathbf{R}^3_+\times\mathbf{R}^3_+$ with $p_2$ or $p_3$ or $q_2$ or $q_3>0$. Therefore, as in the first situation, $\omega((p,q))\subset\{(x,y)=(x_1,x_2,x_3,y_1,y_2,y_3)\in\mathbf{R}^3_+\times\mathbf{R}^3_+|x_1=y_1=0\}$. It is clear from our change of coordinates that under $F$ we also have $\omega((p,q))\subset\{(x,y)=(x_1,x_2,x_3,y_1,y_2,y_3)\in\mathbf{R}^3_+\times\mathbf{R}^3_+|x_1=y_1=0\}$. This establishes the result. □

## 6 Extinction of species 2

If, in patch $A$, the growth rate of the sedentary species 1 always exceeds that of species 2 and patch $B$ is not suitable for the survival of species 2, then species 1 could dominate the system and drive species 2 to extinction. In this section, we obtain conditions that guarantee the extinction of species 2.

**Definition 3** In system (1), species 1 dominates 2 if there exists a positive constant $k\in[1,\infty)$ such that $a_1^k>\max\{a_2,a_3\}$, $b_1^k>\max\{b_2,b_3\}$, $s_1^k>\max\{s_2,s_3\}$ and $[g_1(u_1(x,y))]^k>\max\{dg_2(u_2(x,y))+(1-d)g_3(\alpha_{33}x_3+y_3),\ (1-d)g_2(u_2(x,y))+dg_3(\alpha_{33}x_3+y_3)\}$ for all points $(x,y)=(x_1,x_2,x_3,y_1,y_2,y_3)\in T$, where $u_i(x,y)=\sum_{j=1}^{2}\{\alpha_{ij}x_j+\beta_{ij}y_j\}$ for each $i\in\{1,2\}$.

In Definition 3, $a_1^k>\max\{a_2,a_3\}$ (respectively, $s_1^k>\max\{s_2,s_3\}$) means the fraction of the small (respectively, large) size individuals of species 1 that remain in the same class after one generation, raised to the power $k$, is bigger than the fraction of the small (respectively, large) size individuals of species 2. The condition $b_1^k>\max\{b_2,b_3\}$ means the fraction of the small size individuals of species 1 that move into the large size class after one generation raised to the power $k$, is bigger than the fraction of the small size individuals of species 2 that move into the large size class. If, in addition, the growth rate of species 1 raised to the power $k$ is always bigger than a linear combination of the growth rate of species 2 in patch $A$ and patch $B$, then species 1, the sedentary species, drives species 2 to extinction (Theorem 6.1). In comparison to Definition 2, Definition 3 reverses the roles as to which species dominates.

**Theorem 6.1** *In system (1), if species 1 dominates 2, then $p_1>0$ or $q_1>0$ at the point $(p,q)=(p_1,p_2,p_3,q_1,q_2,q_3)$ in $\mathbf{R}^3_+\times\mathbf{R}^3_+$ implies that $\omega((p,q))\subset\{(x,y)=(x_1,x_2,x_3,y_1,y_2,y_3)\in\mathbf{R}^3_+\times\mathbf{R}^3_+|x_2=x_3=y_2=y_3=0\}$. Hence, species 1 drives 2 to extinction.*

**Proof** To prove Theorem 6.1, proceed exactly as in the proof of Theorem 5.2 with the Lyapunov function $V:\overline{T}\to\mathbf{R}_+$ defined by

$$V(x,y)=\max\left\{\frac{(1-d)x_2+dx_3}{x_1^k},\ \frac{dx_2+(1-d)x_3}{x_1^k},\ \frac{y_2}{y_1^k},\ \frac{y_3}{y_1^k}\right\},$$

where $x_1,\ y_1>0$ and $k\in[1,\infty)$ is given by the fact that species 1 dominates 2. Then

$$V(\overline{F}(x,y))=\max\left\{\frac{(1-d)\overline{F}_2(x,y)+d\overline{F}_3(x,y)}{[\overline{F}_1(x,y)]^k},\right.$$

$$\left.\frac{d\overline{F}_2(x,y)+(1-d)\overline{F}_3(x,y)}{[\overline{F}_1(x,y)]^k},\ \frac{\overline{F}_5(x,y)}{[\overline{F}_4(x,y)]^k},\ \frac{\overline{F}_6(x,y)}{[\overline{F}_4(x,y)]^k}\right\}.$$

Using the dominance condition with $(x,y)\in\overline{T}$ and the fact that the constant $k\in[1,\infty)$ we obtain

$$\frac{(1-d)\overline{F}_2(x,y)+d\overline{F}_3(x,y)}{[\overline{F}_1(x,y)]^k}$$
$$\leq \frac{(1-d)b_2y_2g_2(\overline{u}_2(x,y))+(1-d)a_2x_2+db_3y_3g_3(\overline{u}_3(x,y))+da_3x_3}{b_1^ky_1^k[g_1(\overline{u}_1(x,y))]^k+a_1^kx_1^k}$$
$$< \frac{a_1^k[(1-d)x_2+dx_3]+b_1^k[(1-d)y_2g_2(\overline{u}_2(x,y))+dy_3g_3(\overline{u}_3(x,y))]}{a_1^kx_1^k+b_1^ky_1^k[\max\{D_1\,,\ D_2\}]},$$

where $D_1 = dg_2(\overline{u}_2(x,y)) + (1-d)g_3(\overline{u}_3(x,y))$ and $D_2 = (1-d)g_2(\overline{u}_2(x,y)) + dg_3(\overline{u}_3(x,y))$.

On using Lemma 5.1, we obtain that

$$\frac{(1-d)\overline{F}_2(x,y)+d\overline{F}_3(x,y)}{[\overline{F}_1(x,y)]^k} < \max\left\{\frac{(1-d)x_2+dx_3}{x_1^k}, \frac{dx_2+(1-d)x_3}{x_1^k}, \frac{y_2}{y_1^k}, \frac{y_3}{y_1^k}\right\}.$$

Proceeding exactly as with $\frac{(1-d)\overline{F}_2(x,y)+d\overline{F}_3(x,y)}{[\overline{F}_1(x,y)]^k}$, we obtain

$$\frac{d\overline{F}_2(x,y)+(1-d)\overline{F}_3(x,y)}{[\overline{F}_1(x,y)]^k} < \max\left\{\frac{(1-d)x_2+dx_3}{x_1^k}, \frac{dx_2+(1-d)x_3}{x_1^k}, \frac{y_2}{y_1^k}, \frac{y_3}{y_1^k}\right\}.$$

Also, $\frac{\overline{F}_5(x,y)}{[\overline{F}_4(x,y)]^k} = \frac{(1-d)x_2+dx_3+s_2y_2}{[x_1+s_1y_1]^k} \leq \frac{(1-d)x_2+dx_3+s_2y_2}{x_1^k+s_1^k y_1{}^k}$. On using Lemma 5.1 and the dominance condition, we obtain

$$\frac{\overline{F}_5(x,y)}{[\overline{F}_4(x,y)]^k} < \max\left\{\frac{(1-d)x_2+dx_3}{x_1^k}, \frac{dx_2+(1-d)x_3}{x_1^k}, \frac{y_2}{y_1^k}, \frac{y_3}{y_1^k}\right\}.$$

Similarly,

$$\frac{\overline{F}_6(x,y)}{[\overline{F}_4(x,y)]^k} < \max\left\{\frac{(1-d)x_2+dx_3}{x_1^k}, \frac{dx_2+(1-d)x_3}{x_1^k}, \frac{y_2}{y_1^k}, \frac{y_3}{y_1^k}\right\}.$$

Therefore, $V(F(x,y)) < V(x,y)$. To complete the proof, proceed exactly as in the proof of Theorem 5.2. □

## 7 Application

In the literature on population dynamics, exponential functions are used to provide a fit for a wide variety of data ([**3**], [**4**], [**5**] and [**6**], [**7**], [**8**], [**9**], [**11**], [**12**], and [**13**], [**17**], [**18**] [**19**], [**20**], and [**21**]). In this section, we use a specific example with exponential functions as growth rates to illustrate applications of our theorems. The example is system (1) with each growth function being exponential.

$$\left.\begin{aligned}
x_1(n+1) &= y_1(n)\exp\Big(p_1 - q_1\Big[\sum_{j=1}^{2}\{\alpha_{1j}x_j(n)+\beta_{1j}y_j(n)\}\Big]\Big) + a_1x_1(n)\\
x_2(n+1) &= y_2(n)\exp\Big(p_2 - q_2\Big[\sum_{j=1}^{2}\{\alpha_{2j}x_j(n)+\beta_{2j}y_j(n)\}\Big]\Big) + a_2x_2(n)\\
x_3(n+1) &= y_3(n)\exp(p_3 - q_3[\alpha_{33}x_3(n)+y_3(n)]) + a_3x_3(n)\\
y_1(n+1) &= b_1x_1(n) + s_1y_1(n)\\
y_2(n+1) &= b_2x_2(n) - d[b_2x_2(n) - b_3x_3(n)] + s_2y_2(n)\\
y_3(n+1) &= b_3x_3(n) + d[b_2x_2(n) - b_3x_3(n)] + s_3y_3(n).
\end{aligned}\right\} \quad (4)$$

**Example 1** Set the following parameter values in system (4). For each $i \in \{1,2,3\}$, $q_i = 1$, $a_i = 0.5$, $b_i = 0.4$, $s_i = 0.4$, $\beta_{ii} = 1$, $p_1 = 1$, $p_2 = 1.2$, $p_3 = 1.1$ and $d = 0.5$.

In Example 1, $b_i g_i(0) = 0.4\ e^{p_i} > 1 - s_i^2 = 0.84$. By Corollary 4.4, if in addition the interspecific competition coefficients $\alpha_{11}, \alpha_{12}, \alpha_{21}, \alpha_{22}, \alpha_{33}, \beta_{12}$ and $\beta_{21}$ are sufficiently small, then the two species continue to persist in system (4). With the above choice of constants, if $\alpha_{11} = \alpha_{12} = \alpha_{21} = \alpha_{22} = \alpha_{33} = \beta_{12} = \beta_{21} = 0$, then system (4) has a stable interior fixed point at (1.9315, 2.2662, 2.0506, 1.2877, 1.4389, 1.4389). Our computer experiments confirm the persistence of the two species with every positive initial condition converging to the fixed point under iteration. On increasing the competition coefficients from zero so that $\alpha_{11} = \alpha_{12} = \alpha_{21} = \alpha_{22} = \alpha_{33} = \beta_{12} = \beta_{21} = 1.2$, a dramatic bifurcation occurs. The system changes from the persistence of all the competing species to the extinction of species 1, the species with the smaller carrying capacity. By our numerical calculations, the system has a stable equilibrium fixed point at (0, 0.7904, 0.7504, 0, 0.5136, 0.5136).

To show that the $\omega$-limit set is not necessarily a fixed point whenever exclusionary dynamics occurs, we fix all the parameters at their current values and increase the parameters $p_2$ and $p_3$ to $p_2 = 4.5$ and $p_3 = 5.5$. With our new choice of constants, the system has a stable period 2 orbit that attracts positive population densities. The points in the period 2 orbit are (0, 2.1483, 4.3034, 0, 1.9314, 1.9314) $\to$ (0, 2.9876, 2.5434, 0, 2.0629, 2.0629). The period 2 orbit illustrates that, because the growth functions are exponential functions, by suitable choice of $p_i$ the $\omega$-limit set can be made periodic.

**Example 2** Set the following parameter values in system (4): For each $i,\ j \in \{1,2\}$, $\alpha_{ij} = \beta_{ij} = 1$, $\alpha_{33} = 1$, $a_1 < \min\{a_2, a_3\}$, $b_1 < \min\{b_2, b_3\}$, $s_1 < \min\{s_2, s_3\}$, $q_1 = q_2 = 1$, $d = \frac{1}{2}$ and $e^{p_1} < \frac{1}{2} e^{p_2}$.

In Example 2, for each point $(x, y) = (x_1, x_2, x_3, y_1, y_2, y_3)$ in $T$ we have that

$$\begin{aligned} g_1(u_1(x,y)) &\equiv \exp\left(p_1 - q_1\left[\sum_{j=1}^{2}\{\alpha_{1j}x_j + \beta_{1j}y_j\}\right]\right) \\ &< \frac{1}{2}\exp\left(p_2 - q_2\left[\sum_{j=1}^{2}\{\alpha_{2j}x_j + \beta_{2j}y_j\}\right]\right). \end{aligned}$$

Thus, $g_1(u_1(x,y)) < \frac{1}{2}[g_2(u_2(x,y)) + g_3(\alpha_{33}x_3 + y_3)]$. Hence, species 2 dominates 1. By Theorem 5.2, in Example 2, every initial population density with species 2 present leads to the extinction of species 1.

**Example 3** Set the following parameter values in system (4). For each $i,\ j \in \{1,2\}$, $\alpha_{ij} = \beta_{ij} = 1$, $a_1 > \max\{a_2, a_3\}$, $b_1 > \max\{b_2, b_3\}$, $s_1 > \max\{s_2, s_3\}$, $p_1 = p_2$, $q_1 = q_2$ and $\min\limits_{(x,y) \in T} e^{p_2 - q_2\left(\sum_{j=1}^{2}\{\alpha_{2j}x_j + \beta_{2j}y_j\}\right)} > e^{p_3}$.

In Example 3, for each point $(x, y) = (x_1, x_2, x_3, y_1, y_2, y_3)$ in $T$, we have that

$$e^{p_1 - q_1\left(\sum_{j=1}^{2}\{\alpha_{1j}x_j+\beta_{1j}y_j\}\right)} = e^{p_2 - q_2\left(\sum_{j=1}^{2}\{\alpha_{2j}x_j+\beta_{2j}y_j\}\right)}$$

$$> \frac{1}{2}\,[e^{p_2 - q_2\left(\sum_{j=1}^{2}\{\alpha_{2j}x_j+\beta_{2j}y_j\}\right)} + e^{p_3}]$$

$$\geq \tfrac{1}{2}\,[e^{p_2 - q_2(\sum_{j=1}^{2}\{\alpha_{2j}x_j+\beta_{2j}y_j\})} + e^{p_3 - q_3(\alpha_{33}x_3+y_3)}].$$

Thus, $g_1(u_1(x,y)) > \frac{1}{2}[g_2(u_2(x,y)) + g_3(\alpha_{33}x_3 + y_3)]$. Hence, species 1 dominates 2. By Theorem 6.1, in Example 3, every initial population density with species 1 present leads to the extinction of species 2.

## References

[1] Butler, G. F., Freedman, H. I., Waltman, P. [1986], *Uniformly Persistent Systems*, Proc. Amer. Math. Soc. **96**, 425-430.

[2] Butler, G. F. and Waltman, P. [1986], *Persistence in Dynamical Systems*, J. Diff. Eqns. **65**, 111-138.

[3] Cushing, J. M. [1992], *A size-structured Model for Cannibalism*, Theor. Popl. Biol. **42**, 347-361.

[4] Cushing, J. M. and Li. J. [1989], *On Ebenman's Model for the Dynamics of a Population With Competing Juveniles and Adults*, Bulletin of Math. Biol. **51**, 687-713.

[5] Cushing, J. M. and Li, J. [1991], *Juvenile Versus Adult Competition*, J. Math. Biol. **29**, 457-473.

[6] Cushing, J. M. and Li, J. [1992], *Intra-Specific Competition and Density Dependent Juvenile Growth*, Bulletin of Math. Biol. **54**, 503-519.

[7] Ebenman, B. [1987], *Niche Differences Between Age Classes and Intraspecific Competition in Age-Structured Populations*, J. Theor. Biol. **124**, 25-33.

[8] Ebenman, B. [1988], *Competition Between Age Classes and Population Dynamics,* J. Theor. Biol. **131**, 389-400.

[9] Ebenman, B. and Persson, L. [1988], *Size-structured Populations: Ecology and Evolution,* Springer, Berlin, New York.

[10] Fonda, A. [1988], *Uniformly Persistent Semi-Dynamical Systems*, Proc. Amer. Math. Soc. **104**, 111-116.

[11] Franke, J. E. and Yakubu, A.-A. [1996-A], *Extinction of Species in Age-Structured Discrete Non-Cooperative Systems*, J. Math. Biol. **34**, 442-454.

[12] Franke, J. E. and Yakubu, A.-A. [1996-B], *Extinction and Persistence of Species in Discrete Competitive Systems With a Safe Refuge*, J. Math. Anal. Appl. **203**, 746-761.

[13] Franke, J. E. and Yakubu, A.-A. [1998], *Exclusionary Population Dynamics in Dynamics in Size-Structured, Discrete Competitive Systems*, J. Difference Eqns. Appl., in press.

[14] Freedman, H. I. and So J. W.-H. [1989], *Persistence in Discrete Semidynamical Systems*, SIAM J. Math. Anal. **20**, 930-938.

[15] Hastings, A. [1993], *Complex Interactions Between Dispersal and Dynamics: Lessons from Coupled Logistic Equations*, Ecology **74**, 1362-1372.

[16] Hofbauer, J. and So, J. W.-H. [1987], *Uniform Persistence and Repellors for Maps*, AMS Proc. **107**, 1137-1142.

[17] Hughes, T. P. [1984], *Population Dynamics Based on Individual Size Rather Than Age, a General Model With a Reef Coral Example*, Amer. Natu. **123**, 778-795.

[18] May R. M. [1974], *Biological Populations with Nonoverlapping Generations: Stable Points, Stable Cycles, and Chaos*, Science **186**, 645-647.

[19] May R. M. [1976], *Simple Mathematical Models with very Complicated Dynamics*, Nature **261**, 159-161.

[20] Nisbet, R. M. and Gurney, W. S. C. [1982], *Modelling Fluctuating Populations,* Wiley-Interscience Publication.

[21] Ricker, W. E. [1954], *Stock and Recruitment*, J. Fish. Res. Bd. Can. **11**, 559-623.

[22] Takeuchi, Y. [1986], *Global Stability is Generalized Lotka-Volterra Diffusion Systems*, J. Math. Anal. Appl. **116**, 209-221.

[23] Takeuchi, Y. [1991], *Diffusion-Mediated Persistence In Three-Species Competition Models With Heteroclinic Cyclies*, Math. Biosci. **106**, 111-128.

[24] Takeuchi, Y. [1992], *Refuge-Mediated Global Coexistence of Multiple Competitors on a Single Resource,* Recent Trends in Differential Equations (R. P. Agarwal, ed.), World Scientific, River Edge, NJ, 531-541.

Fields Institute Communications
Volume **21**, 1999

# Global Attractivity in a Nonlocal Reaction-Diffusion Model

**H. I. Freedman**
Department of Mathematical Sciences
University of Alberta
Edmonton, Alberta, Canada T6G 2G1
hfreedma@vega.math.ualberta.ca

**Xiao-Qiang Zhao**
Institute of Applied Mathematics
Academia Sinica
Beijing 100080, China
xzhao@amath6.amt.ac.cn

**Abstract.** A threshold dynamics for a reaction-diffusion model for the dispersal of a population among islands is established in terms of the principal eigenvalue of a coupled nonlocal elliptic system.

## 1 Introduction

Over the last several years, there has been some interest shown in population models which diffuse in a patchy environment. The diffusion can occur in several ways. Models were considered where the population disperse, i.e., they move from one homogeneous patch to another (see, e.g., [**6, 8, 9**]). There were models of diffusion of populations in a single nonhomogeneous patch (see, e.g., [**2, 3, 20, 21, 22**]). Finally, there were models of a single population diffusing among discrete patches with different carrying capacities [**5, 7, 10**], or diffusing in a continuously varying environment [**14**]. The dispersal models involved systems of ordinary differential equations, whereas the diffusion models contained systems of parabolic (reaction-diffusion) partial differential equations.

In the present paper, we extend the diffusion models to consider the case where the diffusion is among islands separated from each other by some distance. The population diffusion may be thought of as airborne and may come from any point on one island and land at any point on the other island, after which it may diffuse on the second island. Since the total population arriving at a given point on the second island is the sum total of such subpopulations from the first island, the model consists of a system of integro-partial differential equations.

1991 *Mathematics Subject Classification.* 35K55, 58F39, 92D25.

HIF Research partially supported by the Natural Sciences and Engineering Research Council of Canada, Grant NSERC OGP4823.

XQZ Research partially supported by the National Science Foundation of China.

The organization of the paper is as follows. In the next section we state some definitions and known results that we require for our analyses, and in Section 3 we introduce our model and discuss its global attractivity.

## 2 Preliminaries

Let $(E, P)$ be an ordered Banach space with positive cone $P$ having nonempty interior int $(P)$. For $x, y \in E$, we write

$$x \geq y \quad \text{if} \quad x - y \in P$$

$$x > y \quad \text{if} \quad x - y \in P \backslash \{0\}$$

$$x \gg y \quad \text{if} \quad x - y \in \text{ int}\,(P).$$

Let $S : P \to P$ be a continuous map. $S$ is said to be *monotone* if for any $x, y \in P$, $x \geq y$ implies $S(x) \geq S(y)$, and *strongly monotone* if for any $x, y \in P$, $x > y$ implies $S(x) \gg S(y)$. $S$ is said to be *subhomogeneous* if for any $x \in P$, $\lambda \in [0, 1]$, $S(\lambda x) \geq \lambda S(x)$, *strictly subhomogeneous* if for any $x \in$ int $(P)$, $\lambda \in (0, 1)$, $S(\lambda x) > \lambda S(x)$, and *strongly subhomogeneous* if for any $x \in$ int $(P)$, $\lambda \in (0, 1)$, $S(\lambda x) \gg \lambda S(x)$. Notice that "subhomogeneous" is also referred to as "sublinear" (see, e.g., [**19**] and [**21**]). Moreover, $S : P \to P$ is said to be *order-compact* if for any $x, y \in P$ with $x < y$, $S([x, y]_E)$ is relatively compact in $E$.

Let $K \in L(E)$ be a linear operator. We call $K$ *strongly positive* if $K(P\backslash\{0\}) \subset$ int $(P)$. If $K$ is compact and strongly positive, we denote its spectral radius by $r(K)$, the existence and positivity of which are guaranteed by the Krein-Rutman Theorem (see, e.g., [**1**], Theorem 3.2, or [**12**], Theorem 7.3).

Let $T(t)$, $t \geq 0$, be the analytic semigroup generated by a linear operator $A : D(A) \to E$. Consider the abstract evolution equation

$$\frac{du}{dt} = Au + g(u), \quad t > 0 \tag{2.1}$$

where $g : E \to E$ is a locally Lipschitz map. For any $\varphi \in E$, let $u(t, \varphi)$ be the unique solution of the integral equation corresponding to (2.1) satisfying $u(0, \varphi) = \varphi$.

**Theorem A** ([**13**], Theorem 3.1) *Assume that for any $\varphi \in E$, $u(t, \varphi)$ exists globally on $[0, \infty)$. Let $\Phi_t : E \to E$ be the solution semiflow defined by $\Phi_t(\varphi) = u(t, \varphi)$, $t \geq 0$. Suppose that*

(1) *$T(t) : E \to E$ is monotone for all $t \geq 0$;*
(2) *$g$ is locally quasimonotone, i.e., for any $u \in E$ there exists a real number $c > 0$ such that $g + cI$ is monotone in some neighborhood of $u$, where $I$ is the identity map on $E$.*

*Then $\Phi_t : E \to E$ is monotone for all $t \geq 0$. Further, if*

(3) *$g : E \to E$ is subhomogeneous (strongly subhomogeneous) and $g(0) \geq 0$,*

*then $\Phi_t : P \to P$ is subhomogeneous (strongly subhomogeneous) for all $t > 0$.*

**Theorem B** ([**13**], Theorem 5.5) *Assume that $S : P \to P$ is monotone and strongly subhomogeneous and suppose there exists a compact nonempty invariant set $K \subset$ int$(P)$. Then $K$ consists of a single fixed point $p \gg 0$, and every trajectory with compact closure in int $(P)$ converges to $p$.*

**Theorem C** ([**21**], Theorem 2.2) *Assume that $S : P \to P$ is monotone and order-compact and suppose that*

(1) $S(0) = 0$, *the Fréchet derivative* $DS(0)$ *exist and is compact and strongly positive, and* $r(DS(0)) \leq 1$;
(2) $S(u) < DS(0)u$ *for any* $u \gg 0$.

*Then* $u = 0$ *is globally attractive for* $S$ *in* $P$.

**Remark 2.1** In ([**21**], Theorem 2.2), the strong monotonicity of $S$ is assumed. However, from the proof of ([**21**], Theorem 2.2), it suffices to assume that for any $u > 0$, $S(u) \gg 0$, which is implied by the strong positivity of $DS(0)$. Indeed, for any fixed $u > 0$, let $\varepsilon \in (0,1]$, then $S(\varepsilon u) = \varepsilon\big[DS(0)u + \frac{0(\varepsilon)}{\varepsilon}\big]$. Since $DS(0)$ is strongly positive, $DS(0)u \gg 0$. Then there exists an $\varepsilon_0 \in (0,1]$ such that for any $\varepsilon \in (0, \varepsilon_0)$, $DS(0)u + \frac{0(\varepsilon)}{\varepsilon} \gg 0$ and hence, $S(u) \geq S(\varepsilon u) = \varepsilon\big[DS(0)u + \frac{0(\varepsilon)}{\varepsilon}\big] \gg 0$.

**Remark 2.2** By ([**21**], Lemma 1), it follows that if $S(0) = 0$ and $DS(0)$ exists, then the strict subhomogeneity and hence strong subhomogeneity implies condition (2) in Theorem C.

## 3 Global attractivity

Consider the island diffusion model with nonlocal reactions

$$\begin{cases} \dfrac{\partial u_1(x_1,t)}{\partial t} & = d_1\Delta u_1(x_1,t) + u_1(x_1,t)g_1(x_1,u_1(x,t)) - h_1(x_1)u_1(x_1,t) \\ & \quad + \displaystyle\int_{\Omega_2} P_{21}(x_1,x_2)h_2(x_2)u_2(x_2,t)dx_2, x_1 \in \Omega_1, \quad t > 0 \\ \dfrac{\partial u_2(x_2,t)}{\partial t} & = d_2\Delta u_2(x_2,t) + u_2(x_2,t)g_2(x_2,u_2(x_2,t)) - h_2(x_2)u_2(x_2,t) \\ & \quad + \displaystyle\int_{\Omega_1} P_{12}(x_1,x_2)h_1(x_1)u_1(x_1,t)dx_1, x \in \Omega_2, \quad t > 0 \\ \dfrac{\partial u_i(x_i,t)}{\partial \nu_i} & = 0, \qquad x_i \in \partial\Omega_i, \quad t > 0, \quad i = 1,2, \end{cases} \tag{3.1}$$

where $d_i > 0$, $i = 1, 2$, $\Delta$ is the Laplacian operator, $\Omega_1$ and $\Omega_2$ represent two islands between which the single species can disperse; $u_i(x_i,t)$ represents the density of the $i^{\text{th}}$ species at point $x_i \in \Omega_i$ and time $t \geq 0$; $g_i(x_i,u_i)$ denotes the specific growth rate of the species in the $i^{\text{th}}$ patch; $h_i(x_i)$ reflects the pressure or desire for the species to leave position $x_i$ in the $i^{\text{th}}$ patch and seek another patch in the total environment; $P_{21}(x_1,x_2)$ denotes the probability that the species, having left point $x_2 \in \Omega_2$, will arrive safely at point $x_1 \in \Omega_1$, and $P_{12}(x_1,x_2)$ denotes the probability that the species, having left from point $x_1 \in \Omega_1$, will arrive safely at point $x_2 \in \Omega_2$. We suppose $\Omega_i$ is a domain in $R^{N_i}$ with smooth boundary $\partial\Omega_i$ and let $\frac{\partial}{\partial \nu_i}$ denote the derivative along the outward normal direction $\nu_i$. Then the Neumann boundary condition $\frac{\partial u_i}{\partial \nu_i} = 0$ is interpreted to mean that there is no diffusion of the species across the boundary $\partial\Omega_i$, $i = 1, 2$.

Due to limited resources, $g_i(x_i,u_i)$ is a decreasing function of $u_i$, eventually becoming negative since the $i^{\text{th}}$ patch can support only a finite population. This is modeled by the following assumption.

($H_1$) $g_i \in C^1(\overline{\Omega}_i \times R^+, R)$, $\frac{\partial g_i}{\partial u_i} < 0$, and there exists $K_i > 0$ such that $g_i(x_i, K_i) \leq 0$, $i = 1, 2$.

We further assume that

($H_2$) $P_{21},\ P_{12} \in C(\overline{\Omega}_1 \times \overline{\Omega}_2, R^+)$, $\int_{\Omega_1} P_{21}(x_1, x_2)dx_1 \leq 1$, and $\int_{\Omega_2} P_{12}(x_1, x_2) dx_2 \leq 1$.

($H_3$) $h_i \in C^1(\overline{\Omega}_i, R^+)$, $i = 1, 2$, and there exist $\overline{x}_i \in \overline{\Omega}_i$, $i = 1, 2$, such that

$$\int_{\Omega_1} P_{12}(x_1, \overline{x}_2)h_1(x_1)dx_1 > 0 \quad \text{and} \quad \int_{\Omega_2} P_{21}(\overline{x}_1, x_2)h_2(x_2)dx_2 > 0.$$

Let $X_i = C(\overline{\Omega}_i)$, and let $X_i^+ = \{\varphi_i \in X_i,\ \varphi_i(x_i) \geq 0,\ x_i \in \overline{\Omega}_i\}$, $i = 1, 2$. For any $\varphi = (\varphi_1, \varphi_2) \in X = X_1 \times X_2$, let $\|\varphi\| = \|\varphi_1\| + \|\varphi_2\|$. Then $X$ is an ordered Banach space with positive cone $X^+ = X_1^+ \times X_2^+$. Clearly, $\operatorname{int}(X^+) = \operatorname{int}(X_1^+) \times \operatorname{int}(X_i^+) \neq \emptyset$. For each $i = 1, 2$, let $A_i^0$ be the differential operator $A_i^0 u_i = d_i \Delta u_i$ defined on the domain

$$D(A_i^0) = \{u_i \in C^2(\Omega_i) \cap C^1(\overline{\Omega}_i) : A_i^0 u_i \in C(\overline{\Omega}_i),\ \frac{\partial u_i}{\partial \nu_i} = 0,\ x_i \in \partial\Omega_i\}.$$

Then the closure $A_i$ of $A_i^0$ in $X_i$ generates an analytic semigroup of bounded linear operator, $T_i(t)$, for $t \geq 0$ such that $u_i(t) = T_i(t)\varphi_i$ is the solution of the abstract linear differential equation in $X_i$ given by

$$\frac{du_i(t)}{dt} = A_i u_i(t), \quad u_i(0) = \varphi_i \in D(A_i).$$

Let $T(t) : X \to X$ be defined by $T(t) = \big(T_1(t), T_2(t)\big)$. Then $T(t)$ is a semigroup of operators on $X$ generated by the operator $A = (A_1, A_2)$ defined on $D(A) = D(A_1) \times D(A_2)$. Then for each $t > 0$, $T(t) : X \to X$ is compact and monotone (see, e.g., [**18**], Section 7.1 and Corollary 7.2.3).

Define $f = (f_1, f_2) : X^+ \to X^+$ by

$$\begin{cases} f_1(\varphi_1, \varphi_2)(\cdot) &= \varphi_1(\cdot)g_1\big(\cdot, \varphi_1(\cdot)\big) - h_1(\cdot)\varphi_1(\cdot) \\ &\qquad + \displaystyle\int_{\Omega_2} P_{21}(\cdot, x_2)h_2(x_2)\varphi_2(x_2)dx_2 \\ f_2(\varphi_1, \varphi_2)(\cdot) &= \varphi_2(\cdot)g_2\big(\cdot, \varphi_2(\cdot)\big) - h_2(\cdot)\varphi_2(\cdot) \\ &\qquad + \displaystyle\int_{\Omega_1} P_{12}(x_1, \cdot)h_1(x_1)\varphi_1(x_1)dx_1. \end{cases} \tag{3.2}$$

Then $f(0, 0) = (0, 0)$. It follows that Equation (3.1) can be viewed as the abstract semilinear ordinary differential equation in $X^+$ given by

$$\begin{cases} \dfrac{du}{dt} = Au + f(u), \quad t > 0 \\ u(0) = \varphi \in X^+. \end{cases} \tag{3.3}$$

Let $B = (B_1, B_2) : X^+ \to X^+$ be defined by

$$B_i(\varphi_1, \varphi_2)(\cdot) = \varphi_i(\cdot)g_i\big(\cdot, \varphi_i(\cdot)\big) - h_i(\cdot)\varphi_i(\cdot),\ i = 1, 2,$$

and let $D = (D_1, D_2) : X^+ \to X^+$ be defined by

$$D_1(\varphi_1, \varphi_2)(\cdot) = \int_{\Omega_2} P_{21}(\cdot, x_2) h_2(x_2) \varphi_2(x_2) dx_2$$

and

$$D_2(\varphi_1, \varphi_2)(\cdot) = \int_{\Omega_1} P_{12}(x_1, \cdot) h_1(x_1) \varphi_1(x_1) dx_1.$$

Clearly, $f(\varphi) = B(\varphi) + D(\varphi), \varphi \in X^+$, and $DX^+ \subset X^+$. We claim that $f : X^+ \to X^+$ is quasipositive in the sense that for any $\varphi \in X^+$, $\lim_{h\to 0^+} \frac{1}{h} d(\varphi + hf(\varphi); X^+) = 0$. Indeed, for any $\varphi \in X^+$ and $h > 0$, since $hD(\varphi) \in X^+$, we have $d(\varphi + hf(\varphi); X^+) = d(\varphi + hB(\varphi) + hD(\varphi); X^+) \leq d(\varphi + hB(\varphi); X^+)$. By the definition of $B : X^+ \to X^+$, we further have $\lim_{h\to 0^+} \frac{1}{h} d(\varphi + hB(\varphi); X^+) = 0$, see, e.g., the proof of ([**16**], Theorem 1), and hence $\lim_{h\to 0^+} \frac{1}{h} d(\varphi + hf(\varphi); X^+) = 0$. By ([**16**], Corollary 4) (see also [**15**], Theorem 8.2.1), it then follows that for every $\varphi \in X^+$, (3.3) admits a unique noncontinuable solution $u(t, \varphi)$ satisfying $u(0, \varphi) = \varphi$ on the maximal interval $[0, \sigma_\varphi)$ and $u(t, \varphi) \in X^+$ for any $t \in [0, \sigma_\varphi)$. Furthermore, $\big(u_1(x_1, t), u_2(x_2, t)\big) = \big(u_1(t, \varphi)(x_1), u_2(t, \varphi)(x_2)\big)$ is also a classical solution of (3.1) for $t > 0$.

We consider the linearized system of (3.1) at its trivial solution $u = (0, 0)$

$$\begin{cases} \dfrac{\partial u_1}{\partial t} &= d_1 \Delta u_1 + \big(g_1(x_1, 0) - h_1(x_1)\big) u_1 \\ & \quad + \displaystyle\int_{\Omega_2} P_{21}(x_1, x_2) h_2(x_2) u_2(x_2, t) dx_2, \\ & \qquad x_1 \in \Omega_1, \quad t > 0 \\ \dfrac{\partial u_2}{\partial t} &= d_2 \Delta u_2 + \big(g_2(x_2, 0) - h_2(x_2)\big) u_2 \\ & \quad + \displaystyle\int_{\Omega_1} P_{12}(x_1, x_2) h_1(x_1) u_1(x_1, t) dx_1, \\ & \qquad x_2 \in \Omega_2, \quad t > 0 \\ \dfrac{\partial u_i}{\partial \nu_i} &= 0, \qquad x_i \in \partial\Omega_i, \quad t > 0, \quad i = 1, 2. \end{cases} \tag{3.4}$$

Let $f^0 = (f_1^0, f_2^0) : X \to X$ be defined by

$$\begin{cases} f_1^0(\varphi_1, \varphi_2)(\cdot) &= \big(g_1(\cdot, 0) - h_1(\cdot)\big)\varphi_1(\cdot) \\ & \quad + \displaystyle\int_{\Omega_2} P_{21}(\cdot, x_2) h_2(x_2) \varphi_2(x_2) dx_2 \\ f_2^0(\varphi_1, \varphi_2)(\cdot) &= \big(g_2(\cdot, 0) - h_2(\cdot)\big)\varphi_2(\cdot) \\ & \quad + \displaystyle\int_{\Omega_1} P_{12}(x_1, \cdot) h_1(x_1) \varphi_1(x_1) dx_1. \end{cases}$$

It easily follows that $f^0 : X \to X$ is a bounded linear operator and satisfies the local quasimonotonicity condition in Theorem A. By the abstract equation of the linear system (3.4) given by

$$\begin{cases} \dfrac{dz}{dt} = Az + f^0(z) \\ \\ z(0) = \varphi \end{cases} \tag{3.5}$$

and Theorem A, $\Phi^0(t)\varphi = z(t,\varphi)$, $t \geq 0$, is a monotone semiflow. Then $\Phi^0(t) : X \to X$ is a positive semigroup, that is, $\Phi^0(t)X^+ \subset X^+$, $t \geq 0$, and by ([**17**], Proposition 3.1.4), $\Phi^0(t)$ is compact for $t > 0$. We further have the following result.

**Lemma 3.1** *$\Phi^0(t) : X \to X$ is strongly positive for each $t > 0$.*

**Proof** Let $\varphi = (\varphi_1, \varphi_2) \in X^+\backslash\{0\}$. Without loss of generality, we assume $\varphi_1 \in X_1^+\backslash\{0\}$. Let $u(t,\varphi) = \big(u_1(t,\varphi),\ u_2(t,\varphi)\big) = \Phi^0(t)\varphi$. Then $u_1(t,\varphi)(x_1) \geq 0$, $u_2(t,\varphi)(x_2) \geq 0$, $t \geq 0$. Clearly, $u_1(x_1,t,\varphi) = u_1(t,\varphi)(x_1)$ satisfies

$$\begin{cases} \dfrac{\partial u_1}{\partial t} \geq d_1 \Delta u_1 + \big(g_1(x_1,0) - h_1(x_1)\big)u_1, x_1 \in \Omega_1, t > 0 \\ \\ \dfrac{\partial u_1}{\partial \nu_1} = 0, x_1 \in \partial\Omega_1, t > 0 \\ \\ u_1(\cdot,0,\varphi) = \varphi_1 > 0. \end{cases} \tag{3.6}$$

By the standard parabolic comparison theorem and maximal principle, it follows that $u_1(x_1,t,\varphi) > 0$ for all $t > 0$ and $x_1 \in \overline{\Omega}_1$, i.e., for each $t > 0$, $u_1(\cdot,t,\varphi) \gg 0$ in $X_1$. We need to prove $u_2(x_2,t,\varphi) = u_2(t,\varphi)(x_2) > 0$ for all $t > 0$ and $x_2 \in \overline{\Omega}_2$. By an argument similar to that for $u_1(x_1,t,\varphi)$, it suffices to prove that for all $t > 0$, $u_2(\cdot,t,\varphi) > 0$ in $X_2$. Assume that, by contradiction, there exists $t_1 > 0$ such that $u_2(\cdot,t_1,\varphi) \equiv 0$. Then we have

$$\frac{\partial u_2(x_2,t_1)}{\partial t} = \int_{\Omega_1} P_{12}(x_1,x_2)h_1(x_1)u_1(x_1,t_1)dx_1.$$

Since $u_1(x_1,t_1) > 0$ for all $x_1 \in \overline{\Omega}_1$, by our assumption ($H_3$),

$$\frac{\partial u_2(\overline{x}_2,t_1)}{\partial t} \geq \min_{x_1 \in \overline{\Omega}_1} u_1(x_1,t_1) \cdot \int_{\Omega_1} P_{12}(x_1,\overline{x}_2)h_1(x_1)dx_1 > 0.$$

On the other hand, since $u_2(\overline{x}_2,t) \geq 0$ for all $t \geq 0$ and $u_2(\overline{x}_2,t_1) = 0$, $\frac{\partial u_2(\overline{x}_2,t_1)}{\partial t} \leq 0$. Now we arrive at a contradiction.

This completes the proof. □

We then consider the eigenvalue problem associated with (3.4)

$$\begin{cases} \lambda w_1(x_1) = d_1\Delta w_1 + \big(g_1(x_1,0) - h_1(x_1)\big)w_1(x_1) \\ \qquad\qquad + \displaystyle\int_{\Omega_2} P_{21}(x_1,x_2)h_2(x_2)w_2(x_2)dx_2, \quad x_1 \in \Omega_1 \\ \lambda w_2(x_2) = d_2\Delta w_2 + \big(g_2(x_2,0) - h_2(x_2)\big)w_2(x_2) \\ \qquad\qquad + \displaystyle\int_{\Omega_1} P_{12}(x_1,x_2)h_1(x_1)w_1(x_1)dx_1, \quad x_2 \in \Omega_2 \\ \dfrac{\partial w_i}{\partial \nu_i} = 0, \qquad x_i \in \partial\Omega_i, \quad i = 1,2. \end{cases} \tag{3.7}$$

By Lemma 3.1, $\Phi^0(t)$ is a compact, strongly positive semigroup on $X^+$. By an argument similar to the proof of ([**18**], Theorem 7.6.1), we have the following result.

**Theorem 3.2** *There exists a real eigenvalue $\lambda_0$ of (3.7) and a corresponding eigenvector $w^0 = (w_1^0, w_2^0) \in int(X^+)$. If $\lambda$ is any other eigenvalue of (3.7), then the real part of $\lambda$, $\mathrm{Re}(\lambda)$, satisfies $\mathrm{Re}(\lambda) < \lambda_0$. Moreover, $\lambda_0$ is an eigenvalue of algebraic multiplicity one and any other nonnegative eigenvector of (3.7) is a positive multiple of $w^0$.*

As usual, we call $\lambda_0$ in Theorem 3.2 the principal eigenvalue of (3.7) and its associated eigenvector is called the principal eigenvector of (3.7).

Now we are in a position to prove our main result on the global dynamics of (3.1).

**Theorem 3.3** *For any $\varphi \in X^+$, the unique solution $u(t,\varphi)$ of (3.1) satisfying $u(0,\varphi) = \varphi$ exists globally on $[0,\infty)$. Furthermore, the following threshold dynamics holds:*

(a) *If $\lambda_0 \le 0$, then for any $\varphi \in X^+$, $\lim_{t\to\infty} u(t,\varphi) = 0$ in $X$, i.e., $\lim_{t\to\infty} u_i(t,\varphi)(x_i) = 0$ uniformly for $x_i \in \overline{\Omega}_i$, $i = 1,2$.*
(b) *If $\lambda_0 > 0$, then (3.1) admits a unique positive steady state $\varphi^* \in int(X^+)$ such that for any $\varphi \in X^+\setminus\{0\}$, $\lim_{t\to\infty} u(t,\varphi) = \varphi^*$ in $X$, i.e., $\lim_{t\to\infty} u_i(t,\varphi)(x_i) = \varphi_i^*(x_i)$ uniformly for $x_i \in \overline{\Omega}_i$, $i = 1,2$.*

**Proof** For any $\varphi \in X^+$, let $u(t,\varphi) = \big(u_1(t), u_2(t)\big)$, $0 \le t < \sigma_\varphi$. As aforementioned, we then have $u_i(t)(x_i) \ge 0$, $x \in \overline{\Omega}_i$, $t \in [0,\sigma_\varphi)$. Let

$$S(t) = \int_{\Omega_1} u_1(t)(x_1)dx_1 + \int_{\Omega_2} u_2(t)(x_2)dx_2, t \in [0,\sigma_\varphi).$$

Then, by system (3.1) and assumption ($H_2$), for all $t \in [0, \sigma_\varphi)$,

$$\begin{aligned}\frac{dS(t)}{dt} &= \int_{\Omega_1} \frac{\partial u_1(x_1,t)}{\partial t}\, dx_1 + \int_{\Omega_2} \frac{\partial u_2(x_2,t)}{\partial t}\, dx_2 \\ &= \int_{\Omega_1} u_1 g_1(x_1,u_1)dx_1 + \int_{\Omega_2} g_2(x_2,u_2)dx_2 \\ &\quad -\int_{\Omega_1} h_1(x_1)u_1dx_1 - \int_{\Omega_2} h_2(x_2)u_2dx_2 \\ &\quad +\int_{\Omega_1}\Big(\int_{\Omega_2} P_{21}(x_1,x_2)h_2(x_2)u_2dx_2\Big)dx_1 \\ &\quad +\int_{\Omega_2}\Big(\int_{\Omega_1} P_{12}(x_1,x_2)h_1(x_1)u_1dx_1\Big)dx_2 \\ &= \int_{\Omega_1} u_1 g_1(x_1,u_1)dx_1 + \int_{\Omega_2} g_2(x_2,u_2)dx_2 \\ &\quad -\int_{\Omega_1} h_1(x_1)u_1dx_1 - \int_{\Omega_2} h_2(x_2)u_2dx_2 \\ &\quad +\int_{\Omega_2}\Big(\int_{\Omega_1} P_{21}(x_1,x_2)dx_1\Big)h_2(x_2)u_2dx_2 \\ &\quad +\int_{\Omega_1}\Big(\int_{\Omega_2} P_{12}(x_1,x_2)dx_2\Big)h_1(x_1)u_1dx_1 \\ &\le \int_{\Omega_1} u_1 g_1(x_1,u_1)dx_1 + \int_{\Omega_2} u_2 g_2(x_2,u_2)dx_2.\end{aligned}$$

By assumption ($H_1$), for each $i = 1, 2$, there exist $M_i > 0$ and $b_i > 0$ such that

$$g_i(x_i,u_i) \le -b_i \quad \text{for all} \quad u_i \ge M_i,\ x_i \in \overline{\Omega}_i.$$

Then there exists $L_i > 0$ such that

$$u_i\Big[g_i(x_i,u_i) + \frac{b_i}{2}\Big] \le L_i \quad \text{for all} \quad u_i \ge 0,\ x_i \in \overline{\Omega}_i.$$

Therefore we have

$$\begin{aligned}\frac{dS(t)}{dt} &\le -\frac{b_1}{2}\int_{\Omega_1} u_1(x_1,t)dx_1 + L_1|\Omega_1| \\ &\quad -\frac{b_2}{2}\int_{\Omega_2} u_2(x_2,t)dx_2 + L_2|\Omega_2| \\ &\le -bS(t) + L,\end{aligned}$$

where $|\Omega_i|$ is the volume of $\Omega_i$, $i = 1, 2$, $b = \min\left(\frac{b_1}{2}, \frac{b_2}{2}\right)$ and $L = L_1|\Omega_1| + L_2|\Omega_2|$. By the standard comparison theorem, it follows that

$$S(t) \le \Big(S(0) - \frac{L}{b}\Big)e^{-bt} + \frac{L}{b}\,. \tag{3.8}$$

Then we have

$$\int_{\Omega_1} u_1(x_1,t)dx_1 + \int_{\Omega_2} u_2(x_2,t)dx_2$$
$$\le \Big(\int_{\Omega_1} \varphi_1(x_1)dx_1 + \int_{\Omega_2} \varphi_2(x_2)dx_2 - \frac{L}{b}\Big)e^{-bt} + \frac{L}{b}, \tag{3.9}$$
$$0 \le t < \sigma_\varphi.$$

Let $\|\varphi_i\|_{L^1} = \int_{\Omega_i} |\varphi_i(x_i)|dx_i$, $i = 1, 2$. Then from (3.9), we have

$$\|u_1(t,\varphi)\|_{L^1} + \|u_2(t,\varphi)\|_{L^1} \le \Big(|\Omega_1|\, \|\varphi_1\| + |\Omega_2|\, \|\varphi_2\| - \frac{L}{b}\Big)e^{-bt} + \frac{L}{b}, \quad 0 \le t \le \sigma_\varphi. \tag{3.10}$$

Clearly, (3.10) implies $u_i(t,\varphi)$ is bounded in $L^1$ norm. For fixed $u_j(x_j,t)$, from (3.1), we see that $u_i(x_i,t)$ $(i \ne j)$ satisfies

$$\frac{\partial u_i}{\partial t} = d_i \Delta u_i + F(x_i, u_i) \tag{3.11}$$

with $|F_i(x_i,u_i)| \le C_1^i |u_i| + C_2^i$, $x_i \in \overline{\Omega}_i$, $C_1^i > 0$, $C_2^i > 0$. By arguments similar to those in ([**4**], Proposition 2.4 and Theorem 2.5), it follows that $u(t,\varphi)$ exists globally on $[0,\infty)$ and there exist a positive continuous function $K_1(\|\varphi\|)$ and a positive constant $K_2$, independent of the initial data $\varphi$, such that

$$\|u_i(t)\| \le K_1(\|\varphi\|), \quad t \ge 0, \tag{3.12}$$

and

$$\limsup_{t\to\infty} \|u_i(t)\| \le K_2. \tag{3.13}$$

Consider the solution semiflow $\Phi(t) : X^+ \to X^+$ defined by $\Phi(t)\varphi = u(t,\varphi)$, $t \ge 0$. By (3.12), it follows that $\Phi(t) : X^+ \to X^+$ is compact for each $t > 0$ (see, e.g., [**18**], Theorem 7.3.1(e)), and by (3.13), $\Phi(t) : X^+ \to X^+$ is point dissipative. Let $f = (f_1, f_2) : X^+ \to X^+$ be defined as in (3.2). Then it is easy to verify that $f$ is locally quasimonotone. By assumption $(\mathrm{H}_1)$, it also easily follows that $f$ is strongly subhomogeneous. Then, by Theorem A, $\Phi(t) : X^+ \to X^+$ is monotone and strongly subhomogeneous for each $t > 0$.

For any given $t_0 > 0$, let $S = \Phi(t_0) : X^+ \to X^+$. Then $S$ is a monotone, compact and strongly subhomogeneous map with $S(0) = 0$. By a standard argument, $DS(0) = \Phi^0(t_0)$, where $\Phi^0(t)$ is the solution semiflow generated by (3.4), and hence, $DS(0)$ is compact and strongly positive. Moreover, by the definition of $\lambda_0$, it follows that $r\big(DS(0)\big) = e^{\lambda_0 t_0}$ (see, e.g., [**18**], Proof of Theorem 7.6.1). In the case where $\lambda_0 \le 0$, we have $r\big(DS(0)\big) \le 1$. By Theorem C and Remark 2.2, for any $\varphi \in X^+$, $\lim_{n\to\infty} S^n(\varphi) = \Phi(nt_0)\varphi = 0$, and hence $\lim_{t\to 0} (\Phi(t)\varphi - \Phi(t)0) = \lim_{t\to\infty} \Phi(t)\varphi = 0$.

In the case where $\lambda_0 > 0$, we have $r\big(DS(0)\big) > 1$. Let $e$ be the principal eigenvector of $DS(0)$ corresponding to the principal eigenvalue $r\big(DS(0)\big)$. By the point dissipativity, compactness and monotonicity of $S$, it easily follows that, (see, e.g., [**21**], Proof of Theorem 2.1), there exist $\varepsilon_0 > 0$ and $q = S(q) \in \operatorname{int}(X^+)$ such that for any $\varepsilon \in (0, \varepsilon_0]$, $S(\varepsilon e) > \varepsilon e$ and $\lim_{n\to\infty} S^n(\varepsilon e) = q$. For any $\varphi \in X^+ \backslash \{0\}$, $S(\varphi) \gg 0$, and hence, there exists some $\varepsilon_1 \in (0, \varepsilon_0]$ such that $S(\varphi) \ge \varepsilon_1 e$. Since $S$ is monotone, $S^{n+1}(\varphi) \ge S^n(\varepsilon_1 e) \ge \varepsilon_1 e$, $n \ge 1$. Then the omega limit set $\omega_S(\varphi)$ of $\varphi$ with respect to $S$ satisfies $\omega_S(\varphi) \ge q \gg 0$ and hence $\omega_S(\varphi) \subset \operatorname{int}(X^+)$.

By the compactness and invariance of $\omega_S(\varphi)$ and Theorem B, there exists $p = S(p) \gg 0$ such that for any $\varphi \in X^+\backslash\{0\}$, $\lim_{n\to\infty} S^n(\varphi) = \lim_{n\to\infty} \Phi(nt_0)\varphi = p$, and hence $\lim_{t\to\infty} (\Phi(t)\varphi - \Phi(t)p) = 0$. Therefore the omega limit set $\omega(\varphi)$ of $\varphi$ with respect to $\Phi(t)$, $t \geq 0$, is a $t_0$-periodic orbit $\Phi(t)p$. By the arbitrariness of $t_0 > 0$, it follows that $\omega(\varphi)$ is an equilibrium $\varphi^*$ of $\Phi(t)$, and hence $p = \varphi^*$. Therefore $\lim_{t\to\infty} \big(\Phi(t)\varphi - \varphi^*\big) = 0$.

This completes the proof. □

Let $\lambda_0^{(i)}$ be the principal eigenvalue of the scalar elliptic eigenvalue problem

$$\begin{cases} \lambda w_i = d_i \Delta w_i + \big(g_i(x_i, 0) - h_i(x_i)\big) w_i, & x \in \Omega_i \\ \dfrac{\partial w_i}{\partial v_i} = 0, & x_i \in \partial\Omega_i. \end{cases} \tag{3.14}$$

Then we have the following result on the uniform persistence of the species at a positive steady state.

**Theorem 3.4** *Let* $\mu_0 = \max(\lambda_0^{(1)}, \lambda_0^{(2)})$. *If* $\mu_0 > 0$, *then the conclusion in Theorem 3.3*(b) *holds.*

**Proof** Let $\lambda_0$ be the principal eigenvalue of (3.7). By Theorem 3.3, it suffices to prove $\lambda_0 \geq \mu_0$. Let $\Phi^0(t) : X \to X$ be the solution semiflow generated by (3.4) and, for each $i = 1, 2$, let $\Phi_i^0(t) : X_i \to X_i$ be the solution semiflow generated by the scalar linear parabolic equation

$$\begin{cases} \dfrac{\partial u_i}{\partial t} = d_i \Delta u_i + \big(g_i(x_i, 0) - h_i(x_i)\big) u_i, & x_i \in \Omega_i,\ t > 0 \\ \dfrac{\partial u_i}{\partial v_i} = 0, & x_i \in \partial\Omega_i,\ t > 0. \end{cases} \tag{3.15}$$

Let $w^0 = (w_1^0, w_2^0)$ be the principal eigenvector corresponding to $\lambda_0$, and let $(u_1(t), u_2(t)) = \Phi^0(t) w^0$. Since $u_i(x_i, t) = u_i(t)(x_i) \geq 0$, $x_i \in \overline{\Omega}_i$, $t \geq 0$, $i = 1, 2$, we have

$$\begin{cases} \dfrac{\partial u_i(x_i, t)}{\partial t} \geq d_i \Delta u_i + \big(g_i(x_i, 0) - h_i(x_i)\big) u_i, & x_i \in \Omega_i,\ t > 0 \\ \dfrac{\partial u_i}{\partial v_i} = 0, & x_i \in \partial\Omega_i,\ t > 0 \\ u_i(\cdot, 0) = w_i^0. \end{cases} \tag{3.16}$$

By the standard comparison theorem,

$$u_i(t) \geq \Phi_i^0(t) w_i^0, \quad t \geq 0, \quad i = 1, 2.$$

That is, $\Phi^0(t) w^0 \geq \big(\Phi_1^0(t) w_1^0, \Phi_2^0(t) w_2^0\big)$, $t \geq 0$. By the definition of $\lambda_0$ and $w^0$, $\Phi^0(t) w^0 = e^{\lambda_0 t} w^0$, $t \geq 0$. Then

$$e^{\lambda_0 t} w_i^0 - \Phi_i^0(t) w_i^0 \geq 0, \quad t \geq 0, \quad i = 1, 2.$$

Since $w_i^0 \gg 0$ and $\Phi_i^0(t) : X_i \to X_i$ is compact and strongly positive in $X_i^+$ for each $t > 0$, by the Krein-Rutman theorem (see, e.g., [**12**], Theorem 7.3),

$$e^{\lambda_0 t} \geq r\big(\Phi_i^0(t)\big) = e^{\lambda_0^{(i)} t}, \quad t > 0, \quad i = 1, 2.$$

Therefore $\lambda_0 \geq \lambda_0^{(i)}$, $i = 1, 2$, i.e., $\lambda_0 \geq \mu_0 = \max(\lambda_0^{(1)}, \lambda_0^{(2)})$.

This completes the proof. □

**Remark 3.1** Let $m_i = \min_{x_i \in \overline{\Omega}_i} \big(g_i(x_i, 0) - h_i(x_i)\big)$. Then $\lambda_0^{(i)} \geq m_i$, $i = 1, 2$. By Theorem 3.4, if either $m_1 > 0$ or $m_2 > 0$, the species will arrive eventually at a unique positive steady state.

**Remark 3.2** By a similar monotone dynamical system approach, one can also discuss the global dynamics of our model under the Robin or Dirichlet boundary conditions.

## References

[1] Amann, H. [1976], *Fixed point equations and nonlinear eigenvalue problems in ordered Banach spaces*, SIAM Review **18**, 620-709.

[2] Cantrell, R. S. and Cosner, C. [1989], *Diffusive logistic equations with indefinite weights: population models in disrupted environments*, Proc. Roy. Soc. Edinburgh **112**, 293-318.

[3] Cantrell, R. S. and Cosner, C. [1991], *Diffusive logistic equations with indefinite weights: population models in disrupted environments II*, SIAM J. Math. Anal. **22**, 1043-1064.

[4] Dung, L. and Smith, H. L. [1996], *A parabolic system modeling microbial competition in an unmixed bio-reactor*, J. Diff. Eqs. **130**, 59-91.

[5] Freedman, H. I. and Krisztin, T. [1992], *Global stability in models of population dynamics with diffusion I. Patchy environments*, Proc. Roy. Soc. Edinburgh **122A**, 69-84.

[6] Freedman, H. I., Rai, B. and Waltman, P. [1986], *Mathematical models of population interactions with dispersal II: Differential survival in a change of habitat*, J. Math. Anal. Appl. **115**, 140-154.

[7] Freedman, H. I., Shukla, J. B. and Takeuchi, Y. [1989], *Population diffusion in a two-patch environment*, Math. Biosci. **95**, 111-123.

[8] Freedman, H. I. and Takeuchi, Y. [1989], *Global stability and predator dynamics in a model of prey dispersal in a patchy environment*, Nonlinear Analysis - TMA **13**, 993-1002.

[9] Freedman, H. I. and Waltman, P. [1977], *Mathematical models of population interactions with dispersal I: Stability of two habitats with and without a predator*, SIAM J. Appl. Math. **32**, 631-648.

[10] Freedman, H. I. and Wu, J. [1992], *Steady state analysis in a model for population diffusion in a multi-patch environment*, Nonlinear Analysis - TMA **18**, 517-542.

[11] Hale, J. K. [1988], *Asymptotic Behavior of Dissipative Systems*, Math. Surveys and Monograph **25**, Amer. Math. Soc., Providence, R.I.

[12] Hess, P. [1991], *Periodic-parabolic Boundary Value Problems and Positivity*, Pitman Research Notes in Math., Series 247, Longman Scientific and Technical, U.K.

[13] Hirsch, M. W. [1994], *Positive equilibria and convergence in subhomogeneous monotone dynamics*, in Comparison Methods and Stability Theory (X. Liu and D. Siegel, eds.), Lecture Notes in Pure and Applied Math. **162**, Marcel Dekker, New York, 169-187.

[14] Krisztin, T. and Freedman, H. I. [1994], *Global stability in models of population dynamics with diffusion II. Continuously varying environments*, Rocky Mountain J. Math. **24**, 155-163.

[15] Martin, R. H. [1976], *Nonlinear Operators and Differential Equations in Banach Spaces*, John Wiley and Sons, New York.

[16] Martin, R. H. and Smith, H. L. [1990], *Abstract functional differential equations and reaction-diffusion systems*, Trans. Amer. Math. Soc. **321**, 1-44.

[17] Pazy, A. [1983], *Semigroups of Linear Operators and Applications to Partial Differential Equations*, Springer-Verlag, New York.

[18] Smith, H. L. [1995], *Monotone Dynamical Systems, An Introduction to the Theory of Competitive and Cooperative Systems*, Math. Surveys and Monographs **41**, Amer. Math. Soc., Providence, R. I.

[19] Takáč, P. [1990], *Asymptotic behavior of discrete-time semigroups of sublinear, strongly increasing mappings with applications in biology*, Nonlinear Analysis - TMA **14**, 35-42.

[20] Zhao, X.-Q. [1995], *Uniform persistence and periodic coexistence states in infinite-dimensional periodic semiflows with applications*, Canadian Appl. Math. Quarterly **3**, 473-495.

[21] Zhao, X.-Q. [1996], *Global attractivity and stability in some monotone discrete dynamical systems*, Bull. Austral. Math. Soc. **53**, 305-324.

[22] Zhao, X.-Q. [1997], *Global asymptotic behavior in a periodic competitor-competitor-mutualist parabolic system*, Nonlinear Analysis - TMA **29**, 551-568.

Fields Institute Communications
Volume **21**, 1999

# Nonlocal Reaction–Diffusion Equations

**Pedro Freitas**
Departamento de Matemática
Instituto Superior Técnico
Av. Rovisco Pais
1096 Lisboa, Codex, Portugal
pfreitas@math.ist.utl.pt

**Abstract.** We present a survey of results and techniques related to nonlocal reaction–diffusion equations. In particular, we describe the complexity of behaviour both in time and in space that results from the introduction of nonlocal terms in this type of equations, with emphasis on the similarities and differences between the two cases. Some open problems and conjectures are mentioned.

## 1 Introduction

There are several biological and physical phenomena that can be modeled by scalar differential equations of the form

$$u_t(x,t) = Eu(x,t) + f(x, u(x,t)),$$

where the space variable $x$ lies in a bounded domain $\Omega$ of $\mathbb{R}^n$ and $E$ denotes a linear elliptic operator of second order. This differential equation is then supplemented with boundary and initial conditions. In these problems, the relation between the unknown $u$ and its derivatives is local in space in the sense that in the equation these functions are all taken at the same point $x$.

There are, however, some cases where a stronger, more global, spatial coupling is present in the phenomenon portrayed and has to be incorporated in the model. One possible way of doing this is by means of an integral term taken over a region in space. Such models have a long history in the mathematical literature going at least as far back as Liouville who published a study of the equation

$$u_t = u_{xx} - b^2 x \int_0^1 x u_t dx$$

in connection with problems in thermo–mechanics [**46**]. According to this paper, the same problem had previously received the attention of Duhamel.

1991 *Mathematics Subject Classification.* Primary 35K57; Secondary 45K05.
Partially supported by FCT (Portugal), under projects PBIC/C/MAT/2139/95 and PRAXIS XXI Project 2/2.1/MAT/199/94.

Other models of this type have been proposed and studied since then, one of the most famous being perhaps the original equation proposed as a model for turbulence in [**9**]:

$$\left\{\begin{array}{l} \dfrac{du}{dt} = P - \dfrac{1}{R}u - \displaystyle\int_0^1 v^2 dx \\ v_t + 2vv_x = \dfrac{1}{R}v_{xx} + uv. \end{array}\right. \tag{1.1}$$

Here $u$ denotes the velocity in a channel due to some applied force $P$, while $v$ stands for the turbulent perturbation of the motion and $R$ is the Reynolds number associated with the viscosity of the fluid. This system has been eclipsed by the better known Burgers equation (which is actually obtained as a simplification of (1.1)), but some of its aspects such as the stability of nontrivial stationary solutions are still not completely understood [**19, 40**].

A scalar equation that combines the better known Burgers equation with a nonlocal term was considered in [**18**]. It takes the form

$$u_t = u_{xx} + \epsilon u u_x + \frac{1}{2}(a\bar{u} + b)u,$$

where the nonlocal term $\bar{u}$ is a power of the $L^p$ norm of $u$.

Over the last fifteen years or so, there have been several papers in the literature dedicated to the study of this type of problem, where quite often the model considered is of the form

$$u_t = \Delta u + f(x, u, \bar{u}), \tag{1.2}$$

where $\Delta$ denotes the Laplacian operator and

$$\bar{u} = \int_\Omega g(y, u(y,t))dy$$

is a nonlocal term that represents the type of strong coupling mentioned above – the case where $g = g(x, y, u)$ has also received some attention (see [**27**] and [**28**]).

These problems include models from biological evolution [**11**], population dynamics ([**10**] and [**30**]), the modelling of electrical devices for which the dependence of the resistivity on the temperature cannot be ignored ([**32**], [**15**], [**43**], [**44**], and [**7**]), models in combustion theory [**6**], laser modelocking [**36**], phase separation in binary mixtures ([**12**], [**24**] and [**52**]), buckling of beams [**4**] and the nonlocal Schrödinger equation arising in quantum mechanics ([**33**] and [**49**]). Other papers that have addressed nonlocal problems in one way or another include [**5**], [**8**] and [**48**].

Another way in which nonlocal terms may arise is when some simplification is introduced in a local problem, which then leads to the appearence of an integral term. In this context, perhaps the most usual case is the concept of a shadow system for systems of reaction–diffusion equations [**37**]. Here a new set of (nonlocal) equations is constructed from the original one, in such a way that the attractors of both systems are in some sense close. More precisely, consider the following system of (local) reaction–diffusion equations

$$\left\{\begin{array}{l} u_t = \Delta u + f(u, v) \\ \delta v_t = D\Delta v + h(u, v) \end{array}\right.$$

together with homogeneous Neumann boundary conditions for $v$. Then, provided that $D$ is large enough and that $f$ and $h$ are such that a global attractor actually exists, the results in [**37**] imply that the global attractor of this system is close to that of the corresponding shadow system given by

$$\begin{cases} u_t = \Delta u + f(u, \bar{v}) \\ \delta \bar{v}_t = \dfrac{1}{|\Omega|} \displaystyle\int_\Omega h(u, \bar{v}) dx, \end{cases}$$

where $\bar{v}$ denotes the spatial average of $v$ over $\Omega$, that is,

$$\bar{v}(t) = \frac{1}{|\Omega|} \int_\Omega v(x,t) dx.$$

Also, if for instance $h(u, v) = g(u) - v$ and one lets $\delta$ go to zero, then this gives (formally) that, on the attractor,

$$\bar{v} = \frac{1}{|\Omega|} \int_\Omega g(u) dx,$$

and substitution in the equation for $u$ yields a scalar nonlocal equation of the form of (1.2).

In spite of its *pedigree* in a wide range of scientific areas, it was only quite recently that a more systematic approach to these equations was tried, in particular by exploring the differences and similarities with the better–studied local reaction–diffusion equation of the form

$$u_t = \Delta u + f(x, u). \tag{1.3}$$

This equation has been widely studied and it is known that its special form imposes severe restrictions on the possible asymptotic behaviour of its solutions, both in time and in space (for a general reference, see, for instance, [**38**]). Two of the main features that restrict the behaviour of solutions in time are the existence of a maximum principle and of a Lyapunov functional.

In the first instance, this implies that the system is monotone in the sense that if $u_0$ and $v_0$ are two distinct initial conditions and $u_0 \geq v_0$, then the corresponding solutions $u$ and $v$ satisfy $u > v$ as long as both solutions exist. Besides being a very useful tool in the study of global existence and blow–up of solutions, this also implies that, generically, bounded trajectories converge to an equilibrium [**50**].

In the second case, the fact that the functional

$$\mathcal{L} = \int_\Omega \frac{1}{2} |\nabla|^2 - \int_0^s f(x, s) ds dx$$

is strictly decreasing along trajectories other than equilibria, implies that solutions either blow–up or converge to a set of equilibria.

Other general properties associated with Equation (1.3) which help in giving it a rich structure refer to the corresponding linear operator $A$ that is obtained in the linearization around a stationary solution. This is a self–adjoint operator which furthermore satisfies the following positivity condition. Let $h$ be a nonnegative function not identically zero and assume that $A$ is invertible. Then the solution $u$ of the linear equation $(A + \alpha I)u = h$ is strictly positive in $\Omega$ for all large enough $\alpha$. This, together with the fact that in this case $A$ has a compact inverse, implies via the Krein–Rutman Theorem the existence of a first eigenvalue which is simple and associated with a positive eigenfuntion.

In some special cases, it is possible to say more about the behaviour of solutions of (1.3). If, for instance, homogenerous Neumann boundary conditions are considered, there is no explicit dependence of the function $f$ in the space variable and either the domain $\Omega$ or $f$ itself are convex, then it is known that the only possible stable stationary solutions are constant [**13**]. For Dirichlet boundary conditions on balls, it is know that the only possible stable stationary solutions do not change sign in $\Omega$ [**45**].

Another important case is when the domain $\Omega$ is a one–dimensional interval on the real aline. In this case, the flux generated by Equation (1.3) is of Morse–Smale type ([**39**] and [**2**]), and for dissipative $f$ the attractor can be described in detail ([**17**] and [**21**]).

Up until 1981, in most if not all of the cases studied where a nonlocal term of the type of $\bar{u}$ was included, either the system enjoyed some of the properties indicated above for Equation (1.3), which made it behave in a similar fashion to that equation, or the analysis carried out was not taken far enough to unveil the possible differences.

Although some slight differences between the local and nonlocal linear operators had been noticed in [**14**], the fact that Equation (1.2) can exhibit a different qualitative behaviour from that of (1.3), seems to have been pointed out first in [**15**], where it was shown that even when the functions $f$ and $g$ do not depend explicitly on the space variable it is possible to obtain nontrivial stable stationary solutions in the Neumann case. The main tools used in the proof were linearization around a trivial solution and bifurcation theory. In this instance, the mechanisms that allow for this behaviour are related to the fact that the linearized operator obtained, although remaining self–adjoint, may no longer have a constant eigenfunction corresponding to the first eigenvalue. Curiously enough, the equation studied by Chafee went back to the beginning of the century, when it had been introduced in an electrical model [**32**].

Similar results for the Dirichlet case were later obtained in [**23**], where it was shown that there could exist stable sign-changing solutions on the interval and that any solution that changed sign an even number of times had to be unstable.

These differences in behaviour referred to the spatial structure of the possible stable equilibria and, in some sense, it could be anticipated as we are introducing a strong spatial coupling. However, nearly a decade later, Fiedler and Poláčik showed that the differences also extended to the behaviour in time and that Equation (1.2) with

$$f(x,u,\bar{u}) = a(x)u + h(x,u) + b(x)\bar{u}, \tag{1.4}$$

$$\bar{u} = \int_0^1 c(\xi)u(\xi)d\xi$$

and homogeneous Dirichlet boundary conditions, could have stable $\omega-$limit sets of arbitrarily high dimension [**20**]. Again this was shown by means of the same tools as in the previous example, and now relied on the fact that the linear operator obtained is no longer self–adjoint in this case. The proof also relied quite heavily on the explicit dependence on the space variable. The existence of a Hopf bifurcation in the case where such dependence is not present was given in [**24**]. This was obtained from the analysis of a specific equation, and there was only numerical evidence that the resulting periodic orbit was stable. Later [**51**], an example was constructed for which it was shown that it was possible for a stable periodic orbit to exist under these circumstances.

In most if not all of the cases arising from specific examples, the function $g = g(x, y, \bar{u})$ used in the definition of $\bar{u}$ does not depend explicitly on the variable $x$. In this case the linear integro–differential operator arising via linearization has a separable kernel and its general form is

$$Lu = \Delta u + a(x)u + b(x)\int_{\Omega} c(\xi)u(\xi)d\xi.$$

The main feature of this operator which still imposes some structure is the fact that it is obtained as a bounded rank–one perturbation of a self–adjoint operator. By taking advantage of this fact in the one–dimensional case, the structure of this operator was studied in [**22**], where, in particular, conditions were given that ensure that the spectrum remains real even when $L$ is no longer self–adjoint. These results were then extended to the $n-$dimensional case in [**28**], where it was also shown that when Equation (1.2) is monotone then its qualitative behaviour is quite similar to that of the local case. In particular, an instability result similar to that of Casten and Holland for the local case still holds.

Positivity conditions for the operator $L$ were given in [**1**], and these were extended in [**27**] to the general case where the kernel in the integral term is no longer separable.

From all this, it is clear that the introduction of a nonlocal term of this type will, in general, destroy most if not all of the properties that are present in the local reaction–diffusion Equation (1.3), and which are responsible for the rich structure found in such equations. In fact, much that is taken for granted both in terms of techniques and of results by people working in scalar local reaction–diffusion equations is lost in the nonlocal case. All of this can be summarized as follows:

1. In general, the equations no longer satisfy a maximum principle, and, as a consequence, the time–dependent problem is not monotone;
2. There is no longer a Lyapunov functional which applies to the general form of Equation (1.2) and, for specific equations of that form, it is known that no such functional exists at all;
3. The linearized operator around a stationary solution will not, in general, be self–adjoint.

The lack of these properties does not, by itself, necessarilly imply that the dynamical behaviour for these equations has to be different. However, it does make the study of simple properties of Equation (1.2) such as global existence/blow–up of solutions, convergence to equilibria, etc., harder, and, from the results mentioned above, we know that in fact this behaviour can be more complicated:

1. Existence of stable nontrivial stationary solutions in the case of Neumann boundary conditions in one dimension;
2. Existence of stable sign–changing stationary solutions for the Dirichlet problem in one dimension;
3. Existence of stable complicated asymptotic behaviour if explicit dependence in the space variable is allowed;
4. Existence of stable periodic orbits even when no explicit dependence on the space variable is allowed.

However, there is still some structure left and it is possible to devise some general techniques to the approach of this type of problem. So far, these rely mostly on general tools such as linearization and bifurcation theory, and take advantage of the already mentioned fact that, in general, we are led to a linear operator that is a rank–one perturbation of a self–adjoint elliptic operator.

The purpose of this paper is to give a general overview of what is known for Equation (1.2), the type of results that one can hope to obtain when faced with a specific equation, and the techniques available to obtain these properties. We also point out some open problems and conjectures and possible directions for future research.

In Section 2 we give conditions for some of the properties that are satisfied in the local case to carry over to the nonlocal equation. These include a weak version of the nonincrease of the lap–number of a solution, and conditions for the time–dependent problem to be monotone. Section 3 describes the structure of the linear nonlocal operator. The results given there are, in part, derived from general perturbation theory. The question of stationary solutions and their stability and bifurcations is treated in Section 4, where the main known stabilization mechanisms are described, and some examples are given. Finally, Section 5 describes some of the known possible complicated dynamical behaviour of solutions, and in the last Section we make some general considerations about this type of problem.

## 2 General results

From the studies that have been carried out over the last decade in the field of scalar nonlocal reaction–diffusion equations it has emerged that it is not possible to treat them in such a general way as in the local case. In the latter, properties such as the existence of a general Lyapunov function, maximum principles, nondecrease of the lap–number and so on, ensured that the behaviour of solutions is quite restricted and simultaneously provided us with tools to study this behaviour. Although most of these properties no longer hold for the general Equation (1.2), the fact that we are still dealing with a scalar equation still imposes some restrictions on its dynamical behaviour. In this Section we mention some of the properties that still hold and under which conditions this happens.

**2.1 Nonincrease of the lap–number.** Consider the non-autonomous scalar reaction–diffusion equation

$$\begin{cases} u_t = u_{xx} + f(t,u) \\ u(x,0) = \psi(x) \end{cases} \tag{2.1}$$

on a bounded real interval $J$, together with either Dirichlet or Neumann homogeneous boundary conditions. Roughly, the classical result on the nonincrease of the lap–number of solutions of this equation in the case of Neumann boundary conditions states that the number of *bumps* of a solution cannot increase with time – note that $f$ does not depend explicitly on the space variable. A similar result, although not as strong, holds for Dirichlet boundary conditions, provided that some further conditions are assumed.

In order to be more precise, we define the lap–number of a function $w$ on $J$ and with continuous derivative on that interval following [**47**] as

$$l(w) = \begin{cases} z(w_x)+1, & \text{if } w_x \not\equiv 0 \\ 0, & \text{otherwise.} \end{cases}$$

Here $z(w)$ denotes the number of sign changes of $w$ on $J$. Then Matano's result for the local case can be stated as follows

**Theorem 2.1** ([**47**]) *Let $\psi(x)$ be a continuous function on $J$ which satisfies the boundary conditions. Assume also that a solution of (2.1) exists and is defined on the interval $(0,T)$, for some positive $T$. Then, if $f$ is a $\mathcal{C}^1$ function on $(0,T)\times J$, one has the following*

(i) *(Neumann b.c.)* $l(w(\cdot,t_1)) \geq l(w(\cdot,t_2))$ *for all* $t_1, t_2 \in [0,T)$ *such that* $t_1 < t_2$.

(ii) *(Dirichlet b.c.) Assume that at least one of the following conditions is satisfied*

$$(a)\ f(t,0) \equiv 0$$
$$(b)\ u \geq 0 \text{ or } u \leq 0 \ \ for\ \ all\ \ t \in [0,T).$$

*Then the statement in (i) still holds, provided that* $l(u(\cdot,t_1)) \neq 0$. *If* $f(t,0) \geq 0$ *or* $f(t,0) \leq 0$, *then* $l(w(\cdot,t_1)) + 2 \geq l(w(\cdot,t_2))$ *for all* $t_1, t_2 \in [0,T)$ *such that* $t_1 < t_2$.

Some of the hypothesis above can somehow be relaxed, but these are sufficient for the present purposes. For the proof (which relies heavily on the Jordan curve theorem on the plane) and other related results, see [**47**] (see also [**53**]).

The fact that a similar result might still hold for (1.2) when the function $f$ does not depend explicitly on the space variable was conjectured in [**30**]. This is indeed the case, as was shown in [**23**]

**Proposition 2.2** *Theorem 2.1 still holds in the nonlocal case provided that* $f$ *does not depend explicitly on the space variable.*

The proof of this fact is just the observation that if a solution $u$ of (1.2) exists on the interval $(0,T)$ for some positive number $T$, then this function is also a solution of an equation of the form (2.1), as $\bar{u} = \bar{u}(t)$.

This is, however, a weak result as it refers only to the decrease of the lap–number along a single trajectory and will not extend in general to the difference of two solutions as in the local case. This last property, which is related to the existence of a maximum principle, is fundamental in the local case in the proof of the fact that Equation (1.3) is of Morse–Smale type ([**39**] and [**2**]).

**2.2 Conditions for the existence of a comparison principle.** In general, an equation of the form of (1.2) will not satisfy a maximum principle. As this is one of the most important tools used in the study of the local case, it is important to know when it can still be applied here. In many instances this will depend on a parameter in the equations or on the form of the reaction terms involved and it has happened quite frequently in the literature that for a specific equation it is only for a region in the parameter space or functions of a special form that results are obtained.

One of the implications of the maximum principle is roughly that if a solution starts above another, then they remain that way as long as both exist. This statement can be made in several ways. Here we mention a possible version.

**Proposition 2.3** *Assume that* $f$ *and* $g = g(y,u)$ *are* $C^1$ *functions of* $\bar{u}$ *and* $u$, *respectively, and let* $I_1 = (c_1, c_2)$ *and* $I_2 = (c_3, c_4)$ *be two intervals (possibly unbounded) and such that*

$$\frac{\partial f(x,u,p)}{\partial p}\frac{\partial g(x,u)}{\partial u} > 0 \tag{2.2}$$

*for all* $(x,u,p) \in \Omega \times I_1 \times I_2$. *Assume also that, in the case of Dirichlet boundary conditions,* $0 \in I_1$. *Then, if* $u$ *and* $v$ *are two solutions of (1.2) satisfying*

$$c_1 < u(x,0) \leq v(x,0) < c_2,\ u(x,0) \not\equiv v(x,0)$$

*and*

$$\bar{u}(0), \bar{v}(0) \in I_2,$$

*it follows that* $u(x,t) < v(x,t)$ *as long as both solutions exist and* $u(x,t) > c_1$, $v(x,t) < c_2$ *and* $\bar{u}, \bar{v} \in I_2$.

The proof follows in the standard way, by considering the equation satisfied by the difference of the two solutions. If condition (2.2) does not hold, then it is always possible to build two solutions $u$ and $v$ such that $u(x,0) \leq v(x,0)$ but $u(x_0,t_0) > v(x_0,t_0)$ for some $x_0$ in $\Omega$ and some positive $t_0$ [**28**]. In Section 4.3 we shall mention some of the restrictions imposed when Proposition 2.3 holds.

One example of application of this result is the equation considered by [**43**]:

$$u_t = u_{xx} + \frac{\mu f(u)}{\left[\int_0^1 f(u(y,t))dy\right]^2},$$

with $f$ a positive $\mathcal{C}^1$ function. When $f$ is decreasing then Proposition 2.3 applies and it is possible to use this comparison principle to study the blow–up and convergence of solutions.

## 3 The linear problem

One of the few techniques that is widely used in the local case to study the stability of stationary solutions and that can still be used for the nonlocal equation is linearization (see [**15**] and [**28**]). Here we consider the linear nonlocal operator in $X = L^2(\Omega)$ defined by

$$Lu = Au + b(x)\int_\Omega c(y)u(y)dy, \tag{3.1}$$

where $A = \Delta + a$ and the functions $a, b$ and $c$ are assumed to be continuous in $(\bar{\Omega})$. The operator $A$ is a densely defined, closed, self–adjoint operator with compact resolvent. Furthermore, its spectrum is real, bounded from above and consists entirely of isolated eigenvalues of finite multiplicity which will be denoted by $\gamma_k$, $k = 0, 1, \cdots$. The corresponding eigenfunctions, which can be chosen in such a way that they form an orthonormal system, will be denoted by $v_k$.

With the exception of self–adjointness (and thus the fact that the spectrum is real), the other properties carry over to $L$. The main idea to study the spectrum of $L$ is to consider it as a bounded rank$-1$ perturbation of $A$. To this end, consider the one–parameter family of linear operators defined by

$$L_\epsilon u = Au + \epsilon b(x)\int_\Omega c(y)u(y)dy, \tag{3.2}$$

where $\epsilon$ is a real number, not necessarily small. From standard results in perturbation theory, the eigenvalues of $L_\epsilon$ are continuous functions of the parameter and we shall denote them by $\lambda_k(\epsilon)$, where the index $k$ indicates that $\lambda_k(0) = \gamma_k$. Furthermore, each of the functions $\lambda_k$ can be chosen in such a way that it will have a continuous derivative when $\epsilon$ is zero [**42**]. It should be stressed, however, that this will not be the case for all values of $\epsilon$.

In this way, and for each $\epsilon$, the spectrum of $L_\epsilon$ may be divided into two disjoint sets $\Sigma^f(L_\epsilon)$ and $\Sigma^m(L_\epsilon)$ in the following way. We shall say that a complex number $\lambda$ belongs to $\Sigma^f(L_\epsilon)$ if it is an eigenvalue of $L_\epsilon$ for all real $\epsilon$ and that it belongs to $\Sigma^m(L_\epsilon)$ otherwise. The first of these sets will be called the fixed spectrum of $L_\epsilon$ and its elements the fixed eigenvalues of $L_\epsilon$, while the second, which depends on the particular value of $\epsilon$ considered, is the moving spectrum of $L_\epsilon$ and its elements the moving eigenvalues.

We now present some of the basic properties that can be derived for the spectrum of $L_\epsilon$. The proofs of these results in the one–dimensional case can be found in [**22**]. For the $n$–dimensional case the proofs are similar, except that now one has to take into account that the eigenvalues of $A$ are not necessarilly simple [**28**]. We begin with some conditions for a point to belong to the fixed spectrum.

**Theorem 3.1** *Let $b_k$ and $c_k$ be defined by*

$$b_k = \int_\Omega b(x) v_k(x) dx \text{ and } c_k = \int_\Omega c(x) v_k(x) dx.$$

*Then we have that $\lambda_k(\epsilon)$ belongs to $\Sigma^f(L_\epsilon)$ if and only if either $b_k$ or $c_k$ is zero. Also, if $\gamma$ is an eigenvalue of $A$ which has multiplicity $m$ greater than one then this point belongs to $\Sigma^f$ and it is an eigenvalue of $L_\epsilon$ of multiplicity at least $m-1$.*

Regarding the moving spectrum, we have the following,

**Theorem 3.2** *Let $\epsilon_1$ and $\epsilon_2$ be two distinct real numbers and assume that there exist integers $k$ and $l$ (not necessarilly distinct) such that $\lambda_k(\epsilon_1) = \lambda_l(\epsilon_2) = \gamma$. Then $\gamma$ belongs to $\Sigma^f$.*

The importance of this result is that it implies that the paths of two moving eigenvalues cannot intersect each other for different values of the parameter except at an eigenvalue of $A$ and, in particular, this means that a moving eigenvalue cannot turn back on its path. This is particularly important for simple real eigenvalues, as they must remain real until they collide with another moving eigenvalue. To complete the picture, it is necessary to know in what direction do moving eigenvalues start going at $\epsilon = 0$.

**Theorem 3.3** *The sign of the derivative of an eigenvalue $\lambda_k$ at $\epsilon = 0$ is the same as that of the product $b_k c_k$.*

**Remark 3.1** If $\gamma$ is a multiple eigenvalue of $A$ with multiplicity $m$, then the corresponding eigenfunctions can be chosen such that they are orthogonal and $m-1$ of these are such that the corresponding coefficients $c_k$ vanish. In this way, the sign of the product $b_k c_k$ is well–defined for the remaining eigenfunction.

Below we shall make use of these results to understand the changes in stability of stationary solutions caused by the introduction of nonlocal terms. For an example of the application to a specific equation, see [**24**].

## 4 Stationary solutions

Stationary solutions of the nonlocal problem (1.2) are time–independent solutions of this problem, that is, solutions for which $u_t$ is identically zero. Knowledge of this type of solution and their stability properties is fundamental to the understanding of Equation (1.2).

A function $w$ will be a stationary solution of (1.2) if and only if it satisfies

$$\Delta w + f(x, w, \bar{w}) = 0, \tag{4.1}$$

together with appropriate boundary conditions. If $\bar{w}$ is taken to be of the form

$$\bar{w} = \int_\Omega g(x, w) dx,$$

then this is constant and it is readily seen that $w$ will be a solution of (4.1) if and only if it satisfies the local problem

$$\Delta w_\alpha + f(x, w_\alpha, \alpha) = 0 \tag{4.2}$$

with the same boundary conditions, and where

$$\alpha = \int_\Omega g(x, w_\alpha)dx. \tag{4.3}$$

Thus, it follows that the set of stationary solutions of the nonlocal problem is contained in the set of stationary solutions of some local problem. More precisely, stationary solutions of the nonlocal problem must also be solutions of the local problem (4.2), while satisfying the restriction (4.3). In particular, for the case of $f = f(u, \bar{u})$ and homogeneous Dirichlet boundary conditions, the usual results for local problems still apply, namely, positive solutions on balls have radial symmetry [**31**] and, in the one–dimensional case, solutions which change sign an even number of times are even about the midpoint of the interval considered.

In spite of this relation between the stationary solutions of both problems, and mainly due to the lack of a maximum principle for (4.1), there are several tools which are commonly used in the local case but are no longer applicable in general now. These include, for instance, the use of sub– and super–solutions and variational methods. Thus, again different techniques from those used in the local case are required.

One general method that suggests itself is to consider the local elliptic problem (4.2), treating $\alpha$ as a real parameter, and to obtain the solutions of this equation as a function of $\alpha$. Substituting this into (4.3) and solving the resulting equation with respect to $\alpha$ then yields the solutions of (4.1) (see [**5**], for instance). Although this method works in principle it will sometimes be very difficult to apply to specific problems, as it requires a very thorough knowledge of the set of solutions of the associated local problem, together with solving Equation (4.3). A possible strategy will be to use this, while at the same time exploring the particular form of the equation.

For the one–dimensional case, a method based on Picard–Fuchs equations can be found in [**26**] and in [**34**]. For some specific problems, examples of different techniques can be found, for instance, in [**43**] and [**44**] and in [**18**].

Here we shall concentrate more on the stabilization mechanisms that may turn a stationary solution which is unstable as a solution of the local problem into a stable solution of the nonlocal problem.

**4.1 Stabilization mechanisms I: pitchfork bifurcations.** The first way to be discovered by which it was possible to obtain stable nontrivial stationary solutions for the Neumann case was via a pitchfork bifurcation [**15**]. Chafee considered an equation of the form

$$u_t = u_{xx} - g(u) + \frac{\mu r(u)}{\left[\int_0^1 r(u(x,t))dx\right]^2}$$

on the interval $(0, 1)$. This corresponds to a model for an electric ballast resistor in the case where the voltage is kept constant and where the functions $g$ and $r$ satisfy certain conditions that ensure that solutions are well–defined for all positive $t$ and that a unique homogeneous equilibrium exists. By choosing the functions $f$ and $r$ appropriately, it is then possible to make a nontrivial solution bifurcate from this homogeneous equilibrium in such a way that it will be stable. This idea has been used again in the case of Neumann boundary conditions for a model in population dynamics [**30**] and in the Dirichlet case [**23**]. We shall now sketch its main ingredients in this last situation. To this end we consider Equation (1.2) on the interval $(-1,1)$ with homogeneous Dirichlet boundary conditions and where $f = f(u,\bar{u})$ and $g = g(u)$ are assumed to be smooth enough and satisfy $f(0,0) = g(0) = 0$.

These conditions ensure that 0 is a stationary solution of (1.2). Denoting $D_1 f(0,0)$ and $D_2 f(0,0)g'(0)$ by $\mu$ and $\epsilon$ respectively, we have that the linearized operator around this solution is of the form

$$L_\epsilon u = Au + \epsilon \int_{-1}^{1} u dx$$

with $Au = u'' + \mu u$. Since $\mu$ and $\epsilon$ can be chosen independenty by an appropriate choice of the functions $f$ and $g$, we are led to the two–parameter self–adjoint eigenvalue problem

$$u'' + \mu u + \epsilon \int_{-1}^{1} u dx = \lambda u.$$

In the local case ($\epsilon = 0$), the eigenvalues are given by $\gamma_k = \mu - (k+1)^2\pi^2/4$, $k = 0, 1, \cdots$ and thus, by increasing $\mu$ it is possible to obtain a sequence of pitchfork bifurcations from the trivial solution. However, in a neighbourhood of the bifurcation point the only that can lead to stable solutions (depending on the higher order terms of $f$ and $g$) is the first, as for all the others the operator already has at least one eigenvalue which is positive. As the eigenfunction corresponding to the first eigenvalue is positive, it follows that the only possible stable solutions that can be obtained in this way do not change sign. The idea to produce a stable solution which changes sign is to use the integral term to shuffle the spectrum in such a way that the first eigenvalue to bifurcate now corresponds to an eigenfunction that changes sign. To see how this can happen, we use the results and notation of Section 3.

A simple calculation gives that in this instance

$$b_k c_k = \begin{cases} \dfrac{16}{(k+1)^2\pi^2}, & k \text{ even} \\ 0, & k \text{ odd}, \end{cases}$$

so that, by Theorem 3.3, $\Sigma^f = \{\gamma_k \in \Sigma(A) : k \text{ is odd}\}$. From the sign of the derivative for the first eigenvalue at $\epsilon = 0$ and the fact that the paths of different eigenvalues do not overlap, it is possible to see [**23**] that, as far as the bifurcation from the first eigenvalue goes, for positive $\epsilon$ the picture is not different from that studied in [**17**].

For negative $\epsilon$ the eigenvalue $\lambda_0$ is to the left of $\gamma_0$ and for sufficiently large values of the parameter it is actually possible to make it go past the fixed eigenvalue $\gamma_1$. As the eigenfunction corresponding to this eigenvalue does not change with $\epsilon$, it is then possible to obtain a stable solution bifurcating from zero and which changes sign once by choosing the higher order terms appropriately.

**Theorem 4.1** *Let $f$ and $g$ be of the form*

$$f(u,\bar{u}) = \alpha_1 u + \alpha_2 \bar{u} + \alpha_3 u^2 + \alpha_4 u\bar{u} + \alpha_5 \bar{u}^2 + \alpha_6 u^3 + \alpha_7 u^2\bar{u} + \alpha_8 u\bar{u}^2 + \alpha_9 \bar{u}^3 + \ h.o.t.$$

*and*

$$g(u) = \beta_1 u + \beta_2 u^2 + \beta_3 u^3 + \text{ h.o.t.},$$

*where* $\alpha_1 = \mu$ *and* $\alpha_2\beta_1 = \epsilon$*. If for* $\epsilon < -\pi^2/2$ *there exists a half–neighbourhood of* $\pi^2$*,* $U_\delta = (\pi^2, \pi^2 + \delta)$ *for some* $\delta$ *small enough, such that*

$$\frac{(\alpha_4\beta_1 + \alpha_3)(2\alpha_2\beta_2 + \alpha_3)}{\pi^2 + 2\epsilon} - \frac{\alpha_3^2}{6\pi^2} - \frac{3}{4}\alpha_6 - \alpha_4\beta_2 > 0,$$

*then Equation (1.2) has a stable stationary solution that changes sign once on* $(-1,1)$*, for all* $\mu \in U_\delta$*.*

The natural question to ask now is, of course, how far one can go and whether it is possible or not to have solutions that change sign more than once which are stable. This is related to the spatial complexity that is possible to obtain in nonlocal equations and so far only incomplete results are available.

To begin with, it should be taken into account that due to Theorem 2.1 the spatial complexity of the $\omega-$limit of a trajectory is highly restricted by the spatial complexity of the initial condition. On the other hand, due mainly to symmetry constraints it is possible to obtain the following result [**23**]

**Theorem 4.2** *Consider Equation (1.2) on a bounded interval* $J$ *of the real line with* $f = f(u, \bar{u})$ *and* $g = g(u)$*. Then if* $w$ *is a stationary solution of (1.2) which changes sign* $m$ *times on* $J$ *with* $m$ *even, the dimension of the unstable manifold of* $w$ *is at least* $m/2$*.*

Some partial results are given in the same paper concerning the stability of stationary solutions which change sign an odd number of times $m$ on $J$ with $m$ larger than 1, but so far it is not known whether or not these have to be unstable in general.

**4.2 Stabilization mechanisms II: saddle–node bifurcations .** As was mentioned at the beginning of this section, stationary solutions of (1.2) must also be stationary solutions of an associated local problem. However, the way in which these solutions depend on a parameter, for instance, might not be the same in both problems. This might bring about a change in the stability of solutions by, for instance, introducing or removing saddle–points. One such example is given by the following model for electrical devices where there is a strong dependence of the resistance with the temperature (see [**43**] for a derivation of the model):

$$\begin{cases} u_t = u_{xx} + \dfrac{\mu r(u)}{\left[\displaystyle\int_{-1}^{1} r(u)dx\right]^2} \\ u(\pm 1, t) = 0 \\ u(x,0) = \psi(x), \end{cases} \tag{4.4}$$

where the function $r$ is taken to be positive. In this case, the stationary solutions solve the equation

$$u_{xx} + \alpha r(u) = 0, \tag{4.5}$$

where

$$\alpha = \frac{\mu}{\left[\displaystyle\int_{-1}^{1} r(u)dx\right]^2}.$$

Although the set of stationary solutions of (4.4) as $\mu$ varies along the positive real numbers is the same as that of Equation (4.5) for positive values of $\alpha$, the dependence of these solutions on the two parameters can be quite different. While in the local case there are positive functions $r$ for which there are multiple solutions, in the nonlocal case there exists at most one stationary solution of (4.4) for each value of the parameter $\mu$. If, furthermore, it is assumed that

$$\int_{-1}^{1} r(s)ds = \infty,$$

then for each value of $\mu$ there exists one and only one stationary solution (see [**43**]). One such case where this happens is for $r(u) = e^u$, where there exists a value of $\alpha$, say $\alpha^*$, such that there are two solutions of the local problem for $0 < \alpha < \alpha^*$, precisely one for $\alpha = \alpha^*$, and none for $\alpha^* > \alpha$. Thus these two solutions appear from a saddle–node bifurcation and we have that the smallest is locally asymptotically stable, while the largest is unstable ([**29**] and [**41**]). The introduction of the nonlocal term has the effect of removing the turning point in the curve representing solutions as a function of the parameter and, as for $r(u) = e^u$ the linear operator obtained by linearization is self–adjoint, it is possible to show that this gives a branch of stable solutions for all values of $\mu$ [**22**].

If it is assumed that the device considered is in series with a resistor, then the denominator in (4.4) becomes of the form

$$\left[a + \int_{-1}^{1} r(u)ds\right]^2,$$

where $a$ is a positive constant. Although for small enough $a$ the bifurcation diagram is still similar to that of the previous example, if it becomes large enough then the turning point will still exist and the way in which stabilization takes place is now by the introduction of a new turning point in such a way that as one moves along the branch of solutions these change from stable to unstable and then back to stable again [**26**].

From these examples we see that the introduction of a nonlocal term may, while not altering the set of stationary solutions of an equation, change their stability by changing the way in which these solutions depend on a parameter.

**4.3 The monotone case.** When the nonlocal reaction term in (1.2) is such that the equation is monotone, then the behaviour of this equation becomes quite like that of Equation (1.3). In this instance the results in [**50**] can be used, and it is known that most trajectories converge to equilibria and any possible higher–dimensional $\omega$–limit sets are unstable. We now show that under these conditions the type of stationary solutions that may be stable is also restricted and again quite like that in the local case. This is done using the results given in Section 3 – for another method which applies to the more general case where $g = g(x, y, u)$ see [**28**].

The main result is that if the functions $b$ and $c$ which appear in the linearized operator in (3.1) both have the same sign, then a stationary solution of (1.2) can only be stable if it is also stable as a stationary solution of a corresponding local equation of the form

$$u_t = \Delta u + f(x, u, \alpha),$$

where $\alpha$ is the constant given by Equation (4.3). This comes as a corollary of the following

**Theorem 4.3** *Let $w$ be a stationary solution of* (1.2) *for which the linearized local problem has $p$ positive eigenvalues. Then, if any of these eigenvalues has multiplicity greater than* 1, *this solution will be an unstable solution of the nonlocal problem. If all the positive eigenvalues are simple, then a necessary condition for $w$ to be a stable solution of the nonlocal problem is that*

$$\begin{cases} b_k c_k < 0, & k \text{ even} \\ b_k c_k > 0, & k \text{ odd} \end{cases}$$

*for all $k = 0, 1, \cdots, p-1$.*

The proof can be found in [**28**], and is based on the fact that because the paths of the eigenvalues of the family of operators (3.2) cannot intersect themselves as $\epsilon$ varies, in order to move all the positive eigenvalues to the left–half of the complex plane it is necessary for them to collide, become complex, and then cross the imaginary axis. As a simple corollary we have

**Corollary 4.4** *Let $w$ be a stationary solution of Equation* (1.2) *and assume that $b$ and $c$ are of the same sign. Then, if $w$ is an unstable stationary solution of the corresponding local problem, it is also an unstable stationary solution of the nonlocal problem.*

In particular, it is possible to derive a Casten and Holland–type result [**13**] in the monotone case and Neumann boundary conditions – see [**28**].

## 5 Asymptotic behaviour

It is to be expected that the introduction of a spatially nonlocal term will affect the spatial asymptotic behaviour of solutions, that is, the type of solutions that may be stable, just like the introduction of nonlocal terms in time in delay equations may cause the asymptotic behaviour (in time) to be more complicated. However, the type of stationary solutions that may be stable in at least some classes of delay–diffusion equations is not altered by the introduction of these terms (see [**25**]). In some sense, it might be said that time and space are, in this instance, kept apart. However, it turns out that in the case of a spatial nonlocal term the asymptotic behaviour in time is also affected. So far, the main result along these lines can be found in [**20**] where it was shown that it is possible to place an arbitrarily high (even) number of eigenvalues of the linearization around a stationary solution on the imaginary axis, while keeping the other eigenvalues on the left–hand side of the complex plane. This last fact ensures that if the higher–order terms are chosen carefully, then a bifurcation from this stationary solution arising in this setting will give rise to stable invariant tori of arbitrarily high–dimension. The main result in this paper is the following

**Theorem 5.1** ([**20**]) *Consider Equation* (1.2) *with Dirichlet boundary conditions on the interval* $(0,1)$ *with $f$ as in* (1.4). *There exists a residual subset $\mathcal{G}$ in $\mathcal{C}[0,1]$ such that for all $a$ in $\mathcal{G}$ the following holds. Let $\omega_1, \cdots, \omega_m$ be any $m$–tuple of mutually distinct positive real numbers. Let $V$ be any polynomial vector field on $\mathbb{R}^{2m}$ of degree $N \geq 2$ such that $V(0) = V'(0) = 0$. Then there exist $b, c$ in $\mathcal{C}[0,1]$ and $h(x,u)$ in $\mathcal{C}^{0,\infty}$ satisfying $h(x,0) = D_u h(x,0) \equiv 0$ such that*

(i) *The spectrum of the linearization $Au = -u_{xx} - a(x)u - b(x)\bar{u}$ is given by the algebraically simple eigenvalues $\pm i\omega_j$, $j = 1, \cdots, m$, and $\gamma_k$, $k = 2m+1, 2m+2, \cdots$, where $\gamma_k$ denote the eigenvalues of the local problem.*

(ii) *Let $H$ be the reduced vector field representing the flow on a $C^N$ centre manifold of the equilibrium $u = 0$ for Equation* (1.2) *with $f$ as in* (1.4). *Then, in real coordinates, the Taylor expansion of $H$ at 0 coincides with $V$ for orders 2 through $N$.*

The proof of this result relies on a result from control theory known as the pole–assignment Lemma, and on a centre–manifold reduction.

Later, in [**51**], the existence of stable periodic orbits bifurcating from a stationary solution was shown to be possible also in the case where there is no explicit dependence on the space variable. Previously [**24**] it had been shown for a specific example that it was possible to have Hopf bifurcations under these circumstances, but only numerical evidence had been given towards the fact that the resulting periodic orbits were stable.

## 6 Comments and open problems

Although the great majority of the work carried out in the field of nonlocal reaction–diffusion equations has been the study of particular equations, the last few years have seen a greater attempt at a more systematic approach. This has centred around the study of the possible behaviours that solutions of such equations may display both in time and in space, and on the comparison with the local case. Along the way, some new techniques have been devised to deal with the problems raised by the loss of structure mentioned in the Introduction. Loosely speaking, it can be said that such problems display a complexity which lies halfway between that of the local scalar reaction–diffusion equation and that of a system of two equations of that type. As was seen, this analogy is strengthened by the fact that nonlocal equations can indeed arise as simplifications of the latter.

When the nonlocal equation is monotone, then its behaviour is in fact quite close to that of the local equation. One question that remains open is how far in the other direction we can go, that is, how close to the behaviour displayed by the second order system can we be. In particular, this is not clear when it comes to the spatial complexity that can be displayed by stable stationary solutions. In [**23**] it was conjectured that for the one–dimensional problem the only solutions that could be stable were those which changed sign at most once. A generalization of this can be stated as

**Conjecture 1** *Let $w$ be a stationary solution of* (1.2). *Then it can only be stable if when considered as a stationary solution of the local problem* (1.3) *the corresponding linearized operator has at most one positive eigenvalue.*

In general, questions related to the stability of stationary solutions that have been addressed in the literature refer only to the case where the system is monotone. This is caused mainly by the fact that then the standard techniques that are used in the local case can extend to the nonlocal problem. However, in general this is not the situation and new techniques should be developed to deal with the situation. One example of a problem that is in some sense typical is that of Equation (4.4). If the function $r$ is decreasing, then the equation satisfies a comparison principle of the type considered in Section 2.2. This has been used in [**43**] to show global asymptotic stability of the only equilibrium. For any other positive function $r$ it is not possible to apply these techniques and only two more cases have been studied: $r(u) = e^u$ and $r(u) = au + b$ with positive $a$ and $b$, for which it was shown that the only equilibrium is still stable [**22**]. This is, in some sense, a typical situation in this field and we believe that developing techniques that will allow for the study of this problem will be important in the understanding of this type of equation. In this respect, we make the following conjencture.

**Conjecture 2** *The single equilibrium of Equation* (4.4) *is stable for all positive functions* $r$.

Another aspect that has received little attention is the question of existence of Lyapunov functionals for equations of this type. That this is not a trivial problem can be seen by two examples. For Equation (4.4), for instance, the only case for which a Lyapunov function is known is when $r(u) = e^{\alpha u}$. Even for the case of general decreasing $r$, for which it is known that if an equilibrium $w$ exists then all solutions converge to $w$ (a condition that implies the existence of a Lyapunov functional [**38**]) explicit forms for this functional are not known.

On the other hand, the existence or not of such a functional might depend on a real parameter in the equation. For the equation

$$u_t = u_{xx} + f(u) - \alpha \int_0^1 f(u)dx$$

with Neumann boundary conditions which was studied in [**24**], it was shown that

$$\mathcal{L}(u) = \int_0^1 \left[ \frac{u_x^2}{2} - \int_0^u f(s)dx \right] dx$$

is a Lyapunov function for $\alpha$ smaller than or equal to 1, but that, under some conditions in $f$, periodic orbits exist for $\alpha$ greater than 6. Note that, in general, this equation is not monotone for any nonzero $\alpha$.

These examples show that, as was to be expected, it is difficult to find Lyapunov functionals for specific equations. However, at the same time they also show that in many cases where the equations are no longer monotone solutions still converge to an equilibrium. This, together with the fact that the number of specific equations studied in the literature which are not monotone is very reduced, clearly means that new (global) techniques have to be developed if we want to be able to understand these problems.

## References

[1] Allegretto, W. and Barabanova, A. [1996], *Positivity of Solutions of Elliptic Equations with Nonlocal Terms*, Proc. Roy. Soc. Edinburgh **126A**, 643–663.

[2] Angenent, S. B. [1986], *The Morse–Smale Property for a Semi–linear Parabolic Equation*, J. Differential Equations **62**, 427–442.

[3] Angenent, S. B. [1988], *The Zero Set of a Solution of a Parabolic Equation*, J. Reine Angew. Math. **390**, 79–96.

[4] Ball, J. [1973], *Stability Theory for an Extensible Beam*, J.Differential Equations **14**, 399–418.

[5] Bazley, N. W. and Weyer, J. [1988], *Explicitly Resolvable Equations with Functional Nonlinearities*, Math. Methods Appl. Sci. **10**, 477–485.

[6] Bebernes, J., Bressan, A. and Lacey, A. [1988], *Total Blow-Up vs. Single Point Blow-Up*, J.Differential Equations **73**, 30–44.

[7] Bebernes, J. and Lacey, A. [1997], *Global Existence and Tinite–time Blow–up for a Class of Nonlocal Parabolic Problems*, preprint.

[8] Budd, C., Dold, B. and Stewart, A. [1993], *Blowup in a Partial Differential Equation with Conserved First Integral*, SIAM J. Appl. Math. **53**, 718–742.

[9] Burgers, J. M. [1939], *Mathematical Examples Illustrating Relations Occuring in the Theory of Turbulent Fluid Motion*, Trans. Roy. Neth. Acad. Sci. Amsterdam **17**, 1–53.

[10] Calsina, A., Perelló, C. and Saldaña, J. [1994], *Nonlocal Reaction–Diffusion Equations Modelling Predator–Prey Coevolution*, Publ. Math. **38**, 315–325.

[11] Calsina, A., Perelló, C. [1995], *Equations for Biological Evolution*, Proc. Roy. Soc. Edinburgh **125A**, 939–958.

[12] Carr, J., Gurtin, E. and Slemrod, M. [1984], *Structured Phase Transitions on a Finite Interval*, Arch. Rat. Mech. Anal **86**, 317–351.

[13] Casten, R. and Holland, C. [1978], *Instability Results for Reaction-Diffusion Equations with Neumann Boundary Conditions*, J. Diff. Equations **27**, 266–273.

[14] Catchpole, E. A. [1972], *A Cauchy Problem for an Ordinary Integro-Differential Equation*, Proc. Roy. Soc. Edinburgh **72A**, 39–55.

[15] Chafee, N. [1981], *The Electric Balast Resistor: Homogeneous and Nonhomogeneous Equilibria*, in Nonlinear Differential Equations: Invariance, Stability and Bifurcation (P. de Mottoni and L. Salvadori, eds.), 97–127, Academic Press, New York.

[16] Chafee, N. [1975], *Asymptotic Behaviour for Solutions of a One–Dimensional Parabolic Equation with Homogeneous Neumann Boundary Conditions*, J. Differential Equations **18**, 111–134.

[17] Chafee, N. and Infante, E. F. [1974], *A Bifurcation Problem for a Nonlinear Partial Differential Equation of Parabolic Type*, Applicable Analysis **4**, 17–37.

[18] Deng, K., Kwong, M. K. and Levine, H. A. [1992], *The Influence of Nonlocal Nonlinearities on the Long Time Behaviour of Solutions of Burger's Equation*, Quart. Appl. Math. **L**, 173–200.

[19] Eckhaus, W. [1965], *Studies in Nonlinear Stability Theory*, Springer Tracts in Natural Philosophy **6**, Springer, Berlin.

[20] Fiedler, B. and Poláčik, P. [1990], *Complicated Dynamics of Scalar Reaction Diffusion Equations with a Nonlocal Term*, Proc. Roy. Soc. Edinburgh **115A**, 167–192.

[21] Fiedler, B. and Rocha, C. [1998], *Orbit Equivalence of Global Attractors of Semilinear Parabolic Differential Equations*, Trans. Amer. Math. Soc., to appear.

[22] Freitas, P. [1994] *A Nonlocal Sturm–Liouville Eigenvalue Problem*, Proc. Roy. Soc. Edinburgh **124**, 169–188.

[23] Freitas, P. [1994], *Bifurcation and Stability of Stationary Solutions of Nonlocal Scalar Reaction–Diffusion Equations*, J. Dynam. Differential Equations **6**, 613–629.

[24] Freitas, P. [1995], *Stability of Stationary Solution of a Nonlocal Reaction–Diffusion Equation*, Quart. J. Mech. Appl. Math. **48**, 558–582.

[25] Freitas, P. [1997], *Some results on the stability and bifurcation of stationary solutions of delay–diffusion equations*, J. Math. Anal. Appl. **206**, 59–82.

[26] Freitas, P. and Grinfeld, M. [1994], *Stationary Solutions of an Equation Modelling Ohmic Heating*, Appl. Math. Lett. **7**, 1–6.

[27] Freitas, P. and Sweers, G. [1997], *Positivity Results for a Nonlocal Elliptic Equation*, Proc. Roy. Soc. Edinburgh. Sec. A., to appear.

[28] Freitas, P. and Vishnevskii, M.P. [1997], *Stability of Stationary Solutions of Nonlocal Reaction–diffusion Equations in m–dimensional Space*, Diff. Int. Equations, to appear.

[29] Fujita, H. [1969], *On the Nonlinear Equations* $\Delta u + e^u = 0$ *and* $\partial v/\partial t = \Delta v + e^v$, Bull. Amer. Math. Soc. **75**, 132–135.

[30] Furter, J. and Grinfeld, M. [1989], *Local vs. Non–local Interactions in Population Dynamics*, J. Math. Biol. **27**, 65–80.

[31] Gidas, B, Ni, W.–M. and Nirenberg, L. [1979], *Symmetry and Related Properties via the Maximum Principle*, Coomun. Math. Phys. **68**, 109–143.

[32] Gifford, R. P. and Page, N. C. [1902], *Experiments with Ballasts of a Nernst Lamp*, B. S. Thesis, MIT, Cambridge, Mass.

[33] Gonze, X., Stumpf, R. and Scheffler, M. [1991], *Analysis of Separable Potentials*, Phys. Rev. B **44**.

[34] Grinfeld, M. and Novick–Cohen, A. [1995], *Counting the Stationary States of the Cahn–Hilliard Equation by Transversality Arguments*, Proc. Roy. Soc. Edinburgh **125A**, 351–370.

[35] Gurtin, M. E. and Matano, H. [1988], *On The Structure of Equilibrium Phase Transitions Within the Gradient Theory of Fluids*, Quart. Appl. Math. **XLVI**, 301–317.

[36] Hagelstein, P. L. [1978], *Solution of a Second-Order Integro-Differential Equation which occurs in Laser Modelocking*, J. Math. Phys. **19**, 1594–1599.

[37] Hale, J. K. and Sakamoto, K. [1989], *Shadow Systems and Attractors in Reaction-Diffusion Equations*, Appl. Anal. **32**, 287-303.

[38] Henry, D. [1981], *Geometric Theory of Semilinear Parabolic Equations*. Springer, New York.

[39] Henry, D. [1985], *Some Infinite–Dimensional Morse–Smale Systems Defined by Parabolic Differential Equations*, J. Differential Equations **59**, 165–205.

[40] Horgan, C. O. and Olmstead, W. E. [1978], *Stability and Uniqueness for a Turbulence Model of Burgers*, Quart. Appl. Math. XXXVI, 121-127.

[41] Joseph, D. D. and Lundgren, T. S. [1973], *Quasilinear Dirichlet Problems Driven by Positive Sources*, Arch. Rational Mech. Anal. **49**, 241–269.

[42] Kato, T. [1980], *Perturbation Theory for Linear Operators*, Springer, New York.

[43] Lacey, A. [1995], *Thermal Runaway in a Non–local Problem Modelling Ohmic Heating. I. Model Derivation and Some Special Cases*, European J. Appl. Math. **6**, 127–144.

[44] Lacey, A. [1995], *Thermal Runaway in a Non–local Problem Modelling Ohmic Heating. II. General Proof of Blow-up and Asymptotics of Runaway*, European J. Appl. Math. **6**, 201–224.

[45] Lin, C.–S. and Ni, W.–M. [1995], *On Stable States of Semilinear Diffusion Equations*, preprint.

[46] Liouville, J. [1837], *Solution Nouvelle d'un Problème d'Analyse, Relatif aux Phénomènes Thermo–mécaniques*, Journal des Mathématiques Pures et Appliquées **2**, 439–456.

[47] Matano, H. [1982], Nonincrease of the Lap-Number of a Solution for a One-Dimensional Semilinear Parabolic Equation, J. Fac. Sci. Univ. Tokyo, Sec. 1A **29**, 401–441.

[48] Miletta, P. D. [1988], *An Evolution Equation with Nonlocal Nonlinearities: Existence, Uniqueness and Asymptotic Behaviour*, Math. Methods in the Applied Sciences **10**, 407–425.

[49] Nasr, A. H. [1984], *A Study of One–Dimensional Schrodinger Equation with Nonlocal Potential*, SIAM J. Math. Anal. **15**, 1984.

[50] Poláčic, P. [1989], *Domains of Attraction of Equilibria and Monotonicity Properties of Convergent Trajectories in Parabolic Systems Admitting Strong Comparison Principle*, J. Reine Angew. Math. **400**, 32–56.

[51] Poláčic, P. and Šošovička, V. [1996], *Stable Periodic Solutions of a Spatially Homogeneous Nonlocal Reaction–diffusion Equation*, Proc. Roy. Soc. Edinburgh **126A**, 867–884.

[52] Rubinstein, J. and Sternberg, P. [1992], *Nonlocal Reaction–Diffusion Equations and Nucleation* IMA J. Appl. Math. **48**, 249–264.

[53] Walter, W. [1970], *Differential and Integral Inequalities*. Springer, Berlin.

Fields Institute Communications
Volume **21**, 1999

# Practical Persistence in Stochastic Population Models

**Thomas C. Gard**
Department of Mathematics
The University of Georgia
Athens, GA 30602, USA
gard@math.uga.edu

**Abstract.** We apply a result due to Miyahara on ultimate boundedness for stochastic systems to obtain practical persistence for some stochastic population interaction models. The assumptions required appear to be less restrictive than those needed to obain recurrence.

## 1 Introduction

To what extent is stability, for example persistence, preserved when random or stochastic fluctuations are included in high level models of population interactions? Persistence has emerged as possibly the most relevant stability concept for population models [**8, 15, 22**]. Practical persistence, which refers to determining specific estimates for persistence in terms of model data, is being recognized at least in the setting of deterministic models [**3, 4, 7, 14**], as a meaningful indicator of long-term population survival. For stochastic systems, sometime ago Ludwig in [**17**], suggested mean exit times from subregions at fixed distances away from the boundary of the feasible region as a characterizaton of persistence and he showed how to calculate such exit times in terms of model parameters. The latter would correspond to estimates of practical persistence. Here we consider a simpler characterization of persistence in the stochastic case, and by applying a theorem due to Miyahara [**18**] on uniform asymptotic boundedness, we show how to obtain practical persistence estimates. Miyahara's result involves construction of a Lyapunov function. Whether or not the approach we are suggesting here actually leads to practical results depends on the choice of the function as well as the sharpness of certain bounds in the construction.

The class of multi-population interaction models we consider can be written as systems of Ito stochastic differential equations (in Kolmogorov form)

$$dX = D(X)[f(X)dt + G(X)dW]. \tag{1.1}$$

In (1.1)

$$X = X(t, \omega) = \{X_i(t, \omega)\}_{i=1}^n$$

1991 *Mathematics Subject Classification.* 92D25, 60H10.

where $X_i$ represents the $i$th - species population density,

$$W = W(t, \omega) = \{W_j(t, \omega)\}_{j=1}^{m}.$$

$W_j$ denoting independent (scalar) Brownian motion processes, $t \geq 0$, $\omega$ is a sample variable in a probability space, and

$$D(X) = \operatorname{diag}\{X_1, X_2, \ldots, X_n\}.$$

Also in (1.1)

$$f(x) = \{f_i(x)\}_{i=1}^{n} \text{ and } G(x) = \{g_{ij}(x)\}_{i,j=1}^{n,m}$$

are $C^1$ functions on the positive cone

$$R_+^n = \{x : x_i > 0,\ i = 1, \ldots, n\}$$

which is the feasible region, and (1.1) can be written in component form as

$$dX_i = X_i[f_i(X)dt + \sum g_{ij}(X)dW_j] \quad i = 1, ..., n.$$

Due to the difficulties in analysis which arise when random or stochastic fluctuations are included in such models, especially if some description of long term qualitative features such as persistence is sought, one can impose a specific form such as (1.1) for how to account for the random aspects which is sufficiently mathematically tractable. A crucial step in the modelling process then is justifying the chosen form from the applications perspective. The above class of models has emerged in this way as a balance between applications relevance and mathematical tractability. Indeed, models of the form (1.1) may be derived (from basic demographic assumptions on the population interactions, e.g., see Abundo [**1**] for a predator-prey model) as a diffusion process limit of a sequence of discrete stochastic models:

$$E([X(t) - X(s)] \mid X(s) = x)/(t - s) \to D(x)f(x)$$

and

$$E([X(t) - X(s)][X(t) - X(s)]^T \mid X(s) = x)/(t - s) \to D(x)G(x)G^T(x)D(x)$$

as $t \to s^+$.

Here $E(\cdot \mid X(s) = x)$ denotes the conditional expectation given $X(s) = x$. Formally, (1.1) also arises if one models the impact of environmental stochasticity by allowing the intrinsic per capita net growth rates of the populations to be random noise fluctuations

$$f(x) + G(x)N$$

about some average values $f(x)$ with intensities $G(x)$ both (possibly) dependent on population size. Here

$$N = \{N_j\}_{j=1}^{m}, N_j = N_j(t, \omega),$$

the $N_j$ specifying independent (scalar) white noises; i.e., for each component species

$$\frac{1}{x_i}(dx_i/dt) = f_i(x) + \sum_{j=1}^{m} g_{ij}(x)N_j.$$

Since the above system is not integrable in the usual way, and since, at least formally

$$N = \frac{dW}{dt}$$

the Ito version of the integrated form of (1.1) gives a natural tractable interpretation. (Stephanopoulos, Aris, and Fredrickson [**21**] give an example of such a model for two competing species in a chemostat in which the stochastic effects destabilize the corresponding deterministic interaction.)

Suppose $x_i(t)$, representing the population density of the $i$th of a certain collection of species at time $t$, is the $i$th component of the solution $x = x(t)$ of the deterministic differential equation

$$\frac{dx}{dt} = D(x)f(x). \tag{1.2}$$

Uniform persistence of (1.2) means that there is a positive number $\delta$ such that if, for each $i, x_i(0) > 0$, then

$$\delta \leq \liminf_{t\to\infty} x_i(t). \tag{1.3}$$

Further, if there is a constant $M$ such that, for all such $x_i(t)$,

$$\limsup_{t\to\infty} x_i(t) \leq M \tag{1.4}$$

then (1.2) is called dissipative (or uniformly ultimately bounded). If both (1.3) (and (1.4)) hold, then (1.2) is called (uniformly) persistent (and permanent). For specific models of real systems, relative sizes of the bounds are important – small values for $\delta$ may not be very good indicators of survival. Practical persistence (resp. permanence) then refers to obtaining specific estimates for $\delta$ (and $M$) in terms of the model data

$$\delta = \delta(f)(\text{and } M = M(f)). \tag{1.5}$$

This is a general theme implicit in using Lyapunov functions to obtain persistence, for example, in the persistence result for simple food chains, in [**9**]. Recently there has been some interest in extending the idea of practical persistence to some PDE [**3, 4**] and discrete population interaction models [**14**], following some remarks in our work with Cao on persistence for models with time delays [**5, 6**]. The main point of this paper is to try to further extend the discussion of practical persistence to stochastic population interaction models at least of the Ito type (1.1). The most natural analogues of (1.3) and (1.4) for stochastic models (1.1) are respectively

$$\delta \leq \liminf_{t\to\infty} EX_i(t) \tag{1.6}$$

$$\limsup_{t\to\infty} EX_i(t) \leq M. \tag{1.7}$$

Practical persistence (permanence) is then described by giving explicit functions

$$\delta = \delta(f,G)(\text{and } M(f,G)). \tag{1.8}$$

There are some shortcomings of using (1.6) and (1.7) for permanence – there may be large deviations from mean values. At least (1.6) - (1.8) allow us to give some crude estimates for the probabilities of such deviations in terms of model data. In the next section we state Miyahara's theorem on ultimate boundedness for stochastic systems and indicate how this result yields practical stochastic permanence for systems having Ito form. Section 3 treats some population interaction examples.

## 2 Miyahara's theorem and practical persistence

It is known (e.g., see [**13**]) that permanence for the deterministic multi-population interaction model

$$\frac{dx}{dt} = D(x)f(x) \tag{2.1}$$

implies the existence of a positive equilibrium $x^*$:

$$x^* \in R^n_+ \text{ such that } f(x^*) = 0.$$

We can then consider the change of variables

$$y_i = ln(x_i/x_i^*) \tag{2.2}$$

which transforms Equation (2.1) into

$$dy/dt = \tilde{f}(y) = f(D(x^*)\exp(y)), \quad y \in R^n, \tag{2.3}$$

where, for $y \in R^n$

$$\exp(y) = (\exp(y_1), \dots, \exp(y_n))^T.$$

The positive equilibrium $x^*$ of (2.1) corresponds to the equilibrium $y^* = 0$ of (2.3) and permanence for (2.1) in $R^n_+$ is equivalent to dissipativeness (or ultimate boundedness) for (2.3) in $R^n$:

$$\limsup_{t\to\infty} \| y(t) \| < M. \tag{2.4}$$

Applying (2.2) to the stochastic Equation (1.1) yields

$$dY = \tilde{f}^G(Y)dt + \tilde{G}(Y)dW \tag{2.5}$$

where $\tilde{f}^G = \{\tilde{f}_i^G\}$ and $\tilde{G} = \{\tilde{g}_{ij}\}$, with, due to Ito's Formula,

$$\tilde{f}_i^G(Y) = \left[ f_i - \frac{1}{2}\sum_j g_{ij}^2 \right] (D(x^*)\exp(Y))$$

and

$$\tilde{g}_{ij}(Y) = g_{ij}(D(x^*)\exp(Y)).$$

If (2.5) has a stable invariant distribution, i.e., a probabiilty measure $\mu$ defined on the Borel sets $\mathcal{B}$ in $R^n$ such that, for each $B \in \mathcal{B}$ and $y \in R^n$

$$\begin{gathered} P\{Y(t) \in B \mid Y(0) = y\} \to \mu(B) \text{ as } t \to \infty, \\ \int_{R^n} P\{Y(t) \in B \mid Y(0) = y\}\mu(dy) = \mu(B), \end{gathered} \tag{2.6}$$

then what might be called stochastic permanence follows: for any $\epsilon > 0$, there is an $M = M(\epsilon) > 0$ such that

$$\limsup_{t\to\infty} P\{\|Y(t)\| \le M\} \ge 1 - \epsilon.$$

Indeed, if $\mu$ has density $\rho = \rho(y)$

$$P\{\|Y(t)\| \le M\} \underset{t\to\infty}{\longrightarrow} \mu(\{\|y\| \le M\}) = \int_{\{\|y\|\le M\}} \rho(y)dy. \tag{2.7}$$

The main relevant results [**2, 10, 11, 16, 19, 20**] state that if (2.3) is globally stable and the noise intensity $G$ is sufficiently small, then (2.5) is positive recurrent which implies that $\mu$ exists. In this setting, practical stochastic permanence then would amount to estimating

$$M = M(\epsilon, \tilde{f}^G, \tilde{G}) \tag{2.8}$$

and this could be accomplished if one could estimate $\mu$ or $\rho$ in (2.6) or (2.7) respectively. The latter may not be easy, however when it exists, $\rho$ solves the equilibrium Fokker-Planck equation

$$\sum_i \frac{\partial}{\partial y_i}[\tilde{f}_i \rho] - \frac{1}{2}\sum_{,j} \frac{\partial^2}{\partial y_i \partial y_j}\left[\left(\sum_k \tilde{g}_{ik}\tilde{g}_{jk}\right)\rho\right] = 0. \tag{2.9}$$

The following theorem (and proof) of Miyahara [**18**], on the other hand, provides a direct way of obtaining (2.8) and seems not to require assumptions as strong as invoked in the above results.

**Theorem 2.1** *(Miyahara) Suppose that $p$ is a positive number and $V = V(y)$ is a $C^2$ function defined on $R^n$ which satisfies, for all $y \in R^n$*

(i) *there are constants $a_1$ and $c_1(> 0)$ such that*

$$V(y) \geq -a_1 + c_1\|y\|^p,$$

(ii) *there are constants $a_2$ and $c_2(> 0)$ such that*

$$\mathcal{L}V(y) \leq a_2 - c_2 V(y),$$

*where*

$$\mathcal{L}v(y) = \sum_i \tilde{f}_i^G \frac{\partial V}{\partial y_i} + \frac{1}{2}\sum_{i,j}\left(\sum_k \tilde{g}_{ik}\tilde{g}_{jk}\right)\frac{\partial^2 V}{\partial y_i \partial y_j},$$

*then there is a constant $K$ such that, if $Y(t)$ is any solution of (2.5)*

$$\limsup_{t\to\infty} E_y\|Y(t)\|^p \leq K, \tag{2.10}$$

*where $E_y(\cdot) = E(\cdot \mid Y(0) = y)$ the conditional expectation given $Y(0) = y$.*

It follows from Miyahara's Theorem that if $0 < \epsilon < \delta$ and $M = (K+\delta)^{1/p}$

$$P(\|Y(t)\| \leq M \mid Y(0) = y) \geq 1 - \frac{K+\epsilon}{M^p} \tag{2.11}$$

for $t$ sufficiently large. What turns (2.10) into a practical persistence result is the explicit formula for the bound $K$, given in Miyahara's proof

$$K = \frac{a_1}{c_1} + \frac{a_2}{c_1 c_2} \tag{2.12}$$

and the fact that the constants $a_1, a_2, c_1$ and $c_2$ can be determined from $\tilde{f}^G, \tilde{G}$, and the choice of $V$.

**Example 1** Suppose (2.5) has the form

$$dY_i = \tilde{f}_i(Y)dt + \tilde{g}_i(Y)dW_i \quad (i = 1, \ldots, n) \tag{2.13}$$

where the functions $\tilde{f} = \{\tilde{f}_i\}$ and $\tilde{g} = \{\tilde{g}_i\}$ satisfy, for some constant $p \geq 2$ and some positive constants $R$, $k_1$ and $k_2$.

$$y^T \tilde{f}(y) \leq -k_1\|y\|^2 \tag{2.14}$$

and

$$\|\tilde{g}(y)\|^2 \leq k_2/\|y\|^{p-2} \tag{2.15}$$

both for $\|y\| \geq R$. To apply Miyahara's result, we take

$$V(y) = \|y\|^p = \left(\sum_{i=1}^{n} y_i^2\right)^{p/2}. \tag{2.16}$$

In this case, $a_1 = 0$ and $c_1 = 1$ and so

$$K = a_2/c_2. \tag{2.17}$$

Further, we get

$$\mathcal{L}V(y) = p\|y\|^{p-2} \sum_i \left\{ y_i \tilde{f}_i(y) + \frac{1}{2}\left(1 + \frac{(p-2)y_i^2}{\|y\|^2}\right) \tilde{g}_i^2(y) \right\} \leq a_2 - c_2\|y\|^p \tag{2.18}$$

where

$$a_2 = \frac{k_2 p(p-1)}{2} \text{ and } c_2 = k_1 p$$

if $\|y\| \geq R$. Now let

$$k_3 = \max\{\mathcal{L}V(y) : \|y\| \leq R\}$$

and replace $a_2$ by

$$a_2 = \max\left\{\frac{k_2 p(p-1)}{2}, k_3 + k_1 p R^p\right\}.$$

Then (2.18) holds for all $y$ and we get

$$\limsup_{t \to \infty} E_y \|Y(t)\|^p \leq K = \frac{a_2}{k_1 p}. \tag{2.19}$$

Consequently for $0 < \epsilon < \delta$, and all $t$ sufficiently large,

$$P(\|Y(t)\| \leq (K+\delta)^{1/p} \mid Y(0) = y) \geq \frac{1 - \epsilon/\delta}{1 + K/\delta}. \tag{2.20}$$

The conditions (2.14) and (2.15) imposed here are somewhat restrictive for population interaction models. In the next section, we give some more relevant examples.

## 3 Population interaction examples

In this section we apply Miyahara's result making use of the Lyapunov function

$$V(y) = \exp\{U(y)\} \tag{3.1}$$

where, with $\tilde{\alpha}_i = \alpha_i x_i^*$,

$$U(y) = \sum_{i=1}^{n} \tilde{\alpha}_i [e^{y_i} - y_i - 1] = \sum_{i=1}^{n} \alpha_i \left[ x_i - x_i^* - x_i^* ln\left(\frac{x_i}{x_i^*}\right)\right] = V_0(x) \tag{3.2}$$

which is Volterra's Lyapunov function. If we take the 1-norm

$$\|y\| = \sum_{i=1}^{n} \mid y_i \mid, \tag{3.3}$$

then it follows that

$$U(y) \geq -\tilde{\beta} + \tilde{\alpha}\|y\| \tag{3.4}$$

where

$$\tilde{\beta} = \sum_{i=1}^{n} \tilde{\alpha}_i \text{ and } \tilde{\alpha} = \min \tilde{\alpha}_i \tag{3.5}$$

and we have

$$V(y) \geq \exp\{-\tilde{\beta} + \tilde{\alpha}\|y\|\} \geq e^{-\tilde{\beta}}(1 + \tilde{\alpha}\|y\|). \tag{3.6}$$

Condition (i) of Theorem 2.1 holds with $a_i$ an arbitrary positive number and

$$c_1 = \tilde{a}e^{-\tilde{\beta}}. \tag{3.7}$$

We can take $a_1 = 0, c_1$ given by (3.7) and hence to apply the theorem it remains to obtain $a_2$ and $c_2$ so that condition (ii) holds. We must determine bounds on $\mathcal{L}V$ in terms of the model data.

**Example 2** (Lotka-Volterra) (In the first part of this example, we essentially follow Polansky [**19**].) Suppose (1.1) has the form

$$dX_i = X_i[(h_i + \sum_{j=1}^{n} b_{ij}X_j)dt + g_i dW_i] \quad (i = 1, \ldots, n), \tag{3.8}$$

where the $h_i, b_{ij}$ and $g_i$ are constants, and assume there is $x^* = \{x_i^*\} \in R_+^n$ such that

$$h_i + \sum_{j=1}^{n} b_{ij}x_i^* = 0 \tag{3.9}$$

and that there is a positive diagonal matrix $A = \operatorname{diag}\{\alpha_1, \ldots, \alpha_n\}$ such that the matrix

$$AB + B^T A \text{ is negative definite} \tag{3.10}$$

where $B = \{b_{ij}\}$. Conditions (3.9) and (3.10) guarantee that there is a positive equilibrium $x^*$ of the corresponding deterministic Lotka-Volterra system

$$\frac{dx_i}{dt} = x_i(h_i + \sum_{j=1}^{n} b_{ij}x_j) \quad (i = 1, \ldots, n) \tag{3.11}$$

which is globally (relative to the positive cone $R_+^n$) asymptotically stable - the latter verifiable by using Volterra's Lyapunov function $V_0(x)$ given in (3.2) with coefficients $\alpha_i$ satisfying (3.10). The transformation $Y_i = ln(X_i/x_i^*)$ applied to (3.8) yields, by Ito's formula,

$$dY_i = \left(\tilde{h}_i + \sum_{j=1}^{n} \tilde{b}_{ij} \exp(Y_j)\right) dt + g_i dW_i \quad (i = 1, \ldots, n) \tag{3.12}$$

where, for each $i$ and $j$,

$$\tilde{h}_i = h_i - \frac{g_i^2}{2} \text{ and } \tilde{b}_{ij} = b_{ij}x_i^*. \tag{3.13}$$

Taking

$$U(y) = V_0(x) \tag{3.14}$$

where $y_i = ln(x_i/x_i^*)$ and, with $\tilde{A} = \operatorname{diag}\{\tilde{\alpha}_1, \ldots, \tilde{\alpha}_n\}$, we can calculate

$$\mathcal{L}U(y) = \frac{1}{2}\{[e^y - 1](\tilde{A}B + B^T\tilde{A})[e^y - 1]^T + \gamma\} \tag{3.15}$$

where

$$\gamma = \sum \tilde{\alpha}_i g_i^2. \tag{3.16}$$

If

$$V(y) = \exp\{U(y)\}, \tag{3.17}$$

then, if $\gamma$ is sufficiently small, there is a positive number $\mu$ such that

$$\mathcal{L}V(Y) = [\mathcal{L}U(y) + \frac{1}{2}\sum \tilde{\alpha}_i^2 g_i^2 [e^{y_i} - 1]^2]V \leq -\mu V \tag{3.18}$$

if $y$ is outside some set $\Omega$. Indeed

$$\mathcal{L}U(y) + \frac{1}{2}\sum \tilde{\alpha}_i^2 g_i^2 [e^{y_i} - 1]^2 = \frac{1}{2}\{\gamma + [e^y - 1](\tilde{A}B + B^T\tilde{A} + \tilde{A}GG^T\tilde{A})[e^y - 1]^T\}. \tag{3.19}$$

If $\gamma$ is sufficiently small, the matrix

$$\tilde{A}B + B^T\tilde{A} + \tilde{A}GG^T\tilde{A} \text{ is negative definite} \tag{3.20}$$

and the set

$$\Omega_0 = \{y : \gamma + [e^y - 1](\tilde{A}B + B^T\tilde{A} + \tilde{A}GG^T\tilde{A})[e^y - 1]^T \geq 0\}$$

which includes the origin, is bounded. By continuity, the same is true for

$$\Omega = \Omega_\mu = \{y : \gamma + [e^y - 1](\tilde{A}B + B^T\tilde{A} + \tilde{A}GG^T\tilde{A})[e^y - 1]^T \geq -\mu\} \tag{3.21}$$

for sufficiently small $\mu > 0$. Now $\mathcal{L}V$ and $V$ are bounded on $\Omega$, say by the numbers $M_1$ and $M_2$ respectively. In particular we can choose $a_2 > M_1$ so large that

$$\mathcal{L}V(y) \leq M_1 \leq a_2 - \mu M_2 \leq a_2 - \mu V \tag{3.22}$$

on $\Omega$. Summarizing, from (3.14), (3.15), (3.18) and (3.22), we have the conditions (i) and (ii) of Miyahara's Theorem satisfied with

$$a_1 = 0, c_1 = \tilde{\alpha}e^{-\tilde{\beta}}, a_2 = M_1 + \mu M_2, \text{ and } c_2 = \mu. \tag{3.23}$$

Then recalling the conclusion (and that $p = 1$ here), we have

$$\limsup_{t\to\infty} E_y\|Y(t)\| \leq K \tag{3.24}$$

with $K$ given by

$$K = \frac{a_1}{c_1} + \frac{a_2}{c_1 c_2} = \frac{(M_1 + \mu M_2)e^{\tilde{\beta}}}{\tilde{\alpha}\mu}. \tag{3.25}$$

**Example 3** (Gause-Type Predator-Prey) Suppose (1.1) has the form

$$\begin{aligned} dX_1 &= [X_1 h(X_1) - p(X_1)q(X_2)]dt + X_1 g_1(X_1, X_2)dW_1 \\ dX_2 &= [n(X_2)(k(X_1) - r(X_2))]dt + X_2 g_2(X_1, X_2)dW_2, \end{aligned} \tag{3.26}$$

where $h$ is a decreasing continuous function on $R_+, h(0) > 0, h(\overline{x}) = 0$, some $\overline{x} > 0$, and $k, n, p, q$, and $r$ are increasing continuous functions with

$$k(0) < 0 \text{ and } n(0) = p(0) = q(0) = r(0) = 0.$$

For the corresponding deterministic model

$$\begin{aligned} dx_1/dt &= x_1 h(x_1) - p(x_1)q(x_2) \\ dx_2/dt &= n(x_2)[k(x_1) - r(x_2)], \end{aligned} \tag{3.27}$$

if there is a stable positive equilibrium $x^* = (x_1^*, x_2^*) \in R_+^2$, Harrison's [**12**] Lyapunov function

$$H(x_1, x_2) = \int_{x_1^*}^{x_1} \frac{k(u) - k(x_1^*)}{p(u)} du + \int_{x_2^*}^{x_2} \frac{q(v) - q(x_2^*)}{n(v)} dv \tag{3.28}$$

can be used to verify global stability of $x^*$. Ragsdale [**20**] has recently used (3.28) to determine some sufficient conditions for positive recurrence of (3.26). With the transformation

$$Y_i = ln(X_i/x_i^*)$$

(3.26) becomes

$$\begin{aligned} dY_1 &= \left( h(x_1^* \exp Y_1) - \frac{p(x_1^* \exp Y_1) q(x_2^* \exp Y_2)}{x_1^* \exp Y_1} - \frac{\tilde{g}_1^2}{2} \right) dt + \tilde{g}_1 dW_1 \\ dY_2 &= \left( \frac{n(x_2^* \exp Y_2)[k(x_1^* \exp Y_1) - r(x_2^* \exp Y_2)]}{x_2^* \exp Y_2} - \frac{\tilde{g}_2^2}{2} \right) dt + \tilde{g}_2 dW_2 \end{aligned} \tag{3.29}$$

with

$$\tilde{g}_i = \tilde{g}_i(Y_1, Y_2) = g_i(x_1^* \exp Y_1, x_2^* \exp Y_2).$$

We can apply Theorem 2.1 to (3.29) using

$$\begin{aligned} V(y_1, y_2) &= \exp\{H(x_1, x_2)\} \\ &= \exp\left\{ x_1^* \int_0^{y_1} \frac{k(x_1^* \exp w) - k(x_1^*)}{p(x_1^* \exp w)} \exp w \, dw \right. \\ &\quad \left. + x_2^* \int_0^{y_2} \frac{q(x_2^* \exp z) - q(x_2^*)}{n(x_2^* \exp z)} \exp z \, dz \right\} \end{aligned} \tag{3.30}$$

where

$$w = ln(u/x_1^*) \quad \text{and} \quad z = ln(v/x_2^*).$$

We now indicate how to determine $\mu$, $M_1$ and $M_2$ for the special case of Michaelis-Menten dynamics, i.e.,

$$k(x) = -d + fx/(m + x), \quad p(x) = cx/(m + x)$$

and

$$h(x) = a - bx, n(x) = q(x) = x, r(x) = sx,$$

where $a, b, c, d, f, s$ and $m$ denote positive constants, and where we also suppose $g_1 = 0$ and $g_2 = g$, a constant. Then (3.29) becomes

$$\begin{aligned} dY_1 &= \left( a - bx_1^* \exp Y_1 - \frac{cx_2^* \exp Y_2}{m + x_1^* \exp Y_1} \right) dt \\ dY_2 &= \left( -d + \frac{fx_1^* \exp Y_1}{m + x_1^* \exp Y_1} - sx_2^* \exp Y_2 - \frac{g^2}{2} \right) dt + g dW \end{aligned} \tag{3.31}$$

and $H$ reduces to a function of the classical Volterra type, so that $V$ has the same form as (3.8)

$$\begin{aligned} V(y_1, y_2) &= \exp\left\{ \frac{x_1^* f}{c} \int_0^{y_1} \exp w - 1 \, dw + x_2^* \int_0^{y_2} \exp z - 1 \, dz \right\} \\ &= \exp\left\{ \frac{x_1^* f}{c} [e^{y_1} - y_1 - 1] + x_2^* [e^{y_2} - y_2 - 1] \right\}. \end{aligned}$$

and

$$\tilde{\alpha} = \left\{ \frac{x_1^* f}{c}, x_2^* \right\}, \quad \tilde{\beta} = \frac{x_1^* f}{c} + x_2^*. \tag{3.32}$$

Since $V$ is the same as in Example 2, and the $g_i$ are also constant here, we have at least the equality part of (3.18)

$$\mathcal{L}V(y) = \{\mathcal{L}H(y) + \frac{1}{2} x_2^{*2} g^2 [e^{y_2} - 1]^2\} V. \tag{3.33}$$

Here, as in [**20**], it can be shown that

$$\begin{aligned}
\mathcal{L}H(y) &= \frac{x_1^* f}{c} [e^{y_1} - 1] \left( a - b x_1^* e^{y_1} - \frac{c x_2^* e^{y_2}}{m + x_2^* e^{y_1}} \right) \\
&\quad + x_2^* [e^{y_2} - 1] \left( -d + \frac{f x_1^* e^{y_1}}{m + x_1^* e^{y_1}} - s x_2^* e^{y_2} - \frac{g^2}{2} \right) + \frac{1}{2} x_2^* e^{y_2} g^2 \\
&= \frac{f}{c} \left( \frac{x_1^* e^{y_1}}{m + x_1^* e^{y_1}} - \frac{x_1^*}{m + x_1^*} \right) \times \\
&\qquad ((a - b x_1^* e^{y_1})(m + x_1^* e^{y_1}) - (a - b x_1^*)(m + x_1^*)) \\
&\qquad\qquad - s[x_2^* e^{y_2} - x_2^*]^2 + \frac{x_2^* g^2}{2}.
\end{aligned} \tag{3.34}$$

Thus from (3.33) and (3.34) we have

$$\begin{aligned}
\mathcal{L}H(y) + \frac{1}{2} x_2^{*2} g^2 [e^{y_2} - 1]^2 &= \frac{f}{c} \left( \frac{x_1^* e^{y_1}}{m + x_1^* e^{y_1}} \frac{x_1^*}{m + x_1^*} \right) \times \\
&\qquad ((a - b x_1^* e^{y_1})(m + x_1^* e^{y_1}) - (a - b x_1^*)(m + x_1^*)) \\
&\qquad - \left( s - \frac{g^2}{2} \right) [x_2^* e^{y_2} - x_2^*]^2 + \frac{x_2^* g^2}{2}.
\end{aligned} \tag{3.35}$$

If $a \leq bm$ and $g < \sqrt{2s}$, the first two terms on the right in (3.35) define a negative definite function of $y$ in $R^2$. For $g$ and $\mu$ sufficiently small, the set

$$\Omega = \Omega_\mu = \{y : \mathcal{L}H(y) + \frac{1}{2} x_2^{*2} g^2 [e^{y_2} - 1]^2 \geq -\mu\}$$

is a compact set in $R^2$ whch contains the origin. Now $M_1$ and $M_2$ can be determined relative to the set $\Omega$, and taking into account (3.32), $K$ can be calculated. For example, if $a = m = 2, b = d = 1, c = 3, f = 4, g = 1/5$ and $s = 1/3$, then

$$x_1^* = 1 = x_2^*$$

and $(x_1^*, x_2^*)$ is stable for the corresponding deterministic system. From (3.32),

$$\tilde{\alpha} = \min \left\{ \frac{x_1^* f}{c}, x_2^* \right\} = 1,$$

$$\tilde{\beta} = \frac{x_1^* f}{c} + x_2^* = \frac{7}{3}.$$

We can take $\mu = .1$ here, and then it can be seen that $\Omega = \Omega_{.1}$ is the bounded region

$$200\left(\frac{e^{y_1}}{2+e^{y_1}} - \frac{1}{3}\right)(e^{2y_1} - 1) + 47[e^{y_2} - 1]^2 \leq 18.$$

For this $\Omega$, one can estimate $M_2 = 1.9, M_1 \leq \frac{(.2)^2}{2} M_2 = .038$ and, from (3.25),

$$K = \frac{e^{\tilde{\beta}}}{\tilde{\alpha}}\left(\frac{M_1}{\mu} + M_2\right) = e^{7/3}(.38 + 1.9) \approx 23.5.$$

## References


[1] Abundo, M. [1991], *A stochastic model for predator-prey systems: basic properties, stability, and computer simulation*, J. Math. Biol. **29**, 495-511.

[2] Barra, M., Del Grosso, G., Gerardi, A., Koch, G. and Marchetti, F. [1979], *Some basic properties of stochastic population models*, in Sys. Theory in Immunology (C. Bruni, ed.), Proc., Rome; Lect. Notes in Biomath. **32**, Springer-Verlag, Berlin, 155-164.

[3] Cantrell, R. S. and Cosner, C., *Practical persistence in ecological models via comparison methods*, preprint.

[4] Cantrell, R. S. and Cosner, C., *Practical persistence in diffusive food chain models*, preprint.

[5] Cao, Y. and Gard, T. C. [1993], *Uniform persistence for population models with time delay using multiple Lyapunov functions*, Differential and Integral Equations **6**, 883-898.

[6] Cao, Y. and Gard, T. C. [1995], *Uniform persistence and net functions*, J. Dynamics and Differential Equations **7**, 491-520.

[7] Cao, Y. and Gard, T. C., *Practical persistence for differential delay models of population interactions*, preprint.

[8] Freedman, H. I. and Moson, P. [1990], *Persistence definitions and their connections*, Proc. Amer. Math. Soc. **109**, 1025-1033.

[9] Gard, T. C. [1987], *Uniform persistence in multispecies population models*, Math. Biosci. **85**, 93-104.

[10] Gard, T. C., *Recurrence for stochastic multi-population models*, in Advances in Mathematical Population Dynamics: Molecules, Cells and Man (O. Arino, D. Axelrod and M. Kimmel, eds.), World Scientific, Singapore, accepted for publication.

[11] Gard, T. C. and Ragsdale, K. B. [1995], *Recurrence for stochastic predator-prey models*, in Differential Equations and Applications to Biology and to Industry (M. Martelli, K. Cooke, E. Cumberbatch, B. Tang and H. Thieme, eds.), World Scientific, Singapore, 117-125.

[12] Harrison, G. [1979], *Global stability of predator-prey interactions*, J. Math. Biol. **8**, 159-171.

[13] Hofbauer, Josef and Sigmund, Karl [1988], *The Theory of Evolution and Dynamical Systems*, Cambridge University Press, Cambridge, England.

[14] Hutson, V. and Mischaikow, K., *An approach to practical persistence*, preprint.

[15] Hutson, V. and Schmidt, K. [1992], *Permanence in dynamical systems*, Math. Biosci. **111**, 1-71.

[16] Kesten, H. and Ogura, Y. [1981], *Recurrence properties of Lotka-Volterra models with random fluctuations*, J. Math. Soc. Japan **32**, 335-366.

[17] Ludwig, D. [1975], *Persistence of dynamical systems under random perturbations*, SIAM Rev. **17**, 605-639.

[18] Miyahara, Y. [1972], *Ultimate boundedness of the systems governed by stochastic differential equations*, Nagoya Math. J. **47**, 111-144.

[19] Polansky, P. [1979], *Invariant distributions for multi-population models in random environments*, Theor. Pop. Biol. **14**, 25-34.

[20] Ragsdale, K. B. [1991], *Qualitative theory of stochastic prey-predator models*, Ph.D. Dissertation, The University of Georgia, Athens, GA.

[21] Stephanopoulos, G., Aris, R. and Fredrickson, A. G. [1979], *A stochastic analysis of the growth of competing microbial populations in a continuous biochemical reactor*, Math. Biosci. **45**, 99-135.

[22] Waltman, P. [1991], *A brief survey of persistence in dynamical systems*, in Delay Differential Equations and Dynamical Systems (S. Busenberg and M. Martelli, eds.), Lect. Notes in Math. **1475**, Springer-Verlag, Berlin, 31-40.

Fields Institute Communications
Volume **21**, 1999

# Structure and Dynamics of Artificial Neural Networks

**Tomáš Gedeon**
Department of Mathematics
Montana State University
Bozeman, MT 59715, USA
gedeon@poincare.math.montana.edu

**Abstract.** We discuss how the structure of connections affects dynamics of artificial neural networks. We review known results and then generalize result of B. Fiedler and T. Gedeon to show that dynamics is quasi-convergent for a certain class of networks. In the process we generalize a theorem of J. Maybee on symmetrization of combinatorially symmetric, irreducible matrices.

## 1 Introduction

In this paper we consider an analog neural network

$$\dot{x}_i = -x_i + \sum_{j=1}^{n} w_{ij} f_j(x_j) \tag{1.1}$$

$i = 1, \dots, n$, where without loss of generality we assume $w_{ii} = 0$. Also, we consider a generalized Lotka-Volterra system

$$\dot{x}_i = x_i(c_i - \sum_{j=1}^{n} w_{ij} f_j(x_j)) \tag{1.2}$$

defined on the positive orthant $\mathbf{R}^{n+} = \{x \in \mathbf{R}^n \mid x_i \geq 0 \text{ for all } i\}$, where $c_i$, $i = 1, \dots, n$ are constants. In fact both of these systems belong to the class of equations

$$\dot{x}_i = a_i(\mathbf{x})(\gamma_i(x_i) - \sum_{j=1}^{n} w_{ij} f_j(x_j)) \tag{1.3}$$

where $w_{ii} = 0$. We assume that

$$\frac{\partial f_j}{\partial x_j} > 0 \text{ for all } j. \tag{\textbf{H1}}$$

and the functions $a_i(\mathbf{x})$ are nonnegative,

---

1991 *Mathematics Subject Classification.* 34C37, 68T10, 92B20.

This research was partially supported by MONTS grant MONTS-291725 and NSF grant DMS-291222.

$$a_i(\mathbf{x}) \geq 0 \qquad \text{for all } \mathbf{x} \in \mathbf{R}^n \text{ and all } i. \qquad \textbf{(H2)}$$

The main goal of this paper is to discuss correspondence between the structure of the matrix $W := [w_{ij}]$ and the dynamics of the system (1.3). We first describe motivation of the problem, then review the history and related results. We also want to point out related results in combinatorial matrix theory.

System of the type (1.1) can be considered as a model of neural network consisting of $n$ neurons. Each $x_i$ represents an activation level of the corresponding neuron. The stable equilibria of the system (1.1) represent stored memory. The trajectories, starting sufficiently close to an asymptotically stable equilibrium of (1.1), will converge to that equilibrium. From the point of view of neural networks this models retrieval of a stored memory pattern upon presentation of a partial, corrupted pattern and preservation of this pattern in the short term memory. This interpretation then leads to applications of systems (1.1) to parallel memory storage, content-addressable memory and the global pattern formation. Models of short term memory and global pattern formation using systems of type (1.1) and (1.3) go back to work in [**9, 11**]. These papers focus on proving general convergence theorems for certain classes of additive neural networks. Indeed, when trying to design a system which could serve as a memory storage, the fundamental question is whether the dynamics of the system is convergent or at least quasi-convergent [**15**]. The system is *convergent*, if every trajectory of the system converges to an equilibrium. To show that the system is convergent, one must prove that equilibria are isolated. Since this is usually a nontrivial problem, we need a more general notion of convergence. We say that the system is *quasi-convergent*, if every trajectory of the system converges to the set of equilibria. The system is *almost quasi-convergent* if trajectory of almost every initial value converges to the set of equilibria. Only after the question of convergence has been resolved can one actually try to design the system so that the equilibria are at the desired places, or, again using neural networks terminology, to design a system, which stores desired memory patterns.

Convergence theorems were proved for various generalizations of additive neural networks by Grossberg in [**6, 7, 8**] and for a system analogous to system (1.3), in [**1**]. Our result below is a generalization of their result.

The *structure matrix* of the system (1.3) is matrix $W$. To every $n \times n$ matrix $W$ we associate a digraph $D(W)$ containing $n$ vertices and a directed edge $e_{ij}$ from point $i$ to point $j$ if, and only if, $a_{ij} \neq 0$, $i \neq j$. Paths and cycles in $D(W)$ are defined in the usual way.

**Definition 1.1** Matrix $W$ is combinatorially symmetric, if $w_{ij} \neq 0$ implies $w_{ji} \neq 0$.

**Definition 1.2** Matrix $W$ is sign symmetric, if $w_{ij} \neq 0$ implies $w_{ij}w_{ji} > 0$.

Observe that if $W$ is combinatorially symmetric, $D(W)$ gives rise to an undirected graph $G(W)$ by identifying the edges $e_{ij}$ and $e_{ji}$.

**Definition 1.3** The system (1.3) is *cooperative* if for any $i \neq j$, $w_{ij} \geq 0$ and *competitive* if for all $i \neq j$, $w_{ij} \leq 0$.

Properties of the structure matrix $W$, which have nontrivial consequences for dynamics of (1.3) are symmetry, non-negativity and a simple structure of matrix $W$ (i.e., tridiagonality). We shall illustrate this in several examples and then generalize some of these results.

**Example 1.4** ([**1**]) Consider system (1.3) on $\mathbf{R}^{n+}$. Assume that the function $a_i(\mathbf{x})$ is a function of first variable only, $a_i(\mathbf{x}) = a_i(x_i)$, and $a_i(x_i) > 0$ for $x_i > 0$. Assume also (H1) and that functions $f_j \geq 0$. Furthermore, let $W$ be symmetric and nonnegative.

Then the system (1.3) admits a Lyapunov function and every bounded trajectory converges to the set of equilibria.

**Example 1.5** ([**12, 13**]) Consider system (1.1). Assume (H1), and that $W$ is sign symmetric. Assume further that $W$ has a *loop property*, which means that along every closed loop in graph $G(W)$ the product of matrix elements corresponding to the edges is positive.

Then the system is almost quasi-convergent.

We want to remark that this result is a combination of a result of [**12**] and [**13**], which states that cooperative systems are almost quasi-convergent, and an abstract combinatorial result, which says that for every sign symmetric matrix $W$ with the loop property, there is a change of variables which makes $W$ a nonnegative matrix [**2**]. The formulation of loop property is due to [**14**] and [**24**].

**Example 1.6** ([**23**], [**25**]) Consider system (1.1). Assume (H1), and that $W$ is sign symmetric and tridiagonal, i.e., $w_{ij} = 0$ unless $j = i + 1$ or $j = i - 1$. Furthermore, assume that $W$ is irreducible, i.e., the graph $G(W)$ has one component.

Then (1.1) is convergent.

We mention briefly, that there are related results about *cyclic feedback systems*. These are systems of the form (1.3) where $w_{ij} = 0$ unless $j = i - 1$. Interested readers can find these results in [**19**], [**4**] and [**5**].

## 2 Main results

We present our results in two forms. We present a dynamical result for the system (1.3) and then a related abstract combinatorial result for matrix $W$. We first list additional assumptions.

There is a spanning tree $T$ of $G(W)$ such that if $e_{ij} \in T$
then $w_{ij}w_{ji} > 0$. **(H3)**

Observe that this is slightly weaker assumption then sign symmetricity of $W$.

A *chord cycle* $C_T$, corresponding to a chord of tree $T$ in the graph $G(W)$, is a collection of edges $e_{i_1i_2}, e_{i_2i_3}, \ldots, e_{i_ki_1}$ such that all edges, except one, are in $T$. Note, that every cycle $C$ in $G(W)$ corresponds to two cycles $C^1 = \vec{e}_{i_1i_2}, \vec{e}_{i_2i_3}, \ldots, \vec{e}_{i_ki_1}$ and $C^2 = \overleftarrow{e}_{i_1i_2}, \overleftarrow{e}_{i_2i_3}, \ldots, \overleftarrow{e}_{i_ki_1}$ in the oriented graph $D(W)$, where the arrows indicate the orientation of the edges. To every cycle $C_T \in G(W)$ we associate two numbers:

$$p(C_T) := \Pi_{C_T^1} w_{ij} \qquad \text{and} \qquad \bar{p}(C_T) := \Pi_{C_T^2} w_{ji}.$$

We assume that

$$p(C_T) = \bar{p}(C_T) \qquad \text{for all chord cycles } C_T. \qquad \textbf{(H4)}$$

Following is a generalization of the result in [**3**].

**Theorem 2.1** *Consider system (1.3) with assumptions (H1-H4). Then there exists a Lyapunov function* $V : \mathbf{R}^n \to \mathbf{R}$, *non-increasing along the trajectories and strictly decreasing along all non-equilibrium trajectories of (1.3).*

Observe that this result replaces the symmetry hypothesis on the structure matrix $W$ in [**1**] (see also [**17, 18**]). Considerable effort was spent trying to remove or weaken this hypothesis. That this hypothesis cannot be removed completely is known for some time. Consider the model in [**20**] of the voting paradox

$$\begin{aligned}\dot{x}_1 &= x_1(1 - x_1 - ax_2 - bx_3)\\ \dot{x}_2 &= x_2(1 - bx_2 - x_2 - ax_3)\\ \dot{x}_3 &= x_3(1 - ax_1 - bx_2 - x_3).\end{aligned} \tag{2.1}$$

Grossberg in [**10**] and Schuster et al. in [**26**] considered the case $b > 1 > a$ and $a+b > 0$. They proved that all positive trajectories, except the uniform trajectories $x_1(0) = x_2(0) = x_3(0)$, persistently oscillate as $t \to \infty$. Observe that the matrix of coefficients

$$\begin{pmatrix} 1 & a & b \\ b & 1 & a \\ a & b & 1 \end{pmatrix}$$

can be chosen arbitrarily close to a symmetric matrix and still exhibits persistent oscillations. Therefore a Lyapunov function in the sense of Theorem 2.1 cannot exist for this system.

We replaced the symmetry assumption by assumptions (H3) and (H4). These two are obviously satisfied if the matrix $W$ is symmetric. Observe that in the system (2.1) the condition (H4) is not satisfied; along the cycle $C := e_{12}, e_{23}, e_{31}$ we have

$$p(C) = \beta_{12}\beta_{23}\beta_{31} = a^3 \neq b^3 = \beta_{21}\beta_{32}\beta_{13} = \bar{p}(C).$$

Observe that the assumption (H4) is always satisfied in the absence of cycles, i.e., when the graph $G(W)$ is a tree.

We want to note that in the assumption (H4) it is required to check certain conditions only for chord cycles of the graph. Fiedler and Gedeon in [**3**] required this condition for all cycles in the graph. This makes a difference in a graph with a lot of edges. If the graph of interactions is a full graph with $n$ vertices, then one needs to check $(n^2 - 3n + 2)/2$ chord cycles to verify assumption (H4), while there is $O(n!)$ cycles in such graph.

To illustrate this advantage, we consider a system (1.1), where we assume that (H1) and (H2) hold and that $w_{ij}w_{ji} > 0$ for $|i - j| = 1$. This implies that the system satisfies (H3) with a tree generated by the edges $e_{ij}$ with $|i - j| = 1$.

Assume further that $w_{kl} = w_{lk} = 0$, for some $l > k + 1$, and the rest of the system satisfies assumption (H4). Then it is always possible to choose weights $w_{kl} \neq 0$ and $w_{lk} \neq 0$ so that (H4) is satisfied. To see that, observe that we only need to check whether the weights along the chord cycle $e_{kl}e_{l,l-1}e_{l-1,l-2}\cdots e_{k+1,k}$ satisfy (H4). It follows that it is sufficient to chose $w_{kl}$ and $w_{lk}$ so that

$$\frac{w_{kl}}{w_{lk}} = \frac{w_{k,k+1}w_{k+1,k+2}\cdots w_{l-1,l}}{w_{k+1,k}w_{k+2,k+1}\cdots w_{l,l-1}}.$$

We say that system (1.1) is *tridiagonal* if $w_{ij} = 0$ for $|i-j| > 1$. As a consequence of these considerations we get that given a tridiagonal system which satisfies (H1-H3), we can add an arbitrary number of connections with appropriate weights so that the resulting system satisfies the assumptions of Theorem 2.1.

In order to show that the system is quasi-convergent we need to show that forward trajectories are bounded. To this end, a natural assumption is the *point dissipativeness* assumption

all trajectories eventually enter a positively invariant bounded set $B$. **(D)**

**Remark 2.2** Point dissipativeness is guaranteed, for example, by the following set of assumptions:

Assume that all $f_i$ are bounded, $a_i(\mathbf{x}) > 0$ for sufficiently large $|\mathbf{x}|$ and $\gamma_i(x_i)x_i \to -\infty$ for $|x_i| \to \infty$. It is easy to see that under these assumptions $\dot{x}_i x_i < 0$, for every $i$, if $|x_i|$ is sufficiently large.

We also remark that in the context of neural networks it is customary to assume that the functions $f_j$ are bounded.

**Corollary 2.3** *Consider system (1.3) with (H1-H4) and (D). Then the system (1.3) is quasi-convergent. If the set of equilibria is finite, then the system is convergent.*

This result has an interesting consequence for Lotka-Volterra systems. Since in this case the physically meaningful region of the phase space is the nonnegative orthant $\mathbf{R}^{n+}$, the assumption (H2) is automatically satisfied.

**Corollary 2.4** *Consider Lotka-Volterra system (1.2) on $\mathbf{R}^{n+}$ with the graph $G(A)$ of interactions being a tree. Assume (H1) and (H3). Then all bounded trajectories converge to the set of equilibria. Moreover, heteroclinic cycles are excluded, even on the boundary of $\mathbf{R}^{n+}$.*

To see that assumption (H3) is sharp, consider a two dimensional predator-prey Lotka-Volterra model

$$\begin{aligned} \dot{x}_1 &= x_1(1 - x_2) \\ \dot{x}_2 &= x_2(-1 + x_1). \end{aligned}$$

This system satisfies (H1) and the graph of interactions is a tree. The condition (H3) is not satisfied and, as it is well known (see for instance [**16**]), all solutions starting in the interior of the positive orthant are periodic.

## 3 Proofs

We first prove an abstract combinatorial result which may be of independent interest to researchers in combinatorial matrix theory. It generalizes theorems of [**22**] and that of [**21**].

**Theorem 3.1** *Let $W$ be a real, combinatorially symmetric, irreducible matrix. Then there is a positive diagonal matrix $D$ such that $DW$ is a symmetric matrix if, and only if,*

1. *there is a spanning tree $T$ of the graph $G(W)$ such that for every edge $e_{ij} \in T$ we have $w_{ij}w_{ji} > 0$ and*
2. *if $C_T$ is a chord cycle in the graph $G$ with respect to tree $T$, then*

$$p(C_T) = \bar{p}(C_T).$$

Observe, that 1. and 2. are precisely assumptions (H3) and (H4).

**Proof** Assume 1. and 2. We seek to define numbers $v_i$, $i = 1, \dots n$ such that

$$\frac{w_{ij}}{w_{ji}} = exp(v_i - v_j). \tag{3.1}$$

To construct the collection $\{v_i\}_{i=1}^n$ we pick $v_1$ and compute $v_i$ for all $i$ with the property that $e_{1i} \in T$, using (3.1). In this way we construct all $v_i$ and, obviously, (3.1) is satisfied along all the edges of $T$. Let us consider an edge $e_{kl} \in (G \setminus T)$. Since $T$ is a spanning tree of $G(W)$, there is a path $P$ in $T$ from vertex $k$ to vertex $l$. Thus $C = P \cup e_{kl}$ forms a chord cycle and by assumption 2., along this cycle we have $p(C) = \bar{p}(C)$, which we write as

$$\Pi_C \frac{w_{ij}}{w_{ji}} = 1.$$

Since (3.1) holds along edges in $P$ this implies

$$\begin{aligned} 1 &= \Pi_C w_{ij} / \Pi w_{ji} \\ &= \frac{w_{kl}}{w_{lk}} \Pi_P \, exp(v_i - v_j). \end{aligned}$$

Simplifying the right-hand side we get

$$\frac{w_{kl}}{w_{lk}} = exp(v_k - v_l).$$

Since $e_{kl}$ was arbitrary, collection $\{v_i\}$ satisfies (3.1) for all edges in $G(W)$.

Let $d_i := exp(-v_i)$ and let $D = diag(d_i)$. Let $Q = DW$. Observe that

$$\begin{aligned} q_{ij}/q_{ji} &= \frac{d_i}{d_j} \frac{w_{ij}}{w_{ji}} \\ &= exp(-v_i) exp(v_j) exp(v_i - v_j) = 1 \end{aligned}$$

and so $Q$ is a symmetric matrix.

If we assume that $Q = DW$ is symmetric, 1. and 2. follow easily. □

**Proof of Theorem 2.1** Observe that (H3) and (H4) are assumptions 1. and 2. in Theorem 3.1. We assume without loss of generality that $W$ is irreducible. In the opposite case we consider every irreducible component separately. We use the symmetric matrix $Q = [q_{ij}]$ and the diagonal matrix $D = [d_{ij}]$ from Theorem 3.1. Let

$$b_i := 2d_i \int^{x_i} (\frac{df_i}{d\zeta_i} \gamma_i)(\zeta_i) d\zeta_i.$$

We define the function $V : \mathbf{R}^n \to \mathbf{R}$

$$V(\mathbf{x}) := -\sum_{i=1}^n \left( b_i(x_i) - \sum_{j=1}^n q_{ij} f_i(x_i) f_j(x_j) \right). \tag{3.2}$$

We calculate the derivative of $V$, abbreviating $f_i' = \frac{df_i}{dx_i}$ to simplify notation.

$$\begin{aligned} \dot{V} &= -\sum_{i=1}^n \left( \frac{db_i}{dx_i} \dot{x}_i - \sum_{j=1}^n q_{ij} (f_i' f_j(x_j) \dot{x}_i + f_j' f_i(x_i) \dot{x}_j) \right) \\ &= -\sum_{i=1}^n \left( 2d_i f_i' \gamma_i \dot{x}_i - \sum_{j=1}^n (q_{ij} f_i' f_j(x_j) \dot{x}_i + q_{ji} f_i' f_j(x_j) \dot{x}_i) \right) = (*) \end{aligned}$$

since, by rearranging the summations,

$$\sum_{i=1}^n \sum_{j=1}^n q_{ij} f_i' f_j(x_j) \dot{x}_i + q_{ij} f_j' f_i(x_i) \dot{x}_j = \sum_{i=1}^n \sum_{j=1}^n q_{ij} f_i' f_j(x_j) \dot{x}_i + q_{ji} f_i' f_j(x_j) \dot{x}_i.$$

We continue our computation

$$(*) \quad = \quad -\sum_{i=1}^{n}\left(2d_i f_i' \dot{x}_i(\gamma_i - \sum_{j=1}^{n} \frac{q_{ij}+q_{ji}}{2d_i} f_j(x_j))\right)$$

$$= \quad -\sum_{i=1}^{n}\left(2d_i f_i' \dot{x}_i(\gamma_i - \sum_{j=1}^{n} \frac{2q_{ij}}{2d_i} f_j(x_j))\right)$$

$$= \quad -\sum_{i=1}^{n} 2d_i f_i' a_i(\mathbf{x})[\gamma_i - \sum_{j=1}^{n} w_{ij} f_j(x_j)]^2$$

where we used equations (1.3) and that $w_{ij} = q_{ij}/d_i$. To finish the computation we observe that $d_i$ are positive constants, $f_i' > 0$ by the assumption (H1), and $a_i(\mathbf{x}) \geq 0$ by assumption (H2). Thus

$$\dot{V} \leq 0.$$

Since $2d_i f_i'$ is strictly positive, $\dot{V} = 0$ if, and only if, $a_i(\mathbf{x})[\gamma_i - \sum_{j=1}^{n} w_{ij} f_j(x_j)]^2 = 0$ for every $i$. This is equivalent to

$$a_i(\mathbf{x})(\gamma_i - \sum_{j=1}^{n} w_{ij} f_j(x_j)) = 0$$

for every $i$ and, consequently, $\dot{x}_i = 0$ for all $i$. Hence $\dot{V}(x) = 0$ if, and only if, $x$ is an equilibrium. This finishes the proof of Theorem 2.1. □

It is interesting to note, that we do not know whether the result of [**21**] (below), which we generalize in our Theorem 3.1, can also be used to construct a Lyapunov function $V$ and prove Theorem 2.1.

**Theorem 3.2** ([**21**]) *Let $W$ be a combinatorially symmetric, irreducible matrix. Then there is a real diagonal matrix $D$ such that $Q := D^{-1}WD$ is symmetric if, and only if, (H3) and (H4) hold.*

We remark that from the proof of this Theorem in [**21**] it is clear that if (H3) and (H4) hold then matrix $D$ can be made to be a positive matrix.

**Proof of Corollary 2.4** On the closed positive orthant $\mathbf{R}^{n+}$ the assumption (H2) is satisfied automatically. The graph of interactions $G(W)$ is a tree and there is a Lyapunov function $V$ on the closed positive orthant by Theorem 2.1. Thus every bounded trajectory converges to the set of equilibria. Since the Lyapunov function is defined also on the boundary of the positive orthant we can exclude the existence of the heteroclinic cycles on the boundary. This proves the Corollary. □

## Acknowledgments

We want to thank P. van den Driessche for calling our attention to the result of J. Maybee.

## References

[1] Cohen, M. A. and Grossberg, S. [1983], *Absolute stability of global pattern formation and parallel memory storage by competitive neural networks*, IEEE Transaction on Systems, Man and Cybernetics, **SMC-13**, No. 5, 815-826.

[2] Eschenbach, C. A. and Johnson, C. R. [1990], *A combinatorial converse to the Perron-Frobenius theorem*, Linear Algebra and its Applications **136**, 173-180.

[3] Fiedler, B. and Gedeon, T. [1998], *A class of convergent neural network dynamics*, Physica D **111**, 288-294.
[4] Gedeon, T. [1998], *Cyclic feedback systems*, Memoirs of AMS, to appear.
[5] Gedeon, T. and Mischaikow, K. [1995], *Structure of the global attractor of cyclic feedback systems*, J. Dynamics and Diff. Eq. **7**, 141-190.
[6] Grossberg, S. [1969], *On the global limits and oscillations of a system of nonlinear differential equations describing a flow of a probabilistic network*, J. Diff. Eq. **5**, 531-563.
[7] Grossberg, S. [1969], *On learning and energy-entropy dependence in recurrent and non-recurrent signed networks*, J. Stat. Phys. **5**, 319-350.
[8] Grossberg, S. [1971], *Pavlovian pattern learning by nonlinear neural networks*, Proc. Natl. Acad. Sci. **68**, 828-831.
[9] Grossberg, S. [1973], *Contour enhancement, short term memory, and constancies in reverberating neural networks*, Studies in Appl. Math. **52**, 217-257.
[10] Grossberg, S. [1978], *Decisions, patterns, and oscillations in the dynamics of competitive systems with applications to Volterra-Lotka systems*, J. Theoret. Biol. **73**, 101-130.
[11] Grossberg, S. [1978], *Competition, decision, and consensus*, J. Math. Anal. Appl. **66**, 470-493.
[12] Hirsch, M. [1982], *Systems of differential equations that are competitive or cooperative, I: Limit sets*, SIAM J. Math. Anal., **13**, 167-179.
[13] Hirsch, M. [1985], *Systems of differential equations that are competitive or cooperative. II: Convergence almost everywhere*, SIAM J. Math. Anal., **16**, 423-439.
[14] Hirsch, M. [1987], *Convergence in neural nets*, Proceedings of 1987 IEEE conference on Neural Networks.
[15] Hirsch, M. [1989], *Convergent activation dynamics in continuous time networks*, Neural Networks **2**, 331-349.
[16] Hale, J. and Koçak, H. [1991], *Dynamics and bifurcations*, Springer-Verlag.
[17] Hopfield, J. J. [1982], *Neural networks and physical systems with emergent collective computational abilities*, Proc. Natl. Acad. Sci., **79**, 2554-2558.
[18] Hopfield, J. J. [1984], *Neurons with graded response have collective computational properties like those of two-state neurons*, Proc. Natl. Acad. Sci., **81**, 3088-3092.
[19] Mallet-Paret, J. and Smith, H. [1990], *The Poincaré-Bendixon theorem for monotone feedback systems*, J. Dynamics and Diff. Eq. **2**, 367-421.
[20] May, R. M. and Leonard, W. J. [1975], *Nonlinear aspects of competition between three species*, SIAM J. Appl. Math. **29**, 243-253.
[21] Maybee, J. S. [1974], *Combinatorially symmetric matrices*, Linear Algebra and its Applications **8**, 529-537.
[22] Parter, S. V. and Youngs, J. W. T. [1962], *The symmetrization of matrices by diagonal matrices*, J. Math. Anal. Appl. **4**, 102-110.
[23] Smilie, J. [1984], *Competitive and cooperative tridiagonal systems of differential equations*, SIAM J. Math. Anal. **15**, No. 3, 530-534.
[24] Smith, H. [1988], *Systems of ordinary differential equations which generate an order preserving flow. A survey of results*, SIAM Review **30**, No. 1, 87-113.
[25] Smith, H. [1991], *Periodic tridiagonal competitive and cooperative systems of differential equations*, SIAM J. Math. Anal. **22**, No.4, 1102-1109.
[26] Schuster, P., Sigmund, K. and Wolff, R. [1979], *On $\omega$-limits for competition between three species*, SIAM J. Appl. Math. **37**, 49-54.

Fields Institute Communications
Volume **21**, 1999

# Continuity of Resetting Curves for FitzHugh-Nagumo Equations on the Circle

**Tomáš Gedeon**
Department of Mathematics
Montana State University
Bozeman, MT 59715, USA
gedeon@poincare.math.montana.edu

**Leon Glass**
Department of Physiology
McGill University
3655 Drummond Street
Montreal, Quebec, Canada H3G 1Y6
glass@cnd.mcgill.ca

**Abstract.** We prove that the resetting curve in a ring of excitable media, modeled by the FitzHugh-Nagumo equation on a circle, is continuous provided that a perturbation, delivered at any phase of a limit cycle oscillation leaves the state point in the basin of attraction of the limit cycle. This is one of the ingredients in the proof of a general conjecture [**2, 9**] that there is always a stimulus which leads to an annihilation of the wave of excitation, travelling around a ring of excitable media. Our formulation of the continuity theorem is quite general and applicable to other models of excitable media.

## 1 Introduction

Abnormal cardiac rhythms are a major cause of illness and death. In tachycardias, the heartbeat is too rapid. Even though tachycardias can arise in different regions of the heart, the underlying mechanisms of these rhythms may be similar. One mechanism for tachycardia assumes that the excitation follows a reentrant path. This means that the period of the oscillation is determined by the length of time it takes for an excitation to follow a circuitous path, rather than the period of the normal pacemaker of the heart. An early proposal by Mines was that reentrant cardiac activity could be modeled by excitation traveling in a ring of cardiac tissue – a viewpoint that has remarkably formed the basis for modern analyses of some

1991 *Mathematics Subject Classification.* 35B10, 35B20, 92C20.

This research was partially supported by grants MONTS-291725 and NSF-DMS-291222(T.G.) and by NSERC (L.G.) .

arrhythmias [**7**]. [1] The hypothesis that an excitation is traveling in a ring of tissue has an important implication to the cardiologist. If the ring is cut, the anatomical substrate for the rhythm is eliminated and the tachycardia does not reoccur. The development of ablation methods, that can often be employed to destroy abnormal cardiac tissue and cure the patient, provides a partial support for the theoretical model. Of course, the real heart is a complex three-dimensional structure and the cardiologist's notion of excitation traveling on a ring is really a topological notion. Thus, though they have usually not studied the mathematics, at least some cardiologists approach their field using a topological perspective.

The current paper is motivated by a set of numerical studies and conjectures generated by Glass and colleagues ([**2**], [**9**]). Reentrant tachycardia was modeled by excitation traveling on a one dimensional ring. The model had the following properties:

1. The system is an *excitable medium*. This means that it supports traveling waves, called excitation waves, similar to what is observed in nerve and heart. Following an excitation the tissue is refractory and it cannot support another excitation, for a length of time called the refractory time. As a consequence, two excitations in a one dimensional ring traveling towards each other in opposite senses will annihilate each other. One of the accepted models of an excitable medium is described by the FitzHugh-Nagumo equations.
2. The ring is sufficiently large that a traveling wave will travel around the ring at a constant velocity.
3. A stimulus delivered to the ring at a fixed location of the ring far outside of the refractory time will lead to initiation of two waves traveling around the ring in opposite directions.

Glass and Josephson in [**2**] proposed the following conjecture:

**Conjecture 1.1** *Given an excitable medium in a one-dimensional ring supporting reentrant excitation, stimuli can be found which would lead to annihilation of the reentrant wave.*

The conjecture is consistent with cardiological practice. Cardiologists often deliver single or multiple stimuli to terminate reentrant tachycardia. Cardiologists often consider the ability to terminate a tachycardia by pacing a diagnostic feature of reentrant tachycardia [**7**]. Finally, a class of medical devices (anti-tachycardia pacers) have been designed based on it [**10**].

The current paper was motivated by an attempt to prove the conjecture using following strategy. Definitions of unfamiliar terms will be presented below.

1. Excitable media in the geometry of a one-dimensional ring, support a stable circulating pulse, i.e., an asymptotically stable limit cycle in the associated partial differential equation.
2. Continuity Theorem: If a perturbation delivered at any phase of a limit cycle oscillation leaves the state point in the basin of attraction of the asymptotically stable limit cycle, then the resetting curves characterizing the effects of the stimuli, will be continuous.
3. Resetting curves for stably circulating pulses on a one dimensional ring are discontinuous.

[1] Reentrant excitation can also have other geometries such as spiral waves, scroll rings, or other complex topologies, but these notions will not occupy us here.

4. It follows from 2. and 3. that there is a critical stimulus (or stimuli) that will lead to a point in the phase space outside of the basin of attraction of the stable periodic orbit.

In ordinary differential equations, the continuity theorem follows from arguments by Guckenheimer in [**3**]. The current manuscript extends the continuity theorem to partial differential equations, in particular, to the FitzHugh-Nagumo equations on the ring. Proofs for 1. and 3. above have not yet been found, although an argument for 3. was presented in [**2**]. Another goal of this paper is to provide a theoretical framework for discussion of continuity of resetting curves in other models, for instance, using functional differential equations [**1**]. This is the reason why we state and prove the continuity theorem in a rather abstract setting.

The paper is organized as follows. In Section 2, we define the resetting function for a semiflow on a complete metric space and prove the continuity theorem. The assumptions of the theorem are abstract and the rest of the paper is devoted to verification of these assumptions in various settings. In Section 3 we show that the assumptions of continuity theorem are verified for flows on finite dimensional spaces. In Section 4 we deal with infinite dimensional problems. We briefly mention main differences with the finite dimensional problems and then, following [**6**], we verify the assumptions of the continuity theorem for semilinear parabolic equations, where the linear operator is sectorial.

Finally, in Section 5 we apply the results of Section 4 to the FitzHugh-Nagumo equations on a circle.

## 2 Continuity theorem

We start with a few definitions which lead to a concept of a phase resetting function and the continuity theorem.

Let $X$ be a complete metric space and $\mathbf{R}^+ = [0, \infty)$. A family of mappings $T(t) : X \to X$, is said to be a $C^r$-semigroup, $r \geq 0$, provided that

(i) $T(0) = I$, the identity map,
(ii) $T(t+s) = T(t)T(s)$, $t \geq 0$, $s \geq 0$,
(iii) $T(t)x$ is continuous in $t, x$ together with Fréchet derivatives in $x$ up through order $r$ for $(t, x) \in \mathbf{R}^+ \times X$.

We shall also use notation $\Phi(t, x) := T(t)x$ for the semiflow generated by the semigroup $\{T(t)\}_{t \geq 0}$.

**Definition 2.1** A compact invariant set $S$ is a *local attractor* if there is an $\delta$-neighborhood $B_\delta(S)$ of $S$ such that

$$dist(\Phi(t, B_\delta(S)), S) \to 0 \text{ as } t \to \infty.$$

**Definition 2.2** A compact invariant set $S$ is a *local attractor with asymptotic phase* if it is a local attractor and there is a $b$ such that for any $x \in B_b(S)$ there is a $y \in S$ such that

$$|\Phi(t, x) - \Phi(t, y)| \to 0 \text{ as } t \to \infty.$$

If $S$ is a periodic orbit $\Gamma = \{\gamma(t), 0 \leq t \leq p\}$, this is equivalent to saying that there is a real number $\Theta(x)$ such that

$$|\Phi(t + \Theta(x), x) - \gamma(t)| \to 0 \text{ as } t \to \infty. \tag{2.1}$$

It follows from these definitions that if a periodic orbit $\Gamma$ is a local attractor with asymptotic phase, and $W$ is the stable manifold of $\Gamma$, there is a map $\Theta : W \to S^1$ satisfying (2.1) for all $x \in W$.

**Definition 2.3** Let $f : W \to W$ be a continuous map. The *phase resetting function* $g_f : S^1 \to S^1$ is a function with the property

$$\Theta \circ f|_\Gamma = g_f \circ \Theta|_\Gamma,$$

where $\Theta|_\Gamma$ and $f|_\Gamma$ are restrictions of the corresponding functions to periodic orbit $\Gamma$.

Observe that the function $\Theta$, restricted to periodic orbit $\Gamma$, is a bijection from $\Gamma$ to $S^1$. Therefore there is an inverse $\Theta_\Gamma^{-1} : S^1 \to \Gamma$ and $g_f = \Theta \circ f|_\Gamma \circ \Theta_\Gamma^{-1}$. The function $\Theta_\Gamma^{-1}$ maps an angle $\theta \in S^1$ to a point $\gamma(\theta) \in \Gamma$. The perturbation function $f$ maps $\gamma(\theta)$ to a point $f(\gamma(\theta)) \in W$. Finally, $\Theta(f(\gamma(\theta)))$ is the asymptotic phase shift (see (2.1)) of the trajectory starting at $f(\gamma(\theta))$.

**Theorem 2.4 (Continuity Theorem)** *Let $X$ be a complete metric space with a semiflow $\Phi(t, x)$. Let $\Gamma$ be a periodic orbit of the semiflow $\Phi(t, x)$, which is a local attractor with asymptotic phase. Let $\Theta(x)$ be defined as above for $x \in W$, the stable manifold of $\Gamma$. Assume $\Theta(x)$ is continuous.*

*Then the phase resetting function $g_f$ is continuous for every continuous function $f : W \to W$.*

**Proof** Observe, $\Theta|_\Gamma$ is a bijection between $\Gamma$ and $S^1$. Thus this map has a continuous inverse $\Theta_\Gamma^{-1}$. Therefore

$$g_f = \Theta \circ f|_\Gamma \circ \Theta_\Gamma^{-1} = \Theta \circ f \circ \Theta_\Gamma^{-1}.$$

Since all maps on the right-hand side are continuous by assumption, $g_f$ is continuous. □

## 3 Finite dimension

In this section we verify the assumptions of Theorem 2.4 for the finite dimensional case. These results are well known.

**Theorem 3.1 (Theorem VI.2.1, [4])** *Consider $\dot{x} = f(x)$ an ordinary differential equation on a finite dimensional manifold. Let $\Gamma$ be a hyperbolic periodic orbit, which has simple multiplier one and all other multipliers have modulus less than one.*

*Then $\Gamma$ is a local attractor with asymptotic phase.*

If the function $\Theta(x)$ exists, then the stable manifold $W$ is divided into sets $\Theta^{-1}(\theta), \theta \in [0, p]$, where $p$ is the period of $\gamma$. If these sets vary continuously throughout $W$, then $\Theta(x)$ will be continuous. This idea is captured in the following definition.

**Definition 3.2** A surjection $\pi : E \to B$ between topological spaces is *a fiber bundle,* if $B$ has an open cover $\{U_\alpha\}$ such that there are fiber preserving homeomorphisms

$$\phi_\alpha : E|_{U_\alpha} \to U_\alpha \times F$$

and the transition functions

$$g_{\alpha\beta} = \phi_\alpha \circ \phi_\beta^{-1}|_{\{x\}} \times F$$

are continuous. The set $\pi^{-1}(\{x\})$ is called a *fiber at $x$.*

In our problem we have $E = W$, $B = \Gamma$ and $\pi = \Theta$.

**Theorem 3.3 (Theorem A, [3])** *Let $\Gamma$ be a periodic orbit of a flow on a finite dimensional manifold, which is local attractor with asymptotic phase. Then the stable manifold $W$ of $\Gamma$ is a fiber bundle $\pi : W \to S^1$.*

These two Theorems verify assumptions of Theorem 2.4. Therefore we have

**Corollary 3.4** *Consider $\dot{x} = f(x)$ an ordinary differential equation on a finite dimensional manifold. Let $\Gamma$ be a hyperbolic periodic orbit, which has a simple multiplier one and all other multipliers have modulus less than one.*

*Then the phase resetting function $g_f$ is continuous for every continuous function $f : W \to W$.*

## 4 Infinite dimensional problems

We begin by pointing out the differences from the finite dimensional case and then proceed with a particular case of the FitzHugh-Nagumo equation on the circle.

Consider

$$u_t = -Au + f(u)$$

with $u(0) = u_0 \in X$, a Banach space. Here $A$ is a linear operator and $f(u)$ is a nonlinearity. A book [**5**] is a great source for results about infinite dimensional problems. We follow Section 3.10 to discuss two major problems with Poincaré maps for periodic orbits in infinite dimensional problems.

The first problem is that in order to define a Poincaré section, a periodic orbit $\Gamma = \{\gamma(t) \mid 0 \le t \le p\}$ must be $C^1$. This is true, provided $\Gamma$ lies on the global attractor (see [**5**]) and $|f'(u)|$ is bounded in the neighborhood of $\Gamma$.

The second problem is the smoothness of the Poincaré map. Given a section $S$ of the semiflow $\Phi(t, x)$, one defines Poincaré map on $S$ by $\Phi(t_r(x), x)$, where $t_r(x)$ is the return time to $S$ for the trajectory starting at $x$.

However, the map $x \to t_r(x)$ is not necessarily smooth and thus a Poincaré map need not be smooth. Thus we cannot express stability of a periodic orbit in terms of the eigenvalues of the derivative of the Poincaré map. Instead, we consider a linear variational equation about $\gamma(t)$

$$v_t = -Av + f'(\gamma(t))v. \tag{4.1}$$

Let $U(t, \sigma)$ be a solution operator of (4.1). Then the characteristic multipliers of (4.1) are the elements of the point spectrum of $U(\sigma + p, \sigma)$, where $p$ is the period of $\gamma(t)$. If we assume that the radius of the essential spectrum is $< 1$ and $\Gamma$ is a $C^1$-manifold then one is always a multiplier, since $\dot{\gamma}(t)$ is a solution of (4.1).

**Definition 4.1** $\Gamma$ is hyperbolic, if one is a simple multiplier of (4.1) and no other multiplier has modulus one.

**4.1 Semilinear parabolic equations.** In this subsection we follow the work of [**6**] on semilinear parabolic equations. We verify the assumptions of the continuity Theorem in this class of equations, provided that the linear operator is sectorial. As in the finite dimensional case we need two results. First, that the hyperbolic attracting periodic orbit is a local attractor with asymptotic phase, and second, that the associated function $\Theta(x)$ is continuous. The first result can be found in [**6**]. The second result essentially follows from his proof, but we need to add some details.

**Definition 4.2 ([6])** A linear operator $A$ in a Banach space $X$ is *sectorial,* if it is a closed densely defined operator such that, for $\varphi \in (0, \pi/2)$ and some $M \geq 1$ and a real $a$, the sector

$$S_{a,\varphi} = \{\lambda \mid \varphi \leq |arg(\lambda - a)| \leq \pi, \lambda \neq a\} \tag{4.2}$$

is in the resolvent set of $A$ and

$$||(\lambda - A)^{-1}|| \leq M/|\lambda - a| \text{ for all } \lambda \in S_{a,\varphi}. \tag{4.3}$$

A sectorial operator is a generalization of an elliptic operator $\Delta$. Given a Banach space $X$ and a sectorial operator $A$ one can define fractional powers $A^\alpha$ of the operator $A$ and for every $\alpha \geq 0$ one can define corresponding fractional power spaces $X^\alpha$. It can be shown (Theorem 1.4.8 [**6**]) that $X^0 = X$ and for $\alpha \geq \beta \geq 0$, $X^\alpha$ is a dense subset of $X^\beta$ with continuous inclusion.

We consider

$$u_t = -Au + f(u) \tag{4.4}$$

and assume:

**Assumptions 4.3** Operator $A$ is sectorial on a Banach space $X$ and $f$ is continuously differentiable from $X^\alpha$ to $X$. Assume that $\gamma(t)$ is a nonconstant periodic solution of period $p$ of (4.4) and that $t \to f_x(\gamma(t)) \in L(X^\alpha, X)$ is Hölder continuous.

**Lemma 4.4 (Lemma 8.2.2, [6])** *Assume (4.3). Then the linear variational equation (4.1) about $\gamma(t)$ has one as a characteristic multiplier.*

**Theorem 4.5 (Theorem 8.2.3, [6])** *Assume (4.3) and that one is an isolated simple eigenvalue of the period map, and that the remainder of the spectrum lies in $\{|\mu| < e^{-\beta p}\}$ for some $\beta > 0$.*

*Then $\Gamma$ is a local attractor with asymptotic phase.*

**Remark 4.6** From the proof of Theorem 4.5 it is clear that $\Theta(x)$ is well defined in a small neighborhood of $\Gamma$ since only there one can guarantee existence of a solution for a sufficiently large time. Although the function $\Theta(x)$ is defined on $W$ since every solution converging to $\Gamma$ must pass through this small neighborhood of $\Gamma$, $W$ does not have to be a "nice" neighborhood of $\Gamma$. Thus the assumption that the perturbation function $f$ in the continuity theorem maps $W$ to $W$ is crucial and may be difficult to verify outside of a small neighborhood of $\Gamma$.

In order to verify the assumptions of the continuity theorem, it remains to show that the function $\Theta(x)$ (which exists because of (4.5)) is continuous.

**Theorem 4.7** *Assume (4.3), that one is an isolated simple eigenvalue of the period map, and that the remainder of the spectrum lies in $\{|\mu| < e^{-\beta p}\}$ for some $\beta > 0$. Then the function $\Theta(x)$ is continuous on $W$.*

**Proof** The proof of the Theorem follows from the proof of Theorem 4.5 in [**6**]. Since it is not explicitly spelled out we lead the reader through the argument and show how it can be changed to yield the proof of Theorem 4.7.

The preliminary step in [**6**] is to construct a family of functions $z(t, a)$, $t \in [0, \infty)$, depending on a parameter $a$, which is related to an initial value of $z(t, a)$ at $t = t_0$. These functions are such that $x^*(t, a) = \gamma(t) + z(t, a)$ are solutions of (4.4) for $t > t_0$ and converge to $\gamma(t)$ exponentially.

Then consider a solution $x(t,\zeta)$ of (4.4) such that

$$x(t_0 - p, \zeta) = \zeta. \tag{4.5}$$

In order to show convergence with asymptotic phase it is enough to show that there is $\theta^*$ real and $a^*$ such that

$$x(t_0 + \theta^*, \zeta) = x^*(t_0, a^*). \tag{4.6}$$

Indeed, uniqueness implies that $x(t,\zeta) = x^*(t - \theta^*, a^*)$ and by the construction of $x^*(t,a^*)$ the later function converges exponentially to $\gamma(t - \theta^*)$.

To show that (4.6) has a solution, we must confine ourselves to a small neighborhood of $\Gamma$ to ensure that $x(t,\zeta)$ exists for $t_0 - p \le t \le t_0 + p$.

Then we can recast solvability of (4.6) in terms of a operator equation

$$a - \theta\dot{\gamma}(t_0) = G(\theta, a; \zeta) \tag{4.7}$$

where $G(0,0;\gamma(t_0)) = 0$. (For details one should consult [**6**].) The Equation (4.7) has a solution $(\theta^*, a^*)$ if and only if (4.6) has.

The left-hand side is linear in $\theta$ and $a$ and thus can be thought of as a linear operator $L$ applied to pair $(\theta, a)$. Henry shows that this operator is invertible. However, no bound on the norm of the inverse $||L^{-1}||$ is given. In other words we try to solve

$$(\theta, a) = L^{-1}G(\theta, a; \zeta)$$

where $L$ is the aforementioned linear operator. Estimates in [**6**] show that the Lipschitz constant of $G$ can be made arbitrarily small in a neighborhood of $\Gamma$. Thus $L^{-1}G$ is a contraction and its unique fixed point $(\theta^*, a^*)$ solves (4.6).

We want to focus on another aspect of (4.7) and that is that it represents a family of equations parameterized by the initial condition $\zeta$. Furthermore, the estimate of the Lipschitz constant does not depend on $\zeta$, i.e., it is uniform in $\zeta$. Thus one can apply the Contraction Mapping Theorem with parameters [**4**] to conclude that the function

$$\zeta \to (\theta^*, a^*)$$

is continuous. This map is defined so far for $||\zeta - \gamma(t_0)||$ sufficiently small.

However, we can replace (4.5) with $x(t_0 - T, \zeta) = \zeta$ for any $T \in [0,p]$, repeat the argument above and conclude the same result. Thus one can define a continuous function $F$ defined on a sufficiently small neighborhood $N$ of $\Gamma$

$$F : \zeta \to (\theta^*, a^*).$$

Observe that for any $\zeta, \eta \in W$ there is a finite time $\tau$ such that both $x(\tau, \zeta)$ and $x(\tau, \eta)$ are in $N$. Thus we can extend $F$ to $W$, and, by the continuous dependence on initial conditions, $F$ is continuous on $W$.

The projection of the continuous function $F$ to the first variable is the function $\Theta(\zeta)$, which is therefore continuous. □

The only place in the above argument where the sectoriality of $A$ is used is in the estimates which show that operator $G$ has a Lipschitz constant uniformly and arbitrarily small.

Since Theorem 4.5 and Theorem 4.7 verify the assumptions of Theorem 2.4, we have:

**Corollary 4.8** *Consider* $u_t = -Au + f(u)$ *where* $A$ *is a sectorial operator on a Banach space* $X$ *and* $f$ *is continuously differentiable from* $X^\alpha$ *to* $X$. *Let* $\gamma(t)$ *be a nonconstant periodic solution, such that one is an isolated simple eigenvalue of the period map, and that the remainder of the spectrum lies in* $\{|\mu| < e^{-\beta p}\}$ *for some* $\beta > 0$. *Assume that* $t \to f_x(\gamma(t)) \in L(X^\alpha, X)$ *is Hölder continuous.*

*Then the phase resetting function* $g_f$ *is continuous for every continuous perturbation function* $f : W \to W$.

## 5 FitzHugh-Nagumo equation on a circle

As a model for the reentrant wave on a circle of excitable media we consider the FitzHugh-Nagumo equations on interval $[0, 2\pi]$ with periodic boundary conditions ([**2**], [**9**]). Consider

$$\begin{aligned} u_t &= Du_{xx} + f(u) - w + I \\ w_t &= c(u - dw) \end{aligned} \tag{5.1}$$

on a space $X := H_p^2(0, 2\pi) \times L_p^2(0, 2\pi)$ with periodic boundary conditions

$$u(0,t) = u(2\pi, t), u'(0,t) = u'(2\pi, t), w(0,t) = w(2\pi, t).$$

Function $f \in C^2(\mathbf{R}, \mathbf{R})$ and $c$ and $d$ are constants. The norm on the space $H$ is $||(u, w)|| = ||u||_2 + ||w||$ where the first norm is the norm in $H_p^2(0, 2\pi)$ and the second in $L_p^2(0, 2\pi)$.

$H_p^2(0, 2\pi)$ is a Hilbert space and a dense subset of $L_p^2(0, 2\pi)$ which can be characterized by (see [**11**])

$$H_p^2(0, 2\pi) = \left\{ u | u = \sum_{n \geq 0} a_n \sin(nx) + \sum_{n \geq 0} b_n \cos(nx), \sum_{n \geq 0} n^4 (|a_n|^2 + |b_n|^2) < \infty \right\}. \tag{5.2}$$

In agreement with the outline in the Introduction we assume that system (5.1) admits an hyperbolic periodic orbit, one is an isolated simple eigenvalue of the period map, and that the remainder of the spectrum lies in $\{|\mu| < e^{-\beta p}\}$ for some $\beta > 0$.

The goal of this subsection is to verify the rest of the assumptions of Corollary 4.8. This shows that the resetting function is continuous, provided that the perturbation function $f$ maps $\Gamma$ into $W$. We first show that the linear operator associated with (5.1) is sectorial.

We rewrite (5.1) as

$$v_t = -Av + F(v)$$

where $v = (u, w)^T$, $F(v) = (f(u) - \frac{\partial f}{\partial u}|_{u=0}, 0)$ is the nonlinear part, and

$$A = \begin{bmatrix} -D\frac{\partial^2}{(\partial x)^2} - \frac{\partial f}{\partial u}|_{u=0} & 1 \\ -c & cd \end{bmatrix}.$$

We compute the spectrum of $A$. The spectrum $\sigma(A) = \sigma_p(A) \cup \sigma_r(A) \cup \sigma_c(A)$ decomposes ([**8**]) to point spectrum $\sigma_p(A)$, residual spectrum $\sigma_r(A)$ and continuous spectrum $\sigma_c(A)$.

First we consider a point spectrum $\sigma_p(A)$. The collection

$$\left\{ \frac{1}{\sqrt{\pi}} \sin(nx), \frac{1}{\sqrt{\pi}} \cos(nx) \right\}_{n=1}^{\infty},$$

together with constant functions, forms an orthonormal basis of $L_p^2(0, 2\pi)$. We set $v = (a_n^1, a_n^2)^T \frac{1}{\sqrt{\pi}} \sin(nx)$ where $(\cdot)^T$ denotes the transpose vector. We compute the eigenvalues from

$$(A - \lambda I)v = 0.$$

We shall omit analogous argument with $v = (b_n^1, b_n^2)^T \frac{1}{\sqrt{\pi}} \cos(nx)$. The characteristic equation is

$$\lambda^2 - \lambda(cd + Q) + Qcd + c = 0 \tag{5.3}$$

where $Q = Dn^2 - \frac{\partial f}{\partial u}|_{u=0}$.

Equation (5.3) has complex roots if and only if

$$(cd - Q)^2 - 4c < 0. \tag{5.4}$$

Since $c, d$ and $\frac{\partial f}{\partial u}|_{u=0}$ are constants and

$$(cd - Q)^2 \to \infty \text{ as } n \to \infty,$$

there are only finitely many $n$ for which (5.4) is satisfied. Therefore, there are only finitely many eigenvalues with a nonzero complex part and it follows that there is a sector $S_{a,\varphi}$ (see (4.2)) such that $\mathbf{C} \setminus S_{a,\varphi}$ contains the point spectrum $\sigma_p(A)$.

We need to study the continuous and residual spectrum of $A : X \to L := L_p^2(0, 2\pi) \times L_p^2(0, 2\pi)$.

To that end we study properties of the resolvent operator $(A - \lambda I)^{-1}$ on $L$. Consider

$$(A - \lambda I)v = F \tag{5.5}$$

for $F = (f_1, f_2) \in L$ and $v = (u, w)$. Then the second equation in (5.5) reads

$$-cu + cdw = f_2 + \lambda w$$

which can be solved for $w$

$$w = -\frac{cu + f_2}{\lambda - cd}$$

if and only if

$$\lambda \neq cd. \tag{5.6}$$

The first equation in (5.5) now yields

$$-\frac{\partial^2 u}{(\partial x)^2} - au - \lambda u - \frac{c}{\lambda - cd} u = f_1 + f_2/(\lambda - cd)$$

where $a = \frac{\partial f}{\partial u}|_{u=0}$. Observe that the right-hand side, $h := f_1 + f_2/(\lambda - cd) \in L_p^2(0, 2\pi)$.

We solve for $u$ by setting

$$u = \sum_{n \geq 0} a_n \sin(nx) + \sum_{n \geq 0} b_n \cos(nx)$$

and $h = \sum_{n \geq 0} c_n \sin(nx) + \sum_{n \geq 0} d_n \cos(nx)$. We only show the calculation for $\sin(nx)$ modes since the calculation for $\cos(nx)$ modes is analogous. The equation for the $n$-th mode is

$$\left(n^2 - a - \lambda - \frac{c}{\lambda - cd}\right) a_n = c_n$$

and so

$$a_n = c_n / \left( n^2 - a - \lambda - \frac{c}{\lambda - cd} \right). \tag{5.7}$$

Analogously, for $\cos(nx)$ modes we have

$$b_n = d_n / \left( n^2 - a - \lambda - \frac{c}{\lambda - cd} \right). \tag{5.8}$$

Obviously, $a_n$ and $b_n$ are not defined when the denominator vanishes. However, an easy calculation shows that $n^2 - a - \lambda - \frac{c}{\lambda - cd} = 0$ if and only if $\lambda \in \sigma_p(A)$, the point spectrum of $A$. If $a_n$ and $b_n$ are defined then

$$\sum_{n \geq 0} n^4 (|a_n|^2 + |b_n|^2) < \infty$$

and so $u \in H_p^2(0, 2\pi)$ by (5.2). This implies, that if $\lambda \in \Lambda := C \setminus \{\sigma_p(A) \cup \{cd\}\}$, then

$$(A - \lambda I)^{-1} : L \to X.$$

Since $X$ is a dense subset of $L$, the set $\Lambda$ does not intersect the residual spectrum $\Lambda \cap \sigma_r(A) = \emptyset$.

For every fixed $\lambda \in \Lambda$ let $k_\lambda$ be the minimum distance of $\lambda$ to $\sigma_p(A)$ in the complex plane. Then the norm of the operator $(A - \lambda I)^{-1}$ is given by

$$\sup \left\{ \sum_{n \geq 0} |a_n|^2 + |b_n|^2 \right\}$$

where $a_n$ and $b_n$ satisfy (5.7) and (5.8) and the supremum is taken over all sequences $c_n, d_n$ with $\sum_{n \geq 0} |c_n|^2 + |d_n|^2 = 1$. It is easy to see that

$$\sum_{n \geq 0} |a_n|^2 + |b_n|^2 \leq \frac{1}{k_\lambda} \left( \sum_{n \geq 0} |c_n|^2 + |d_n|^2 \right) = \frac{1}{k_\lambda}. \tag{5.9}$$

Hence $(A - \lambda I)^{-1}$ is continuous and the set $\Lambda$ does not intersect the continuous spectrum $\Lambda \cap \sigma_c(A) = \emptyset$.

A short calculation shows that if $\lambda = cd$ then

$$(A - \lambda I)^{-1}(f_1, f_2) = (-f_2/c, f_1 + f_2/c(\lambda + a) - \frac{\partial^2 f_2}{(\partial x)^2}/c).$$

Observe that this operator is defined for $(f_1, f_2) \in L_p^2(0, 2\pi) \times H_p^2(0, 2\pi)$ which is a dense subset of $L$. However, as we can find $f_2 \in H_p^2(0, 2\pi)$ with the norm $|| \frac{\partial^2 f_2}{(dx)^2} ||_2$ arbitrarily large, the operator $(A - \lambda I)^{-1}$ is not bounded. Hence $\lambda = cd$ is in the continuous spectrum $\sigma_c(A)$.

This shows that spectrum of $A$, $\sigma(A) = \sigma_p(A) \cup \{cd\}$. Since $cd$ is real, the set $\mathbf{C} \setminus S_{a,\varphi}$ which contains point spectrum $\sigma_p(A)$, also contains $cd$ and thus $\sigma(A)$. Therefore the sector $S_{a,\varphi}$ is contained in the resolvent set of $A$. The Equation (5.9) verifies (4.3) and so $A : X \to L$ is a sectorial operator.

To verify the remaining assumptions of Corollary 4.8, we need to show that the nonlinearity

$$F := (f_1(u), 0) = \left( f(u) - \frac{\partial f}{\partial u}|_{u=0}, 0 \right)$$

is continuously differentiable, $F : X^\alpha \to X$, for some $\alpha$. We first identify $X^\alpha$. Recall that [**6**] if $u = \sum c_i e_i$ is the expansion of $u$ in basis $e_i$ of $X$, then $Au = \sum \lambda_i c_i e_i$, where $\lambda_i$ are eigenvalues of A. We define fractional power of $A$ by $A^\alpha u = \sum \lambda_i^\alpha c_i e_i$. The space $X^\alpha$ is the domain of $A^\alpha$, $X^\alpha = D(A^\alpha)$, with the graph norm $||x||_\alpha = ||A^\alpha x||$, $x \in X^\alpha$.

It follows from (5.3) and (5.4) that

$$Re(\lambda_n) = O(n^2) \text{ as } n \to \infty. \tag{5.10}$$

Using (5.2) we see that

$$X = \left\{ u | u = \sum_{n \geq 0} (a_n^1, a_n^2)^T \sin(nx) + \sum_{n \geq 0} (b_n^1, b_n^2)^T \cos(nx), \right.$$
$$\left. \sum_{n \geq 0} n^4 (|a_n^1|^2 + |b_n^1|^2) < \infty, \sum_{n \geq 0} |a_n^2|^2 + |b_n^2|^2 < \infty \right\}$$

and, by definition of $X^\alpha$ and (5.10),

$$X^\alpha = \left\{ u | u = \sum_{n \geq 0} (a_n^1, a_n^2)^T \sin(nx) + \sum_{n \geq 0} (b_n^1, b_n^2)^T \cos(nx), \right.$$
$$\left. \sum_{n \geq 0} n^{4\alpha} (|a_n^1|^2 + |b_n^1|^2) < \infty, \sum_{n \geq 0} |a_n^2|^2 + |b_n^2|^2 < \infty \right\}.$$

From the embedding theorems [**6**] follows that if $\alpha > 1/4$ then $X^\alpha \subset C(0, 2\pi) \times L^2(0, 2\pi)$. Since the nonlinearity $F$ involves only a function of the first component $u$, which is a continuous function, it is easy to see that $F : X^\alpha \to X$ is continuously differentiable for any $\alpha > 1/4$. Choose $\alpha = 1/2$. It follows now that the problem (5.1) is locally well posed in $X^{1/2}$. To show that the solutions exist in $X^{1/2}$ for all time, we use a Lyapunov function $V = \frac{1}{2}||A^{1/2}u|| + \int_0^{2\pi} F(u)$. From the standard argument (see [**5**] Section 4.3.), it follows that solutions of (5.1) form a semigroup, which is analytic by Theorem 1.3.4 of [**6**]. Therefore, since $t \to \gamma(t)$ is analytic and $\gamma(t) \to F_x(\gamma(t))$ is continuous, $t \to F_x(\gamma(t))$ is continuous.

This verifies assumptions of Corollary 4.8.

## 6 Discussion

The goal of this paper was to prove continuity theorem. It is one of the ingredients in the proof of a general conjecture ([**2**] and [**9**]) that there is always a stimulus which leads to annihilation of the wave of excitation, traveling around a ring of excitable media. We proved continuity theorem in a general setting of semiflow on a complete metric space. Thus the theorem is applicable to a wide variety of models, ranging from ordinary differential equations, partial differential equations to functional differential equations.

We pay the price for the simplicity and generality of the theorem and its proof when we verify the assumptions of the theorem in the concrete setting of the FitzHugh-Nagumo equations on the circle. Since the phase space for the semi-flow generated by these equations is infinite dimensional, one must work harder in order to avoid problems mentioned in Section 4. Though calculations in Section 5 are tedious, they are straightforward and commonly used in the literature [**5**].

## Acknowledgment

T. G. would like to thank Jack Dockery for his help with calculations in Section 5.

## References

[1] Glass, L. [1997], *The topology of phase resetting and the entrainment of limit cycles*, in: Case Studies in Mathematical Modeling – Ecology, Physiology, and Cell Biology, (H. G. Othmer, F. R. Adler, M. A. Lewis, C. Dallon, eds.), Prentice-Hall, New York, 255-276.

[2] Glass, L. and Josephson, M. E. [1995], *Resetting and annihilation of reentrant abnormally rapid heartbeat*, Phys. Rev. Lett. **75**, 2059-2062.

[3] Guckenheimer, J. [1975], *Isochrons and Phaseless Sets*, J. Math. Biol. **1**, 259-273.

[4] Hale, J. [1980], *Ordinary Differential Equations*, 2nd edition, Robert E. Krieger, Inc.

[5] Hale, J. [1988], *Asymptotic Behavior of Dissipative Systems*, Mathematical Surveys and Monographs N. 25, AMS Providence, Rhode Island.

[6] Henry, D. [1981], *Geometric theory of semilinear parabolic equations*, Lecture Notes in Mathematics, Springer-Verlag.

[7] Josephson, M. E., Callans, D., Almendral, J. M. et al. [1993], *Resetting and Entrainment of Ventricular Tachycardia Associated with Infraction: Clinical and Experimental Studies*, in Tachycardia: Mechanisms and Management (M. E. Josephson and H. J. J. Wellens, eds.) Futura, Mount Kisco, pp. 505-536.

[8] Naylor, A. and Sell, G. [1982], *Linear Operator Theory in Engineering and Science*, Springer-Verlag, New York, Heidelberg, Berlin.

[9] Nomura, T. and Glass, L. [1996], *Entrainment and termination of reentrant wave propagation in a periodically stimulated ring of excitable media*, Phys. Rev. E **53**, 6353-6360.

[10] Saksena, S. and Goldschlager, N. [1990], *Electrical Therapy for Cardiac Arhythmias: Pacing, Antitachycardia Devices, Catheter Ablation*, W. B. Saunders, Philadelphia.

[11] Temam, R. [1983], *Navier-Stokes Equations and Nonlinear Functional Analysis*, CBMS-NSF Regional Conference Series in Applied Mathematics **41**, SIAM, Philadelphia.

Fields Institute Communications
Volume **21**, 1999

# An Epidemic Model with Recruitment-Death Demographics and Discrete Delays

**Lynne Genik**
Department of Mathematics and Statistics
University of Victoria
Victoria, B. C., Canada V8W 3P4
Current address: Nortel Wireless Networks
P.O. Box 3511, Station C
Ottawa, Ontario, Canada K1Y 4H7
lgenik@nortel.ca

**P. van den Driessche**
Department of Mathematics and Statistics
University of Victoria
Victoria, B. C., Canada V8W 3P4
pvdd@smart.math.uvic.ca

**Abstract.** A new SEIRS disease transmission model is formulated for a variable size population that is divided into susceptible, exposed, infective, and recovered classes. The model has recruitment-death demographics and includes a disease-related death rate. Standard incidence, a constant latent period, and a constant period of immunity are assumed. The model is formulated as an integro-differential equation system, which is shown to be equivalent to a delay differential equation system with two discrete delays. This latter system is analysed. A reproduction number $\mathsf{R}_0$ is identified; if $\mathsf{R}_0 < 1$ the disease dies out of the population, and a unique endemic equilibrium exists if and only if $\mathsf{R}_0 > 1$. Periodic solutions about the endemic equilibrium are possible for a range of parameter values when $\mathsf{R}_0 > 1$. The limiting case for diseases with no latent period, resulting in an SIRS model, is analysed in detail. Periodic solutions are found for a range of parameter values, which is algorithmically determined and related to data on *Pasteurella muris*.

## 1 Introduction

When modelling disease transmission, the population is often divided into classes of individuals, giving a compartmental model. In general, a greater number of classes allows for the inclusion of more detail and is more realistic, although the

1991 *Mathematics Subject Classification.* Primary 92D30; Secondary 34K20.
PvdD research was partially supported by an NSERC research grant and the University of Victoria Committee on Research and Travel.

models are usually more complicated and may not yield qualitatively different results than a simpler model with fewer classes.

The model considered here is referred to as an SEIRS type model. The population is divided into susceptible, exposed, infective, and recovered classes, with population numbers in each class denoted as functions of time by $S(t), E(t), I(t)$, and $R(t)$, respectively. The population is assumed to be uniform and homogeneously mixing. The total population number is denoted by $N(t)$, thus

$$N(t) = S(t) + E(t) + I(t) + R(t). \tag{1.1}$$

Upon an effective contact with an infective individual, a susceptible individual enters the exposed (or latent) class, and is infected but not yet infective. After a latent period an exposed individual then enters the infective class. The latent period is the period from the point of infection to the beginning of the state of infectiousness. (This should not be confused with the incubation period, which is the period from the point of infection to the appearance of symptoms of the disease.) Upon recovery, an individual enters the recovered class with temporary immunity. After a period of immunity, an individual becomes susceptible once again. This type of model is appropriate for diseases such as influenza and whooping cough (where the cycling back to the susceptible class is due to waning immunity of immunization; see [**3**], p. 348). The model also includes important limiting cases such as the SIRS model for diseases with no latent period, e.g., *Pasteurella muris*, analysed in Section 3.

A recruitment-death demographic structure is assumed. Individuals enter the susceptible class at rate $A > 0$, the constant recruitment (or immigration) rate. The natural death rate is assumed to be proportional to the population number in each class, with rate constant $d > 0$. In addition, there is a disease-related death rate in the infective class, which is assumed to be proportional to the population number in that class, with rate constant $\epsilon \geq 0$. Thus, the total population number changes according to the ordinary differential equation

$$N'(t) = A - dN(t) - \epsilon I(t). \tag{1.2}$$

In the absence of the disease, or if the disease causes no deaths, the population size approaches $A/d$ as $t \to \infty$.

The incidence is assumed to be of the standard form $\lambda SI/N$ where $\lambda > 0$ is the average number of effective contacts of one infective individual per unit time, so that $\lambda S/N$ is the average number of effective contacts of one infective with susceptibles per unit time. In some models in the literature (see, e.g., [**1**]) the incidence is assumed to be of the mass action form $\beta SI$ where $\beta$ is the transmission coefficient. Note that in a variable population these two forms of incidence are different. From data on human and animal diseases, the standard incidence seems to be the better approximation (see [**8**] and references therein).

The model is assumed to have a constant finite latent period $\omega > 0$ and a constant finite period of immunity $\tau > 0$. Thus, the probability that an individual remains in the exposed class at least $t$ time units before becoming infective is given by the step function with value 1 for $0 \leq t \leq \omega$ and value 0 for $t > \omega$. Similarly, the probability that an individual remains in the recovered class at least $t$ units before cycling back into the susceptible class is given by the step function with value 1 for $0 \leq t \leq \tau$ and value 0 for $t > \tau$. This gives $\tau$ as the expected time in the recovered class, with death-adjusted expected time $(1 - e^{-d\tau})/d$. The probability that an individual remains infective $t$ time units after having become infective is assumed to be $e^{-\gamma t}, \gamma > 0$. Thus, the recovery rate is proportional to the number in the infective class, and the mean death-adjusted time in the infective class is $1/(\gamma + \epsilon + d)$. The two choices of the probability function used in this model are extreme cases; see, e.g., [**4**]. For more details on the SEIRS model and for other SEIRS models and limiting cases with recruitment-death demographics, see [**9**].

For a disease with no latent period, i.e., in the limit as $\omega \to 0^+$, an SIRS model results, which will be analysed in Section 3. For a disease that confers no immunity, i.e., in the limit when $\tau \to 0^+$, an SEIS model results, which has been analysed in ([**10**], Section 4). For diseases with permanent immunity, i.e., in the formal limit when $\tau \to \infty$, an SEIR model results. This type of model is appropriate for childhood diseases, such as measles, and has been analysed in ([**9**], Section 5.4.1). In the limit when both lags approach zero, i.e., $\omega, \tau \to 0^+$, an o.d.e. SIS model results, which has been analysed for a more general case with $\lambda(N)$ in ([**17**], Section 2). In each of these latter three models, no Hopf bifurcation is possible. For the SIS model, there is a sharp threshold with the disease dying out below the threshold and tending to a globally stable endemic equilibrium above the threshold [**17**]. The SEIS and SEIR models exhibit this same behaviour, although in both models only local stability of the endemic equilibrium is proved [**9, 10**].

The presence of a discrete delay in the recovered class is known to give rise to oscillations in some SIRS models, e.g., constant population models [**12, 16**] and a variable population model with exponential demographics [**6**]. Ordinary differential equation SIRS models similar to these cited models, but with an exponentially distributed period of immunity, do not exhibit oscillations. (Note that an o.d.e. model with at least 3 removed subclasses, causing a "delay", can exhibit oscillations ([**12**], Section 4).) In particular, o.d.e. SIRS models with recruitment-death demographics and standard or mass action incidence do not have periodic solutions ([**14**], Sections 2 and 3). In Section 3 of this paper, we show that a delay in the recovered class in the variable population SIRS model can give rise to oscillations. We give an algorithm to determine the parameter ranges where oscillations occur and relate this to data on *Pasteurella muris*.

We remark that our SEIRS model is similar to that of Cooke and van den Driessche [**6**], but they assume an exponential demographic structure, i.e., input into the susceptible class is assumed to be proportional to $N$, rather than the constant immigration rate, $A$. Their ordinary differential equation for the total population number is different from (1.2) and in the absence of the disease $N(t)$ either grows or decays exponentially.

## 2 SEIRS model

The model assumptions in the introduction lead to the following transfer diagram.

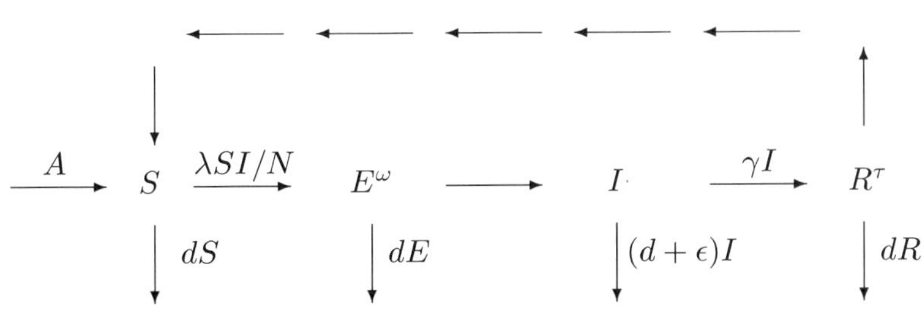

For $t > r = max\{\omega, \tau\}$, the model equations are

$$S'(t) = A - dS(t) - \lambda \frac{S(t)I(t)}{N(t)} + \gamma I(t-\tau)e^{-d\tau}, \tag{2.1}$$

$$E(t) = \int_{t-\omega}^{t} \frac{\lambda S(u)I(u)}{N(u)} e^{-d(t-u)} du, \tag{2.2}$$

$$I'(t) = \frac{\lambda S(t-\omega)I(t-\omega)}{N(t-\omega)} e^{-d\omega} - (\gamma + \epsilon + d)I(t), \tag{2.3}$$

$$R(t) = \int_{t-\tau}^{t} \gamma I(u) e^{-d(t-u)} du, \tag{2.4}$$

with $N(t)$ given by (1.1).

The equations can be determined directly from the transfer diagram. For example, consider (2.1). The rate of change of the population number in the susceptible class is given by the immigration rate, minus the natural death rate and the rate at which individuals become exposed, plus the recovery rate. The rate at which individuals recover is the term $\gamma I(t-\tau)e^{-d\tau}$, which is the product of the recovery rate constant and the number of individuals who were in the recovered class $\tau$ time units before and have not died. The integral in (2.2) represents the summation over the interval $[t-\omega, t]$ of those individuals who became exposed at time $u \geq 0$ and have neither become infective nor died. Shifting time back by $r$, the above equations hold for the new time $t > 0$. The initial data can be specified as $S(t) \geq 0, E(t) \geq 0, I(t) \geq 0$, and $R(t) \geq 0$, with $N(t) > 0$ on $[-r, 0]$ and satisfying (1.1). The system (2.1)-(2.4) is an integro-differential equation system.

To analyse this SEIRS model, it is convenient to find an equivalent delay differential equation system so that standard theorems may be applied. Differentiating (2.2) and (2.4) gives

$$E'(t) = \frac{\lambda S(t)I(t)}{N(t)} - \frac{\lambda S(t-\omega)I(t-\omega)}{N(t-\omega)} e^{-d\omega} - dE(t), \tag{2.5}$$

$$R'(t) = \gamma I(t) - \gamma I(t-\tau)e^{-d\tau} - dR(t). \tag{2.6}$$

Note that adding (2.1), (2.3), (2.5), (2.6) and using (1.1) gives (1.2).

**Theorem 2.1** (see [**6**], Theorem 2.1). *A solution of the integro-differential equation system (2.1)-(2.4) with $N(t)$ given by (1.1) satisfies (2.5) and (2.6). Conversely, let $S(t)$, $E(t)$, $I(t)$, and $R(t)$ be a solution of the delay differential equation system (2.1), (2.3), (2.5), and (2.6) with $N(t)$ given by (1.1) and the initial conditions on $[-r, 0]$ stated above. If, in addition,*

$$E(0) = \int_{-\omega}^{0} \frac{\lambda S(u)I(u)}{N(u)} e^{du} du \quad \text{and} \quad R(0) = \int_{-\tau}^{0} \gamma I(u) e^{du} du,$$

*then this solution satisfies the integro-differential equation system (2.1)-(2.4).*

The model system (1.2), (2.1), (2.3), (2.5), and (2.6) is a nonlinear system of delay differential equations in five variables. However, in the analysis of the model it is sufficient to consider four variables, since the fifth is uniquely determined by (1.1).

Setting $\mathbf{v} = (S, I, R, N)^T$, the system can be written as the functional differential equation

$$\begin{bmatrix} S' \\ I' \\ R' \\ N' \end{bmatrix} = \mathbf{v}'(t) = \mathbf{F}(\mathbf{v}_t) = \begin{bmatrix} A - dS(t) - \lambda \frac{S(t)I(t)}{N(t)} + \gamma I(t-\tau)e^{-d\tau} \\ \frac{\lambda S(t-\omega)I(t-\omega)}{N(t-\omega)} e^{-d\omega} - (\gamma + \epsilon + d)I(t) \\ \gamma I(t) - \gamma I(t-\tau)e^{-d\tau} - dR(t) \\ A - dN(t) - \epsilon I(t) \end{bmatrix}, \tag{2.7}$$

where $\mathbf{v}_t = \mathbf{v}(t+\theta)$ for $-r \le \theta \le 0$ and $t \ge 0$. The initial conditions can be written as $\mathbf{v}_0 \ge 0$ on $[-r, 0]$, with $N(t) > 0$ on this interval.

To show that the model is well-posed, define the regions

$$D_+ = \{S, I, R, N | S \ge 0, I \ge 0, R \ge 0, N \ge S + I + R, N > 0\}$$

and

$$D = \{S, I, R, N | S \ge 0, I \ge 0, R \ge 0, N \ge S + I + R, A/d \ge N \ge A/(d + \epsilon)\}.$$

**Theorem 2.2** *The solution $\mathbf{v}(t) = (S(t), I(t), R(t), N(t))^T$ of the system (2.7) subject to the specified initial conditions exists and is unique for all $t \ge 0$, with $S(t) \ge 0$, $I(t) \ge 0$, $R(t) \ge 0$, and $N(t) \ge S(t) + I(t) + R(t)$, $N(t) > 0$. For the model system (2.7), the region $D$ is positively invariant and all solutions starting in $D_+$ approach, enter, or stay in $D$.*

**Proof** For the delay differential equation system (2.7), the local existence, uniqueness, and continuation results follow from standard results (see [**13**], pp. 19-20, and [**11**], pp. 43-44). Regardless of which of the two lags $\omega$ or $\tau$ is larger, by taking time steps of $\omega$ with the initial data in $D_+$ on the interval $[-r, 0]$, the solution can be shown to stay in $D_+$ as long as it exists. Since $N(t) \ge I(t)$, then $A - (d+\epsilon)N(t) \le N'(t) \le A - dN(t)$ implies that $N(t)$ is bounded and all solutions starting in $D_+$ approach, enter, or stay in $D$. For given initial data the solution starting in $D_+$ exists and is unique on a maximum time interval. Furthermore, since the solution is bounded and $\mathbf{F}(\mathbf{v}_t)$ is continuous, the maximum time interval is $[0, \infty)$. Thus, the solution is unique and exists for all $t \ge 0$. Standard results also imply that the solution depends continuously on the parameters and initial data. □

In fact, solutions enter or stay in $D$ except when $N(t) > A/d$ and $I(t) = 0$ on $[-r, 0]$, in which case $N(t)$ approaches $A/d$. Therefore, it suffices to consider solutions starting in $D$.

The following theorem identifies the equilibrium with all individuals susceptible, and stability is determined by the basic reproduction number defined as

$$\mathsf{R}_0 = \frac{\lambda e^{-d\omega}}{\gamma + \epsilon + d}, \tag{2.8}$$

which is the product of the contact rate constant $\lambda$, the fraction of individuals surviving the exposed class $e^{-d\omega}$, and the average death-adjusted infective period $1/(\gamma + \epsilon + d)$. The parameter $\mathsf{R}_0$ can be interpreted as the expected number of secondary cases produced, in a completely susceptible population, by a typical infective individual during its entire period of infectiousness [**2, 7**].

**Theorem 2.3** *The model system (2.7) always has the disease free equilibrium $\mathbf{v} = (A/d, 0, 0, A/d)^T$. If $\mathsf{R}_0 < 1$, then all solutions starting in $D$ approach the disease free equilibrium. If $\mathsf{R}_0 > 1$, this equilibrium is unstable.*

**Proof** Using (2.7) at equilibrium gives $I = 0$ as one solution and substituting this yields $S = A/d$, $R = 0$, and $N = A/d$.

From (2.3), $I'(t) \leq \lambda I(t-\omega)e^{-d\omega} - (\gamma + \epsilon + d)I(t)$. Using a Razumikhin-type theorem ([**13**], Theorem 6.1, p. 39), with $V(I) = I$, as in ([**10**], Theorem 4.1), then $I(t) \to 0$ as $t \to \infty$ when $\mathsf{R}_0 < 1$. Using (2.6) implies that $R(t) \to 0$ as $t \to \infty$, and (2.1) and (1.2) imply that $S(t) \to A/d$ and $N(t) \to A/d$ as $t \to \infty$. Thus, the disease free equilibrium is globally asymptotically stable if $\mathsf{R}_0 < 1$.

Linearization of (2.7) near the disease free equilibrium gives a characteristic quasi-polynomial with a triple root $z = -d$, and solutions $z$ of the quasipolynomial $\lambda e^{-(z+d)\omega} - (z + \gamma + \epsilon + d) = 0$. This quasipolynomial has a positive root if $\mathsf{R}_0 > 1$ (see [**6**], Theorem A2). □

Using (2.7) at equilibrium also gives an endemic equilibrium, i.e., $I_e > 0$, if $\mathsf{R}_0 > 1$.

**Theorem 2.4** *If* $\mathsf{R}_0 > 1$, *the model system (2.7) has a unique endemic equilibrium in D, namely* $\mathbf{v}_e = (S_e, I_e, R_e, N_e)^T$, *where* $S_e = N_e/\mathsf{R}_0$, $I_e = \frac{A(\mathsf{R}_0 - 1)}{\lambda - (\epsilon + \gamma \mathsf{R}_0 e^{-d\tau})}$, $R_e = \gamma(1 - e^{-d\tau})I_e/d$, *and* $N_e = \frac{A}{d}\left[\frac{\lambda - \mathsf{R}_0(\epsilon + \gamma e^{-d\tau})}{\lambda - (\epsilon + \gamma \mathsf{R}_0 e^{-d\tau})}\right]$. *If* $\mathsf{R}_0 \leq 1$, *then the disease free equilibrium is the only equilibrium in D.*

Linearization near the endemic equilibrium gives a double root $z = -d$, and the remaining characteristic equation

$$\begin{aligned} 0 \;=\; & (z + d + \lambda I_e/N_e)(z + \gamma + \epsilon + d) - [(z+d)\lambda S_e/N_e + \lambda \epsilon S_e I_e/N_e^2]e^{-(z+d)\omega} \\ & -\lambda\gamma I_e/N_e e^{-(z+d)(\omega+\tau)}. \end{aligned} \tag{2.9}$$

The characteristic equation (2.9) is complicated since it contains the two lags $\omega$ and $\tau$ in both the exponential terms and the coefficients, and stability of the endemic equilibrium is an open problem.

Numerical simulations of (2.7) with the specified initial conditions using XP-PAUT indicate that there is a transcritical bifurcation at $\mathsf{R}_0 = 1$, where the stability is transferred from the disease free equilibrium to the endemic equilibrium. As $\mathsf{R}_0$ further increases there are periodic solutions for a range of parameter values when $\omega, \tau > 0$. In the limit when $\tau \to 0^+$, the model reduces to an SEIS model; in this limit the endemic equilibrium is locally asymptotically stable and, thus, there are no periodic solutions arising from a Hopf bifurcation ([**10**], Section 4). The next section deals with another limiting case, namely $\omega \to 0^+$.

## 3 SIRS model

For diseases that have no latent period but confer temporary immunity, e.g., *Pasteurella muris* in laboratory mice [**1**], the limiting case $\omega \to 0^+$ of the SEIRS model is appropriate. In this case $E(t) \equiv 0$, and the SIRS model considered here results. This model is also suitable as a simple model for influenza, since the latent period is short ([**3**], p. 247), although a more realistic model should incorporate the three different types of the influenza virus.

Defining $\tilde{\mathbf{v}} = (S, I, N)^T$, the system can be written from (2.7) as

$$\begin{bmatrix} S' \\ I' \\ N' \end{bmatrix} = \tilde{\mathbf{v}}'(t) = \begin{bmatrix} A - dS(t) - \lambda \frac{S(t)I(t)}{N(t)} + \gamma I(t-\tau)e^{-d\tau} \\ \lambda \frac{S(t)I(t)}{N(t)} - (\gamma + \epsilon + d)I(t) \\ A - dN(t) - \epsilon I(t) \end{bmatrix}. \tag{3.1}$$

The initial conditions are given as $\tilde{\mathbf{v}}_0 \geq 0$ on $[-\tau, 0]$, with $N(t) > 0$ on the interval, and the invariant region is

$$\tilde{D} = \{S, I, N | S \geq 0, I \geq 0, N \geq S + I, A/d \geq N \geq A/(d+\epsilon)\}.$$

The equilibria and the stability of the disease free equilibrium can be determined using the theorems from Section 2 in the limiting case $\omega \to 0^+$. The basic reproduction number $\mathsf{R}_0$ is given by (2.8) with $\omega = 0$, namely

$$\mathsf{R}_0 = \frac{\lambda}{\gamma + \epsilon + d}, \tag{3.2}$$

and is now the product of the contact rate constant and the average death-adjusted infective period.

**Theorem 3.1** *The model system (3.1) always has the disease free equilibrium* $\tilde{\mathbf{v}} = (A/d, 0, A/d)^T$. *With* $\mathsf{R}_0$ *given by (3.2), if* $\mathsf{R}_0 < 1$, *then all solutions of the system starting in* $\tilde{D}$ *approach the disease free equilibrium. If* $\mathsf{R}_0 \leq 1$, *this is the only equilibrium in* $\tilde{D}$. *If* $\mathsf{R}_0 > 1$, *then the disease free equilibrium is unstable and there is a unique endemic equilibrium in* $\tilde{D}$, *namely* $\tilde{\mathbf{v}}_e = (S_e, I_e, N_e)^T$, *where* $S_e, I_e$, *and* $N_e$ *are given in Theorem 2.4. If* $d^2 + (\lambda I_e/N_e)^2 - 2\lambda I_e/N_e[\gamma + \epsilon(1 - 1/\mathsf{R}_0)] \geq 0$, *the endemic equilibrium is locally asymptotically stable.*

**Proof** All statements except the last result from Theorems 2.3 and 2.4 in the limiting case $\omega \to 0^+$. Linearizing near the endemic equilibrium gives a root $z = -d$ and solutions $z$ of the quasipolynomial that results from (2.9) with $\omega = 0$, namely

$$z^2 + (d + \lambda I_e/N_e)z + \lambda I_e/N_e[\gamma + d + \epsilon(1 - S_e/N_e)] - \lambda\gamma I_e/N_e e^{-(z+d)\tau} = 0. \tag{3.3}$$

When $\mathsf{R}_0 > 1$ and $\tau = 0$, this reduces to a quadratic with both roots having $\Re(z) < 0$ and the endemic equilibrium is locally asymptotically stable. When $\mathsf{R}_0 > 1$ and $\tau > 0$, then $z = 0$ is not a root of (3.3). Without loss of generality, assuming $z = iy$ with $y > 0$, taking absolute values, and squaring gives

$$\begin{aligned} &y^4 + [d^2 + (\lambda I_e/N_e)^2 - 2\lambda I_e/N_e(\gamma + \epsilon(1 - S_e/N_e))]y^2 \\ &\quad + (\lambda I_e/N_e)^2[(\gamma + d + \epsilon(1 - S_e/N_e))^2 - \gamma^2 e^{-2d\tau}] = 0. \end{aligned}$$

Setting $Y = y^2$, the above equation becomes a quadratic. The constant term is positive if $\mathsf{R}_0 > 1$, and the coefficient of the $Y$ term is nonnegative if

$$d^2 + (\lambda I_e/N_e)^2 - 2\lambda I_e/N_e[\gamma + \epsilon(1 - 1/\mathsf{R}_0)] \geq 0. \tag{3.4}$$

Therefore, if $\mathsf{R}_0 > 1$ and (3.4) is satisfied, all coefficients of the quadratic equation are nonnegative. Thus, a root $Y$ cannot be positive and no purely imaginary root $z = iy$ can exist. Since a root can only appear in the right half plane by moving across the imaginary axis, then if $\mathsf{R}_0 > 1$ and (3.4) is satisfied, all roots have $\Re(z) < 0$, and the endemic equilibrium is locally asymptotically stable. □

Sufficient conditions have been identified for the endemic equilibrium to be locally asymptotically stable. Now the possibility of a Hopf bifurcation, leading to periodic solutions, is investigated.

Let $Z = z\tau, \tau > 0$, so that the time lag in the new variable is scaled to 1. Multiplying (3.3) by $\tau^2$ gives

$$\begin{aligned} &Z^2 + (d\tau + \lambda\tau I_e/N_e)Z + \lambda\tau I_e/N_e[d\tau + \epsilon\tau(1 - S_e/N_e)] \\ &\quad + \lambda\tau\gamma\tau I_e/N_e(1 - e^{-(Z+d\tau)}) = 0. \end{aligned} \tag{3.5}$$

Without loss of generality, assuming $Z = i\mu$ with $\mu > 0$ and equating the imaginary parts gives

$$(d\tau + \lambda\tau I_e/N_e)\mu + \lambda\tau\gamma\tau I_e/N_e e^{-d\tau} sin(\mu) = 0. \tag{3.6}$$

In order for (3.6) to be satisfied, it is necessary that $sin(\mu) < 0$ so that $\mu \in ((2k-1)\pi, 2k\pi)$ for $k \in \{1, 2, ...\}$.

Rewriting (3.5), dividing by $Z + d\tau$, and setting $Z = i\mu$ with $\mu > 0$ gives

$$i\mu + \lambda\tau I_e/N_e + \lambda\tau\epsilon\tau I_e/N_e(1 - S_e/N_e)\frac{d\tau - i\mu}{\mu^2 + d^2\tau^2} + \lambda\tau\gamma\tau I_e/N_e\frac{s - iq}{\mu^2 + d^2\tau^2} = 0,$$

where $q = \mu(1 - e^{-d\tau}cos(\mu)) - d\tau e^{-d\tau}sin(\mu)$ and $s = d\tau(1 - e^{-d\tau}cos(\mu)) + \mu e^{-d\tau}sin(\mu)$. Equating the real parts and solving for $\gamma\tau$ gives

$$\gamma\tau = -\left[1 + \epsilon\tau d\tau\frac{1 - S_e/N_e}{\mu^2 + d^2\tau^2}\right]\left[\frac{\mu^2 + d^2\tau^2}{s}\right], \tag{3.7}$$

and equating the imaginary parts gives

$$\mu = \lambda\tau I_e/N_e\left[\frac{\epsilon\tau\mu(1 - S_e/N_e) + \gamma\tau q}{\mu^2 + d^2\tau^2}\right]. \tag{3.8}$$

The smallest value of $\mu > 0$ that satisfies (3.6) and also the real part of (3.5) with $Z = i\mu$ occurs when $\mu \in (\pi, 2\pi)$, and gives the first pair of imaginary roots. The curve generated with $\mu \in (\pi, 2\pi)$ is referred to as the lowest imaginary root curve. Other imaginary root curves can be found by taking $\mu \in ((2k-1)\pi, 2k\pi)$ where $k \in \{2, 3, ...\}$.

With $\epsilon = 0$, i.e., no disease-related deaths, Equations (3.7) and (3.8) are equivalent to those of [**16**], in which a similar model with a constant population number was analysed. In this case a Hopf bifurcation is possible. Also with $\epsilon = 0$, (3.1) is a special case of a model with arbitrary distributed removal rate analysed by Castillo-Chavez and Thieme in [**5**]. They proved that when $\mathsf{R}_0 > 1$ and $\gamma\tau < -(\mu^2 + d^2\tau^2)/s$ (see (3.7) with $\epsilon = 0$), then given a positive number of initial infectives all solutions approach the endemic equilibrium for $t \to \infty$.

Consider the case of disease-related deaths, i.e., $\epsilon > 0$. Substituting $S_e/N_e = (\gamma\tau + \epsilon\tau + d\tau)/(\lambda\tau)$ into (3.7) and solving for $\lambda\tau$ gives

$$\lambda\tau = \frac{\epsilon\tau d\tau(\gamma\tau + \epsilon\tau + d\tau)}{\mu^2 + d^2\tau^2 + \epsilon\tau d\tau + \gamma\tau s}. \tag{3.9}$$

Calculating $\lambda\tau I_e/N_e$ in terms of the parameters gives

$$\lambda\tau I_e/N_e = \frac{d\tau[\lambda\tau - (\gamma\tau + \epsilon\tau + d\tau)]}{\gamma\tau(1 - e^{-d\tau}) + d\tau}. \tag{3.10}$$

Substituting (3.9) for $\lambda\tau$ into the right-hand sides of (3.10) and of $S_e/N_e = (\gamma\tau + \epsilon\tau + d\tau)/(\lambda\tau)$, then substituting for $\lambda\tau I_e/N_e$ and $S_e/N_e$ into (3.8) and simplifying gives

$$\begin{aligned}&\mu[\gamma\tau(1 - e^{-d\tau}) + d\tau]\,(\mu^2 + d^2\tau^2 + \epsilon\tau d\tau + \gamma\tau s)\\ &\qquad -(\gamma\tau + \epsilon\tau + d\tau)(\mu^2 + d^2\tau^2 + \gamma\tau s)(\mu + \gamma\tau e^{-d\tau}sin(\mu)) = 0,\end{aligned} \tag{3.11}$$

where $s$ and $q$ depend on $\mu$ and $d\tau$. Equation (3.11) is a cubic equation in $\gamma\tau$.

The algorithm is applied by fixing a $d\tau$ and an $\epsilon\tau$ and choosing $\mu \in (\pi, 2\pi)$. Then MAPLE can be used to numerically solve (3.11) for $\gamma\tau$. If $\gamma\tau > 0$, it is substituted into (3.9) to give $\lambda\tau$, and finally $\mathsf{R}_0$ is calculated from (3.2). Only those points with $\mathsf{R}_0 > 1$ are considered. Varying $\mu$, it is possible to plot the bifurcation curve $\mathsf{R}_0$

versus $\gamma\tau$. By examining the output, it can be seen that a fixed $\mu$ may produce either one or two roots with $\gamma\tau > 0$. If there are two positive roots produced, then only one of these two roots gives $R_0 > 1$. Since two different values of $\mu$ may yield the same $\gamma\tau$, but a different $\lambda\tau$, there are two values of $R_0$ for most values of $\gamma\tau$ generated. For example, for $d\tau = 0.252$ and $\epsilon\tau = 2.52$, $\mu \approx 3.85$ gives $\gamma\tau \approx 9.84$ and $R_0 \approx 1.67$, while $\mu \approx 5.43$ gives $\gamma\tau \approx 9.84$ and $R_0 \approx 4.25$ (see Figure 1). This lowest imaginary root curve or bifurcation curve indicates the boundary of a region where periodic solutions may occur.

The assumptions of the Hopf bifurcation theorem ([**15**], Theorem 2.1), have not been proven. Due to the complicated nature of the characteristic Equation (3.3), numerics are now used to investigate periodic solutions. Since the first purely imaginary roots occur on the lowest imaginary root curve generated, it follows that, in the region below the curve, solutions are locally asymptotically stable (see Figure 1). Therefore, periodic solutions, if they exist, will occur in the region above the curve.

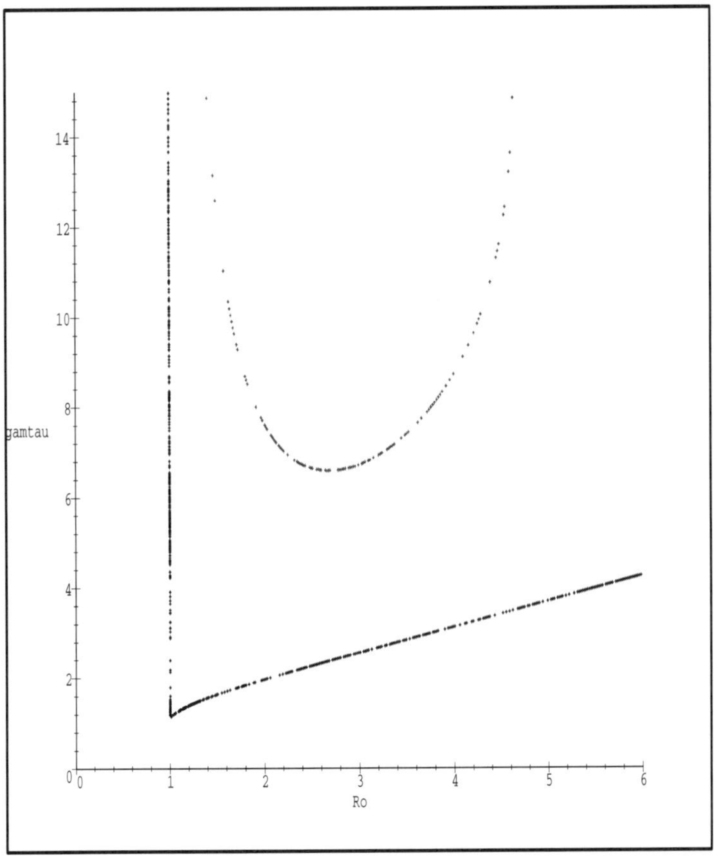


**Figure 1** The Bifurcation and Stability Estimate Curves in Terms of $R_0$ Versus $\gamma\tau$ for $d = 0.006, \epsilon = 0.06$, and $\tau = 42$.

The lowest imaginary root curve is the upper U-shaped curve. The stability curve estimate is given by $d^2 + (\lambda I_e/N_e)^2 - 2\lambda I_e/N_e[\gamma + \epsilon(1 - 1/R_0)] = 0$.

Anderson and May in ([**1**], p. 363), give crude estimates of parameters for a SIRS model with mass action incidence and an exponentially distributed period of immunity for the virus *Pasteurella muris* in laboratory mice. The natural death rate constant, disease-related death rate constant, and recovery rate constant are given as $d = 0.006, \epsilon = 0.06$, and $\gamma = 0.04$, respectively, in units of $\text{day}^{-1}$. The parameters are based on an experiment in which mice were introduced at rates ranging from 0.33 to 6 mice per day. Thus, $A$ is chosen as 0.33 mice per day. The loss of immunity rate constant is $0.021 \text{ day}^{-1}$. To get an estimate for $\tau$, the expected death-adjusted time in the recovered class for each model is equated. In the case of the Anderson and May ordinary differential equation model, this is given by one divided by the sum of the loss of immunity rate constant and the natural death rate constant. In this SIRS model, the expected death-adjusted time in the recovered class is $(1 - e^{-d\tau})/d$. Equating the two quantities gives an estimate for $\tau$ as 42 days. Anderson and May use the mass action incidence and estimate the transmission coefficient to be $0.0056 \text{ mice}^{-1}\text{day}^{-1}$. An estimate for the contact rate constant is $\lambda = 0.0056N$, where $N$ is taken to be the disease free equilibrium value, $A/d$. Thus, the estimate is $\lambda = 0.308 \text{ day}^{-1}$.

Using the estimates for the natural death rate constant $d = 0.006$, the disease-related death rate constant $\epsilon = 0.06$, and the constant period of immunity $\tau = 42$, programs written in MAPLE can be used to generate the lowest imaginary root curves and the stability estimate curves (given by (3.4) with the inequality replaced by equality); see Figure 1. Inequality (3.4) gives the region below the curve as an estimate of where solutions are locally stable. As $\epsilon$ increases, the range of $\mathsf{R}_0$ for the bifurcation curve decreases. Also, as $\epsilon$ increases, the stability estimate curve has higher values of $\gamma\tau$ for the same $\mathsf{R}_0$, which means that the estimated region of stability is larger.

Figures 2 and 3 are numerical simulations of (3.1) done in XPPAUT. For each of the two figures, the solution curve for short time $t > 0$ depends strongly on the initial values taken on $[-\tau, 0]$. For a small number of initial infectives introduced into a susceptible population, $I(t)$ first increases if $\mathsf{R}_0 > 1$; see (3.1). The parameters for Figure 2 are given above and based on the Anderson and May estimates. For these parameters, the solution approaches the endemic equilibrium. The parameters for Figure 3 are found by keeping the values of $d\tau, \epsilon\tau$, and $\mathsf{R}_0$ from Figure 2 fixed and choosing a value of $\gamma\tau$ slightly above the lowest imaginary root curve. The solution is periodic for these parameters, with a period of about $275/5 \approx 55$ days. This period agrees with the estimate given in the Hopf bifurcation theorem ([**15**], Theorem 2.1), namely $2\pi\tau/\mu \approx 84\pi/4.75$. Additional numerical simulations performed (but not shown), with parameters in the region above the bifurcation curve, gave periodic solutions in each variable.

## 4 Discussion

The main features of the new SEIRS disease transmission model formulated and analysed here are the recruitment-death demographics with disease-related deaths, the standard incidence, and the constant latent and immunity periods. A reproduction number $\mathsf{R}_0$ has been identified in (2.8). For $\mathsf{R}_0 < 1$ the disease dies out, whereas for $\mathsf{R}_0 > 1$ there is a unique endemic equilibrium. This threshold criterion suggests a possible means of controlling the disease, i.e., forcing the basic reproduction number to be less than one. Perhaps the contact rate constant $\lambda$ could be decreased and the recovery rate constant $\gamma$ increased by early diagnosis and

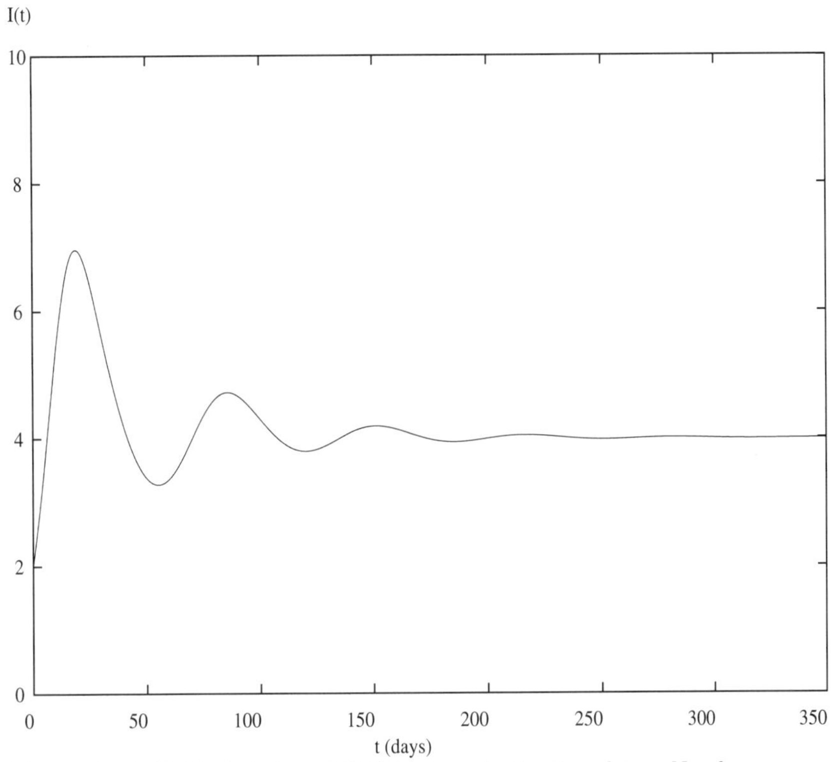


**Figure 2** A Stable Solution of (3.1) in the Infective Population Number.
Numerical simulations of (3.1) using XPPAUT with $S(t) = 15, I(t) = 2, N(t) = 20$ on $[-\tau, 0]$, and parameters $A = 0.33, d = 0.006, \lambda = 0.308, \gamma = 0.04, \epsilon = 0.06$, and $\tau = 42$. For these parameter values, $\mathsf{R}_0 = 2.9, S_e = 5.2, I_e = 4.0$, and $N_e = 15.1$.

treatment of the disease. Numerical simulations indicate that for some parameter values the SEIRS model can exhibit periodic solutions when $\mathsf{R}_0 > 1$.

The SIRS model with a constant finite period of immunity has been shown to exhibit periodic solutions for a range of parameter values when $\mathsf{R}_0 > 1$. The algorithm given in Section 3 was used to identify this parameter range. For parameter values for *Pasteurella muris*, solutions approach the endemic equilibrium, but when $\gamma\tau$ is sufficiently increased the infective population number oscillates about the endemic equilibrium. This periodicity is in contrast to the ordinary differential equation model ([**14**], Section 3), where periodic solutions arising by Hopf bifurcation are not possible. Thus, as has been found in other disease transmission models (see the Introduction), assuming a constant (finite) period of immunity, as opposed to an exponentially distributed period, qualitatively alters the solution behaviour. A Hopf bifurcation cannot occur in the SEIS model with a constant latent period [**10**], or the SEIR model with a constant latent period and permanent immunity [**9**]. This supports the assertion that it is the discrete delay in the removed class that gives rise to periodic solutions. Anderson and May in ([**2**], p. 59), remark that adding an exposed class does not add qualitatively new features to the dynamics in most cases. This appears to also be the case here.

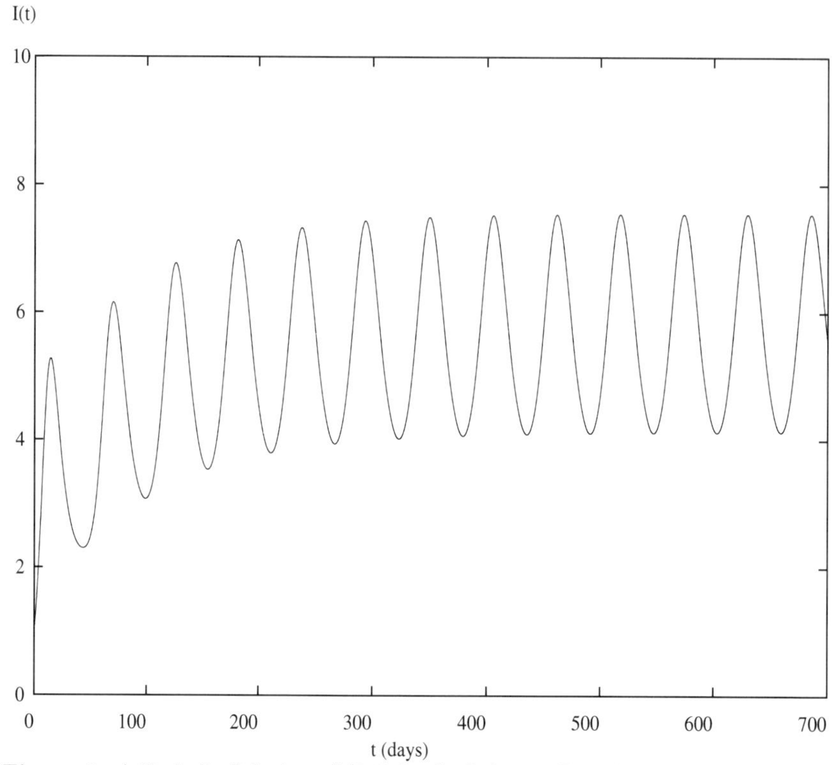


**Figure 3** A Periodic Solution of (3.1) in the Infective Population Number.
Numerical simulations of (3.1) using XPPAUT with $S(t) = 13, I(t) = 1, N(t) = 20$ on $[-\tau, 0]$, and parameters $A = 0.70$, $d = 0.006$, $\lambda = 0.655$, $\gamma = 0.160$, $\epsilon = 0.06$, and $\tau = 42$. For these parameter values, $\mathsf{R}_0 = 2.9, S_e = 20.7, I_e = 5.7$, and $N_e = 60.0$.

## Acknowledgement

We thank Dr. Xingfu Zou for discussion on Theorem 2.2.

## References

[1] Anderson, R. M. and May, R. M. [1979], *Population Biology of Infectious Diseases: Part I*, Nature **280**, 361-367.

[2] Anderson, R. M. and May, R. M. [1991], *Infectious Diseases of Humans*, Oxford University Press, Oxford, New York, Tokyo.

[3] Benenson, A. S. [1995] (ed.), *Control of Communicable Diseases in Man*, Sixteenth Edition, American Public Health Association.

[4] Brauer, F. and Castillo-Chavez, C. [1995], *Basic Models in Epidemiology*, in Ecological Time Series (T. M. Powell and J. H. Steele, eds.), Chapman and Hall, pp. 410-447.

[5] Castillo-Chavez, C. and Thieme, H.R. [1995], *Asymptotically Autonomous Epidemic Models*, in Mathematical Population Dynamics: Analysis of Heterogeneity (O. Arino, D. Axelrod, M. Kimmel, M. Langlais, eds.), Wuerz Publishing Ltd., Winnipeg, Canada, pp. 33-50.

[6] Cooke, K. L. and van den Driessche, P. [1996], *Analysis of an SEIRS Epidemic Model with Two Delays*, Journal of Mathematical Biology **35**, 240-260.

[7] Diekmann, O., Heesterbeek, J. A. P. and Metz, J. A. J. [1990], *On the Definition and the Computation of the Basic Reproduction Ratio* $\mathsf{R}_0$ *in Models for Infectious Diseases in Heterogeneous Populations*, Journal of Mathematical Biology **28**, 365-382.

[8] Gao, L. Q., Mena-Lorca, J. and Hethcote, H. W. [1995], *Four SEI Endemic Models with Periodicity and Separatrices*, Mathematical Biosciences **128**, 157-184.

[9] Genik, L. [1997], *An SEIRS Epidemic Model with Recruitment-Death Demographics, Standard Incidence, and Two Discrete Delays*, M.Sc. Thesis, University of Victoria.

[10] Genik, L. and van den Driessche, P. [1998], *A Model for Diseases Without Immunity in a Variable Size Population*, Canadian Applied Mathematics Quarterly **6**, to appear.

[11] Hale, J. K. and Verduyn Lunel, S. M. [1993], *Introduction to Functional Differential Equations*, Springer Verlag, New York.

[12] Hethcote, H. W., Stech, H. W. and van den Driessche, P. [1981], *Nonlinear Oscillations in Epidemic Models*, SIAM Journal of Applied Mathematics **40**, 1-9.

[13] Kuang, Y. [1993], *Delay Differential Equations with Applications in Population Dynamics*, Academic Press, Inc., Boston.

[14] Mena-Lorca, J. and Hethcote, H. W. [1992], *Dynamic Models of Infectious Diseases as Regulators of Population Sizes*, Journal of Mathematical Biology **30**, 693-716.

[15] Stech, H. W. [1985], *Hopf Bifurcation Calculations for Functional Differential Equations*, Journal of Mathematical Analysis and Applications **109**, 472-491.

[16] van den Driessche, P. [1983], *A Cyclic Epidemic Model with Temporary Immunity and Vital Dynamics*, in Lecture Notes in Biomathematics (H. I. Freedman, C. Strobeck eds.), Springer-Verlag, pp. 433-440.

[17] Zhou, J. and Hethcote, H. W. [1994], *Population Size Dependent Incidence in Models for Diseases without Immunity*, Journal of Mathematical Biology **32**, 809-834.

Fields Institute Communications
Volume **21**, 1999

# Nonlinear Propagation in Reaction Transport Systems

**K. P. Hadeler**
Biomathematics and Department of Mathematics
University of Tübingen
D-72076 Tübingen, Germany
k.p.hadeler@uni-tuebingen.de

**Abstract.** Optimal characterizations for travelling fronts of semilinear reaction transport equations are derived by reduction of the problem of heteroclinic orbits to the corresponding problem for Fisher's equation. Particular attention is given to the problem whether the minimal speed can be obtained from the linearization at the trivial solution.

## 1 Travelling fronts in transport systems

Since the seminal work of R. A. Fisher in [**5**] and of Kolmogorov et al. in [**20**], spatial spread caused by the joint effects of diffusion and reaction is a well-known phenomenon that has been extensively studied (see, e.g., [**4**]). Sometimes the phenomenon has been described, in rather crude terms, as a linear principle: if matter is diffusing with rate $D$ and reproducing with rate $a$ then the bulk of matter (i.e., the boundary of a level set) is moving forward as a front with speed $2\sqrt{Da}$. This principle disregards nonlinear effects. In general, the number $2\sqrt{Da}$ is only a lower bound for the speed of propagation.

The reaction diffusion equation is based on the diffusion model of spatial spread which shows the unrealistic effect of infinitely fast propagation of small amounts of matter. This effect is closely connected to the parabolic character of the diffusion equation. More realistic models for the spread of matter are reaction transport systems (see [**1, 21, 18, 11**]) or semilinear Cattaneo systems [**17, 11**]. These systems are hyperbolic, particles have uniformly bounded speeds, and the speeds of fronts depend on these bounds. In the simplest case of space dimension one and constant particle speed, transport systems as well as Cattaneo systems are correlated random walk systems.

The existence problem for travelling fronts (for equations without non-local terms) amounts to finding heteroclinic orbits for certain systems of ordinary differential equations. For hyperbolic systems arising from correlated random walks the problem of travelling fronts has been studied in [**8, 9, 10, 6**]. Throughout that

1991 *Mathematics Subject Classification.* Primary 35K57, 35L60; Secondary 34C37, 92D25.
Work supported in part by DFG Sonderforschungsbereich 382.

investigation there has been a discrepancy between the direct approach (which appears very natural) and a seemingly awkward reduction to the parabolic case (which nevertheless produces rather general results). Here we finally show that this second approach is the natural one and that it leads to results which are optimal in a mathematical sense and intuitively convincing from a modelling point of view.

Our starting point is Fisher's equation

$$u_t = Du_{xx} + f(u) \tag{1.1}$$

where $u = u(t,x)$ is the time-dependent density of particles, $D > 0$ is the diffusion coefficient, and the nonlinearity $f$ comprises the effects of production and annihilation of particles. We assume $f \in C^1[0,1]$, $f(0) = f(1) = 0$. There are two standard types of functions $f$ leading to saddle-node or saddle-saddle connections, respectively.

I: $f'(0) > 0$, $f'(1) < 0$, $f(u) > 0$ for $0 < u < 1$.

II: There is an $\alpha \in (0,1)$, $f(\alpha) = 0$, $f'(0) < 0$, $f'(\alpha) > 0$, $f'(1) < 0$, $f(u) < 0$ for $0 < u < \alpha$, $f(u) > 0$ for $\alpha < u < 1$.

A travelling front solution of Fisher's equation (1.1) is a solution $u(t,x) = u(x - ct)$ with $u(-\infty) = 1$, $u(+\infty) = 0$, and $0 < u(x) < 1$ for $x \in \mathbb{R}$ (in fact $u$ turns out to be strictly decreasing). The number $c$ is the velocity or speed of the front. The following results are standard (see, e.g., [**12**]). In case I there is a positive minimal speed $c_P$, depending on $f$ and $D$. For every $c \geq c_P$ there is a travelling front, unique up to translation. In case II there is a unique velocity $c_P$ for a travelling front, and $c_P > 0$ if and only if $\int_0^1 f(u)du > 0$.

Now we consider reaction transport systems based on correlated random walks, (see [**7, 19, 11**]). Let $u^+(t,x)$, $u^-(t,x)$ be the time-dependent density of particles moving right or left, respectively. Let $\mu/2 > 0$ be the turning rate, $\gamma > 0$ be the particle speed, $m(u)$ and $g(u)$ be the production rate and annihilation rate, both depending on the total particle density $u = u^+ + u^-$. Then $f(u) = m(u)u - g(u)u$ is the net production. The system to be studied is the semilinear reaction transport system

$$\begin{aligned} u_t^+ + \gamma u_x^+ &= \frac{1}{2}\mu(u^- - u^+) + \frac{1}{2}m(u)u - g(u)u^+, \\ u_t^- - \gamma u_x^- &= \frac{1}{2}\mu(u^+ - u^-) + \frac{1}{2}m(u)u - g(u)u^-. \end{aligned} \tag{1.2}$$

This system is the natural hyperbolic analogue of Fisher's equation (1.1). Mathematically, apart from the physical interpretation, the nonlinearity in (1.2) contains terms depending on $u^+, u^-$ separately and others depending only on the sum. This feature becomes more transparent if we introduce the total particle density $u = u^+ + u^-$ and the particle flow $v = u^+ - u^-$ as dependent variables,

$$\begin{aligned} u_t + \gamma v_x &= f(u), \\ v_t + \gamma u_x &= -h(u)v, \end{aligned} \tag{1.3}$$

where $h(u) = \mu + g(u)$. For the connection of the system (1.3) to Cattaneo systems and reaction telegraph equations, see [**11, 17**].

For the system (1.2) we assume that $m, g \in C^2[0,1]$ are positive functions and that $f(u) = m(u)u - g(u)u$ has the properties I or II stated for the parabolic case. For the system (1.3), when studied in its own right, we assume that $f, h \in C^2[0,1]$, that $f$ has the properties I or II, and that $h$ is positive. A travelling front solution of the system (1.2) is a solution $u^+ = u^+(x - ct)$, $u^- = u^-(x - ct)$ with speed $c$, $u^+(-\infty) = u^-(-\infty) = 1/2$, $u^+(+\infty) = u^-(+\infty) = 0$, and $u^+(x) > 0$, $u^-(x) > 0$, $u^+(x) + u^-(x) < 1$ for $x \in \mathbb{R}$. A travelling front of the system (1.3) is a solution $u = u(x - ct)$, $v = v(x - ct)$ with speed $c$ and $u(-\infty) = 1$, $u(+\infty) = 0$, $v(\pm\infty) = 0$, and $0 < u(x) < 1$ of $x \in \mathbb{R}$. Once a front solution of (1.3) has been found, one has still to show $u^+ > 0$, $u^- > 0$ to get a front solution for the system (1.2).

Assume a travelling front $(u(x - ct), v(x - ct))$ of the system (1.3). The shape functions $u, v$ satisfy the ordinary differential equations

$$\begin{aligned} -c\dot{u} + \gamma\dot{v} &= f(u) \\ -c\dot{v} + \gamma\dot{u} &= -h(u)v, \end{aligned} \tag{1.4}$$

where the parameter $c$ is unknown. We are looking for fronts which move slower than the particle speed, i.e., we require $|c| < \gamma$. Thus, we can solve for $\dot{u}, \dot{v}$ and obtain the system

$$\begin{aligned} (\gamma^2 - c^2)\dot{u} &= cf(u) - \gamma h(u)v \\ (\gamma^2 - c^2)\dot{v} &= \gamma f(u) - ch(u)v. \end{aligned} \tag{1.5}$$

In the following we shall repeatedly employ a rescaling of the time variable: if we multiply the vector field (the right-hand side) by a positive factor (possibly depending on $u$, $v$) then we do not change the trajectories. We multiply the right hand side of (1.5) by $(\gamma^2 - c^2)/(\gamma h(u))$. We introduce the normalized speed

$$\kappa = c/\gamma \tag{1.6}$$

and the nonlinearity

$$F(u) = \frac{f(u)}{h(u)}. \tag{1.7}$$

Then we get the system

$$\begin{aligned} \dot{u} &= \kappa F(u) - v \\ \dot{v} &= F(u) - \kappa v \end{aligned} \tag{1.8}$$

where $\kappa$ ranges in $0 < \kappa < 1$. The system (1.8) has been derived in [**10**]. It is not difficult to establish the existence of some heteroclinic orbits for this system. However, in general a trajectory for one value of $\kappa$ does not form the boundary of a positively invariant domain for larger values of $\kappa$. Thus, it is difficult to show that the set of speeds is connected. Such difficulties do not arise in the following reduction to the parabolic case.

For this reduction we need the hypothesis

$$F'(u) < 1 \quad \text{for} \quad 0 \leq u \leq 1. \tag{1.9}$$

We call condition (1.9) a damping condition. In fact, the inequality (1.9) is exactly the condition which ensures the positivity of the damping term in corresponding reaction telegraph equations or semilinear wave equations, see [**13, 14, 15, 16**]. It might appear that the inequality (1.9) is too restrictive and that more general results can be obtained from the direct treatment of (1.8). However, if $F'(0) > 1$ then the stationary point $(0, 0)$ is a repellor and there will be no heteroclinic orbits at all. Thus, (1.9) seems to be a very natural condition. Recall that in terms of the original problem (1.2) condition (1.9) is satisfied whenever the turning rate $\mu$ is sufficiently large.

We apply a sequence of transformations to this system. We differentiate the first equation with repect to time, replace $\dot{v}$ from the second equation, and replace again $v$ from the first equation. Thus, we get a second order equation for the function $u$,

$$\ddot{u} = -\kappa(1 - F'(u))\dot{u} - (1 - \kappa^2)F(u). \tag{1.10}$$

Now we return to a first order system by setting $\dot{u} = z$,

$$\begin{aligned} \dot{u} &= z \\ \dot{z} &= -\kappa(1 - F'(u))z - (1 - \kappa^2)F(u). \end{aligned} \tag{1.11}$$

The sign of the damping term $1 - F'(u)$ plays an essential role. We require (1.9). Then we can define a diffeomorphism $G : [0,1] \to [0,1]$ by

$$G(u) = u - F(u). \tag{1.12}$$

We define a new dependent variable $w = G(u)$. Then $u = G^{-1}(w)$ and

$$\dot{w} = (1 - F'(u))\dot{u} = (1 - F'(u))z.$$

Thus, we obtain a system for the variables $w$ and $z$,

$$\begin{aligned} \dot{w} &= (1 - F'(u))z \\ \dot{z} &= -\kappa(1 - F'(u))z - (1 - \kappa^2)F(u) \end{aligned} \tag{1.13}$$

where $u = G^{-1}(w)$. We multiply the vector field by $1/(1 - F'(u))$ which amounts to another rescaling of the time variable. The final system is

$$\begin{aligned} \dot{w} &= z \\ \dot{z} &= -\kappa z - (1 - \kappa^2)H(w) \end{aligned} \tag{1.14}$$

where

$$H(w) = \frac{F(G^{-1}(w))}{1 - F'(G^{-1}(w))}. \tag{1.15}$$

Since all the steps so far can be inverted, we have the following proposition.

**Proposition 1.1** Assume the damping condition (1.9) holds. Then the systems (1.4) and (1.14) are equivalent. Moreover, if the function $f$ is of type I (type II) then the function $H$ is of type I (type II, respectively).

Now we proceed as follows. The system

$$\begin{aligned} \dot{w} &= z \\ \dot{z} &= -\zeta z - H(w) \end{aligned} \tag{1.16}$$

is the system for travelling fronts $w = w(x - \zeta t)$ of the parabolic equation $w_t = w_{xx} + H(w)$. As described earlier, we have rather complete information on (1.16). We can use this information to handle (1.14).

**Theorem 1.2** *Assume $f$ is of type I and the damping condition (1.9) holds. Then the system (1.3) has a family of travelling front solutions. There is a minimal speed $c_H$, with $0 < c_H < \gamma$. For every $c \in [c_H, \gamma)$ there is a travelling front solution, unique up to translation.*

**Proof** Since $f$ is of type I, the function $H$ has the same property. For the system (1.16) the set of speeds of travelling fronts is a half-line $[c_P, \infty)$ where the minimal speed $c_P$ satisfies $c_P \geq 2\sqrt{H'(0)}$. If $H(w)$ is multiplied by the positive factor $1 - \kappa^2$ then the minimal speed changes to $\zeta(\kappa) = \sqrt{1 - \kappa^2} c_P$. According to (1.14) we have to look for solutions of the equation $\zeta(\kappa) = \kappa$. Thus, the minimal $\kappa$ is $\kappa_H = c_P/\sqrt{1 + c_P^2}$. Then use (1.6) and put $c_H = \kappa\gamma$. □

**Theorem 1.3** *Assume $f$ is of type II and (1.9) holds. Then the system has, up to translation, a unique travelling front solution with speed $c_H$. The speed $c_H$ is positive if $\int_0^1 F(u)du > 0$.*

**Proof** Since $f$ is of type II, the function $H$ has the same property. The system (1.16) has a unique travelling front solution with speed $c_P$ which may be of either sign. If the function $H$ is multiplied by $1-\kappa^2$ then $c_P$ changes to $c_P\sqrt{1-\kappa^2}$. Thus, the problem (1.14) has a unique travelling front solution for $\kappa = \kappa_H = c_P/\sqrt{1+c_P^2}$. The sign of $\kappa_H$ is the same as that of $c_P$. But the sign of $c_P$ is the same as that of $\int_0^1 H(w)dw = \int_0^1 F(u)du$. □

In the parabolic case we know that the "subtangential condition" on the function $f$ of type I ensures that the minimal speed can be determined from the linearized problem as $c_* = 2\sqrt{Df'(0)}$, see (2.1) below. The following theorem gives the corresponding condition for the hyperbolic problem. In condition iii) we recover a result from ([**9**], Theorem 4). Whereas in the parabolic case the subtangential condition, i.e., the comparison with a linear problem, seems a distinguished and natural hypothesis (though not optimal), we did not find such distinguished criterion in the the hyperbolic case. The concavity property iii) is satisfied in many practical examples, though.

**Theorem 1.4** *Assume that the function $f$ is of type I and (1.9) holds. Assume that $F$ satisfies one of the following conditions.*

(i)

$$\frac{F(u)}{1-F'(u)} \le \frac{F'(0)}{(1-F'(0))^2}(u-F(u)) \quad \textit{for} \quad 0<u<1.$$

(ii) *$F'(u) \le F'(0)$ and $F(u) \le F'(0)u$ for $0 \le u \le 1$.*

(iii) *$F(u)$ is concave.*

*Then*

$$c_H = \frac{2\sqrt{F'(0)}}{1+F'(0)}\gamma. \tag{1.17}$$

**Proof** For the parabolic problem (1.16) we know that the inequality $H(w) \le H'(0)w$ implies $c_P = 2\sqrt{H'(0)}$. Using (1.15), (1.12), we find $H'(0) = F'(0)/(1-F'(0))^2$ and then, using again (1.12), the condition i). Simple arguments show (iii) $\to$ (ii) $\to$ (i). □

**Theorem 1.5** *Assume the hypotheses of Theorem 1.2 or 1.3. Let $(u,v)$ be a travelling front solution of the system (1.3). Let $u^+ = (u+v)/2$, $u^- = (u-v)/2$ be the corresponding solution of the system (1.2). Then $u^+(x) > 0$, $u^-(x) > 0$ for $x \in \mathbb{R}$.*

**Proof** From (1.14), by construction, we have that $w \in (0,1)$ and $z = \dot{w} < 0$. Thus, in (1.10) we have $u \in (0,1)$, $u$ decreasing, $\dot{u} < 0$. Therefore, in (1.8) $\kappa F(u) - v < 0$, i.e., $v > \kappa F(u) > 0$. Thus, $u^+ = (u+v)/2 > 0$. From the second equation in (1.2) we find

$$\dot{u}^- - \frac{1}{\gamma+c}g(u)u^- = -\frac{1}{\gamma+c}(\tfrac{1}{2}\mu v + \tfrac{1}{2}m(u)u).$$

Thus, the function $\exp\{-(\gamma+c)^{-1}\int_0^t g(u(s))ds\}u^-(t)$ is decreasing. Since the function goes to zero for $t \to +\infty$, it is positive for finite $t$. Hence also $u^- > 0$. □

## 2 The subtangential condition

As announced above, we address the question whether for the parabolic problem and a function of type I the minimal speed of travelling fronts can be obtained from the linearization. We assume $D = 1$. Let $c_* = 2\sqrt{f'(0)}$. We know $c_P \geq c_*$. It is well-known that the "sub-tangential condition"

$$f(u) \leq f'(0)u \quad \text{for} \quad 0 \leq u \leq 1 \tag{2.1}$$

is sufficient to ensure $c_P = c_*$. On the other hand, the example of the general cubic nonlinearity shows that (2.1) is not necessary (see [**12**]), where the minimal speed has been determined explicitly: for $f(u) = u(1-u)(1+\nu u)$ we have $c_0 \equiv 2$ for $-1 \leq \nu \leq 2$, but the subtangential condition is satisfied only for $-1 \leq \nu \leq 1$. Now we go a few steps in the direction of characterizing those functions $f$ for which $c = c_*$ holds. The following proposition is a direct consequence of the usual existence proof via invariant domains.

**Proposition 2.1** The following statements are equivalent.

(i) $c_P = c_*$.

(ii) There is $\varphi \in C^1[0,1]$, $\varphi(0) = 0$, $\varphi(u) > 0$ for $0 < u < 1$, such that

$$\frac{f(u)}{\varphi(u)} + \varphi'(u) \leq c_* \quad \text{for} \quad 0 < u < 1. \tag{2.2}$$

(iii) There is $\varphi \in C^1[0,1]$, $\varphi(0) = 0$, $\varphi(u) > 0$ for $0 < u < 1$, $\varphi(1) = 0$, such that (2.2) holds.

(iv) There is $\varphi \in C^1[0,1]$, $\varphi(0) = 0$, $\varphi(u) > 0$ for $0 < u < 1$, $\varphi(1) = 0$, such that

$$\frac{f(u)}{\varphi(u)} + \varphi'(u) \equiv c_*. \tag{2.3}$$

Of course, the function $\varphi$ in (iv) is the heteroclinic orbit.

We can normalize the problem by assuming $f'(0) = 1$, $\varphi'(0) = 1$, $c_* = 2$. Then the key inequality (2.2) becomes

$$f(u) \leq (2 - \varphi'(u))\varphi(u), \quad 0 < u < 1. \tag{2.4}$$

Thus, the problem of finding all functions $f$ for which $c_P = c_*$ amounts to solving the inequality (or equality!) (2.4). The problem is somewhat awkward for several reasons: i) In typical situations (say $\varphi' > 0$) the two factors on the right-hand side are monotone in opposite ways. ii) Although the problem is monotone in the sense that $c_P(f) = c_*(f) = 2$ and $\tilde{f} \leq f$, $\tilde{f}'(0) = 1$ implies $c_P(\tilde{f}) = c_*(\tilde{f})$, there is no maximal $f$. With some effort one can convince oneself that functions $f$ with the property $c_P = c_* = 2$ can have very different shapes and usually two functions with this property do not have a joint majorant with the same property.

If we put $\varphi(u) = u$ in (2.4) and require equality then we get $f(u) = u$. This is again (2.1). Thus, the linear function seems to be distinguished. But the following result shows again that the subtangential condition is not necessary.

**Proposition 2.2** Assume

$$f(u) \leq f'(0)(u + u^2/8), \quad 0 \leq u \leq 1. \tag{2.5}$$

Then $c_P = c_*$.

**Proof** In (2.2) choose $\varphi(u) = u - bu^2$ with $b \leq 1/4$.

□

**Proposition 2.3** Let $\alpha > 0$ and

$$f(u) \leq f'(0)\frac{1}{\alpha}(2 - 3e^{-\alpha u} + e^{-2\alpha u}), \quad 0 \leq u \leq 1. \tag{2.6}$$

Then $c_P = c_*$ holds.

**Proof** Let $f_\alpha(u) = (2 - 3e^{-\alpha u} + e^{-2\alpha u})/\alpha$ and $\varphi(u) = (e^\alpha u - 1)/\alpha$. Then $f_\alpha(0) = 0$, $f'_\alpha(0) = 1$, $f_\alpha(u) > 0$ for $0 < u < 1$, and $f_\alpha(u) = (2 - \varphi'(u))\varphi(u)$. □

The family of functions $f_\alpha$ does not have a maximal element. However, for $0 < \alpha < \log(2/\sqrt{3})$ the expression $f_\alpha(u)$ is increasing in $\alpha$ for all $u \in (0, 1)$. For small $\alpha > 0$ Proposition 2.3 is stronger than Proposition 2.2.

## References


[1] Alt, W. [1980], *Biased random walk models for chemotaxis and related diffusion approximations*, J. Math. Biol. **9**, 147–177.

[2] Dunbar, S. [1988], *A branching random walk evolution and a nonlinear hyperbolic equation*, SIAM J. Appl. Math. **48**, 1510–1526.

[3] Dunbar, S. and Othmer, H. [1974], *On a nonlinear hyperbolic equation describing transmission lines, cell movement, and branching random walks*, in: Nonlinear Oscillations in Biology and Chemistry (H. G. Othmer, ed.), Lecture Notes in Biomathematics **66**, Springer Verlag.

[4] Fife, P. C. [1979], *Mathematical aspects of reacting and diffusing systems*, Lecture Notes in Biomathematics **28**, Springer Verlag, New York.

[5] Fisher, R. A. [1937], *The advance of advantageous genes*, Annals of Eugenics **7**, 355–361.

[6] Gallay T. and Raugel, G. [1997], *Stability of travelling waves for a damped hyperbolic equation*, Z. Angew. Math. Phys. **48**, 451–479.

[7] Goldstein, S. [1951], *On diffusion by discontinuous movements and the telegraph equation*, Quart. J. Mech. Appl. Math. **4**, 129–156

[8] Hadeler, K. P. [1988], *Hyperbolic travelling fronts*, Proc. Edinburgh Math. Soc. **31**, 89–97.

[9] Hadeler, K. P. [1994], *Travelling fronts for correlated random walks*, Canadian Appl. Math. Quart. **2**, 27–43.

[10] Hadeler, K. P. [1995], *Travelling fronts in random walk systems*, Forma (Japan) **10** (Special Issue on Travelling Waves in Biology, Chemistry, Ecology and Medicine), 223–233.

[11] Hadeler, K. P. [1996], *Reaction telegraph equations and random walk systems*, in: Stochastic and spatial structures of dynamical systems (S. van Strieen, S. Verduyn Lunel, eds.), Roy. Acad. of the Netherlands. North Holland, Amsterdam, p. 133–161.

[12] Hadeler, K. P. and Rothe, F. [1975], *Travelling fronts in nonlinear diffusion equations*, J. Math. Biol. **2**, 251–263.

[13] Hillen, T. [1996], *A Turing model with correlated random walk*, J. Math. Biol. **35**, 49–72.

[14] Hillen, T. [1996], *Nonlinear hyperbolic systems describing random motion and their application to the Turing model*, Dissertation Summaries in Math **1**, 121–128.

[15] Hillen, T. [1996], *Qualitative analysis of hyperbolic random walk systems*, Department of Mathematics, University of Tübingen, SFB 382, Preprint No.43

[16] Hillen, T. [1997], *Invariance principles for hyperbolic random walk systems* J. Math. Anal. Appl. **210**, 360–374.

[17] Hillen, T. [1998], *Qualitative analysis of semilinear Cattaneo equations*, Math. Models Methods Appl. Sci. **3**, to appear.

[18] Holmes, E. E. [1993], *Are diffusion models to simple? A comparison with telegraph models of invasion*, American Naturalist **142**, 779–795.

[19] Kac, M. [1956], *A stochastic model related to the telegrapher's equation*, reprinted in Rocky Mtn. Math. J. **4** (1974), 497–509.

[20] Kolmogorov, A. N., Petrovskij, I. and Piskunov, N. [1937], *Etude de l'équation de la diffusion avec croissance de la quantité de la matière et son application a une problème biologique*, Bull. Univ. Moscou Ser. Int. Sec A **1,6**, 1–25.

[21] Othmer, H. G., Dunbar, S. R. and Alt, W. [1988] *Models for dispersal in biological systems*, J. Math. Biol. **26**, 263–298.

[22] Schwetlick, H. [1997], *On the minimal wave speed of a semilinear reaction transport equation*, Department of Mathematics, University of Tübingen, SFB 382, Report 73.

Fields Institute Communications
Volume **21**, 1999

# On the Dynamics of Nonlinear Feedback Loops in Cell Membrane, Neural and Sensory Systems

**Uwe an der Heiden**
Department of Mathematics
University of Witten/Herdecke
D-58448 Witten, Germany
adheiden@uni-wh.de

**Abstract.** Feedback is a very old topic with a very rich literature. In this paper we demonstrate that there are still many features in the dynamics of feedback loops which either have been detected only recently or are still to be detected. In the context of examples from physiology we study the effects of (i) multiple loops, (ii) mixed feedback in conjunction with delays, (iii) non-smooth feedback in first and second order delay systems. Applications range from mathematical models for signal processing in cell membranes (Section 2), for excitatory-inhibitory neuronal interactions in context of epileptic seizures (Section 3) to mathematical models of the pupil-light-reflex and its manipulation by a non-invasive control loop with non-smooth feedback (Section 4). Non-smooth feedback is also considered in second order delay differential equations (Section 5). Here rather complicated structures of periodic solutions and the existence of chaos can be derived analytically.

## 1 Introduction

For the engineers, feedback is essentially a control instrument helping to keep some variable constant or at least in a bounded domain of variation. Since the deviations from the stationary level should be small most often a linear approach, and thus essentially linear differential equations (from the mathematical point of view), are sufficient to deal with this problem.

However in biology, cells, organs, organisms, and populations often operate far from equilibrium and develop rich spatio-temporal structures and dynamics. This has been deliberately worked out in the last decades in many areas of mathematical biology, and for some overview the reader is referred to the books [**16**] and [**2**].

In this paper we will demonstrate that nonlinear feedback is one way how biological systems generate particularly rich temporal dynamics. Though feedback is ubiquitous in the realm of the living we choose three particular examples:

---

1991 *Mathematics Subject Classification.* Primary 34; Secondary 39, 92.
This work has been supported by Deutsche Forschungsgemeinschaft (DFG).

(i) In the first one we show that molecular signalling across cell membranes is a highly nonlinear processing and not simply a 1-1-transmission, the mathematical model being a system of three ordinary differential equations representing two feedback loops as distilled from the experimental literature. A variety of adaptation, desensitization and resensitization phenomena can be derived from this model.

(ii) Second, a mathematical model in terms of a first order differential-delay equation is presented describing a feedback loop existing in neural tissue of the hippocampus brain area where pyramidal cells excite basket cells which in turn inhibit pyramidal cells. This model has been developed in [**12**] in an effort to explain experimentally induced epileptic seizures in brain tissues from guinea pigs. A rather detailed sequence of bifurcations ranging from increasingly complex periodic patterns to chaotic neural activity shows up from both the mathematical and the animal model.

(iii) Higher order differential equations with delays come into play in an experimental setup developed in [**8**] and [**14**] for inducing large amplitude oscillations of the pupil: A body-external feedback loop is installed by turning on or off a thin light beam falling onto the retina depending on the diameter of the pupil. Here the feedback is "non-smooth" as a consequence of the light being either "on" or "off". Equations with non-smooth feedback functions are particularly amenable to constructing solutions analytically such that we can prove existence of high-frequency periodic solutions and chaos.

## 2 A mathematical model for G-protein induced signal processing across cell membranes

Structure and function of cells crucially depend on signals (hormones, neurotransmitters, metabolites, drugs) arriving at their surface. Some of these signals ("first messenger") do not permeate the cell membrane but bind to more or less specific receptors. In order that the signal has some effect on the interior of the cell it is then necessary that some "second messenger" inside the cell becomes triggered by some membrane bridging mechanism.

One of the most ubiquitous second messsengers is cAMP which is important, e.g., in neuromodulatory processes, glyco-genolysis, gluconeogenesis, and cardiac contraction. However, the triggering of cAMP is only possible by recruiting it from adenosinetriphosphate (ATP). The reaction ATP $\rightarrow$ cAMP has to be catalyzed by the membrane enzyme adenylate cyclase. Still more difficult, the adenylate cyclase is not activated directly by the signal-receptor complex, but only via the interference of stimulatory heterotrimeric GTPase, briefly called G-protein. However, not the GDP-bound, but only the GTP-bound G-protein is able to activate the adenylate cyclase. The binding of GTP to the G-protein is effectively enhanced in the presence of the signal-receptor complex.

Nauroschat & an der Heiden in [**18**] developed a detailed mathematical model in terms of basic biomolecular reactions to account for the just sketched rather complicated signal-processing mechanisms. The primary model, consisting of a system of eleven ordinary differential equation, can be equivalently transformed, by exploiting conservation laws and quasi-steady state conditions, into the following system of four ordinary differential equations:

$$dr/dt = -(h(t)+c_1)r + c_2x + c_3 \tag{2.1}$$
$$dx/dt = h(t)r - c_4x - F(x)(1-a) - Q_1(a)Q_2(x) \tag{2.2}$$
$$da/dt = c_2F(x)(1-a) - c_6a \tag{2.3}$$
$$dm/dt = G(a) - c_7m. \tag{2.4}$$

Here $c_1, \ldots, c_7$ are positive constants. The time-dependent extracellular signal is represented by $h(t)$ (e.g., concentration of a hormone or drug). The variable $r$ represents the fraction of cell surface receptors not bound to a signal molecule. The variable $x$ represents the fraction of signal-coupled, non-phosphorylated cell surface receptors. The variable $a$ measures the fraction of active G-protein. Finally, the concentration of the second messenger (cAMP) is represented by $m$, and, as is obvious from (2.4), only depends on $a$. The nonlinear reaction functions $F$, $G$, $Q_1$, $Q_2$ have the following properties: They are nonnegative, monotone increasing and bounded on their domain $\mathbb{R}_+$; the value at 0 is always zero (thus they are essentially hyperbolic).

System (2.1) - (2.4) reflects that the inner signal (second messenger $m$) is not simply a mirror image of the outer signal (first messenger $h$). The linear pathway $h \to r \to x \to a \to m$ is modified by several mechanisms with "reversed arrays", thus creating several feedback loops, e.g., the term $-Q_1(a)Q_2(x)$ reflects the inhibition of coupling between signal-receptor complex and inactive G-protein by receptor kinase whose activity depends on $a$. The term $F(x)(1-a)$ in Equation (2.3) shows that the transduction from the level $x$ to the level $a$ depends very much on the load in this "channel". And the term $-c_6a$ includes the GTP-ase activity of the active G-protein.

Note that Equation (2.4) is decoupled from the first three equations, thus allowing us to concentrate on the three-dimensional systems (2.1), (2.2), (2.3). It has been shown, see [**18**], that there is a compact domain (of prismatic shape) in the $(r, x, a)$-state space which is invariant under the flow of the system and contains the physiologically relevant trajectories $(0 \leq r + x \leq c_3/c_1, 0 \leq a \leq 1)$. Moreover, for constant input $(h(t) \equiv \bar{h} \geq 0)$ there is a unique, locally asymptotically stable equilibrium $(\bar{r}, \bar{x}, \bar{a}, \bar{m})$. Whereas $\bar{r}$ strictly decreases as a function of $\bar{h}$, $0 \leq \bar{h} < \infty$, the steady state values $\bar{x}, \bar{a}, \bar{m}$ are strictly increasing functions of $\bar{h} \in \mathbb{R}_+$. Since $(\bar{r}, \bar{x}, \bar{a})$ lies for all $\bar{h}$ in the described prism the model exhibits clearly a saturation effect of the steady states.

However, more interesting is the dynamic behaviour of the model. Let us consider the situation where at time $t = 0$ the system is at the resting state $(\bar{r}_0, \bar{x}_0, \bar{a}_0, \bar{m}_0)$, i.e., the steady state for $\bar{h} = 0$. We are interested in the transitory behaviour occuring when at $t = 0$ the signal strength is suddenly increased to another constant level $\bar{h} = h_1 > 0$. It is easy to calculate that whereas $dr(0)/dt < 0$ and $dx(0)/dt > 0$ we have $da(0)/dt = dm(0)/dt = 0$; one can even calculate $d^2a(0)/dt^2 > 0$ and $d^2m(0)/dt^2 = 0$! The conclusion is that a sudden, jump-like change of the external signal is responded to very smoothly by the internal second messenger, at least at the onset of the reaction. All our numerical simulations indicate that finally the system reaches the new steady state $(\bar{r}(h_1), \bar{x}(h_1), \bar{a}(\bar{h}_1), \bar{m}(h_1))$. However, in a broad range of parameters the approach to this steady state is not monotone, but the variables $x, a$ and $m$ first go beyond the new steady state levels ("overshooting") and only after this excursion do they approach monotonously the steady state, for illustration see Figure 1.

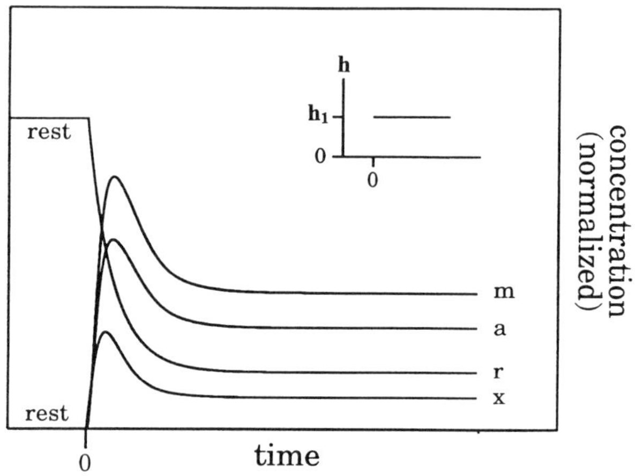


**Figure 1** Responses of the free receptors ($r$), the bounded receptors ($x$), the active G-protein (a) and cAMP (m) to a sudden increase of the extracellular signal (h) from the level $h = 0 (t \leq 0)$ to a level $h_1 > 0 (t > 0)$, as predicted from the model Equations (2.1)-(2.4). Redrawn after [**18**].

This result can be interpreted in two ways:

(i) when the cell is at the resting state its sensitivity is increased ("hyperresponsiveness"): signal changes are more strongly responded than when the cell is already in an elevated, active state;

(ii) the system adapts under steady stimulation: the discriminative capacity of the signalling system is decreased ("desensitization"). A consequence of this dynamic feature is that if two equally sized signal pulses follow each other in a brief temporal distance then the response to the second pulse is significantly lower than to the first pulse.

In some of our numerical calculations we could even observe a certain paradoxon: Though both the initial response and the steady state response are increasing with the strength of the input level $h$, there may be an intermediate period of time during which the response to the stronger stimulus ($h_3$) is weaker than the response to the weaker stimulus ($h_2$), for illustration see Figure 2 ("exhaustion").

Further implications and certain extensions of this membrane signalling system are currently under investigation [**19**].

## 3 A mathematical model of the feedback between pyramidal and basket cells: on the way to understand epilepsy

The following model has been developed in [**12**]. For the experimental background the reader is referred to that paper. Activity in the hippocampus is mostly dominated by three populations of neurons: the presynaptic *mossy fibres*, originating in other brain areas, the CA 3 *pyramidal cells*, and the interneural *basket cells*. In modelling these neuron types the activity variables (either membrane potentials or impulse frequencies) are averages over many neurons since (i) we are interested in the overall behaviour of the cells and (ii) each neuron is connected to many other neurons such that the individual neuron appears to be not very important.

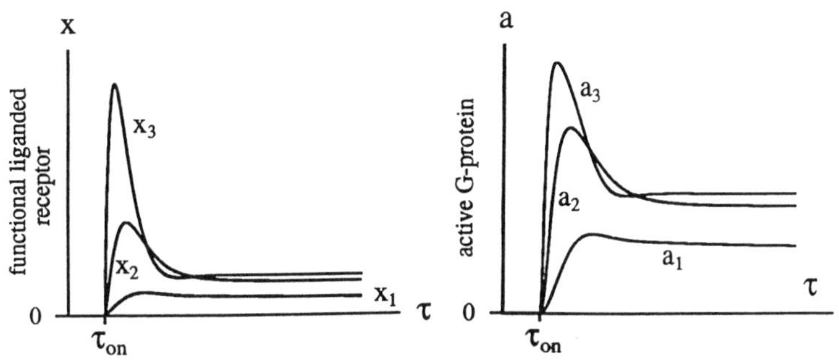


**Figure 2** Responses of the bounded receptors $(x_1, x_2, x_3)$ and the active G-protein $(a_1, a_2, a_3)$ as in Figure 1, however for three different stimulus levels $h_1 < h_2 < h_3$ resp. Temporarily, the response to a stronger stimulus may be weaker than the response to a weaker stimulus ("exhaustion"). Redrawn after [**17**].

Let $E(t)$ denote the average postsynaptic potential at time t generated in the pyramidal cells by the excitatory input from mossy fibres. Additional to this input the pyramidal cells are inhibited via the transmitter GABA by the basket cells; the average amount of this inhibition is denoted by $I(t)$. The balanced input to the pyramidal cells is thus given by

$$E(t) - I(t).$$

The output of the pyramidal cells, their axonal impulse frequency $F(t)$, is determined by the potential-to-frequency-characteristic

$$F(t) = \kappa\vartheta\left(E(t) - I(t) - \Theta\right), \tag{3.1}$$

where $\Theta$ is a constant threshold, $\kappa > 0$,and

$$\vartheta(x) = 0 \quad \text{for} \quad x \leq 0 \quad \text{and} \quad \vartheta(x) = x \quad \text{for} \quad x \geq 0.$$

The pyramidal cells excite the basket cells, and it is assumed that their firing frequency $F^*$ is proportional to $F$:

$$F^*(t) = \alpha F(t - \tau), \tag{3.2}$$

however with a delay $\tau(> 0)$ due to time for impulse conduction and synaptic transmission. The basket cells in turn act inhibitorily on the pyramidal cells across GABA-ergic synapses. The effect of $F^*$ on the inhibitory postsynaptic potential is proportional to the number $G$ of not yet activated (unbound) postsynaptic GABA-receptors. A quasi-steady state receptor kinetics leads to the expression

$$G = TK/(K + [C]^n), \tag{3.3}$$

where $T$ is the total number of receptors/cell, $[C]$ the concentration of the transmitter, $n$ the number of transmitter molecules required to activate one receptor, and $K$ the equilibrium constant for transmitter-receptor reaction. Assuming $[C] = mF^*$

with a proportionality constant $m$, we arrive at the rate equation for the inhibitory potential $I$:

$$\frac{dI}{dt} = V_m\ T \frac{K}{K + (mF^*)^n}\ F^* - \gamma I, \tag{3.4}$$

$\gamma$ the membrane time constant, $V_m$ the inhibitory potential due to activation of one receptor. Finally, with (3.2) the total model is

$$\frac{dI}{dt} = \alpha\ T\ V_m\ F(t-\tau) \frac{K}{K + (m\alpha F(t-\tau)^n)} - \gamma I \tag{3.5}$$

in conjunction with

$$F(t) = \kappa\vartheta(E(t) - I(t) - \Theta). \tag{3.6}$$

Assuming a constant input $E(t) = e$, it is shown in [**12**] that Equation (3.5) in combination with (3.6) may have one, two or three steady states depending on the parameters. Moreover numerical simulations exhibited undamped periodic solutions, period doubling bifurcations and apparently chaotic solutions.

Plugging the expression (3.6) for $F$ into Equation (3.5) one arrives at an equation of the following form:

$$\frac{dx}{dt}(t) = f(x(t-\tau)) - ax(t). \tag{3.7}$$

The most important aspect here is that the feedback function $f$ is not monotone, meaning that we have neither positive nor negative feedback in their pure form but so-called *mixed feedback* [**5**]. First in 1977, M. C. Mackey and L. Glass discovered by numerically solving Equation (3.7) with

$$f(\xi) = \alpha \frac{\xi}{\Theta + \xi^n}, \quad n \geq 2 \tag{3.8}$$

that mixed feedback may imply very complicated types of solutions including periodic solutions arising through a cascade of period doubling bifurcations and also deterministic chaos. Similarly, [**12**] could show that the neural model (3.5) exhibits a bifurcating sequence of more and more complex solutions resulting from a successive lowering of the number $T$ of postsynaptic receptors. This lowering of the number of postsynaptic receptors is assumed to mimic certain experiments where slices of hippocampus tissue, taken from guinea pigs, were infused with penicillin. Different levels of applied penicillin led to very different firing patterns in the tissue which are sometimes nearly periodic and becoming more and more irregular with increasing amounts of penicillin. The average impulse frequency also increases with the amount of penicillin giving the impression that epileptic seizures take place ([**20**], see also [**13**]). A possible explanation is that penicillin blocks GABA-receptors thus in effect diminishing $T$ and making inhibition less effective. Our model not only explains the increase of activity but also why apparently a sequence of bifurcations of very different activity patterns do occur in the tissue.

It appears difficult to prove the existence of chaos analytically for equations like (3.7) with nonlinearities $f$ of type (3.8). Without repeating the results in detail we mention that proofs have been found for cases where $f$ is piecewise constant or, if smooth, at least near to a piecewise constant function, compare [**6**] and [**4**]. We will take up this discussion again in the following section.

## 4 A non-invasive delayed feedback loop in the pupil-light-reflex: non-smooth feedback

The pupil-light-reflex which controls pupil diameter in accordance to the amount of light falling onto the retina is also connected to deeper areas inside the brain since the diameter of the pupil may also depend on emotional states. This fact has already been exploited for a long time in a diagnostic tool for detecting neurological disorders (e.g. early states of multiple sclerosis): A beam of light is thrown just onto the boundary of the pupil and it is observed what kinds of oscillations of the pupil diameter will thereupon arise. People with certain neurological disorders show different oscillation patterns than healthy persons. Recently Milton and Longtin in [**8**] and [**14**] developed a more sophisticated method by applying a thin light beam onto the center of the pupil. This light beam is either turned on or turned off depending on whether a certain threshold of pupil diameter is passed from below or from above respectively. However, peculiarily, this turning on or off takes place not just at the time $t$ of threshold passsing but at a certain delay time $\tau$ after $t$. The controlling apparatus is built in a way such that $\tau$ can be chosen arbitrarily. The simplest mathematical model to account for the observed phenomena (e.g., nearly periodic oscillations of the pupil with amplitude and frequency depending on the threshold diameter $\Theta$ and the delay $\tau$; see [**8**], [**14**]) assumes that after a step change of light the pupil area approaches exponentially (with a time constant $\alpha > 0$) a new stationary level:

$$\frac{dx}{dt} + \alpha x(t) = f(x(t-\tau)). \tag{4.1}$$

This is Equation (3.7). However, here the function $f$ is piecewise constant with only two values corresponding to the two states "light on" or "light off":

$$f(\xi) = \begin{cases} a & \text{if} \quad \xi < \Theta \\ b & \text{if} \quad \xi > \Theta \end{cases} \tag{4.2}$$

with $a > b$ because the stationary area of the pupil is smaller when the light is "on" compared to "off". Since $f$ is monotone decreasing we have a negative feedback situation.

Solutions to Equation (4.1) with $f$ given by (4.2) are piecewise composed of exponential functions either approaching the level $a/\alpha$ or the level $b/\alpha$. Depending on the valves of $\Theta$ and $\tau$ there exist in fact periodic solutions with amplitudes and frequencies satisfactorily close to those in the experiment. (It is easy to calculate these periodic solutions, and their amplitudes and frequencies analytically).

Longtin and Milton also performed experiments with two thresholds $\Theta_1 < \Theta_2$ where the light is only on when the pupil's diameter is between $\Theta_1$ and $\Theta_2$, again just a time $\tau$ after these thresholds have been passed. (Interestingly they did this knowing the results on deterministic chaos from the work in [**5**] and [**6**] where piecewise constant functions $f$ with two thresholds were used.) Also for this *mixed feedback* situation there is good coincidence between theory and experiment. In particular, the pupil could be driven into an apparently chaotic motion. However, there is discrepancy in that solutions of (4.1) with piecewise constant $f$ are not differentiable at times $t$ such that at time $t - \tau$ a threshold has been passed (at such a time $t$ the solution is only continuous), whereas the experimental curves appear to be very smooth.

## 5 Higher order delay-differential equations: periodic solutions, bifurcations, and chaos

These last findings and experimental investigation of the gain of the pupil-light-reflex suggest that a higher order low pass filter is more adequate than a first order one used in model (4.1), see [**9**]. Thus it has been suggested that a more general ansatz like

$$\frac{d^n x}{dt^n}(t) + a_{n-1}\frac{d^{n-1}x}{dt^{n-1}}(t) + \ldots + a_1\frac{dx}{dt}(t) + a_0 x(t) = f(x(t-\tau)) \qquad (5.1)$$

could be more realistic [**7**]. Measurements suggest $n = 2$ or $3$ [**9**].

In the following we report analytic results obtained for (5.1) in case of $n = 2$, however under the not very realistic restriction $a_1 = 0$. Thus, we consider the second order delay equation

$$d^2x(t)/dt^2 + x(t) = f(x(t-\tau)), \qquad (5.2)$$

first under the condition that

$$f(\xi) = \begin{cases} a & \text{for} \quad \xi \leq \Theta \\ b & \text{for} \quad \xi > \Theta, \end{cases} \qquad (5.3)$$

$a > b$, $\Theta \in \mathbb{R}$, corresponding to negative feedback (note that $a_0 = 1$ without loss of generality). Equation (5.2) may be transformed into the system

$$\begin{aligned} dx(t)/dt &= y(t) \\ dy(t)/dt &= f(x(t-\tau)) - x(t). \end{aligned} \qquad (5.4)$$

An initial condition of (5.4) is a pair $(\varphi, y_0) \in C^1([-\tau, 0], \mathbb{R}) \times \mathbb{R}$ such that $\varphi'(0) = y_0$. Solutions $(x(t), y(t))$, $t \geq 0$, of (5.4) can be represented as continuous trajectories $t \longmapsto (x(t), y(t))$ in the x-y-plane ($\mathbb{R}^2$). If $f$ is of type (5.3) then these trajectories are piecewise composed of arcs (or sectors) of circles having centers at either $(a, 0)$ or $(b, 0)$, e.g., the center is $(a, 0)$ for all $t \in [t_1, t_2]$ if for all $t \in [t_1 - \tau, t_2 - \tau]$ the inequality $x(t) \leq \Theta$ is satisfied. Note that the angular length of the arc associated with $[t_1, t_2]$ is just $t_2 - t_1$, hence the angular velocity of the trajectory point $(x(t), y(t))$ is always 1, independent of the radii of the arcs.

It can be shown that (5.4) in conjunction with (5.3) has a very large variety of periodic solutions. More precisely we have the following theorem on the existence of high frequency periodic solutions:

**Theorem 5.1** *Let $f$ be given by (5.3) and let $a = 1/2$ and $b = -1/2$ (without loss of generality). Let $\Theta \in [0, 1/2]$. Then the following holds:*

(A) *For each $n \in \mathbb{N}$ and for each $\tau \in (0, 2n\pi)$ System (5.4) has a periodic solution with minimal period $\tau/n$.*

(B) *For each $n \in \mathbb{N}$, $n$ odd, and for each $\tau \in (n\pi, 2n\pi)$ System (5.4) has periodic solutions with minimal period $2\tau/n$.*

The trajectories in the x-y-plane corresponding to the periodic solutions of (A) and (B) consist exactly of two arcs of circles. In [**7**] it is shown that there exist also periodic orbits composed of more than two arcs. However, their domain of existence in parameter space is not well known. [**1**] contains a proof of Theorem 5.1 and shows the bifurcation structure of these and also of other types of periodic orbits with two arcs.

Moreover it contains some results on the stability of such orbits. System (5.4) shows *multistability* in that for some of the parameter values there is coexistence of several stable limit cycles (sometimes more than 2 of them).

A proof showing the existence of chaos for Equation (5.2), i.e., System (5.4), in case of a non-smooth mixed feedback nonlinearity has just been given [**3**]:

**Theorem 5.2** *Let $\Theta$ be a positive constant. Let $f(\xi) = 1/2$ for $\xi \in [0, \Theta]$, $f(\xi) = -1/2$ otherwise. Then there is an open set in the $(\tau, \Theta)$-parameter space such that for each element of this set System (5.4) has infinitely (countably) many periodic solutions of different minimal period and, moreover, uncountably many asymptotically aperiodic orbits such that for any two of these aperiodic solutions $(x_1, y_1)$ and $(x_2, y_2)$ the relations*

$$\liminf_{t\to\infty} \| (x_1(t), y_1(t)) - (x_2(t), y_2(t)) \| = 0$$

*and*

$$\limsup_{t\to\infty} \| (x_1(t), y_1(t)) - (x_2(t), y_2(t)) \| > 0$$

*hold.*

These last limit conditions tell that the aperiodic solutions come arbitrarily close to each other again and again, but nevertheless get substantially far away from each other, thereby reflecting one of the hallmarks of deterministic chaos, the critical dependence of solutions on initial conditions. The proof of Theorem 5.2 relies on identifying a one-dimensional subset of the state space $C([-\tau, 0], \mathbb{R}^2)$ where the dynamics can be hormeomorphically mapped onto that of a difference equation obeying a snap back repeller property.

## References


[1] Bayer, W., an der Heiden, U. [1998], *Oscillations types and bifurcations of a nonlinear second order differential equation*, J. Dynamics and Diff. Equations **10** , 303-326.

[2] Edelstein-Keshet, L. [1988], *Mathematical Models in Biology*, Random House, New York.

[3] an der Heiden, U. and Bayer, W. [1998], *Chaos proved for $d^2(x(t))/dt^2 + x(t) = f(x(g - \tau))$ with non-smooth $f$*, in preparation.

[4] an der Heiden, U. and Mackey, M. C. [1982], *The dynamics of production and destruction: analytic insight into complex behaviour*, J. Math. Biol. **16**, 75-101.

[5] an der Heiden, U. and Mackey, M. C. [1987], *Mixed feedback – A paradigm for regular and irregular oscillations*, in Temporal Disorder in Human Oscillatory Systems (L. Rensing, U. an der Heiden, M. C. Mackey, eds.), pages 30-47, Springer-Verlag, Berlin, Heidelberg, New York.

[6] an der Heiden, U. and Walther, H. O. [1983], *Existence of chaos in control systems with delayed feedback*, J. Diff. Eqs. **47**, 273-295.

[7] an der Heiden, U., Longtin, A., Mackey, M. C., Milton, J. G., and Scholl, R. [1990], *Oscillatory modes in a nonlinear second-order differential equation with delay*, J. Dynamics and Diff. Equations **2**, 423-449.

[8] Longtin, A. and Milton, J. [1988], *Complex oscillations in the human pupil light reflex with "mixed" and delayed feedback*, Math. Biosci. **90**, 183-199.

[9] Longtin, A. and Milton, J. G. [1989], *Insight into the transfer function, gain and oscillation onset for the pupil light reflex using nonlinear delay-differential equations*, Biol. Cybern **61**, 51-58.

[10] Mackey, M. C. and Glass, L. [1977], *Oscillation and chaos in physiological control systems*, Science **197**, 287-289.

[11] Mackey, M. C. and an der Heiden, U. [1982], *Dynamical diseases and bifurcations: understanding functional disorders in physiological systems*, Funkt. Biol. Med. **1**, 156-164.

[12] Mackey, M. C. and an der Heiden, U. [1984], *The dynamics of recurrent inhibition*, J. Math Biol. **19**, 211-225.

[13] Mackey, M. C. and Milton, J. [1987], *Dynamical diseases*, Annals New York Acad. Sci. **504**, 16-32.

[14] Milton, J. and Longtin, A. [1990], *Clamping the pupil light reflex with external feedback: Evaluation of constriction and dilation from cyclic measurements*, Vision Res. **30**, 515-525.

[15] Milton, J., Bayer, W. and an der Heiden, U. [1998], *Modelling of the pupil-light-reflex with differential delay equations*, Z. Angew. Math. Mech., in press.

[16] Murray, J. D. [1989], *Mathematical Biology*, Springer–Verlag, Berlin, Heidelberg, New York.

[17] Nauroschat, J. and an der Heiden, U. [1996], *Mathematical aspects of signal-processing via G-proteins*, in Computation in Cellular and Molecular Biological Systems (R. Cuthbertson, M. Holcomb, R. Paton, eds.), pages 57-68, World Scientific, London.

[18] Nauroschat, J. and an der Heiden, U. [1997], *A theoretical approach towards G-protein modulation of cellular responsiveness*, J. Math. Biol. **35**, 609-627.

[19] Nauroschat, J. and an der Heiden, U. [1998], *Involvement of receptor-kinase in the biochemistry upstream from cAMP synthesis in cells*, in Information processing in cells and tissues (R. C. Paton and M. Holcombe, eds.), Plenum Publ. Corp., in press.

[20] Prince, D. A. [1978], Ann. Rev. Neurosci. **1**, 395.

Fields Institute Communications
Volume **21**, 1999

# Skew Product Flows of Quasi-Processes and Stabilities

**Yoshiyuki Hino**
Department of Mathematics and Informatics
Chiba University
1-33 Yayoi-cho, Inage-ku, Chiba 263, Japan
hino@math.s.chiba-u.ac.jp

**Satoru Murikami**
Department of Applied Mathematics
Okayama University of Science
1-1 Ridai-cho, Okayama 700, Japan
murakami@xmath.ous.ac.jp

**Abstract.** The concept of quasi-processes which is a generalization of processes in the sense of Dafermos is introduced, and some equivalence relations concerning with stability properties for quasi-processes are studied in connection with stability properties for limiting quasi-processes. Results are applied to stability problems in functional differential equations with infinite delay.

## 1 Introduction

The concept of processes discussed in [**1**], [**2**], [**3**] and [**5**] is a useful tool in the study of mathematical analysis for some phenomena whose dynamics is described by equations that contain the derivative with respect to the time variable. Indeed, Hale in [**2**] derived some stability properties for processes and applied those to get stability results for some kinds of equations, including functional differential equations, partial differential equations and evolution equations. These equations have a common property: the space of initial functions for these equations is identical with the phase space and solutions take values which vary continuously in the phase space with time $t$. On the other hand, in the theory of functional differential equations with infinite delay as well as integrodifferential equations, the space BC which consists of all bounded continuous functions on $(-\infty, 0]$ is one of the important classes for the space of initial functions. Clearly, the space BC is different from the phase space of these equations, because solutions do not take values which vary continuously in BC with time $t$. In a recent paper [**6**], Murakami and Yoshizawa

1991 *Mathematics Subject Classification.* Primary 34K15; Secondary 34K20.

YH was supported in part by Grant-in-Aid for Scientific Research (C), No.09640156, Japanese Ministry of Education, Science, Sports and Culture.

SM was supported in part by Grant-in-Aid for Scientific Research (C), No.09640235, Japanese Ministry of Education, Science, Sports and Culture.

gave the following result under a mild assumption that the BC-uniform asymptotic stability is equivalent to the $\rho$-uniform asymptotic stability, where $\rho$-stability means that the solution remains small with respect to the metric $\rho$, which induces the compact open topology, if the initial function is small with respect to $\rho$. The $\rho$-stability is a useful tool in the study of the existence of almost periodic solutions for almost periodic systems (see [**6**]). As will be seen by an example in Section 3, however, the concept of processes does not fit in with the study of $\rho$-stability in the stability problems for functional differential equations with infinite delay.

To overcome the difficulty stated above, in this article we generalize the concept of processes and get a more extended concept which is called quasi-processes. In Section 2, we discuss some stability properties for quasi-processes and obtain some equivalence relations concerning stabilities for quasi-processes in connection with those for "limiting" quasi-processes (Theorem 2.1), which is a generalization of Theorem 3.7.4 in [**2**]. In Section 3, we treat functional differential equations with infinite delay defined on the fading memory space $C_g^0$ and, corresponding to $\rho$-metric, we construct quasi-processes associated with functional differential equations with infinite delay (Propositions 3.2 and 3.3). Finally, establishing that quasi-processes fit in with the analysis of $\rho$-stability problems, we obtain an equivalence relationship between these stabilities for the original equation and those for limiting equations (Theorem 3.4) and an equivalence relationship between the BC-stability and $\rho$-stabilities (Corollary) which is a generalization of Theorem 2 in [**6**].

## 2 Quasi-processes and some stability properties

In this section, we shall introduce the concept of quasi-processes which is a generalization of processes and deduce some results on stability properties for quasi-processes. Suppose $\mathcal{X}$ is a metric space with metric $d$ and let $w : R^+ \times R^+ \times \mathcal{X} \mapsto \mathcal{X}, R^+ := [0, \infty)$, be a function satisfying the following properties for all $t, \tau, s \in R^+$ and $x \in \mathcal{X}$

(p1) $w(0, s, x) = x$;

(p2) $w(t + \tau, s, x) = w(t, \tau + s, w(\tau, s, x))$.

Let $\mathcal{Y}$ be a nonempty closed set in $\mathcal{X}$. We call the mapping $w$ *a* $\mathcal{Y}$*-quasi-process* on $\mathcal{X}$ or simply *a quasi-process* on $\mathcal{X}$, if $w$ satisfies the condition,

(p3) the restricted mapping $w : R^+ \times R^+ \times \mathcal{Y} \mapsto \mathcal{X}$ is continuous,

together with the conditions (p1) and (p2). In case of $\mathcal{Y} = \mathcal{X}$, the concept of quasi-processes is identical with that of processes investigated in [**1**], [**2**], [**3**] and [**5**]. We emphasize that the concept of processes does not fit in with the study of $\rho$-stabilities in functional differential equations in contrast with the concept of quasi-processes, as will be seen in the next section.

Denote by $W$ the set of all quasi-processes on $\mathcal{X}$. For $\tau \in R^+$ and $w \in W$, we define the translation $\sigma(\tau)w$ of $w$ by

$$(\sigma(\tau)w)(t, s, x) = w(t, \tau + s, x), \qquad (t, s, x) \in R^+ \times R^+ \times \mathcal{X},$$

and set $\gamma_\sigma^+(w) = \bigcup_{t \geq 0} \sigma(t)w$. Clearly $\gamma_\sigma^+(w) \subset W$. We denote by $H_\sigma(w)$ all functions $\chi : R^+ \times R^+ \times \mathcal{X} \mapsto \mathcal{X}$ such that for some sequence $\{\tau_n\} \subset R^+$, $\{\sigma(\tau_n)w\}$ converges to $\chi$ pointwise on $R^+ \times R^+ \times \mathcal{X}$, that is, $\lim_{n \to \infty}(\sigma(\tau_n)w)(t, s, x) = \chi(t, s, x)$ for any $(t, s, x) \in R^+ \times R^+ \times \mathcal{X}$. The set $H_\sigma(w)$ is considered as a topological space with the pointwise convergence, and it is called the hull of $w$.

Consider a $\mathcal{Y}$-quasi-process $w$ on $\mathcal{X}$ satisfying (p4) $H_\sigma(w) \subset W$.

Clearly, $H_\sigma(w)$ is invariant with respect to the tanslation $\sigma(\tau), \tau \in R^+$. For any $t \in R^+$, we consider a function $\pi(t) : \mathcal{X} \times H_\sigma(w) \mapsto \mathcal{X} \times H_\sigma(w)$ defined by

$$\pi(t)(x, \chi) = (\chi(t, 0, x), \sigma(t)\chi)$$

for $(x, \chi) \in \mathcal{X} \times H_\sigma(w)$. $\pi(t)$ is called *the skew product flow of the quasi-process* $w$, if the following property holds true.

(p5) $\pi(t)(y, \chi)$ is continuous in $(t, y, \chi) \in R^+ \times \mathcal{Y} \times H_\sigma(w)$.

The skew product flow $\pi(t)$ is said to be $\mathcal{Y}$*-strongly asymptotically smooth* if, for any nonempty, closed bounded set $B \subset \mathcal{Y} \times H_\sigma(w)$, there exists a compact set $J \subset \mathcal{Y} \times H_\sigma(w)$ with the property that $\{\pi(t_n)(y_n, \chi_n)\}$ has a subsequence which approaches to $J$ whenever sequences $\{t_n\} \subset R^+$ and $\{(y_n, \chi_n)\} \subset B$ satisfy $\lim_{n\to\infty} t_n = \infty$ and $\pi(t)(y_n, \chi_n) \in B$ for all $t \in [0, t_n]$. In case of $\mathcal{Y} = \mathcal{X}$, the $\mathcal{Y}$-strong asymptotic smoothness of $\pi(t)$ implies the asymptotic smoothness of $\pi(t)$ introduced in [**2, 3**]. In the next section, we shall see that the $\mathcal{Y}$-strong asymptotic smoothness of $\pi(t)$ is ensured when $w$ is a quasi-process generated by some functional differential equations with infinite delay.

Now we suppose that $H_\sigma(w)$ is sequentially compact and we discuss relationships between some stability properties for the quasi-process $w$ and those for the "limiting" quasi-processes $\chi \in \Omega_\sigma(w)$; here $\Omega_\sigma(w)$ denotes the $\omega$-limit set of $w$ with respect to the translation semigroup $\sigma(t)$. A continuous function $\mu : R^+ \mapsto \mathcal{Y}$ is called *an integral* of the quasi-process $w$ on $R^+$, if $w(t, s, \mu(s)) = \mu(t+s)$ for all $t, s \in R^+$ (cf., [**2**], p.80). In the following, we suppose that there exists an integral $\mu : R^+ \mapsto \mathcal{Y}$ of the quasi-process $w$ on $R^+$ such that the set $O^+(\mu) = \{\mu(t) : t \in R^+\}$ is contained in $\mathcal{Y}$ and is relatively compact in $\mathcal{X}$.

From (p5) we see that $\pi(\delta)(\mu(s), \sigma(s)w) = (w(\delta, s, \mu(s)), \sigma(s+\delta)w) = (\mu(s+\delta), \sigma(s+\delta)w)$ tends to $(\mu(s), \sigma(s)w)$ as $\delta \to 0^+$ uniformly for $s \in R^+$; consequently, the integral $\mu$ must be uniformly continuous on $R^+$. From Ascoli-Arzéla's theorem and the sequential compactness of $H_\sigma(w)$, it follows that for any sequence $\{\tau'_n\} \subset R^+$, there exist a subsequence $\{\tau_n\}$ of $\{\tau'_n\}$, a $\chi \in H_\sigma(w)$ and a function $\nu : R^+ \mapsto \mathcal{Y}$ such that $\lim_{n\to\infty} \sigma(\tau_n)w = \chi$ and $\lim_{n\to\infty} \mu(t+\tau_n) = \nu(t)$ uniformly on any compact interval in $R^+$. In this case, we write as

$$(\mu^{\tau_n}, \sigma(\tau_n)w) \to (\nu, \chi) \text{ compactly}$$

for simplicity. Denote by $H_\sigma(\mu, w)$ the set of all $(\nu, \chi)$ such that $(\mu^{\tau_n}, \sigma(\tau_n)w) \to (\nu, \chi)$ compactly for some sequence $\{\tau_n\} \subset R^+$. In particular, we denote by $\Omega_\sigma(\mu, w)$ the set of all $(\nu, \chi) \in H_\sigma(\mu, w)$ for which one can choose a sequence $\{\tau_n\} \subset R^+$ so that $\lim_{n\to\infty} \tau_n = \infty$ and $(\mu^{\tau_n}, \sigma(\tau_n)w) \to (\nu, \chi)$ compactly. We easily see that $\nu$ is an integral of the quasi-process $\chi$ on $R^+$ whenever $(\nu, \chi) \in H_\sigma(\mu, w)$.

For any $x_0 \in \mathcal{X}$ and $\varepsilon > 0$, we set $V_\varepsilon(x_0) = \{x \in \mathcal{X} : d(x, x_0) < \varepsilon\}$. We shall give the definition of stabilities for the integral $\mu$ of the quasi-process $w$.

**Definition** The integral $\mu : R^+ \mapsto \mathcal{Y}$ of the quasi-process $w$ is said to be:

(i) $\mathcal{Y}$*-uniformly stable* ($\mathcal{Y}$-US) if for any $\varepsilon > 0$, there exists a $\delta := \delta(\varepsilon) > 0$ such that $w(t, s, \mathcal{Y} \cap V_\delta(\mu(s))) \subset V_\varepsilon(\mu(t+s))$ for $(t, s) \in R^+ \times R^+$;

(ii) $\mathcal{Y}$*-uniformly asymptotically stable* ($\mathcal{Y}$-UAS), if it is $\mathcal{Y}$-US and there exists a $\delta_0 > 0$ with the property that for any $\varepsilon > 0$, there is a $t_0 > 0$ such that $w(t, s, \mathcal{Y} \cap V_{\delta_0}(\mu(s))) \subset V_\varepsilon(\mu(t+s))$ for $t \geq t_0$, $s \in R^+$;

(iii) $\mathcal{Y}$-*attractive* in $\Omega_\sigma(w)$ if there is a $\delta_0 > 0$ such that for $y \in \mathcal{Y} \cap V_{\delta_0}(\nu(0))$ and $(\nu, \chi) \in \Omega_\sigma(\mu, w)$, $d(\chi(t, 0, y), \nu(t)) \to 0$ as $t \to \infty$.

We assume the following property on $\mathcal{Y}$, $w$ and $\mu$:

(p6) There is a $\delta_1 > 0$ such that for any $s \in R^+$ and $t_0 > 0$, $w(t_0, s, y) \in \mathcal{Y}$ whenever $y \in \mathcal{Y}$ and $w(t, s, y) \in V_{\delta_1}(\mu(t+s))$ for all $t \in (0, t_0]$.

If $\mathcal{Y} = \mathcal{X}$, then (p6) is clearly satisfied. In the next section, we will give a nontrivial example for which (p6) is satisfied.

**Theorem 2.1** *Let $\mathcal{Y}$ be a closed set in a metric space $\mathcal{X}$ and suppose that $w$ is a $\mathcal{Y}$-quasi-process on $\mathcal{X}$ for which $H_\sigma(w)$ is sequentially compact and that the skew product flow $\pi(t) : \mathcal{X} \times H_\sigma(w) \mapsto \mathcal{X} \times H_\sigma(w)$ of the quasi-process $w$ is $\mathcal{Y}$-strongly asymptotically smooth. Also, suppose that $\mu : R^+ \mapsto \mathcal{Y}$ is an integral of $w$ on $R^+$ such that $O^+(\mu)$ is a relative compact subset of $\mathcal{Y}$ and that* (p4) *and* (p6) *are satisfied. Then the following statements are equivalent:*

(i) *The integral $\mu$ is $\mathcal{Y}$-UAS.*
(ii) *The integral $\mu$ is $\mathcal{Y}$-US and $\mathcal{Y}$-attractive in $\Omega_\sigma(w)$.*

In the case that $\mathcal{Y} = \mathcal{X}$, $\mu \equiv 0$ and $\mathcal{X}$ is complete, the equivalence (i) $\Leftrightarrow$ (ii) is a direct consequence of Theorem 3.7.4 in [**3**], because the $\mathcal{Y}$-strong asymptotic smoothness of $\pi(t)$ implies the asymptotic smoothness of $\pi(t)$. Following the argument employed in Theorem 3.7.4 in [**3**], we can see that the equivalence mentioned above holds true even if $\mathcal{Y} \neq \mathcal{X}$ or $\mu \not\equiv 0$ or $\mathcal{X}$ is not necessarily complete; we omit the details.

## 3 Quasi-processes generated by functional differential equations with infinite delay

In this section, we shall treat functional differential equations on a fading memory space and show that quasi-processes are naturally generated by these functional differential equations.

We first explain some notation and convention employed throughout this section. Let $R^n$ be the $n$-dimensional real linear space with norm $|\cdot|$. For any interval $J \subset R := (-\infty, \infty)$, we denote by $\mathrm{BC}(J; R^n)$ the space of all bounded and continuous functions mapping $J$ into $R^n$. Clearly $\mathrm{BC}(J; R^n)$ is a Banach space with the norm $|\cdot|_{\mathrm{BC}(J;R^n)}$ defined by $|\phi|_{\mathrm{BC}(J;R^n)} = \sup\{|\phi(t)| : t \in J\}$. If $J = R^- := (-\infty, 0]$, then we simply write $\mathrm{BC}(J; R^n)$ and $|\cdot|_{\mathrm{BC}(J;R^n)}$ as BC and $|\cdot|_{\mathrm{BC}}$, respectively. For any function $u : (-\infty, a) \mapsto R^n$ and $t < a$, we define a function $u_t : R^- \mapsto R^n$ by $u_t(s) = u(t+s)$ for $s \in R^-$. Let $g : R^- \mapsto [1, \infty)$ be any continuous, nonincreasing function such that $g(0) = 1$ and $g(s) \to \infty$ as $s \to -\infty$. We set

$$C_g^0 := \{\phi : R^- \mapsto R^n \text{ is continuous with } \lim_{s \to -\infty} |\phi(s)|/g(s) = 0\}.$$

Then the space $C_g^0$ equipped with the norm

$$|\phi|_g = \sup_{s \le 0} \frac{|\phi(s)|}{g(s)}, \quad \phi \in C_g^0$$

is a separable Banach space.

Throughout the remainder of this article, let $C(R^+ \times C_g^0; R^n)$ be the space of all continuous functions on $R^+ \times C_g^0$ with values in $R^n$, which is equipped with the compact open topology.

Now we consider the following functional differential equation

$$\frac{du}{dt} = F(t, u_t), \tag{3.1}$$

where $F(t,\phi) \in C(R^+ \times C_g^0; R^n)$. In what follows, we shall show that (3.1) generates a quasi-process on an appropriate space under some conditions and deduce equivalence relationships between some stability properties of (3.1) and those of its limiting equations as an application of Theorem 2.1.

We assume the following conditions on $F$:

(H1) $F(t,\phi)$ is uniformly continuous on $R^+ \times \mathcal{K}$ for any compact set $\mathcal{K}$ in $C_g^0$.

(H2) For any $H > 0$, there is an $L(H) > 0$ such that $|F(t,\phi)| \leq L(H)$ for all $t \in R^+$ and $\phi \in C_g^0$ such that $|\phi|_g \leq H$.

For $\tau \in R^+$, we denote the $\tau$-translation $F^\tau$ of $F(t,\phi)$ by

$$F^\tau(t,\phi) = F(t+\tau,\phi), \quad (t,\phi) \in R^+ \times C_g^0.$$

Clearly, $F^\tau$ is in $C(R^+ \times C_g^0; R^n)$, too. Set

$$H(F) = \overline{\{F^\tau; \tau \in R^+\}},$$

where $\overline{\{F^\tau; \tau \in R^+\}}$ denotes the closure of $\{F^\tau; \tau \in R^+\}$ in $C(R^+ \times C_g^0; R^n)$. The subspace $H(F)$ of $C(R^+ \times C_g^0; R^n)$ is called the hull of $F$. It is known that $H(F)$ is metrizable, see Proposition 8.1.3 in [**4**]. Clearly, the hull $H(F)$ is invariant with respect to the $\tau$-translation, that is, $G^\tau \in H(F)$ whenever $G \in H(F)$ and $\tau \in R^+$. Moreover, from (H1) it follows that $H(F)$ is a compact set in $C(R^+ \times C_g^0; R^n)$, see Theorem 7.1.4 in [**4**]. If $G \in H(F)$, one can choose a sequence $\{\tau_n\} \subset R^+$ so that $F^{\tau_n}$ tends to $G$ in $C(R^+ \times C_g^0; R^n)$, that is, $F(t+\tau_n,\phi) \to G(t,\phi)$ as $n \to \infty$ uniformly on any compact subset of $R^+ \times C_g^0$. It is easy to see that each $G \in H(F)$ satisfies (H1) and (H2) (with the common $L(\cdot)$) when $F$ does. We denote by $\Omega(F)$ the set of all elements $G$ in $H(F)$ for which one can choose a sequence $\{\tau_n\} \subset R^+$ so that $\{\tau_n\} \to \infty$ as $n \to \infty$ and $F^{\tau_n} \to G$ in $C(R \times C_g^0; R^n)$. If $G \in H(F)$, the system

$$\frac{du}{dt} = G(t, u_t), \qquad t \in R^+, \tag{3.2}$$

is called an equation in the hull of System (3.1). In particular, if $G \in \Omega(F)$, then it is called a limiting equation of (3.1).

In the following, we always assume the following condition,

(H3) For any $G \in H(F)$ and $(\sigma,\phi) \in R^+ \times C_g^0$, Equation (3.2) has a *unique* solution through $(\sigma,\phi)$ which exists for *all* $t \geq \sigma$.

Consider a mapping $\Phi : R^+ \times R^+ \times C_g^0 \times H(F) \mapsto C_g^0$ defined by

$$\Phi(t,s,\phi,G) = u_{t+s}(s,\phi,G) \in C_g^0, \qquad (t,s,\phi,G) \in R^+ \times R^+ \times C_g^0 \times H(F),$$

where $u(\cdot,s,\phi,G)$ is the solution of (3.2) through $(s,\phi)$.

**Proposition 3.1** (cf., Theorem 8.1.5 in [**4**]) *Assume that conditions* (H1)–(H3) *are satisfied. Then the mapping* $\Phi$ *is continuous.*

Suppose the following condition:

(H4) Equation (3.1) has a bounded solution $\bar{u}(t)$ defined on $R^+$ such that $\bar{u}_0 \in \mathrm{BC}$.

We see that the set $\{\bar{u}_t : t \in R^+\}$ is relatively compact in $C_g^0$ and that $\bar{u}_t \in C_g^0$ is uniformly continuous in $t \in R^+$, see [**6**]. Therefore, for any sequence $\{\tau_n'\} \subset R^+$ one can choose a subsequence $\{\tau_n\} \subset \{\tau_n'\}$, $\bar{v} \in C(R; R^n)$ and $G \in H(F)$ such that $\lim_{n\to\infty} F^{\tau_n} = G$ in $C(R^+ \times C_g^0; R^n)$ and $\lim_{n\to\infty} |\bar{u}_{t+\tau_n} - \bar{v}_t|_g = 0$ uniformly on any compact interval in $R^+$. In this case, we write as

$$(\bar{u}^{\tau_n}, F^{\tau_n}) \to (\bar{v}, G) \text{ compactly}$$

for simplicity. Denote by $H(\bar{u}, F)$ the set of all $(\bar{v}, G) \in C(R; R^n) \times H(F)$ such that $(\bar{u}^{\tau_n}, F^{\tau_n}) \to (\bar{v}, G)$ compactly for some sequence $\{\tau_n\} \subset R^+$. In particular, we denote by $\Omega(\bar{u}, F)$ the set of all elements $(\bar{v}, G)$ in $H(\bar{u}, F)$ for which one can choose a sequence $\{\tau_n\} \subset R^+$ so that $\lim_{n\to\infty} \tau_n = \infty$ and $(\bar{u}^{\tau_n}, F^{\tau_n}) \to (\bar{v}, G)$ compactly. We can easily see that $\bar{v}$ is a solution of (3.2) whenever $(\bar{v}, G) \in H(\bar{u}, F)$.

Next we shall construct a quasi-process with $\mathcal{X} = \mathrm{BC}_\rho$ associated with (3.2); here and hereafter, $\mathrm{BC}_\rho$ denotes the space BC which is equipped with the metric $\rho$ defined by

$$\rho(\phi, \psi) = \sum_{n=0}^{\infty} \frac{1}{2^n} \frac{|\phi - \psi|_{\mathrm{BC}([-n,0];R^n)}}{1 + |\phi - \psi|_{\mathrm{BC}([-n,0];R^n)}}, \qquad \phi,\ \psi \in \mathrm{BC}_\rho.$$

It is well known that the topology induced by the metric $\rho$ is equivalent to the compact open topology in BC.

We first provide an example which shows that a process on $\mathrm{BC}_\rho$ cannot be always constructed for functional differential equations with infinite delay.

**Example** Consider a scalar delay equation

$$\dot{x}(t) = \sum_{n=1}^{\infty} (1/n^3) x(t-n), \tag{3.3}$$

which is a special case of (3.1) with $g(s) = s + 1$ and $F(t, \phi) = \sum_{n=1}^{\infty} (1/n^3)\phi(-n)$. It is clear that the conditions (H1)–(H3) are satisfied for this equation. Consider a sequence $\{\phi^k\} \subset \mathrm{BC}$ defined by $\phi^k(\theta) = 0$ if $-k \le \theta \le 0$, $k^4$ if $\theta \le -k-1$ and linear if $-k-1 \le \theta \le -k$. Clearly $\phi^k \to 0$ in $\mathrm{BC}_\rho$. Let denote by $x(t, s, \phi)$ the solution of (3.3) through $(s, \phi)$. Then

$$\begin{aligned} x(1, 0, \phi^k) &= \int_0^1 \sum_{n=1}^{\infty} (1/n^3)\phi^k(s-n)ds \\ &\ge 1/(k+2)^3 \int_0^1 \phi^k(s-k-2)ds \\ &\ge k^4/(k+2)^3 \ge 1 \end{aligned}$$

for $k \ge 10$. Note that $x(t, 0, 0) \equiv 0$ and $x_1(0, \phi^k) \not\to x_1(0, 0)$ in $\mathrm{BC}_\rho$. Hence the associated mapping $w : R^+ \times R^+ \times \mathrm{BC}_\rho \mapsto \mathrm{BC}_\rho$ defined by $w(t, s, \phi) = x_{t+s}(s, \phi)$, $(t, s, \phi) \in R^+ \times R^+ \times \mathrm{BC}_\rho$, is not continuous on $R^+ \times R^+ \times \mathrm{BC}_\rho$.

From the above example, we see that the concept of processes does not fit in with the study of the $\rho$-stabilities in functional differential equations with infinite delay. In what follows, we shall consider a subset $\mathcal{Y}$ of $\mathrm{BC}_\rho$ and construct a $\mathcal{Y}$-quasi-process on $\mathrm{BC}_\rho$ associated with (3.2) to overcome the above difficulty.

Let $U$ be a closed, bounded subset of $R^n$ whose interior contains the closure of the set $\{\bar{u}(t) : t \in R\}$, where $\bar{u}$ is the one in (H4). Set

$$\mathrm{BC}_\rho^U = \{\phi \in \mathrm{BC}_\rho : \phi(\theta) \in U \text{ for all } \theta \in R^-\}.$$

It is clear that $\mathrm{BC}_\rho^U$ is a nonempty, closed subset of $\mathrm{BC}_\rho$. With $\mathcal{X} = \mathrm{BC}_\rho$ and $\mathcal{Y} = \mathrm{BC}_\rho^U$, we shall construct the quasi-process associated with (3.2). Consider a function $w_\rho^G : R^+ \times R^+ \times \mathrm{BC}_\rho \mapsto \mathrm{BC}_\rho$ defined by

$$w_\rho^G(t, s, \phi) = u_{t+s}(s, \phi, G), \qquad (t, s, \phi) \in R^+ \times R^+ \times \mathrm{BC}_\rho.$$

**Proposition 3.2** *$w_\rho^G$ is a $\mathrm{BC}_\rho^U$-quasi-process on $\mathrm{BC}_\rho$.*

**Proof** From (H3) we easily see that $w_\rho^G$ satisfies (p1) and (p2) with $\mathcal{X} = \mathrm{BC}_\rho$. We shall show that $w_\rho^G$ satisfies (p3) with $\mathcal{X} = \mathrm{BC}_\rho$ and $\mathcal{Y} = \mathrm{BC}_\rho^U$. Suppose the condition (p3) is not satisfied for $w_\rho^G$. Then there exist a point $(\bar{t}, \bar{s}, \bar{\phi}) \in R^+ \times R^+ \times \mathrm{BC}_\rho$ and sequences $\{t_n\} \subset R^+$, $\{s_n\} \subset R^+$ and $\{\phi^n\} \subset \mathrm{BC}_\rho^U$ such that $(t_n, s_n, \phi^n) \to (\bar{t}, \bar{s}, \bar{\phi})$ in $R^+ \times R^+ \times \mathrm{BC}_\rho$ as $n \to \infty$ and that $\inf_n \rho(u_{t_n+s_n}(s_n, \phi^n, G), u_{\bar{t}+\bar{s}}(\bar{s}, \bar{\phi}, G))$ $> 0$. Then there exists an integer $l > 0$ such that $\inf_n |u_{t_n+s_n}(s_n, \phi^n, G) - u_{\bar{t}+\bar{s}}(\bar{s}, \bar{\phi}, G)|_{\mathrm{BC}([-l,0];R^n)} > 0$, and hence there exists a sequence $\{\tau_n\} \subset [-l, 0]$ such that

$$\inf_n |u(t_n + s_n + \tau_n, s_n, \phi^n, G) - u(\bar{t} + \bar{s} + \tau_n, \bar{s}, \bar{\phi}, G)| > 0. \tag{3.4}$$

Since $u(t, \bar{s}, \bar{\phi}, G)$ is continuous in $t \in R$, we get $\inf_n |u(t_n+s_n+\tau_n, s_n, \phi^n, G) - u(t_n + \bar{s} + \tau_n, \bar{s}, \bar{\phi}, G)| > 0$. Therefore it must hold that $t_n + \tau_n \geq 0$ for all sufficiently large $n$, because of $\rho(\phi^n, \bar{\phi}) \to 0$ as $n \to \infty$. Thus we can assume that $\lim_{n\to\infty} \tau_n = \bar{\tau}$ for some $\bar{\tau} \in [-l, 0]$ with $\bar{t} + \bar{\tau} \geq 0$. Since $\lim_{n\to\infty} |\phi^n - \bar{\phi}|_g = 0$ as easily can be seen, it follows from Proposition 3.1 that $\lim_{n\to\infty} |u_{t_n+s_n+\tau_n}(s_n, \phi^n, G) - u_{\bar{t}+\bar{s}+\bar{\tau}}(\bar{s}, \bar{\phi}, G)|_g = 0$; which implies that $\lim_{n\to\infty} |u(t_n+s_n+\tau_n, s_n, \phi^n, G) - u(\bar{t}+\bar{s}+\bar{\tau}, \bar{s}, \bar{\phi}, G)| = 0$. Therefore $\lim_{n\to\infty} |u(t_n+s_n+\tau_n, s_n, \phi^n, G) - u(\bar{t}+\bar{s}+\tau_n, \bar{s}, \bar{\phi}, G)| = 0$, which is a contradiction to (3.4).

Now we consider the $\mathrm{BC}_\rho^U$-quasi-process $w_\rho^F$ on $\mathrm{BC}_\rho$ generated by (3.1). (H3) yields the relation $u(t + \tau, s + \tau, \phi, F) = u(t, s, \phi, F^\tau)$ for $t \in R$ and $(s, \tau, \phi) \in R^+ \times R^+ \times C_g^0$. Hence $(\sigma(\tau)w_\rho^F)(t, s, \phi) = w_\rho^F(t, \tau + s, \phi) = u_{t+\tau+s}(\tau + s, \phi, F) = u_{t+s}(s, \phi, F^\tau) = w_\rho^{F^\tau}(t, s, \phi)$ for $(t, s, \phi) \in R^+ \times R^+ \times C_g^0$; in other words,

$$\sigma(\tau)w_\rho^F \equiv w_\rho^{F^\tau}.$$

Furthermore, we see that $H_\sigma(w_\rho^F) = \{w_\rho^G : G \in H(F)\}$ and $\Omega_\sigma(w_\rho^F) = \{w_\rho^G : G \in \Omega(F)\}$. Moreover, we see that $H_\sigma(w_\rho^F)$ is sequentially compact and the $\mathrm{BC}_\rho^U$-quasi-process satisfies the condition (p4). For $t \in R^+$, consider a mapping $\pi_\rho(t) : \mathrm{BC}_\rho \times H_\sigma(w_\rho^F) \mapsto \mathrm{BC}_\rho \times H_\sigma(w_\rho^F)$ defined by

$$\pi_\rho(t)(\phi, w_\rho^G) = (u_t(0, \phi, G), \sigma(t)w_\rho^G), \qquad (\phi, w_\rho^G) \in \mathrm{BC}_\rho \times H_\sigma(w_\rho^F).$$

Notice that $\lim_{n\to\infty} |\phi^n - \phi|_g = 0$ whenever $\{\phi^n\} \subset \mathrm{BC}_\rho^U$ satisfies $\lim_{n\to\infty} \rho(\phi^n, \phi)$ $= 0$. Therefore, repeating almost the same argument as in the proof of Proposition

3.2, one can see that $\pi_\rho(t)$ satisfies the condition (p5) with $\mathcal{Y} = \mathrm{BC}_\rho^U$, and hence $\pi_\rho(t)$ is the skew product flow of $w_\rho^F$.

**Proposition 3.3** *The skew product flow $\pi_\rho(t)$ is $\mathrm{BC}_\rho^U$-strongly asymptotically smooth.*

**Proof** It suffices to show that for the set $\mathrm{BC}_\rho^U \times H_\sigma(w_\rho^F)$ there exists a compact set $J \subset \mathrm{BC}_\rho^U \times H_\sigma(w_\sigma^F)$ with the property that $\{\pi_\rho(t_n)(\phi^n, w_\rho^{G_n})\}$ approaches to $J$ whenever sequences $\{t_n\} \subset R^+$ and $\{(\phi^n, w_\rho^{G_n})\} \subset \mathrm{BC}_\rho^U \times H_\sigma(w_\rho^F)$ satisfy $\lim_{n\to\infty} t_n = \infty$ and $\pi_\rho(t)(\phi^n, w_\rho^{G_n}) \subset \mathrm{BC}_\rho^U \times H_\sigma(w_\rho^F)$ for all $t \in [0, t_n]$. Since the set $U$ is bounded in $R^n$, (H2) implies that there exists a constant $L > 0$ such that $|F(t,\phi)| \le L$ for $t \in R^+$ and $\phi \in \mathrm{BC}_\rho^U$. Consider the set $J_U$ of all elements $\phi$ in $\mathrm{BC}_\rho^U$ with the proerty that $|\phi(s_1) - \phi(s_2)| \le L|s_1 - s_2|$ for all $s_1, s_2 \in R^-$, and set $J = J_U \times H_\sigma(w_\rho^F)$. Then $J$ is a compact set in $\mathrm{BC}_\rho^U \times H_\sigma(w_\rho^F)$. Since $u_t(0,\phi^n,G_n) \in \mathrm{BC}_\rho^U$ for all $t \in [0,t_n]$, we get $|u(t_n+s_1,0,\phi^n,G_n) - u(t_n+s_2,0,\phi^n,G_n)| \le L|s_1 - s_3|$ for $s_1, s_2 \in [-t_n, 0]$. Let $y^n$ be a function defined by $y^n(t) = u(t+t_n,0,\phi^n,G_n)$ if $t \ge -t_n$, and $y^n(t) = \phi^n(0)$ if $t < -t_n$. Then $y_0^n \in J_U$ and $\rho(y_0^n, u_{t_n}(0,\phi^n,G_n)) \to 0$ as $n \to \infty$, and hence $\pi_\rho(t_n)(\phi^n, w_\rho^{G_n}) \to J$ as $n \to \infty$.

For any function $\xi : R \mapsto R^n$ such that $\xi_0 \in \mathrm{BC}_\rho^U$ and $\xi$ is continuous on $R^+$, we define a continuous function $\mu_\rho^\xi : R^+ \mapsto \mathrm{BC}_\rho$ by

$$\mu_\rho^\xi(t) = \xi_t, \qquad t \in R^+.$$

It follows from (H4) that $\mu_\rho^{\bar u}$ is an integral of the quasi-process $w_\rho^F$ on $R^+$. Let $\eta > 0$ be chosen so that the interior of $U$ contains the $\eta$-neighborhood of the set $\{\bar u(t) : t \in R\}$. Then we easily see that (p6) is satisfied with $\delta_1 := \eta$ as $\mathcal{Y} = \mathrm{BC}_\rho^U$, $w = w_\rho^F$ and $\mu = \mu_\rho^{\bar u}$ because of the inequality $|u(t+s,s,\phi,F) - \bar u(t+s)| \le \rho(w_\rho^F(t,s,\phi), \mu_\rho^{\bar u}(t+s))$. Also, we get $H_\sigma(\mu_\rho^{\bar u}, w_\rho^F) = \{(\mu_\rho^{\bar v}, w_\rho^G) : (\bar v, G) \in H(\bar u, F)\}$ and $\Omega_\sigma(\mu_\rho^{\bar u}, w_\rho^F) = \{(\mu_\rho^{\bar v}, w_\rho^G) : (\bar v, G) \in \Omega(\bar u, F)\}$.

The $\mathrm{BC}_\rho^U$-stabilities of the integral $\mu_\rho^{\bar u}$ of the quasi-process $w_\rho^F$ yield the $\rho$-stabilities with respect to $U$ for the solution $\bar u(t)$ of (3.1). For example, the solution $\bar u(t)$ of (3.1) is $\rho$-uniformly stable with respect to $U$ in $\Omega(F)$($\rho$-US with respect to $U$ in $\Omega(F)$), if for any $\varepsilon > 0$ there exists a $\delta(\varepsilon) > 0$ such that $\rho(u_t(s,\phi,G), \bar v_t) < \varepsilon$ for $t \ge s \ge 0$ whenever $(\bar v, G) \in \Omega(\bar u, F)$, $\rho(\phi, \bar v_s) < \delta(\varepsilon)$ and $\phi(s) \in U$ for all $s \in R^-$. The other $\rho$-stabilities with respect to $U$ for $\bar u(t)$ are given in a similar way; we omit the details. □

The following result is a direct consequence of Theorem 2.1 and Propositions 3.2 and 3.3.

**Theorem 3.4** *Suppose that the conditions* (H1)–(H4) *are satisfied. Also, let $U$ be a closed and bounded subset of $X$ whose interior contains the closure of the set $\{\bar u(t) : t \in R\}$. Then the following statements are equivalent:*

(i) *The solution $\bar u(t)$ of* (3.1) *is $\rho$-UAS with respect to $U$.*

(ii) *The solution $\bar u(t)$ of* (3.1) *is $\rho$-US with respect to $U$ and $\rho$-attractive with respect to $U$ in $\Omega(F)$.*

On the other hand, practical phenomena are intimately related to the BC-stabilities, where, for example, the solution $\bar u$ of (3.1) is BC-uniformly stable (BC-US) if for any $\varepsilon > 0$ there exists a $\delta(\varepsilon) > 0$ such that $|\bar u(t) - u(t,s,\psi,F)| < \varepsilon$ for $t \ge s \ge 0$ whenever $|\bar u_s - \psi|_{\mathrm{BC}} < \delta(\varepsilon)$ and $\psi \in \mathrm{BC}$. The other BC-stabilities are given in a similar way. We omit the details. Thus the BC-stabilities would seem to be more usual than the $\rho$-stabilities.

The following result is given in [**6**].

**Theorem 3.5** *Under the conditions of Theorem 3.4, the following statements are equivalent:*

(i) *The solution* $\bar{u}(t)$ *of* (3.1) *is* $\rho$-*UAS with respect to* $U$.
(ii) *The solution* $\bar{u}(t)$ *of* (3.1) *is* BC-*UAS.*

Hence, by combining Theorem 3.4 with Theorem 3.5, we have the following result.

**Corollary 3.6** *Under the conditions of Theorem 3.4, the following statements are equivalent:*

(i) *The solution* $\bar{u}(t)$ *of* (3.1) *is* $\rho$-*UAS with respect to* $U$.
(ii) *The solution* $\bar{u}(t)$ *of* (3.1) *is* $\rho$-*US with respect to* $U$ *and* $\rho$-*attractive with respect to* $U$ *in* $\Omega(F)$.
(iii) *The solution* $\bar{u}(t)$ *of* (3.1) *is* BC-*UAS.*

## References

[1] Dafermos, C. M. [1971], *An invariance principle for compact processes,* J. Differential Equations **9**, 239–252.

[2] Hale, J. K. [1977], *Theory of Functional Differential Equations,* Applied Math. Sciences **3**, Springer-Verlag, New York.

[3] Hale, J. K. [1988], *Asymptotic Behavior of Dissipative Systems,* American Mathematical Society, Providence, Rhode Island.

[4] Hino, Y., Murakami, S. and Naito, T. [1991], *Functional Differential Equations with Infinite Delay,* Lecture Notes in Math. **1473**, Springer-Verlag, Berlin -Heidelberg.

[5] Kato, J., Marttynyuk, A. A. and Shestakov, A. A. [1996], *Stability of Motion of Nonautonomous Systems,* Gordon and Breach Publishers.

[6] Murakami, S. and Yoshizawa, T. [1992], *Relationships between BC-stabilities and $\rho$-stabilities in functional differential equations with infinite delay,* Tohoku Math. J. **44**, 45–57.

Fields Institute Communications
Volume **21**, 1999

# Periodic Solutions to System of Globally Coupled Oscillators

**Ying Sue Huang**
Department of Mathematics
Pace University
Pleasantville, NY 10570, USA
huang@risc.dac.pace.edu

**Abstract.** In this paper we study the existence of certain types of periodic solutions. We will state the results on existence of single wave form running periodic solutions, including the in-phase, discrete rotating wave and clustered discrete rotating wave solutions. Then we study the existence of semirotor solutions. In such solutions, the oscillators are divided into two groups. The first group will follow the same periodic pattern while the second group will follow the same running periodic pattern. Finally we study conditions under which semirotor solutions do not exist.

## 1 Introduction

In practice, an oscillator could be a simple pendulum, a biological oscillator, a Josephson junction in physics and so on. We consider the following system of $N$-globally coupled oscillators

$$\beta\frac{d^2\phi_j}{dt^2}+\frac{d\phi_j}{dt}+\sin\phi_j+\frac{dQ}{dt}=I,\quad j=1,2,\cdots,N, \tag{1.1}$$

$$L\frac{d^2Q}{dt^2}+R\frac{dQ}{dt}+C^{-1}Q=\frac{1}{N}\sum_{k=1}^{N}\frac{d\phi_k}{dt}, \tag{1.2}$$

where $\beta, L, R,\ C^{-1}$ and $I$ are all nonnegative constants and $C^{-1}=0$ if $C=\infty$. In this system, the oscillators are coupled globally with equal weights through $Q$. There are two special cases of this system that have been studied by many authors. When $L=0$ and $C^{-1}=0$, system (1.1) and (1.2) can be reduced to

$$\frac{\beta d^2\phi_j}{dt^2}+\frac{d\phi_j}{dt}+\sin\phi_j+\frac{R^{-1}}{N}\sum_{k=1}^{N}\frac{d\phi_k}{dt}=I,\quad j=1,2,\cdots,N. \tag{1.3}$$

1991 *Mathematics Subject Classification.* 34C15, 34C05, 34C25, 47H10, 58C15, 92C15, 92C20, 94C05.

When $L = 0$ and $R = 0$, then system (1.1) and (1.2) can be reduced to

$$\frac{\beta d^2\phi_j}{dt^2} + \frac{d\phi_j}{dt} + \sin\phi_j + \frac{C}{N}\sum_{k=1}^{N}\frac{d^2\phi_k}{dt^2} = I, \quad j = 1, 2, \cdots, N. \tag{1.4}$$

In both cases, the $Q$ equation can actually be eliminated and therefore should be much easier to study than system (1.1) and (1.2).

These systems arise in the study of Josephson junctions in physics. However they may also be used in studying the biological oscillators and neural nets. They are of mathematical interest since the dynamics of these systems are very rich, in particular, they are permutation invariant (see [**1**]). These systems have been extensively studied both numerically and analytically, especially in the special cases of systems (1.3) and (1.4). For example, they were studied numerically in [**6**], [**9**], [**10**] and [**11**], and analytically in [**1**], [**2**], [**3**], [**4**], [**5**], [**7**] and others.

In this paper we study the existence of certain periodic solutions to system (1.1) and (1.2), including the running periodic solutions and semirotor solutions.

This article is organized as follows. First, in Section 2 we give the definition of running periodic solutions, including the in-phase running and discrete rotating wave solutions. Then we summarize the known results on the existence of such type of periodic solutions for system (1.1) and (1.2). In Section 3, we define and then study the existence of the semirotor solutions to system (1.1) and (1.2). Finally, we study the conditions in which semirotor solutions do not exist. We give conditions that semirotor solutions do not exist to systems (1.3) and (1.4). We believe the results could be generalized to system (1.1) and (1.2).

## 2 Existence of running periodic solutions

A function $\phi : \mathbf{R} \to \mathbf{R}$ is said to be a *periodic function* of *period* $T$ if there is a minimum $T > 0$ such that

$$\phi(t + T) = \phi(t) \text{ for all } t \in \mathbf{R}.$$

A function $\phi : \mathbf{R} \to \mathbf{R}$ is said to be a *running function* of *period* $T$ if there is a minimum $T > 0$ such that

$$\phi(t + T) = \phi(t) + 2\pi \text{ for all } t \in \mathbf{R}.$$

Note that a running solution on $\mathbf{R}$ is a $2\pi$-periodic function on the circle $S^1$.

A *running solution* of period $T$ to the system (1.1) and (1.2) is an $(N+1)$-tuple

$$\{\phi_1, \phi_2, \cdots, \phi_N, Q\}$$

in which each $\phi_j$ ( $j = 1, 2, \cdots, N$) is a running function of period $T$.

If there exist positive numbers $\rho_1 \leq \rho_2 \cdots \leq \rho_N$ such that

$$\phi_j(t) = \phi(t - \rho_j T), \quad j = 1, 2, \cdots, N$$

for some running function $\phi$, then the running solution is called a single wave form solution. If there exist integers $l$ and $M$ such that $N = lM$ and

$$\rho_j = \frac{h}{N}T, \quad j = hl + 1, \cdots, (h+1)l \text{ and } h = 0, 1, \cdots, M - 1,$$

we call the single wave form solution an $(l, M, N)$ type single wave form solution of period $T$ and $\phi$ is called the wave form of the solution. In particular, if $l = N$ and $M = 1$, then

$$\phi_1 = \phi_2 = \cdots = \phi_N.$$

We also call the $(N, 1, N)$ type single wave form solution an in-phase running solution. If $l = 1$ and $M = N$, then

$$\phi_j(t) = \phi(t - \frac{j-1}{N}T), \quad j = 1, 2, \cdots, N.$$

We call the $(1, N, N)$ type single wave form solution a discrete rotating wave solution. An $(l, M, N)$ type single wave form solution other than in-phase or discrete rotating wave solution is often referred to as a clustered discrete rotating wave form solution.

Existence and behavior of the running solutions to system (1.1) and (1.2) have been extensively studied especially in the special cases to systems (1.3) and system (1.4). In [**1**] it is proved that there exist in-phase running periodic solutions to systems (1.3) and (1.4). For the same two systems, in [**2**], it is proved that there exist discrete rotating wave solutions and clustered discrete rotating wave solutions. These results were generalized to system (1.1) and (1.2) in [**3**]. When $R + C^{-1} > 0$, based on homotopy argument and degree theory, it is proved that there exist $(l, M, N)$ type of single wave form solutions to system (1.1) and (1.2). As a special case, in-phase running solution and discrete rotating wave solution both exist to system (1.1) and (1.2). One can prove that for a single wave form solution as we defined, the $\rho_j$'s have to be rationally related to each other. It is also conjectured that single wave form solutions are generically those $(l, M, N)$ type of single wave form solutions. The discrete rotating wave solutions, when $N$ goes to infinity, are proved to converge to a certain special wave form exponentially in [**3**] and [**7**].

There is also an extensive literature devoted to the case $\beta = 0$; see, e.g., [**11**] or [**9**] for further information and references.

The results on existence of single wave form solutions can be summarized in the following theorem.

**Theorem 2.1** ([**7**]) *Suppose $\beta \geq 0$, $I > 1$, $L, R$ and $C$ are all nonnegative with $R + C^{-1} > 0$. If there exist integers $l$ and $M$ such that $N = lM$, then there exist $(l, M, N)$ type of single wave form solutions to system (1.1) and (1.2).*

When $N$ is large, the behavior of the discrete rotating wave solutions satisfy the following theorem.

**Theorem 2.2** ([**3**]) *For system (1.1) and (1.2), suppose $\beta \geq 0$, $L, R$ and $C$ are all nonnegative with $R + C^{-1} > 0$. Let $\phi_N$ be the wave form of the discrete rotating wave solution with period $T_N$. Then there exist functions $\phi$ and $Q$ and a positive constant $T$ such that*

$$\lim_{N\to\infty} T_N = T, \quad \lim_{N\to\infty} Q_N(t) = Q(t), \quad \lim_{N\to\infty} \phi_N(t) = \phi(t).$$

*Furthermore, the limiting wave form $\phi$ is a running solution of period $T$ to the pendulum equation*

$$\beta\phi'' + \phi' + \sin\phi + \rho\frac{2\pi}{T} = I$$

*where $\rho = 0$ if $C^{-1} \neq 0$, and $\rho = R^{-1}$ if $C^{-1} = 0$ but $R \neq 0$.*

## 3 The existence of semirotor solutions

Semirotor solutions were first observed in [**1**]. In their numerical experiments, Aronson et al. found a different type of periodic solutions, the so called semirotor solutions.

A *semirotor solution* of period $T$ to system (1.1) and (1.2) is an $(N+1)$-tuple

$$\{\phi_1, \phi_2, \cdots, \phi_N, Q\}$$

such that there exists a $k$ and

$$\phi_j(t) = \psi_1(t) \text{ for } j = 1, 2, \cdots, k$$

and

$$\phi_j(t) = \psi_2(t) \text{ for } j = k+1, k+2, \cdots, N$$

for some running periodic function $\psi_1$ and a periodic function $\psi_2$ of period $T$. Functions $\psi_1$ and $\psi_2$ are called the running periodic wave form and periodic wave form, respectively, of a semirotor solution. If the running periodic form $\psi_1$ satisfies that $\psi'(t) > 0$ for all $t$, then the semirotor solution is called a *simple semirotor solution.* It is proved that such semirotor solutions exist to systems (1.3) and (1.4) when $\beta$ is sufficiently large and $I$ satisfies certain conditions in [**5**]. It is of interest to generalize their result to more general cases.

If $\{\phi_1, \phi_2, \cdots, \phi_N, Q\}$ is a semirotor solution with the running periodic wave form $\psi_1$ and periodic wave form $\psi_2$, respectively, then $\psi_1$ and $\psi_2$ will satisfy

$$\beta\frac{d^2\psi_j}{dt^2} + \frac{d\psi_j}{dt} + \sin\psi_j + \frac{dQ}{dt} = I, \quad j = 1, 2, \tag{3.1}$$

$$L\frac{d^2Q}{dt^2} + R\frac{dQ}{dt} + C^{-1}Q = \rho\frac{d\psi_1}{dt}(t) + (1-\rho)\frac{d\psi_2}{dt}(t), \tag{3.2}$$

where

$$\rho = \frac{k}{N}.$$

Suppose $R \neq 0$ and $L + C^{-1} \neq 0$. Under certain nonresonance conditions, we show that when $I$ is in a certain range of values, system (1.1) and (1.2) has two simple semirotor solutions depending smoothly on $\beta$. We will use an implicit function theorem to prove the existence of such semirotor solutions. Our proof is similar to the method used in [**5**] but much more complicated. As a matter of fact, for the two special systems (1.3) and (1.4), the $Q$ equation is eliminated in system (1.1) and (1.2), which makes it much easier to study. We will prove the following theorem.

**Theorem 3.1** *Suppose $R^2 - 4LC^{-1} \neq 0$. For every fixed $k \in \{1, 2, \cdots, N-1\}$, when $C^{-1} \neq 0$ and $I < 1$ or when $C^{-1} = 0$ and $I < 1 + \frac{k}{NR}$, there exists $\beta(I)$ such that for every $\beta > \beta(I)$, system (1.1) and (1.2) has at least two simple semirotor solutions depending smoothly on $\beta$.*

**Proof** To prove the theorem, it suffices to show that there is a solution $(\psi_1, \psi_2)$ to system (3.1) and (3.2) such that $\psi_1$ is running periodic and $\psi_2$ is periodic with the same period $T$. Let $\epsilon = 1/\beta$. We divide both sides of the Equations (3.1) and

(3.2) by $\beta$ and rewrite the system as a system with first order equations.

$$\begin{aligned} \frac{d\psi_1}{dt} &= \xi, \\ \frac{d\xi}{dt} &= \epsilon(I - q - \sin\psi_1 - \xi), \\ \frac{d\psi_2}{dt} &= \zeta, \\ \frac{d\zeta}{dt} &= \epsilon(I - q - \sin\psi_2 - \zeta), \\ \frac{dQ}{dt} &= q, \\ L\frac{dq}{dt} &+ Rq + C^{-1}Q = \rho\xi + (1-\rho)\zeta. \end{aligned} \tag{3.3}$$

We assume that $\frac{d\psi_1}{dt} > 0$ and let $s = \psi_1(t)$. Then

$$\frac{ds}{dt} = \frac{d\psi_1}{dt} = \xi.$$

Furthermore, all functions can be reduced as functions in $s$. For the rest of the section, we use the following notations

$$\psi_2 = \psi, \qquad ' = \frac{d}{ds}.$$

Then system (3.3) can be written as

$$\begin{aligned} \xi\xi' &= \epsilon(I - q - \sin s - \xi), \\ \xi\psi' &= \zeta, \\ \xi\zeta' &= \epsilon(I - q - \sin\psi - \zeta), \\ \xi Q' &= q, \\ L\xi q' &= -Rq - C^{-1}Q + \rho\xi + (1-\rho)\zeta. \end{aligned} \tag{3.4}$$

We prove the theorem in two cases.

**Case I** $C^{-1} \neq 0$.

Motivated by the numerical result of [**1**] and the results of [**5**], we will show that there exist periodic solutions of form

$$\begin{aligned} \psi &= A + \epsilon X, \\ \xi &= B + \epsilon Y, \\ \zeta &= \epsilon Z, \\ Q &= \rho C B + \epsilon U, \\ q &= \epsilon V. \end{aligned} \tag{3.5}$$

To show that system (3.4) has a solution of form (3.5), it suffices to show that there exist periodic solutions to the following system (with new order in $X, Y, Z, U, V$.)

$$\begin{aligned} (B+\epsilon Y)X' &= Z, \\ (B+\epsilon Y)Y' &= I - B - \epsilon V - \epsilon Y - \sin s, \\ (B+\epsilon Y)Z' &= -\epsilon V - \epsilon Z + I - \sin(A + \epsilon X), \\ (B+\epsilon Y)U' &= V, \\ L(B+\epsilon Y)V' &= -RV - C^{-1}U + \rho Y + (1-\rho)Z. \end{aligned} \tag{3.6}$$

To show that we can find a periodic solution to system (3.6) when $\epsilon$ is small enough, first we look at the case that $\epsilon = 0$. When $\epsilon = 0$, system (3.6) can be reduced as

$$\begin{aligned} BX' &= Z, \\ BY' &= I - B - \sin s, \\ BZ' &= I - \sin A, \\ BU' &= V, \\ LBV' &= -RV - C^{-1}U + \rho Y + (1-\rho)Z. \end{aligned} \tag{3.7}$$

In order that system (3.7) has a periodic solution, it suffices that $I = B = \sin A$. Since $0 < I \leq 1$, such $A$ and $B$ do exist and the solution can be expressed as

$$\begin{aligned} X &= 0, \\ Y &= \cos s / I, \\ Z &= 0, \end{aligned} \tag{3.8}$$

$U$ is the unique periodic solution to

$$LI^2U'' + RIU' + C^{-1}U = \rho \frac{\cos s}{I}$$

and $V = IU'$. Since $R^2 - 4LC^{-1} \neq 0$, we can get

$$U = a \cos s + b \sin s, \tag{3.9}$$

$$V = -aI \sin s + bI \cos s \tag{3.10}$$

with

$$a = \frac{\frac{\rho}{I}(-LI^2 + C^{-1})}{(-LI^2 + C^{-1})^2 + (RI)^2}, \tag{3.11}$$

$$b = \frac{\rho R}{(-LI^2 + C^{-1})^2 + (RI)^2}. \tag{3.12}$$

System (3.6) can then be written as

$$\begin{aligned} (I + \epsilon Y)X' &= Z, \\ (I + \epsilon Y)Y' &= -\epsilon V - \epsilon Y - \sin s, \\ (I + \epsilon Y)Z' &= -\epsilon V - \epsilon Z + \sin A - \sin(A + \epsilon X), \\ (I + \epsilon Y)U' &= V, \\ L(I + \epsilon Y)V' &= -RV - C^{-1}U + \rho Y + (1-\rho)Z. \end{aligned} \tag{3.13}$$

Dividing both sides by $I + \epsilon Y$ and then expanding in terms of $\epsilon$, System (3.13) can be further reduced as follows

$$\begin{aligned} X' &= \frac{Z}{I} + \frac{\epsilon}{I} g_1(X, Y, Z, U, V, s, \epsilon), \\ Y' &= \frac{-\sin s}{I} + \frac{\epsilon}{I} g_2(X, Y, Z, U, V, s, \epsilon), \\ Z' &= \frac{\epsilon}{I} g_3(X, Y, Z, U, V, s, \epsilon), \\ U' &= \frac{V}{I} + \frac{\epsilon}{I} g_4(X, Y, Z, U, V, s, \epsilon), \\ V' &= \frac{1}{LI}(-RV - C^{-1}U + \rho Y + (1-\rho)Z) \\ &\quad + \frac{\epsilon}{LI} g_5(X, Y, Z, U, V, s, \epsilon), \end{aligned} \tag{3.14}$$

where

$$g_1(X,Y,Z,U,V,s,\epsilon) = -\frac{YZ}{I} + O(\epsilon),$$

$$g_2(X,Y,Z,U,V,s,\epsilon) = \frac{Y\sin s}{I} - V - Y + O(\epsilon)$$

$$g_3(X,Y,Z,U,V,s,\epsilon) = -X\cos A - V - Z + O(\epsilon),$$

$$g_4(X,Y,Z,U,V,s,\epsilon) = -\frac{VY}{I} + O(\epsilon),$$

$$g_5(X,Y,Z,U,V,s,\epsilon) = -\frac{Y}{I}(-RV - C^{-1}U + \rho Y + (1-\rho)Z)) + O(\epsilon).$$

We are looking for conditions on initial values $X_0 = X(0), Y_0 = Y(0), Z_0 = Z(0), U_0 = U(0), V_0 = V(0)$ so that system (3.14) has a periodic solution.

From the equations in system (3.14) one can get that

$$Z = Z_0 + O(\epsilon) \tag{3.15}$$

$$Y = Y_0 - \frac{1}{I} + \frac{\cos s}{I} + O(\epsilon) \tag{3.16}$$

$$X = X_0 + \frac{Z_0}{I}s + O(\epsilon) \tag{3.17}$$

and $U$ and $V$ satisfies

$$LI^2U'' + RIU' + C^{-1}U = \rho\left(Y_0 - \frac{1}{I}\right) + \rho\frac{\cos s}{I} + (1-\rho)Z_0 + O(\epsilon).$$

Let $\lambda_1, \lambda_2$ be the two solutions to the characteristic equation

$$LI^2\lambda^2 + RI\lambda + C^{-1} = 0.$$

Then

$$U = c_1e^{\lambda_1 s} + c_2e^{\lambda_2 s} + C\rho(Y_0 - \tfrac{1}{I}) + C(1-\rho)Z_0 + a\cos s + b\sin s + O(\epsilon), \tag{3.18}$$

$$V = c_1\lambda_1e^{\lambda_1 s} + c_2\lambda_2e^{\lambda_2 s} - aI\sin s + bI\cos s + O(\epsilon), \tag{3.19}$$

where

$$c_1 = \frac{\lambda_2(U_0 - C\rho(Y_0 - 1/I) - C(1-\rho)Z_0 - a) - (V_0 - bI)}{\lambda_2 - \lambda_1},$$

$$c_2 = \frac{-\lambda_1(U_0 - C\rho(Y_0 - 1/I) - C(1-\rho)Z_0 - a) + (V_0 - bI)}{\lambda_2 - \lambda_1}$$

and $a$ and $b$ are given in (3.11) and (3.12). In order that system (3.14) has a periodic solution with initial condition $(X_0, Y_0, Z_0, U_0, V_0)$, by integrating both sides of the equations of (3.14) over $[0, 2\pi]$, we obtain the following set of equations

$$\int_0^{2\pi} Z ds + \epsilon \int_0^{2\pi} g_1(X, Y, Z, U, V, s, \epsilon) ds = 0, \tag{3.20}$$

$$\int_0^{2\pi} g_2(X, Y, Z, U, V, s, \epsilon) ds = 0, \tag{3.21}$$

$$\int_0^{2\pi} g_3(X, Y, Z, U, V, s, \epsilon) ds = 0, \tag{3.22}$$

$$\int_0^{2\pi} V ds + \epsilon \int_0^{2\pi} g_4(X, Y, Z, U, V, s, \epsilon) ds = 0, \tag{3.23}$$

$$\begin{aligned}\int_0^{2\pi} &(-RV - C^{-1}U + \rho Y + (1-\rho)Z) ds \\ &+ \epsilon \int_0^{2\pi} g_5(X, Y, Z, U, V, s, \epsilon) ds = 0.\end{aligned} \tag{3.24}$$

One can see from (3.23) that

$$\int_0^{2\pi} V(s) ds = -\epsilon \int_0^{2\pi} g_4(X, Y, Z, U, V, s, \epsilon) ds. \tag{3.25}$$

Let $F_j(X_0, Y_0, Z_0, U_0, V_0, \epsilon)$ denote respectively the left-hand sides of (3.20)-(3.22) for $j = 1, 2, 3$, (3.23) multiplied by $R^{-1}$ for $j = 4$ and (3.24) multiplied by $C$ for $j = 5$. Then $F_j$'s are operators from $R^5$ to $R$. Using (3.25), then equations (3.20)-(3.24) are equivalent to the following equations

$$\begin{aligned}
F_1(X_0, Y_0, Z_0, U_0, V_0, \epsilon) &= 2\pi Z_0 + O(\epsilon) = 0, \\
F_2(X_0, Y_0, Z_0, U_0, V_0, \epsilon) &= 2\pi \left( Y_0 - \frac{1}{I} \right) + O(\epsilon) = 0, \\
F_3(X_0, Y_0, Z_0, U_0, V_0, \epsilon) &= -2\pi X_0 \cos A - \frac{2\pi^2 Z_0 \cos A}{I} - 2\pi Z_0 I + O(\epsilon) = 0, \\
F_4(X_0, Y_0, Z_0, U_0, V_0, \epsilon) &= c_1(e^{2\pi\lambda_1} - 1) + c_2(e^{2\pi\lambda_2} - 1) + O(\epsilon) = 0, \\
F_5(X_0, Y_0, Z_0, U_0, V_0, \epsilon) &= c_1(e^{2\pi\lambda_1} - 1)/\lambda_1 + c_2(e^{2\pi\lambda_2} - 1)/\lambda_2 + O(\epsilon) = 0.
\end{aligned} \tag{3.26}$$

The solution of equation

$$F(X_0, Y_0, Z_0, U_0, V_0, \epsilon) = (F_1, F_2, F_3, F_4, F_5)(X_0, Y_0, Z_0, U_0, V_0, \epsilon) = 0 \tag{3.27}$$

gives the initial condition that there exist periodic solutions to system (3.14). When $\epsilon = 0$,

$$F_1(X_0, Y_0, Z_0, U_0, V_0, 0) = 2\pi Z_0,$$

$$F_2(X_0, Y_0, Z_0, U_0, V_0, 0) = 2\pi \left(Y_0 - \frac{1}{I}\right),$$

$$F_3(X_0, Y_0, Z_0, U_0, V_0, 0) = -2\pi X_0 \cos A - \frac{2\pi^2 Z_0 \cos A}{I} - 2\pi Z_0 I,$$

$$F_4(X_0, Y_0, Z_0, U_0, V_0, 0) = c_1(e^{2\pi\lambda_1} - 1) + c_2(e^{2\pi\lambda_2} - 1),$$

$$F_5(X_0, Y_0, Z_0, U_0, V_0, 0) = c_1(e^{2\pi\lambda_1} - 1)/\lambda_1 + c_2(e^{2\pi\lambda_2} - 1)/\lambda_2.$$

One can see from the definition of $c_1$ and $c_2$ that when $\epsilon = 0$, operator equation (3.27) has a unique solution

$$(X_0, Y_0, Z_0, U_0, V_0) = \left(0, \frac{1}{I}, 0, a, Ib\right).$$

From system (3.26), we can show that

$$\det \frac{\partial F}{\partial(X_0, Y_0, Z_0, U_0, V_0)} \left(0, \frac{1}{I}, 0, a, Ib, 0\right) = -\frac{8\pi^3 \cos A(e^{2\pi\lambda_1} - 1)(e^{2\pi\lambda_2} - 1)}{\lambda_1\lambda_2}.$$

Since $\lambda_1, \lambda_2 \neq 0$, the above expression is not zero for the two solutions of $A$ with $\sin A = I$. The implicit function theorem implies that when $\epsilon$ is small enough, there exist $(X_0(\epsilon), Y_0(\epsilon), Z_0(\epsilon), U_0(\epsilon), V_0(\epsilon))$ which is the initial condition of the periodic solution to system (3.13) such that (3.27) holds.

When $\epsilon = 0$, we have that

$$\begin{aligned}
X &= 0,\\
Y &= \frac{\cos s}{I},\\
Z &= 0,\\
U &= a\cos s + b\sin s,\\
V &= -aI\sin s + bI\cos s,
\end{aligned}$$

where, $a$ and $b$ are given in (3.11) and (3.12). As a result, using the expressions in (3.5) and (3.4), we get

$$\begin{aligned}
\psi_1 &= It + \frac{\epsilon}{I^2}\sin(It) + O(\epsilon^2),\\
\psi_2 &= A + O(\epsilon^2),\\
T &= \frac{2\pi}{I} + O(\epsilon),\\
Q &= C\rho B + \epsilon(a\cos(It) + b\sin(It)) + O(\epsilon^2),
\end{aligned} \tag{3.28}$$

where $I = B = \sin A$.

**Case II** $C^{-1} = 0$ and $R \neq 0$.

If we replace $\frac{dQ}{dt}$ by $q$, then equation (3.2) can be written as

$$L\frac{dq}{dt} + Rq = \rho\frac{d\psi_1}{dt}(t) + (1-\rho)\frac{d\psi_2}{dt}(t).$$

As a result, system (3.3) can be written as

$$\begin{aligned} \xi\xi' &= \epsilon(I - q - \sin s - \xi), \\ \xi\psi' &= \zeta, \\ \xi\zeta' &= \epsilon(I - q - \sin\psi - \zeta), \\ L\xi q' &= -Rq + \rho\xi + (1-\rho)\zeta. \end{aligned}$$

We will show that there exist periodic solutions of form

$$\begin{aligned} \psi &= A + \epsilon X, \\ \xi &= B + \epsilon Y, \\ \zeta &= \epsilon Z, \\ q &= R^{-1}\rho B + \epsilon V. \end{aligned}$$

As we did in Case I, first we look at the case that $\epsilon = 0$. When $\epsilon = 0$, we get

$$\begin{aligned} BX' &= Z, \\ BY' &= I - B - R\rho B - \sin s, \\ BZ' &= I - R\rho B - \sin A, \\ LBV' &= -RV + \rho Y + (1-\rho)Z. \end{aligned}$$

In order that this system has a periodic solution, it suffices that $I = B + R\rho B = \sin A + R\rho B$. We need

$$B = \sin A = \frac{I}{1+\rho R}.$$

It suffices that $0 < I \leq 1 + \rho R$ holds. The other parts of the proof are similar to Case I with the use of the implicit function theorem argument.

When $\epsilon = 0$, we have that

$$\begin{aligned} X &= 0, \\ Y &= \frac{\cos s}{I}, \\ Z &= 0, \\ V &= a_1 \sin s + a_2 \cos s \end{aligned}$$

with

$$\begin{aligned} a_1 &= \frac{L\rho}{(LI)^2 + R^2}, \\ b_1 &= \frac{R\rho}{I[(LI)^2 + R^2]}. \end{aligned}$$

Thus

$$\begin{aligned} \psi_1 &= It + \frac{\epsilon}{I^2}\sin(It) + O(\epsilon^2), \\ \psi_2 &= A + O(\epsilon^2), \\ T &= \frac{2\pi}{I} + O(\epsilon), \\ Q' &= q = R^{-1}\rho B + \epsilon(a_1\sin(It) + b_1\cos(It)) + O(\epsilon^2), \end{aligned} \tag{3.29}$$

where $I = B = \sin A$. □

From the proof of the theorem and the expressions in (3.28) and (3.29), we have the following result.

**Theorem 3.2** *Let $\psi_1^j$ and $\psi_2^j$, $j = 1, 2$ be, respectively, the running periodic wave-forms and the periodic wave-forms of the two simple semirotor solutions found in Theorem 3.1. As $\beta \to \infty$, the curve $\{\psi_1^j(t), {\psi_1^j}'(t)\}$ approaches the horizontal line $y = I$ and the curve $\{\psi_2^j(t), {\psi_2^j}'(t)\}$ approaches a single point $(A, 0)$.*

## 4 Nonexistence of semirotor solutions

We have shown that when $R^2 - 4LC^{-1} \neq 0$, when $C^{-1} \neq 0$ and $I < 1$ or when $C^{-1} = 0$ and $I < 1 + \frac{k}{NR}$, semirotor solutions for system (1.1) and (1.2) exist for $\beta$ sufficiently large. Here we study conditions that semirotor solutions do not exist to systems (1.3) and (1.4). Even though we will establish our result only for systems (1.3) and (1.4), we believe similar results can be established for other cases.

To get our results, we need the following lemma.

**Lemma 4.1** *For equation*

$$\beta y'' + \alpha y' = w(t), \tag{4.1}$$

*where both $\alpha$ and $\beta$ are positive, if $M_1 < w(t) < M_2$ for some constants $M_1$ and $M_2$, then for any $\epsilon > 0$, there exists $t(\epsilon)$ such that when $t > t(\epsilon)$, the following holds:*

$$\frac{M_1}{\alpha} - \epsilon \leq y'(t) \leq \frac{M_2}{\alpha} + \epsilon.$$

**Proof** From (4.1), one can get

$$y'(t) = e^{-\frac{\alpha}{\beta}t}\left(y'(0) + \int_0^t e^{\frac{\alpha}{\beta}s} w(s)ds\right).$$

Thus,

$$y'(0)e^{-\frac{\alpha}{\beta}t} + \frac{M_1}{\alpha}(1 - e^{-\frac{\alpha}{\beta}t}) \leq y'(t) \leq y'(0)e^{-\frac{\alpha}{\beta}t} + \frac{M_2}{\alpha}(1 - e^{-\frac{\alpha}{\beta}t}).$$

Since $\alpha$ and $\beta$ are both positive, we can see that the lemma holds. □

We can prove the following result:

**Theorem 4.2** *There exist constants $I(R)$ and $I(\beta, C)$, such that there is no semirotor solution to system (1.3) or system (1.4) when $I > I(R)$ or $I > I(\beta, C)$, respectively.*

**Proof** It suffices to show that $\phi'(t) > 0$ when $t$ is large enough. Denote

$$g(t) = \frac{1}{N}\sum_{k=1}^{N} \phi_k(t)$$

and

$$h(t) = \frac{1}{N}\sum_{k=1}^{N} \sin\phi_k(t).$$

First, we consider system (1.3). By adding all the equations together and then divide by $N$, we get

$$\beta g'' + (1 + R^{-1})g'(t) = I - h(t). \tag{4.2}$$

Since

$$I - 1 \leq I - h(s) \leq I + 1,$$

using Lemma 4.1, for any $\epsilon$, there exists a $t(\epsilon)$ when $t > t(\epsilon)$,

$$\frac{I-1}{1+R^{-1}} - \epsilon \leq |g'(t)| \leq \frac{I+1}{1+R^{-1}} + \epsilon. \tag{4.3}$$

From System (1.3), we get

$$\beta\phi_j'' + \phi_j' = I - \sin\phi_j - R^{-1}g'. \tag{4.4}$$

Using (4.3), we then get

$$\begin{aligned} I - 1 - \frac{I+1}{1+R^{-1}}R^{-1} - \epsilon &\leq I - \sin\phi_j - R^{-1}g' \\ &\leq I + 1 + \frac{I+1}{1+R^{-1}}R^{-1} + \epsilon. \end{aligned} \tag{4.5}$$

Using (4.5) and applying Lemma 4.1 to Equation (4.4), there exists a $t_1(\epsilon)$, when $t > t_1(\epsilon)$, we have

$$\phi'(t) \geq I - 1 - \frac{I+1}{1+R^{-1}}R^{-1} - \epsilon - \epsilon.$$

Thus, when $I > 2R^{-1} + 1$ and $t$ is sufficiently large, we have $\phi'(t) > 0$. Thus, there is no semirotor solution in this case.

Next, we consider system (1.4). As we did for system (1.3), we add all the equations together and then divide both sides by $N$. We then get

$$(\beta + C)g''(t) + g'(t) = I - h(t), \tag{4.6}$$

using Lemma 4.1, there is a $t(\epsilon)$, such that when $t > t(\epsilon)$,

$$I - 1 - \epsilon < g'(t) < I + 1 + \epsilon.$$

From (4.6),

$$g''(t) = \frac{I - h(t) - g'(t)}{\beta + C}.$$

From system (1.4), we get

$$\beta\phi_j'' + \phi_j' = I - \sin\phi_j - Cg''. \tag{4.7}$$

Since

$$\begin{aligned} I - \sin\phi_j - Cg'' &= I - \sin\phi_j - C\frac{I - h(t) - g'(t)}{\beta + C} \\ &\geq I - 1 - C\frac{I + 1 - I + 1 + \epsilon}{\beta + C} \\ &\geq I - 1 - \frac{2C + \epsilon C}{\beta + C}, \end{aligned} \tag{4.8}$$

using (4.8) and applying Lemma 4.1 to (4.7), we know that for any $\epsilon$, there exist $t(\epsilon)$ such that when $t > t(\epsilon)$,

$$\phi_j'(t) \geq I - 1 - \frac{2C + \epsilon C}{\beta + C} - \epsilon.$$

When $I > 1 + \frac{2C}{\beta + C}$, if $t$ is large enough, then $\phi_j'(t) > 0$ for all $j = 1, 2, \cdots, N$. We thus get the result. □

## References

[1] Aronson, D. G., Golubitsky, M. and Krupa, M. [1991], *Coupled arrays of Josephson junctions and bifurcation of maps with $S_N$ symmetry*, Nonlinearity **4**, 861-902.

[2] Aronson, D. G., Golubitsky, M. and Mallet-Paret, J. [1991], *Ponies on a merry-go-round in large arrays of Josephson junctions*, Nonlinearity **4**, 903-910.

[3] Aronson, D. G. and Huang, Y. S. [1994], *Limits and uniqueness of discrete rotating waves in large arrays of Josephson junctions*, Nonlinearity **7**, 777-804.

[4] Aronson, D. G. and Huang, Y. S. [1997], *Single waveform solutions for linear arrays of Josephson junctions*, Physica D **101**, 157-177.

[5] Aronson D. G., Krupa, M. and Ashwin, P. [1996], *Semirotors in the Josephson junction equations*, J. Nonlinear Sci. **6**, 85-103.

[6] Hadley, P., Beasley, M. R. and Wiesenfeld, K. [1988], *Phase locking of Josephson junction series arrays*, Phys. Rev. B **38**, 8712-8719.

[7] Huang, Y. S. and Aronson, D. G. [1996], *Discrete rotating wave solutions for system of globally coupled Josephson junctions*, Int. J. Bif. & Chaos. **10**, 1789-1797.

[8] Mirollo, R. E. [1992], *Splay-phase orbits for equivariant flows on tori*, SIAM J. Math. Anal. **25**, 1176-1180.

[9] Strogatz, S. H. and Mirollo, R. E. [1993], *Splay states in globally coupled Josephson arrays: Analytic prediction of Floquet multipliers*, Phys. Rev. E **47**, 220-227.

[10] Tsang, K. Y., Mirollo, R. E., Strogatz, S. H. and Wiesenfeld, K. [1991], *Dynamics of globally coupled oscillator array*, Physica D **48**, 102-112.

[11] Watanabe, S. and Strogatz, S. H. [1994], *Constants of motion for superconductor arrays*, Physica D **74**, 197-253.

Fields Institute Communications
Volume **21**, 1999

# Bifurcation Structure of Stationary Solutions of a Lotka-Volterra Competition Model with Density-Dependent Diffusion

**Yukio Kan-on**
Department of Mathematics
Faculty of Education
Ehime University
Matsuyama 790-8377, Japan
kanon@ed.ehime-u.ac.jp

**Abstract.** In this paper, we consider a Lotka-Volterra competition model with density-dependent diffusion which describes the dynamics of the population of two competing species, and establish the bifurcation structure of positive stationary solutions of the model. To do this, the comparison principle and the bifurcation theory are employed, and the spatial profile of positive stationary solutions is studied.

## 1 Introduction

In order to discuss the mechanism of the coexistence for two competing species, Shigesada, Kawasaki and Teramoto in [**7**] proposed a model of reaction-diffusion equations with density-dependent diffusion

$$\begin{cases} u_t = \varepsilon\,[(1+\alpha\,v)\,u]_{xx} + (1-u-c\,v)\,u, \\ v_t = \varepsilon\,d\,[(1+\beta\,u)\,v]_{xx} + (a-b\,u-v)\,v, \quad x\in(0,1), \quad t>0, \\ u_x = 0, \quad v_x = 0, \quad x = 0,1, \quad t>0, \\ u(0,x)\ge 0, \quad v(0,x)\ge 0, \quad x\in[0,1], \end{cases} \tag{1.1}$$

where $\alpha$ and $\beta$ are nonnegative constants, and the others are positive ones. The terms $1+\alpha\,v$ and $1+\beta\,u$ represent the virtual diffusivity depending on the population density, and mean that each species moves under the influence of the random movement and the repulsive force due to the uncomfortableness. As $\mathbf{u} = (u,v)$ indicates the population density of two competing species, we restrict our discussion to positive solutions of (1.1), where we say that $(u,v)$ is *positive* if $u>0$ and $v>0$ hold.

1991 *Mathematics Subject Classification.* 35B32.

This research was partially supported by Grant-in-Aid for Encouragement of Young Scientists (No. 09740150) from the Ministry of Education, Science, Sports and Culture, Japan.

When $\alpha = 0$ and $\beta = 0$ are satisfied, the model (1.1) falls into a class of competition-diffusion equations, and the asymptotic behavior of the solution $\mathbf{u}(t,x)$ of (1.1) with positive initial data is classified into the following (for example, see [**2**] and [**4**]).

(I) If $a < \min(b, 1/c)$, then $\lim_{t\to+\infty} \mathbf{u}(t,x) = (1,0)$.

(II) If $b < a < 1/c$, then $\lim_{t\to+\infty} \mathbf{u}(t,x) = \left(\dfrac{1-a\,c}{1-b\,c}, \dfrac{a-b}{1-b\,c}\right)$.

(III) If $1/c < a < b$, then $\mathbf{u}(t,x)$ generically converges to either $(1,0)$ or $(0,a)$ as $t \to +\infty$. Furthermore, every spatially inhomogeneous positive stationary solution of (1.1) is unstable.

(IV) If $a > \max(b, 1/c)$, then $\lim_{t\to+\infty} \mathbf{u}(t,x) = (0,a)$.

We should note here that (1.1) has no spatially inhomogeneous positive stationary solutions for the cases (I), (II) and (IV). Furthermore, the bifurcation theory says that (1.1) has spatially inhomogeneous positive stationary solutions with small amplitude for a certain value of parameters such that the case (III) is satisfied. On the other hand, when $\alpha > 0$ and $\beta = 0$ are satisfied, Mimura et al. [**5**] and [**6**] employed the singular perturbation technique, and studied the bifurcation structure of positive stationary solutions of (1.1) for sufficiently small $d > 0$. Summarizing the above facts, we see that (1.1) has many spatially inhomogeneous positive stationary solutions for the case (III).

We now address the following problem. Suppose that all parameters are suitably fixed so that the case (III) is satisfied. How many positive stationary solutions does the model (1.1) have? To obtain an answer of the problem, we assume the following:

$$a = 1, \quad b = c > 1, \quad d = 1, \tag{A.1}$$

$$0 \le \alpha = \beta < \gamma(b) \equiv \min\left\{\frac{b-1}{2}, \frac{1}{b}\right\}. \tag{A.2}$$

The above assumption says that if $\mathbf{u}(x)$ is a stationary solution of (1.1), then $T\,\mathbf{u}(x)$ is also a stationary solution of (1.1), where

$$T = \begin{pmatrix} 0 & 1 \\ 1 & 0 \end{pmatrix}.$$

We set

$$X = \{\mathbf{u} \in C^2([0,1], \mathbf{R}^2) \mid \mathbf{u}_x(0) = 0 = \mathbf{u}_x(1)\}, \quad \bar{u} = \frac{1}{1+b},$$

$$\bar{\mathbf{u}} = (\bar{u}, \bar{u}), \quad \varepsilon_0 = +\infty, \quad \varepsilon_n = \frac{b-1}{n^2\,\pi^2\,(b+1)} \ (n \in \mathbf{N}),$$

and define the order relations $\preceq_s$ and $\preceq_o$ in the following manner:

$$(u_1, v_1) \preceq_s (u_2, v_2) \iff u_1 \le u_2, \ v_1 \le v_2,$$
$$(u_1, v_1) \preceq_o (u_2, v_2) \iff u_1 \le u_2, \ v_1 \ge v_2.$$

We denote by $\prec_s$ and $\prec_o$ the relations obtained from the above definition by replacing $\le$ with $<$. The following is the main result of this paper.

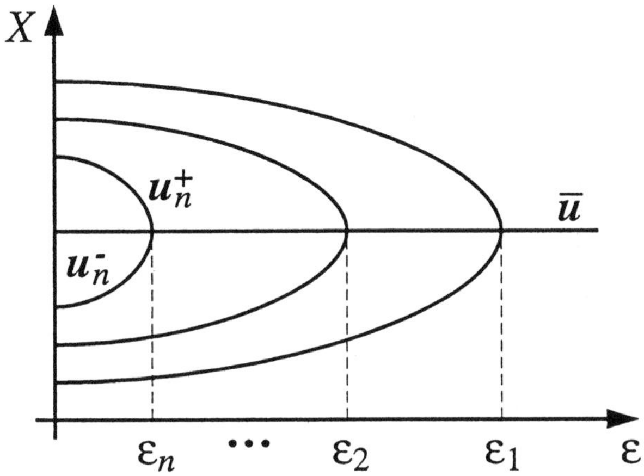


**Figure 1** Bifurcation structure of postive stationary solutions

**Theorem 1.1** *Under the assumptions* (A.1) *and* (A.2), *the following holds for each* $n \in \mathbf{N}$:

(i) *For any* $\varepsilon \in (0, \varepsilon_n)$, *there exists a pair of spatially inhomogeneous positive stationary solutions* $\mathbf{u}_n^{\pm}(x, \varepsilon)$ *of* (1.1) *such that*

$$(-1)^j\, \mathbf{u}_{nx}^{-}(x, \varepsilon) \prec_o 0 \prec_o (-1)^j\, \mathbf{u}_{nx}^{+}(x, \varepsilon)$$

*is satisfied for each* $0 \leq j < n$ *and* $x \in (j/n, (j+1)/n)$.

(ii) *Both* $\mathbf{u}_n^-(., \varepsilon)$ *and* $\mathbf{u}_n^+(., \varepsilon)$ *are* $C^1$*-class functions from* $(0, \varepsilon_n)$ *to* $X$.

(iii) *The model* (1.1) *has no positive stationary solutions other than* $\bar{\mathbf{u}}$ *and* $\mathbf{u}_j^{\pm}(x, \varepsilon)$ $(1 \leq j < n)$ *for any* $\varepsilon \in [\varepsilon_n, \varepsilon_{n-1})$.

The above theorem gives us information on the bifurcation structure of positive stationary solutions of (1.1), as shown in Figure 1. We note here that $\varepsilon = \varepsilon_n$ is a bifurcation point at which positive stationary solutions of (1.1) bifurcate from $\mathbf{u} = \bar{\mathbf{u}}$, and that the secondary bifurcation of stationary solutions never occurs.

To prove the above theorem, our strategy is stated as follows:

(i) Determine the geometrical position of the curve of positive stationary solutions of (1.1) bifurcating from $\mathbf{u} = \bar{\mathbf{u}}$ at $\varepsilon = \varepsilon_n$ for each $n \in \mathbf{N}$.

(ii) Study spatial profiles for an arbitrary positive stationary solution of (1.1) by using the comparison principle.

(iii) Show that the linearized operator of (1.1) around the spatially inhomogeneous positive stationary solution does not have the eigenvalue 0.

According to the strategy, we prove the theorem in the following sections.

## 2 Preliminary

Setting

$$\varphi(\mathbf{u}) = (\varphi_1, \varphi_2)(\mathbf{u}), \quad \varphi_1(\mathbf{u}) = (1 + \alpha\, v)\, u, \quad \varphi_2(\mathbf{u}) = (1 + \alpha\, u)\, v,$$

we easily have $\varphi(T\,\mathbf{u}) = T\,\varphi(\mathbf{u})$ for any $\mathbf{u}\in \mathbf{R}^2$. Since

$$\det \varphi_{\mathbf{u}}(\mathbf{u}) = \det \begin{pmatrix} 1+\alpha\, v & \alpha\, u \\ \alpha\, v & 1+\alpha\, u \end{pmatrix} = 1+\alpha\, u+\alpha\, v > 0$$

is satisfied for any $\mathbf{u} \succeq_s 0$, it is obvious that $\varphi(\mathbf{u})$ is bijective from the first quadrant to itself. We denote by $\Phi(\mathbf{w}) = (\Phi_1, \Phi_2)(\mathbf{w})$ the inverse function of $\varphi(\mathbf{u})$ on the first quadrant, where $\mathbf{w} = (w, z)$. By definition, we have

$$\Phi(T\,\mathbf{w}) = T\,\Phi(\mathbf{w}), \quad \Phi(\mathbf{w}) \preceq_s \mathbf{w} \quad \text{for any } \mathbf{w} \succeq_s 0. \tag{2.1}$$

By the change of variables $\mathbf{w} = \varphi(\mathbf{u})$, we can rewrite the stationary problem of (1.1) as

$$\begin{cases} 0 = \varepsilon\, \mathbf{w}_{xx} + \mathbf{f}(\mathbf{w}), & x \in (0,1), \\ \mathbf{w}_x = 0, & x = 0,1, \end{cases} \tag{2.2}$$

where $\mathbf{f}(\mathbf{w}) = (f_1, f_2)(\mathbf{w})$, $f_1(\mathbf{w}) = w\, f_1^0(\mathbf{w})$, $f_2(\mathbf{w}) = z\, f_2^0(\mathbf{w})$,

$$f_1^0(\mathbf{w}) = \frac{1-\Phi_1(\mathbf{w}) - b\,\Phi_2(\mathbf{w})}{1+\alpha\,\Phi_2(\mathbf{w})}, \quad f_2^0(\mathbf{w}) = \frac{1-b\,\Phi_1(\mathbf{w}) - \Phi_2(\mathbf{w})}{1+\alpha\,\Phi_1(\mathbf{w})}.$$

It is easy to check $f_2^0(\mathbf{w}) = f_1^0(T\,\mathbf{w})$ and $f_2(\mathbf{w}) = f_1(T\,\mathbf{w})$ for any $\mathbf{w} \succeq_s 0$. From $\mathbf{w} = \varphi(\mathbf{u})$, we obtain

$$\mathbf{u}_x = \frac{1}{1+\alpha\,\Phi_1(\mathbf{w}) + \alpha\,\Phi_2(\mathbf{w})} \begin{pmatrix} 1+\alpha\,\Phi_1(\mathbf{w}) & -\alpha\,\Phi_1(\mathbf{w}) \\ -\alpha\,\Phi_2(\mathbf{w}) & 1+\alpha\,\Phi_2(\mathbf{w}) \end{pmatrix} \mathbf{w}_x$$

for any $\mathbf{w} \succeq_s 0$, which implies that if a positive solution $\mathbf{w}(x)$ of (2.2) satisfies $\mathbf{w}_x(\tau) \succ_o 0$ (resp. $\mathbf{w}_x(\tau) \prec_o 0$), $\mathbf{u}(x) = \Phi(\mathbf{w}(x))$ is a positive stationary solution of (1.1) satisfying $\mathbf{u}_x(\tau) \succ_o 0$ (resp. $\mathbf{u}_x(\tau) \prec_o 0$). Hence we may prove the theorem for the problem (2.2) instead of the model (1.1).

After simple calculations, we obtain $f_1^0(0, 1/b) = 0$, $f_1^0(1,0) = 0$,

$$\begin{aligned} f_{1w}^0(\mathbf{w}) &= -\frac{1+\alpha\,(w+z) - \alpha\,(b+\alpha)\,\Phi_2(\mathbf{w})}{(1+\alpha\,\Phi_1(\mathbf{w}) + \alpha\,\Phi_2(\mathbf{w}))\,(1+\alpha\,\Phi_2(\mathbf{w}))^2}, \\ f_{1z}^0(\mathbf{w}) &= -\frac{b+\alpha-2\,\alpha\,\Phi_1(\mathbf{w})}{(1+\alpha\,\Phi_1(\mathbf{w}) + \alpha\,\Phi_2(\mathbf{w}))\,(1+\alpha\,\Phi_2(\mathbf{w}))}. \end{aligned} \tag{2.3}$$

Furthermore if $\mathbf{w}$ satisfies $\mathbf{w} \succeq_s 0$ and $f_1^0(\mathbf{w}) = 0$, then we have $f_{1z}^0(\mathbf{w}) < 0$ and

$$f_{1w}^0(\mathbf{w}) = -\frac{1-\alpha+2\,\alpha\,\Phi_1(\mathbf{w})}{(1+\alpha\,\Phi_1(\mathbf{w}) + \alpha\,\Phi_2(\mathbf{w}))\,(1+\alpha\,\Phi_2(\mathbf{w}))} < 0$$

because of (A.2) and $0 \le \Phi_1(\mathbf{w}) \le 1$. We see from the implicit function theorem that there exists a $C^1$-class function $p(z)$ defined on $[0, 1/b]$ such that $w = p(z)$ is a unique zero-level curve of $f_1^0(\mathbf{w})$ and satisfies $p(0) = 1$, $p(1/b) = 0$ and $0 \le p(z) \le 1$ for any $z \in [0, 1/b]$. By (A.2) and (2.1), we have

$$p_z(z) = \frac{2\,\alpha\,\Phi_1(p(z), z) - b - \alpha}{2\,\alpha\,\Phi_1(p(z), z) + 1 - \alpha} \le \frac{\alpha - b}{\alpha + 1} < -1$$

for any $z \in [0, 1/b]$, which implies

$$\{\,\mathbf{w}\,|\,\mathbf{w} \succ_s 0,\ f_1^0(\mathbf{w}) = 0\,\} \subset \Omega \equiv \{\,\mathbf{w}\,|\,\mathbf{w} \succ_s 0,\ 1/b < w+z < 1\,\}. \tag{2.4a}$$

From $f_2^0(\mathbf{w}) = f_1^0(T\,\mathbf{w})$ and $\Omega = T\,\Omega$, we also have

$$\{\,\mathbf{w}\,|\,\mathbf{w} \succ_s 0,\; f_2^0(\mathbf{w}) = 0\,\} \subset \Omega. \tag{2.4b}$$

By (2.4) and the comparison principle, we have the following lemma.

**Lemma 2.1** *Every positive solution* $\mathbf{w}(x)$ *of* (2.2) *satisfies* $\mathbf{w}(x) \in \Omega$ *for any* $x \in [0,1]$.

By the above lemma, we may restrict our discussion to the region $\Omega$. From (A.2), (2.1) and (2.3), we easily have $f_{1z}^0(\mathbf{w}) < 0$ and

$$1 + \alpha\,(w+z) - \alpha\,(b+\alpha)\,\Phi_2(\mathbf{w}) > 1 + \frac{\alpha}{b} - \alpha\,(b+\alpha) > 0$$

for any $\mathbf{w} \in \Omega$, which implies $f_{1\mathbf{w}}^0(\mathbf{w}) \prec_s 0$ for any $\mathbf{w} \in \Omega$. Since $f_{2\mathbf{w}}^0(\mathbf{w}) = f_{1\mathbf{w}}^0(T\,\mathbf{w})\,T$ and $\Omega = T\,\Omega$ hold, we have $f_{2\mathbf{w}}^0(\mathbf{w}) \prec_s 0$ for any $\mathbf{w} \in \Omega$.

**Lemma 2.2** $f_{1\mathbf{w}}^0(\mathbf{w}) \prec_s 0$ *and* $f_{2\mathbf{w}}^0(\mathbf{w}) \prec_s 0$ *are satisfied for any* $\mathbf{w} \in \Omega$.

## 3 Bifurcation from spatially homogeneous solution

In this section, we calculate spatially inhomogeneous positive solutions of (2.2) bifurcating from $\mathbf{w} = \varphi(\bar{\mathbf{u}})$ at $\varepsilon = \varepsilon_n$ for each $n \in \mathbf{N}$. To do it, we set

$$\hat{\varepsilon}_n = \varepsilon_n + \tilde{\varepsilon}_n(\mu)\,\mu^2, \quad \hat{\mathbf{w}}_n(x,\mu) = \varphi(\bar{\mathbf{u}}) + \mu\,\mathbf{e}\,\cos(n\,\pi\,x) + \mu^2\,\tilde{\mathbf{w}}_n(x,\mu), \tag{3.1}$$

where $\mathbf{e} = (1,-1)$. Substituting $\varepsilon = \hat{\varepsilon}_n(\mu)$ and $\mathbf{w} = \hat{\mathbf{w}}_n(x,\mu)$ into (2.2), we have

$$\tilde{\varepsilon}_n(0) = -\frac{\{(11\,b-5)\,(b+1) + 4\,\alpha\,(5\,b-3)\}\,(b-1)}{2\,\{(5\,b-3)\,(b+1) + 8\,\alpha\,(b-1)\}\,n^2\,\pi^2} < 0$$

because of $b > 1$ and $\alpha \geq 0$. We have the following lemma by the bifurcation theory.

**Lemma 3.1** ([1]) *For each* $n \in \mathbf{N}$, *there exist* $\delta_n > 0$, $\mu_n > 0$ *and* $C^1$*-class functions* $\hat{\varepsilon}_n(\mu)$, $\hat{\mathbf{w}}_n(.,\mu)$ *defined on* $(-\mu_n, \mu_n)$ *such that the following property holds:*

*(i)* $\hat{\mathbf{w}}_n(x,\mu)$ *is a spatially inhomogeneous positive solution of* (2.2) *with* $\varepsilon = \hat{\varepsilon}_n(\mu)$ *for each* $0 < |\,\mu\,| < \mu_n$.
*(ii)* $\hat{\varepsilon}_n(0) = \varepsilon_n$ *and* $\hat{\varepsilon}_n(\mu) < \varepsilon_n$ *for any* $0 < |\,\mu\,| < \mu_n$ *are satisfied.*
*(iii) Suppose that* $\mathbf{w}(x)$ *is a spatially inhomogeneous positive solution of* (2.2) *for* $\varepsilon = \eta$. *If* $|\,\eta - \varepsilon_n\,| < \delta_n$ *and* $\|\,\mathbf{w} - \varphi(\bar{\mathbf{u}})\,\|_X < \delta_n$ *hold, then there exists* $\rho \in (-\mu_n, \mu_n)$ *such that* $\eta = \hat{\varepsilon}_n(\rho)$ *and* $\mathbf{w}(.) = \hat{\mathbf{w}}_n(.,\rho)$ *are satisfied.*

## 4 Property of positive stationary solutions

Let $\mathbf{w}_0(x) = (w_0, z_0)(x)$ be an arbitrary spatially inhomogeneous positive solution of (2.2) for $\varepsilon > 0$. It is obvious from Lemma 3.1 that such a solution exists for some $\varepsilon > 0$. In this section, we study spatial profiles of $\mathbf{w}_0(x)$, which will be employed in order to discuss the possibility of the secondary bifurcation in the next section.

From Lemma 2.1, we obtain $\mathbf{w}_0(x) \in \Omega$ for any $x \in [0,1]$. By the boundary condition, we can regard $\mathbf{w}_0(x)$ as a periodic function with period 2 which satisfies $\mathbf{w}(x) = \mathbf{w}(-x)$ for any $x \in \mathbf{R}$. We set

$$\mathbf{u}_0(x)\,(= (u_0, v_0)(x)) = \Phi(\mathbf{w}_0(x)), \quad q_0(x) = u_0(x) + v_0(x).$$

It is obvious that $\mathbf{u}_0(x)$ is positive and spatially inhomogeneous. By (2.1), we have $0 < q_0(x) \le w_0(x) + z_0(x) < 1$ for any $x \in [0,1]$. Furthermore, we find that $h_0(x) \equiv u_0(x) - v_0(x)$ satisfies

$$\begin{cases} 0 = \varepsilon\, h_{xx} + (1 - q_0(x))\, h, & x \in (0,1), \\ h_x(0) = 0, \quad h_x(1) = 0. \end{cases} \tag{4.1}$$

If $h_0(x) \gtrless 0$ holds for any $x \in [0,1)$, then we have

$$0 = \int_0^1 (1 - q_0(x))\, h_0(x)\, dx \gtrless 0.$$

This contradiction implies that there exists $x_0 \in [0,1)$ such that $h_0(x_0) = 0$ and $h_0(x) \ne 0$ for any $x \in [0, x_0)$ are satisfied.

We consider the case $x_0 = 0$. We obtain $h_0(x) = 0$ for any $x \in \mathbf{R}$ by virtue of uniqueness, and then find that $u_0(x)$ satisfies

$$\begin{cases} 0 = \varepsilon\, [(1 + \alpha\, u)\, u]_{xx} + u\,\{1 - (1+b)\, u\}, & x \in (0,1), \\ u_x(0) = 0, \quad u_x(1) = 0. \end{cases} \tag{4.2}$$

In a similar manner with the above argument, we see from

$$\int_0^1 u_0(x)\,\{1 - (1+b)\, u_0(x)\}\, dx = 0$$

that there exists $x_1 \in [0,1)$ such that $u_0(x_1) = \bar{u}$ and $u_0(x) \ne \bar{u}$ for any $x \in [0, x_1)$ hold. By (4.2), we have

$$0 \le u_{0x}(x_1)^2 = -\frac{u_{0x}(x_1)}{\varepsilon\,(1 + 2\,\alpha\, u_0(x_1))} \int_0^{x_1} u_0(x)\,\{1 - (1+b)\, u_0(x)\}\, dx \le 0,$$

which implies $u_{0x}(x_1) = 0$. By uniqueness, we obtain $u_0(x) = \bar{u}$ for any $x \in \mathbf{R}$. This contradicts that $\mathbf{u}_0(x)$ is spatially inhomogeneous. Hence, we have $x_0 \in (0,1)$ and $h_{0x}(x_0) \ne 0$. If $\varepsilon > 4\,x_0^2/\pi^2$ is satisfied, then we have

$$\begin{aligned} 0 &= \int_0^{x_0} \{\varepsilon\, h_{0xx}(x) + (1 - q_0(x))\, h_0(x)\}\, \cos(\pi\, x/(2\, x_0))\, dx \\ &= \int_0^{x_0} \{1 - q_0(x) - \varepsilon\, \pi^2/(4\, x_0^2)\}\, h_0(x)\, \cos(\pi\, x/(2\, x_0))\, dx \ne 0, \end{aligned}$$

which implies that $\varepsilon$ must satisfy $\varepsilon \le 4\, x_0^2/\pi^2$.

Since $\mathbf{f}(T\,\mathbf{w}) = T\,\mathbf{f}(\mathbf{w})$ is satisfied for any $\mathbf{w} \succeq_s 0$, we see that

$$\mathbf{W}(x)\,(= (W, Z)(x)) = \mathbf{w}_0(x) - T\,\mathbf{w}_0(2\, x_0 - x)$$

is a periodic function with period 2, and that $\mathbf{W}(x)$ satisfies $\mathbf{W}(x_0) = 0$, $W_x(x_0) = Z_x(x_0)$ and

$$0 = \varepsilon\, \mathbf{W}_{xx} + A(x)\, \mathbf{W}, \quad x \in \mathbf{R},$$

where

$$A(x) = \int_0^1 \mathbf{f}_{\mathbf{w}}(\theta\, \mathbf{w}_0(x) + (1-\theta)\, T\, \mathbf{w}_0(2\,x_0 - x))\, d\theta.$$

We assume that $W_x(x_0) \neq 0$ holds. It follows from $\mathbf{W}(x_0+2) = 0$ that there exists $x_2 \in (x_0, x_0+2]$ such that $W(x_2)\, Z(x_2) = 0$ and $W(x)\, Z(x) > 0$ for any $x \in (x_0, x_2)$ hold. If $W(x_2) = 0$ is satisfied, then we have

$$\begin{aligned} 0 &= \int_{x_0}^{x_2} (\varepsilon\, w_{0xx}(x) + f_1(\mathbf{w}_0(x)))\, z_0(2\,x_0 - x)\, dx \\ &= \varepsilon\,(w_0(x_2)\, W_x(x_2) - w_0(x_0)\, W_x(x_0)) \\ &\quad - \int_{x_0}^{x_2} \int_0^1 f^0_{1\mathbf{w}}(\theta\, \mathbf{w}_0(x) + (1-\theta)\, T\, \mathbf{w}_0(2\,x_0 - x))\mathbf{W}(x) \\ &\qquad\qquad \times\, w_0(x)\, z_0(2\,x_0 - x)\, d\theta\, dx \neq 0 \end{aligned}$$

because of Lemma 2.2 and the convexity of $\Omega$. In a similar manner, we can derive a contradiction when $Z(x_2) = 0$ holds. Hence we have $\mathbf{W}_x(x_0) = 0$. By uniqueness, we obtain $\mathbf{w}_0(x) = T\,\mathbf{w}_0(2\,x_0 - x)$ for any $x \in \mathbf{R}$.

**Lemma 4.1** *Let $\mathbf{w}_0(x) = (w_0, z_0)(x)$ be an arbitrary spatially inhomogeneous positive solution of* (2.2) *for $\varepsilon > 0$. Then there exists a unique constant $x_0 \in (0, 1)$ such that $\mathbf{w}_0(x)$ satisfies $w_0(x) \neq z_0(x)$ for any $x \in [0, x_0)$ and*

$$\mathbf{w}_0(x) = \begin{cases} \mathbf{w}_0(r(x)) & \text{if } m(x) \equiv 0 \mod 4, \\ T\,\mathbf{w}_0(x_0 - r(x)) & \text{if } m(x) \equiv 1 \mod 4, \\ T\,\mathbf{w}_0(r(x)) & \text{if } m(x) \equiv 2 \mod 4, \\ \mathbf{w}_0(x_0 - r(x)) & \text{if } m(x) \equiv 3 \mod 4 \end{cases}$$

*for any $x \in \mathbf{R}$, where $m(x) \in \mathbf{Z}$ and $r(x) \in [0, x_0)$ are determined by $x = r(x) + m(x)\, x_0$.*

The linearized operator $\mathcal{L}$ of (2.2) around $\mathbf{w} = \mathbf{w}_0(x)$ can be represented as

$$\mathcal{L}\,\mathbf{w} = \varepsilon\,\mathbf{w}_{xx} + \mathbf{f}_{\mathbf{w}}(\mathbf{w}_0(x))\,\mathbf{w}.$$

Since $\mathbf{w}_{0x}(x)$ is a nontrivial solution of $\mathcal{L}\,\mathbf{w} = 0$ with $\mathbf{w}(0) = 0$, we obtain $\mathbf{w}_{0xx}(0) \neq 0$. We set

$$\begin{pmatrix} f_{11} & f_{12} \\ f_{21} & f_{22} \end{pmatrix}(x) = \mathbf{f}_{\mathbf{w}}(\mathbf{w}_0(x)), \qquad \begin{pmatrix} f^0_{11} & f^0_{12} \\ f^0_{21} & f^0_{22} \end{pmatrix}(x) = \frac{\partial(f^0_1, f^0_2)}{\partial \mathbf{w}}(\mathbf{w}_0(x)).$$

By Lemma 2.2 and the functional form of $\mathbf{f}(\mathbf{w})$, we obtain

$$\begin{gathered} f^0_{11}(x) < 0, \quad f^0_{12}(x) < 0, \quad f^0_{12}(x) < 0, \quad f^0_{22}(x) < 0, \\ f_{11}(x) = f^0_1(\mathbf{w}_0(x)) + w_0(x)\, f^0_{11}(x), \quad f_{12}(x) = w_0(x)\, f^0_{12}(x), \\ f_{21}(x) = z_0(x)\, f^0_{21}(x), \quad f_{22}(x) = f^0_2(\mathbf{w}_0(x)) + z_0(x)\, f^0_{22}(x) \end{gathered} \tag{4.3}$$

for any $x \in \mathbf{R}$.

**Lemma 4.2** *Let $(\phi_1, \phi_2)(x)$ be an arbitrary solution of $\mathcal{L}\,\mathbf{w} = 0$. Suppose that there exist $x_3^-$ and $x_3^+$ $(x_3^- < x_3^+)$ such that $\phi_1(x_3^\pm) = 0$ (resp. $\phi_2(x_3^\pm) = 0$) and $\phi_1(x) > 0$ (resp. $\phi_2(x) > 0$) for any $x \in (x_3^-, x_3^+)$ hold. Then $\phi_2(x) < 0$ (resp. $\phi_1(x) < 0$) is satisfied for some $x \in (x_3^-, x_3^+)$.*

**Proof** We only show the proof for the former case, because the latter can be proved in a similar manner. If $\phi_2(x) \geq 0$ holds for any $x \in (x_3^-, x_3^+)$, then we have

$$\begin{aligned} 0 &= \int_{x_3^-}^{x_3^+} (\varepsilon\,\phi_{1xx}(x) + f_{11}(x)\,\phi_1(x) + f_{12}(x)\,\phi_2(x))\,w_0(x)\,dx \\ &= \varepsilon\,(\phi_{1x}(x_3^+)\,w_0(x_3^+) - \phi_{1x}(x_3^-)\,w_0(x_3^-)) \\ &\qquad + \int_{x_3^-}^{x_3^+} (f_{11}^0(x)\,\phi_1(x) + f_{12}^0(x)\,\phi_2(x))\,w_0(x)^2\,dx < 0. \end{aligned}$$

This contradiction implies that the desired result holds. □

Since $\frac{d}{dx}(w_{0xx}\,z_{0xx})(0) = 0$ and

$$\varepsilon\,\frac{d^2}{dx^2}(w_{0xx}\,z_{0xx})(0) = -f_{12}(0)\,z_{0xx}(0)^2 - f_{21}(0)\,w_{0xx}(0)^2 > 0$$

hold if $w_{0xx}(0)\,z_{0xx}(0) = 0$ is satisfied, it follows from $\mathbf{w}_{0x}(2\,x_0) = 0$ that there exists $x_4 \in (0, 2\,x_0]$ such that $w_{0x}(x_4)\,z_{0x}(x_4) = 0$ and $w_{0x}(x)\,z_{0x}(x) \neq 0$ for any $x \in (0, x_4)$ hold. We consider the case $\mathbf{w}_{0x}(x_4) \neq 0$. It is obvious that

$$\mathbf{W}(x)\,(= (W, Z)(x)) = \mathbf{w}_0(x) - \mathbf{w}_0(2\,x_4 - x)$$

is a periodic function with period 2, and that $\mathbf{W}(x)$ satisfies $\mathbf{W}(x_4) = 0$, $\mathbf{W}_x(x_4) \neq 0$, $W_x(x_4)\,Z_x(x_4) = 0$, $\mathbf{W}_{xx}(x_4) = 0$ and

$$\begin{aligned} &\varepsilon\,W_{xxx}(x_4)\,Z_x(x_4) = -f_{12}(x_4)\,Z_x(x_4)^2 > 0 \quad \text{if } W_x(x_4) = 0, \\ &\varepsilon\,Z_{xxx}(x_4)\,W_x(x_4) = -f_{21}(x_4)\,W_x(x_4)^2 > 0 \quad \text{if } Z_x(x_4) = 0. \end{aligned}$$

Hence, we see that there exists $x_5 \in (x_4, x_4 + 2]$ such that $W(x_5)\,Z(x_5) = 0$ and $W(x)\,Z(x) > 0$ for any $x \in (x_4, x_5)$ hold. We can derive a contradiction in the same manner with the proof of Lemma 4.1. Therefore, we have $\mathbf{w}_{0x}(x_4) = 0$. Since $\mathbf{w}_0(x) = \mathbf{w}_0(2\,x_4 - x)$ holds for any $x \in \mathbf{R}$ because of uniqueness, we obtain $h_0(2\,x_4 - x_0) = 0$. By Lemma 4.1, we have either $x_4 = x_0$ or $x_4 = 2\,x_0$. If $x_4 = x_0$ holds, then we obtain $h_0(x) = 0$ for any $x \in \mathbf{R}$ because of $h_0(x_0) = 0$ and $h_{0x}(x_0) = 0$, which indicates that $\mathbf{w}_0(x)$ is a constant function. This contradiction implies that $x_4 = 2\,x_0$ holds. By Lemma 4.2 and the choice of $x_4$, we have $w_{0x}(x)\,z_{0x}(x) < 0$ for any $x \in (0, 2\,x_0)$.

**Lemma 4.3** *Let $\mathbf{w}_0(x) = (w_0, z_0)(x)$ be an arbitrary spatially inhomogeneous positive solution of* (2.2) *for $\varepsilon > 0$ and let $x_0$ be a unique constant given in Lemma* 4.1. *Then there exists $\ell \in \mathbf{N}$ such that $2\,x_0\,\ell = 1$ holds. Furthermore, $w_{0x}(x)\,z_{0x}(x) < 0$ is satisfied for any $j \in \mathbf{Z}$ and $x \in (j/\ell, (j+1)/\ell)$.*

For any spatially inhomogeneous positive solution $\mathbf{w}(x)$ of (2.2), we denote by $\ell(\mathbf{w})$ the integer which is given by applying Lemma 4.3 to $\mathbf{w}(x)$.

**Corollary 4.4** *Let $n \in \mathbf{N}$ and let $\hat{\mathbf{w}}_n(x, \mu) = (\hat{w}_n, \hat{z}_n)(x, \mu)$ be a $C^1$-class function given in Lemma* 3.1. *Then $\ell(\hat{\mathbf{w}}_n(., \mu)) = n$ holds for any $\mu$ ($\neq 0$) in a neighborhood of $\mu = 0$.*

**Proof** Since

$$(\hat{w}_n(x,\mu) - \hat{z}_n(x,\mu))/\mu = 2\cos(n\pi x) + O(\mu)$$

as $\mu \to 0$ by virtue of (3.1), we see from Lemmas 4.1 and 4.3 that the desired result holds. □

**Lemma 4.5** (Lemma 4.5 in [**3**]) *Let* $0 \le \varepsilon_- < \varepsilon_+$ *and let* $\mathbf{w}(.,\varepsilon)$ *be a continuous function on* $(\varepsilon_-,\varepsilon_+)$ *such that* $\mathbf{w}(x,\varepsilon)$ *is a spatially inhomogeneous positive solution of* (2.2) *for each* $\varepsilon \in (\varepsilon_-,\varepsilon_+)$. *Then* $\ell(\mathbf{w}(.,\varepsilon))$ *is a constant function on* $(\varepsilon_-,\varepsilon_+)$.

## 5 Nonexistence of the eigenvalue 0

Let $\mathbf{w}_0(x) = (w_0, z_0)(x)$ be an arbitrary spatially inhomogeneous positive solution of (2.2) for $\varepsilon > 0$ and let $\mathcal{L}$ be the linearized operator of (2.2) around $\mathbf{w} = \mathbf{w}_0(x)$. We denote by $\sigma(\mathcal{L})$ the set of spectra of $\mathcal{L}$ relative to $X$. In this section, we show that $0 \notin \sigma(\mathcal{L})$ holds.

We assume $0 \in \sigma(\mathcal{L})$ in order to derive a contradiction. Let $x_0 = 1/(2\,\ell(\mathbf{w}_0))$ and let $\phi(x) = (\phi_1, \phi_2)(x)$ be an arbitrary eigenfunction of $\mathcal{L}$ corresponding to the eigenvalue 0. We may assume $\phi_1(0) \ge 0$ without loss of generality. By Lemma 4.1 and the boundary condition, we can regard $\phi(x)$ as a periodic function with period 2 which satisfies $\phi(x) = \phi(-x)$ for any $x \in \mathbf{R}$. In the similar manner with the argument of Section 5 in [**3**], we have the following lemma.

**Lemma 5.1**

(i) $\mathcal{N}(\mathcal{L}) = \mathrm{span}\{\,\phi\,\}$ *is satisfied, where* $\mathcal{N}(\mathcal{L}) = \{\,\mathbf{w} \in X \mid \mathcal{L}\,\mathbf{w} = 0\,\}$.
(ii) $\phi(x) = T\,\phi(2\,x_0 - x)$ *holds for any* $x \in \mathbf{R}$.
(iii) *There exists* $\xi_1 \in (0, 2\,x_0)$ *such that* $\phi_1(x)\,(x - \xi_1) < 0$ *is satisfied for any* $x \in [0,\xi_1) \cup (\xi_1, 2\,x_0]$.

By the above lemma, we see that $\phi(x)$ is a nontrivial solution of

$$\begin{cases} \mathcal{L}\,\mathbf{w} = 0, & x \in (0, 2\,x_0), \\ \mathbf{w}_x(0) = 0, & \mathbf{w}_x(2\,x_0) = 0. \end{cases}$$

We may assume $h_0(x) > 0$ for any $x \in [0, x_0)$. (If not, we use the argument below by replacing $[0, 2\,x_0]$ and $\phi(x)$ with $[2\,x_0, 4\,x_0]$ and $-\phi(x)$, respectively.) By Lemma 4.1 and $h_{0x}(x_0) \ne 0$, we have $h_{0x}(x_0) < 0$ and $h_0(x) < 0$ for any $x \in (x_0, 2\,x_0]$. We set

$$\xi_2 = 2\,x_0 - \xi_1, \quad x_6^- = \min(\xi_1,\xi_2), \quad x_6^+ = \max(\xi_1,\xi_2),$$
$$K(x) = \phi_1(x) + \phi_2(x), \quad H(x) = \phi_1(x) - \phi_2(x).$$

Since $\phi_1(x) = \phi_2(2\,x_0 - x)$ holds for any $x \in \mathbf{R}$, we have

$$K(x) = K(2\,x_0 - x), \quad H(x) = -H(2\,x_0 - x)$$

for any $x \in \mathbf{R}$. By $\phi_2(0) < 0 < \phi_1(0)$, we obtain

$$H(x) \begin{cases} > 0 & \text{for any } x \in [0, x_6^-), \\ < 0 & \text{for any } x \in (x_6^+, 2\,x_0]. \end{cases}$$

Furthermore if $K(x_0) \ne 0$ holds, then we have $K(x_0)\,K(x) > 0$ for any $x \in [x_6^-, x_6^+]$.

Setting $k = u + v + 2\,\alpha\,u\,v$ and $h = u - v$, we have $k = w + z$ and $h = w - z$, where $(w, z) = \varphi(u, v)$. By

$$k = u + v + \frac{\alpha}{2}\,(u+v)^2 - \frac{\alpha}{2}\,h^2.$$

we obtain

$$u + v = Q(k,h) \equiv \frac{-1 + \sqrt{1 + 2\,\alpha\,k + \alpha^2\,h^2}}{\alpha}.$$

It is obvious that $q_0(x) = Q(k_0(x), h_0(x))$ is satisfied for any $x \in \mathbf{R}$, where $k_0(x) = w_0(x) + z_0(x)$. Since the stationary problem of (1.1) can be rewritten as

$$\begin{cases} 0 = \varepsilon\,k_{xx} + \left(1 - \dfrac{b+1}{2}\,Q(k,h)\right)\,Q(k,h) + \dfrac{b-1}{2}\,h^2, \\ 0 = \varepsilon\,h_{xx} + (1 - Q(k,h))\,h, \quad x \in (0,1), \\ k_x = 0, \quad h_x = 0, \quad x = 0, 1, \end{cases}$$

we see that $(K, H)(x)$ is a nontrivial solution of

$$\begin{cases} 0 = \varepsilon\,K_{xx} + a_{11}(x)\,K + a_{12}(x)\,H, \\ 0 = \varepsilon\,H_{xx} + a_{21}(x)\,K + a_{22}(x)\,H, \quad x \in (0,1), \\ K_x = 0, \quad H_x = 0, \quad x = 0, 1, \end{cases} \tag{5.1}$$

where $q_1(x) = Q_k(k_0(x), h_0(x))$,

$$a_{11}(x) = \{(\alpha + b + 1)\,q_1(x) - b - 1\}/\alpha, \quad a_{12}(x) = h_0(x)\,\{(\alpha + b + 1)\,q_1(x) - 2\},$$
$$a_{21}(x) = -q_1(x)\,h_0(x), \quad a_{22}(x) = -\alpha\,q_1(x)\,h_0(x)^2 + 1 - q_0(x).$$

By Lemma 2.1, we have $1/b \le k_0(x) \le 1$ and $|\,h_0(x)\,| \le 1$ for any $x \in \mathbf{R}$. Hence we obtain

$$\frac{1}{1+\alpha} \le q_1(x) = \frac{1}{\sqrt{1 + 2\,\alpha\,k_0(x) + \alpha^2\,h_0(x)^2}} < \sqrt{\frac{b}{b + 2\,\alpha}}$$

for any $x \in \mathbf{R}$. By (A.2), we obtain

$$\begin{aligned} a_{11}(x) &\le \frac{1}{\alpha}\left\{(b + \alpha + 1)\sqrt{\frac{b}{b+2\,\alpha}} - b - 1\right\} \\ &= \frac{b\,\alpha - 2\,b - 2}{(b + 2\,\alpha)\left\{(b + \alpha + 1)\sqrt{\dfrac{b}{b+2\,\alpha}} + b + 1\right\}} < 0, \\ \hat{a}_{12}(x) &\equiv \frac{a_{12}(x)}{h_0(x)} \ge \frac{\alpha + b + 1}{1 + \alpha} - 2 \ge \frac{b - \alpha - 1}{1 + \alpha} > 0 \end{aligned}$$

for any $x \in \mathbf{R}$.

First of all, we consider the case $K(x_0) > 0$. Suppose that $x_7 < 2\,x_0$ holds, where

$$x_7 = \sup\{\,y \in [x_0, 2\,x_0] \mid K(x) \ge 0 \text{ for any } x \in [x_0, y]\,\}\;(> x_6^+).$$

Setting

$$x_8 = \sup\{\,y \in [x_7, 2\,x_0] \mid K(x) \le 0 \text{ for any } x \in [x_7, y]\,\}\;(\le 2\,x_0),$$

we have $K_x(x_8) \geq 0$ because of $K_x(2\,x_0) = 0$. Since $K(x)$ satisfies $K(x_7) = 0$, $K_x(x_7) \leq 0$ and

$$\varepsilon\, K_{xx}(x) = -a_{11}(x)\, K(x) - \hat{a}_{12}(x)\, h_0(x)\, H(x) < 0 \text{ for any } x \in (x_7, x_8),$$

we obtain $x_7 < x_8$ and $K_x(x_8) < 0$. This contradiction implies $K(x) \geq 0$ ($\not\equiv 0$) for any $x \in [0, 2\,x_0]$. It follows from $H(x_0) = 0$ that there exists $x_9 \in (0, x_0]$ such that $H(x_9) = 0$ and $H(x) > 0$ for any $x \in [0, x_9)$ hold. Since $h_0(x)$ is a nontrivial solution of (4.1), we have

$$\begin{aligned} 0 &= \int_0^{x_9} \{\, \varepsilon\, h_{0xx}(x) + (1 - q_0(x))\, h_0(x)\, \}\, H(x)\, dx \\ &= -\, \varepsilon\, h_0(x_9)\, H_x(x_9) + \int_0^{x_9} q_1(x)\, h_0(x)^2\, (K(x) + \alpha\, h_0(x)\, H(x))\, dx > 0. \end{aligned}$$

This is a contradiction.

We consider next the case $K(x_0) < 0$. It is obvious that $K(x) < 0$ holds for any $x \in [x_6^-, x_6^+]$. If there exist $x_{10}^-$ and $x_{10}^+$ ($x_6^- \leq x_{10}^- < x_{10}^+ \leq x_0$) such that $H(x_{10}^{\pm}) = 0$ and $H(x) \leq 0$ for any $x \in [x_{10}^-, x_{10}^+]$ are satisfied, then we have

$$\begin{aligned} 0 &= \int_{x_{10}^-}^{x_{10}^+} \{\, \varepsilon\, h_{0xx}(x) + (1 - q_0(x))\, h_0(x)\, \}\, H(x)\, dx \\ &= \varepsilon\, (h_0(x_{10}^-)\, H_x(x_{10}^-) - h_0(x_{10}^+)\, H_x(x_{10}^+)) \\ &\qquad + \int_{x_{10}^-}^{x_{10}^+} q_1(x)\, h_0(x)^2\, (K(x) + \alpha\, h_0(x)\, H(x))\, dx < 0 \end{aligned}$$

because of $h_0(x_{10}^{\pm}) \geq 0$. This fact and $H(x_0) = 0$ show that $H(x) \geq 0$ holds for any $x \in [0, x_0]$. Since

$$\varepsilon\, K_{xx}(x) = -a_{11}(x)\, K(x) - \hat{a}_{12}(x)\, h_0(x)\, H(x) < 0$$

holds if both $K(x) < 0$ and $0 \leq x < x_0$ are satisfied, we obtain $K(x) < 0$ and $K_x(x) > 0$ for any $x \in [0, x_0)$ because of $K_x(x_0) = 0$. This contradicts $K_x(0) = 0$.

Finally we consider the case $K(x_0) = 0$. We have $x_6^- = x_6^+ = x_0$ because of $K(x_0) = 2\,\phi_1(x_0)$. It is easy to check $K(x_0) = 0$, $K_x(x_0) = 0$ and $H(x_0) = 0$. Since $(K, H)(x)$ is a nontrivial solution of (5.1), we obtain $H_x(x_0) < 0$, $K_{xx}(x_0) = 0$, $K_{xxx}(x_0) = 0$ and

$$\varepsilon\, K_{xxxx}(x_0) = -h_{0x}(x_0)\, \hat{a}_{12}(x_0)\, H_x(x_0) < 0.$$

We can derive a contradiction in a similar manner with the proof for the case $K(x_0) < 0$.

The above contradictions show that the following lemma holds.

**Lemma 5.2** $0 \notin \sigma(\mathcal{L})$ *holds.*

## 6 Proof of Theorem 1.1

Let $n \in \mathbf{N}$ be arbitrarily fixed. It follows from Lemmas 3.1, 4.5, 5.2 and Corollary 4.4 that there exists a maximal extended interval $J_n = (\eta_n, \varepsilon_n)$ such that the following are satisfied:

(i) There exists a pair of $C^1$-class functions $\mathbf{w}_n^{\pm}(.,\varepsilon)$ defined on $J_n$ such that the fact (i) in Theorem 1.1 hold for any $\varepsilon \in J_n$.
(ii) $\mathbf{w}_n^{\pm}(.,\varepsilon_n(\mu)) = \hat{\mathbf{w}}_n(.,\pm\mu)$ holds for any $0 \le \mu < \mu_n$.

By (2.2), we can easily obtain $\int_0^1 \mathbf{f}(\mathbf{w}_n^{\pm}(x,\varepsilon))\,dx = 0$ for each $\varepsilon \in J_n$. Since $\mathbf{w}_n^{\pm}(x,\varepsilon) \in \Omega$ holds for any $x$ and $\varepsilon$ because of Lemma 2.1, we see from the Ascoli-Arzela theorem that there exist solutions $\mathbf{w}_0^-(x)$ and $\mathbf{w}_0^+(x)$ of (2.2) which satisfy $\mathbf{w}_n^{\pm}(.,\varepsilon) \to \mathbf{w}_0^{\pm}$ as $\varepsilon \to \eta_n$ in $X$. Furthermore, we also have $\mathbf{w}_0^{\pm}(x) \in \overline{\Omega}$ for any $x \in [0,1]$.

We consider the case where there exists $x_{11} \in [0,1]$ such that $w_0^+(x_{11}) = 0$ holds, where $\mathbf{w}_0^+(x) = (w_0^+, z_0^+)(x)$. By uniqueness, we obtain $w_0^+(x) = 0$ for any $x \in [0,1]$. Since $z_0^+(x)$ is a nonnegative solution of

$$\begin{cases} 0 = \varepsilon\, z_{xx} + (1-z)\,z, & x \in (0,1), \\ z_x(0) = 0, \quad z_x(1) = 0, \end{cases}$$

we have $z_0^+(x) = 1$ for any $[0,1]$. By

$$f_1^0(\mathbf{w}_n^+(.,\varepsilon)) \to 1 - b\,(<0) \qquad \text{as } \varepsilon \to \eta_n \text{ in } X,$$

we have $\int_0^1 f_1(\mathbf{w}_n^+(.,\varepsilon))\,dx < 0$ in a neighborhood of $\varepsilon = \eta_n$. This contradiction implies that $w_0^+(x) > 0$ holds for any $x \in [0,1]$. In a similar manner, we can prove $z_0^+(x) > 0$ for any $x \in [0,1]$. Hence, we see that $\mathbf{w}_0^+(x)$ is positive. Analogously we can prove that $\mathbf{w}_0^-(x)$ is positive.

We consider the case $\eta_n > 0$. By Corollary 4.4 and Lemma 4.5, we have $\ell(\mathbf{w}_n^{\pm}(.,\varepsilon)) = n$ for any $\varepsilon \in J_n$. By Lemma 5.2 and the implicit function theorem, it follows that $\mathbf{w}_0^-(x)$ and/or $\mathbf{w}_0^+(x)$ must be spatially homogeneous. Lemma 3.1 leads to the fact that $\eta_n = \varepsilon_m$ holds for some $m > n$. By Corollary 4.4, we see that $\ell(\mathbf{w}_n^-(.,\varepsilon)) = m$ or $\ell(\mathbf{w}_n^+(.,\varepsilon)) = m$ holds in a neighborhood of $\varepsilon = \eta_n$. This is a contradiction. Hence, we have $\eta_n = 0$.

Let $\eta > 0$ and let $\mathbf{w}_0(x)$ be an arbitrary spatial inhomogeneous positive solution of (2.2) for $\varepsilon = \eta$. We find from Lemma 5.2 and the implicit function theorem that there exists a maximal extended interval $J = (\varepsilon_-, \varepsilon_+)$ such that the following hold,

(i) $\eta \in \operatorname{Int} J$ is satisfied.
(ii) There exists a $C^1$-class function $\mathbf{w}(.,\varepsilon)$ defined on $J$ such that $\mathbf{w}(x,\varepsilon)$ is a spatially inhomogeneous positive solution of (2.2) for each $\varepsilon \in J$ and satisfies $\mathbf{w}(.,\eta) = \mathbf{w}_0$.

By Lemma 4.5, we have $\ell(\mathbf{w}(.,\varepsilon)) = \ell(\mathbf{w}_0)$ for any $\varepsilon \in J$, which implies $\varepsilon_+ \le 1/(\pi^2\,\ell(\mathbf{w}_0)^2)$. It follows from Lemma 5.2 and the implicit function theorem that $\lim_{\varepsilon \to \varepsilon_+} \mathbf{w}(x,\varepsilon)$ must be spatially homogeneous. By Lemma 3.1, we have $\varepsilon_+ = \varepsilon_{\ell(\mathbf{w}_0)}$, and then obtain $\mathbf{w}_0 = \mathbf{w}^{\nu}_{\ell(\mathbf{w}_0)}(.,\eta)$ for some $\nu \in \{+,-\}$. Thus the proof is completed. $\square$

## References

[1] Crandall, M. G. and Rabinowitz, P. H. [1971] *Bifurcation from simple eigenvalues,* J. Funct. Anal. **8**, 321–340.

[2] de Mottoni, P. [1979] *Qualitative analysis for some quasi-linear parabolic systems,* Inst. Math. Pol. Acad. Sci. Zam. 11/70 **190**.

[3] Kan-on, Y. [1998] *Bifurcation structure of stationary solutions of a Lotka-Volterra competition model with diffusion,* SIAM J. Math. Anal. **29**, 424-426.

[4] Kishimoto, K. and Weinberger, H. F. [1985] *The spatial homogeneity of stable equilibria of some reaction-diffusion systems on convex domains,* J. Differential Equations **58**, 15–21.

[5] Mimura, M. [1982] *Aggregation and segregation phenomena in reaction-diffusion equations,* Lecture Notes in Biomath. **45**, 71–82.

[6] Mimura, M., Nishiura, Y., Tesei, A. and Tsujikawa, T. [1984] *Coexistence problem for two competing species models with density-dependent diffusion,* Hiroshima Math. J. **14**, 425–449.

[7] Shigesada, N., Kawasaki, K. and Teramoto, E. [1979] *Spatial segregation of interacting species,* J. theoret. Biol. **79**, 83–99.

Fields Institute Communications
Volume **21**, 1999

# Stability Domain of a Chemostat System with Delay

**Junji Kato**
Mathematical Institute
Tohoku University
Sendai 981-77, Japan

**Jiaqi Pan**
Department of Mathematics
Northeast Normal University
Changchun, Jilin, China

**Abstract.** In this paper, we consider a chemostat model with a discrete delay proposed by Thingstad and Langeland (J. Theor. Biol. **48**, (1974), 149-159). Stability domain of the positive equilibrium point is determined by analyzing the corresponding characteristic equation.

The system

$$\begin{aligned}\frac{dS(\tau)}{d\tau} &= (S^0 - S(\tau))D - \frac{\mu M S(\tau)}{A + S(\tau)} X(\tau) \\ \frac{dX(\tau)}{d\tau} &= -DX(\tau) + \frac{MS(\tau - T)}{A + S(\tau - T)} X(\tau)\end{aligned} \tag{1.1}$$

is a mathematical model of a simple chemostat with delay, which is one of the most important objects in mathematical biology.

Here $\mu$, $A$ and $M$ are determinate positive constants associated with the species, while $S^0$ and $D$ are controllable positive parameters. For the model refer to [**1**], [**5**] and [**2**].

By the scaling

$$s = \frac{S}{S^0}, \quad x = \frac{\mu X}{S^0}, \quad t = D\tau, \tag{1.2}$$

the system (1.1) will be transformed into

$$\begin{aligned}\dot{s}(t) &= 1 - s(t) - p(s(t))x(t) \\ \dot{x}(t) &= -x(t) + p(s(t - r))x(t),\end{aligned} \tag{1.3}$$

---

1991 *Mathematics Subject Classification.* Primary 34K20; Secondary 92B05.

where

$$p(s) := \frac{ms}{a+s}, \tag{1.4}$$

or, in general, $p$ is a $C^2$-function with the properties $p(0) = 0, \dot{p}(s) > 0$ for $s > 0$ and

$$m = \frac{M}{D}, \quad a = \frac{A}{S^0}, \quad r = DT. \tag{1.5}$$

The system (1.3) defined on the first quadrant is also called a chemostat system and has been extensively studied by many authors, for example, refer to [**3**] and [**7**].

The system (1.3) has a unique interior equilibrium point $E(s_0, x_0)$ if, and only if, $p(1) > 1$ where $s_0 > 0$ is a unique solution of $p(s) = 1$ and $x_0 := 1 - s_0 > 0$, and the corresponding characteristic equation is

$$\lambda^2 + (1+\beta)\lambda + \beta e^{-r\lambda} = 0 \tag{1.6}$$

where $\beta := \dot{p}(s_0)(1 - s_0) > 0$.

The stability and the bifurcation problem is closely related with the location of the pure imaginary roots of (1.6).

Here, $\lambda = i\omega, \omega > 0$, is a solution of (1.6) if, and only if,

$$\omega^2 + \cos[r\omega] - \omega \sin[r\omega] = 0, \tag{1.7}$$

and

$$\beta = \frac{\omega^2}{\cos[r\omega]}. \tag{1.8}$$

In [**3**] and [**7**] it has been shown that for any given $\beta > 0$ there exists a series of numbers $0 < r_0(\beta) < r_1(\beta) < r_2(\beta) < \cdots$ such that:

(i) (1.6) has a pair of pure imaginary roots $\{i\omega_k, -i\omega_k\}$ if $r = r_k(\beta)$;
(ii) (1.3) is asymptotically stable for $r \in (0, r_0(\beta))$ and unstable for $r > r_0(\beta)$;
(iii) The system (1.3) has a Hopf bifurcation at $r = r_k(\beta)$.

Returning to the system (1.1), the above results are expected to be translated for (1.1). However, considering the relation (1.5) both of the corresponding parameters $r$ and $\beta$ depend on the parameter $D$. In fact, $r = D\tau$ and

$$\beta := \frac{m-1-a}{ma}(m-1) = \frac{S^0(M-D)^2 - AD(M-D)}{MAD}. \tag{1.9}$$

This fact shows that the translation mentioned above is not effective for the system (1.1).

In order to see the bifurcation diagram of (1.1) the second author, Pan in [**4**] has analysed this for the Equation (1.3) completely. More precisely he has shown:

(i) Since $\beta > 0$, for each non-negative integer $k$ the equation (1.7) has a solution $\omega_k(r)$, which is analytic on $(2k\pi + \frac{\pi}{2}, \infty)$ and satisfies $\omega_k(r)r \in (2k\pi, 2k\pi + \frac{\pi}{2})$;

(ii) Define the function $\beta_k(r)$ by (1.8) with $\omega = \omega_k(r)$. Then, it is analytic on $(2k\pi + \frac{\pi}{2}, \infty)$ and satisfies

$$\lim_{r\to\infty} \beta_k(r) = 0, \qquad \lim_{r\to 2k\pi+\frac{\pi}{2}} \beta_k(r) = \infty,$$

$$\beta_k(r) < \beta_{k+1}(r) \quad \text{on the common domain.}$$

Moreover, the interior equilibrium point $E$ is asymptotically stable when $r < \frac{\pi}{2}$ or $\beta < \beta_0(r)$ and unstable when $\beta > \beta_0(r)$, and (1.3) has a Hopf bifurcation at $\beta = \beta_k(r)$ for $r > 2k\pi + \frac{\pi}{2}$ and any $k \geq 0$;

(iii) Returning to the system (1.1) the relation (1.9) implies

$$S^0 = \frac{MAD\beta_k(DT) + AD(M-D)}{(M-D)^2} =: f_k(D),$$

which is defined on $(\frac{4k+1}{2T}\pi, M)$ and satisfies

$$\lim_{D\to M} f_k(D) = \infty, \qquad \lim_{D\to\frac{4k+1}{2T}\pi} f_k(D) = \infty,$$

$$f_k(D) < f_{k+1}(D) \quad \text{on the common domain.}$$

And the interior equilibrium point $E(\frac{AD}{M-D}, \frac{\mu(S^0(M-D)-AD)}{M-D})$ exists only when

$$\Sigma : M - D > 0 \text{ and } S^0 > \frac{AD}{M-D},$$

and it is asymptotically stable when $D < \frac{\pi}{2T}$ or $S^0 < f_0(D)$ and unstable when $S^0 > f_0(D)$ on $\Sigma$. Moreover, (1.1) has a Hopf bifurcation at $S^0 = f_k(D)$ for $D \in (\frac{4k+1}{2T}\pi, M)$ and any $k \geq 0$.

This result fairly matches with the graph given in [**6**] by a numerical analysis. However, the chemostat system proposed in [**6**]

$$\begin{aligned} \frac{dS(\tau)}{d\tau} &= (S^0 - S(\tau))D - \frac{\mu MS(\tau)}{A+S(\tau)}X(\tau) \\ \frac{dX(\tau)}{d\tau} &= -DX(\tau) + \frac{MS(\tau-T)}{\theta A + S(\tau - T)}X(\tau) \end{aligned} \tag{1.10}$$

with $\theta > 0$ is slightly different from the system (1.1), which corresponds to the case where $\theta = 1$ in (1.10), but their graph suggests that the behavior will not be essentially different from that for (1.1). A value of $\theta$ different from unity corresponds to a yield in cell number varying with growth rate, though we omit giving the meaning of the parameters.

Here, by the arguments similar to those in [**4**] we shall confirm this about the stability problem.

By the same scaling (1.2) the system (1.10) will be transformed into

$$\begin{aligned} \dot{s}(t) &= 1 - s(t) - p(s(t))x(t) \\ \dot{x}(t) &= -x(t) + q(s(t-\tau))x(t), \end{aligned} \tag{1.11}$$

where $p(t)$ is the one given in (1.4) with the same notations (1.5), while

$$q(s) := \frac{ms}{\theta a + s},$$

and hence the interior equilibrium point $E(s_0, x_0)$ is given by

$$s_0 := \frac{\theta a}{m-1}, \quad x_0 := \frac{1-s_0}{p(s_0)}$$

under the condition $m > 1$ and $m - 1 - \theta a > 0$ for (1.11), or

$$M > D, \quad S^0 > \frac{\theta AD}{M-D} \tag{1.12}$$

for (1.10), and the corresponding characteristic equation is

$$\lambda^2 + (1+\beta)\lambda + \gamma e^{-\lambda r} = 0, \tag{1.13}$$

where

$$\begin{aligned}\beta : &= \frac{\dot{p}(s_0)}{p(s_0)}(1-s_0) = \frac{(m-1)^2 - \theta a(m-1)}{ma} \\ &= \frac{S_0(M-D)^2 - \theta AD(M-D)}{MDA}, \\ \gamma : &= \frac{\dot{q}(s_0)}{q(s_0)}(1-s_0) = K\beta\end{aligned}$$

with

$$K := \frac{p(s_0)\dot{q}(s_0)}{\dot{p}(s_0)} = \frac{m-1+\theta}{m} = \frac{M-(1-\theta)D}{M}. \tag{1.14}$$

Therefore, the corresponding Equations (1.7) and (1.8) become

$$\omega^2 + K\cos[r\omega] - K\omega\sin[r\omega] = 0 \tag{1.15}$$

and

$$\beta = \frac{\omega^2}{K\cos[r\omega]}, \tag{1.16}$$

respectively, and the arguments similar to those in [**4**] may reach analogous results on the parametric analysis for the system (1.11). However, in this case the solution $\omega_k(r)$ of (1.15) and the function $\beta_k(r)$ defined by (1.16) with $\omega = \omega_k(r)$ must be depending on the parameter $K$, and this can not be adequate to provide a parametric analysis in the $(S^0, D)$-space for the system (1.10) because $K$ given in (1.14) also depends on $D$.

Now, we shall rewrite the relation (1.14) into

$$K = 1 - \frac{1-\theta}{M}D = 1 - \frac{1-\theta}{MT}DT = 1 + \epsilon r$$

by setting $\epsilon := \frac{\theta-1}{MT}$, and hence the equations (1.13), (1.15) and (1.16) become

$$\lambda^2 + (1+\beta)\lambda + (1+\epsilon r)\beta e^{-\lambda r} = 0, \tag{1.17}$$

$$\omega^2 + (1+\epsilon r)\cos[r\omega] - (1+\epsilon r)\omega\sin[r\omega] = 0 \tag{1.18}$$

and

$$\beta = \frac{\omega^2}{(1+\epsilon r)\cos[r\omega]}, \tag{1.19}$$

respectively, where we note that $r$ ranges over the set $J$ defined as

$$J := (0,\infty) \text{ if } \epsilon \geq 0 \text{ and } J := (0, -\frac{1}{\epsilon}) \text{ if } \epsilon < 0 \tag{1.20}$$

because $r = DT < MT < \frac{MT}{1-\theta} = -\frac{1}{\epsilon}$ if $0 < \theta < 1$.

Consider the function

$$g(\xi, r) := \xi^2 + r^2(1+\epsilon r)\cos\xi - r(1+\epsilon r)\xi\sin\xi$$

on $I \times J$, where $I := \cup_{k=0}^{\infty} I_k$ with $I_k := (2k\pi, 2k\pi + \frac{\pi}{2})$. Here, obviously $\omega(r)$ is a solution of (1.18) if, and only if, $\xi := r\omega(r)$ is a zero of $g(\xi, r)$.

Modifying the results in [**4**] we can proceed with the following:

The following lemma is similar to the Lemma 2.1 in [**4**] by noting

$$(1+\epsilon r)r\frac{\sin\xi}{\xi} > 1 \text{ if } g(\xi, r) = 0. \tag{1.21}$$

**Lemma 1.1** *If $\xi^*$ is a zero of $g(\xi, r)$, then $g_\xi(\xi^*, r) < 0$ for $r \geq 1$.*

Put

$$\sigma(\epsilon) := \frac{2}{1+\sqrt{1+4\epsilon}}, \quad \rho(\epsilon) := \frac{\pi}{1+\sqrt{1+2\epsilon\pi}},$$

$$\rho_k^*(\epsilon) := \frac{\pi}{1-\sqrt{1+2\epsilon\pi}} \text{ if } \epsilon < 0, \ \rho_k^*(\epsilon) \ := \infty \text{ if } \epsilon \geq 0.$$

In the following $\sigma(\epsilon), \rho(\epsilon)$ and $\rho^*(\epsilon)$ will be simply written as $\sigma, \rho$ and $\rho^*$, respectively. Here, note that $\sigma$ is a solution of $r(1+\epsilon r) = 1$ while $\rho$ and $\rho^*$ are solutions of $r(1+\epsilon r) = \frac{\pi}{2}$.

The following propositions correspond to Propositions 2.3 and 2.4 in [**4**].

**Proposition 1.2** *There exists a positive constant $\epsilon_+$ such that: Suppose that*

$$-\frac{1}{2\pi} < \epsilon < \epsilon_+. \tag{1.22}$$

(i) *If $0 < r \leq \rho$ or $r \geq \rho^*$, then $g(\xi, r)$ has no zero on $I$;*
(ii) *For each $r \in (\rho, \rho^*)$, $g(\xi, r)$ has a unique zero $\xi(r, \epsilon)$, or simply $\xi(r)$, in $I_0 = (0, \frac{\pi}{2})$;*
(iii) *The function $\xi(r)$ is an analytic function on the maximal domain $(\rho, \rho^*)$, and $\omega(r) := \frac{1}{r}\xi(r)$ satisfies $\dot{\omega}(r) < 0$ there. Hence, $\xi \to \frac{\pi}{2}$ as $r \to \rho$ or as $r \to \rho^*$ when $\rho^* < \infty$.*

**Proof** First of all, note that

$$g(0, r) = r^2(1+\epsilon r) > 0,$$

$$\begin{aligned} g(\xi, r) &\geq \xi^2(1 - r(1+\epsilon r)\frac{\sin\xi}{\xi}) + r^2(1+\epsilon r)\cos\xi \\ &> \xi^2(1 - r(1+\epsilon r)) \end{aligned}$$

and

$$g\left(\frac{\pi}{2}, r\right) = \left(\frac{\pi}{2}\right)^2 \left(1 - \frac{2r(1+\epsilon r)}{\pi}\right).$$

Then, by the arguments similar to those in [**4**] the conclusions (i) and (ii) follow from Lemma 1.1 if $\rho > 1$ which follows from $\epsilon < \frac{\pi}{2} - 1$ and if $r \leq \sigma$ or $r \geq 1$ in (i).

Consider the case where $\sigma < r < 1$, which implies $\epsilon > 0$. Then, we have

$$r^2(1+\epsilon r) > \sigma \text{ and } r(1+\epsilon r) < 1+\epsilon,$$

hence,

$$g(\xi, r) > \xi^2 + \sigma\cos\xi - (1+\epsilon)\xi\sin\xi =: h(\xi).$$

Let $\xi_0 \in (1, \frac{\pi}{2})$ be fixed. Then, we can see that if $0 < \epsilon < \xi_0 - 1$ and if $\xi \geq \xi_0$, then

$$h(\xi) \geq \xi^2 - (1+\epsilon)\xi \geq \xi(\xi_0 - (1+\epsilon)) > 0.$$

Choose $\epsilon_+ \in (0, \xi_0 - 1)$ so that

$$\frac{\sigma}{\epsilon} = \frac{2}{\epsilon(1+\sqrt{1+4\epsilon})} > \xi_0\tan\xi_0 \text{ for } 0 < \epsilon < \epsilon_+.$$

Then, for any $\epsilon \in (0, \epsilon_+)$ and $\xi \leq \xi_0$,

$$h(\xi) = \xi^2\left(1 - \frac{\sin\xi}{\xi}\right) + \epsilon\cos\xi\left(\frac{\sigma}{\epsilon} - \xi\tan\xi\right) > 0.$$

Thus, we can conclude that if $\epsilon \in (0, \epsilon_+)$, then $g(\xi, r) > 0$ for any $r \in (\sigma, 1)$ and $\xi \in I$.

To show (iii), put $\omega(r) := \frac{\xi(r)}{r}$, where $\xi(r)$ is the solution of $g(\xi, r) = 0$ such that $\xi(r) \in I_0$. Then,

$$\begin{aligned}
\dot{\omega}(r) &= \frac{1}{r^2}(\dot{\xi}r - \xi) \\
&= -\frac{1}{r^2 g_\xi}\left[(2+3\epsilon r)r^2\cos\xi - (1+2\epsilon r)r\xi\sin\xi + \xi g_\xi\right] \\
&= -\frac{1}{r^2 g_\xi}\left[\frac{2+3\epsilon r}{1+\epsilon r}\{(1+\epsilon r)r\xi\sin\xi - \xi^2\} - (1+2\epsilon r)r\xi\sin\xi\right. \\
&\quad \left. +2\xi^2 - (1+\epsilon r)(r^2+r)\xi\sin\xi - (1+\epsilon r)r\xi^2\cos\xi\right] \\
&= \frac{\xi^2}{rg_\xi}\left[\frac{\epsilon}{1+\epsilon r} + (1+\epsilon r)r\frac{\sin\xi}{\xi} + (1+\epsilon r)\cos\xi\right] \\
&< \frac{\xi^2}{rg_\xi}\left(\frac{\epsilon}{1+\epsilon r} + 1\right) = \frac{\xi^2}{rg_\xi(1+\epsilon r)}(\epsilon + 1 + \epsilon r)
\end{aligned}$$

by (1.21) and the fact $g_\xi < 0$.

Obviously $\dot{\omega}(r) < 0$ if $\epsilon \geq 0$. If $\epsilon < 0$ and if $r \in (\rho, \rho^*)$, then

$$\epsilon + 1 + \epsilon r > \epsilon + 1 + \epsilon\rho^* = \epsilon + \frac{1 - \sqrt{1+2\epsilon\pi}}{2}.$$

Therefore, $\dot{\omega}(r) < 0$ for $r \in (\rho, \rho^*)$, when $\epsilon > -\frac{1}{2\pi}$.

Finally, by the monotonicity of $\omega(r)$, $\xi(r)$ must be convergent to a $\xi^* \in \overline{I_0} = [0, \frac{\pi}{2}]$ as $r \to \rho$ or as $r \to \rho^*$ when $\rho^* < \infty$. Since $\xi(r)$ cannot be extended beyond $r = \rho$ and $r = \rho^*$ (if $\rho^* < \infty$), $\xi^* = 0$ or $\xi^* = \frac{\pi}{2}$. However, it is impossible that $g(0, r) = 0$ for $r = \rho$ or $\rho^*$, hence, $\xi^* = \frac{\pi}{2}$. □

**Proposition 1.3** *Let $\xi(r)$ be the function given in Proposition 1.2, and let $\beta(r, \epsilon)$ or simply $\beta(r)$ be defined by*

$$\beta(r) := \frac{\xi^2(r)}{(1+\epsilon r)r^2\cos\xi(r)}$$

*according to (1.19). Then:*

(i) *The characteristic Equation (1.17) has a pair of pure imaginary roots $\lambda = \pm i\frac{\xi(r)}{r}, \xi(r) \in I_0$ if, and only if, $\beta = \beta(r)$ for $r \in (\rho, \rho^*)$;*

(ii) *For the branch $\lambda = \lambda(r, \beta)$ of the root of (1.17) through $ir\xi(r)$ at $\beta = \beta(r)$ we have $Re\frac{\partial\lambda}{\partial\beta}\,|_{\beta=\beta(r)} > 0$ for $r \in (\rho, \rho^*)$;*

(iii) *$\beta(r)$ is an analytic function defined on $(\rho, \rho^*)$ satisfying*

$$\lim_{r\to\rho}\beta(r) = \infty, \quad \lim_{r\to\rho^*}\beta(r) = \infty \text{ if } \rho^* < \infty.$$

**Proof** Consider the characteristic Equation (1.17)

$$h(\lambda, \beta, r) := \lambda^2 + (1+\beta)\lambda + \beta(1+\epsilon r)e^{-\lambda r} = 0.$$

Then, noting that

$$\lambda = i\frac{\xi}{r} \text{ at } \beta = \beta(r) = \frac{\xi^2}{(1+\epsilon r)r^2\cos\xi} \text{ for } \xi = \xi(r)$$

and

$$\xi^2 + (1+\epsilon r)r^2\cos\xi - (1+\epsilon r)r\xi\sin\xi = 0,$$

we have

$$\begin{aligned} h_\lambda &= 2\lambda + (1+\beta) - \beta(1+\epsilon r)re^{-\lambda r} \\ &= 2i\frac{\xi}{r} + \frac{(1+\epsilon r)r^2\cos\xi + \xi^2}{(1+\epsilon r)r^2\cos\xi} - \frac{\xi^2}{r\cos\xi}(\cos\xi - i\sin\xi) \\ &= \frac{\xi}{r}[\tan\xi - \xi + i(2+\xi\tan\xi)] \end{aligned}$$

and

$$\begin{aligned} h_\beta &= \lambda + (1+\epsilon r)e^{-\lambda r} \\ &= i\frac{\xi}{r} + (1+\epsilon r)(\cos\xi - i\sin\xi) \\ &= (1+\epsilon r)\cos\xi + i\frac{1}{\xi r}(\xi^2 - (1+\epsilon r)r\xi\sin\xi) \\ &= \frac{1}{\xi}(1+\epsilon r)\cos\xi\cdot(\xi - ir). \end{aligned}$$

Therefore,

$$\begin{aligned} Re\frac{\partial\lambda}{\partial\beta}\,|_{\beta=\beta(r)} &= Re(-\lambda\frac{h_\beta}{h_\lambda}) \\ &= -\frac{r}{\xi^2}Re\frac{(1+\epsilon r)\cos\xi\cdot(\xi - ir)}{\tan\xi - \xi + i(2+\xi\tan\xi)} \\ &= -\frac{r}{\xi^2}(1+\epsilon r)\cos\xi\cdot\frac{\xi\tan\xi - \xi^2 - 2r - r\xi\tan\xi}{(\tan\xi-\xi)^2 + (2+\xi\tan\xi)^2} \\ &= \frac{r}{\xi^2}(1+\epsilon r)\cos\xi\cdot\frac{\xi^2 + 2r + \xi(r-1)\tan\xi}{(\tan\xi-\xi)^2 + (2+\xi\tan\xi)^2} > 0 \end{aligned}$$

since $r > \rho > 1$.

Since $\xi(r) \to \frac{\pi}{2}$ as $r \to \rho$ or as $r \to \rho^*$ when $\rho^* < \infty$, we can conclude that

$$\beta(r) \to \infty \textbf{ as } r \to \rho \text{ or as } r \to \rho^* \text{ when } \rho^* < \infty.$$

□

By the same argument in [**4**] we can see that the characteristic Equation (1.17) has no roots with positive real parts for $\beta < \beta(r)$. This fact together with (i) shows that:

**Theorem 1.4** *Under the condition (1.22) the interior equilibrium point $E$ of system (1.11) is asymptotically stable if $\beta > 0$ and $r \in J\backslash(\rho, \rho^*)$, $J$ being defined by (1.20), or if $0 < \beta < \beta(r)$ when $r \in (\rho, \rho^*)$, and it is unstable if $\beta > \beta(r)$ and $r \in (\rho, \rho^*)$.*

Now back to the Equation (1.10). Then, the equilibrium point $E$ becomes $E(S^*, X^*)$, where

$$S^* = \frac{\theta DA}{M-D},\quad X^* = \frac{(S^0(M-D) - \theta DA)(M + (\theta-1)D)}{M\theta(M-D)},$$

which exists only when

$$(S^0, D) \in \Omega := \{(S^0, D) : S^0 > \frac{\theta DA}{M-D},\ 0 < D < M\}.$$

Therefore, by the transformation

$$\tau = \frac{t}{D},\quad S = S^* + S^0 s,\quad X = X^* + \frac{1}{\mu}S^0 x$$

the $E(S^*, x^*)$ is reduced into the origin in the $(s, x)$-space and the corresponding characteristic equation is (1.17) with

$$r = DT, \quad \epsilon = \frac{\theta - 1}{MT} \tag{1.23}$$

and

$$\beta = \frac{S^0(M-D)^2 - \theta AD(M-D)}{\theta AD(M + (\theta - 1)D)}. \tag{1.24}$$

Set

$$f(D) := \frac{\theta AD}{(M-D)^2}\{(M + (\theta-1)D)\beta\left(DT, \frac{\theta-1}{MT}\right) + M - D\},$$

$$\Delta := \frac{1}{T}\rho\left(\frac{\theta-1}{MT}\right) = \frac{\pi\sqrt{M}}{T\sqrt{M} + \sqrt{MT^2 + 2\pi T(\theta-1)}},$$

and

$$\Delta^* := \frac{1}{T}\rho^*(\frac{\theta-1}{MT}) \text{ if } \theta < 1 \text{ and } \Delta := \infty \text{ if } \theta \geq 1.$$

Here, note that under rewriting $r$ and $\epsilon$ by (1.23) $f(D)$ is the solution $S^0$ of (1.24) with $\beta = \beta(r, \epsilon)$, $\Delta$ and $\Delta^*$ are the solution $D$ of $DT = r = \rho(\epsilon)$ and $DT = r = \rho^*(\epsilon)$.

The following proposition immediately follows from Proposition 1.3.

**Proposition 1.5** *Suppose that $M, T, A, \theta$ are positive constants and satisfy*

$$1 - \frac{MT}{2\pi} < \theta < 1 + \epsilon_+ MT \tag{1.25}$$

*with $\epsilon_+$ given in Proposition 1.2, which corresponds to (1.22) under the relation (1.23). Then:*

(i) *The function $f(D)$ is an analytic function defined on $(\Delta, \Delta^*) \cap (0, M)$ and satisfies*

$$\lim_{D\to\Delta} f(D) = \infty \text{ and } \lim_{D\to\Delta^{**}} f(D) = \infty \text{ for } \Delta^{**} = \inf\{\Delta^*, M\};$$

(ii) *Let $\lambda(S^0, D)$ be a branch of the root of (1.17) with (1.23) and (1.24) satisfying $\lambda(S^0, D) = i\frac{\xi(DT)}{DT}$ when $S^0 = f(D)$. Then*

$$\frac{\partial\lambda(S^0, D)}{\partial S^0}|_{S^0=f(D)} > 0 \text{ for } D \in (\Delta, \Delta^*) \cap (0, M).$$

Thus, we can prove the following theorem.

**Theorem 1.6** *Under the condition (1.25) the equilibrium point $E$ of system (1.10) is asymptotically stable if*

$$\begin{aligned}(S^0, D) \in \{(S^0, D) : S^0 > 0 \quad & \text{when } 0 < D < \Delta, \text{ or} \\ 0 < S^0 < f(D) \quad & \text{when } \Delta < D < \Delta^*\} \cap \Omega\end{aligned}$$

*and unstable if*

$$(S^0, D) \in \{(S^0, D) : \Delta < D < \Delta^* \text{ and } S^0 > f(D)\} \cap \Omega.$$

## References

[1] Butler, G. J. and Wolkowicz, G. S. K. [1985], *A mathematical model of the chemostat with a general class of functions describing nutrient uptake,* SIAM J. Applied Math. **45**, 138–151.

[2] Caperon, J. [1969], *Time lag in population growth response of isochrysis galbana to a variable nitrate environment,* Ecology **50**, 188–192.

[3] Freedman, H. I., So, J. W. H. and Waltman, P. [1989], *Coexistence in a model of competition in the chemostat incorporating discrete delays,* SIAM J. Applied Math. **49**, 859–870.

[4] Pan, J., *Parameter analysis of a chemostat system with a delay,* submitted.

[5] Smith, H. L. and Waltman, P. [1995], *The Theory of the Chemostat, Dynamics of Microbial Competition,* Cambridge Univ. Press, London.

[6] Thingstad, T. F. and Langeland, T. I. [1974], *Dynamics of chemostat culture: the effect of a delay in cell response,* J. Theor. Biol. **48**, 149–159.

[7] Zhao, T. [1995], *Global periodic solutions for a differential delay system modeling a microbial population in the chemostat,* J. Math. Anal. Appl. **193**, 329–352.

Fields Institute Communications
Volume **21**, 1999

# Oscillation Criteria for Second Order Half-Linear Differential Equations

**Qingkai Kong**
Department of Mathematical Sciences
Northern Illinois University
DeKalb, IL 60115, USA

**Abstract.** New oscillation criteria are established for equations of the form $(p(y')^{\sigma})' + qy^{\sigma} = 0$. Some of them are extensions of Kamenev's type, and more generally, Philos' type criteria for linear equations, while the others are the so-called interval criteria which are based on the information only on a sequence of subintervals of $[t_0, \infty)$, rather than on the whole half-line.

## 1 Introduction

We consider the oscillation behavior of solutions of the second order nonlinear differential equation

$$(p(t)(y')^{\sigma})' + q(t)y^{\sigma} = 0 \tag{1.1}$$

on the half-line $[t_0, \infty)$ where $\sigma = \alpha/\beta$ for some odd positive integers $\alpha$ and $\beta$, and $1/p, q \in L_{loc}([t_0, \infty), \mathbb{R})$, the set of real valued, locally integrable functions on $[t_0, \infty)$, and $p > 0$ *a.e.* on $[t_0, \infty)$. The existence, uniqueness, and the extension to $[t_0, \infty)$ of solutions to the initial value problem for (1.1) is established, see [**2, 3**]. When $\sigma = 1$, Equation (1.1) becomes the linear equation

$$(p(t)(y'))' + q(t)y = 0. \tag{1.2}$$

Equation (1.1) is said to be *half-linear*, because if $y(t)$ is a solution of (1.1), then $c\,y(t)$ is also a solution of (1.1) for every $c \in \mathbb{R}$, although the set of solutions of (1.1) is not a linear space. Some basic properties for the linear equation (1.2) are still true for the nonlinear equation (1.1). For example, the Sturmian Comparison Theorem and the Sturmian Separation Theorem for (1.2) also hold for (1.1), see [**2, 3**] for details.

A solution of Equation (1.1) is said to be *oscillatory* if it has arbitrarily large zeros. According to the Sturmian Separation Theorem, all solutions of Equation (1.1) are either oscillatory or nonoscillatory, so we may classify Equation (1.1) as oscillatory or nonoscillatory.

1991 *Mathematics Subject Classification.* Primary 34C10.

For a solution $y$ of (1.1) which has no zero in an interval $J$ we can define the Riccati variable $w = -p(y')^\sigma/y^\sigma$, and hence obtain the Riccati equation on $J$

$$w'(t) = q(t) + \sigma p^{-1/\sigma}(t) w^{\frac{\sigma+1}{\sigma}}(t). \tag{1.3}$$

As a powerful tool for oscillation analysis of Equation (1.2), the Riccati technique remains useful to determine the oscillation properties of Equation (1.1). Some results have been found in this direction, see [**7, 8**], and the references therein. In this paper, using the Riccati technique we will obtain Kamenev's type, and more generally, Philos' type criteria for oscillation of (1.1). Furthermore, using the Riccati technique but from a different approach, we obtain some interval criteria for oscillation, that is, criteria given by the behavior of Equation (1.1) (or of $p$ and $q$) only on a sequence of subintervals of $[t_0, \infty)$, see Theorems 3.2, 3.3, and their corollaries for detail. These results are consistent with the Sturmian Separation Theorem which reveals the fact that oscillation of (1.1) is only an interval property, i.e., if there exists a sequence of subintervals $[a_i, b_i]$ of $[t_0, \infty)$, $a_i \to \infty$, such that for each $i$, there exists a solution of Equation (1.1) which has at least two zeros in $[a_i, b_i]$, then every solution of Equation (1.1) is oscillatory, no matter how "bad" Equation (1.1) is (or $p$ and $q$ are) on the remaining parts of $[t_0, \infty)$. These theorems can be applied to some extreme cases such as $\int_{t_0}^\infty q(t)\,dt = -\infty$, and improve Kamenev's type of criteria for many cases.

**Remark** All the results obtained in this paper can be extended to the equation

$$(p(t)|y'|^{\sigma-1}y')' + q(t)|y|^{\sigma-1}y = 0 \tag{1.4}$$

where $\sigma > 1$ is a fixed real number.

## 2 Preliminary results

In the sequel we say that a function $H = H(t,s)$ belongs to a function class $\mathcal{H}$, denoted by $H \in \mathcal{H}$, if $H \in C(D, \mathbb{R}_+)$ where $D = \{(t,s) : -\infty < s \le t < \infty\}$, which satisfies

$$H(t,t) = 0, \qquad H(t,s) > 0 \text{ for } t > s, \tag{2.1}$$

and has partial derivatives $\partial H/\partial t$ and $\partial H/\partial s$ on $D$ such that

$$\frac{\partial H}{\partial t} = h_1(t,s) H(t,s)^{\frac{\sigma}{\sigma+1}} \qquad \text{and} \qquad \frac{\partial H}{\partial s} = -h_2(t,s) H(t,s)^{\frac{\sigma}{\sigma+1}}. \tag{2.2}$$

where $h_1, h_2 \in L_{loc}(D, \mathbb{R})$.

First, we state a result on the maximal value of a function which can be proved by a direct computation.

**Proposition 2.1** *Let $\sigma = \alpha/\beta$ for some odd positive integers $\alpha$ and $\beta$, and let*

$$f(w) = bw - aw^{\frac{\sigma+1}{\sigma}}$$

*where $a > 0$ and $b \in \mathbb{R}$ are constants. Then $f$ assumes its maximal value on $\mathbb{R}$ at $w^* = \dfrac{b\sigma}{a(\sigma+1)}$ and*

$$\max_{w \in \mathbb{R}} f(w) = f(w^*) = \frac{\sigma^\sigma}{(\sigma+1)^{(\sigma+1)}} \frac{b^{\sigma+1}}{a^\sigma}.$$

The following lemmas will be useful to establish oscillation criteria for Equation (1.1).

**Lemma 2.1** *Assume $y$ is a solution of Equation (1.1) such that $y(t) > 0$ on $[c, b)$. Let*

$$w(t) = -p(t)(y')^{\sigma}(t)/y^{\sigma}(t) \tag{2.3}$$

*on $[c, b)$. Then for any $H \in \mathcal{H}$*

$$\int_c^b H(b,s)q(s)\,ds \le -H(b,c)w(c) + \frac{1}{(\sigma+1)^{(\sigma+1)}} \int_c^b p(s)h_2^{\sigma+1}(b,s)\,ds. \tag{2.4}$$

**Proof** From (1.1) and (2.3) we have that for $s \in [c, b)$

$$q(s) = w'(s) - \sigma p^{-1/\sigma}(s) w^{\frac{\sigma+1}{\sigma}}(s). \tag{2.5}$$

Multiplying (2.5) by $H(t, s)$, integrating it with respect to $s$ from $c$ to $t$ for $t \in [c, b)$, and using (2.1), (2.2) we get that

$$\begin{aligned} \int_c^t H(t,s)q(s)\,ds &= \int_c^t H(t,s)w'(s)\,ds - \int_c^t \sigma p^{-1/\sigma}(s)H(t,s)w^{\frac{\sigma+1}{\sigma}}(s)\,ds \\ &= -H(t,c)w(c) + \int_c^t \left(h_2(t,s)H(t,s)^{\frac{\sigma}{\sigma+1}}w(s) - \sigma p^{-1/\sigma}(s)H(t,s)w^{\frac{\sigma+1}{\sigma}}(s)\right) ds. \end{aligned}$$

Using Proposition 2.1 for

$$f(w) = h_2(t,s)H(t,s)^{\frac{\sigma}{\sigma+1}}w - \sigma p^{-1/\sigma}(s)H(t,s)w^{\frac{\sigma+1}{\sigma}}$$

we have that

$$\int_c^t H(t,s)q(s)\,ds \le -H(t,c)w(c) + \frac{1}{(\sigma+1)^{(\sigma+1)}} \int_c^t p(s)h_2^{\sigma+1}(t,s)\,ds.$$

Letting $t \to b^-$ in the above we obtain (2.4). □

**Lemma 2.2** *Assume $y$ is a solution of Equation (1.1) such that $y(t) > 0$ on $(a, c]$. Let $w(t)$ be defined by (2.3) on $(a, c]$. Then for any $H \in \mathcal{H}$*

$$\int_a^c H(s,a)q(s)\,ds \le H(c,a)w(c) + \frac{1}{(\sigma+1)^{(\sigma+1)}} \int_a^c p(s)h_1^{\sigma+1}(s,a)\,ds. \tag{2.6}$$

**Proof** Similar to the proof of Lemma 2.1, we multiply (2.5) by $H(s,t)$, integrate it with respect to $s$ from $t$ to $c$ for $t \in (a, c]$, and use (2.1), (2.2) and Proposition 2.1, then we get that

$$\begin{aligned} \int_t^c H(s,t)q(s)\,ds &= \int_t^c H(s,t)w'(s)\,ds - \int_t^c \sigma p^{-1/\sigma}(s)H(s,t)w^{\frac{\sigma+1}{\sigma}}(s)\,ds \\ &= H(c,t)w(c) + \int_t^c \left(-h_1(s,t)H^{\frac{\sigma}{\sigma+1}}w(s) - \sigma p^{-1/\sigma}(s)H(s,t)w^{\frac{\sigma+1}{\sigma}}(s)\right) ds \\ &\le H(c,t)w(c) + \frac{1}{(\sigma+1)^{(\sigma+1)}} \int_t^c p(s)h_1^{\sigma+1}(s,a)\,ds. \end{aligned}$$

Letting $t \to a^+$ in the above we obtain (2.6). □

As by-products of the lemmas, in Corollary 2.1, we find some necessary conditions for disfocality and disconjugacy of Equation (1.1) which are defined similarly to those for linear equations, see, for example, [**6**] for details.

**Corollary 2.1** *i) Assume Equation (1.1) is right-disfocal on $[c,b)$. Then for any $H \in \mathcal{H}$*

$$\int_c^b H(b,s)q(s)\,ds \le \frac{1}{(\sigma+1)^{(\sigma+1)}}\int_c^b p(s)h_2^{\sigma+1}(b,s)\,ds. \tag{2.7}$$

*ii) Assume Equation (1.1) is left-disfocal on $(a,c]$. then for any $H \in \mathcal{H}$*

$$\int_a^c H(s,a)q(s)\,ds \le \frac{1}{(\sigma+1)^{(\sigma+1)}}\int_a^c p(s)h_1^{\sigma+1}(s,a)\,ds. \tag{2.8}$$

*iii) Assume Equation (1.1) is disconjugate on $(a,b)$. Then for any $H \in \mathcal{H}$ there exists $c \in [a,b]$ such that both (2.7) and (2.8) hold.*

**Proof** i) Let $y$ be a solution of (1.1) satisfying $y'(c)=0$ and $y(c)>0$. Since (1.1) is right-disfocal on $[c,b)$, then $y(t)>0$ on $[c,b)$. So the $w(t)$ defined by (2.3) exists on $[c,b)$, and $w(c)=0$. The inequality (2.7) hence follows from Lemma 2.1.

ii) Similar to i).

iii) For any $c \in [a,b]$ let $y_c(t)$ be the solution of (1.1) satisfying $y(c)=1$ and $y'(c)=0$. Pick any $c_0 \in (a,b)$. Since (1.1) is disconjugate on$(a,b)$, $y_{c_0}(t)$ has at most one zero in $(a,b)$. Without loss of generality we assume $y_{c_0}(t)>0$ on $(a,c_0]$. By the continuity of $y_c(t)$ with respect to $c$, we see that either $y_c(t)>0$ on $(a,c]$ for all $c \in [c_0,b)$, in this case, both (2.8) and (2.7) hold for $c=b$ by part ii); or there exists a $c_1 \in (c_0,b)$ such that $y_{c_1}(t)>0$ on $(a,c_1]$, $y_{c_1}(a)=0$, and $y_c(t)$ has a zero in $(a,c)$ for all $c$ in a right-neighborhood of $c_1$. On the other hand, $y_{c_1}(t)$ cannot have any zeros in $(c_1,b)$, for otherwise, there exists $c_2>c_1$ such that $y_{c_2}(t)$ has a zero in $(c_2,b)$, thus $y_{c_2}(t)$ has two zeros in $(a,b)$, contradicting the disconjugacy of (1.1) on $(a,b)$. Now, Equation (1.1) is both right-disfocal on $[c_1,b)$ and left-disfocal on $(a,c_1]$. Therefore, (2.7) and (2.8) hold for $c=c_1$. □

**Corollary 2.2** *Assume that for some $c \in (a,b)$ and for some $H \in \mathcal{H}$*

$$\frac{1}{H(c,a)}\int_a^c H(s,a)q(s)\,ds + \frac{1}{H(b,c)}\int_c^b H(b,s)q(s)\,ds \tag{2.9}$$

$$> \frac{1}{(\sigma+1)^{(\sigma+1)}}\left(\frac{1}{H(c,a)}\int_a^c p(s)h_1^{\sigma+1}(s,a)\,ds + \frac{1}{H(b,c)}\int_c^b p(s)h_2^{\sigma+1}(b,s)\,ds\right).$$

*Then every solution of Equation (1.1) has at least one zero in $(a,b)$.*

**Proof** (2.9) implies that both (2.4) and (2.6) do not hold for the given $c$, and hence every solution of (1.1) must have a zero either in $[c,b)$ or in $(a,c]$. □

**Corollary 2.3** *Assume that for some $H \in \mathcal{H}$ and for any $c \in [a,b]$ at least one of the following inequalities holds:*

$$\int_a^c H(s,a)q(s)\,ds > \frac{1}{(\sigma+1)^{(\sigma+1)}}\int_a^c p(s)h_1^{\sigma+1}(s,a)\,ds \tag{2.10}$$

*and*

$$\int_c^b H(b,s)q(s)\,ds > \frac{1}{(\sigma+1)^{(\sigma+1)}}\int_c^b p(s)h_2^{\sigma+1}(b,s)\,ds. \tag{2.11}$$

*Then every solution of Equation (1.1) has at least one zero in $(a,b)$.*

**Proof** From (2.10), (2.11) and Corollary 2.1, iii) we see that (1.1) is not disconjugate on $(a,b)$, and hence there exists a solution of (1.1) with 2 zeros in $(a,b)$. Then the conclusion is from the Sturm Separation Theorem for (1.1). □

## 3 Oscillation criteria

The first theorem presents an oscillation criterion for (1.1) which is an analogue of Philos' criterion for (1.2).

**Theorem 3.1** *Equation (1.1) is oscillatory provided for each* $r \geq t_0$

$$\limsup_{t\to\infty} \frac{1}{H(t,r)} \int_r^t [H(t,s)q(s) - \frac{1}{(\sigma+1)^{(\sigma+1)}} p(s) h_2^{\sigma+1}(t,s)]\, ds = \infty. \tag{3.1}$$

**Proof** Assume (1.1) is not oscillatory. Then (1.1) has a solution $y(t) > 0$ for $t \in [r, \infty)$ where $r \geq t_0$. Let $w(t)$ be defined by (2.3). From Lemma 2.1 we have that for all $t \geq r$

$$\int_r^t H(t,s)q(s)\, ds \leq -H(t,r)w(r) + \frac{1}{(\sigma+1)^{(\sigma+1)}} \int_r^t p(s) h_2^{\sigma+1}(t,s)\, ds.$$

Hence

$$\int_r^t [H(t,s)q(s) - \frac{1}{(\sigma+1)^{(\sigma+1)}} p(s) h_2^{\sigma+1}(t,s)]\, ds \leq -H(t,r)w(r).$$

Divide both sides by $H(t,r)$ and take limitsup as $t \to \infty$ we find that

$$\limsup_{t\to\infty} \frac{1}{H(t,r)} \int_r^t [H(t,s)q(s) - \frac{1}{(\sigma+1)^{(\sigma+1)}} p(s) h_2^{\sigma+1}(t,s)]\, ds \leq -w(r) < \infty$$

which contradicts the assumption. This completes the proof. □

The next theorem provides some interval criteria for oscillation which are extensions of those in [**5**] to (1.1).

**Theorem 3.2** *Equation (1.1) is oscillatory provided for each* $T \geq t_0$ *there exists* $H \in \mathcal{H}$ *and either*

*i) there exists* $a, b, c \in \mathbb{R}$ *such that* $T \leq a < c < b$ *and (2.9) holds, or*

*ii) there exist* $a, b \in \mathbb{R}$ *such that* $T \leq a < b$ *and for any* $c \in [a, b]$ *at least one of (2.10) and (2.11) holds.*

**Corollary 3.1** *Assume that for some* $H \in \mathcal{H}$ *and for each* $r \geq t_0$

$$\limsup_{t\to\infty} \int_r^t [H(s,r)q(s) - \frac{1}{(\sigma+1)^{(\sigma+1)}} p(s) h_1^{\sigma+1}(s,r)]\, ds > 0, \tag{3.2}$$

*and*

$$\limsup_{t\to\infty} \int_r^t [H(t,s)q(s) - \frac{1}{(\sigma+1)^{(\sigma+1)}} p(s) h_2^{\sigma+1}(t,s)]\, ds > 0. \tag{3.3}$$

*Then Equation (1.1) is oscillatory.*

**Proof** For any $T \geq t_0$, let $a = T$. In (3.2) we choose $r = a$. Then there exists $c > a$ such that

$$\int_a^c [H(s,a)q(s) - \frac{1}{(\sigma+1)^{(\sigma+1)}} p(s) h_1^{\sigma+1}(s,a)]\, ds > 0. \tag{3.4}$$

In (3.3) we choose $r = c$. Then there exists $b > c$ such that

$$\int_c^b [H(b,s)q(s) - \frac{1}{(\sigma+1)^{(\sigma+1)}} p(s) h_2^{\sigma+1}(b,s)]\, ds > 0. \tag{3.5}$$

Combining (3.4) and (3.5) we obtain (2.9). The conclusion thus comes from Theorem 3.2, i). □

For the case that $H \equiv H(t-s) \in \mathcal{H}$, we have that $h_1(t-s) \equiv h_2(t-s)$ and denote them by $h(t-s)$. The subclass of $\mathcal{H}$ containing such $H(t-s)$ is denoted by $\mathcal{H}_0$. Applying Theorem 3.2, i) to $\mathcal{H}_0$ we obtain

**Theorem 3.3** *Assume that for each $T \geq t_0$, there exists $H \in \mathcal{H}_0$ and $a, c \in \mathbb{R}$ such that $T \leq a < c$ and*

$$\int_a^c H(s-a)[q(s)+q(2c-s)]\,ds > \frac{1}{(\sigma+1)^{(\sigma+1)}} \int_a^c [p(s)+p(2c-s)]h^{\sigma+1}(s-a)\,ds. \tag{3.6}$$

*Then Equation (1.1) is oscillatory.*

**Proof** Let $b = 2c-a$. Then $H(b-c) = H(c-a) = H((b-a)/2)$, and for any $f \in L[a,b]$, we have

$$\int_c^b f(s)\,ds = \int_a^c f(2c-s)\,ds.$$

Hence (3.6) implies (2.9) and hence (1.1) is oscillatory by Theorem 3.2, i). □

Now, we choose $H(t-s) = (t-s)^\lambda$ for $\lambda > \sigma$. Then $H \in \mathcal{H}_0$, and $h(t-s) = \lambda(t-s)^{\frac{\lambda}{\sigma+1}-1}$. Based on the above results we obtain the following oscillation criteria of generalized Kamenev's type.

**Theorem 3.4** *Let $p(t) \equiv 1$. Then Equation (1.1) is oscillatory provided that for each $r \geq t_0$ and for some $\lambda > \sigma$, either*

*i) the following inequality holds:*

$$\limsup_{t\to\infty} \frac{1}{t^\lambda} \int_r^t (t-s)^\lambda q(s)\,ds = \infty; \quad \textit{or} \tag{3.7}$$

*ii) the following two inequalities hold:*

$$\limsup_{t\to\infty} \frac{1}{t^{\lambda-\sigma}} \int_r^t (s-r)^\lambda q(s)\,ds > \frac{\lambda^{\sigma+1}}{(\sigma+1)^{(\sigma+1)}(\lambda-\sigma)} \tag{3.8}$$

*and*

$$\limsup_{t\to\infty} \frac{1}{t^{\lambda-\sigma}} \int_r^t (t-s)^\lambda q(s)\,ds > \frac{\lambda^{\sigma+1}}{(\sigma+1)^{(\sigma+1)}(\lambda-\sigma)}; \quad \textit{or} \tag{3.9}$$

*iii) the following inequality holds:*

$$\limsup_{t\to\infty} \frac{1}{t^{\lambda-\sigma}} \int_r^t (s-r)^\lambda [q(s)+q(2t-s)]\,ds > \frac{2\lambda^{\sigma+1}}{(\sigma+1)^{(\sigma+1)}(\lambda-\sigma)}. \tag{3.10}$$

**Proof** i) It is easy to see that

$$\lim_{t\to\infty} \frac{1}{H(t-t_0)} \int_{t_0}^t h^{\sigma+1}(t-s)\,ds = 0,$$

then the conclusion is clear from Theorem 3.1.

ii) Noting that

$$\lim_{t\to\infty} \frac{1}{t^{\lambda-\sigma}} \int_r^t h^{\sigma+1}(s-r)\,ds = \frac{\lambda^{\sigma+1}}{\lambda-\sigma}, \tag{3.11}$$

from (3.8) and (3.11) we have that

$$\begin{aligned}\limsup_{t\to\infty} \frac{1}{t^{\lambda-\sigma}} \int_r^t [H(s-r)q(s) - \frac{1}{(\sigma+1)^{(\sigma+1)}} h^{\sigma+1}(s-r)]\, ds \\ = \limsup_{t\to\infty} \frac{1}{t^{\lambda-\sigma}} \int_r^t [(s-r)^{\lambda} q(s) - \frac{\lambda^{\sigma+1}}{(\sigma+1)^{(\sigma+1)}} (s-r)^{\lambda-\sigma-1}]\, ds \\ = \limsup_{t\to\infty} \frac{1}{t^{\lambda-\sigma}} \int_r^t (s-r)^{\lambda} q(s)\, ds - \frac{\lambda^{\sigma+1}}{(\sigma+1)^{(\sigma+1)}(\lambda-\sigma)} > 0.\end{aligned}$$

It follows that

$$\limsup_{t\to\infty} \int_r^t [H(s-r)q(s) - \frac{1}{(\sigma+1)^{(\sigma+1)}} h^{\sigma+1}(s-r)]\, ds > 0,$$

i.e., (3.2) holds. Similarly, (3.9) implies that (3.3) holds. By Corollary 3.1, (1.1) is oscillatory.

iii) Combining (3.10) with (3.11) we have that for $a \geq T$

$$\limsup_{t\to\infty} \frac{1}{t^{\lambda-\sigma}} \int_a^t \left\{ (s-a)^{\lambda}[q(s) + q(2t-s)] - \frac{2\lambda^{\sigma+1}}{(\sigma+1)^{(\sigma+1)}} (s-a)^{\lambda-\sigma-1} \right\} ds > 0.$$

It follows that

$$\limsup_{t\to\infty} \int_a^t \left\{ (s-a)^{\lambda}[q(s) + q(2t-s)] - \frac{2\lambda^{\sigma+1}}{(\sigma+1)^{(\sigma+1)}} (s-a)^{\lambda-\sigma-1} \right\} ds > 0.$$

Hence there exists $c > a$ such that

$$\int_a^c (s-a)^{\lambda}[q(s) + q(2c-s)] > \frac{2\lambda^{\sigma+1}}{(\sigma+1)^{(\sigma+1)}} \int_a^c (s-a)^{\lambda-\sigma-1}\, ds,$$

i.e., (3.6) holds. By Theorem 3.3, Equation (1.1) is oscillatory. □

## References


[1] Erbe, L. H., Kong, Q. and Ruan, S. [1993], *Kamenev type theorems for second order matrix differential systems*, Proc. Amer. Math. Soc. **117**, 957-962.

[2] Del Pino, M. and Manasevich, R. [1989], *Some nonlinear Sturmian comparison theory for* $(|u'|^{p-2}u')' + c(t)|u|^{p-2}u = 0$, Lecture notes in Pure and Appl. Math. **118**, Dekker, New York, 183-190.

[3] Del Pino, M. and Manasevich, R. [1988], *Oscillation and nonoscillation for* $(|u'|^{p-2}u')' + a(t)|u|^{p-2}u = 0, p > 1$, Houston J. Math. **14**, 173-177.

[4] Kamenev, I. V. [1978], *Integral criterion for oscillation of linear differential equations of second order*, Mat. Zametki **23**, 249-251.

[5] Kong, Q., *Interval criteria for oscillation of second order linear ordinary differential equations*, preprint.

[6] Kwong, M. K. [1981], *On Lyapunov's inequality for disfocality*, J. Math. Anal. Appl. **83**, 486-494.

[7] Li, H. J. and Yeh, C. C. [1995], *Oscillations of half-linear second order differential equations*, Hiroshima Math. J. **25**, 585-594.

[8] Li, H. J. and Yeh, C. C. [1995], *Sturmian comparison theorem for half-linear second order differential equations*, Proc. Roy. Soc., Edinburgh, **125A**, 1193-1204.

[9] Philos, Ch. G. [1989], *Oscillation theorems for linear differential equations of second order*, Arch. Math. (Basel) **53**, 483-492.

Fields Institute Communications
Volume **21**, 1999

# Rich Dynamics of Gause-Type Ratio-Dependent Predator-Prey System

**Yang Kuang**
Department of Mathematics
Arizona State University
Box 871804, Tempe, AZ 85287-1804, USA
kuang@math.la.asu.edu

**Abstract.** Ratio-dependent predator-prey models are increasingly favored by field ecologists as an alternative or more suitable ones for predator-prey interactions where predation involves the searching process. However, in the past such models were not well studied mathematically. In our recent work, we have shown that such models exhibit much richer dynamics than the traditional ones. This is especially true in boundary dynamics. For example, the ratio-dependent models can exhibit dynamics such as for some parameters and initial conditions, both species can become extinct. In this paper, we consider the global behaviors of solutions of the rather general Gause-type ratio-dependent predator-prey system. In addition, to confirm that Gause-type ratio dependent predator-prey models are rich in boundary dynamics we shall also present very sharp sufficient conditions to assure that if the positive steady state of the ratio-dependent predator-prey system is locally asymptotically stable, then the system has no nontrivial positive periodic solutions. We also give sufficient conditions for each of the possible three steady states to be globally asymptotically stable. We note that for ratio-dependent systems, paradox of enrichment can not occur. In general, local asymptotic stability of the positive steady state does not even guarantee the so-called persistence of the system, and therefore does not imply global asymptotic stability.

## 1 Introduction

A milestone progress in the study of predator-prey interactions was the discovery of the now well known "paradox of enrichment" by Hairston et al. in [**14**] and by Rosenzweig in [**24**], which states that according to the Lotka-Volterra type predator-prey theory (with Michaelis-Menten-Holling type functional response) enriching a predator-prey system will cause an increase in the equilibrium density of the predator but not in that of the prey, and will destabilize the community

1991 *Mathematics Subject Classification.* 34C05, 34C25, 92A15.
Research partially supported by NSF Grant DMS-9306239.

equilibrium. Another similar paradox is the so-called "biological control paradox", which was recently brought into discussion by [**22**], stating that according to the classic predator-prey theory, you can not have both a low and stable prey equilibrium density. Both of these two paradoxes can be easily illustrated by a standard local stability analysis for the following well studied Lotka-Volterra type predator-prey model with Michaelis-Menten (or Michaelis-Menten-Holling) type functional response ([**10**], [**23**], [**25**])

$$\begin{cases} dx/dt = x' = ax(1 - x/K) - cxy/(m + x) \\ dy/dt = y' = y(-d + fx/(m + x)) \\ x(0) > 0, y(0) > 0, \end{cases} \tag{1.1}$$

where $x, y$ stands for prey and predator density, respectively. $a, K, c, m, f, d$ are positive constants that stand for prey intrinsic growth rate, carrying capacity, capturing rate, half capturing saturation constant, conversion rate, predator death rate, respectively. While many of the mathematicians working in mathematical biology may regard these as important contributions that mathematics had for ecology, they are very controversial among ecologists up to this day. Some ecologists may simply view it as a problem ([**1**], [**2**], [**5**] , [**7**], [**11**], [**16**]). Indeed, the above predictions are simply not in line with many field observations ([**2**], [**3**], [**12**]).

The controversial part is, of course, not the mathematical analyses of the model, rather, it is the model itself. Recently, there is growing explicit biological and physiological evidence ([**5**], [**4**], [**13**]) that in many situations, especially when predators have to search for food (and therefore have to share or compete for food), a more suitable general predator-prey theory should be based on the so-called **ratio-dependent** theory, which can be roughly stated as that the per capita predator growth rate should be a function of the ratio of prey to predator abundance, and so should be the so-called predator functional responses (see below). This is strongly supported by numerous field and laboratory experiments and observations ([**2**], [**4**], [**16**]). Generally, a ratio-dependent predator-prey model takes the form

$$\begin{cases} x' = xg(x/K) - yp(x/y) \equiv F(x, y) \\ y' = cyp(x/y) - dy \equiv G(x, y) \\ x(0) > 0, y(0) > 0 \end{cases} \tag{1.2}$$

while the traditional (or **prey-dependent**) model takes the form

$$\begin{cases} x' = xg(x/K) - yp(x) \\ y' = cyp(x) - dy \\ x(0) > 0, y(0) > 0. \end{cases} \tag{1.3}$$

Here $g(x/K)$ is the so-called per capita growth rate, $p(x/y)$ is the so-called predator functional response and $c$ is the conversion rate. $g(x), p(x)$ satisfy the usual properties that make system (1.3) a predator-prey one. Specifically, we shall assume in the rest of this paper that $g(x), p(x)$ are continuously differentiable and

(A1) $g(0) > 0$, there is a $K > 0$ such that $g(x/K)(x-K) < 0$ for $x > 0, x \neq K$. Also, $g'(x) < 0$ for $K \geq x \geq 0$;

(A2) $p(0) = 0, p'(z) > 0$ for $z \geq 0, \lim_{z\to\infty} p(z) = p^* < \infty$, and $\lim_{(x,y)\to(0,0)} yp(x/y) = 0$.

In the following, the constant $K$ is referred as the carrying capacity.

Very recently and independently, [**26**] and [**9**] were able to justify through some very basic (but different) principles that indeed, ratio dependent models are more reasonable choices for many predator-prey interactions. Thieme's approach is an adaption of the so-called complex-formation arguments, which is used for marriage and epidemic models, as well as in the models of enzyme reaction kinetics. Some variations of Holling's classic approach [**17**] can also lead to justification of ratio-dependence predator's functional response. The derivation of [**9**] is based on the principle of mass action with the crucial refinement that the nature of the spatial distribution of predator and/or opportunities for predation are taken into account in some implicit way. For example, they find that if the predators are assumed to have a homogeneous distribution, then the prey dependence functional response is derived. If the predators are assumed to form a dense colony in a single location, then the ratio-dependence one is obtained.

Geometrically, the differences of prey-dependent and ratio-dependent models are obvious, the former has a vertical predator isocline, while the latter has a slanted one. There are even more differences in their prey isoclines. Local stability analyses and simulations ([**2**], [**7**]) show that the ratio-dependent models are capable of producing richer and more reasonable dynamics. It is easy to show that the paradox of biological control is no longer valid for ratio-dependent systems. It can also be shown (in the next section) that the ratio-dependent type models do not produce the so-called paradox of enrichment.

Our recent work [**20**] focused on the ratio-dependent type predator-prey model with Michaelis-Menten type functional response, which takes the form of

$$\begin{cases} x' = ax(1 - x/K) - cxy/(my + x) \\ y' = y(-d + fx/(my + x)) \\ x(0) > 0, y(0) > 0. \end{cases} \tag{1.4}$$

Clearly, the above system is a special case of system (1.2). The objective of that paper was to perform a global qualitative study on system (1.4). The present paper has the same objective for system (1.2). In particular, we shall systematically study the boundary dynamics and give sharp sufficient conditions for the positive steady state to be globally asymptotically stable. This is mainly accomplished by finding conditions for the system to have no nontrivial positive periodic solutions. Our method involves a nontrivial application of the so-called divergency criterion for the stability of periodic solutions in planar systems (see [**15**], Lemma 3.1 on p224) and some critical transformations. This is the same method that was successfully used on system (1.4) [**20**]. We will also present some other related qualitative results for system (1.2). All our results here are either more general or sharper than similar ones for system (1.4) [**20**].

This paper is organized as follows. In the next section, we present results on the boundedness of solutions (or dissipativity of the systems), local stability analysis, and nonexistence of paradox of enrichment. Some permanence results and global stability results of $(K, 0)$ and $(0, 0)$ are given in Section 3. Section 4 contains the main result, namely, sufficient conditions for the nonexistence of nontrivial positive periodic solutions. It also contains some results on the global stability of the positive steady state. The paper ends with a brief discussion section.

## 2 Preliminaries, nonexistence of paradox of enrichment

In this section, we shall present some preliminary results, including the boundedness of solutions, local stability analysis, and show that for the Gause-type ratio-dependent predator-prey model (1.2), the so-called paradox of enrichment can not happen. We shall point out here that although $(0, 0)$ is defined for system (1.2), it usually (see (1.4)) can not be linearized. So, local stability of $(0, 0)$ can not be studied in the usual way. Indeed, this singularity at the origin, while it causes much difficulty in our analysis of the system, contributes significantly to the richness of dynamics of the model.

Clearly, in the interior of the positive cone, the solution exists and is unique. The only possible trouble is when $y = 0$. However, the predator equation gives us $ly < y' < Ly$ for some constants $l, L$. Hence, the solution of system (1.2) is unique even if for $x(0) \geq 0, y(0) \geq 0$.

System (1.2) always has equilibria $(0,0), (K,0)$ and has a unique positive equilibrium $E^* = (x^*, y^*)$ if and only if the following two conditions hold:

(B1) There is a $z^* > 0$ such that $p(z^*) = d/c$, which is equivalent to $cp^* > d$.

(B2) There is a unique $x^* \in (0, K)$ such that $g(x^*/K) = d/[cp^{-1}(d/c)]$.

Here $p^{-1}(s)$ stands for the inverse function of $s = p(x)$. When both conditions are satisfied, we have a unique positive steady state $E^* = (x^*, x^*/z^*)$.

The Jacobian matrix $J_K = J(K, 0)$ of system (1.2) at $(K, 0)$ takes the form of

$$\begin{pmatrix} g'(1) & -p^* \\ 0 & cp^* - d \end{pmatrix}.$$

Clearly, whenever the positive steady state $E^*$ exists, $(K, 0)$ is unstable.

The Jacobian matrix $J^* = J(x^*, y^*)$ of system (1.2) at $E^*$ takes the form of (recall $z^* = x^*/y^*$)

$$\begin{pmatrix} (x^*/K)g'(x^*/K) + p(z^*)/z^* - p'(z^*) & -p(z^*) + z^*p/(z^*) \\ cp'(z^*) & -cz^*p'(z^*) \end{pmatrix}.$$

It is easy to see that the determinant of $J^*$ is

$$\det J^* = -(x^*/K)g'(x^*/K)cz^*p'(z^*) > 0.$$

Its trace is

$$\delta^* \equiv (x^*/K)g'(x^*/K) + p(z^*)/z^* - p'(z^*) - cz^*p'(z^*).$$

Clearly, $E^*$ is locally asymptotically stable if the trace is less than zero. If we let $r = d/c$, than we have $p(z^*) = r$ or $z^* = p^{-1}(r)$. If we let $g^{-1}$ stand for the inverse function of $g$ on $[0, g(0)]$, then $x^*/K = g^{-1}(r/p^{-1}(r)) \equiv k$. Clearly, both $z^*, k$ are independent of $K$. By a straightforward computation, we can have

$$\delta^* = kg'(k) + p(z^*)/z^* - p'(z^*) - cz^*p'(z^*),$$

which is also independent of $K$. This shows that the stability of $E^*$ is independent of $K$. Hence we have shown

**Theorem 2.1** *Increasing the carrying capacity $K$ in system (1.2) will not change the stability of $E^*$. Hence, the paradox of enrichment can not happen to system (1.2).*

The prey isocline inside the positive quadrant is described by

$$xg(x/K) = yp(x/y).$$

By differentiating it implicitly and solving for $y'$, we obtain

$$(p(z) - zp'(z))y' = (x/K)g'(x/K) + p(z)/z - p'(z).$$

Since $zp'(z) > 0$ for $z > 0$, we have the following simple geometric criterion for the stability of positive equilibrium $E^*$. In applications, this can be very convenient to use, since all we need to know is the prey isocline shape (slope) near the equilibrium, not the precise analytic expression (which is impossible to obtain).

**Corollary 2.1** *In system (1.2), assume that* $p(z) > zp'(z)$ *for* $z > 0$. *Let* $y = f(x)$ *be the prey isocline inside the positive quadrant. If* $f'(x^*) \leq 0$, *then* $E^*$ *is locally asymptotically stable.*

Standard and simple arguments show that solutions of system (1.2) always exist and stay positive. Indeed, it is obvious that for system (1.2), we have

$$\limsup_{t\to\infty} x(t) \leq K \tag{2.1}$$

and

$$y(t) < y(0)e^{(cp^*-d)t}.$$

We consider now the boundedness of $y(t)$ for system (1.2). Clearly, if $d \geq cp^*$, then $\lim_{t\to\infty} y(t) = 0$. So we assume below that $cp^* > d$. In the following, we denote by $y_M$ the unique value of $y$ such that $cp(K/y) = d$. We shall show that

**Theorem 2.2** *For system (1.2), if* $cp^* > d$,

$$\limsup_{t\to\infty} y(t) \leq y_M. \tag{2.2}$$

**Proof** Observe that there exists a $T > 0$ such that, for $t > T$, $x(t) < K$ (otherwise, from the prey equation and the fact that $-dy \leq y' \leq (cp^* - d)y$, we have $\lim_{t\to\infty} x(t) = K$, $\lim_{t\to\infty} y(t) = 0$; while from the predator equation we see that $\lim_{t\to\infty} y(t) = +\infty$, a contradiction). From the predator equation, we see that, for $t > T$, we have

$$y'(t) \leq y(cp(K/y) - d).$$

A standard comparison argument shows that

$$\limsup_{t\to\infty} y(t) \leq y_M.$$

□

Therefore we have shown that system (1.2) is dissipative.

## 3 Permanence and rich boundary dynamics

In this section, we shall present some results on permanence, and boundary dynamics (the global stability of boundary equilibria $(K, 0)$ and $(0, 0)$) for system (1.2). We also would like to point out that some of the results in this section are modified from similar ones in [**6**].

For convenience, we let

$$p_1 \equiv \sup\{p(x)/x : x > 0\}.$$

It is easy to see that, for system (1.4), if $g(0) > p_1$, then

$$x' = x[g(x/K) - p(x/y)/(x/y)] \geq x(g(x/K) - p_1). \tag{3.1}$$

If we let $\underline{x}$ be the root of $g(x/K) - p_1 = 0$, then we see that the above inequality implies (again, by standard comparison argument) that $\liminf_{t\to+\infty} x(t) \geq \underline{x}$. Hence, for large $t$, $x(t) > \underline{x}/2$, and

$$y' \geq y[-d + cp(\underline{x}/(2y))].$$

Let $\underline{y}$ be the root of $-d + cp(\underline{x}/(2y)) = 0$. Then, we obtain

$$\liminf_{t\to+\infty} y(t) \geq \underline{y}.$$

The above arguments and Theorem 2.2 imply that:

**Theorem 3.1** *If $cp^* > d$ and $g(0) > p_1$, then system (1.2) is permanent.*

Recall that system (1.3) is permanent if and only if it has a positive steady state. The above theorem is not nearly as sharp (see (B1), (B2)). The same can be said for most results in the rest of this section.

Assume below that

$$p'(0) > g(0) + d$$

in system (1.4). Then there is an $\alpha > 0$, such that for all $s \in (0, \alpha]$,

$$p(s)/s \geq g(0) + d.$$

Let $\delta = x(0)/y(0) < \alpha$. We claim that for all $t > 0$, $x(t)/y(t) < \alpha$ and $\lim_{t\to+\infty} x(t) = 0$. Otherwise, there is a first time $t_1$, $x(t_1)/y(t_1) = \alpha$ and for $t \in [0, t_1)$, $x(t)/y(t) < \alpha$. Then for $t \in [0, t_1]$, we have

$$x'(t) = x[g(x/K) - (y/x)p(x/y)] \leq x[g(0) - p(x/y)/(x/y)] \leq -dx(t)$$

which implies that $x(t) \leq x(0)e^{-dt}$. However, for all $t \geq 0$,

$$y'(t) \geq -dy(t)$$

which implies that $y(t) \geq y(0)e^{-dt}$. This shows that for $t \in [0, t_1]$,

$$x(t)/y(t) \leq x(0)/y(0) = \delta < \alpha,$$

a contradiction to the existence of $t_1$, proving the claim. This in turn implies that $x(t) \leq x(0)e^{-dt}$ for all $t \geq 0$. That is, $\lim_{t\to+\infty} x(t) = 0$. Hence we have established the following result (recall that system (1.2) is said to be *not persistent* if $\min(\liminf_{t\to+\infty} x(t), \liminf_{t\to+\infty} y(t)) = 0$ for some of its positive solutions).

**Theorem 3.2** *If $p'(0) > g(0) + d$, then system (1.2) is not persistent.*

In fact, we also have

**Theorem 3.3** *If $p'(0) > g(0)+d$, then there exist positive solutions $(x(t), y(t))$ of system (1.2) such that $\lim_{t\to+\infty}(x(t), y(t)) = (0, 0)$.*

**Proof** If $cp^* \leq d$, then this is obvious from the previous argument that leads to Theorem 3.2. Assume now that $cp^* > d$. Again, the argument leading to Theorem 3.2 shows that $\lim_{t\to+\infty} x(t) = 0$ and for $t \geq 0$, $x(t)/y(t) \leq \delta$, provided that $\delta = x(0)/y(0) < \alpha$, where $\alpha$ is as described in the proof of Theorem 3.2. Let $(x(t), y(t))$ be the solution of (1.2), with $x(0)/y(0) < \alpha$. Since $y(t)$ is bounded, we have

$$0 \leq s_1 \equiv \limsup_{t\to+\infty} y(t) < +\infty, \qquad 0 \leq s_2 \equiv \liminf_{t\to+\infty} y(t) < +\infty.$$

If $s_2 > 0$, then we see that for large $t$, $y'(t) < -dy(t)/2$, which leads to $\lim_{t\to+\infty} y(t) = 0$. This is a contradiction. So we must have $s_2 = 0$. Assume below that $s_1 > 0$. Let $k_1$ be the root of $cp(x/(s_1/2)) - d = 0$. Since $\lim_{t\to+\infty} x(t) = 0$, we see that (i) there is a $t_1$, such that for $t > t_1$, we have $x(t) < k_1$.

From the assumption $s_1 > 0$, we see that there is a $t_2 > t_1$ such that $y(t_2) > s_1/2$, and $y'(t_2) > 0$. However, we note that $y'(t_2) > 0$ leads to

$$cp(x(t_2)/(s_1/2)) > cp(x(t_2)/y(t_2)) > d,$$

which implies that $x(t_2) > k_1$. This is a contradiction to (i). Hence we must have $s_2 = 0$. This completes the proof. □

Note that under the assumption $p'(0) > g(0) + d$, system (1.2) may have a positive steady state. This shows that system (1.2) can have both positive steady state and positive solutions that tend to the origin. **In particular, we can see that if $cp^*$ is just slightly greater than $d$ and $p'(0) > g(0) + d$, then $E^*$ is locally asymptotically stable but not globally asymptotically stable.**

Theorem 3.3 generalizes Theorem 2.6 in [**6**]. Clearly, Theorem 3.3 implies that if $p'(0) > g(0) + d$, then the boundary equilibrium $(K, 0)$ of system (1.2) is not globally asymptotically stable. Note that under the condition $p'(0) > g(0) + d$, system (1.2) can have no positive steady state and at the same time, $(K, 0)$ is locally stable (just add the assumption $cp^* \le d$). In such case, some solutions tend to $(K, 0)$ and some tend to $(0, 0)$. Hence the above theorem shows that system (1.2) can exhibit behavior similar to bistability. However, we can also show that

**Theorem 3.4** *If $p_1 < g(0)$ and $cp^* \le d$, then $(K, 0)$ is globally asymptotically stable for system (1.2).*

**Proof** It is easy to see that $\lim_{t\to+\infty} y(t) = 0$, and $\liminf_{t\to+\infty} x(t) \ge \underline{x} > 0$, where $g(\underline{x}/K) = p_1$. Then, for any $\varepsilon \in (0, a)$, there exists $T = T(\varepsilon)$, such that for $t > T$,

$$x(t)(g(x(t)/K) - \varepsilon) \le x'(t) \le x(t)g(x(t)/K).$$

This clearly shows that

$$\lim_{t\to+\infty} x(t) = K.$$

This proves the theorem. □

Our last theorem of this section shows that under some conditions, the origin can be globally attractive!

**Theorem 3.5** *Assume that $p(z)/z$ is nonincreasing for $z > 0$ and $p_1 > g(0)$. Let $z_1$ be the root of $p(z)/z - g(0) = 0$. If $z_1 \ge p^{-1}(d/c)$, then $(0, 0)$ is globally asymptotically stable for system (1.2).*

**Proof** This theorem can be proved by a simple flow analysis. The prey isocline in the positive quadrant is the curve described by $g(x/K) = p(z)/z, z = x/y$. Since $p(z)/z$ is nonincreasing, we see that the least value that z can possibly attain is when $x = 0$, which is the value $z_1$. Note also that the predator isocline is simply the line $x/y = p^{-1}(d/c)$. Since $z_1 \ge p^{-1}(d/c)$, we see that the whole predator isocline lies above the prey isocline. So, we can divide the positive cone into three regions, $I, II, III$.

$$III = \{(x, y) : x \ge 0, y \ge 0, F(x, y) \ge 0, G(x, y) > 0\};$$

$$II = \{(x, y) : x \ge 0, y \ge 0, F(x, y) < 0, G(x, y) \ge 0\};$$

$$I = \{(x, y) : x \ge 0, y \ge 0, F(x, y) < 0, G(x, y) < 0\}.$$

Clearly, solutions that start in region $III$ will enter region $II$ by crossing the prey isocline vertically (upwards) in finite time, since there is no steady state in $III$. And then, for the same reason, the solutions will enter region $I$ in finite time by crossing the predator isocline horizontally (left wards). Once in region $I$, it is easy to see that solutions are trapped (cannot cross the predator isocline from left to right) and we must have $(x(t), y(t)) \to (0, 0)$ as $t \to +\infty$. This proves the theorem. □

## 4 Global stability of positive equilibrium and more on permanence

In this section we shall obtain some sufficient conditions for the positive equilibrium of system (1.2) to be globally attractive. In addition, we will present another permanence result. Obviously, we shall assume that the positive steady state $E^*$ exists for the rest of this section. This is equivalent to assuming (B1)-(B2). In particular, this implies that throughout this section, we assume that $cp^* > d$. If we let $z = x/y$, and

$$f(x,z) \equiv g(x/K) - p(z)/z.$$

Then we can rewrite system (1.2) as

$$\begin{cases} x' = xf(x,z) \equiv F_1(x,z) \\ z' = z[f(x,z) - cp(z) + d] \equiv F_2(x,z). \end{cases} \tag{4.1}$$

System (4.1) has $(0,0), (x^*, z^*)$, and possibly $(0, z_*)$ and/or $(x_*, 0)$ as its equilibria, where $z_*$ is any positive root of

$$f(0,z) - cp(z) + d = 0$$

and $x_*$ is the any positive root of

$$g(x/K) - p'(0) = 0.$$

The Jacobian matrix $J(0,0)$ of system (4.1) at $(0,0)$ takes the form of

$$J(0,0) = \begin{pmatrix} g(0) - p'(0) & 0 \\ 0 & g(0) - p'(0) + d \end{pmatrix}.$$

Clearly, $(0,0)$ is a repeller if $g(0) > p'(0)$. The Jacobian matrix $J(0, z_*)$ of system (4.1) at $(0, z_*)$ takes the form of

$$J(0,z_*) = \begin{pmatrix} g(0) - p(z_*)/z_* & 0 \\ (z_*/K)g'(0) & -cz_*p'(z_*) - p'(z_*) + p(z_*)/z_* \end{pmatrix}.$$

If $g(0) > p(z_*)/z_*$, then $(0, z_*)$ is unstable. Clearly, if $g(0) > p_1$, then both $(0,0), (0, z_*)$ are unstable. The Jacobian matrix $J(x_*, 0)$ of system (4.1) at $(x_*, 0)$ takes the form of

$$J(x_*,0) = \begin{pmatrix} x_*g'(x_*/K) & -x_*p''(0)/2 \\ 0 & d \end{pmatrix}.$$

Clearly, when it exists, $(x_*, 0)$ is always a saddle, and is stable along the $x$ direction.

Using a method involving a nontrivial application of the so-called divergency criterion for the stability of periodic solution in planar systems (see [**15**], Lemma 3.1 on p224) and some critical transformations, Kuang and Beretta in [**20**] succeeded in showing that whenever $E^*$ is locally stable, there exists no nontrivial positive periodic solutions in system (1.4). This approach, however, seems formidable to be applied to the more general setting of system (1.2). Nevertheless, applying Dulac criterion, we can obtain some simple yet practical conditions for the positive equilibrium $E^*$ to be globally asymptotically stable.

Let $b_1(x,z) \equiv 1/(xz)$ and

$$D_1 \equiv \frac{\partial(b_1F_1)}{\partial x} + \frac{\partial(b_1F_2)}{\partial z}.$$

Then we have

$$D_1 = g'(x/K)/(zK) + [p(z) - p'(z)z - cp'(z)z^2]/(xz^2).$$

Recall that $g'(x/K) \leq 0$ for $x \in [0, K]$. It is easy to see that any nontrivial positive periodic solution must surround $E^*$ and be contained in the strip $\{(x, z) : x \in [0, K], z > 0\}$. By Dulac criterion, we arrive at:

**Theorem 4.1** *If the predator response function $p(z)$ satisfies*

$$\Delta \equiv p(z) - p'(z)z - cp'(z)z^2 \leq 0, \tag{4.2}$$

*then system (1.2) has no nontrivial positive periodic solutions. If, in addition, that $g(0) > p_1$, then $E^*$ is globally asymptotically stable.*

**Proof** From Dulac criterion, we see that system (1.2) has no nontrivial positive periodic solution. $\Delta \leq 0$ implies that $E^*$ is locally asymptotically stable, while $cp^* > d$ (since $E^*$ exists) and $g(0) > p_1$ ensure that the system is permanent (Theorem 3.1). These together yield the conclusion. □

An alternative proof of Theorem 4.1 can bypass Theorem 3.1 by simply making use of the facts that when $g(0) > p_1$, both $(0, 0)$ and $(0, z_*)$(if any) are unstable plus that $(K, 0)$ is a saddle in the original system (1.2).

As a simple example for Theorem 4.1. Consider the case that

$$p(z) = \frac{bz}{a + z}.$$

Then, condition (4.2) is reduced to $ca \geq 1$. Notice that we assume only that $g(x/K)$ is decreasing.

Let $b_2(x, z) \equiv 1/(xp(z))$ and

$$D_2 \equiv \frac{\partial(b_2 F_1)}{\partial x} + \frac{\partial(b_2 F_2)}{\partial z}.$$

Then we have

$$D_2 = g'(x/K)/(p(z)K) + [(z/p(z))'g(x/K) - c + d(z/p(z))']/x.$$

Hence, $D_2 \leq 0$, if $(z/p(z))'(g(x/K) + d) \leq c$. This is equivalent to

$$cp(z)^2 \geq (g(x/K) + d)(p(z) - zp'(z)). \tag{4.3}$$

Again, by Dulac criterion, we obtain:

**Theorem 4.2** *If the predator response function $p(z)$ satisfies $p(z) \geq zp'(z)$, and*

$$cp(z)^2 \geq (g(0) + d)(p(z) - zp'(z)), \tag{4.4}$$

*then system (1.2) has no nontrivial positive periodic solution. If, in addition, that $E^*$ is locally asymptotically stable and $g(0) > p_1$, then $E^*$ is globally asymptotically stable.*

**Proof** Clearly condition (4.4) implies (4.3). The rest of the proof is the same as that for Theorem 4.1. □

As applications of Theorem 4.2, we again consider

$$p(z) = \frac{bz}{a + z}.$$

We can easily show that condition (4.4) is reduced to

$$cb \geq g(0) + d, \tag{4.5}$$

instead of $ca \geq 1$ from the previous theorem.

We can obtain the following somewhat general and practically useful result (independent of Theorems 4.1 and 4.2) by a simple flow analysis.

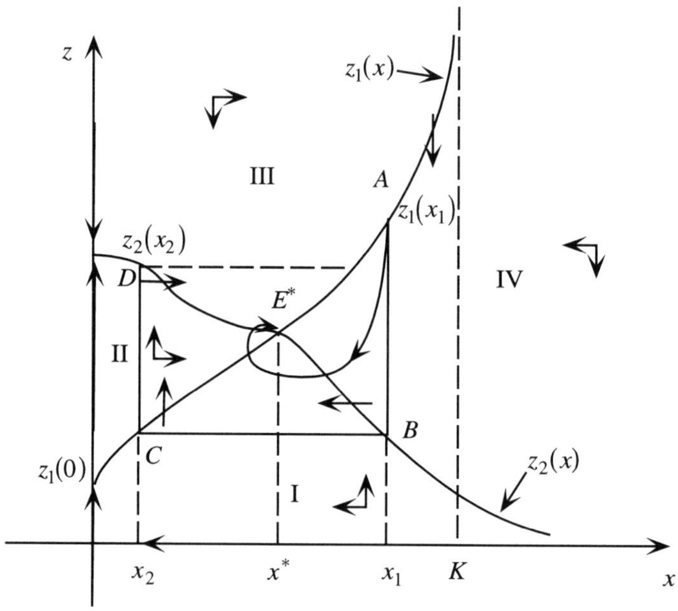


**Figure 1** A flow diagram of system (4.1)

**Theorem 4.3** *Let $x_*, x_u$ be two constants such that $0 \le x_* < K < x_u$. Suppose that $z_1(x)$ is the prey isocline defined on $[x_*, K)$ and $z_2(x)$ is the z isocline defined on $[0, x_u)$ in the interior of the first quadrant of $x - z$ plane. Assume that $z_2(x) > z^* > z_1(x)$ for $x < x^*$ and $z_2(x) < z^* < z_1(x)$ for $x > x^*$. Assume also that $p(z)/z$ is strictly decreasing, $p_1 < g(0) + d$. For any $x_1 \in (x^*, K)$, let $x_2 \in [0, x^*)$ be such that $z_1(x_2) = z_2(x_1)$. If for all $x_1 \in (x^*, K)$*

$$z_1(x_1) > z_2(x_2), \tag{4.6}$$

*then system (1.2) has no nontrivial positive periodic solutions and $E^*$ is globally asymptotically stable.*

**Proof** Since $p(z)/z$ is strictly decreasing, we see that $z_1(x)$ is strictly increasing. So the value of $x_2$ in the theorem is uniquely defined.

We divide the interior of the positive cone into four regions (see Figure 1) $I, II, III, IV$.

$$I = \{(x, z) : x > 0, z > 0, F_1(x, z) < 0, F_2(x, z) > 0\};$$

$$II = \{(x, z) : x > 0, z > 0, F_1(x, z) > 0, F_2(x, z) > 0\};$$

$$III = \{(x, z) : x > 0, z > 0, F_1(x, z) > 0, F_2(x, z) < 0\};$$

$$IV = \{(x, z) : x > 0, z > 0, F_1(x, z) < 0, F_2(x, z) < 0\}.$$

Let $A = (x_1, z_1(x_1)), B = (x_1, z_2(x_1)), C = (x_2, z_1(x_2)), D = (x_2, z_2(x_2))$. If we follow the solution starting at $A = (x_1, z_1(x_1))$, we see that it is trapped inside the polygon $ABCDA$. The solution either tends to $E^*$ before returning to the upper portion of the $x$ isocline $z_1(x > x^*)$, or will reach it in finite time at, say $(x_3, z_1(x_3))$. We define a function $h(x)$ on $(x^*, K)$ as follows. Let $h(x_1) = x_3$ if $x_3$ exists, otherwise we set $h(x_1) = x^*$. Clearly $h(x)$ is well defined and strictly monotone decreasing. If we continue to follow the flow and repeat this simple argument, we see that the solution must have $E^*$ as the limit or there is a monotone decreasing sequence $\{x_{2n+1}\}_{n=1}^{n=+\infty}$. If $\lim_{n\to\infty} x_{2n+1} = x_0 \neq x^*$, then we must have $h(x_0) = x_0$, a contradiction to assumption (4.6). Similar arguments can be applied to all other positive solutions. This proves the theorem. □

As a consequence of the local stability analysis at the beginning of this section and the thought contained in the proof of the previous theorem, we can obtain the following permanence result, which is quite different from Theorem 3.1.

**Theorem 4.4** *Assume $E^*$ and $z_*$ exist, $p'(0) < g(0) + d$ and $g(0) > p(z_*)/z_*$. Then system (1.2) is permanent.*

**Proof** Clearly, $x$ is eventually bounded above by $K+1$. Again, we divide the interior of the positive quadrant into four separate regions as in the proof of the previous theorem. Since $z_*$ exists and $p'(0) < g(0)+d$, we see that for any positive solution $(x(t), z(t))$, we have

$$\limsup_{t\to\infty} z(t) \leq z_M,$$

where $z_M \equiv \sup\{z_2(x) : x \geq 0.\}$.

Assume first that $g(0) < p_1$. Notice that as a result of $g(0) < p_1$, we have $x_* = 0$ and $z_1(0) > 0$. Also, $(0,0)$ is a saddle due to $g(0) < p_1 < g(0) + d$. For the same reason, we see that at a given point $(x, z)$, if $0 < z < z_2(x)$, then $z' = z[f(x,z) - cp(z) + d] > 0$, while if $z > z_2(x)$, then $z' = z[f(x,z) - cp(z) + d] > 0$. This indicates that $-cz_*p'(z_*) - p'(z_*) + p(z_*)/z_* \leq 0$ where $z_* = z_2(0)$. More specifically, we can see that $z_*$ is locally attractive for the scalar equation

$$z' = z[g(0) - p(z)/z - cp(z) + d].$$

Observe that $g(0) = p(z_1(0))/z_1(0)$. Since $z_2(0) > z_1(0)$, and $p(z)/z$ is strictly decreasing, we see that $g(0) - p(z_2(0))/z_2(0) > 0$. Thus, we have shown that $(0, z_2(0))$ is also a saddle. If $g(0) \geq p_1$, then $x_* \geq 0$ and $(x_*, 0)$ is an equilibrium. We can show that if $z_* > 0$, then $(0,0)$ is an unstable node and $(0, z_*)$ remains a saddle. While if $z_* = 0$, then $(0,0) = (0, z_*)$ is a saddle. As a consequence, in all the above possible cases, we see that no positive solution can tend to either $(0,0)$, or $(0, z_*)$, or $(x_*, 0)$. This, together with the fact that $x, z$ are eventually bounded above by $K+1, z_M+1$ respectively, implies (by standard permanence theory) that system (4.1) is permanent. This is equivalent to say that the original system (1.2) is also permanent. □

## 5 Discussion

In this paper, we have conducted a preliminary qualitative study on the global dynamics of Gause-type ratio-dependent predator-prey models. We have shown that compared to the traditional predator-prey system (1.3), the ratio-dependent predator-prey system (1.2) has much richer boundary dynamics. For ratio-dependent predator-prey models, there can be no paradox of enrichment or paradox of biological control. Even when there is a positive steady state, both prey and predator can become extinct. Such extinction (the collapse of the system) occurs in two cases. In one case, both species become extinct regardless of the initial data. In the other case, both species will die out only if the initial prey/predator ratio is too low (otherwise, only the predator goes extinct, while prey tends to carrying capacity). The first case is often the result of the intrinsic high efficiency of the predator species (e.g., too efficient in catching and/or converting the prey biomass). The second case is most subtle and interesting in application. It indicates that over harvesting (or catching) certain prey species, or stocking too much predator species at a given time may lead to the collapse of the whole system (the extinction of both species). These are all clearly reasonable outcomes that are frequently observed (e.g., [**11**]) in nature. However, traditional predator-prey systems fail to exhibit these. Simulation work that illustrates these and other dynamics for system (1.4) can be found in [**20**].

While the above facts do not provide enough evidence to discredit the traditional predator-prey models, it suggests the traditional predator-prey models have a number of limitations. Our conviction is that the ratio-dependent predator-prey models are more appropriate for predator-prey interactions where predation involve serious searching processes (like animals hunting for animals), while the traditional ones may be more suitable to those which involve no or a random searching process (like chemostat situation, plant-herbivore interaction, etc.).

There are still many interesting and challenging mathematical questions need to be studied for system (1.2) (in fact, even for system (1.4)). For example, most of our results in Sections 3 and 4 have much room for improvement. However, significant improvements appear to be difficult at this moment. Some obviously important questions that are left untouched here are the existence and uniqueness of the limit cycle of system (1.2) when $E^*$ becomes unstable. These should be some of the worthy objectives of future work. Nevertheless, with standard phase plane analysis, one can easily show that whenever $E^*$ is unstable, system (1.2) always has a cycle (limit cycle or heterclinic cycle connecting $(0,0)$ to $(K,0)$). In addition if the system is permanent (sufficient conditions are given in Theorem 3.1 and 4.4), then the system has a nontrivial positive limit cycle. At this moment, we are still unable to show that system (1.4) has a unique positive limit cycle, when $E^*$ exists and is unstable. This is, however, known to be true for system (1.1) (see [**21**] and the reference cited).

Finally, we would like to point out that almost nothing has been done to the qualitative analyses of food chains or multi-species or patchy interaction models based on the ratio-dependent approach. It is easy to conceive that such models should possess much richer and more reasonable dynamics (such as the collapse of the whole ecosystem in a closed environment) than the traditional counterparts can provide. And most surely, they pose more interesting and challenging questions both mathematically and biologically.

## References

[1] Arditi, R. and Berryman, A. A. [1991], *The biological control paradox*, Trends in Ecology and Evolution, **6**, 32.

[2] Arditi, R. and Ginzburg, L. R. [1989], *Coupling in predator-prey dynamics: ratio-dependence*, J. Theoretical Biology, **139**, 311–326.

[3] Arditi, R., Ginzburg, L. R. and Akcakaya, H. R. [1991], *Variation in plankton densities among lakes: a case for ratio-dependent models*, American Naturalist, **138**, 1287–1296.

[4] Arditi, R., Perrin, N. and Saiah, H. [1991], *Functional responses and heterogeneities: an experimental test with cladocerans*, OIKOS, **60**, 69–75.

[5] Arditi, R. and Saiah, H. [1992], *Empirical evidence of the role of heterogeneity in ratio-dependent consumption*, Ecology, **73**, 1544–1551.

[6] Beretta, E. and Kuang, Y. [1998], *Global analyses in some delayed ratio-dependent predator-prey systems*, Nonlinear Analysis - TMA **32**, 381-408. in press.

[7] Berryman, A. A. [1992], *The origins and evolution of predator-prey theory*, Ecology, **73**, 1530–1535.

[8] Butler, G. J., Hsu, S. B. and Waltman, P. [1983], *Coexistence of competing predators in a chemostat*, J. Math. Biol., **17**, 133–151.

[9] Cosner, C., DeAngelis, D. L., Ault, J. S. and Olson, D. B. [1997], *Effects of spatial grouping on the functional response of predators*, preprint.

[10] Freedman, H. I. [1980] *Deterministic Mathematical Models in Population Ecology*, Marcel Dekker, New York.

[11] Getz, W. M. [1984], *Population dynamics: a per-capita resource approach*, J. Theoretical Biology, **108**, 623–643.

[12] Ginzburg, L. R. and Akcakaya, H. R. [1992], *Consequences of ratio-dependent predation for steady state properties of ecosystems*, Ecology, **73**, 1536–1543.

[13] Gutierrez, A. P. [1992], *The physiological basis of ratio-dependent predator-prey theory: a metabolic pool model of Nicholson's blowflies as an example*, Ecology, **73**, 1552–1563.

[14] Hairston, N. G., Smith, F. E. and Slobodkin, L. B. [1960], *Community structure, population control and competition*, American Naturalist, **94**, 421–425.

[15] Hale, J. K. [1980], *Ordinary Differential Equations*, Krieger Publishing Co., Malabar.

[16] Hanski, I. [1991], *The functional response of predator: worries about scale*, TREE, **6**, 141–142.

[17] Holling, C. S. [1966], *The functional response of invertebrate predators to prey density*, Mem. Ent. Soc. Canada **48**.

[18] Hsu, S.-B. and Huang, T.-W. [1995], *Global stability for a class of predator-prey systems*, SIAM J. Appl. Math., **55**, 763–783.

[19] Kuang, Y. [1990], *Global stability of Gause-type predator-prey systems*, J. Math. Biol., **28**, 463–474.

[20] Kuang, Y. and Beretta, E. [1996], *Global qualitative analysis of a ratio-dependent predator-prey system*, J. Math. Biol., in press.

[21] Kuang, Y. and Freedman, H. I. [1988], *Uniqueness of limit cycles in Gause-type predator-prey systems*, Math. Biosci., **88**, 67–84.

[22] Luck, R. F. [1990], *Evaluation of natural enemies for biological control: a behavior approach*, Trends in Ecology and Evolution, **5**, 196–199.

[23] May, R. M. [1974], *Stability and Complexity in Model Ecosystems*, Princeton Univ. Press.

[24] Rosenzweig, M. L. [1969], *Paradox of enrichment: destabilization of exploitation systems in ecological time*, Science, **171**, 385–387.

[25] Smith, J. Maynard [1974], *Models in Ecology*, Cambridge Univ. Press.

[26] Thieme, H. [1997], *Mathematical Biology, An Introduction via Selected Topics*, lecture note at Arizona State University.

Fields Institute Communications
Volume **21**, 1999

# Differentiability with Respect to Delays for a Neutral Differential-Difference Equation

**Luiz A. C. Ladeira**
Departamento de Matemática
Instituto de Ciências Matemáticas de São Carlos
Universidade de São Paulo
CEP 13560-970 São Carlos, SP, Brazil
ladeira@icmsc.sc.usp.br

**Abstract.** Regularity of solutions with respect to delays for neutral differential-difference equation is studied for the case of several delays. Equations with delays depending on the time are also studied.

## 1 Introduction

Let $r \geq 0$, $a \geq 0$ be given real numbers. Denote by $C([-r,a],\mathbb{R}^n)$ the Banach space of continuous functions mapping the interval $[-r,a]$ into $\mathbb{R}^n$, with the norm $\|\phi\|^0 = \sup_{-r\leq\theta\leq 0}|\phi(\theta)|$, by $W^{1,\infty}([-r,a],\mathbb{R}^n)$ the Banach space of absolutely continuous functions from $[-r,a]$ into $\mathbb{R}^n$ having essentially bounded derivatives, with norm $\|\phi\|^{1,\infty}_{[-r,a]} = |\phi(-r)| + \operatorname{ess\,sup}_{\theta\in[-r,a]}|\dot\phi(\theta)|$; in $W^{1,\infty}([-r,a],\mathbb{R}^n)$ we can also define the (noncomplete) norms $\|\phi\|^0_{[-r,a]}$ and $\|\phi\|^{1,1}_{[-r,a]} = |\phi(-r)| + \int_{-r}^{a}|\dot\phi(s)|\,ds$. If $a = 0$ we denote simply $C = C([-r,0],\mathbb{R}^n)$, $W^{1,\infty} = W^{1,\infty}([-r,0],\mathbb{R}^n)$, and denote the above norms simply by $\|\phi\|^0$, $\|\phi\|^{1,1}$ and $\|\phi\|^{1,\infty}$. Let $\tau$, $\sigma \in [-r,0]$, $\sigma > 0$, $\phi \in C$, and let $f\colon \mathbb{R}^{2n} \to \mathbb{R}^n$ and $g\colon \mathbb{R}^n \to \mathbb{R}^n$ be Lipschitz continuous functions. Consider the initial value problem

$$\begin{cases} \dfrac{d}{dt}[x(t) - g(x(t-\sigma))] = f(x(t), x(t-\tau)), & \text{if } t > 0 \\ x(t) = \phi(t), & \text{if } -r \leq t \leq 0. \end{cases} \tag{1.1}$$

A *solution* of (1.1) is a continuous function $x\colon [-r,a) \to \mathbb{R}^n$ (for some $a > 0$) such that $x(t) = \phi(t)$ for $-r \leq t \leq 0$, $x(t) - g(x(t-\sigma))$ is continuously differentiable and satisfies (1.1) for $t \geq 0$. It can be shown that, for each such $\phi$, $\tau$, $\sigma$, the initial value problem (1.1) has a unique solution $x(t) = x(\phi,\tau,\sigma)(t)$, defined on a maximal interval $[-r,b)$, for some $b > 0$ (see, for example [**6**]). We are concerned with the smoothness properties of $x(\phi,\tau,\sigma)$ with respect to $(\phi,\tau,\sigma)$. This problem has been considered in [**7**] for retarded differential delay equations (that is, when $g = 0$), and in [**8**] for retarded differential delay equations with a diffusion term. In this paper

---

1991 *Mathematics Subject Classification.* 34K05.

Partially supported by FAPESP, Proc. 95/6922-7 and CAPES, Proc. AEX 0605/97-2.

we prove the results of [**7**] for a class of equations of neutral type. As noted in [**7**], to study smoothness with respect to delays, we must take initial functions in $W^{1,\infty}$. It follows that the solution will be absolutely continuous and the differential equation in (1.1) can be rewritten as $\dot{x}(t) = f(x(t), x(t-\tau)) + Dg(x(t-\sigma))\dot{x}(t-\sigma)$. Thus, we shall be working in the space $W^{1,\infty}([-r,a],\mathbb{R}^n)$, with the norm $\|\cdot\|_a^{1,1}$ instead of its natural (and complete) norm $\|\cdot\|_a^{1,\infty}$, since it imposes severe restrictions to the equations even in the study of continuous dependence of solutions (see [**2**], [**9**], [**10**] and [**11**] for useful remarks and examples). We shall also use the space $W^{k,\infty}([-r,0],\mathbb{R}^n)$ (which we shall denote simply $W^{k,\infty}$) of the functions from $[-r,0]$ into $\mathbb{R}^n$ whose $k^{th}$ order derivatives are essentially bounded; in this space we shall use the norm $\|\phi\|^{k,1} = |\phi(-r)| + \int_{-r}^0 |\phi^{(k)}(s)|\,ds$. The basic tool to handle this type of problem is a convenient uniform contraction principle. In [**7**], the following extension of that principle has been used to prove the differentiability result for retarded equations (in the statement of the next lemma and in everything that follows, the expressions open, continuous, $C^k$, etc., should be understood with respect to the norm $\|\cdot\|^{1,1}$).

**Lemma 1.1** ([**7**], Theorem 2.1) *Suppose $Y$ is a normed space, $V \subset Y$ and $U \subset W^{1,\infty}([a,b],\mathbb{R}^n)$ are open subsets, and $T\colon \overline{U} \times V \to \overline{U}$ satisfies:*

(a) *$T$ is a uniform contraction (with respect to both norms) in $x \in \overline{U}$ (that is, there exists $q \in [0,1)$ such that $\|T(w,y) - T(x,y)\|_{[a,b]}^{1,1} \le q\|w-x\|_{[a,b]}^{1,1}$ and $\|T(w,y) - T(x,y)\|_{[a,b]}^{1,\infty} \le q\|w-x\|_{[a,b]}^{1,\infty}$ for all $w, x \in \overline{U}$ $y \in V$);*

(b) *for each $\rho > 0$ there exists $R > 0$ such that $T((U \cap B_{R,\|\cdot\|^{1,\infty}}) \times V_R) \subset \overline{U} \cap B_{R,\|\cdot\|^{1,\infty}}$, where $V_R = \{y \in V;\ \|y\| < R\}$, and $B_{R,\|\cdot\|_{[a,b]}^{1,\infty}} = \{\psi \in W^{1,\infty}([a,b],\mathbb{R}^n);\ \|\psi\|_{[a,b]}^{1,\infty} < R\}$,*

(c) *$T$ is $C^k$ on $\overline{U} \times V$, $k \ge 1$.*

*Then, for each $y \in V$, there exists a unique fixed point $g(y)$ of $T(\cdot,y)$ in $\overline{U}$, and the map $g$ is $C^k$ in $V$.*

Actually, Theorem 2.1 in [**7**] is more general than the above lemma. We present here only a special version for $W^{1,\infty}([a,b],\mathbb{R}^n)$ (the space we will be working in) to avoid new definitions. Another point to be remarked is about hypothesis (b), which is not adequate when we deal with neutral equation due to the presence of the derivative term $\dot{x}(t-\sigma)$. Since the only reason to assume is to insure the existence of the fixed point (which we do not need here, since it is already proved in [**2**] (see also [**9**]), we shall replace it in the next theorem by a more convenient hypothesis. As can be easily seen, the proof for this result is contained in [**7**]. Also, since the dependence of the solution $x(\phi,\tau,\sigma)$ on $\sigma$ is not as good as its dependence on $\tau$, we shall state our conclusions in two theorems. We shall need the following version of the uniform contraction principle.

**Theorem 1.2** *Suppose $Y$ is a normed space, $\Lambda$ is a topological space, $V \subset Y$ and $U \subset W^{1,\infty}([a,b],\mathbb{R}^n)$ are open sets, and $T\colon \overline{U} \times V \times \Lambda \to U$ satisfy:*

(a) *For each $(y,\lambda) \in V \times \Lambda$ there exists a unique $x = g(y,\lambda) \in U$ such that $g(y,\lambda) = T(g(y,\lambda),y,\lambda)$;*

(b) *$T(x,y,\lambda)$ is a contraction in $x$ uniformly with respect to $(y,\lambda) \in V \times \Lambda$ (that is, there is $q < 1$ such that $\|T(x,y,\lambda) - T(x',y,\lambda)\|_{[a,b]}^{1,1} \le q\|x - x'\|_{[a,b]}^{1,1}$ for all $(y,\lambda) \in V \times \Lambda$);*

(c) *For each* $x \in g(V \times \Lambda)$, *the map* $T(x, \cdot, \cdot)$ *is continuous (resp., Lipschitz continuous in* $\lambda$*) in* $V \times \Lambda$.

*Then* $g$ *is a continuous (resp., Lipschitz continuous in* $\lambda$*) map on* $V \times \Lambda$. *If, in addition, we have*

(d) *For each* $\lambda \in \Lambda$, $T(\cdot, \cdot, \lambda)$ *is* $C^k$ *in* $\overline{U} \times V$.

(e) $T(x, y, \lambda)$ *is a contraction with respect to the norm* $\|\cdot\|^{1,\infty}_{[a,b]}$ *in* $x$ *uniformly with respect to* $(y, \lambda) \in V \times \Lambda$ *(that is, for all* $(y, \lambda)$ *in* $V \times \Lambda$, *we have* $\|T(x, y, \lambda) - T(x', y, \lambda)\|^{1,\infty}_{[a,b]} \leq q\|x - x'\|^{1,\infty}_{[a,b]}$*),*

*then, for each* $\lambda \in \Lambda$, *the partial map* $g(\cdot, \lambda) \colon V \to U$ *is* $C^k$ *in* $V$.

The first part of the above theorem (on the continuity of the map $g$) is proven in [**5**]; the proof of the second part is contained in the proof of Theorem 2.1 of [**7**] (see also [**6**]). The next lemma will be very useful in the computations.

**Lemma 1.3** *Let* $\phi \in W^{1,\infty}([a, b], \mathbb{R}^n)$. *Then,*

(a) $|\phi(t) - \phi(s)| \leq K_\phi |t - s|$ *for* $s, t \in [a, b]$, *where* $K_\phi = \|\phi\|^{1,\infty}_{[a,b]}$;

(b) $\|\phi\|^{0}_{[a,b]} \leq \|\phi\|^{1,1}_{[a,b]}$;

(c) $\displaystyle\int_a^b |\phi(s + h) - \phi(s)|\, ds \leq \|\phi\|^{1,1}_{[a,b]} |h|$;

(d) $\displaystyle\int_a^b |\dot\phi(s + h) - \dot\phi(s)|\, ds \to 0, \quad$ *as* $h \to 0$,

*where, in (c) and (d) we define* $\phi(t) = \phi(a)$ *for* $t < a$ *and* $\phi(t) = \phi(b)$ *for* $t > b$.

**Proof** The inequalities in (a) and (b) follow from the relation $\phi(t) = \phi(s) + \int_s^t \dot\phi(u)\, du$. To prove (c) we write

$$\int_a^b |\phi(s + h) - \phi(s)|\, ds = \int_a^b \Big| \int_0^1 \dot\phi(s + th) h\, dt \Big|\, ds \leq \|\phi\|^{1,1}_{[a,b]} |h|.$$

The proof of (d) is immediate when $\dot\phi$ is continuous. If $\dot\phi \in L^\infty$, we note that, for any $\varepsilon > 0$, there is a continuous function $g$ such that $\int_a^b |\dot\phi(s) - g(s)|\, ds < \varepsilon$, and then (c) follows. ☐

## 2 Differentiability of the solutions

The following notation will be used in this section: for each $r > 0$, $\alpha > 0$, $\beta > 0$, denote:

$W^{1,\infty} := W^{1,\infty}([-r, 0], \mathbb{R}^n)$, and $\|\phi\|^{1,1} := |\phi(-r)| + \int_{-r}^0 |\dot\phi(s)|\, ds$;

$W^{1,\infty}_{\alpha,0} := \{\xi \in W^{1,\infty}([-r, \alpha], \mathbb{R}^n);\ \xi(t) \equiv 0 \text{ on } [-r, 0]\}$ with the norms

$\|\xi\|^{1,1}_\alpha = \int_0^\alpha |\dot\xi(s)|\, ds$ and $\|\xi\|^{1,\infty}_\alpha = \operatorname{ess\,sup}_{s\in[0,\alpha]} |\dot\xi(s)|$ for $\xi \in W^{1,\infty}_{\alpha,0}$;

$A(\alpha, \beta) := \{\xi \in W^{1,\infty}_{\alpha,0};\ \|\xi\|^{1,1}_\alpha < \beta\}$; $\quad B_\beta := \{\psi \in W^{1,\infty};\ \|\psi\|^{1,1} < \beta\}$.

Clearly $W^{1,\infty}_{\alpha,0}$ is a closed subspace of $W^{1,\infty}([-r, \alpha], \mathbb{R}^n)$ and $A(\alpha, \beta)$ is an open subset of $W^{1,\infty}_{\alpha,0}$.

For each $\alpha > 0$, $\phi \in W^{1,\infty}$ we define the element $\tilde\phi \in W^{1,\infty}([-r, \alpha], \mathbb{R}^n)$ by $\tilde\phi(t) := \phi(t)$, if $-r \leq t \leq 0$, and $\tilde\phi(t) := \phi(0)$, if $0 \leq t \leq \alpha$. It is easy to see that a function $x(t)$ is a solution of (1.1) if and only if $x(t)=\tilde{\phi}(t)+z(t)$ and $z(t)$ satisfies $z(t)=0$, for $-r\le t\le 0$ and

$$\begin{aligned} z(t) = \;& g(\tilde{\phi}(t-\sigma)+z(t-\sigma))-g(\phi(-\sigma)) \\ & +\int_0^t f(\phi(0)+z(s),\tilde{\phi}(s-\tau)+z(s-\tau))\,ds\,,\text{for } t\ge 0. \end{aligned}$$

For any $\alpha>0$, $z\in W^{1,\infty}_{\alpha,0}$, $\phi\in W^{1,\infty}$, $\tau,\sigma\in[0,r]$ with $\sigma>0$, we define the function $T(z,\phi,\tau,\sigma)(t)$ by: $T(z,\phi,\tau,\sigma)(t)=0$, if $-r\le t\le 0$ and

$$\begin{aligned} T(z,\phi,\tau,\sigma)(t) &= g(\tilde{\phi}(t-\sigma)+z(t-\sigma))-g(\phi(-\sigma)) \\ & +\int_0^t f(u(s),v(s))\,ds\,, \end{aligned} \tag{2.1}$$

if $0\le t\le\alpha$, where, to shorten notation, we denote $u(s)=\phi(0)+z(s)$, $v(s)=\tilde{\phi}(s-\tau)+z(s-\tau)$. Clearly $T(z,\phi,\tau,\sigma)\in W^{1,\infty}_{\alpha,0}$, and a function $x$ is a solution of (1.1) on $[0,\alpha]$ if and only if $x=\tilde{\phi}+z$ and $z=T(z,\phi,\tau,\sigma)$. Also, since we consider only $\sigma>0$, we shall take $\alpha<\sigma$, which yields $z(t-\sigma)=0$, for $0\le t\le\alpha$.

For Equation (1.1) we have the following result on continuous dependence of the solutions (actually, very general results on the continuous dependence of solutions of neutral equations are given in [**2**] and [**11**]). We include the proof in order to introduce some notation to be used throughout this paper.

**Theorem 2.1** *Suppose $f\colon\mathbb{R}^{2n}\to\mathbb{R}^n$ and $g\colon\mathbb{R}^n\to\mathbb{R}^n$ are continuously differentiable functions, and the derivative $Dg$ is locally Lipschitz continuous. Then, for $t$ on each compact interval $[-r,\alpha]\subset[-r,b)$, the solution $x(\phi,\tau,\sigma)(t)$ of (1.1) is continuous in $(\phi,\tau,\sigma)$; for each fixed $\sigma^0>0$, $x(\phi,\tau,\sigma^0)(t)$ is locally Lipschitz continuous with respect to $(\phi,\tau)$, and for each fixed $\phi^0$ such that $\dot{\phi}^0$ is of bounded variation, the solution $x(\phi,\tau,\sigma)(t)$ is locally Lipschitz continuous with respect to $(\tau,\sigma)$.*

**Proof** It suffices to prove the Theorem for $t$ on a small interval $[0,\alpha]$. Fix $(\phi^0,\tau^0,\sigma^0)$, $\phi^0\in W^{1,\infty}$, $\tau^0,\sigma^0\in[-r,0]$, with $\sigma^0>0$, and take $\beta>0$ such that $\|\phi^0\|^1<\beta$. Choose $L(=L(\beta))>0$ such that

$$\begin{gathered} |f(x,y)|\le L,|g(x)|\le L,|Dg(x)|\le L,\;|f(x,y)-f(x',y')|\le L(|x-x'|+|y-y'|), \\ |g(x)-g(x')|\le L|x-x'|,\;|Dg(x)-Dg(x')|\le L|x-x'| \end{gathered} \tag{2.2}$$

for $|x|,|y|,|x'|,|y'|\le 3\beta$. Choose $\alpha<\min\{\frac{\sigma^0}{2},\frac{\beta}{(2\beta+1)L}\}$, and define $q=2\alpha L$ (clearly, we have $q<1$, and $z(t-\sigma)=0$, for all $0\le t\le\alpha$). For any $\xi\in\bar{A}(\alpha,\beta)$, $\phi\in B_\beta$, $\tau,\sigma\in[-r,0]$ we have $|\phi(0)+\xi(s)|\le 2\beta$, $|\tilde{\phi}(s-\tau)+\xi(s-\tau)|\le 2\beta$, $|\tilde{\phi}(s-\sigma)+\xi(s-\sigma)|\le 2\beta$, so that inequalities (2.2) hold when $x$ is replaced by either $\phi(0)+\xi(s)$ or $\tilde{\phi}(s-\sigma)+\xi(s-\sigma)$, and $y$ is replaced by $\tilde{\phi}(s-\tau)+\xi(s-\tau)$.

Then, for $z, w \in \bar{A}(\alpha,\beta)$, $\phi,\psi \in B_\beta$, $\tau,\rho \in [0,r]$, $\sigma,\nu \in (\frac{\sigma^0}{2},\ r)$, we have

$$\|T(z,\phi,\tau,\sigma)\|_\alpha^{1,1} \le \int_0^\alpha |Dg(\tilde\phi(t-\sigma))\dot{\tilde\phi}(t-\sigma)|\,dt + \int_0^\alpha |f(u(s),v(s))|\,dt$$

$$\le (\beta+1)L\alpha < \beta,$$

$$\|T(z,\phi,\tau,\sigma) - T(w,\phi,\tau,\sigma)\|_\alpha^{1,1}$$

$$= \int_0^\alpha |f(u(s),v(s)) - f(\phi(0)+w(s),\tilde\phi(s-\tau)+w(s-\tau))|\,ds$$

$$\le L\int_0^\alpha (|z(s)-w(s)| + |z(s-\tau)-w(s-\tau)|)\,ds$$

$$\le 2L\alpha\|z-w\|_\alpha^{1,1} = q\|z-w\|_\alpha^{1,1},$$

$$\|T(z,\psi,\rho,\nu) - T(z,\phi,\tau,\sigma)\|_\alpha^{1,1}$$

$$\le \int_0^\alpha \left|Dg(\tilde\psi(t-\nu))\dot{\tilde\psi}(t-\nu) - Dg(\tilde\phi(t-\sigma))\dot{\tilde\phi}(t-\sigma)\right|dt$$

$$+ \int_0^\alpha |f(\psi(0)+z(t),\tilde\psi(t-\rho)+z(t-\rho)) - f(u(t),v(t))|\,dt$$

$$\le \int_0^\alpha \left|Dg(\tilde\psi(t-\nu))\right|\left|\dot{\tilde\psi}(t-\nu) - \dot{\tilde\phi}(t-\sigma)\right|dt$$

$$+ \int_0^\alpha \left|Dg(\tilde\psi(t-\nu)) - Dg(\tilde\phi(t-\sigma))\right|\left|\dot{\tilde\phi}(t-\sigma)\right|dt +$$

$$+L\int_0^\alpha \left[|\psi(0)-\phi(0)| + |\tilde\psi(t-\rho)-\tilde\phi(t-\tau)|\right]dt$$

$$\le L\big[(1+2\alpha+\|\phi\|^{1,1})\|\psi-\phi\|^{1,1} + (\|\phi\|^{1,1}+\|z\|_\alpha^{1,1})|\rho-\tau|$$

$$+ \int_0^\alpha |\dot{\tilde\phi}(t-\nu) - \dot{\tilde\phi}(t-\sigma)|\,dt + K_\phi\|\phi\|^{1,1}|\nu-\sigma|\big].$$

This shows that $T(z,\phi,\tau,\sigma)$ is a contraction in $z$ uniformly with respect to $(\phi,\tau,\sigma)$, is continuous in $(\phi,\tau,\sigma)$ and, for each fixed $\sigma > 0$, is locally Lipschitz with respect to $(\phi,\tau)$. Now the conclusion follows from Theorem 1.2. □

We now analyze the differentiability of the solution $x(\phi,\tau,\sigma)$ of (1.1). First, we observe that the dependence of the solution $x$ with respect to $\sigma$ is somewhat complicated; even for a very simple equation, the solution is not differentiable with respect to $\sigma$, as the next example shows. Consider the initial value problem

$$\begin{cases} \dot x(t) = \dot x(t-\sigma), & \text{for } t > 0 \\ x(t) = \phi(t), & \text{for } -2 \le t \le 0, \end{cases}$$

where $\phi(\theta) = 2+\theta$ if $\theta \in [-2,-1]$ and $\phi(\theta) = 1$ if $\theta \in [-1,0]$. Clearly, for any $\sigma_1 \le 1$, the solution $x(\sigma_1)$ of this initial value problem is $x(\sigma_1)(t) \equiv 1$ which implies

$D_\sigma x(1^-) = 0$. On the other hand, if $\sigma_2 > 1$, the solution $x(\sigma_2)$ is given on the interval $[0, \sigma_2 - 1]$ by $x(\sigma_2)(t) = \phi(0) + \phi(t - \sigma_2) - \phi(-\sigma_2) = 1 + t$, so, by Lemma 1.3, $\|x(\sigma_2) - x(1)\|_{[0,2]}^{1,1} \geq |x(\sigma_2)(\sigma_2 - 1) - x(1)(\sigma_2 - 1)| = \sigma_2 - 1$, and it follows that the solution $x(\sigma)(t)$ does not have a derivative (with respect to $\sigma$) at $\sigma = 1$. Actually, this example shows that the solution $x(\sigma)$ may not be even Lipschitz continuous with respect to $\sigma$. If we take the initial function $\phi \in W^{1,\infty}$ such that $\dot{\phi}$ is not of bounded variation, then $\int_{-r}^{0} |\dot{\phi}(s-h) - \dot{\phi}(s)|\, ds$ is not of order $O(h)$ as $h \to 0$ (see, for example, [**1**], Theorem 4.4, page 124); this implies that the solution is not Lipschitz continuous in $\sigma$. This example also shows that $D_\phi x(\phi, \tau, \sigma)$ can be discontinuous with respect to $\sigma$. For any $\sigma \geq 1$ we have $D_\phi x(0, \sigma)h(t) = h(t - \sigma) + h(0) - h(-\sigma)$ if $0 \leq t \leq 1$, so $\|D_\phi x(0, \sigma)h - D_\phi x(0, 1)h\|_{[0,1]}^{1,1} = \int_0^1 |\dot{h}(t - \sigma) - \dot{h}(t - 1)|\, dt$. By taking $\sigma_n = 1 + \frac{1}{n}$ and $h_n(t) = 0$, for $-2 \leq t \leq -\sigma_n$, $h_n(t) = n(t + \sigma_n)$, for $-\sigma_n \leq t \leq -1$, $h_n(t) = 1$, for $-1 \leq t \leq 0$, we have $\|h_n\|^{1,1} = 1$ and we can easily see that, $\|D_\phi x(0, \sigma_n)h_n - D_\phi x(0, 1)h_n\|_{[0,1]}^{1,1} = 1$, which shows that $D_\phi x(0, \sigma)h$ is not continuous on $\sigma$ uniformly with respect to $h$, for $h$ varying in the unit ball $\{h \in W^{1,\infty}[-2, 0] \;:\; \|h\|^{1,1} \leq 1\}$ (that is, $D_\phi x(0, \sigma)$ is not continuous in the operator norm). Because of the above example, we will keep $\sigma$ fixed and search smoothness of the solution $x(\phi, \tau, \sigma)$ of (1.1) only with respect to $(\phi, \tau)$. By Theorem 1.2, we need to analyze the differentiability properties of the operator $T$ defined by (2.1).

**Lemma 2.2** *Suppose $f\colon \mathbb{R}^{2n} \to \mathbb{R}^n$ is a $C^k$ function and $g\colon \mathbb{R}^n \to \mathbb{R}^n$ is $C^{k+1}$, $k \geq 1$. Then:*

(a) *$T(z, \phi, \tau, \sigma)$ is a contraction with respect to the norm $\|\cdot\|_\alpha^{1,\infty}$ in $x$ uniformly with respect to $(\phi, \tau, \sigma) \in B_\beta \times [0, r] \times (\frac{\sigma^0}{2}, r)$;*

(b) *For any fixed $\sigma \in (\frac{\sigma^0}{2}, r)$, the map $T(z, \phi, \tau, \sigma)$ is continuously differentiable, it is $C^k$ with respect to $(z, \phi)$, and for any $\phi \in W^{k,\infty}$, the map $T$ is $C^k$ with respect to $\tau$, for $z \in A(\alpha, \beta)$, $\phi \in B_\beta$, $|\tau - \tau^0| < \delta$.*

**Proof** In order to simplify the estimates below, we assume further that the derivatives $D^k f$ and $D^{k+1} g$ are Lipschitz continuous. It can be easily seen from the proof that the mere continuity of the derivatives $D^k f$ and $D^{k+1} g$ is sufficient to conclude that $T$ is $C^k$. Using the notation introduced in Theorem 1.2, we have, for $(z, \phi, \tau, \sigma)$, $(w, \phi, \tau, \sigma)$ in $\bar{A}(\alpha, \beta), \times B_\beta \times [0, r] \times (\frac{\sigma^0}{2}, r)$,

$$\begin{aligned}
\|T(z, \phi, \tau, \sigma) &- T(w, \phi, \tau, \sigma)\|_\alpha^{1,\infty} \\
&= \operatorname{ess\,sup}_{t \in [0,\alpha]} |f(u(t), v(t)) - f(\phi(0) + w(t), \phi(t - \tau) + w(t - \tau))| \\
&\leq L \sup_{t \in [0,\alpha]} (|z(t) - w(t)| + |z(t - \tau) - w(t - \tau)|) \\
&\leq 2L\alpha \|z - w\|_\alpha^\infty,
\end{aligned}$$

which shows (a). Since $f$ and $g$ have continuous second order partial derivative, their first order partial derivatives are locally Lipschitz continuous: we keep denoting by $L$ the bounds and Lipschitz constants of these functions and their derivatives in the set $\{x \in \mathbb{R}^n\,;\ |x| \leq 3\beta\}$, that is, for $|x|, |y|, |x'|, |y'| \leq 3\beta : |D^2 g(x)| \leq L$, $|D_x f(x, y) - D_x f(x', y')| \leq L(|x - x'| + |y - y'|)$, $|D_y f(x, y) - D_y f(x', y')| \leq L(|x - x'| + |y - y'|)$, $|D^2 g(x) - D^2 g(x')| \leq L|x - x'|$.

The candidates to partial derivatives of $T$ are the linear operators defined by:

$$D_zT(z,\phi,\tau,\sigma)h = D_\phi T(z,\phi,\tau,\sigma)h = D_\tau T(z,\phi,\tau,\sigma)h = 0 \text{ on } [-r,0],$$

and

$$\big(D_zT(z,\phi,\tau,\sigma)h\big)(t) = \int_0^t D_xf(u(s),v(s))h(s)\,ds + \int_0^t D_yf(u(s),v(s))h(s-\tau)\,ds,$$

$$\big(D_\phi T(z,\phi,\tau,\sigma)h\big)(t) = Dg(\tilde{\phi}(t-\sigma))\tilde{h}(t-\sigma) - Dg(\tilde{\phi}(-\sigma))h(-\sigma) + \int_0^t D_xf(u(s),v(s))h(0)\,ds + \int_0^t D_yf(u(s),v(s))\tilde{h}(s-\tau)\,ds,$$

$$\big(D_\tau T(z,\phi,\tau,\sigma)\big)(t) = -\int_0^t D_yf(u(s),v(s))(\dot{\tilde{\phi}}(s-\tau)+\dot{z}(s-\tau))\,ds,$$

for $0 \le t \le \alpha$, where $u(s) = \phi(0) + z(s)$ and $v(s) = \tilde{\phi}(s-\tau) + z(s-\tau)$. Now we show that these operators are in fact the partial derivatives of $T$ and they are continuous. For $z \in A(\alpha,\beta)$, $h \in W^{1,\infty}_{\alpha,0}$, $\phi \in B_R$, $\tau \in [-r,0]$, $\sigma \in (\frac{\sigma^0}{2}, r)$, we have

$$\begin{aligned}
\|T(z+h,\phi,\tau,\sigma) &- T(z,\phi,\tau,\sigma) - D_zT(z,\phi,\tau,\sigma)\|^{1,1}_\alpha \\
&= \int_0^\alpha \big|f(u(s)+h(s),v(s)+h(s-\tau)) - f(u(s),v(s)) \\
&\quad - D_xf(u(s),v(s))h(s) - D_yf(u(s),v(s))h(s-\tau)\big|\,ds \\
&\le \int_0^\alpha \sup_{0\le\lambda_s\le1} \big|D_xf(u(s)+\lambda_sh(s),v(s)+\lambda_sh(s-\tau)) \\
&\quad - D_xf(u(s),v(s))\big||h(s)|\,ds \\
&\quad + \int_0^\alpha \sup_{0\le\lambda_s\le1} \big|D_yf(u(s)+\lambda_sh(s),v(s)+\lambda_sh(s-\tau)) \\
&\quad - D_yf(u(s),v(s))\big||h(s-\tau)|\,ds \\
&\le L\int_0^\alpha (|h(s)|+|h(s-\tau)|)^2\,ds \\
&\le 4L\alpha(\|h\|^{1,1}_\alpha)^2,
\end{aligned}$$

and this implies $T$ has a partial derivative with respect to $z$. Now we show $D_zT$ is continuous. Since for $(z,\phi,\tau,\sigma)$, $(w,\psi,\rho,\nu) \in A(\alpha,\beta) \times B_R \times (0,r) \times (\frac{\sigma^0}{2}, r)$ and $h \in W^{1,\infty}_{\alpha,0}$, we have (for typographical reasons we denote $\hat{u}(s) = \psi(0)+w(s)$, $\hat{v}(s) =$

$\psi(s-\rho)+w(s-\rho))$,

$$
\begin{aligned}
&\big\|D_zT(w,\psi,\rho,\nu)h - D_zT(z,\phi,\tau,\sigma)h\big\|_\alpha^{1,1}\\
&\qquad\le \int_0^\alpha \big|D_xf(\hat u(s),\hat v(s))h(s) - D_xf(u(s),v(s))h(s)\big|\,ds\\
&\qquad\quad+\int_0^\alpha \big|D_yf(\hat u(s),\hat v(s))h(s-\rho) - D_yf(u(s),v(s))h(s-\tau))\big|\,ds\\
&\qquad\le L\int_0^\alpha \big(|\psi(0)-\phi(0)| + |w(s)-z(s)| + |\psi(s-\rho)-\phi(s-\tau)|\\
&\qquad\quad+|w(s-\rho)-z(s-\tau)|\big)\big(|h(s)|+|h(s-\tau)|\big)\,ds\\
&\qquad\quad+L\int_0^\alpha \big|D_yf(u(s),v(s))\big|\big|h(s-\tau)-h(s-\rho)|\big)\,ds\\
&\qquad\le L\big[4\alpha(\|\psi-\phi\|^{1,1}+\|w-z\|_\alpha^{1,1})\big]\,\|h\|_\alpha^{1,1}\\
&\qquad\quad+L\big[(2\|\phi\|^{1,1}+2\|z\|_\alpha^{1,1}+1)|\rho-\tau|\big]\,\|h\|_\alpha^{1,1},
\end{aligned}
$$

it follows that $D_zT$ is continuous. An induction argument shows that $D_z^kT$ exists and is continuous.

To show that, for each fixed $\sigma>0$, the map $T(z,\phi,\tau,\sigma)$ has a continuous derivative with respect to $\phi$, it is convenient to define $G(\phi,\sigma)(t) = g(\tilde\phi(t-\sigma)) - g(\phi(-\sigma))$ and $F(z,\phi,\tau) = T(z,\phi,\tau,\sigma) - G(\phi,\sigma)$. We omit the computations for the map $F$, since they are similar to the ones above. The candidate to $D_\phi G(\phi,\sigma)$ is the linear operator defined by $D_\phi G(\phi,\sigma)h(t) = Dg(\tilde\phi(t-\sigma))\tilde h(t-\sigma) - Dg(\phi(-\sigma))h(-\sigma)$. We can write

$$
\begin{aligned}
\|G(\phi+h,\sigma)\ &-G(\phi,\sigma) - D_\phi G(\phi,\sigma)h\|_\alpha^{1,1}\\
&= \int_0^\alpha \Big|\tfrac{d}{dt}\{g(\tilde\phi(t-\sigma)+\tilde h(t-\sigma)) - g(\tilde\phi(t-\sigma))\\
&\quad -Dg(\tilde\phi(t-\sigma))\tilde h(t-\sigma)\}\Big|\,dt\\
&\le \int_0^\alpha |Dg(\tilde\phi(t-\sigma)+\tilde h(t-\sigma)) - Dg(\tilde\phi(t-\sigma))||\tilde h(t-\sigma)|\,dt\\
&\quad+\int_0^\alpha |Dg(\tilde\phi(t-\sigma)+\tilde h(t-\sigma)) - Dg(\tilde\phi(t-\sigma))\\
&\quad -D^2g(\tilde\phi(t-\sigma))\tilde h(t-\sigma)||\dot{\tilde\phi}(t-\sigma)|\,dt\\
&\le L(1+\|\phi\|^{1,1})(\|h\|^{1,1})^2,
\end{aligned}
$$

and this implies $T$ has a partial derivative with respect to $\phi$. The continuity is a consequence of the inequalities

$$
\begin{aligned}
\|D_\phi G(\psi,\sigma)h - & D_\phi G(\phi,\sigma)h\|_\alpha^{1,1} \\
&= \int_0^\alpha \left|\tfrac{d}{dt}\{Dg(\tilde\psi(t-\sigma))\tilde h(t-\sigma) - Dg(\tilde\phi(t-\sigma))\tilde h(t-\sigma)\}\right| dt \\
&\le \int_0^\alpha \left|D^2 g(\tilde\psi(t-\sigma))[\dot{\tilde\psi}(t-\sigma) - \dot{\tilde\phi}(t-\sigma)]\tilde h(t-\sigma)\right| dt \\
&\quad + \int_0^\alpha \left|[D^2 g(\tilde\psi(t-\sigma)) - D^2 g(\tilde\phi(t-\sigma))]\dot{\tilde\phi}(t-\sigma)\tilde h(t-\sigma)\right| dt \\
&\quad + \int_0^\alpha \left|[Dg(\tilde\psi(t-\sigma)) - Dg(\tilde\phi(t-\sigma))]\dot{\tilde h}(t-\sigma)\right| dt \\
&\le L(2 + \|\phi\|^{1,1})\|\psi - \phi\|^{1,1}\|h\|^{1,1}.
\end{aligned}
$$

An induction argument shows that $D_\phi^k T$ exists and is continuous. Now we show that, for each fixed $\sigma > 0$, $T$ has a partial derivative with respect to $\tau$. Since we are interested in the differentiability for $\tau > 0$, we assume that $\tau \in [\delta, r]$, for some $\delta > 0$. Then, we can choose $0 < \alpha < \delta$, so that, for $0 \le t \le \alpha$, we have $z(t-\tau) = 0$ and $\tilde\phi(t-\tau) = \phi(t-\tau)$ in the above expressions of $T$ and $D_\tau T$. Defining $\Delta\phi(= \Delta\phi(s))\colon = \phi(s-\tau-h) - \phi(s-\tau)$, we can write $\Delta\phi = -\dot\phi(s-\tau)h + h\omega_h(s)$, a.e. in $s$, and $\omega_h(s) \to 0$, a. e. in $s$, as $h \to 0$, and then

$$
\begin{aligned}
\|T\ (z,\phi,\tau+h,\sigma) &- T(z,\phi,\tau,\sigma) - D_\tau T(z,\phi,\tau,\sigma)h\|_\alpha^{1,1} \\
&= \int_0^\alpha |f(u(s),v(s)+\Delta\phi) - f(u(s),v(s)) + D_y f(u(s),v(s))\dot\phi(s-\tau)h|ds \\
&\le \int_0^\alpha |f(u(s),v(s)+\Delta\phi) - f(u(s),v(s)) - D_y f(u(s),v(s))\Delta\phi)|ds \\
&\quad + \int_0^\alpha |D_y f(u(s),v(s))||\omega_h(s)||h|\, ds \\
&\le L\int_0^\alpha |\Delta\phi|^2\, ds + L|h|\int_0^\alpha |\omega_h(s)|\, ds \\
&\le L\alpha(K_\phi)^2|h|^2 + L|h|\int_0^\alpha |\omega_h(s)|\, ds.
\end{aligned}
$$

Since $|\omega_h(s)| \le 2K_\phi$ and $|\omega_h(s)| \to 0$ a.e., as $h \to 0$, it follows from Lebesgue Dominated Convergence that $\int_0^\alpha |\omega_h(s)|\, ds \to 0$, as $h \to 0$. Thus $T$ has a partial derivative with respect to $\tau$.

Now let $z, w \in A(\alpha,\beta),\ \phi,\psi \in B_R,\ \tau,\rho \in (\delta, r), \sigma \in (\frac{\sigma^0}{2}, r)$

$$\left\|D_\tau T\ (w,\psi,\rho,\sigma) - D_\tau T(z,\phi,\tau,\sigma)\right\|_\alpha^{1,1}$$

$$= \int_0^\alpha \left|D_y f(\hat{u}(s),\hat{v}(s))\dot{\psi}(s-\rho) - D_y f(u(s),v(s))\dot{\phi}(s-\tau)\right| ds$$

$$\leq L\int_0^\alpha \big(|\psi(0)-\phi(0)| + |w(s)-z(s)| + |\psi(s-\rho)-\phi(s-\tau)|\big)\cdot$$

$$\cdot|\dot{\phi}(s-\rho)|\, ds + L\int_0^\alpha |\dot{\psi}(s-\rho)-\dot{\phi}(s-\tau)|\, ds$$

$$\leq L\big[(1+2\|\phi\|^{1,1})\|\psi-\phi\|^{1,1} + \|\phi\|^{1,1}\|w-z\|_\alpha^{1,1}$$

$$+K_\phi\|\phi\|^{1,1}|\rho-\tau|\big] + L\int_0^\alpha |\dot{\phi}(s-\rho)-\dot{\phi}(s-\tau)|\, ds.$$

From Lemma 1.3 we have $\int_0^\alpha |\dot{\phi}(s-\rho)-\dot{\phi}(s-\tau)|\, ds \to 0$, as $\rho\to\tau$, and hence $D_\tau T$ is continuous at $(\phi,\tau,\sigma)$. Similar computations show that if ($\sigma$ is fixed and) $\phi \in W^{k,\infty}$, then $D_\tau^k T$ exists and is continuous. □

From Theorem 1.2 and Lemma 2.2, we can conclude:

**Theorem 2.3** *Suppose $f\colon \mathbb{R}^{2n} \to \mathbb{R}^n$ is $C^k$ and $g\colon \mathbb{R}^n \to \mathbb{R}^n$ is $C^{k+1}$, $k \geq 2$. Then, for fixed $\sigma > 0$, the solution $x(\phi,\tau,\sigma)(t)$ of Equation (1.1) is continuously differentiable in $(\phi,\tau) \in W^{1,\infty} \times (0,r)$, and is $C^k$ in $\phi \in W^{1,\infty}$, for $t$ in each compact subinterval of $[0,b)$; furthermore, if $\phi^0$ is any fixed initial condition in $W^{k,\infty}$, then $x(\phi^0,\tau)$ is $C^k$ in $\tau \in (0,r)$.*

When $g(x) \equiv 0$, Equation (1.1) becomes an equation of retarded type; Theorem 2.3 gives one of the main results of [**7**] (actually, the result present here is slightly modified, with a few minor changes in the statement).

**Corollary 2.1** ([**7**], Theorem 3.1) *Suppose $f\colon \mathbb{R}^{2n} \to \mathbb{R}^n$ is a $C^k$ function, $k \geq 1$ and its $k^{th}$ order derivative $D^k f$, is locally Lipschitz continuous. Then, the unique solution $x(\phi,\tau)(t)$ of*

$$\begin{cases} \dot{x}(t) = f(x(t), x(t-\tau)) & \textit{if}\ t > 0 \\ x(t) = \phi(t) & \textit{if}\ -r \leq t \leq 0 \end{cases} \tag{2.3}$$

*is continuous in $(\phi,\tau)$; $x(\phi,\tau)(t)$ is continuously differentiable in $(\phi,\tau)$ in $W^{1,\infty} \times (0,r)$, and is $C^k$ in $\phi$, for $t$ in each compact subinterval of $[0,b)$; furthermore, if $\phi^0 \in W^{k,\infty}$, then $x(\phi^0,\tau)(t)$ is $C^k$ in $\tau \in (0,r)$.*

## 3 Time dependent delay

The conclusions of Theorem 2.3 still remain true for equations with delays depending on $t$; in this case, the delay is treated as an element of a convenient open set in a Banach space of continuously differentiable functions. In this section we point out the changes in the above estimates in order to get the differentiability for this case. In order to avoid excessive complications in notation, we shall restrict the delays to a neighborhood of a constant one. As it can be easily seen from the computations below we could make the delays to vary in a larger open set. For any $a > 0$, consider the Banach space $C^1([0,a],\mathbb{R})$ of continuously differentiable functions

$\sigma\colon [0,a] \to \mathbb{R}$ with norm $\|\sigma\|_{C^1} = \max\{\sup\{|\sigma(t)|\,;\, 0 \le t \le a\}, \sup\{|\dot\sigma(t)|\,;\, 0 \le t \le a\}\}$. Take constant functions $\sigma^0, \tau^0 \in C^1([0,a],\mathbb{R})$, with $0 < \sigma^0(t), \tau^0(t) < r$, numbers $c > 0$, $c < \min\{\sigma^0(0), \tau^0(0), 1-\sigma^0(0), 1-\tau^0(0)\}$, and consider the balls $B_c(\sigma^0) = \{\sigma \in C^1([0,a],\mathbb{R})\,;\, \|\sigma - \sigma^0\|_{C^1} < c\}$ and $B_c(\tau^0) \subset C^1([0,a],\mathbb{R})$. As in Section 2, for smooth functions $f, g$, delays $\sigma \in B_c(\sigma^0)$, $\tau \in B_c(\tau^0)$, and initial condition $\phi \in W^{1,\infty}$, denote by $x(\phi,\tau,\sigma)$ the solution of the initial value problem

$$\begin{cases} \frac{d}{dt}[x(t) - g(x(t-\sigma(t)))] = f(x(t), x(t-\tau(t))) & if\ \ t > 0 \\ x(t) = \phi(t) & if\ \ -r \le t \le 0. \end{cases} \tag{3.1}$$

Define, as in Section 2, the operator $T(z,\phi,\tau,\sigma) = F(z,\phi,\tau) + G(\phi,\sigma)$, where $F, G$ are defined by $F(z,\phi,\tau)(t) = G(\phi,\sigma)(t) = 0$, if $-r \le t \le 0$ and

$$G(\phi,\sigma)(t) = g(\tilde\phi(t-\sigma(t))) - g(\tilde\phi(\sigma(0)))$$
$$F(z,\phi,\tau)(t) = \int_a^t f(\phi(0) + z(s), \tilde\phi(s-\tau(s)) + z(t-\tau(s)))\,ds\,,$$

for $0 \le t \le \alpha$. The operator $T$ satisfies estimates similar to those in Section 2. The estimates showing that $T$ is a contraction with respect to either norms $\|\cdot\|^{1,1}$ and $\|\cdot\|^{1,\infty}$ are exactly the same as before. The inequalities to show that the derivatives $D_zT$, $D_\phi T$ and $D_\tau T$ exist are also the same. To prove that $D_\tau T$ exists, we define, for $h \in B_c(\sigma^0)$, $\xi_h(s) = \omega_{h(s)}(s)$ (where $\omega_h$ is defined in the proof of Lemma 2.2); then we have $\xi_{h(s)}(s) \to 0$ a.e. as $\|h\|_{C^1} \to 0$, and the remaining of the proof is the same.

The other inequalities become, for $z, w \in B(\alpha,\beta)$, $\phi, \psi \in B_\beta$, $\tau, \rho \in B_c(\tau^0)$, $\nu, \sigma \in B_c(\sigma^0)$, with $\|\rho - \tau\|_{C^1} < \varepsilon < 1 - c$,

$$\|D_zT\ (w,\psi,\rho,\nu)h - D_zT(z,\phi,\tau,\sigma)h\|_\alpha^{1,1} \le 4\alpha L\|\psi - \phi\|^{1,1}\|h\|_\alpha^{1,1}$$
$$+\big[4\alpha L\|w - z\|_\alpha^{1,1} + \tfrac{L}{1-c-\varepsilon}(1 + 2\|\phi\|^{1,1} + 2\|z\|_\alpha^{1,1})\|\rho - \tau\|_{C^1}\big]\|h\|_\alpha^{1,1},$$

$$\|D_\phi G\ (\psi,\sigma)h - D_\phi G(\phi,\sigma)h\|_\alpha^{1,1} \le L(2 + \|\phi\|^{1,1})\|\psi - \phi\|^{1,1}\|h\|^{1,1},$$

$$\|D_\tau T\ (w,\psi,\rho,\sigma)h - D_\tau T(z,\phi,\tau,\sigma)h\|_\alpha^{1,1}$$
$$\le L\|h\|_{C^1}\int_0^\alpha |\dot\phi(s-\rho) - \dot\phi(s-\tau)|\,ds + \tfrac{L}{1-c}\|\phi\|^{1,1}\|w - z\|_\alpha^{1,1}$$
$$+\tfrac{L}{1-c}\big[(1 + 2\|\phi\|^{1,1})\|\psi - \phi\|^{1,1}\|h\|_{C^1} + K_\phi\|\phi\|^{1,1}\|\rho - \tau\|_{C^1}\big]\|h\|_{C^1}.$$

The proof of these estimates is similar to the corresponding ones in the previous section, by using the following version of Lemma 1.3.

**Lemma 3.1** *Suppose $\phi, \psi \in W^{1,\infty}$, $h \in W_{0,\alpha}^{1,\infty}$, $\tau, \rho \in B_c(\sigma^0)$ and $0 < \varepsilon < 1 - c$. Then:*

(a) $\displaystyle\int_0^\alpha |\dot{\tilde\psi}(t-\tau(t)) - \dot{\tilde\phi}(t-\tau(t))|\,dt \le \tfrac{1}{1-c}\|\psi - \phi\|^{1,1}$;

(b) $\displaystyle\int_0^\alpha |\tilde\psi(t-\tau(t)) - \tilde\phi(t-\tau(t))||\dot h(t-\tau(t))|\,dt \le \tfrac{\|h\|_\alpha^{1,1}}{1-c}\|\psi - \phi\|^{1,1}$;

(c) $\displaystyle\int_0^\alpha |\tilde\phi(t-\rho(t)) - \tilde\phi(t-\tau(t))|\,dt \le \tfrac{\|\phi\|^{1,1}}{1-c-\varepsilon}\|\rho - \tau\|_{C^1}$, *if* $\|\rho - \tau\|_{C^1} < \varepsilon$;

(d) $\displaystyle\int_0^\alpha |\dot{\tilde\phi}(t-\tau(t)-\mu(t)) - \dot{\tilde\phi}(t-\tau(t))|\,dt \to 0$, *as* $\|\mu\|_{C^1} \to 0$.

**Proof** To prove (a) we note that $|1-\dot\tau(t)| > 1-c$, for all $t \in [0,\alpha]$, and then

$$\begin{aligned}\int_0^\alpha |\dot{\tilde\psi}(t-\tau(t)) - \dot{\tilde\phi}(t-\tau(t))|\,dt &\\ &\le \tfrac{1}{1-c}\int_0^\alpha |\dot{\tilde\psi}(t-\tau(t)) - \dot{\tilde\phi}(t-\tau(t))|(1-\dot\tau(t))\,dt \\ &\le \tfrac{1}{1-c}\|\psi-\phi\|^{1,1}.\end{aligned}$$

The inequality in (b) is an immediate consequence of $|\tilde\psi(t-\tau(t)) - \tilde\phi(t-\tau(t))| \le \sup_{-r\le t\le\alpha}|\tilde\psi(t-\tau(t)) - \tilde\phi(t-\tau(t))| \le \|\psi-\phi\|^{1,1}$. The inequality in (c) follows from

$$\begin{aligned}\int_0^\alpha |\tilde\phi(t-\rho(t)) - \tilde\phi(t-\tau(t))|\,dt &\\ &\le \int_0^\alpha \Big|\int_0^1 \dot{\tilde\phi}(t-\tau(t)+u(\tau(t)-\rho(t)))(\tau(t)-\rho(t))\,du\Big|\,dt \\ &\le \tfrac{\|\tau-\rho\|_{C^1}}{1-c-\varepsilon}\int_0^1\int_0^\alpha |\dot{\tilde\phi}(t-\tau(t)+u(\tau(t)-\rho(t)))|\cdot \\ &\qquad \cdot(1-\dot\tau(t)+u(\dot\tau(t)-\dot\rho(t))\,dt\,du \\ &\le \tfrac{\|\phi\|^{1,1}}{1-c-\varepsilon}\|\tau-\rho\|_{C^1}.\end{aligned}$$

To prove (d) let $\varepsilon > 0$. There exists a continuous function $\xi$ such that $\int_0^\alpha |\dot{\tilde\phi}(s) - \xi(s)|\,ds < \frac{\varepsilon(1-c)}{6}$. From the continuity of $\xi$, there is a $\delta > 0$ such that $|h| < \delta$ implies $|\xi(s+h)-\xi(s)| < \frac{\varepsilon}{2\alpha}$, for all $s \in [-r,\alpha]$. Therefore, if $\|\mu\|_{C^1} < \min(\delta, \frac{1-c}{2})$, we have

$$\begin{aligned}\int_0^\alpha |\tilde\phi\,(t-\tau(t)-\mu(t)) - \tilde\phi(t-\tau(t))|\,dt &\\ &\le \tfrac{2}{1-c}\int_0^\alpha |\tilde\phi(t-\tau(t)-\mu(t)) - \xi(t-\tau(t)-\mu(t))|(1-\dot\tau(t)-\dot\mu(t))\,dt \\ &+\int_0^\alpha |\xi(t-\tau(t)-\mu(t)) - \xi(t-\tau(t))|\,dt \\ &+\int_0^\alpha |\xi(t-\tau(t)) - \tilde\phi(t-\tau(t))|\,dt < \frac{3}{1-c}\,\frac{\varepsilon(1-c)}{6} + \frac{\varepsilon}{2} = \varepsilon.\end{aligned}$$

□

We can thus state the following result.

**Theorem 3.2** *Suppose $f\colon \mathbb{R}^{2n} \to \mathbb{R}^n$ is $C^k$, $g\colon \mathbb{R}^n \to \mathbb{R}^n$ is $C^{k+1}$, $k \ge 2$ and $a > 0$. Then, for each fixed function $\sigma$ in the ball $B_c(\sigma^0) \subset C^1([0,a],\mathbb{R})$, ($\sigma^0$ is a constant function), the solution $x(\phi,\tau,\sigma)(t)$ of Equation (1.1) is continuously differentiable in $(\phi,\tau) \in W^{1,\infty} \times B_c(\tau^0)$ and is $C^k$ in $\phi$; furthermore, if $\phi \in W^{k,\infty}$, then $x(\phi,\tau,\sigma)(t)$ is $C^k$ in $\tau$.*

## 4 Remarks

**1.** Similar results hold if in $W^{1,\infty}_{\alpha,0}$ we use the norm $\|\cdot\|^{1,p}_{\alpha}$, $p \geq 1$, instead of $\|\cdot\|^{1,1}_{\alpha}$, where, for each $z \in W^{1,\infty}_{\alpha,0}$, we define $\|z\|^{1,p}_{\alpha} = \left(\int_0^{\alpha} |z(t)|^p\, dt\right)^{1/p}$.

**2.** The significance of the results of Theorems 2.3 and 3.2 above is that they remain valid for equations with finitely many delays. Suppose $f$, $g$ are smooth functions. In order to simplify the notation below, we consider only constant delays and denote $\hat{\tau} = (\tau_1, \ldots, \tau_p)$, $\hat{\sigma} = (\sigma_1^0, \ldots, \sigma_s^0)$, $X(t-\hat{\tau}) = (x(t-\tau_1), \ldots, x(t-\tau_p))$, $X(t-\hat{\sigma}) = (x(t-\sigma_1), \ldots, x(t-\sigma_m))$, and by $x(\phi, \hat{\tau}, \hat{\sigma}) = x(\phi, \tau_1, \ldots, \tau_p, \sigma_1^0, \ldots, \sigma_s^0)$ the unique solution of the initial value problem

$$\begin{cases} \dfrac{d}{dt}\left[x(t) - g(X(t-\hat{\sigma}))\right] = f(x(t), X(t-\hat{\tau})), & \text{if } t > 0 \\ x(t) = \phi(t), & \text{if } -r \leq t \leq 0. \end{cases} \tag{4.1}$$

Let $\hat{\tau} = (\tau_1^0, \ldots, \tau_p^0)$, $\hat{\sigma} = (\sigma_1^0, \ldots, \sigma_m^0)$ be fixed. For each $j$, $1 \leq j \leq p$, the proof of Lemma 2.2 applies to the corresponding operator

$$U(z, \phi, \tau_j) = T(z, \phi, \tau_1^0, \ldots, \tau_{j-1}^0, \tau_j, \tau_{j+1}^0, \ldots, \tau_p^0, \hat{\sigma}^0).$$

Then we can state a result corresponding to Theorem 2.3: for each fixed $\hat{\sigma}^0$, the solution $x(\phi, \hat{\tau}, \hat{\sigma}^0)$ is continuously differentiable in $(\phi, \hat{\tau})$, it is $C^k$ in $\phi$, and if $\phi \in W^{k,\infty}$, then $x(\phi, \hat{\tau}, \hat{\sigma}^0)$ is $C^k$ in $\hat{\tau}$.

**3.** When $g(x) \equiv 0$ Equation (1.1) becomes an equation of retarded type (2.3). This class of equations have a very important smoothing property, namely, its solutions become smoother at each interval of length $r$. Thus, if a solution of (2.3) is defined and is bounded in $\mathbb{R}$, then it is as smooth in $t$ as $f$ is. So, the initial function of (2.3) for such a solution is $C^k$, if $f$ is $C^k$, and therefore, the solution will be $C^k$ in $\tau$. In the case $f$ is dissipative, Equation (2.3) has an attractor. If $f$ is analytic, then the elements of the attractor are analytic functions of $t$. Then we can conclude from the above results that $x(\phi, \tau)$ is $C^\infty$ in $(\phi, \tau)$. This property is not true in general for Eq. (1.1), as we can easily see from the example in Section 2. However, we state below a very interesting result in this direction (see [**6**], Theorem 6.2, p. 294). In the statement of the theorem we use the following concept: the difference operator $S\colon C \to \mathbb{R}^n$ defined by $Sx_t = x(t) - g(x(t-r))$ is said to be *stable*, if the zero solution of the difference equation $x(t) = g(x(t-r))$ is asymptotically stable.

**Theorem** ([**6**], Theorem 6.2) *If $f\colon \Omega \to \mathbb{R}^n$ ($\Omega \subset C$ is open) has derivatives through order $k$ continuous and bounded on $\Omega$, and the operator $Sx_t = x(t) - g(x(t-r))$ is stable, then any solution $x(t)$ which is bounded on $(-\infty, a]$ must be $C^{k+1}$. Also, the $\omega-$limit set of a bounded orbit consists of $C^{k+1}$ functions such that, for each $j \leq k+1$, the family of functions consisting of the $j$th derivative of each element of the $\omega-$limit set is uniformly bounded and equicontinuous.*

For this class of neutral equations (that is, when the operator $S$ is stable) we can state similar conclusions as for retarded equations.

## References

[1] Benedetto, J. [1976], *Real Variable and Integration*, B. G. Teubner, Stuttgart.

[2] Driver, R. [1965], *Existence and continuous dependence of solutions of a neutral functional differential equation*, Arch. Rational Mech. Anal. **19**, 149-166.

[3] Driver, R. [1975], *Topologies for equations of neutral type and classical electrodynamics*, Univ. of Rhode Island Technical Report No. 60, in Differential Equations with Deviating Argument, Izdat. Nauk. Dumka, Kiev, 1977, 113-127.

[4] Hale, J. K. [1969], *Ordinary Differential Equations*, Wiley.

[6] Hale, J. K. [1977], *Theory of Functional Differential Equations*, Springer-Verlag, New York/Berlin.

[7] Hale, J. K. and Ladeira, L. A. C. [1991], *Differentiability with respect to delays*, J. Differential Equations **92**, 14-26.

[8] Hale, J. K. and Ladeira, L. A. C. [1993], *Differentiability with respect to delays for a retarded reaction-diffusion equation*, Nonlinear Analysis - TMA **20**, 793-801.

[9] Melvin, W. [1972], *A class of neutral functional differential equations*, J. Differential Equations **12**, 524-534.

[10] Melvin, W. [1973], *Topologies for neutral functional differential equations*, J. Differential Equations **13**, 24-31.

[11] Melvin, W., *Bounded* $\omega_1^*$*-dynamical systems*, preprint.

Fields Institute Communications
Volume **21**, 1999

# Renormalizations for Trimodal Maps

**J. P. Lampreia**
Departamento de Matemática
Faculdade de Ciências e Tecnologia
Universidade Nova de Lisboa
2825 Monte da Caparica, Portugal

**R. Severino**
Departamento de Matemática
Universidade do Minho, Campus de Gualtar
4710 Braga, Portugal
ricardo@math.uminho.pt

**J. Sousa Ramos**
Departamento de Matemática
Instituto Superior Técnico
Universidade Técnica de Lisboa
Av. Rovisco Pais
1096 Lisboa Codex, Portugal
sramos@math.ist.utl.pt

**Abstract.** We study a family $f_a$ of trimodal maps for which the topological entropy varies in the interval $[0,1]$. We identify the possible renormalization types occuring in the family from the point of view of symbolic dynamics.

## 1 Introduction

Modern biology has sought, for the last decades, to understand the basic constituents of living organisms and the mechanism leading to its extraordinary diversity. Since then, it was recognized that genes are made of long chains of molecules, the nucleic acids, whose components were identified as four distinct pieces, the nucleotide bases, namely adenine, thymine, guanine and cytosine. One extremely important characteristic of these chains of molecules is the absolute lack of regularity of the bases position in their structure, being its precise sequence is important genetic information (one believes it is its exact sequence along a length of the DNA molecule that determines the structure and purpose of a particular organism). Thus, we can think of the nucleotide bases as the four letters — A, T, G and C — of an alphabet used to spell the total information of an organism genetic

1991 *Mathematics Subject Classification.* Primary 34C35; Secondary 58F03, 54H20.
Partially supported by CMA-UNL and Praxis XXI Project 2/2.2/Mat/132/94.

code, that is, its genoma. *Breaking the code* has been undoubtedly one of the most challenging problems of our time.

One aspect, already known, of the nucleotide bases, is how they join together to form amino acids. Indeed, the vast diversity among proteins can be attributed to a particular sequence of only twenty different amino acids, which compose their constituent polypeptide chains. Furthermore, it is known that each amino acid is coded by triplets of bases, called codons. However, since there are sixty four possible triplets but only twenty amino acids to be coded, we have that two or more triplets give equivalent descriptions of the same amino acid (for example, the lysine amino acid is specified by both AAA and AAG codons). There are also a few codons which do not represent any amino acid but have the meaning of signals to start and stop reading the genetic code.

One-dimensional discrete maps on the interval have had a key role for the understanding of many important properties and phenomena found in nonlinear dynamics and complex systems. In the context of the present work, we choose a one-dimensional family of maps $f_{abe}$ on the interval, with three critical points (trimodal maps), whose symbolic codification corresponds to an alphabet of four symbols, one for each subinterval of monotonicity. It is that analogy with the genetic code we want to start exploring. However, the space of the symbolic sequences obtained from such family $f_{abe}$ of maps is extremely complex, whence we restrict our study to the properties of those maps related only with the topological entropy. To accomplish that, we introduce a one-parameter subfamily of trimodal maps $f_a$ characterized by the topological entropy, for which we show that the topological entropy is monotone nondecreasing as a function of the parameter $a$. Within the subintervals of constant entropy, we also show the existence of self-similarities and renormalization properties for $f_a$, introducing a symbolic product between trimodal sequences.

In this work, we put a particular emphasis on the analysis of sequences generated by four different symbols and their dependence with the topological entropy. From the point of view of dynamical systems, this entropy is easily computed, e.g., from the spectral radius of the Markov matrix transition associated to the kneading sequence and, since the concept of entropy gives us a measure of the complexity of a biological system, we think it can be useful in the search for a mathematical model that allows the understanding of the complexity of the nucleotide bases sequences.

## 2 Preliminaries

This section presents some of the definitions, notations and results related to symbolic dynamics of trimodal maps on the interval. Consider a 3-parameter family of trimodal maps $f_{abe}$ on the interval $[c_0, c_4]$, satisfying the following conditions:

i) $f_{abe}(c_0) = f_{abe}(c_4) = c_0, \quad f'_{abe}(c_0) > 0;$
ii) $f'_{abe}(x)$ has 3 real roots $c_1, c_2$ and $c_3$ in the interval $[c_0, c_4]$;
iii) the Schwarzian derivative of $f_{abe}$ is negative, that is

$$S(f_{abe})(x) = \frac{f'''_{abe}(x)}{f'_{abe}(x)} - \frac{3}{2}\left(\frac{f''_{abe}(x)}{f'_{abe}(x)}\right)^2 < 0,$$

for all $x \in [c_0, c_4]$, differs from $c_1$, $c_2$ and $c_3$;
iv) $f_{abe}(c_1) \leq c_4, f_{abe}(c_2) \geq c_0$ and $f_{abe}(c_3) \leq c_4$.

Note that the condition of negative Schwarzian derivative is extremely simple and useful to reduce the topological study of these families to the orbits of the critical points [**3**].

Under these assumptions, we are able to use Milnor-Thurston's kneading theory to classify and understand the topological structure of such a family of maps. The strategy in this case involves the following three steps: first, for each value of the parameters $(a, b, e)$, consider the orbits of the (critical) points $c_1, c_2$ and $c_3$, i.e., the sets of points,

$$\mathcal{O}_{abe}(c_i) = \{x_j^{(i)} \ : \ x_j^{(i)} = f_{abe}^j(c_i),\ j \in \mathbb{N}\}, \quad i = 1, 2, 3.$$

Next, associate to each monotone subinterval $[c_i, c_{i+1}]$, with $i = 0, 1, 2, 3$, and each critical point $c_k$, with $k = 1, 2, 3$, the symbols $L$,$M$,$N$,$R$, and $A$,$B$,$C$, respectively. Finally, to each numerical orbit $\mathcal{O}_{abe}(c_i)$ associate a sequence of symbols, $S = S_1\, S_2\, \ldots$, given by:

$$\begin{array}{lll} S_j = L & \text{if} & f_{abe}^j(c_i) < c_1, \\ S_j = A & \text{if} & f_{abe}^j(c_i) = c_1, \\ S_j = M & \text{if} & c_1 < f_{abe}^j(c_i) < c_2, \\ S_j = B & \text{if} & f_{abe}^j(c_i) = c_2, \\ S_j = N & \text{if} & c_2 < f_{abe}^j(c_i) < c_3, \\ S_j = C & \text{if} & f_{abe}^j(c_i) = c_3, \\ S_j = R & \text{if} & f_{abe}^j(c_i) > c_3. \end{array}$$

Thus, we get a classification theorem (see [**5**] for details) depending on these three symbolic sequences associated with the critical points of $f_{abe}$.

Now, we introduce an ordering $\prec$ between symbolic sequences related with the order of the interval. Indeed, it can be easily seen that the order of the real line induces the order on the symbols $L \prec A \prec M \prec B \prec N \prec C \prec R$. Then, given two symbolic sequences $U = U_1 \ldots U_i U_{i+1} \ldots$ and $V = V_1 \ldots V_i V_{i+1} \ldots$ such that $U_j = V_j$, for $j \leq i$, and $U_{i+1} \neq V_{i+1}$, we say that $U \prec V$ if either

i) $\rho(U_1 \ldots U_i) = \rho(V_1 \ldots V_i)$ is even and $U_{i+1} \prec V_{i+1}$, or
ii) $\rho(U_1 \ldots U_i) = \rho(V_1 \ldots V_i)$ is odd and $V_{i+1} \prec U_{i+1}$,

where $\rho(U_1 \ldots U_i)$, the parity of $U_1 \ldots U_i$, denotes the number of symbols $M$ and $R$ present in the common subsequence $U_1 \ldots U_i$. As we shall see later, this order relation will allow a nice characterization of the triples of four symbols dynamically interesting.

The infinite sequences of symbols associated to the orbits of the critical points $c_1, c_2$ and $c_3$ are usually called kneading sequences and its triple, the kneading data, $K(f_{abe})$, of the map $f_{abe}$. Observe that when an orbit $\mathcal{O}_{abe}(c_i)$ is periodic, of period $p$, then the corresponding symbolic sequence is also periodic, with the same period $p$, and it can be written as $S = (S_1 \ldots S_p)^\infty$. Thus, we shall adopt the abbreviated notation $S = S_1 \ldots S_p = S^{(p)}$ for these periodic symbolic sequences. Hence, when $f_{abe}$ is such that the orbits of the three critical points $c_1, c_2$ and $c_3$ are periodic (of periods $p, q$ and $s$, respectively), we get that its kneading data can be written as

$$K(f_{abe}) = (P^{(p)}, Q^{(q)}, S^{(s)}) = (P_1 \ldots P_{p-1}A, Q_1 \ldots Q_{q-1}B, S_1 \ldots S_{s-1}C).$$

We shall use a similar abbreviated notation when the orbit of one critical point includes another critical point, such as, $P_1 \ldots P_{p-1}BQ_1 \ldots Q_{q-1}A$. Observe that, in this case, the sequence $P_1 \ldots P_{p-1}BQ_1 \ldots Q_{q-1}A$ describes the symbolic orbits of both $c_1$ and $c_2$ (the latter being $Q_1 \dots Q_{q-1}AP_1 \dots P_{p-1}B$). Then, we simplify the notation and write the kneading data as a pair of sequences

$$K(f_{abe}) = (P_1 \dots P_{p-1}BQ_1 \dots Q_{q-1}A, S_1 \dots S_{s-1}C).$$

## 3 One-parameter family of trimodal maps

The topological classification of a generic family $f_{abe}$ of trimodal maps depends on three topological invariants, making its study extremely difficult. Hence, we need to consider some additional assumptions to restrict our study to a subfamily of trimodal maps depending only on one topological invariant — the well known topological entropy $h_t(f_{abe}) = \log_4 s(f_{abe})$, where $s(f_{abe})$ is the growth number of $f_{abe}$, that is, $s(f_{abe}) = \lim_{k\to\infty} N(f^k)^{1/k}$, where $N(f^k)$ is the number of fixed points of $f^k$. As we shall prove, this problem can be solved considering trimodal maps as a second iterate of a one-parameter unimodal family of maps, such as

$$g_a(x) = -ax^2 - 1 + a, \qquad x \in [-1, 1]$$

with the parameter $a \in [1, 2]$. Then, a family of trimodal maps that satisfies the condition above is given by

$$f_a(x) = g_a^2(x) = -a^3x^4 + 2a^2(a-1)x^2 - (a^3 - 2a^2 + 1).$$

Since any unimodal map is totally characterized, up to self-similarities, by its topological entropy (see [**4**]), we immediately ensure that $f_a$ does not depend on the other two invariants. One implication of our construction is the following result:

**Proposition 3.1** *The topological entropy $h_t(f_a)$ is a monotone non-decreasing function of the parameter $a$, with $h_t(f_a) \in [0, 1]$.*

**Proof** Since $s(f_a) = s(g_a^2) = s(g_a)^2$ and $1 \le s(g_a) \le 2$, the result follows directly from Theorem 13.1 and Corollary 13.2 of [**6**]. □

Observe that $f_a$ is a full family of maps for the entropy, in the sense that every value in the interval $[0, 1]$ occurs as $h_t(f_a)$, for some $a \in [1, 2]$, where the interval $[0, 1]$ is known to be the full range of $h_t(f_{abe})$, with $f_{abe}$ a generic family of trimodal maps.

Now, we are going to characterize the kneading sequences obtained from the family of maps $f_a$. Being $f_a$ the second iterate of the family $g_a$ of unimodal maps, we can easily rewrite the kneading sequences of $g_a$ as kneading sequences of $f_a$ following the translation table given below:

| unimodal | $\rightarrow$ | trimodal |
|---|---|---|
| $ll$ | | $L$ |
| $lc$ | | $A$ |
| $lr$ | | $M$ |
| $cr$ | | $B$ |
| $rr$ | | $N$ |
| $rc$ | | $C$ |
| $rl$ | | $R$ |

where $l, c, r$ are the symbols defined in the symbolic codification of the unimodal maps $g_a$ (we associate the symbol $l$ if $g_a(x) < 0$, $c$ if $g_a(x) = 0$ and $r$ if $g_a(x) > 0$). Thus, to each kneading sequence $K(g_a)$ there is a corresponding kneading data $K(f_a)$, with the following properties: if the length of $K(g_a)$ is even, that is, if $K(g_a) = s_1 \dots s_{2k-1}c$, we have that $K(f_a) = (P_1 \dots P_{k-1}E, Q_1 \dots Q_{k-1}B,\ P_1 \dots P_{k-1}E)$, with $E = A$ or $C$, depending on the symbol $s_{2k-1}$ (note that only one of these possibilities can actually occur). Notice that under these circumstances, the orbits of both critical points $c_1$ and $c_3$ are equal, which motivates the abbreviated notation $K(f_a) = (P_1 \ldots P_{k-1}E, Q_1 \ldots Q_{k-1}B)$. On the other hand, if the length of $K(g_a)$ is odd, that is, if $K(g_a) = s_1 \ldots s_{2k}c$, we get, in a similar abbreviated notation, $K(f_a) = P_1 \ldots P_{k-1}BQ_1 \ldots Q_{k-2}E$, with $E = A$ or $C$, depending on the symbol $s_{2k}$.

Let $\mathcal{T}_H$ denote the set of kneading data $K(f_a)$. The following result establishes the characterization of a triple of sequences as an element of $\mathcal{T}_H$.

**Proposition 3.2** *Given two sequences $P$ and $Q$ of the symbols $L, A, M, B, N, C$ and $R$, then we have $(P, Q) = K(f_a)$, for some value of the parameter $a$, if and only if*

$$Q \preceq \sigma^k(Q), \sigma^k(P) \preceq P$$

*where $\sigma$ denotes the usual shift operator ($\sigma(P_1P_2P_3...) = P_2P_3...$) and $\sigma^k$ is the k-fold iterate of $\sigma$ .*

**Proof** Follows as a simple application of the results for unimodal maps (see [**1**]). □

Considering the order relation $\prec$ between sequences, given above, we can present $\mathcal{T}_H$ with a tree structure, as seen in Figure 1:

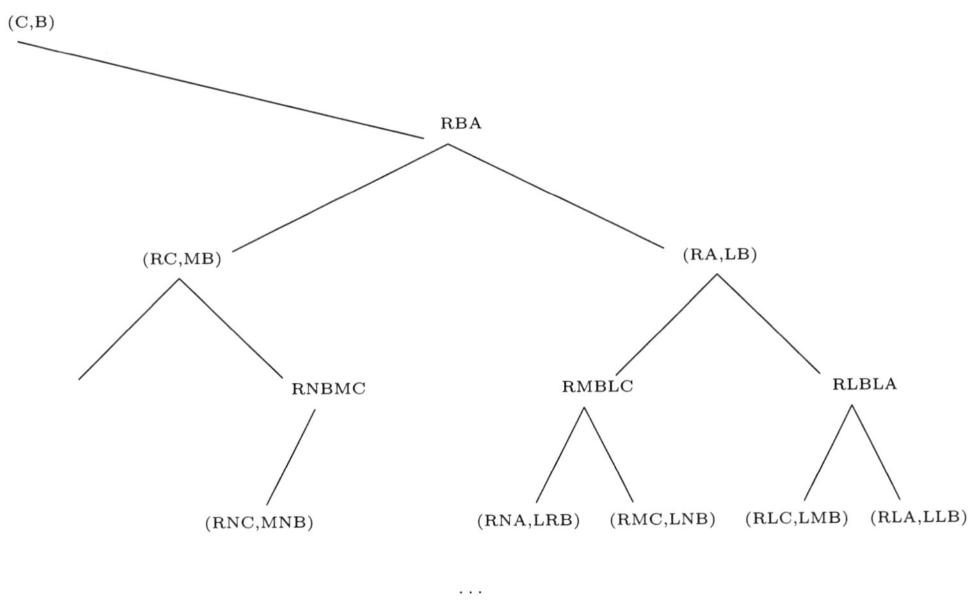


**Figure 1** Tree $\mathcal{T}_H$ of the kneading data $K(f_a)$, where each level corresponds to a new iteration and the unlabeled vertices correspond to non-admissible itineraries of the critical points.

The tree structure of $\mathcal{T}_H$ is useful since it gives an insight of the relations between its elements.

## 4 Product of symbolic sequences

This section generalizes the $*$-product introduced by [**2**] for unimodal maps to the family $f_a$ of trimodal maps. We remember that the universal periodic doubling for unimodal maps can be given by the $*$-product $rc \to rc * rc = rlrc \to rlrc * rc = rlrrrlrc \to \ldots$ .

The generalization of this $*$-product to symbolic sequences obtained from a generic family $f_{abe}$ of trimodal maps will be a key step for the understanding of the topological structure of the set of symbolic sequences with four symbols. Besides the mathematical interest of this generalization, we believe that it could help to give an insight of the relation between this topological structure and the genetic code table.

As we just refereed, there are two kinds of kneading data $K(f_a)$, depending on whether the length of $K(g_a)$ is even or odd. Therefore, it will be necessary to consider four different situations in order to define a product between such symbolic sequences. It should be noted that, since it is possible to characterize the result of the product by one kneading sequence of the second factor's kneading data, we are going to simplify the notation, writing only that kneading sequence.

**Type I:**

$$\begin{aligned}(P_1 \ldots P_{p-1}E, Q_1 \ldots Q_{p-1}B) * X_1 \ldots X_{x-1}D \\ = (P_1 \ldots P_{p-1}X_{11}P_1 \ldots P_{p-1}X_{12}P_1 \ldots P_{p-1} \ldots D_1 P_1 \ldots P_{p-1}D_2, \\ Q_1 \ldots Q_{p-1}\widetilde{X}_{11}Q_1 \ldots Q_{p-1}\widetilde{X}_{12}Q_1 \ldots Q_{p-1} \ldots B_1 Q_1 \ldots Q_{p-1}B_2).\end{aligned}$$

**Type II:**

$$\begin{aligned}(P_1 \ldots P_{p-1}\, E, Q_1 \ldots Q_{p-1}B) * X_1 \ldots X_{x-1}B \\ = (P_1 \ldots P_{p-1}X_{11}P_1 \ldots P_{p-1}X_{12}P_1 \ldots P_{p-1} \ldots X_{x-1,2}P_1 \ldots P_{p-1}E, \\ Q_1 \ldots Q_{p-1}\widetilde{X}_{11}Q_1 \ldots Q_{p-1}\widetilde{X}_{12}Q_1 \ldots Q_{p-1} \ldots \widetilde{X}_{x-1,2}Q_1 \ldots Q_{p-1}B),\end{aligned}$$

where $E$, $D = A$ or $C$, $(X_{i1}, X_{i2}) = \Phi_E^{\rho(P)}(X_i)$ and $(\widetilde{X}_{i1}, \widetilde{X}_{i2}) = \Psi^{\rho(Q)}(X_i)$ (see table below).

$(X_{i1}, X_{i2}) = \Phi_E^{\rho(P)}(X_i)$

| $X_i$ | $E$ | $\rho(P)=+1$ | $\rho(P)=-1$ |
|---|---|---|---|
| L | A | ( L, L) | ( M, M) |
| L | C | ( N, N) | ( R, R) |
| A | A | ( L, A) | ( M, A) |
| A | C | ( N, C) | ( R, C) |
| M | A | ( L, M) | ( M, L) |
| M | C | ( N, R) | ( R, N) |
| B | A | A | A |
| B | C | C | C |
| N | A | ( M, M) | ( L, L) |
| N | C | ( R, R) | ( N, N) |
| C | A | ( M, A) | ( L, A) |
| C | C | ( R, C) | ( N, C) |
| R | A | ( M, L) | ( L, M) |
| R | C | ( R, N) | ( N, R) |

$(\widetilde{X}_{i1}, \widetilde{X}_{i2}) = \Psi^{\rho(Q)}(X_i)$

| $X_i$ | $\rho(Q)=+1$ | $\rho(Q)=-1$ |
|---|---|---|
| L | ( M, M) | ( N, N) |
| A | ( M, B) | ( N, B) |
| M | ( M, N) | ( N, M) |
| B | B | B |
| N | ( N, N) | ( M, M) |
| C | ( N, B) | ( M, B) |
| R | ( N, M) | ( M, N) |

**Type III:**

$$
\begin{aligned}
&(P_1 ... P_{p-1}\, BQ_1 ... Q_{p-2}E) * X_1 ... X_{x-1}D \\
&\qquad = (P_1 ... P_{p-1}X_{11}Q_1 ... Q_{p-2}X_{12}P_1 ... P_{p-1} ... D_1 Q_1 ... Q_{p-2}D_2, \\
&\qquad\quad Q_1 ... Q_{p-2}\widetilde{X}_{11}P_1 ... P_{p-1}\widetilde{X}_{12}Q_1 ... Q_{p-2} ... B_1 P_1 ... P_{p-1}B_2).
\end{aligned}
$$

**Type IV:**

$$
\begin{aligned}
&(P_1 ... P_{p-1}\, BQ_1 ... Q_{p-2}E) * X_1 ... X_{x-1}BX_{x+1} ... X_{2x-2}D \\
&\qquad = (P_1 ... P_{p-1}X_{11}Q_1 ... Q_{p-2}X_{12}P_1 ... P_{p-1} ... X_{x-1,2}P_1 ... P_{p-1}B_1 \\
&\qquad\quad Q_1 ... Q_{p-2}B_2 P_1 ... P_{p-1}X_{x+1,1}Q_1 ... Q_{p-2}X_{x+1,2}P_1 ... P_{p-1} ... D_1 \\
&\qquad\quad Q_1 ... Q_{p-2}D_2),
\end{aligned}
$$

where $E$, $D = A$ or $C$, $(X_{i1}, X_{i2}) = \Theta_E^{\rho(P)}(X_i)$ and $(\widetilde{X}_{i1}, \widetilde{X}_{i2}) = \Lambda_E^{\rho(Q)}(X_i)$ (see table below).

$(X_{i1}, X_{i2}) = \Theta_E^{\rho(P)}(X_i)$

| $X_i$ | $E$ | $\rho(P) = +1$ | $\rho(P) = -1$ |
|---|---|---|---|
| L | A | ( N, M) | ( M, L) |
| L | C | ( N, N) | ( M, R) |
| A | A | ( N, A) | ( M, A) |
| A | C | ( N, C) | ( M, C) |
| M | A | ( N, L) | ( M, M) |
| M | C | ( N, R) | ( M, N) |
| B | A | ( B, L) | ( B, M) |
| B | C | ( B, R) | ( B, N) |
| N | A | ( M, L) | ( N, M) |
| N | C | ( M, R) | ( N, N) |
| C | A | ( M, A) | ( N, A) |
| C | C | ( M, C) | ( N, C) |
| R | A | ( M, M) | ( N, L) |
| R | C | ( M, N) | ( N, R) |

$(\widetilde{X}_{i1}, \widetilde{X}_{i2}) = \Lambda_E^{\rho(Q)}(X_i)$

| $X_i$ | $E$ | $\rho(Q) = +1$ | $\rho(Q) = -1$ |
|---|---|---|---|
| L | A | ( L, M) | ( M, N) |
| L | C | ( N, N) | ( R, M) |
| A | A | ( L, B) | ( M, B) |
| A | C | ( N, B) | ( R, B) |
| M | A | ( L, N) | ( M, M) |
| M | C | ( N, M) | ( R, N) |
| B | A | ( A, N) | ( A, M) |
| B | C | ( C, M) | ( C, N) |
| N | A | ( M, N) | ( L, M) |
| N | C | ( R, M) | ( N, N) |
| C | A | ( M, B) | ( L, B) |
| C | C | ( R, B) | ( N, B) |
| R | A | ( M, M) | ( L, N) |
| R | C | ( R, N) | ( N, M) |

The following examples are intended to illustrate these definitions:

**Type I:**

$$(RA, LB) * RA = (RLRMRMRA, LMLNLNLB).$$

**Type II:**

$$(RA, LB) * RBA = (RLRMRA, LMLNLB).$$

**Type III:**

$$RNBMC * RMA = (RNMMNRNNMRRNNMC, MRRNNMNRNMMNRNB).$$

**Type IV:**

$$RMBLC * RNBMC = RMNLRRMNLNRMBLNRMMLNRMNLC.$$

The importance of this $*$-product will be clear with the following results:

**Theorem 4.1** *Given $U$ and $W \in \mathcal{T}_H$, then $U * W \in \mathcal{T}_H$.*

**Proof** The admissibility of the product $U*W$ is a consequence of the properties of the unimodal $*$-product and the rewrite rules given above. The arguments which shows the admissibility of the different product types are similar, therefore we shall prove the result only for the type I.

Let $U = (P_1 \dots P_{p-1}E, Q_1 \dots Q_{p-1}B)$ and $W = X_1 \dots X_{x-1}D$. From the rewrite rules, we obtain the corresponding unimodal sequences $u = p_1 \dots p_{2p-1}c$ and $w = x_1 x_2 \dots x_{2x-1}c$, and its unimodal $*$-product

$$u * w = p_1 \dots p_{2p-1}\widehat{x}_1 p_1 \dots p_{2p-1}\widehat{x}_2 p_1 \dots p_{2p-1} \cdots p_1 \dots p_{2p-1}\widehat{x}_{2x-1} p_1 \dots p_{2p-1}c$$

where $\widehat{l} = r$ and $\widehat{r} = l$, if $u$ is odd and $\widehat{l} = l, \widehat{r} = r$ if $u$ is even.

Now, we are going to show that the kneading data corresponding to $u * w$ is the product $U * W$, according to the definitions given above.

Without lost of generality, since the other situation can be proven by similar arguments, suppose $\rho(P_1 \dots P_{p-1}E) = +1$. Then, the $r$-parity of the unimodal sequence $u$ is also $+1$ and we can write

$$u * w = P_1 \dots P_{p-1} p_{2p-1} \widehat{x}_1 P_1 \dots P_{p-1} p_{2p-1} \widehat{x}_2 P_1 \dots P_{p-1} \dots$$
$$P_1 \dots P_{p-1} p_{2p-1} \widehat{x}_{2x-1} P_1 \dots P_{p-1} p_{2p-1} c$$
$$= P_1 \dots P_{p-1} X_{11} P_1 \dots P_{p-1} X_{12} P_1 \dots P_{p-1} \dots P_1 \dots P_{p-1} D_1 P_1 \dots P_{p-1} D_2,$$

that is, the only thing yet to be determined are the symbols $X_{i1}$ and $X_{i2}$, with $i = 1, \dots, p-1$. Now, suppose $E = A$; then $p_{2p-1} = l$ and we have the following different situations:

$$\text{if } X_i = L \;\Rightarrow\; X_i = x_{2i-1}x_{2i} = ll \;\Rightarrow\; \begin{cases} p_{2p-1}\widehat{x}_{2i-1} = l\widehat{l} = ll \Rightarrow X_{i1} = L \\ p_{2p-1}\widehat{x}_{2i} = l\widehat{l} = ll \Rightarrow X_{i2} = L \end{cases}$$
$$\text{if } X_i = A \;\Rightarrow\; X_i = x_{2i-1}x_{2i} = lc \;\Rightarrow\; \begin{cases} p_{2p-1}\widehat{x}_{2i-1} = l\widehat{l} = ll \Rightarrow X_{i1} = L \\ p_{2p-1}\widehat{x}_{2i} = l\widehat{c} = lc \Rightarrow X_{i2} = A \end{cases}$$
$$\text{if } X_i = M \;\Rightarrow\; X_i = x_{2i-1}x_{2i} = lr \;\Rightarrow\; \begin{cases} p_{2p-1}\widehat{x}_{2i-1} = l\widehat{l} = ll \Rightarrow X_{i1} = L \\ p_{2p-1}\widehat{x}_{2i} = l\widehat{r} = lr \Rightarrow X_{i2} = M \end{cases}$$
$$\text{if } X_i = N \;\Rightarrow\; X_i = x_{2i-1}x_{2i} = rr \;\Rightarrow\; \begin{cases} p_{2p-1}\widehat{x}_{2i-1} = l\widehat{r} = lr \Rightarrow X_{i1} = M \\ p_{2p-1}\widehat{x}_{2i} = l\widehat{r} = lr \Rightarrow X_{i2} = M \end{cases}$$
$$\text{if } X_i = C \;\Rightarrow\; X_i = x_{2i-1}x_{2i} = rc \;\Rightarrow\; \begin{cases} p_{2p-1}\widehat{x}_{2i-1} = l\widehat{l} = lr \Rightarrow X_{i1} = M \\ p_{2p-1}\widehat{x}_{2i} = l\widehat{c} = lc \Rightarrow X_{i2} = A \end{cases}$$
$$\text{if } X_i = R \;\Rightarrow\; X_i = x_{2i-1}x_{2i} = rl \;\Rightarrow\; \begin{cases} p_{2p-1}\widehat{x}_{2i-1} = l\widehat{l} = lr \Rightarrow X_{i1} = M \\ p_{2p-1}\widehat{x}_{2i} = l\widehat{r} = ll \Rightarrow X_{i2} = L \end{cases}$$

from which we determine the values of the function $\Phi_E^{\rho(P)}$, i.e., that $(X_{i1}, X_{i2}) = \Phi_E^{\rho(P)}(X_i)$, with, in this particular case, $E = A$ and $\rho(P) = +1$. As stated before, the other cases follow in the same way. □

**Definition 4.1** *Given $U \in \mathcal{T}_H$, denote by $\mathcal{T}_U$ the subset of kneading sequences given by*

$$\mathcal{T}_U = \{V \in \mathcal{T}_H : V = U * W, \;\; W \in \mathcal{T}_H\}.$$

One implication of this construction is that, to each vertex $U \in \mathcal{T}_H$, there exists an isomorphism between the subtree $\mathcal{T}_U$ and $\mathcal{T}_H$, but where the elements of the subtree $\mathcal{T}_U$, are located in multiple $u$-levels of $\mathcal{T}_H$ (see Figure 2, for a partial view of the subtree $\mathcal{T}_{(RA,LB)}$). The next result states that there are no overlapping of subtrees.

**Proposition 4.2** *Any subtree $\mathcal{T}_U$ consists of all the admissible sequences in $\mathcal{T}_H$ located between its extreme branches, that is, between the sequences $U * L^\infty$ and $U * RL^\infty$.*

**Proof** From [4], the unimodal $*$-product satisfies a similar result. Since the trimodal $*$-product is induced by it, both order and admissibility structures are preserved. Thus, these properties of the unimodal family $g_a$ are naturally transfered to the trimodal family $f_a$. □

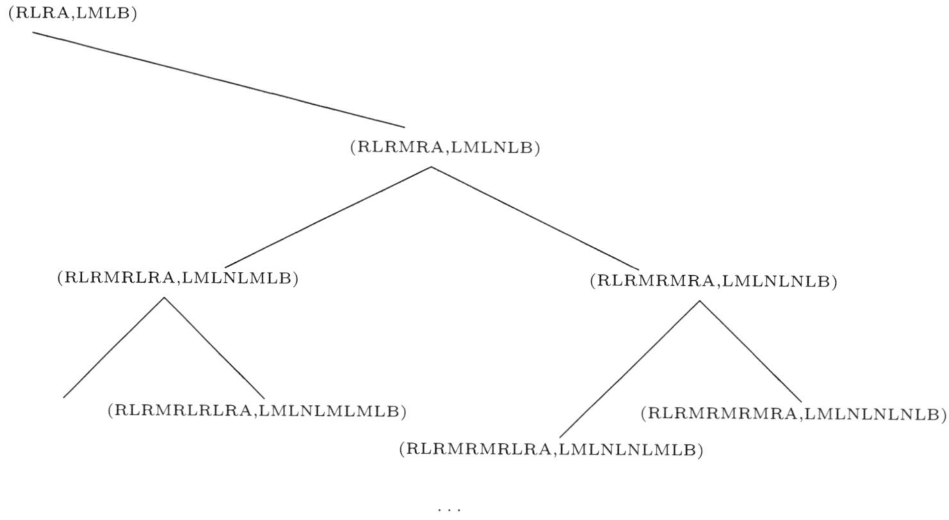


**Figure 2** Subtree of kneading sequences $\mathcal{T}_{(RA,LB)}$.

Our final result characterizes the variation of the topological entropy in a subtree. But, to simplify the notation, we are going to write $h_t(U)$ for the topological entropy of the map $f_a$ whose kneading data $K(f_a)$ is $U$.

**Proposition 4.3** *Let $U \in \mathcal{T}_H$. Then, given $V \in \mathcal{T}_U$, that is, $V = U * W$, for some $W \in \mathcal{T}_H$, we have the following results concerning the topological entropy:*

$$\begin{array}{ll} h_t(V) = \frac{1}{2} h_t(W) & \text{if } U = (C,B) \\ h_t(V) = h_t(U) & \text{if } U \neq (C,B). \end{array}$$

**Proof** Follows from similar results for the unimodal maps (see [**1**]). □

## Acknowledgements

We are grateful to the referee for his suggestions.

## References

[1] Collet, P. and Eckmann, J. P. [1980], *Iterated maps on the interval as dynamical systems,* Birkhauser: Basel and Boston.

[2] Derrida, B., Gervois, A. and Pomeau, Y. [1978], *Iteration of endomorphisms on the real axis and representation of numbers,* Ann. Inst. Henri Poincaré **A XXIX**, 305-356.

[3] Guckenheimer, J. [1979], *Sensitive dependence on initial conditions for one dimensional maps,* Comm. Math. Phys. **70**, 133-160.

[4] Lampreia, J. P., Rica da Silva, A. and Sousa Ramos, J. [1986], *Subtrees of unimodal maps tree,* Boll. U. M. I., Anal. Funz. Appl., serie VI, **V-C**, 159-167.

[5] Melo, W. and van Strein, S. [1993], *One-Dimensional Dynamics.* Springer-Verlag.

[6] Milnor, J. and Thurston, W. [1988], *On iterated maps of the interval,* in Proceedings Univ. Maryland 1986-1987 (J. C. Alexander, ed.), Lect. Notes in Math. n. 1342, Springer-Verlag, 465-563.

Fields Institute Communications
Volume **21**, 1999

# Statistical Stability of Strongly Perturbed Dynamical Systems

**Andrzej Lasota**
Institute of Mathematics
Silesian University
ul. Bankowa 14
40–007 Katowice, Poland
lasota@gate.us.edu.pl

**Michael C. Mackey**
Departments of Physiology, Physics, and Mathematics
Centre for Nonlinear Dynamics in Physiology and Medicine
McGill University
3655 Drummond
Montreal, Quebec, Canada H3G 1Y6
mackey@cnd.mcgill.ca

**Abstract.** In this note we outline the study of the asymptotic behaviour of randomly perturbed dynamical systems from a statistical point of view, using four different techniques. The results are formulated in the language of Markov operator theory. Each technique is applied to a typical biological model. We also show an extension of the last technique – the Foguel alternative – to continuous time systems and its application to the logistic equation with parametric noise.

## 1 Introduction

It has been known for many years that randomly perturbed dynamical systems can be effectively studied by the use of Markov operators. Recently discovered methods in the theory of Markov operators facilitate this study. The purpose of this paper is to present a few of these techniques which, in our opinion, are of great potential value, and to illustrate their use with applications to different biological models.

In Section 2 we start with a description of discrete time stochastically perturbed systems, and derive the corresponding transition operators. Then in Section 3 we review some basic facts from the theory of Markov operators, and outline four techniques which allow one to determine the asymptotic behaviour of perturbed

---

1991 *Mathematics Subject Classification.* Primary 60J05, 92D25; Secondary 45D05, 47B38.

This research was partially supported by grants from the State Committee for Scientific Research (Poland), the Natural Sciences and Research Council (NSERC, Canada), and Le Fonds pour la Formation de Chercheurs et l'Aide à la Recherche (FCAR, Québec).

systems from a statistical point of view. Sections 4 and 5 are devoted to the lower bound function technique and its application to mathematical models of the cell cycle. In Section 6 we present the asymptotic decomposition theorem, and show how it can be used to describe dynamical systems with additive and multiplicative perturbations. In Section 7 we study integral operators having a positive invariant density and we show an application to the Tyson–Hannsgen model of the cell cycle. Finally in Sections 8 and 9 we present a new version of the Foguel alternative and its application to the parametrically perturbed logistic equation. We also formulate an open problem concerning the existence of a subinvariant function for integral operators of Volterra type with advanced argument.

Most of the results discussed here are relatively new (e.g., [**1, 9, 13**]) or were especially prepared for this note. In particular we present new proofs of the asymptotic stability for the Tyson & Hannsgen and Tyrcha models of the cell cycle and also show new applications of the Foguel alternative.

## 2 Perturbed dynamical systems

We consider a stochastically perturbed discrete time dynamical system of the form

$$x_{n+1} = T(x_n, \xi_n), \quad n = 0, 1, \dots , \tag{2.1}$$

where $x_n$ is the state variable of the system at time $t = n$ and the $\xi_n$ are independent random variables which represent the perturbations. Throughout we make the following assumptions:

1. The transformation $T : X \times W \to X$ is defined on a set $X \times W$ where $X \subset \mathbb{R}^d$ is a closed set and $W \subset \mathbb{R}^k$ is Borel measurable. For every fixed $w \in W$ the function $T(x, w)$ is continuous in $x$ and for every fixed $x \in X$ it is Borel measurable in $w$.
2. The random variables $\xi_0, \xi_1, \dots$ with values in $W$ are independent and have the same distribution,
$$\nu(B) = \text{prob}\,(\xi_n \in B), \quad B \subset W, \quad B \text{ Borelian.}$$
3. The initial vector $x_0$ with values in $X$ is independent of $\{\xi_n\}$.

Our goal is to study the statistical properties of the trajectories $\{x_n\}$. Thus we define the corresponding sequence of distributions

$$\mu_n(A) = \text{prob}\,(x_n \in A), \qquad A \subset X, \quad A \text{ Borelian,}$$

and look for a recurrence relation that will give $\mu_{n+1}$ in terms of $\mu_n$.

Let $h$ be a real valued bounded Borel measurable function defined on $X$. Then the mathematical expectation of $h(x_{n+1})$ is given by the formula

$$E(h(x_{n+1})) = \int_X h(x)\mu_{n+1}(dx). \tag{2.2}$$

On the other hand, using (2.1), we have $h(x_{n+1}) = h(T(x_n, \xi_n))$ and consequently

$$E(h(x_{n+1})) = \int_X \int_W h(T(x, w))\mu_n(dx)\nu(dw). \tag{2.3}$$

From Equations (2.2), (2.3), in the special case where $h = 1_A$ is the indicator function of a set $A$, we obtain

$$\mu_{n+1}(A) = \int_X \int_W 1_A(\,T(x, w)\,)\mu_n(dx)\nu(dw). \tag{2.4}$$

This is the desired recurrence condition. If the $\mu_n$ are absolutely continuous measures with densities $f_n(x) = d\mu_n/dx$ then Equation (2.4) also allows us to calculate $f_{n+1}$ from $f_n$. In this case we write $f_{n+1} = Pf_n$ and we call $P$ the *transition operator* corresponding to the dynamical system (2.1). We illustrate this situation by two important examples.

**Example 2.1** Additive Perturbations. Assume that $X = W = \mathbb{R}^d$ and that $T(x,w)$ is linear in $w$, so $T(x,w) = S(x) + w$ with a continuous $S : \mathbb{R}^d \to \mathbb{R}^d$. In this case (2.1) reduces to

$$x_{n+1} = S(x_n) + \xi_n, \qquad n = 0, 1, \dots , \tag{2.5}$$

and (2.4) takes the form

$$\mu_{n+1}(A) = \int_{\mathbb{R}^d}\int_{\mathbb{R}^d} 1_A(S(x) + w)\mu_n(dx)\nu(dw).$$

If in addition we assume that the measure $\nu$ is absolutely continuous with a density $g$, then

$$\mu_{n+1}(A) = \int_{\mathbb{R}^d}\Big\{\int_{\mathbb{R}^d} 1_A(S(x) + w)g(w)dw\Big\}\mu_n(dx).$$

Substituting $x = u$, $S(x) + w = y$ and changing the order of integration we obtain

$$\mu_{n+1}(A) = \int_A\Big\{\int_{\mathbb{R}^d} g(y - S(u))\mu_n(du)\Big\}dy.$$

The function inside the braces is integrable with respect to $y$ and consequently $\mu_{n+1}$ is an absolutely continuous measure. Therefore the densities $f_n = d\mu_n/dx$ exist for $n \geq 1$ and satisfy the recurrence relation $f_{n+1} = Pf_n$ with the transition operator

$$Pf(x) = \int_{\mathbb{R}^d} g(x - S(u))f(u)du. \tag{2.6}$$

**Example 2.2** Multiplicative Perturbations. Assume that $X = W = [0, \infty)$ and that $T(x,w) = wS(x)$ with a continuous $S : [0,\infty) \to (0,\infty)$. Then

$$x_{n+1} = \xi_n S(x_n), \qquad n = 0, 1, \dots \tag{2.7}$$

and

$$\mu_{n+1}(A) = \int_0^\infty\int_0^\infty 1_A(wS(x))\mu_n(dx)\nu(dw). \tag{2.8}$$

Again assuming that $\nu$ is absolutely continuous with $d\nu/dx = g(x)$ and substituting $x = u$, $wS(x) = y$ we obtain

$$\mu_{n+1}(A) = \int_A\Big\{\int_0^\infty g\Big(\frac{y}{S(u)}\Big)\frac{\mu_n(du)}{S(u)}\Big\}dy.$$

Thus for $n \geq 1$ the $\mu_n$ are absolutely continuous and the densities $f_n = d\mu_n/dx$ satisfy the relation $f_{n+1} = Pf_n$ where

$$Pf(x) = \int_0^\infty g\Big(\frac{x}{S(u)}\Big)f(u)\frac{du}{S(u)}. \tag{2.9}$$

In these two special cases the transition operator $P$ was derived under the mild assumption that the distribution $\nu$ of perturbations is absolutely continuous. An analogous calculation can be carried out for the general form of $T(x,y)$. However, in this general situation the derivation of $P$ is less transparent and requires some additional assumptions concerning the derivatives $\partial T/\partial w$. In the next section we will discuss the transition operators from a more abstract point of view.

## 3 Markov operators

Let $(X, \mathcal{A}, \mu)$ be a $\sigma$–finite measure space. By $D$ we denote the set of all normalized densities, i.e.,

$$D = \{f \in L^1 : \ f \geq 0, \ \|f\| = 1\},$$

where $\|\cdot\|$ stands for the norm in $L^1 = L^1(X, \mathcal{A}, \mu)$. A linear mapping $P : L^1 \to L^1$ is called a *Markov operator* if $P(D) \subset D$.

An important class of Markov operators is given by stochastic kernels. We say that a measurable function $k : X \times X \to \mathbb{R}$ is a stochastic kernel if

$$k(x, z) \geq 0 \quad \text{and} \quad \int_X k(u, z)\mu(du) = 1, \qquad x, z \in X. \tag{3.1}$$

Having $k$, we define the corresponding *integral Markov* operator by the formula

$$Pf(x) = \int_X k(x, u)f(u)\mu(du). \tag{3.2}$$

Conditions (3.1) imply that $P$ given by (3.2) is in fact a Markov operator.

The transition operators (2.6) and (2.9) are typical examples of integral Markov operators. Conversely, it is interesting that every integral Markov operator is a transition operator of some stochastically perturbed dynamical system of the form (2.1). We may easily demonstrate this fact for operators defined on $L^1(\mathbb{R})$. Thus assume that an integral Markov operator

$$Pf(x) = \int_{-\infty}^{+\infty} k(x, u)f(u)du, \tag{3.3}$$

is given. Assume moreover that $g \in D(\mathbb{R})$ is an arbitrary positive ($g(x) > 0$ a.e.) density. Having $k$ and $g$ we define the function $T : \mathbb{R} \times \mathbb{R} \to \mathbb{R}$ as follows. For every fixed $(x, w) \in \mathbb{R}^2$ the value $T(x, w)$ is given as the smallest solution of the equation.

$$\int_{-\infty}^{T(x,w)} f(u, x)du = \int_{-\infty}^{w} g(u)du. \tag{3.4}$$

Since the integral on the right-hand side of (3.4) has values in the open interval $(0, 1)$ and $k$ is a stochastic kernel the solution $T(x, w)$ always exists. An elementary calculation shows that the dynamical system $x_{n+1} = T(x_n, \xi_n)$, in which the $\xi_n$ are independent random variables having the same density distribution function $g$, has a transition operator of the form (3.3). Formula (3.4) was proposed by K. Łoskot.

From the above considerations it follows that in many cases the study of the asymptotic behaviour of the perturbed dynamical system (2.1) can be replaced by the examination of the iterates $\{P^n\}$ of its transition operator $P$.

We introduce two simple definitions which describe the typical behaviour of $\{P^n\}$, assuming that a $\sigma$–finite measure space $(X, \mathcal{A}, \mu)$ is given.

A Markov operator $P : L^1 \to L^1$ is called *asymptotically stable* if there exists a density $f_*$ such that $Pf_* = f_*$ and

$$\lim_{n \to \infty} \|P^n f - f_*\| = 0 \quad \text{for} \quad f \in D. \tag{3.5}$$

A density $f_*$ satisfying $Pf_* = f_*$ is called stationary. Evidently (3.5) implies that for an asymptotically stable operator there exists exactly one stationary density.

Let a Markov operator $P : L^1 \to L^1$ and a subfamily of measurable subsets $\mathcal{A}_* \subset \mathcal{A}$ be given. We say that $P$ is *sweeping* with respect to $\mathcal{A}_*$ (or, simply, sweeping) if

$$\lim_{n\to\infty} \int_A P^n f(x)\mu(dx) = 0 \quad \text{for} \quad A \in \mathcal{A}_*, \quad f \in D. \tag{3.6}$$

We say that the family $\mathcal{A}_*$ is *regular* if there exists a sequence $A_n \in \mathcal{A}_*$ such that $X = \bigcup_n A_n$. Evidently an operator which is sweeping with respect to a regular family $\mathcal{A}_*$ cannot be asymptotically stable.

In the following sections we will show four effective criteria for the determination of asymptotic stability and sweeping. Namely we will discuss:

1. the lower bound function theorem,
2. asymptotic decomposition of constrictive operators,
3. integral operators satisfying a transitivity property,
4. the Foguel alternative.

Each criterion will be illustrated by an application to a dynamical system. In particular we will examine some mathematical models of the cell cycle and the logistic growth equation with parametric noise.

## 4 Lower bound function

Let $(X, \mathcal{A}, \mu)$ be a $\sigma$–finite measure space and $P : L^1 \to L^1$ a Markov operator. We say that a function $h \in L^1$ is a *lower bound function* if for every $f \in D$ there is a sequence $\varepsilon_n(f) \in L^1$ such that

$$P^n f \geq h - \varepsilon_n(f) \quad \text{and} \quad \|\varepsilon_n(f)\| \to 0. \tag{4.1}$$

A lower bound function $h$ is nontrivial if $h \geq 0$ and $\|h\| > 0$.

At first glance conditions (4.1) look much weaker than condition (3.5) appearing in the definition of the asymptotic stability. We have, however, the following theorem from [**11**].

**Theorem 4.1** *A Markov operator $P$ is asymptotically stable if and only if it has a nontrivial lower bound function.*

An efficient condition to prove the asymptotic stability of integral Markov operators can be formulated by a use of a Liapunov function. Let $G \subset \mathbb{R}^d$ be an unbounded Borel measurable set. A continuous function $V : G \to \mathbb{R}$ is called a Liapunov function if

$$V(x) \geq 0 \quad \text{for} \quad x \in G \quad \text{and} \quad \lim_{|x|\to\infty} V(x) = \infty.$$

Consider on $L^1(G)$ an integral Markov operator

$$Pf(x) = \int_G k(x,y) f(y)dy. \tag{4.2}$$

From Theorem 4.1 it is easy to derive the following corollary which is a slight generalization of a result of [**14**].

**Corollary 4.2** *Assume that there exists a Liapunov function $V$ and constants $\delta \geq 0$, $\gamma < 1$ such that*

$$\int_G V(x) Pf(x) dx \leq \delta + \gamma \int_G V(x) f(x) dx \quad \textit{for} \quad f \in D. \tag{4.3}$$

*Assume moreover that*

$$\int_G \inf_{u\in C} k(x,u)dx > 0 \tag{4.4}$$

*for every compact set* $C \subset G$. *Then the operator* $P$ *defined by* (4.2) *is asymptotically stable.*

## 5 A cell cycle model

As an example of the applicability of the lower bound function technique we consider a model of cell cycle proposed by [**14**] which generalizes a model of [**8**] and the tandem model of [**15**].

In the Tyrcha model it is assumed that the cell cycle consists of two phases $A$ and $B$. Phase $A$ begins at birth and lasts until the occurrence of a critical event which is necessary for mitosis. Then the cell enters phase $B$. The end of phase $B$ coincides with cell division. The duration $t_B$ of phase $B$ is constant while the length $t_A$ of phase $A$ is random. More precisely the probability that the critical moment occurs in the interval $[t, t+\Delta t]$ is

$$\text{prob}\,(t \le t_A \le t+\Delta t \mid t_A \ge t) = \varphi(x(t))\Delta t + o(\Delta y), \tag{5.1}$$

where $x(t)$ is the size (or amount of mitogen) of the cell at time $t$ and $\varphi$ is a given nonnegative function. Further it is assumed that the cell size grows according to the equation

$$\frac{dx}{dt} = g(x), \qquad x(0) = r, \tag{5.2}$$

where $g(x)$ for $x > 0$ and $g(0) = 0$. Denote by $x_n$ the initial size of a cell in the $n$–th generation. Evidently $x_n$ can be considered as a random variable. Using the above assumptions it can be shown (see [**10**]) that

$$x_{n+1} = \lambda^{-1}\{Q^{-1}[\,Q(x_n) + \xi_n\,]\}, \tag{5.3}$$

where

$$Q(x) = \int_0^x \frac{\varphi(y)}{g(y)}dy, \qquad \lambda(x) = \pi(-t_B, 2x)$$

and $\pi(t,x)$ is the solution of Equation (5.2). The random variables $\xi_n$ are independent and have the common density distribution function $g(x) = \mathrm{e}^{-x}$. An elementary calculation shows that the transition operator for the dynamical system (5.3) has the form

$$Pf(x) = -\int_0^{\lambda(x)} \Big\{\frac{\partial}{\partial x}\exp[Q(y) - Q(\lambda(x))]\Big\} f(y)dy. \tag{5.4}$$

The asymptotic properties of operator (5.4) can be studied under the quite general assumptions that $Q$ and $\lambda$ are absolutely continuous, nondecreasing and

$$\lim_{x\to\infty} Q(x) = \lim_{x\to\infty}\lambda(x) = \infty, \qquad \lambda(0) = 0. \tag{5.5}$$

These conditions imply by a straightforward calculation that $P$ is an integral Markov operator on $L^1([0,\infty))$. The asymptotic behaviour of $P$ is described by the following.

**Theorem 5.1** *If*

$$\liminf_{x\to\infty}\,[\,Q(\lambda(x)) - Q(x)\,] > 1, \tag{5.6}$$

*then $P$ is asymptotically stable, and if*

$$\limsup_{x \to \infty} [\, Q(\lambda(x)) - Q(x) \,] < 1, \tag{5.7}$$

*then $P$ is sweeping with respect to compact subsets of* $[0, \infty)$.

**Proof** (outline). First assume (5.6) setting $V(x) = \exp(\alpha Q(\lambda(x)))$, $0 < \alpha < 1$, so using (5.4) it is easy to verify that

$$\int_0^\infty V(x) Pf(x) dx = \frac{1}{1-\alpha} \int_0^\infty f(x) \mathrm{e}^{\alpha Q(x)} dx. \tag{5.8}$$

According to (5.6), there exists an $x_0 > 0$ and $\varrho > 1$ such that $Q(x) \leq Q(\lambda(x)) - \varrho$ for $x \geq x_0$. Using this and (5.8) we obtain

$$\int_0^\infty V(x) Pf(x) dx \leq \frac{1}{1-\alpha} \int_0^{x_0} f(x) \mathrm{e}^{\alpha Q(x)} dx + \frac{\mathrm{e}^{-\varrho\alpha}}{1-\alpha} \int_{x_0}^\infty V(x) f(x) dx$$

or

$$\int_0^\infty V(x) Pf(x) dx \leq \delta(\alpha) + \gamma(\alpha) \int_0^\infty V(x) f(x) dx,$$

where $\delta = \mathrm{e}^{\alpha Q(x_0)}(1-\alpha)^{-1}$ and $\gamma(\alpha) = \mathrm{e}^{-\varrho\alpha}(1-\alpha)^{-1}$. The inequality $\varrho > 1$ implies that $\gamma(\alpha) < 1$ for some $\alpha \in (0,1)$ and consequently condition (4.3) of Corollary 4.1 is satisfied. The verification of (4.4) is elementary due to the specific (exponential) form of the kernel in (5.4).

Now assume (5.7) and define $V(x) = w(\, Q(\lambda(x)) \,)$ where

$$w(x) = \begin{cases} \mathrm{e}^{-\alpha x_0} & \text{for } x \leq x_0 \,, \\ \mathrm{e}^{-\alpha x} & \text{for } x > x_0 \,. \end{cases}$$

Choosing $x_0$ sufficiently large and $\alpha > 0$ sufficiently small it can be shown that

$$\int_0^\infty V(x) Pf(x) dx \leq \gamma \int_0^\infty V(x) f(x) dx, \quad \text{for} \quad f \in D, \tag{5.9}$$

where $\gamma < 1$ is a constant. From (5.9) it follows

$$\lim_{n \to \infty} \int_0^\infty V(x) P^n f(x) dx = 0$$

which implies sweeping. □

The detailed proof is given in [**4**]. The proof presented here is much shorter due to another choice of the Liapunov function $V(x)$.

## 6 Asymptotic decomposition theorem

In order to apply the lower bound function technique it is necessary to evaluate the sequence $\{P^n f\}$ from below. The asymptotic decomposition theorem also describes the behaviour of $\{P^n f\}$ in the case when the sequence $\{P^n f\}$ has an integrable upper bound.

Let $(X, \mathcal{A}, \mu)$ be a $\sigma$–finite measure space. A Markov operator $P : L^1 \to L^1$ is called *constrictive* if there exists constant $\kappa < 1$, $\delta > 0$ and a set $C \in \mathcal{A}$ with the following property: For every $f \in D$ there is an integer $n_0(f)$ such that

$$\int_{B \cup (X \setminus C)} P^n f(x) \mu(dx) \leq \kappa \quad \text{for } n \geq n_0(f), \;\; \mu(B) \leq \delta, \;\; B \subset C. \tag{6.1}$$

The constrictiveness rules out the possibility that for large $n$ the densities $P^n f$ are either concentrated on a set of very small measure or dispersed throughout the entire space. In particular $P$ is constrictive if there exists an integrable function $h \geq 0$ such that

$$P^n f \leq h + \varepsilon_n(f) \quad \text{and} \quad \lim_{n\to\infty} \|\varepsilon_n(f)\| = 0.$$

The asymptotic behaviour of constrictive operators is described by the following theorem due to [**6**].

**Theorem 6.1** *The iterates of a constrictive operator $P$ can be written in the form*

$$P^n f = \sum_{i=1}^{r} \lambda_i(f) g_{\alpha^n(i)} + Q_n f, \qquad \text{for } f \in L^1, \tag{6.2}$$

*where:*

1. *$g_1, \dots, g_r$ are densities with disjoint supports;*
2. *$\lambda_1, \dots \lambda_r$ are linear functionals on $L^1$;*
3. *$\alpha$ is a permutation of numbers $1, \dots, r$ such that $Pg_i = g_{\alpha(i)}$ and $\alpha^n$ denotes the $n^{\text{th}}$ iterate of $\alpha$;*
4. *$Q_n$ is a sequence of operators such that $\lim_{n\to\infty} \|Q_n f\| = 0$ for $f \in L^1$.*

The terms under the summation in (6.2) are just permuted with each application of $P$ and since $r$ is finite this sum is periodic with a period smaller than or equal to $r!$. Thus for every $f \in L^1$ the sequence $\{P^n f\}$ is asymptotically periodic. From Theorem 6.1 it is easy to derive a simple criterion for asymptotic stability.

**Corollary 6.2** *Let $P$ be a constrictive Markov operator. Assume there is a set $A \in \mathcal{A}$ of positive measure with the following property: For every $f \in D$ there exists an integer $n_0(f)$ such that*

$$P^n f(x) > 0 \quad \text{for } x \in A \ \text{ a.e. and } \ n \geq n_0. \tag{6.3}$$

*Then $P$ is asymptotically stable.*

Theorem 6.1 and Corollary 6.2 can be easily applied to dynamical systems with additive and multiplicative perturbations. We show this application in the following two examples.

**Example 6.3** Again consider the dynamical system (2.5) and the corresponding transition operator (2.6). Assume that there is a number $M > 0$ such that for every initial vector $x_0$ the sequence $\{x_n\}$ satisfies

$$\limsup_{n\to\infty} E(|x_n|) < M. \tag{6.4}$$

Then by the Chebyshev inequality

$$\int_{|x|\geq 2M} P^n f(x)dx \leq \frac{E(|x_n|)}{2M} \leq \frac{1}{2}$$

for $f \in D$ and $n$ sufficiently large, say $n \geq n_0(f)$. Moreover for $f \in D$ we have

$$\begin{aligned}\int_B P^n f(x)dx &= \int_B \Big\{ \int_{\mathbb{R}^d} g(x - S(u)) P^{n-1} f(u) du \Big\} dx \\ &\leq \sup_{z\in\mathbb{R}^d} \int_{B+z} g(x)dx.\end{aligned}$$

Since $g$ is Lebesgue integrable, there exists a $\delta > 0$ such that the last integral is smaller than $\varepsilon = 1/4$ whenever $\mu(B) \leq \delta$. Setting $C = \{x : |x| \leq 2M\}$ we have

$$\int_{B\cup(\mathbb{R}^d\setminus C)} P^n f(x)dx \leq \int_B P^n f(x)dx + \int_{|x|\geq 2M} P^n f(x)dx \leq \frac{3}{4}$$

for $f \in D$ and $n \geq n_0(f)$ which implies constrictivness. Thus, using Theorem 6.1 we obtain the following conclusion: Every dynamical system with additive perturbations is either asymptotically unbounded ((6.4) is not satisfied with any $M$ independent of $x_0$) or the sequence of densities $\{P^n f\}$ is asymptotically periodic. If (6.4) is satisfied and in addition $g(x) > 0$ for $x \in \mathbb{R}^d$, then condition (6.3) is also fullfilled and $\{P^n f\}$ converges to a unique stationary density.

**Example 6.4** Now consider the dynamical system (2.7) with the transition operator (2.9). Assume that the transformation $S$ is linearly bounded, i.e.,

$$0 < S(x) \leq \alpha x + \beta \quad \text{for} \quad x \geq 0, \tag{6.5}$$

and that the perturbations $\xi_n$ have a common density distribution function $g$ with a finite first moment

$$m = \int_0^\infty xg(x)dx < \infty. \tag{6.6}$$

Horbacz in [**5**] has shown that for $\alpha m < 1$ the operator $P$ given by (2.9) is constrictive. A simplified version of her proof goes as follows. Setting $V(x) = x$ it is easy to verify that

$$\begin{aligned}\int_0^\infty V(x)Pf(x)dx &= m\int_0^\infty f(x)S(x)dx \\ &\leq \alpha m\int_0^\infty V(x)f(x)dx + \beta m,\end{aligned}$$

which shows that condition (4.3) of Corollary 4.1 is satisfied. This, by the Chebyshev inequality, implies

$$\int_0^c P^n f(x)dx \geq \frac{3}{4} \quad \text{for} \quad n \geq n_0(f),$$

where $c$ is a sufficiently large number. Since $g$ is integrable there must be a $\delta_1 > 0$ such that

$$\int_A g(x)dx \leq \frac{1}{4} \quad \text{for} \quad \mu(A) \leq \delta_1.$$

Define $\delta = \delta_1 \min_{0\leq u\leq c} S(u)$. Then for $\mu(B) \leq \delta$ and $0 \leq u \leq c$ we have $\mu(B/S(u)) \leq \delta_1$. Consequently

$$\begin{aligned}\int_B P^n f(x)dx &= \int_0^\infty P^{n-1}f(u)\Big\{\int_B g\Big(\frac{x}{S(u)}\Big)\frac{dx}{S(u)}\Big\}du \\ &\leq \int_c^\infty P^{n-1}f(u)du + \int_0^c P^{n-1}f(u)\Big\{\int_B g\Big(\frac{x}{S(u)}\Big)\frac{dx}{S(u)}\Big\}du \\ &\leq \frac{1}{4} + \frac{1}{4}\int_0^c P^{n-1}f(u)du \leq \tfrac{1}{2} \qquad \text{for} \quad n \geq n_0(f) + 1.\end{aligned}$$

Setting $X = [0, \infty)$, $C = [0, c]$ we have $X \setminus C = [c, \infty)$ and

$$\int_{B\cup(X\setminus C)} P^n f(x)dx \leq \int_B P^n f(x)dx + \int_0^\infty P^n f(x)dx \leq \frac{1}{2} + \frac{1}{4}$$

which again shows that $P$ is constrictive. If in addition $g(x) > 0$ for $x > 0$ then $Pf(x) > 0$ for $x > 0$ and condition (6.3) is automatically satisfied. In this case $P$ is asymptotically stable.

In the above argument, the conditions $S(x) > 0$ for $x \geq 0$ and $g(x) > 0$ for $x > 0$ play an important role. A more sophisticated argument given in [**5**] shows that $P$ remains asymptotically stable if

$$S(0) = 0, \quad S(x) > 0 \ \text{ for } \ x > 0, \quad g(x) > 0 \ \text{ for } \ x \ \text{ sufficiently large.}$$

However in this case it is assumed not only that $\alpha m < 1$ but also that

$$\int_0^\infty \frac{g(x)}{(\gamma x)^\lambda} < 1, \qquad \gamma = S'(0) \tag{6.7}$$

for some $\lambda > 0$. The Horbacz result can be applied to the "bottleneck" model of annual plants with "seedbank" studied in [**16**] and [**2**]. The dynamical system describing this model can be written in the form

$$x_{n+1} = \frac{\xi_n x_n}{(1 + (k+p)x_n)^\alpha (1 + px_n)^{1-\alpha}},$$

where $\alpha$, $k$, $p$ are positive constants and $x_n$ is the size of the population in the $n^{\text{th}}$ generation. According to Horbacz the bottleneck model is asymptotically stable if the density $g(x)$ of perturbation $\xi_n$ is positive for large $x$ and

$$\int_0^\infty \frac{g(x)}{x^\lambda} dx < 1$$

for some $\lambda > 0$.

## 7 Transitive integral operators

In this section we discuss Markov operators defined by the integral equation (3.2), i.e.,

$$Pf(x) = \int_X k(x,u) f(u) \mu(du), \tag{7.1}$$

where, as before, $(X, \mathcal{A}, \mu)$ is a $\sigma$–finite measure space and $k : X \times X \to \mathbb{R}$ is a stochastic kernel. Integral Markov operators have some specific properties which simplify the conditions for the convergence of iterates. Namely, the existence of an invariant density and a simple transition property imply asymptotic stability. To formulate this criterion precisely, recall that in the theory of Markov operators the support of an $f \in L^1$ is defined by

$$\operatorname{supp} f = \{x \in X : \ f(x) \neq 0\}.$$

We say that an operator $P : L^1 \to L^1$ *overlaps supports* if for every $f, g \in D$ there is an $n_0(f, g)$ such that

$$\mu(\operatorname{supp} P^{n_0} f \cap \operatorname{supp} P^{n_0} g) > 0. \tag{7.2}$$

Baron and Lasota in [**1**] have proved the following

**Theorem 7.1** *An integral Markov operator which overlaps supports and has a positive stationary density ($f_*(x) > 0$ a.e.) is asymptotically stable.*

Using Theorem 7.1 it is possible to find a sharp sufficient condition for the asymptotic stability of the operator

$$Pf(x) = \int_0^{\lambda(x)} \left\{ -\frac{\partial}{\partial x} H(\, Q(\lambda(x)) - Q(u)\,) \right\} f(u)du \tag{7.3}$$

which appears in the general model of irregular biological events proposed in [**10**]. Formula (7.3) represents the transition operator for dynamical system of the form (5.3) in which the perturbations $\xi_n$ have the common survival function

$$\text{prob}\,(\xi_n > x) = H(x).$$

This application is, however, quite complicated and we are not going to present it here. For the sake of simplicity we show how Theorem 7.1 can be applied to the Tyson & Hannsgen model of the cell cycle. In this model the transition operator $P : L^1([\sigma,\infty)) \to L^1([\sigma,\infty))$ has the form

$$Pf(x) = \int_\sigma^{x/\sigma} K(x,u) f(u)du, \tag{7.4}$$

where

$$K(x,u) = \begin{cases} (\alpha/\sigma)(x/\sigma)^{-1-\alpha} & \text{for } \sigma \le u \le 1, \\ (\alpha/\sigma)(x/\sigma)^{-1-\alpha}u^\alpha & \text{for } 1 < \sigma \le x/\sigma. \end{cases} \tag{7.5}$$

and $\alpha > 0$, $0 < \sigma < 1$ are constants. The operator (7.4) evidently overlaps supports, since for every $f \in D$ the values $Pf(x)$ are positive if $x$ is sufficiently large. Further an immediate calculation shows that the function

$$f_*(x) = \frac{c}{x^{\beta+1}}\,, \qquad x \ge \sigma \tag{7.6}$$

is invariant with respect to $P$ if $\beta$ is a solution to the transcendental equation

$$w(\beta) := \sigma^\beta + \beta/\alpha = 1. \tag{7.7}$$

Such a solution $\beta > 0$ exists if $w'(0) < 1$ or

$$\alpha > -\frac{1}{\ln \sigma}. \tag{7.8}$$

In this case (7.6) with a properly chosen $c$ is a positive invariant density on $[\sigma,\infty)$ and according to Theorem 7.1 the operator $P$ defined by (7.4), (7.5) is asymptotically stable.

The fact that condition (7.8) implies asymptotic stability of $P$ was already predicted in [**15**]. They also found the stationary density (7.6) and proved the asymptotic stability under more restrictive condition $\alpha > 1/(1-\sigma)$. The first proof that (7.8) actually implies asymptotic stability was given in [**14**]. Another proof follows from the result of [**4**]. This short history of stability conditions for operator (7.4), (7.5) shows the utility of Theorem 7.1. Once a positive stationary density is found the proof of stability is easy. It remains only to verify the "overlapping support" property.

In the next section we present a theorem which can be used not only to prove asymptotic stability, but also sweeping.

## 8 The Foguel alternative

Let a $\sigma$–finite measure space and a regular family $\mathcal{A}_* \subset \mathcal{A}$ be given. In order to formulate the main result of this section – the Foguel alternative – we need to introduce a few notions.

A measurable function $f : X \to \mathbb{R}$ is called *locally* integrable if

$$\int_A f(x)\mu(dx) < \infty \qquad \text{for } A \in \mathcal{A}_*.$$

A Markov operator $P$ is called *expanding* if

$$\lim_{n\to\infty} \mu(A \setminus \operatorname{supp} P^n f) = 0 \qquad \text{for} \qquad f \in D, \quad \mu(A) < \infty. \tag{8.1}$$

Finally, a measurable nonnegative function $h : X \to \mathbb{R}$ is called *subinvariant* with respect to an integral operator $P$ given by (7.1) if

$$Ph(x) := \int_X k(x,u)h(u)\mu(du) \le h(x) \qquad \text{a.e.}$$

Using the result of [**7**] and [**13**] the following version of the Foguel alternative ([**3, 12**]) can be stated.

**Theorem 8.1** *Assume that $P$ is an expanding integral operator and that there is a locally integrable positive subinvariant function $h$ ($h > 0, Ph \le h$ a.e.). Then either $P$ is asymptotically stable or $P$ is sweeping.*

A simple example of an operator $P$ for which all the assumptions of Theorem 8.1 are satisfied is given by the Tyson & Hannsgen equations (7.4), (7.5). In this case it is natural to assume that $\mathcal{A}_*$ consists of all compact subsets of $[\sigma, \infty)$. Since for every positive $\sigma$ and $\alpha$ equation (7.7) has a solution $\beta = 0$, we may choose $h(x) = 1/x$ as an invariant locally integrable function. Finally, condition (8.1) follows from the fact that $\sigma < 1$ and $K$ given by (7.1) is positive.

The arguments used in the above example can be extended to a large family of Markov operators of the form

$$Pf(x) = \int_a^{\lambda(x)} K(x,u)f(u)du$$

in which $K(x,u) > 0$ for $a < u < \lambda(x)$, $x > a$ and $\lambda(x) > x$ for $x > a$ ($a \ge 0$). The only difficulty lies in the proof of the existence of a positive locally integrable subinvariant function. Thus, it is an important and open problem to characterize the class of kernels $K$ and bounds $\lambda$ for which such a function exists.

## 9 Foguel alternative for continuous time semigroups

The Foguel alternative may be easily extended to continuous time systems and applied to differential equations with stochastic perturbations. We are going to show such applications using the formalism of stochastic semigroups. As before we assume that $(X, \mathcal{A}, \mu)$ and a regular $\mathcal{A}_*$ are given.

A family $\{P_t\}_{t\ge 0}$ of Markov operators is called a (continuous) *stochastic semi-group* if the following conditions are satisfied.

$1^o$ $P_0 f = f$ for $f \in L^1$;

$2^o$ $P_{t+s}f = P_t(P_s f)$ for $t, s \ge 0$, $f \in L^1$;

$3^o$ $\lim_{t\to 0} \|P_t f - f\| = 0$ for $f \in L^1$.

A stochastic semigroup is called *asymptotically stable* if there exists an $f_* \in D$ such that $P_t f_* = f_*$ (stationary density) and if

$$\lim_{t\to\infty} \|P_t f - f_*\| = 0 \quad \text{for} \quad f \in D \ .$$

A semigroup $\{P_t\}_{t\geq 0}$ is called *sweeping* if

$$\lim_{t\to\infty} \int_A P_t f(x)\mu(dx) = 0 \quad \text{for} \quad f \in D, \quad A \in \mathcal{A}_* \ .$$

Stability and sweeping of a stochastic semigroup $\{P_t\}_{t\geq 0}$ and a single operator $P_t$ with a fixed $t$ are closely related. It can be proved ([**9**]) that asymptotic stability of $P_{t_0}$ with $t_0 > 0$ implies the asymptotic stability of $\{P_t\}_{t\geq 0}$ with the same stationary density $f_*$, and analogously that the sweeping of $P_{t_0}$ implies the sweeping of $\{P_t\}_{t\geq 0}$. The inverse implications are obvious. Using this it is easy to derive from Theorem 8.1 the following.

**Corollary 9.1** *Let $\{P_t\}_{t\geq 0}$ be a continuous stochastic semigroup. Assume that for some $t_0 > 0$ the operator $P_{t_0}$ is expanding and given by a stochastic kernel. Assume moreover that there exists for $P_{t_0}$ a positive locally integrable subinvariant function. Then either the semigroup $\{P_t\}_{t\geq 0}$ is asymptotically stable or $\{P_t\}_{t\geq 0}$ is sweeping.*

As an example consider the logistic equation

$$\frac{dx}{dt} = x(a_1 - x), \qquad \text{with} \qquad a_1 = a + \sigma\xi_t, \tag{9.1}$$

where $a, \sigma$ are positive constants and $\xi_t$ is a normalized white noise. The density distribution function $u(t,x)$ of $x(t)$ satisfies the Fokker–Planck equation

$$\frac{\partial u}{\partial t} = \frac{\sigma^2}{2}\frac{\partial^2(x^2u)}{\partial x^2} - \frac{\partial}{\partial x}[x(a-x)u] \tag{9.2}$$

and the corresponding stochastic semigroup is defined by

$$P_t f(x) = u(t,x), \qquad u(0,x) = f(x) \ .$$

In order to study the asymptotic behaviour of $\{P_t\}_{t\geq 0}$ it is not necessary to solve (9.2). It is enough to know that $u(t,x)$ is given by an integral formula

$$u(t,x) = \int_0^\infty \Gamma(t,x,y)f(y)dy$$

with a stochastic kernel (Green's function) $\Gamma$. We obtain the stationary density by solving the stationary equation

$$\frac{\sigma^2}{2}\frac{d^2(x^2u(x))}{dx^2} - \frac{d}{dx}[x(a-x)u(x)] = 0$$

which gives

$$u(x) = f_*(x) = cx^\gamma \mathrm{e}^{-x/\sigma^2}, \tag{9.3}$$

where $\gamma = 2(a/\sigma^2) - 2$. This function is, for every $\gamma$, locally integrable with respect to the family

$$\mathcal{A}_* = \{\, [\varepsilon, \infty) : \ \varepsilon > 0 \,\} \ . \tag{9.4}$$

If $\gamma > -1$ so $a > \frac{1}{2}\sigma^2$, the function (9.3) can be normalized and $\{P_t\}_{t\geq 0}$ has a stationary density. In this case $\{P_t\}_{t\geq 0}$ cannot be sweeping and by the Foguel alternative it is asymptotically stable. If the inverse inequality $a \leq \frac{1}{2}\sigma^2$ holds, then

$\{P_t\}_{t\geq 0}$ has no stationary density and cannot be asymptotically stable. According to the Foguel alternative the semigroup $\{P_t\}_{t\geq 0}$ must be sweeping with respect to (9.4). This means that for large $t$, the density $u(t,x)$ is concentrated in a neighborhood of $x = 0$. A large parametric perturbation can kill the population.

## References


[1] Baron, K. and Lasota, A. [1993], *Asymptotic properties of Markov operators defined by Volterra type integrals*, Ann. Polon. Math. **58**, 161-175.

[2] Ellner, S. [1984], *Asymptotic behaviour of some stochastic difference equation population models*, J. Math. Biology **19**, 169–200.

[3] Foguel, S. R. [1966], *Limit theorems for Markov processes*, Trans. Amer. Math. Soc. **121**, 200–209.

[4] Gacki, H. and Lasota, A. [1990], *Markov operators defined by Volterra type integrals with advanced argument*, Ann. Polon. Math. **51**, 155–166.

[5] Horbacz, K. [1989], *Asymptotic stability of dynamical systems with multiplicative perturbations*, Ann. Polon. Math. **50**, 209–218.

[6] Kormornik, J. and Lasota, A. [1987], *Asymptotic decomposition of Markov operators*, Bull. Polish Acad. Sci. Math. **35**, 321–327.

[7] Komorowski, T. and Tyrcha, J. [1989], *Asymptotic properties of some Markov operators*, Bull. Polon. Acad. Sci. Math. **37**, 221–228.

[8] Lasota, A. and Mackey, M. C. [1984], *Globally asymptotic properties of proliferating cell populations*, J. Math. Biology **19**, 43–62.

[9] Lasota, A. and Mackey, M. C. [1994], *Chaos, Fractals and Noise: Stochastic Aspects of Dynamics*, (second edition), Springer Verlag, New York.

[10] Lasota, A., Mackey, M. C. and Tyrcha, J. [1992], *The statistical dynamics of irregular biological events*, J. Math. Biol. **30**, 775–800.

[11] Lasota, A. and Yorke, J. A. [1982], *Exact dynamical systems and the Frobenius–Perron operator*, Trans. Amer. Math. Soc. **273**, 375–384.

[12] Lin, M. [1971], *Mixing for Markov operators*, Z. Wahrsheinlichkeitstheorie Verw. Gebiete **19**, 231–242.

[13] Malczak, J. [1992], *An application of Markov operators in differential and integral equations*, Rend. Sem. Mat. Univ. Padova **87**, 281–297.

[14] Tyrcha, J. [1988], *Asymptotic stability in a generalized probabilistic/deterministic model of the cell cycle*. J. Math. Biol. **26**, 465–475.

[15] Tyson, J. J. and Hannsgen, K. B. [1986], *Cell growth and division: A deterministic/ probabilistic model of the cell cycle*, J. Math. Biol. **23**, 231–246.

[16] Watkinson, A. R. [1980], *Density dependence in single–species population of plants*, J. Theor. Biol. **82**, 345–357.

Fields Institute Communications
Volume **21**, 1999

# A Generalization of Variation of Parameters and Lyapunov's Method

**Xinzhi Liu**
Department of Applied Mathematics
University of Waterloo
Waterloo, Ontario, N2L 3G1
xzliu@monotone.waterloo.ca

**Abstract.** The method of variation of parameters has been a well-known and useful tool in the investigation of the properties of solutions of differential equations. Since one can employ this method to nonlinear differential equations whose unperturbed parts are linear, it has gained importance in what is known as linearization techniques. It is also well-known that Lyapunov's second method is an important and fruitful technique that has gained increasing significance and has given a decisive impetus for modern development of stability analysis of differential equations. By unifying these two important methods, we develop a fruitful technique which brings out the advantages of both methods. We then establish some stability criteria in this new framework. To illustrate the advantages of this technique, some examples are worked out.

## 1 Introduction

It is well-known that the method of variation of parameters is a powerful tool in the study of nonlinear differential equations whose unperturbed parts are linear or smooth nonlinear. On the other hand, Lyapunov's second method has proven to be an important and fruitful technique in the investigation of nonlinear problems and has given a decisive impetus for modern development of stability analysis of differential equations. Since these two techniques are extremely useful and effective in the investigation of nonlinear problems, it is natural to combine these two approaches in order to utilize the benefits of both methods. Such a unification has recently been achieved and awaits further development [**1, 3, 4, 5, 6**].

In this paper, we shall exploit the ideas introduced in [**1, 3, 4, 5, 6**] to refine the unification from a different perspective. The method of variation of parameters is generalized so that known results become special cases. Moreover, stability criteria are established by using the new framework. It is shown that stability

1991 *Mathematics Subject Classification.* 34D10, 34D20.
Research supported by NSERC-Canada.

properties of unperturbed systems can be improved by utilizing the good character of perturbation compared to the useful Lyapunov method.

## 2 Generalized variation of parameters formula

Consider the differential systems

$$x' = F(t, x), \quad x(t_0) = x_0 \tag{2.1}$$

and

$$y' = f(t, y), \quad y(t_0) = x_0 \tag{2.2}$$

where $F, f \in C[R_+ \times R^n, R^n]$. A special case is when $f(t, y) = A(t)y$, i.e.,

$$y' = A(t)y, \quad y(t_0) = x_0, \tag{2.3}$$

where $A(t) = \dfrac{\partial F(t, \overline{x})}{\partial x}$, $\overline{x}$ being a known solution of system (2.1), and $\dfrac{\partial F}{\partial x}$ being assumed to exist. Let $U(t, t_0)$ be the fundamental matrix solution of (2.3). Then solutions $x(t) = x(t, t_0, x_0)$ of (2.1) can be expressed by

$$x(t) = U(t, t_0)x_0 + \int_{t_0}^{t} U(t, t_0)U^{-1}(s, t_0)R(s, x(s))\, ds\,, \tag{2.4}$$

where $R(t, x) = F(t, x) - A(t)x$. This is known as the linear variation of parameters formula. If $A(t) \equiv A$ is independent of $t$, then $U(t, t_0) = e^{A(t-t_0)}$ and (2.4) reduces to

$$x(t) = e^{A(t-t_0)}x_0 + \int_{t_0}^{t} e^{A(t-s)}R(s, x(s))\, ds\,. \tag{2.5}$$

Now let $y(t, t_0, x_0)$ be the solution of (2.2) such that $\dfrac{\partial y}{\partial t_0}(t, t_0, x_0)\,, \dfrac{\partial y}{\partial x_0}(t, t_0, x_0)$ exist and are continuous on $R_+ \times R^n$. Then setting

$$m(s) = y(t, s, x(s))\,, \quad t_0 \le s \le t\,,$$

where $x(s) = x(s, t_0, x_0)$ is any solution of (2.1) for $s \ge t_0$, we have

$$\begin{aligned} m'(s) &= \frac{\partial y}{\partial t_0}(t, s, x(s)) + \frac{\partial y}{\partial x_0}(t, s, x(s)) \cdot F(s, x(s)) \\ &\equiv G(t, s, x(s))\,. \end{aligned} \tag{2.6}$$

Integrating both sides from $t_0$ to $t$, we get

$$m(t) - m(t_0) = \int_{t_0}^{t} G(t, s, x(s))\, ds$$

which implies

$$x(t) = y(t) + \int_{t_0}^{t} G(t, s, x(s))\, ds, \quad t \ge t_0\,. \tag{2.7}$$

This is the nonlinear variation of parameters formula. If, on the other hand, set

$$m(s) = V(s, y(t, s, x(s))),$$

where $V \in C^1[R_+ \times R^n, R]$, then proceeding as before, we arrive at

$$V(t, x(t)) = V(t_0, y(t)) + \int_{t_0}^{t} \frac{dv}{ds}(s, y(t, s, x(s)))\, ds\,, \tag{2.8}$$

where

$$\frac{dv}{ds}(s, y(t,s,x(s))) = \frac{\partial V}{\partial t}(s, y(t,s,x(s))) + \frac{\partial V}{\partial x}(s, y(t,s,x(s)))G(t,s,x(s)) .$$

The relation (2.8) can be considered as the generalized variation of parameters formula which brings out the advantage of unifying the methods of variation of parameters and Lyapunov-like functions.

If $f(t,y)$ has continuous partial derivatives with respect to $y$, then the solutions $y(t,t_0,x_0)$ of (2.2) are differentiable with respect to $(t_0,x_0)$ and satisfy

$$\begin{cases} \dfrac{\partial y}{\partial t_0}(t,t_0,x_0) = -\Phi(t,t_0,x_0)\cdot f(t_0,x_0) , \\ \dfrac{\partial y}{\partial x_0}(t,t_0,x_0) = \Phi(t,t_0,x_0), \quad t \geq t_0, \end{cases} \tag{2.9}$$

where $\Phi(t,t_0,x_0)$ is the matrix solution of the variational equation

$$z' = f_y(t, y(t,t_0,x_0))z, \quad z(0) = I ,$$

$I$ being the identity matrix. By (2.6) and (2.9), we see that

$$G(t,s,x(s)) = \Phi(t,s,x)[F(s,x) - f(s,x))] . \tag{2.10}$$

Thus if $F(t,x) = f(t,x) + R(t,x)$, $R(t,x)$ being treated as the perturbation, then the relation (2.7) reduces to

$$x(t) = y(t) + \int_{t_0}^{t} \Phi(t,s,x(s))R(s,x(s))\,ds, \quad t \geq t_0 , \tag{2.11}$$

which is the Alekseev's nonlinear variation of parameters formula. Using (2.11) to estimate $|x(t)|$ leads to evaluating $\|\Phi(t,s,x(s))\|$ and $|R(s,x(s))|$, which destroys the good behaviour of $R(t,x)$, if any. As a result, one can at best only preserve the good properties of the unperturbed system (2.2). If, on the other hand, the relation (2.8) is utilized, it is easy to see that the behavior of $R(t,x)$ plays an important role in improving the properties of the unperturbed system. For example, if $V(t,x) = |x|^2$, then we get by (2.8)

$$|x(t)|^2 = |y(t)|^2 + 2\int_{t_0}^{t} y(t,s,x(s))\Phi(t,s,x(s))R(s,x(s))\,ds , \tag{2.12}$$

which is useful in deriving estimates on solutions $x(t)$ of (2.1) in terms of norms since the expression $y(t,s,x(s))\Phi(t,s,x(s))R(s,t(s))$ need not be nonnegative.

## 3 Stability criteria

In this section, we shall establish some stability criteria utilizing the idea developed in the previous section. For this, we shall assume $F(t,0) \equiv 0$ and $f(t,0) \equiv 0$ so that systems (2.1) and (2.2) admit a trivial solution. We shall also assume that solutions $y(t,t_0,x_0)$ of (2.2) exist for $t \geq t_0$, unique and $|y(t,t_0,x_0|$ is locally Lipschitzian in $x_0$. In order to discuss stability results corresponding to Lyapunov's original theorems in the new framework, we need the notion of strict stability of the system (2.2) which is a refinement of the concept of stability in tube-like domains introduced in [**2**].

**Definition 3.1** System (2.2) is said to be

(i) *strictly stable* if for any $\epsilon_1 > 0,\ t_0 \in R_+$, there exists a $\delta_1 = \delta_1(t_0, \epsilon) > 0$ such that

$$|x_0| \leq \delta_1 \quad \text{implies} \quad |y(t, t_0, x_0)\,| < \epsilon_1,\ t \geq t_0\,,$$

and for any $\delta_2 \leq \delta_1$, there exists a $\epsilon_2 < \delta_2$ such that

$$\delta_2 \leq |x_0| \quad \text{implies} \quad |y(t, t_0, x_0)| \geq \epsilon_2,\ \ t \geq t_0\,,$$

where $y(t, t_0, x_0)$ is any solution of (2.2);

(ii) *strictly uniformly stable,* if $\delta_1, \delta_2$ and $\epsilon_2$ in (i) are independent of $t_0$.

Let $V \in C(R_+ \times R^n, R_+]$. We define

$$D^+V(t, s, x) = \lim_{h \to 0^+} \sup \frac{1}{h}\,[V(s+h, y(t, s+h, x+hF(s, x))) - V(s, y(t, s, x))],$$

where $y(t, s, x)$ is any solution of (2.2) such that $y(s, s, x) = x$.

If $f(t, y) \equiv 0$, then $y(t, s, x) \equiv x$,

$$y(t, s+h, x+hF(s, x)) \equiv x + hF(s, x)$$

and the above definition reduces to

$$D^+V(s, x) = \lim_{h \to 0^+} \sup \frac{1}{h}\,[V(s+h, x+hF(s, x)) - V(s, x)]$$

which is the usual definition of the upper right generalized derivative of $V$ along solutions of (2.1).

Let $K$ denote the class of continuous functions $u : R_+ \to R_+$ which are strictly increasing and satisfying $u(0) = 0$. For $\rho > 0$, we define $s(\rho)$ by

$$s(\rho) = \{x \in R^n,\ \ |x| < \rho\}\,.$$

**Theorem 3.1** *Assume that*

(i) $V \in C(R_+ \times R^n, R_+], V(t, x)$ *is locally Lipschitzian in* $x$, *positive definite and*

$$D^+V(s, y(t, s, x)) \leq 0\,,\quad s \in [t_0, t)\,,\quad x \in s(\rho)\,; \tag{3.1}$$

(ii) $V(t, 0) \equiv 0$.

*Then any nonuniform stability property of the trivial solution of system* (2.2) *implies the corresponding nonuniform stability property of the trivial solution of system* (2.1).

**Proof** Since $V(t, x)$ is positive definite, there exists $b \in K$ and $\rho_0 \in (0, \rho)$ such that

$$b(\|x\|) \leq V(t, x)\,,\quad x \in s(\rho_0)\,. \tag{3.2}$$

Let $\epsilon \in (0, \rho)$ and $t_0 \in R_+$ be given. Then by (ii) there exists $\eta = \eta(t_0) > 0$ such that

$$V(t_0, x) < b(\epsilon)\,,\ \ \|x\| < \eta\,. \tag{3.3}$$

Assume that the trivial solution of system (2.2) is stable, then there exists $\delta = \delta(t_0, \eta) > 0$ such that

$$\|x_0\| < \delta \quad \text{implies} \quad \|y(t, t_0, x_0)\| < \eta\,,\ \ t \geq t_0, \tag{3.4}$$

where $y(t, t_0, x_0)$ is any solution of system (2.2). Let $x(t) = x(t, t_0, x_0)$ be any solution of system (2.1) with $\|x_0\| < \delta$. It is easy to see from (3.2) and (3.3) that $\|x(t_0)\| < \epsilon$. If there exists $t_1 > t_0$ to such that

$$\|x(t_1)\| = \epsilon \quad \text{and} \quad \|x(t)\| < \epsilon, \ t \in [t_0, t_1), \tag{3.5}$$

For $t \in [t_0, t_1]$, define $m(s) = V(s, y(t, s, x(s)))$, $s \in [t_0, t]$, where $y(t, s, x(s))$ denotes the solution of system (2.2). Then by condition (i), we get $D^+m(s) \leq 0$, $t_0 \leq s \leq t \leq t_1$. Thus $m(t) \leq m(t_0)$, $t_0 \leq t \leq t_1$, which implies

$$V(t_1, x(t_1)) \leq V(t_0, y(t_1, t_0, x_0)) < b(\epsilon).$$

But by (3.2) and (3.5)

$$V(t_1, x(t_0)) \geq b(\|x(t_1)\|) = b(\epsilon),$$

which is a contradiction. Thus we have $\|x(t)\| < \epsilon, t \geq t_0$, and the trivial solution of (2.1) is stable.

We next assume that the trivial solution of system (2.2) is asymptotically stable. Then it implies stability of the trivial solution of system (2.1). Thus for $t_0 \in R_+$, $\rho_0 > 0$, there exists $\delta^0 = \delta(\rho_0) > 0$ such that

$$\|x_0\| < \delta^0 \quad \text{implies} \quad \|x(t)\| < \rho_0, \ t \geq t_0\,, \tag{3.6}$$

where $x(t) = x(t, t_0, x_0)$ is any solution of system (2.1). By condition (i), we have

$$V(t, x(t)) \leq V(t_0, y(t, t_0, x_0)), \ \ t \geq t_0\,. \tag{3.7}$$

Since the trivial solution of system (2.2) is asymptotically stable, we get $\lim_{t\to\infty} y(t, t_0, x_0) = 0$, which implies, in view of (3.7) and condition (iii) that

$$\lim_{t\to\infty} V(t, x(t)) = 0\,.$$

Thus by (3.2) we derive that $\lim_{t\to\infty} x(t) = 0$, i.e., the trivial solution of system (2.1) is asymptotically stable. □

**Theorem 3.2** *If condition (ii) in Theorem 3.1 is changed to (ii*) $V(t, x)$ is decrescent, then any uniform stability property of the trivial solution of system* (2.2) *implies the corresponding uniform stability property of the trivial solution of system* (2.1).

**Proof** Since $V(t, x)$ is decrescent, there exist $a \in K$ and $\sigma > 0$ such that

$$V(t, x) \leq a(\|x\|), \ \ x \in s(\sigma). \tag{3.8}$$

Choose $\eta > 0$ such that

$$a(\eta) < b(\epsilon)\,. \tag{3.9}$$

Assume that the trivial solution of system (2.2) is uniformly stable. Then there exists $\delta = \delta(\eta) > 0$ such that for any $t_0 \in R_+$

$$\|x_0\| < \delta \quad \text{implies} \quad \|y(t, t_0, x_0)\| < \eta\,, \ \ t \geq t_0, \tag{3.10}$$

where $y(t, t_0, x_0)$ is any solution of system (2.2). Let $x(t) = x(t, t_0, x_0)$ be any solution of system (2.1) with $\|x_0\| < \delta$. Then an argument similar to that used in the proof of Theorem 1 shows that $\|x(t)\| < \epsilon$, $t \geq t_0$. Thus the trivial solution of system (2.1) is uniformly stable.

We next assume that the trivial solution of system (2.2) is uniformly asymptotically stable. Then it implies uniform stability of the trivial solution of system (2.1). Thus for $\rho^* = \min\{\rho_0, \sigma\}$, there exists a $\delta^* = \delta(\rho^*) > 0$ such that $\|x_0\| < \delta^*$

implies $\|x(t,t_0,x_0)\| < \rho^*$, $t \geq t_0$, where $x(t) = x(t,t_0,x_0)$ is any solution of system (2.1). By condition (i) and (3.8) we have

$$V(t,x(t)) \leq V(t_0, y(t,t_0,x_0)) \leq a(\|y(t,t_0,x_0)\|), \quad t \geq t_0 . \tag{3.11}$$

Since the trivial solution of system (2.2) is uniformly asymptotically stable, for any $\epsilon \in (0,\rho^*)$, there exists $T = T(\epsilon) > 0$ such that

$$\|y(t,t_0,x_0)\| < \alpha^{-1}(b(\epsilon)), \, t \geq t_0 + T . \tag{3.12}$$

It thus follows from (3.2), (3.11) and (3.12) that

$$|x(t)| < \epsilon, \quad t \geq t_0 + T ,$$

and hence the trivial solution of system (2.1) is uniformly asymptotically stable, completing the proof. □

**Theorem 3.3** *Assume that*

(i) $V \in C[R_+ \times R^n, R_+], V(t,x)$ *is locally Lipschitzian in* $x$*, positive definite and decrescent;*

(ii) *there exists* $C \in K$ *such that*

$$D^+V(s,y(t,s,x)) \leq -C(\|y(t,s,x)\|), \quad s \in [t_0,t), \, x \in s(\rho) ;$$

(iii) *system* (2.2) *is uniformly strictly stable.*

*Then the trivial solution of* (2.1) *is uniformly asymptotically stable.*

**Proof** Uniform stability of the trivial solution of system (2.1) follows from Theorem 3.2. Thus there exists a $\delta_0 = \delta(\rho^*) > 0$ such that

$$\|x_0\| < \delta_0 \quad \text{implies} \quad \|x(t)\| < \rho^*, \quad t \geq t_0, \tag{3.13}$$

where $\rho^*$ is the same as defined in the proof of Theorem 3.2, and $x(t) = x(t,t_0,x_0)$ is any solution of system (2.1). By condition (i), (3.2) and (3.8) remain valid. Now let $\epsilon \in (0,\rho^*)$ and $\delta = \delta(\epsilon) > 0$ be the same as defined in uniform stability. Since system (2.2) is uniformly strictly stable, there exist $\delta_1 = \delta_1(\rho^*) > 0$ and $\eta = \eta(\delta^*) > 0$, where $\delta^* \in (0, \min\{\delta,\delta_1\})$, such that

$$\delta^* \leq \|x_0\| < \delta_1 \quad \text{implies} \quad \eta < \|y(t,t_0,x_0)\| < \rho^*, \quad t \geq t_0 , \tag{3.14}$$

where $y(t,t_0,x_0)$ is any solution of system (2.2). Choose $\delta = \min\{\delta_0,\delta_1\}$ and $T = \dfrac{1}{c(\eta)} a(\rho^*)$. For any $t_0 \in R_+$, let $x(t) = x(t,t_0,x_0)$ be any solution of system (2.1) with $\|x_0\| < \delta$. We are going to prove

$$\|x(t)\| < \epsilon, \quad t \geq t_0 + T . \tag{3.15}$$

By uniform stability, it is sufficient to show that there exists a $t^* \in [t_0, t_0+T]$ such that $\|x(t^*)\| < \delta$. If this is not true, then there exists a solution $x(t) = x(t,t_0,x_0)$ of (2.1) with $\|x_0\| < \delta$ such that

$$\|x(t)\| \geq \delta, \quad t \in [t_0, t_0+T] . \tag{3.16}$$

Define $m(s) = V(s, y(t,s,x(s)))$, $t_0 \leq s \leq t \leq t_0 + T$. Then by condition (ii) and (3.14), we get

$$D^+m(s) \leq -C(\eta), \quad t_0 \leq s \leq t, \quad t \in [t_0, t_0+T] ,$$

which implies

$$m(t) \leq m(t_0) - C(\eta)(t-t_0), \quad t \in [t_0, t_0+T] .$$

Since $m(t) = V(t, y(t, t, x(t))) = V(t, x(t)), m(t_0) = V(t_0, y(t, t_0, x_0))$, we get in view of (3.8) and the definition of $T$,

$$\begin{aligned} V(t_0 + T, x(t_0 + T)) &\leq V(t_0, y(t, t_0, x_0)) - C(\eta)T \\ &\leq a(\|y(t, t_0, x_0\|) - C(\eta)T \\ &< a(\rho^*) - a(\rho^*) = 0\,, \end{aligned}$$

which is a contradiction. Thus (3.15) is true and the trivial solution of system (2.1) is uniformly asymptotically stable. □

**Theorem 3.4** *Assume that*

(i) $V \in C[R_+ \times R^n, R_+]$, $V(t, x)$ *is locally Lipschitzian in* $x$, *positive definite and* $V(t, 0) \equiv 0$;

(ii) *there exists* $C \in K$ *such that*

$$D^+V(s, y(t, s, x)) \leq -C(\|y(t, s, x)\|), \ s \in [t_0, t), x \in s(\rho)\,;$$

(iii) $F(t, x)$ *is bounded on* $R_+ \times S(\rho)$ *and system* (2.2) *is uniformly strictly stable.*

*Then the trivial solution of* (2.1) *is asymptotically stable.*

**Proof** Since system (2.2) is uniformly strictly stable, there exist $\delta^* = \delta(\rho) > 0$ such that for any $t_0 \in R_+$

$$\|x_0\| < \delta^* \quad \text{implies} \quad \|y(t, t_0, x_0)\| < \rho, \ t \geq t_0 \tag{3.17}$$

and for any $\alpha \in (0, \delta^*)$, there exists $a\ \eta = \eta(\alpha) > 0$ such that for any $t_0 \in R_+$

$$\|x_0\| \geq \alpha \quad \text{implies} \quad \|y(t, t_0, x_0)\,\| > \eta, \ t \geq t_0, \tag{3.18}$$

where $y(t, t_0, x_0)$ is any solution of system (2.2).

By Theorem 3.1, the trivial solution of system (2.1) is stable. Thus for $t_0 \in R_+$ and $\delta^* > 0$, there exist a $\delta_0 = \delta(\delta^*) > 0$ such that

$$\|x_0\| < \delta_0 \quad \text{implies} \quad \|x(t)\| < \delta^*, \ t \geq t_0, \tag{3.19}$$

where $x(t) = x(t, t_0, x_0)$ is any solution of system (3.1). We first claim that

$$\liminf_{t \to \infty} \|x(t))\| = 0 \tag{3.20}$$

where $x(t) = x(t, t_0, x_0)$ satisfies (3.19). If this is not true, then there exists $\alpha \in (0, \delta_0)$ and $T > t_0$ such that

$$\|x(t)\| \geq \alpha, \ t \geq T\,,$$

which implies, in view of (3.18), that there exists a $\eta = \eta(\alpha) > 0$ such that

$$\|y(t, s, x(s))\| > \eta, \ t \geq s \geq T. \tag{3.21}$$

But by condition (ii), we get

$$V(t, x(t)) \leq V(T, y(t, T, x(T))) - C(\eta)(t - T), \ t \geq T,$$

which implies

$$\limsup_{t \to \infty} V(t, x(t)) = -\infty\,.$$

This contradiction shows that (3.20) must be true.

Suppose $\limsup_{t \to \infty} \|x(t)\| \neq 0$. Then there exist $\gamma \in \left(0, \dfrac{\delta_0}{2}\right)$ and a sequence

$$t_0 < a_1 < b_1 < a_2 < b_2 < \cdots < a_k < b_k < \cdots$$

such that $a_k \to \infty$ as $k \to \infty$ and

$$\|x(a_k)\| = r, \ \ \|x(b_k)\| = 2\gamma \quad \text{and} \quad \|x(t)\| \geq \gamma, \ \ t \in [a_k, b_k]. \tag{3.22}$$

By condition (iii), $F(t, x)$ is bounded on $R_+ \times s(\rho)$ by some positive constant $M$. This together with (3.22), implies

$$b_k - a_k \geq \frac{\gamma}{M}, \qquad k = 1, 2, \cdots . \tag{3.23}$$

Since system (2.2) is uniformly strictly stable, it follows that there exist a $\sigma = \sigma(\gamma) > 0$ such that

$$\|y(t, s, x(s))| > \sigma, \quad t \geq s, \quad s \in [a_k, b_k]. \tag{3.24}$$

Now by condition (ii), we get, in view of (3.22) - (3.24)

$$\begin{aligned}
\limsup_{t\to\infty} V(t, x(t)) &\leq V(t_0, y(t, t_0, x_0)) - \lim_{t\to\infty} \int_{t_0}^{t} C(\|y(t, s, x(s))\|)\, ds \\
&\leq V(t_0, y(t, t_0, x_0)) - \lim_{k\to\infty} \sum_{j=1}^{k} \int_{a_j}^{b_j} C(\|y(t, s, x(s))\|)\, ds \\
&\leq V(t_0, y(t, t_0, x_0)) - \lim_{k\to\infty} \sum_{j=1}^{k} C(\sigma)(b_j - a_j) \\
&= -\infty ,
\end{aligned}$$

which is a contradiction. Thus $\limsup_{t\to\infty} \|x(t)\| = 0$ and hence by (3.20) we conclude $\lim_{t\to\infty} x(t) = 0$. Therefore the trivial solution of system (2.1) is asymptotically stable. □

Finally, to illustrate our results, we consider some examples.

**Example 3.1.** Consider systems

$$\begin{cases} x_1' = x_1 \cos t - x_1^3 \sin^2 x_2 , \\ x_2' = x_2 \sin t \ - x_2 \cos^2 x_1 x_2 \end{cases} \tag{3.25}$$

and

$$\begin{cases} y_1' = y_1 \cos t , \\ y_2' = y_2 \sin t . \end{cases} \tag{3.26}$$

The solution of (3.26) is given by

$$y(t, t_0, x_0) = \begin{bmatrix} x_{10} e^{\sin t - \sin t_0} \\ x_{20} e^{-\cos t + \cos t_0} \end{bmatrix},$$

which is uniformly stable. Choose $V = \frac{1}{2}(x_1^2 + x_2^2)$. Then

$$\begin{aligned}
V'(t, s, x) &= \begin{bmatrix} x_1 e^{2(\sin t - \sin s)} \\ x_2 e^{2(-\cos t + \cos s)} \end{bmatrix}^T \begin{bmatrix} -x_1^3 \sin^2 x_2 \\ -x_2 \cos^2 x_1 x_2 \end{bmatrix} \\
&= y_1^2(t, s, x) x_1^2 \sin^2 x_2 - y_2^2(t, s, x) \cos^2(x_1 x_2) \leq 0 .
\end{aligned}$$

Thus, by Theorem 3.2 the trivial solution of (3.25) is uniformly stable.

**Example 3.2.** Consider systems

$$\begin{cases} x_1' = (\sin t - 1)x_1 + x_2 + x_1 x_2^2, \\ x_2' = (\sin t - 1)x_2 - x_1 + x_1^2 x_2, \end{cases} \tag{3.27}$$

$$\begin{cases} y_1' = y_1 \sin t, \\ y_2' = y_2 \sin t. \end{cases} \tag{3.28}$$

The solution of (3.28) is given by

$$y(t, t_0, x_0) = \begin{bmatrix} x_{10}\, e^{-\cos t + \cos t_0} \\ x_{20}\, e^{-\cos t + \cos t_0} \end{bmatrix},$$

which is uniformly strictly stable. Let $V = \frac{1}{2}(x_1^2 + x_2^2)$. Then

$$\begin{aligned} V'(t, s, x) &= e^{-2[\cos t - \cos s]}[-x_1^2 - x_2^2 + 2x_1^2 x_2^2] \\ &\leq -k|y(t, s, x)|^2, \; k = 1 - \rho^2 > 0, \rho < 1 \end{aligned}$$

and $|x| < \rho$.

By Theorem 3.3, the trivial solution of (3.27) is uniformly asymptotically stable.

**Remark** In general, it is not necessary to find suitable known system (2.2). It would be sufficient to choose any function $y \in C[R_+^2 \times R^n, R^n]$ satisfying

(i) $|y(t, s, x)|$ is locally Lipschitzian in $x$ for each $(t, s)$;

(ii) $y(t, t, x) = x$;

(iii) $y(t, t_0, x_0)$ has required stability properties.

## References

[1] Ladde, G. S., Lakshmikantham, V. and Leela, S. [1997], *A new technique in perturbation theory*, Rocky Mountain J. Math. **6**, 133-144.

[2] Lakshmikantham, V. and Leela, S. [1969], *Differential and Integral Inequalities*, Vol. I, Academic Press, New York.

[3] Lakshmikantham, V. and Liu, X. [1993], *Stability Analysis in Terms of Two Measures*, World Scientific, Singapore.

[4] Liu, X. and Vatsala, A. S., *Variations of parameters in terms of Lyapunov-like functions and stability of perturbed systems*, to appear.

[5] Liu, X. and Sivasundaram, S. [1995], *Stability of nonlinear systems under constantly acting perturbations*, International J. Math. and Math. Sci. **18**, 273-278.

[6] Rajalakshmi, S. and Sivasundaram, S. [1993], *Variational Lyapunov second method*, Dyn. Sys. and Appl. **2**, 484-490.

Fields Institute Communications
Volume **21**, 1999

# Persistence in an Epidemiological Model

**Jaroslav Milota**
Matematicko-fyzikalni fakulta UK
Sokolovska 83, 186 00
Praha 8, Czech Republic
milota@karlin.mff.cuni.cz

**Abstract.** Transmission of a sexual disease is considered in a population divided into a core and a noncore group. The model of SIS type leads to an asymptotically autonomous system of ODEs. Global asymptotic stability either of the disease-free equilibrium or of an endemic equilibrium is proved. For the model of SIR type, global asymptotic stability of the disease free equilibrium is proved provided there is no endemic equilibrium. Conditions for the existence of a (unique) endemic equilibrium are found and strong uniform persistence of the disease is proved in this case.

## 1 Introduction

We consider transmission of a sexual disease in a multigroup population (for the sake of simplicity $n = 2$) which differs by various sexual activity. The SIS type was introduced in [**20**], see also the deep study in [**17**], where they supposed that all subpopulations are homogeneous and closed and the incidence functions are given by the simple mass action. The main result, namely that either the disease-free equilibrium is globally asymtotically stable or there exists a unique endemic equilibrium which is globally asymptotically stable, is a consequence of the monotonicity of this dynamical system (see [**18**] or [**23**]).

A general multigroup model was formulated in [**2**], but conditions for global asymptotic stability of an endemic equilibrium are not easily verified (for numerical simulations of an SIR multigroup model (see [**3**] or [**8**], Section 2.3.4.2).

The "constant total population size" assumption is reasonable for quickly spreading diseases or for endemic situations. In modeling the transmission of sexual diseases we should have in mind that the disease-related outflow is not balanced with inflow, e.g., because of AIDS, or, from the opposite side, because of increasing prostitution in East European countries since 1989. Another reason for a varying population size were formulated in [**1**], [**7**], [**14**], [**11**], and [**10**].

1991 *Mathematics Subject Classification.* 92D30.

More generally, nonlinear (e.g., Holling type, or Heesterbeek type [**15**]) or population size dependent incidence functions were considered in e.g., [**4**], [**22**], [**25**], and [**27**] (see also the references given there) and in the survey paper [**16**].

In this paper we will consider a two-group-model for both SIS and SIR types with simple vital dynamics and nonlinear, population size dependent incidence functions, which are time independent. In order to describe a model in the form of four ordinary differential equations, we suppose that the duration of the infection is exponentially distributed and there is no delay in the incidence functions. A similar SIS model with a core group and constant population size has been recently investigated in [**5**]. For a more general (nonconstant population size, arbitrary rate of recovery) two sex SIS model with a local stable endemic equilibrium see [**6**].

The SIS case is more simple because this leads to an asymptotically autonomous system. Applying the results of [**21**] (see also [**24**] and [**9**]) and the Poincaré-Bendixson theory we generalize the above mentioned result of Lajmanovich-Yorke. The SIR case is more complicated. If there is only the disease-free equilibrium then this is globally asymtotically stable as we will prove with the help of the fluctuation lemma [**19**]. We will find threshold parameters for the existence of a unique endemic equilibrium. Its global stability remains an open problem. Instead of that we will give a weaker result, namely that the disease is strongly uniformly persistent. To show this we will first prove weak persistence (via approach due to [**26**]) and then we will use a result of [**12**] (see also [**13**]).

## 2 Description of the model

We consider a population of the size $\Lambda$ which is divided into a very active (core) group of the size $P$ and a less active (non-core) group of the size $N = \Lambda - P$. Both groups consist of susceptible parts of the sizes $P^-$ (in the core group) and $S$ (in the non-core group) and infective parts $P^+$, resp. $I$.

The contact functions $C_i$ (for the non-core group), $D_i$ (for the core group), $i = 1, 2$, are supposed to be continuous functions of $N, P$. The force of infection for the non-core group is therefore given by

$$r_1 D_1(N,P)\frac{I}{N} + r_2 D_2(N,P)\frac{P^+}{P} ,$$

where $r_1, r_2$ are risk factors.

We note that the functions $C_2, D_2$ should fulfill the "conservation law" $C_2(N,P)N = D_2(N,P)P$ .

The whole model is described in the block diagram on the next page.

The dotted lines in the diagram describe the SIS type. Further, $a, b$ denote the inflow into the non-core group and into the core group, respectively. As we mentioned in the Introduction, we suppose simple vital dynamics, i.e., that $a$ and $b$ are constants. The constants $m$ and $\mu$ denote the "mortality" and $\delta$ and $\epsilon$ express extra "mortality" due to the disease.

Under these assumptions and notations a model can be expressed in the form

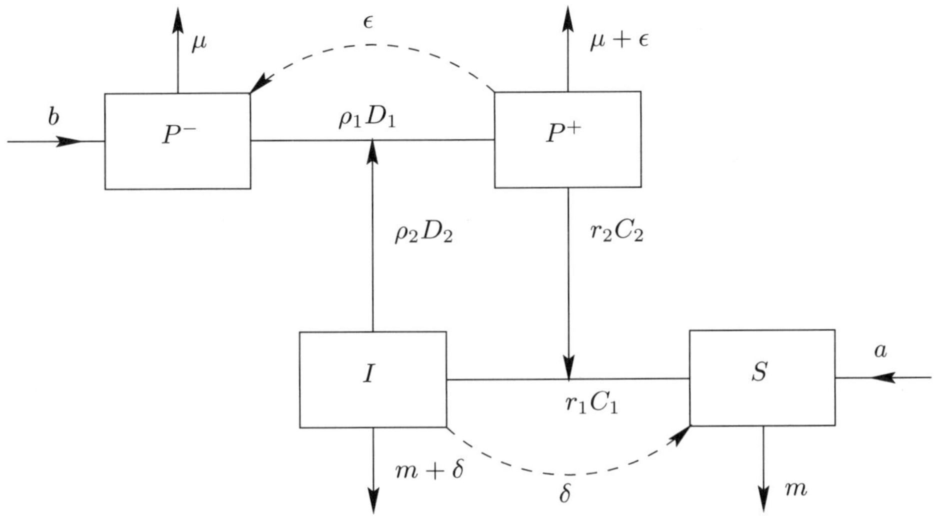


**Figure 1. The diagram for the transmission of a disease.**

of the following system of ordinary differential equations

$$\dot{S} = a - mS - r_1 C_1 \frac{I}{N} S - r_2 C_2 \frac{P^+}{P} S + (\delta I) \tag{1}$$

$$\dot{I} = -(m+\delta) I + r_1 C_1 \frac{I}{N} S + r_2 C_2 \frac{P^+}{P} S \tag{2}$$

$$\dot{P}^- = b - \mu P^- - \rho_1 D_1 \frac{P^+}{P} P^- - \rho_2 D_2 \frac{I}{N} P^- + (\epsilon P^+) \tag{3}$$

$$\dot{P}^+ = -(\mu+\epsilon) P^+ + \rho_1 D_1 \frac{P^+}{P} P^- + \rho_2 D_2 \frac{I}{N} P^- . \tag{4}$$

The terms in parantheses in (1), (3) are considered for the SIS model only.

## 3 Auxiliary results

We will need the so-called fluctuation lemma which was basically proved in [**19**] and we state this in the following form ([**26**], Corollary 2.4):

**Theorem A** *Let $I$ be a bounded interval and $g$ be a bounded uniformly continuous function on $(t_0, +\infty) \times I$. Let $x : (t_0, +\infty) \to I$ be a solution of the equation $\dot{x} = g(t, x)$. Then the inequalities*

$$\liminf_{t\to+\infty} g(t, x^\infty) \le 0 \le \limsup_{t\to+\infty} g(t, x^\infty)$$

$$\liminf_{t\to+\infty} g(t, x_\infty) \le 0 \le \limsup_{t\to+\infty} g(t, x_\infty)$$

*hold with* $x^\infty = \limsup_{t\to+\infty} x(t), \quad x_\infty = \liminf_{t\to+\infty} x(t).$

We recall (see [**12, 13**]) the notion of persistence for a system of differential equations. Consider a system

$$\dot{x} = h(x) \tag{5}$$

in the set $\Omega \subset \mathbb{R}^n$ and suppose that for any $x_0 \in \Omega$ there is a unique solution $\varphi(., x_0), \varphi(0, x_0) = x_0$, defined on the interval $[0, +\infty)$ and taking values in the set $\Omega$.

**Definition** A system (5) is said to be in $\Omega$

1. weakly persistent (WP) if $\limsup_{t \to +\infty} \operatorname{dist}(\varphi(t, x_0), \partial\Omega) > 0$ for any $x_0 \in \Omega$.
2. weakly uniformly persistent (WUP) if there is $\epsilon > 0$ such that $\limsup_{t \to +\infty} \operatorname{dist}(\varphi(t, x_0), \partial\Omega) \geq \epsilon$ for any $x_0 \in \Omega$.
3. strongly uniformly persistent (SUP) if there is $\epsilon > 0$ such that $\liminf_{t \to +\infty} \operatorname{dist}(\phi(t, x_0), \partial\Omega) \geq \epsilon$ for any $x_0 \in \Omega$.

Freedman and Moson in [**12**] proved that (WUP) $\Longrightarrow$ (SUP) and Thieme in [**26**] found conditions for (WP) $\Longrightarrow$ (WUP). The following theorem ([**26**], Theorem 4.5) will be useful for our purposes.

**Theorem B** *Let $X$ be locally compact, and let $X_2$ be compact in $X$ and $X_1$ be forward invariant under the continuous semiflow $\Phi$ on $X$. Assume that $\Omega_2$ defined by*

$$\Omega_2 = \bigcup_{y \in Y_2} \omega(y), \quad Y_2 = \{x \in X_2; \Phi_t(x) \in X_2 \quad \forall t > 0\},$$

*has an acyclic isolated covering $M = \cup_{k=1}^m M_k$. If each part $M_k$ of $M$ is a weak repeller for $X_1$, then $X_2$ is a uniform strong repeller for $X_1$.*

## 4 Results for the SIS type model

In this case we have for the core and non-core subpopulations

$$\dot{N} = a - mN, \quad \dot{P} = b - \mu P, \tag{6}$$

and hence

$$\lim_{t \to +\infty} N(t) = \frac{a}{m} =: \bar{N}, \qquad \lim_{t \to +\infty} P(t) = \frac{b}{\mu} =: \bar{P}. \tag{7}$$

Introducing the new variables $x = \frac{I}{N}, y = \frac{P^+}{P}$, we get

$$\dot{x} = -(m + \delta)x + r_1 C_1 x(1 - x) + r_2 C_2 y(1 - x) - x\frac{\dot{N}}{N}, \tag{8}$$

$$\dot{y} = -(\mu + \epsilon)y + \rho_1 D_1 y(1 - y) + \rho_2 D_2 x(1 - y) - y\frac{\dot{P}}{P}. \tag{9}$$

Because of (7), the system (8), (9) is asymptotically autonomous with the following limit equations

$$\frac{1}{m + \delta}\dot{x} = -x + R_N x(1 - x) + R_{P,N} y(1 - x), \tag{10}$$

$$\frac{1}{\mu + \epsilon}\dot{y} = -y + R_P y(1 - y) + R_{N,P} x(1 - y), \tag{11}$$

where

$$R_N = \frac{r_1 C_1(\bar{N}, \bar{P})}{m + \delta}, \quad R_{P,N} = \frac{r_2 C_2(\bar{N}, \bar{P})}{m + \delta},$$
$$R_P = \frac{\rho_1 D_1(\bar{N}, \bar{P})}{\mu + \epsilon}, \quad R_{N,P} = \frac{\rho_2 D_2(\bar{N}, \bar{P})}{\mu + \epsilon}.$$

The system (10), (11) always has the trivial (i.e., disease-free) equilibrium $(0,0)$. To find another equilibrium we introduce the function $f(x) = \frac{x}{1-x}\frac{1-R_N(1-x)}{R_{P,N}}$ and an analogous function $g(y)$ from the Equation (11). The graphs of these functions show that a (unique) equilibrium in the domain $(0,1) \times (0,1)$ exists if and only if at least one of the folowing conditions is satisfied :

$$R_N > 1, \tag{H1}$$
$$R_P > 1, \tag{H2}$$
$$R_{N,P}R_{P,N} > (1 - R_N)(1 - R_P). \tag{H3}$$

The rectangle $[0,1] \times [0,1]$ is positively invariant for the system (10), (11). In the case of the existence of a nontrivial equilibrium, the Dulac divergence criterion (after division by $xy$) rules out nonconstant periodic solutions. In this case the (central-) stable manifold (if exists) of the trivial equilibrium is oriented outward the positive cone. Therefore all solutions in $[0,1] \times [0,1] \setminus (0,0)$ converge either to a unique nontrivial equilibrium (if this exists) or to the trivial equilibrium (if there is no nontrivial equilibrium in $[0,1] \times [0,1]$). As all solutions of (6), (8), (9) with initial conditions in $\mathbb{R}^2 \times [0,1]^2$ stay in this set and are bounded, we can apply the result of Markus (Theorems 1,2 in [**21**], see also [**24**]) to obtain the following result for the SIS case :

**Theorem 1** *For any solution of* (1) - (4) *with non-negative initial conditions and* $(I(0), P^+(0)) \neq (0,0)$*, the solution converges either to a (unique) endemic equilibrium (provided that at least one of the conditions* (H1) - (H3) *is fulfilled) or to the disease-free equilibrium* $[\frac{a}{m}, 0, \frac{b}{\mu}, 0]$.

**Remark** If we use the cooperative character of the limit system (10) - (11) instead of the Poincaré-Bendixson theory, we can generalize Theorem 1 for a $n$-groups model of the SIS type. But the condition (H3) will be more complicated.

## 5 Results for the SIR type model

To describe the SIR epidemics we consider the system (1) - (4) together with the following two additional equations

$$\dot{R} = \delta I - mR,$$
$$\dot{T} = \epsilon P^- - \mu T.$$

Then for $N = S + I + R, \quad P = P^- + P^+ + T$, we again have the Equations (6) and the new system is also asymptotically autonomous. To simplify the discussion, we will suppose that both subpopulations are in their equilibria, i.e., $N = \bar{N}, P = \bar{P}$. In other words, we will discuss the limit system.

In the new variables $x = \frac{I}{N}, r = \frac{R}{N}, y = \frac{P^+}{P}, s = \frac{T}{P}$, the model for the SIR case can be rewritten in the form

$$\dot{x} = -(m + \delta)x + \bar{C}_1 x(1 - x - r) + \bar{C}_2 y(1 - x - r), \tag{12}$$
$$\dot{r} = \delta x - mr, \tag{13}$$
$$\dot{y} = -(\mu + \epsilon)y + \bar{D}_1 y(1 - y - s) + \bar{D}_2 x(1 - y - s), \tag{14}$$
$$\dot{s} = \epsilon y - \mu s, \tag{15}$$

where $\bar{C}_i = C_i(\bar{N}, \bar{P}), \bar{D}_i = D_i(\bar{N}, \bar{P}), i = 1, 2$. The phase space $\Delta \times \Delta$, where $\Delta = \{(u,v) \in \mathbb{R}^2_+; u + v \le 1\}$, is positively invariant.

A nontrivial equilibrium of this system can be found in the same way as in Section 4, i.e., from the first two equations one gets $y = f(x)$ with $f'(0) = \frac{1-R_N}{R_{P,N}}$. Similarly, the third and fourth equations yield $x = g(y)$ with $g'(0) = \frac{1-R_P}{R_{N,P}}$. Therefore, a (unique) nontrivial equilibrium in $\Delta \times \Delta$ exists if and only if one of the conditions (H1) - (H3) is satisfied.

The system (12) - (15) has always the trivial equilibrium. An easy computation shows that this equilibrium is locally asymptotically stable provided none of the conditions (H1) - (H3) is satisfied.

**Theorem 2** *If the system* (12) - (15) *has a unique equilibrium in* $\Delta \times \Delta$ *(namely the trivial equilibrium), then this is globally asymptotically stable in* $\Delta \times \Delta$.

**Proof** Let $x^\infty := \limsup_{t\to+\infty} x(t)$, $x_\infty := \liminf_{t\to+\infty} x(t)$. By Theorem A and the Equation (12), we have

$$-(m+\delta)x^\infty + \bar{C}_1 x^\infty(1 - x^\infty - r_\infty) + \bar{C}_2 y^\infty(1 - x^\infty - r_\infty) \ge 0$$

since $(x, r) \in \Delta$. As $r_\infty \ge 0$, we obtain that $y^\infty \ge f(x^\infty)$, where the function $f$ is as above, i.e., a strictly increasing convex function on $[0, 1)$ and $f'(0) = \frac{1-R_N}{R_{P,N}} > 0$. Suppose that $x^\infty > 0$. Then $f(x^\infty) > f'(0)x^\infty$. Similarly,

$$x^\infty \ge g(y^\infty) > g(f'(0)x^\infty) \ge g'(0)f'(0)x^\infty.$$

According to the negation of (H3), $g'(0)f'(0) \ge 1$. This contradiction shows that $x^\infty = 0$, i.e., $\lim_{t\to\infty} x(t) = 0$. An analogous computation yields $\lim_{t\to\infty} y(t) = 0$. The Equations (13) - (15) imply $\lim_{t\to\infty} r(t) = \lim_{t\to\infty} s(t) = 0$.

□

We are not able to prove that the endemic equilibrium is globally asymptotically stable provided it exists. So we will give a weaker result, namely that the disease is uniformly persistent in both subpopulations.

**Theorem 3** *Let the system* (12) - (15) *have a nontrivial equilibrium in* $\Delta \times \Delta$. *Then this system is strongly uniformly persistent in* $\Delta \times \Delta$.

**Proof** We will apply Theorem B. Let $X = \Delta \times \Delta$, $X_1 = \mathrm{Int}(\Delta \times \Delta)$, $X_2 = \partial(\Delta \times \Delta)$. The maximal positive invariant subset of $X_2$ is $Y_2 = \{(0, r, 0, s);\ r \in [0, 1],\ s \in [0, 1]\}$. By (13), (15) all solutions in $Y_2$ converge to the origin $\mathcal{O}$, i.e., $\Omega_2 = \{\mathcal{O}\}$. An easy computation shows that the linearization of (12) - (15) at $\mathcal{O}$ has eigenvalues $\lambda_1 = -m$, $\lambda_2 = -\mu$ and $\lambda_-$, $\lambda_+$. Under the assumption of the theorem, at least one of the conditions (H1) - (H3) is satisfied. This means that at least $\lambda_+$ is positive ($\lambda_-$ is also positive iff (H1), (H2) are satisfied and (H3) not). The corresponding eigenfunctions are $e_1 = (0, 1, 0, 0)$, $e_2 = (0, 0, 0, 1)$, $e_+ = (x_+, r_+, y_+, s_+)$, $e_- = (x_-, r_-, y_-, s_-)$, where $e_+$ can be chosen in $\mathbb{R}^4_+$ and $x_- > 0$, $y_- < 0$. If $\lambda_- > 0$ then the stable manifold $W^s(\mathcal{O}) = \mathrm{Lin}\{e_1,\ e_2\}$ and therefore $W^s(\mathcal{O}) \cap X_1 = \emptyset$. If $\lambda_- \le 0$, the same result ( if $\lambda_- = 0$ then for the central-stable manifold) can be obtained as follows. The stable manifold $W^s(\mathcal{O})$ can be locally expressed in the form

$$\alpha e_1 + \beta e_2 + \gamma e_- + f(\alpha, \beta, \gamma)e_+, \quad \alpha, \beta, \gamma \text{ are small reals,}$$

where $f \in C^1$, $f(0,0,0) = 0$, $Df(0,0,0) = 0$. Moreover, $f(\alpha,\beta,0) = 0$ as $\mathrm{Lin}(e_1,e_2)$ is a part of the stable manifold. If $W^s(\mathcal{O}) \cap X_1 \neq \emptyset$, then this intersection contains points in any neighborhood of the origin. Let

$$z_n = \alpha_n e_1 + \beta_n e_2 + \gamma_n e_- + f(\alpha_n,\beta_n,\gamma_n)e_+ \in W^s(\mathcal{O}) \cap X_1$$

and $\alpha_n$, $\beta_n$, $\gamma_n \to 0$. Suppose that $\gamma_n > 0$. The third Euclidean coordinate $\gamma_n y_- + f(\alpha_n,\beta_n,\gamma_n)y_+$ is strictly positive. Since $f \in C^1$, $f(\alpha,\beta,0) = 0$, it follows that

$$\lim \frac{f(\alpha_n,\beta_n,\gamma_n)}{\gamma_n} = 0.$$

This yields the contradiction as we have chosen $y_- < 0$. If $\gamma_n < 0$ then we obtain the same contradiction by considering the first coordinate of $z_n$. This proves that the stable manifold $W^s(\mathcal{O})$ is disjoint with $X_1$ and thus there is no homoclinic orbit of $\mathcal{O}$ in $X_1$, i.e., $\mathcal{O}$ is itself the acyclic covering and $\mathcal{O}$ is also a weak repeller for $X_1$. The statement of the theorem follows now from Theorem B.

□

## 6 Discussion

The threshold parameters given in the conditions (H1) - (H3) could be interpreted in the following way. If $S$ would be zero, then the Equation (2) has the form $\dot{I} = -(m+\delta)I$ and hence $\frac{1}{m+\delta}$ is the average time for a non-core individual to be infective. The coefficient $r_1$ is a risk factor for contacts between the members of the non-core group. Since $\bar{C}_1$ is the number of contacts per unit time of a typical individual if both populations are in stationary states, we can see that $R_N = \frac{r_1\bar{C}_1}{m+\delta}$ is the average number of new cases in the non-core group caused by one infective non-core individual during the whole infective period. Therefore, the condition (H1) for endemicity of the disease in the isolated (i.e., $C_2 = 0$) non-core group is plausible.

The condition (H3) shows that sufficiently high promiscuity between subpopulations produces persistence of the disease even though it would become extinct provided both subpopulations are isolated. The interpretation of the condition (H3) is not so obvious as for (H1), (H2).

The special case when the core group consists of prostitutes can be described in the model choosing $D_1 = 0$. If $R_N$ is small ($< 1$) then the condition (H3) is necessary and sufficient for the existence of an endemic equilibrium. In this case we can also assume that the contact function $C_2$ does not depend on $N$ and is of Holling type II, e.g., $C_2(P) = \frac{\alpha P^2}{\beta^2+P^2}$. By the "conservation law", we have $D_2(N,P) = \frac{\alpha NP}{\beta^2+P^2}$ and $R_{N,P}R_{P,N} = \alpha^2 c\frac{\bar{P}^3\bar{N}}{(\beta^2+\bar{P}^2)^2}$, where $c$ is a constant depending on the mortalities and risk factors. Looking at this expression as the function $\Phi(\bar{P})$, we see that $\Phi(0) = \lim_{\bar{P}\to+\infty} \Phi(\bar{P}) = 0$ and $\Phi$ has a unique maximum. Therefore the disease cannot be endemic if the size $\bar{P}$ of the core group is sufficiently small or large. In other words, there is a window for $\bar{P}$ in which the disease is endemic.

Our SIS and SIR models cannot describe a fatal disease, since susceptibles persist in both subpopulations. This result follows easily from the fluctuation lemma. In the SIR case, the Equation (12), (13) yield $\dot{\sigma} = m(1-\sigma) - \bar{C}_1 x\sigma - \bar{C}_2 y\sigma$ for

$\sigma = \frac{S}{N}$. By Theorem A,

$$\sigma_\infty \geq \frac{m}{m + \bar{C}_1 x^\infty + \bar{C}_2 y^\infty} > 0.$$

The analogous inequality for $(\frac{P^-}{P})_\infty$ follows from (14), (15). For the SIS case, we get

$$\sigma_\infty \geq \frac{1}{1 + R_N x^\infty + R_{P,N} y^\infty}$$

from the Equation (10).

## References


[1] Anderson, R. M. and May, R. M. [1979] *Population biology of infectious diseases I,II,* Nature **280**, 361-367, 455-461.

[2] Beretta, E. and Capasso, V. [1986] *On the general structure of epidemic systems. Global asymptotic stability,* Comp. and Math. with Appl. **12A**, 677-694.

[3] Beretta, E. and Capasso, V. [1988] *Global stability results for a multigroup SIR epidemic model,* in Mathematical Ecology (L. J. Gross, T. G. Hallam and S. A. Levin, eds.), World Scientific, Singapore.

[4] Brauer, F. [1990] *Models for the spread of universally fatal diseases,* J. Math. Biol. **28**, 451-462.

[5] Brauer, F., Castillo-Chavez, C. and Velasco-Hernándes, J. X. [1996] *Recruitment into a core group and its effect on the spread of a sexually-transmitted disease,* preprint.

[6] Brauer, F., Castillo-Chavez, C. and Velasco-Hernándes, J. X. [1996] *Recruitment effects in heterosexually transmitted disease models,* Int. J. Appl. Sci. & Computations **3**, 78-90.

[7] Busenberg, S. and Hadeler K. [1990] *Demography and epidemics,* Math. Biosci. **101**, 41-62.

[8] Capasso, V. [1993] *Mathematical Structures of Epidemic Systems,* Lecture Notes in Biomathematics **97**, Springer, Berlin-Heidelberg-New York.

[9] Castillo-Chavez, C. and Thieme, H. [1995] *Asymptotic autonomous epidemiological models,* in Mathematical Population Dynamics: Analysis of Heterogeneity I. Theory of Epidemics (O. Arino et al., eds.), Wuerz, 33-50.

[10] Derrick, W. R. and van den Driessche, P. [1993] *A disease transmission model in a nonconstant population,* J. Math. Biol. **31**, 495-512.

[11] Gao, L. Q. and Hethcote, H. W. [1992] *Disease transmission models with density - dependent demographics,* J. Math. Biol. **30**, 717-731.

[12] Freedman, H. I. and Moson, P. [1990] *Persistence definitions and their connections, Proc. Amer. Math. Soc.* **109**, 1025-1033.

[13] Freedman, H. I., Ruan, S. and Tang, M. [1994] *Uniform persistence and flows near a closed positively invariant set,* J. Dynamics and Diff. Eqs. **6**, 583-600.

[14] Hadeler, K. and Ngoma, K. [1990] *Homogeneous models for sexually-transmitted diseases,* Rocky Mount. J. Math. **20**, 967-986.

[15] Heesterbeek, J. A. and Metz, J. A. [1993] *The saturating contact rate in marriage- and epidemic models,* J. Math. Biol. **31**, 529-539.

[16] Hethcote, H. W. *A thousand and one epidemic models,* in Frontiers in Theoretical Biology (S. A. Levin, ed.), Lecture Notes in Biomathematics **100**, Springer, Berlin-Heidelberg-New York, 504-515.

[17] Hethcote, H. W. and Yorke, J. A. [1984] *Gonorrhea Transmission Dynamics and Control,* Lecture Notes in Biomathematics **56**, Springer, Berlin-Heidelberg-New York.

[18] Hirsch, W. M. [1984] *The dynamical system approach to differential equations,* Bull. Amer. Math. Soc. **11**, 1-64.

[19] Hirsch, W. M., Hanisch, H. and Gabriel, J. P. [1985] *Differential equation models of some parasitic infections: Methods for the study of asymptotic behavior,* Comm. Pure Appl. Math. **38**, 733-753.

[20] Lajmanovich, A. and Yorke, J. A. [1976] *A deterministic model for gonorrhea in a nonhomogenous population,* Math. Biosci. **28**, 221-236.

[21] Markus, L. [1956] *Asymptotic autonomous differential systems,* in Contributions to the Theory of Nonlinear Oscillations III (S. Lefschetz, ed.), Ann. Math. Stud. **36**, Princeton Univ. Press, Princeton, 17-29.

[22] Pugliese, A. [1991] *An SEI epidemic model with varying population size,* in Differential Equation Models in Biology (S. Busenberg and M. Martelli, eds.), Lecture Notes in Biomathematics **92**, Springer, Berlin-Heidelberg-New York, 121-138.

[23] Smith, H. L. [1988] *Systems of ordinary differential equations which generate an order preserving flow. A survey of results,* SIAM Review **30**, 87-113.

[24] Thieme, H. [1992] *Convergence results and a Poincaré-Bendixson trichotomy for asymptotically autonomous differential equations,* J. Math. Biol. **30**, 755-763.

[25] Thieme, H. [1992] *Epidemic and demographic interaction in the spread of potentially fatal diseases in growing populations,* Math. Biosci **111**, 99-130.

[26] Thieme, H. [1993] *Persistence under relaxed point-dissipativity (with application to an endemic model),* SIAM J. Math. Anal. **24**, 407-435..

[27] Zhou, J. and Hethcote, H. W. [1994] *Population size dependent incidence in models for diseases without immunity,* J. Math. Biol. **32**, 809-834.

Fields Institute Communications
Volume **21**, 1999

# Homoclinic Bifurcations in a Diffusively Coupled Excitable System

**Yoshihisa Morita**
Department of Applied Mathematics and Informatics
Faculty of Science and Technology
Ryukoku University, Seta Ohtsu 520-21, Japan
morita@math.ryukoku.ac.jp

**Mark Pernarowski**
Department of Mathematical Sciences
Montana State University, Bozeman, MT 59717, USA
pernarow@math.montana.edu

**Jack Dockery**
Department of Mathematical Sciences
Montana State University, Bozeman, MT 59717, USA
umsfjdoc@math.montana.edu

**Abstract.** In this paper we consider the diffusively coupled system:

$$\begin{aligned} \dot{u}^1 &= f(u^1, \mu) + \nu D(u^2 - u^1) \\ \dot{u}^2 &= f(u^2, \mu) + \nu D(u^1 - u^2) \end{aligned} \tag{0.1}$$

where $u^j \in \mathbf{R}^2$ $(j = 1, 2)$, $f : \mathbf{R}^2 \times \mathbf{R}^1 \mapsto \mathbf{R}^2$ is sufficiently smooth, $D = \operatorname{diag}(d_1, d_2), d_j \geq 0 (j = 1, 2), d_1 + d_2 > 0$ and $\nu$ is a nonnegative parameter. The corresponding uncoupled system

$$\dot{u} = f(u, \mu) \tag{0.2}$$

models the force-length relationship in muscle cells and is shown to be excitable for $\mu < \mu_H$ and have a homoclinic bifurcation at $\mu = \mu_H$ from which stable periodic solutions emerge and persist for a range of $\mu > \mu_H$. A numerically computed global bifurcation diagram illustrates that, for some parameter values, the coupled system (0.1) has stable asymmetric periodic solutions which continue in some region of $\mu < \mu_H$, where no periodic solutions of (0.2) exist. Some degenerate codimension 2 bifurcations of equilibria are summarized after which we investigate conditions for asymmetric homoclinic and heteroclinic bifurcations to occur in (0.1). The role of such bifurcations in the whole bifurcation structure is discussed.

---

1991 *Mathematics Subject Classification.* Primary 34C23, 92C10; Secondary 34C15, 34C37, 92B99.

MP was supported in part by NSF Grants DMS-94-04-521, DMS-97-04-966 and OSR-93-50-546.

JD was supported in part by NSF Grants DMS-94-04-160 and OSR-93-50-546.

## 1 Introduction

There are a variety of models which describe excitability in different systems. The literature is laden with examples and analyses of excitable systems describing the electrical activity of nerve and enodcrine cells. Some notable examples include the Hodgkin-Huxley model of electrical activity in the squid giant axon [**6**], and its analog version the FitzHugh-Nagumo model [**4, 10**]. Less studied is the excitability exhibited in models based on the force-length relationship in the muscle system.

Here, we examine a minor variant of one such model developed by Murase in [**9**]. First we introduce the model, describing its biological interpretation and bifurcation structure. In subsequent sections we then examine bifurcations of equilibria and homoclinic solutions in a related coupled system. The theorem stated in the last section demonstrates that asymmetric homoclinic solutions of the coupled system can bifurcate from in-phase homoclinic solutions that are asymptotic to nonhyperbolic equilibria.

The original model of Murase is the 2-component system of ordinary differential equations

$$\gamma\frac{dx}{dt} = nF(x) - kx + z, \tag{1.1}$$

$$\frac{dn}{dt} = g(x,n) = \begin{cases} b(1-n) & (x \le x_0) \\ -cn & (x > x_0), \end{cases} \tag{1.2}$$

where

$$F(x) = ax^2(1-x). \tag{1.3}$$

Here, $x$ is the dimensionless sliding displacement of cross-bridges connecting thin and thick muscle filaments, $n$ is the fraction of the active cross-bridges, $z$ is the external shear force and the parameters $\lambda = (a, b, c, k, x_0, \gamma)$ have the interpretations:

$a$ : the force constant of the active cross-bridges,
$b$ : the activation rate constant,
$c$ : the inactivation rate constant,
$k$ : the force constant of the passive elastic component,
$x_0$ : activation region,
$\gamma$ : the internal viscous shear resistance.

Equation (1.1) describes the balance of the external and elastic shear forces (right side), and the viscous force (left side). Equation (1.2) illustrates how the nonlinear shear force $F(x)$ is activated or inactivated depending on whether the cross-bridge displacement is below or above $x_0$. Thus, depending on the sliding displacement, the level of excitation $n$ is driven to either $n = 0$ (a complete lack of excitability) or $n = 1$ (the maximal excitability) (see [**9**] for details).

First, we briefly explain some characteristic features of this model. When $z = 0$, $(x, n) = (0, 0)$ is a rest point which is asymptotically stable. With an appropriate choice of $\lambda$ the system (1.1)-(1.2) exhibits excitability, that is, a sufficiently large stimulus $z$ (of short duration) causes a large excursion of $x$ from and then back to its stable equilibrium. For constant external forces above a threshold, (1.1)-(1.2) exhibits periodic excitation. As $z$ is increased, a bifurcation analysis shows [**9**] that these (stable) periodic solutions emerge through a homoclinic bifurcation. Hence large amplitude periodic motion suddenly appears as $z$ is increased.

For convenience, we modify the model (1.1)-(1.2) by smoothing the discontinuity in $g(x, n)$ as follows:

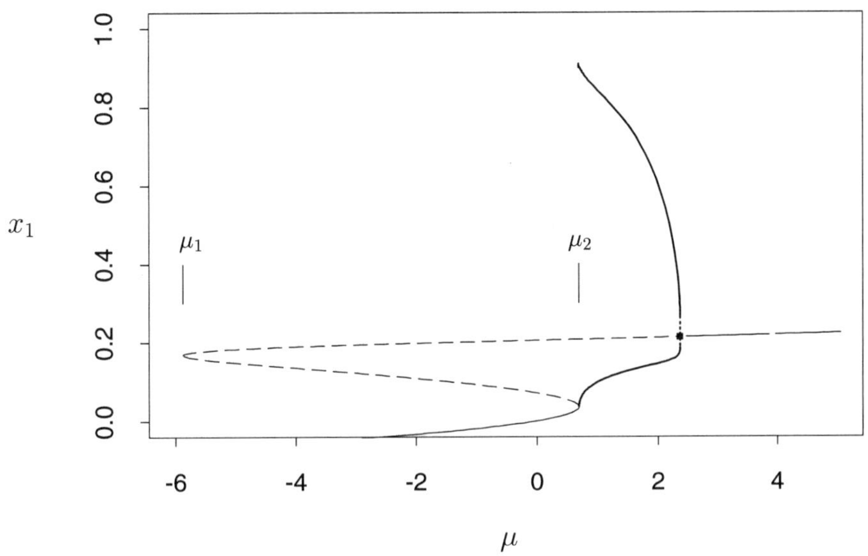


**Figure 1** Bifurcation diagram of the modified Murase model. Thick and thin lines indicate periodic solutions and equilibria, respectively. Solid lines indicate stable solutions whereas dashed lines indicate unstable solutions. Note the emergence of periodic solutions from a Hopf point (solid square) which terminate at a homoclinic bifurcation at $\mu = \mu_H$.

$$\dot{u} = f(u, \mu) = \begin{pmatrix} \gamma^{-1}(nF(x) - kx + \mu/100) \\ G(x) - cn \end{pmatrix}, \tag{1.4}$$

where $u = (x, n)^T \in \mathbf{R}^2$ and

$$G(x) := \frac{b}{2}\left\{1 + \tanh\left(\frac{x_0 - x}{\delta}\right)\right\}. \tag{1.5}$$

With these definitions $G(x) - cn \to g(x, n)$ as $\delta \to 0^+$ if $b = c$. In this manner, the modified system (1.4) is an approximation of the original model provided $0 < \delta << 1$. Henceforth, we fix the parameters $\lambda = (6, 1, 1, 2/5, 1/5, 1/10)$, $\delta = 3/100$, and $b = c = 1$, leaving $\mu$ as a bifurcation parameter in later analyses.

In Figure 1, AUTO [**3**] was used to numerically generate a bifurcation diagram for (1.4). The qualitative features depicted in this figure are identical to those found by Murase in [**9**] for (1.1)-(1.2). Stated precisely, the bifurcation structure is:

1) There are three branches of equilibria $u = u_K(\mu) = (x_K(\mu), n_K(\mu))^T, K = 0, +, -$ of (1.4):

$$\begin{array}{ll} (a) & C_+ := \{u = u_+(\mu) : \mu_1 < \mu\} \\ (b) & C_0 := \{u = u_0(\mu) : \mu_1 < \mu < \mu_2\} \\ (c) & C_- := \{u = u_-(\mu) : \mu < \mu_2\} \end{array}$$

for $\mu_1, \mu_2$ with $\mu_1 < 0 < \mu_2$. A saddle-node bifurcation occurs at $\mu = \mu_1$ and $\mu_2$. Then the two branches $C_+$ (resp. $C_-$) and $C_0$ join at $\mu = \mu_1$ (resp. $\mu = \mu_2$), hence

$$\lim_{\mu \to \mu_1 +} u_+(\mu) = \lim_{\mu \to \mu_1 +} u_0(\mu), \qquad \lim_{\mu \to \mu_2 -} u_-(\mu) = \lim_{\mu \to \mu_2 -} u_0(\mu).$$

2) Equilibria on $C_-$ are asymptotically stable node points, while equilibria on $C_0$ are saddle points. The stability of the equilibria on $C_+$ changes through a Hopf bifurcation at $\mu_c(>\mu_2)$. Indeed, these equilibria are asymptotically stable for $\mu > \mu_c$ and unstable for $\mu \in (\mu_1, \mu_c)$.
3) There is a $\mu_H, \mu_1 < \mu_H < \mu_2$ such that a homoclinic orbit asymptotic to the equilibrium $u_0$ appears at $\mu = \mu_H$. For $\mu > \mu_H$, an asymptotically stable periodic orbit bifurcates from the homoclinic orbit. This branch of periodic solutions terminate at the Hopf bifurcation at $\mu = \mu_c$.

We note that the numerics indicate $\mu_H < \mu_2$ only in closeups of Figure 1 (the numerical values are, to three significant figures, $\mu_H = 0.689, \mu_2 = 0.691$). Here, we have presented a view where the equilibria in the range $\mu \in (\mu_1, \mu_2)$ and the emergence of periodic solutions from the Hopf point are more clearly visible.

In terms of the Jacobians

$$A_K(\mu) := D_u f(u_K(\mu), \mu), \qquad K = 0, +, -, \tag{1.6}$$

the stability criteria embodied in the description above are realized by the conditions

$$\begin{aligned} &\mathrm{tr}A_+(\mu) \begin{cases} > 0 & (\mu_1 < \mu < \mu_c) \\ = 0 & (\mu = \mu_c) \\ < 0 & (\mu_c < \mu), \end{cases} \\ &\det A_+(\mu) > 0 \qquad (\mu_1 < \mu), \\ &\det A_0(\mu) < 0 \qquad (\mu_1 < \mu < \mu_2), \\ &\mathrm{tr}A_-(\mu) < 0, \ \det A_-(\mu) > 0 \qquad (\mu < \mu_2), \end{aligned} \tag{1.7}$$

where it is understood that the limiting values of $detA_K(\mu)$ and $trA_K(\mu)$ are assumed to be the same near the knees at $\mu = \mu_1, \mu_2$.

Moreover, stable periodic solutions can bifurcate from the homoclinic orbit at $\mu_H$ only if

$$\mathrm{tr}A_0(\mu) < 0 \quad \text{at } \mu = \mu_H, \tag{1.8}$$

which in turn implies the modulus of the negative eigenvalue of $A_0(\mu_H)$ is greater than the positive one.

We are now in a position to define a related coupled system

$$\begin{aligned} \dot{u}^1 &= f(u^1, \mu) + \nu D(u^2 - u^1), \\ \dot{u}^2 &= f(u^2, \mu) + \nu D(u^1 - u^2), \end{aligned} \tag{1.9}$$

where $\nu > 0$ and $D = \mathrm{diag}(d_1, d_2), d_1, d_2 \geq 0, d_1 + d_2 > 0$. Taking into account the background of the model, we assume $d_2 = 0$, hence we may set $d_1 = 1$ without any loss of generality. With these values, the parameter $\nu$ can be regarded as an elastic coupling strength between two different regions on the filaments. In later discussions of this system we assume the description 1)-3), the conditions (1.7)-(1.8) (all of which are supported by the numerical results in Figure 1) and these values of $d_i$.

In Sections 2 and 3, we discuss the following problems for the system (1.9):

(i) existence of stable periodic solutions for some region of $\mu < \mu_H$; in other words, the existence of periodic orbits induced by the coupling for which (1.4) exhibits no oscillation;

(ii) classification of degenerate bifurcation points;
(iii) existence of asymmetric homoclinic/heteroclinic orbits.

## 2 Bifurcations in the coupled system

In this section, we consider bifurcations of the coupled system indicating some codimension 2 bifurcations of the equilibria. To begin, we clearly define some terminology. Firstly, by in-phase periodic solutions we shall mean any periodic solution having the form $(u^1, u^2) = (p(t), p(t))$, where $p(t)$ is a periodic solution of (1.4). By anti-phase (periodic) solutions of (1.9) we shall mean a solution which can be written as $(u^1, u^2) = (\tilde{p}(t), \tilde{p}(t + T/2))$ where $\tilde{p}(t)$ is some $T$-periodic function. Note that under the reflection $(u^1, u^2) \mapsto (u^2, u^1)$, both the in-phase and anti-phase solutions are symmetric (up to phase shift for the anti-phase solutions). Lastly, by asymmetric periodic solutions we mean any periodic solution of (1.9) which is neither in-phase nor anti-phase but can be written in the form $(u^1, u^2) = (p^1(t), p^2(t))$ where $p^k(t), k = 1, 2$, are periodic functions of commensurate periods. Note that for any asymmetric periodic solution $(p^1(t), p^2(t))$, there also exists the reflected one $(p^2(t), p^1(t))$.

In Figure 2 we present bifurcation diagrams for (1.9) generated using AUTO. The parameter values for the simulations are as described in the previous section with the coupling strength set at $\nu \simeq 0.089$. As is evident from this figure, the in-phase periodic solution of (1.9) exists for $\mu > \mu_H$ until it disappears at the Hopf bifurcation point of the branch $C_+$. Furthermore, we observe that there are two branches of asymmetric periodic solutions connecting the anti-phase solutions with the in-phase periodic solutions. These asymmetric periodic solutions are certainly stable for an interval containing $\mu_H$ (approximately (0.4, 1.05)). We note that these numerical results include those of the corresponding coupled Murase equations [**12**] and are similar to bifurcation structures found in other excitable systems [**11**].

Next we consider codimension 2 bifurcations of the coupled system (1.9). First note that for any $\nu$ the system (1.9) has three symmetric (homogeneous) equilibria

$$(u_K(\mu), u_K(\mu)), \quad K = 0, +, -.$$

By defining

$$y = \frac{u^1 + u^2}{2}, \qquad z = \frac{u^1 - u^2}{2}, \tag{2.1}$$

the system (1.9) can be transformed into

$$\begin{aligned} \dot{y} &= F_1(y, z, \mu) := \frac{1}{2}\{f(y + z, \mu) + f(y - z, \mu)\}, \\ \dot{z} &= F_2(y, z, \mu, \nu) := \frac{1}{2}\{f(y + z, \mu) - f(y - z, \mu)\} - 2\nu D z. \end{aligned} \tag{2.2}$$

In this form, it is easy to see that a local stability analysis of the symmetric equilibria of (1.9) reduces to finding the eigenvalues of the matrices

$$A_K(\mu) \quad \text{and} \quad A_K(\mu) - 2\nu D, \quad K = 0, +, -.$$

Now, we determine conditions which imply $A_K(\mu) - 2\nu D$ has either a zero eigenvalue or a pair of purely imaginary eigenvalues. Given $d_1 = 1$ and $d_2 = 0$, it follows from

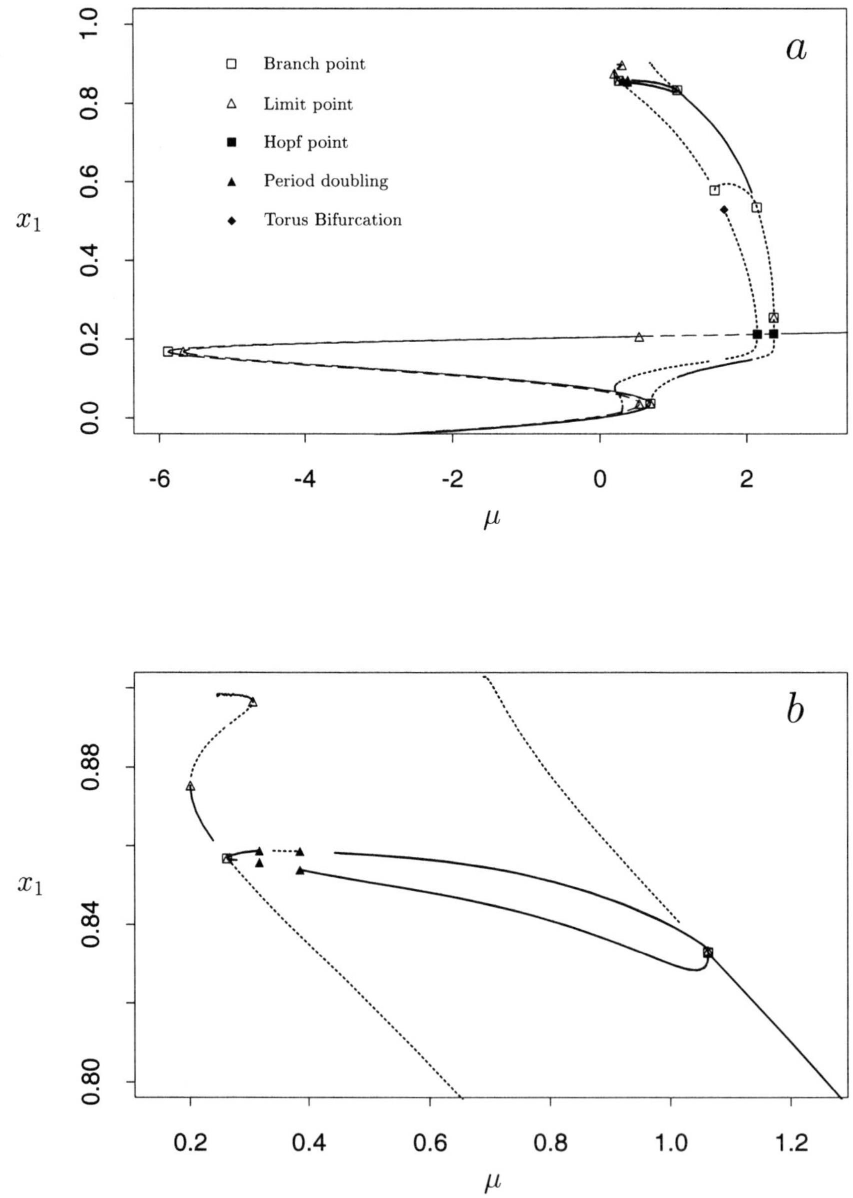


**Figure 2** Bifurcation diagrams for the coupled system: (a) Shows the overall bifurcation structure. Periodic solutions emanating from the right Hopf point are in-phase, (b) shows a closeup of the stable asymmetric periodic solutions connecting periodic branches.

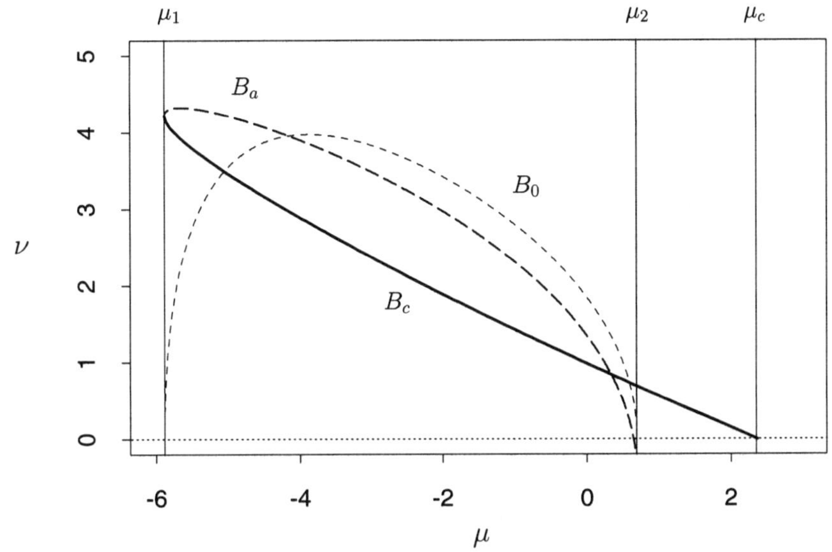


**Figure 3** Bifurcation curves in $(\mu, \nu)$-plane.

(1.4)-(1.6) that

$$\det(A_K(\mu) - 2\nu D) = \det A_K(\mu) + 2\nu c,$$
$$\mathrm{tr}(A_K(\mu) - 2\nu D) = \mathrm{tr} A_K(\mu) - 2\nu. \tag{2.3}$$

Considering the conditions (1.7), there are three curves in the $(\mu, \nu)$-plane on which the eigenvalues of $A_K(\mu) - 2\nu D$ are either zero or $\pm \alpha i$:

$$B_c := \{(\mu, \nu) : \nu = \mathrm{tr} A_+(\mu)/2, \ \mu_1 < \mu < \mu_c\},$$
$$B_0 := \{(\mu, \nu) : \nu = -\tfrac{1}{2c} \det A_0(\mu), \ \mu_1 < \mu < \mu_2\}, \tag{2.4}$$
$$B_a := \{(\mu, \nu) : \nu = \mathrm{tr} A_0(\mu)/2, \ \nu > -\tfrac{1}{2c} \det A_0(\mu)\}.$$

On $B_c$ anti-phase periodic solutions bifurcate from a Hopf point on $C_+$ while on $B_0$ a pair of asymmetric equilibria bifurcate from $u_0(\mu)$. On $B_a$ anti-phase periodic solutions also bifurcate but from $u_0(\mu)$ on $C_0$.

With these definitions, codimension 2 bifurcations can occur at points where any two of the six curves $B_c, B_0, B_a, \mu = \mu_1, \mu = \mu_2$ and $\mu = \mu_c$ intersect. In Figure 3, all six of these curves are plotted in the $(\mu, \nu)$-plane and it is seen that there are two such intersection points $(\bar{\mu}_0, \bar{\nu}_0)$ and $(\mu_1, \nu_1)$:

$$\{(\bar{\mu}_0, \bar{\nu}_0)\} = B_0 \cap B_a, \qquad \{(\mu_1, \nu_1)\} = B_c \cap \{(\mu, \nu) : \mu = \mu_1\},$$

where

$$\bar{\nu}_0 = \mathrm{tr} A_0(\bar{\mu}_0)/2 = -\frac{1}{2c} \det A_0(\bar{\mu}_0), \qquad \nu_1 := \mathrm{tr} A_+(\mu_1)/2.$$

At $(\bar{\mu}_0, \bar{\nu}_0)$, the matrix $A_0(\bar{\mu}_0) - 2\bar{\nu}_0 D$ has a zero eigenvalue with double algebraic multiplicity while $A_0(\bar{\mu}_0)$ has two nonzero eigenvalues with opposite signs on account of (1.7). For the other point $(\mu_1, \nu_1)$ the former matrix has a pair of pure imaginary eigenvalues and the latter one has a simple zero eigenvalue. At both of these points, the corresponding equilibria have one-dimensional unstable directions.

We remark that around each of these singularities, a classification of bifurcations were locally investigated by Guckenheimer and Holmes in [**5**] using normal form calculations. We do not perform these tedious computations here.

## 3 Existence of asymmetric homoclinic orbits

In the previous section we demonstrated the existence of some locally degenerate bifurcations of equilibria occurring in the coupled system. In this section we examine a degenerate bifurcation which depends on global information. Specifically, we consider bifurcations of the in-phase homoclinic solution $(u_h(t), u_h(t))$ for parameters $(\mu, \nu)$ local to the intersecting point $(\mu_H, \nu_H)$ of $\mu = \mu_H$ and $B_0$. At this point the in-phase homoclinic solution is degenerate in the sense it is asymptotic to an equilibrium $(u_0(\mu), u_0(\mu))$ which is not hyperbolic. Under certain generic conditions we can prove that a pair of asymmetric homoclinic (or heteroclinic) solutions bifurcate from $(u_h(t), u_h(t))$ and its respective bifurcation curve in the $(\mu, \nu)$-plane emerges from $(\mu_H, \nu_H)$. We outline our assumptions, then state the main theorem and lastly present numerical evidence for the existence of the aforementioned bifurcation curve in the $(\mu, \nu)$-plane.

First, consider the linear variational equation of the single equation (1.4) about the homoclinic solution $u_h(t)$:

$$\dot{v} = A(t)v, \qquad A(t) := D_u f(u_h(t), \mu_H), \tag{3.1}$$

and its respective adoint equation:

$$\dot{y} = -A(T)^T y, \tag{3.2}$$

Equation (3.1) has a bounded solution $v = \dot{u}_h(t)$ over $\mathbf{R}$. Therefore, the adjoint equation (3.2) also has a bounded solution $q(t)$ for $t \in \mathbf{R}$. Letting $\alpha$ be the modulus of the negative eigenvalue of $A_0(\mu_H)$, it can be shown with the normalization

$$q(t)^T \dot{u}_h(t) = 1 \quad \text{and} \quad \lim_{t\to\infty} q(-t)^T \dot{u}_h(t) e^{2\alpha t} = 1,$$

the condition

$$\alpha_1 = \int_{-\infty}^{\infty} q(t)^T D_\mu f(u_h(t), \mu_H) dt < 0 \tag{3.3}$$

implies a bifurcation of a stable periodic solution from $u_h(t)$ occurs in the direction $\mu > \mu_H$. This is easily computed from the definition of the Melnikov function (see [**2**]). We assume this condition is satisfied.

Next, consider the linear variational equation of (1.9) about the in-phase homoclinic orbit $(u^1, u^2) = (u_h(t), u_h(t))$. By means of the transformation (2.1) this linearization can be decomposed into the system

$$\dot{v}^1 = A(t)v^1, \qquad \dot{v}^2 = (A(t) - 2\nu D)v^2, \tag{3.4}$$

where $v_k \in \mathbf{R}^2$. The equation for $v_1$ is the same as (3.1). The latter equation has a bounded solution when $\nu = \nu_H$. Indeed, if $p_0(t)$ is this bounded solution, its asymptotics as $t \to \pm\infty$ can be characterized in the following way:

$$\lim_{t\to-\infty} p_0(t) = \zeta_0, \quad \lim_{t\to\infty} p_0(t) = \hat{c}\zeta_0, \tag{3.5}$$

where $\zeta_0$ is in the nullspace of $A_0 - 2\nu_H D$ and $\hat{c}$ is some constant.

Given these assumptions and definitions we can now state the main theorem which shows the existence of a pair of asymmetric homoclinic orbits bifurcating from the homogeneous one.

**Theorem 3.1** *Assume (3.3), and that $\hat{c} > 0$ in (3.5). Then there is a curve $(\mu(\varepsilon), \nu(\varepsilon)), 0 < \varepsilon < \varepsilon_1$, emerging from $(\mu_H, \nu_H)$ such that for each $(\mu, \nu) = (\mu(\varepsilon), \nu(\varepsilon))$ Equation (1.9) has a pair of asymmetric homoclinic orbits bifurcating from the in-phase homoclinic solution $(u_h(t), u_h(t))$ and $\lim_{\varepsilon \to 0^+}(\mu(\varepsilon), \nu(\varepsilon)) = (\mu_H, \nu_H)$. Each asymmetric homoclinic orbit is asymptotic to one of the equilibria which bifurcate from the homogeneous equlibrium $(u_0(\mu), u_0(\mu))$. Moreover, $\mu(\epsilon)$ solves a bifurcation equation which is expanded as*

$$H(\mu, \epsilon) = \alpha_1(\mu - \mu_H) + \frac{\beta_1}{2}\epsilon^2 + h.o.t = 0, \tag{3.6}$$

*where the coefficients $\alpha_1$ and $\beta_1$ are given by (3.3) and*

$$\beta_1 := \int_{-\infty}^{\infty} q(t)^T D_{uu} f(u_h(t), \mu_H) \circ (p_0(t), p_0(t)) dt, \tag{3.7}$$

*respectively. Hence if in addition $\beta_1$ is positive (resp. negative), then $\mu(\varepsilon) - \mu_H$ is positive (resp. negative) for $\varepsilon$ sufficiently small.*

The last assertion of the theorem implies that the direction of the bifurcation curve, $\mu(\varepsilon) > \mu_H$ or $\mu(\varepsilon) < \mu_H$, can be determined by the sign of $\alpha_1$ and $\beta_1$ in (3.3) and (3.7).

In Figure 4 we present numerical results where the bifurcation curve $(\mu(\varepsilon), \nu(\varepsilon))$ in the theorem was computed and continued using AUTO [**1**]. Parameter values in the computation were the same as before except with $b = c = 2$ and $a = 4$. The figure indicates a branch of asymmetric (dashed curve) homoclinic solutions of (1.9) bifurcating from the symmetric (solid vertical line) homoclinic points at the point $(\mu_H, \nu_H)$. Note that the theorem predicts such a bifurcation local to $(\mu_H, \nu_H)$. Figure 4 clearly demonstrates that the numerically continued homoclinics away from $(\mu_H, \nu_H)$ are asymmetric.

Finally, we remark that if $\hat{c} < 0$ in Theorem 3.1, asymmetric heteroclinic orbits rather than homoclinic orbits, bifurcate from the in-phase homoclinic solution. See [**9**] for details and the proof of Theorem 3.1.

## 4 Discussion

We conclude with a few remarks addressing the relevance of our results. Firstly, we note that it is possible that asymmetric periodic solutions could bifurcate from the asymmetric homoclinics in the theorem. An example of a similar situation occurring can be found in [**7**]. The stability of any such asymmetric periodic solutions would become especially relevant in the application being examined. If stable, then other periodic mechanical oscillations in muscle fibers may be induced by heterogeneities within the region (through $\nu$).

Secondly, it is often practical to numerically continue homoclinic solutions by finding a large period periodic orbit continued from a Hopf point. For our model this procedure failed to find the asymmetric homoclinic orbit of the theorem. As pointed out in the first remark, periodic solutions may bifurcate from these asymmetric homoclinics. Thus, it may not always be possible to find all periodic motion in coupled muscle fiber (and excitable cell) models by continuing periodic solutions from Hopf points. We point this out as a cautionary reminder to those researchers interested in such problems.

Finally it is a very interesting but a hard problem to reveal the whole bifurcation structure in $(\mu, \nu)$ plane. Indeed, to perform it, we will not only need precise local bifurcation diagrams of all the degenerate equilibria, but also have to investigate

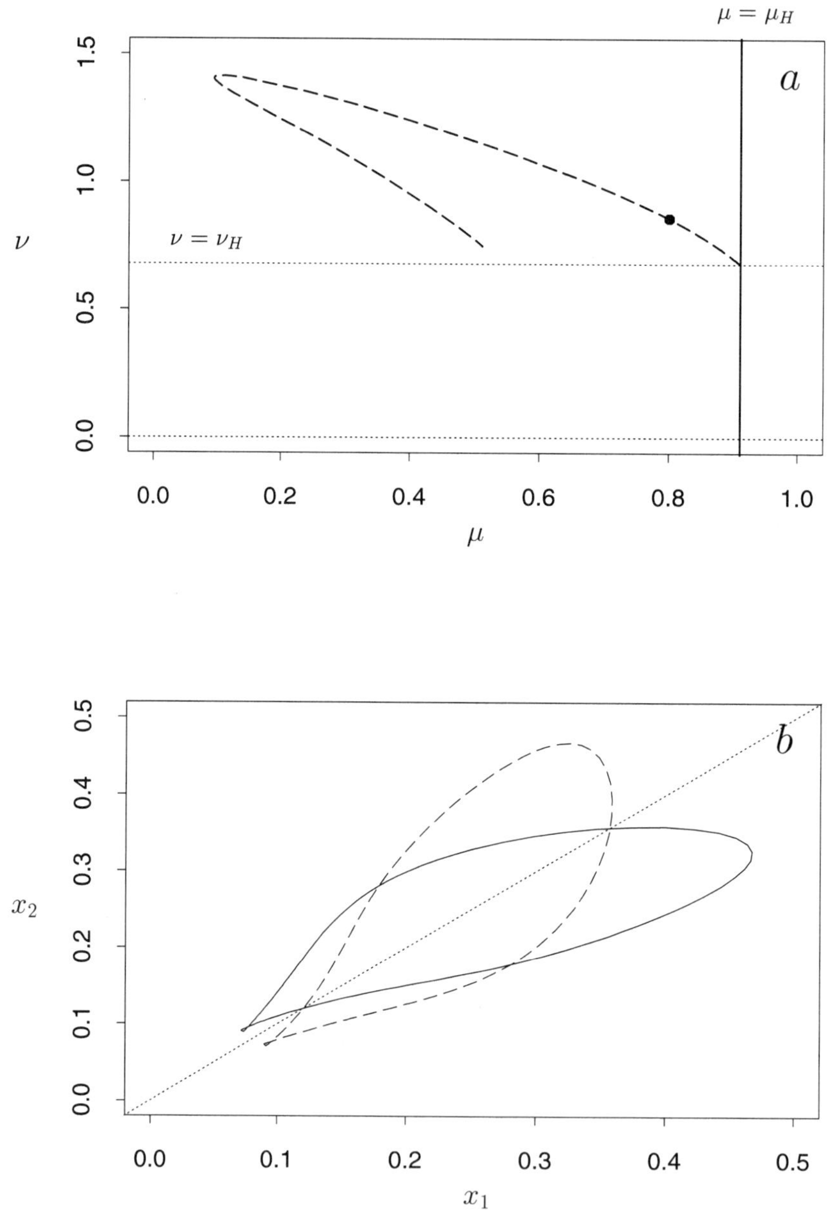


**Figure 4** Bifurcation curves of symmetric and asymmetric homoclinic orbits: a) vertical line $\mu = \mu_H$ and dark dashed curve indicate parameter values where symmetric and asymmetric homoclinics exist, respectively. b) the projection of an asymmetric homoclinic orbit of (1.9) onto $x_1 - x_2$ plane for parameter values indicated by solid dot in a).

the existence of other asymptotic behaviors such as quasi-periodic motion which have not been examined. Those will perhaps be the subject of a later study.

## References


[1] Champneys, A. R., Kuznetsov, Y. A. and Sandstede, B. [1995], *HomCont: an AUTO86 driver for homoclinic bifurcation analysis*, Version 2, Centrum voor Wiskunde en Informatica, Report AM-R9516, Amsterdam, NL.

[2] Chow, S.-N. and Yamashita, M. [1992], *Geometry of the Melnikov vector,* Nonlinear Equations in the Applied Sciences, Academic Press, 79–148.

[3] Doedel, E. J. [1981], *AUTO: A program for the automatic bifurcation and analysis of autonomous systems*, in Proc. 10th Manitoba Conf. on Numer. Math. and Comput., Winnipeg, Canada, Congr. Numer. **30**, 265–284.

[4] FitzHugh, R. [1961], *Impulses and physiological states in theoretical models of nerve membranes*, Biophys. J. **1**, 445–466.

[5] Guckenheimer, J. and Holmes, P. [1983], *Nonlinear Oscillations, Dynamical Systems, and Bifurcations of Vector Fields*, Springer-Verlag.

[6] Hodgkin, A. L. and Huxley, A. F. [1952], *A quantitative description of membrane current and its application to conduction and excitation in nerve*, J. Physiol. (London), **117**, 500–544.

[7] Lin, X.-B. [1996], *Twisted and nontwisted bifurcations induced by diffusion*, J. Dynamics and Diff. Eqn. **8**, 325-372.

[8] Morita, Y., Pernarowski, M. and Dockery, J. [1998], *Symmetry Breaking Homoclinic Bifurcation in Diffusively Coupled Equations*, preprint.

[9] Murase, M. [1992], *Dynamics of Celluar Motility,* John Wiley & Sons Ltd.

[10] Nagumo, J. S., Arimoto, S. and Yoshizawa, S. [1962], *An action pulse transmission line simulating nerve axon*, Proc. IRE, 2061–2071.

[11] Sherman, A. [1994], *Anti-phase, asymmetric and aperiodic oscillations in excitable cells-I. Coupled bursters*, Bulletin of Mathematical Biology **56**, 811–835.

[12] Tsuruzoe, M. [1995], *Numerical computation and structure of exciting patterns in some coupled nonlinear ocsillators* (in Japanese), Masters thesis, Ryukoku University, Japan.

Fields Institute Communications
Volume **21**, 1999

# Compactness and Asymptotic Spatial Homogeneity in Conservation Laws with Memory

**Hana Petzeltová**
Institute of Mathematics AVČR
Žitná 25, 115 67
Praha 1, Czech Republic
petzelt@math.cas.cz

**Abstract.** Compactness of a set of bounded solutions of conservation law with memory

$$(u + K * u)_t + \sum a_i(u)u_{x_i} = 0 \text{ on } (0, \infty) \times R^N$$

is shown provided some nondegeneracy conditions on the kernel $K$ and the nonlinear functions $a_i$, $i = 1, .., N$ are imposed. This implies that any solution of the problem starting from a spatially periodic initial datum converges to a constant (its integral mean), which is also true in the linear case. The results are compared with their analogies in standard and fractional conservation laws.

## 1 Introduction

In this paper, we collect recent results on compactness of bounded solutions and asymptotic homogenization of space periodic solutions to conservation laws when the memory effects come into play.

We consider equations of the form

$$\frac{\partial}{\partial t}\Big(u + \int_{-\infty}^{t} k(t-s)u(s)\, ds\Big) + \sum_{i=1}^{N} a_i(u)\frac{\partial u}{\partial x_i} = 0 \text{ on } (0, \infty) \times R^N, \tag{1.1}$$

$$\frac{\partial}{\partial t}\int_{-\infty}^{t} K(t-s)u(s)\, ds + \sum_{i=1}^{N} a_i(u)\frac{\partial u}{\partial x_i} = 0 \text{ on } (0, \infty) \times R^N, \tag{1.2}$$

where $u = u(t, x)$ is a function of the time $t$ and the spatial variable $x \in R^N$. The "history" of the solution is given by the initial condition,

$$u|_{(-\infty,0]} = u_0 \tag{1.3}$$

---

1991 *Mathematics Subject Classification.* 45K05, 35L65, 35B40.
Supported by grants A 1019703 GA AV ČR and 201/96/0432 GA ČR.

where $u_0$ is a spatially periodic function with the period 1. More precisely, we shall assume

$$u_0 \in C((-\infty,0];L^1(\mathcal{T}^N)) \cap L^\infty((-\infty,0)\times\mathcal{T}^N) \tag{1.4}$$

where the symbol $\mathcal{T}^N \equiv ([0,1]|_{\{0,1\}})^N$ stands for the N-dimensional torus.

As for the kernels $k$, $K$, we shall suppose

(k) $k \in L^1(R^+)$, nonnegative, nonincreasing on $R^+$,

(K) $K \in L^1(R^+)$, nonnegative, nonincreasing and convex on $R^+$, $K(t) \sim t^{-\alpha}$ for $t \to 0_+$.

We will mainly deal with the first equation, comparing the results with the standard conservation laws, i.e., $k \equiv 0$ and the second equation, the so called fractional conservation law. In the latter case, the convolution term may be interpreted as the time derivative of the order $\alpha$.

The problem of propagation, interaction and cancellation of oscillations in solutions to nonlinear conservation laws is closely related to the questions of the long-time behaviour, homogenization and strong convergence of exact and approximate solutions where no uniform bounds on the derivative are available.

Since the pioneering work of L. Tartar in [**23**], the method of compensated compactness has become an important tool to describe the constraints imposed on oscillations of some functions because of the differential equations they satisfy. The obvious success of this approach has been illustrated by the existence results for special classes of nonlinear systems in one spatial dimension endowed with Riemann invariants and a rich family of entropies identified by DiPerna [**4**] and Serre [**22**]. This is more or less related to the fact that the so-called genuine nonlinearities give rise to a multitude of dissipative mechanisms.

The solution of (1.1) necessarily develops shock waves as soon as the datum is "large" and the functions $a_i$ are sufficiently "nonlinear". Such a phenomenon was observed, e.g., by Malek-Madani and Nohel in [**18**]. On the other hand, the existence of global weak solutions was established by Gripenberg in [**13**] and Dafermos in [**3**], etc.

The regularizing effect of singular kernels on solutions of integro - differential equations has been observed, e.g., in [**5**], [**6**]. Gripenberg and Londen in [**14**] considered the Riemann problem for a scalar "fractional derivative" conservation law in one spatial dimension. Assuming monotonicity of the flux function they show that the problem possesses a continuous solution for any positive time. Thus in this particular case, the effect of the singularity of the kernel is similar to that of viscosity in scalar conservation laws. Further examples as well as the corresponding physical background may be found in the monograph [**21**] or [**20**].

There is another potential source of dissipation in the above equation. Experiments indicate that certain materials have memory, e.g., in viscoelasticity the stress may not depend only on the strain at the present time but also on the entire temporal history of the strain. Typically, the memory fades with time, i.e., disturbances that occurred in the distant past have less influence than those more recent, which brings about the dissipation of energy (see, e.g., the monograph [**21**]). Here, the memory effects are represented by the convolution term $k * u$.

The physical intuition suggests that various types of dissipation when acting together should strengthen their effects on the solutions and thus lead to better regularity and smoothness properties. Consequently, we can expect the long-time stabilization of any periodic solution to a constant related to the mean value of its initial data, which is a well known feature of standard one-dimensional conservation laws with quadratic nonlinearity, under much weaker assumptions on the flow. This is indeed true as stated in Theorems 3.1 and 4.1, though the rigorous proof is far from being straightforward (see [**10**] for N=1 and [**11**] for higher dimensions). The main reason is probably that the cooperation of those mechanisms is very subtle and, sometimes, the fact that the material "remembers" its initial shape may slow down the stabilization (and regularization) process rather than to speed it up. For instance, the well-known $L^1 - L^\infty$ regularizing effect for genuinely nonlinear conservation laws in 1-D does not seem to have its counterpart when memory is involved (cf., e.g., [**12**]).

On the other hand, as disturbances are transmitted along wave fronts, it is plausible to expect the dissipation effect of the memory term provided all waves intersect the x-direction transversally, in particular, when $\sigma$ is linear.

The situation is more complicated in higher dimensions. For one thing, a straightforward generalization of the Lax's concept of genuine nonlinearity to more space dimensions would be too restrictive and would practically exclude all the physically interesting cases. One of the first attempts to find the minimum nonlinearity conditions required for two-dimensional scalar conservation laws (without memory) to exhibit typical features of one-dimensional genuinely nonlinear systems, including formation of shocks, time decay of solutions and cancellation of oscillations was made in [**7**]. Using a different approach based on the so-called kinetic formulation of the problem, Lions, Perthame and Tadmor obtained in [**17**] similar results for scalar conservation laws in arbitrary space dimension. In particular, a nondegeneracy (LPT) condition was found which corresponds to a multidimensional extension of the nonlinearity type condition introduced in [**23**].

The concept of the entropy solutions introduced in [**1**] makes it possible to derive the kinetic formulation also for (1.1) and (1.2). This, in turn, together with deep results on Fourier multipliers in $L^p$ spaces enables to prove compactness of bounded solutions to (1.1) provided the (LPT) condition holds true. This condition can be weakened in the case of fractional conservation law (1.2). The compactness results still hold even if weaker entropy conditions on the solution are imposed (see Section 5).

The paper is organized as follows. In Section 2, we recall the basic existence and uniqueness theory of entropy solutions and state a comparison lemma. Section 3 is devoted to asymptotic behaviour of solutions in the one-dimensional case, while the multidimensional equation is treated in Section 4. Compactness results are given in Section 5.

## 2 Entropy solutions and comparison principle

Following Cockburn, Gripenberg and Londen in [**1**] we introduce the concept of the entropy solution, as follows,

**Definition** We shall say that a function $u$,

$$u \in L^\infty((-\infty, T] \times \mathcal{T}^N) \cap C((-\infty, T]; L^1(\mathcal{T}^N)),$$

is an *entropy solution* of the problem (1.1), (1.3) on the time interval $[0, T]$ if $u$ satisfies (1.3) and

$$\frac{\partial}{\partial t}\left(S(u(t)) + \int_{-\infty}^{t} k_s(t-s)S(u(s))\,ds\right) + \tag{2.1}$$

$$\left(k_r(0+)u(t) + \int_{-\infty}^{t} u(s)\, dk_r(t-s)\right) S'(u(t)) + \sum_{i=1}^{N} \frac{\partial}{\partial x_i}\eta_i(u(t)) \leq 0$$

in the sense of distributions for any convex entropy $S$ and the corresponding flux $\eta$,

$$\eta_i(v) = \int_0^v S'(z)a_i(z)\, dz,$$

and for any decomposition of the type

$$k = k_r + k_s,\ k_r, k_s \geq 0,\ k_r, k_s \text{ nonincreasing on } (0,\infty),\ k_r(0+) < \infty.$$

The entropy solution of (1.2) is defined in the same way, just without the first term in (2.1).

The existence and, most importantly, uniqueness of entropy solutions for the problem (1.1), (1.3) was proved in [**1**, Theorem 7].

To prove compactness of bounded solutions (Section 5), a weaker entropy condition is needed, namely, no decomposition of the kernel is required. It is sufficient if the condition (2.1) holds with $k_r = 0,\ k_s = k$.

Our next goal is to show a comparison principle for the entropy solutions.

**Proposition 2.1** *Let $u$, $v$ be two entropy solutions of (1.1), (1.3) on the interval $[0,T]$ with the initial data $u_0$, $v_0$ respectively satisfying (1.4).*
*Then the inequality*

$$\|(u-v)^+(t)\|_{L^1(\mathcal{T}^N)} + \int_{\infty}^{t} k(t-s)\|(u-v)^+(s)\|_{L^1(\mathcal{T}^N)}\, ds \leq$$

$$\|(u-v)^+(0)\|_{L^1(\mathcal{T}^N)} + \int_{\infty}^{0} k(-s)\|(u-v)^+(s)\|_{L^1(\mathcal{T}^N)}\, ds \tag{2.2}$$

*holds for all $t \in [0,T]$.*

For the regular kernel $k$, the proof is based on the Kružkov's definition of entropy solutions to conservation law with the right-hand side and singular kernels can be approximated such that corresponding solutions converge to the unique entropy solution (see [**1**]). Passing to the limit in the formula (2.2) completes the proof for general $k$. For the details see [**10**] and [**11**].

This proposition enables us to prove convergence results for smooth data only, and then approximate general data from above and from below to get rid of the smoothness assumptions.

## 3 Asymptotic homogenization, $N = 1$

In this section, we show that any solution starting from spatially periodic initial datum stabilizes to its integral mean as time goes to infinity under the assumption that $a = a_1$ does not vanish on any open subinterval of the range of the solution. The result reads as follows:

**Theorem 3.1** *Let the kernel $k$ and the function $a$ satisfy the following hypotheses:*

$$k \in L^1(R^+) \text{ is a nonnegative, nonincreasing function, } k(0+) > 0, \tag{3.1}$$

$$a \in C^1(R^1) \text{ such that the set } \{z|\ a(z) \neq 0\} \text{ is dense in } R^1.$$

*Then for any initial datum satisfying (1.4), the (unique) entropy solution $u$ of (1.1), (1.3) stabilizes to the mean value of its initial state, i.e.,*

$$u(t) \to \bar{u} \text{ in } L^1 \text{ as } t \to \infty$$

*where*

$$\bar{u} = \frac{\int_0^1 u_0(0,x)dx + \int_{-\infty}^0 \int_0^1 k(-s)u_0(x,s)\ dxds}{1 + \int_0^\infty k(s)\ ds}.$$

**Sketch of the proof**

We use the modified localization method of [**7**] based on the fact that long time asymptotics of solutions are determined by the local behaviour of the flux in a neighbourhood of the mean of the initial data. This observation along with a comparison lemma will enable us to reduce the original problem to the situation where $a$ is either constant or $a' > 0$ ($a' < 0$). In the former case the desired result is obtained using the Laplace transform technique (see [**15**]) while the latter may be treated essentially by the method of Greenberg [**12**], who showed that in this case solutions of (1.1) constructed by the method employed in [**19**] enjoy similar regularity and stabilization properties as those of the problem without memory.

**Remarks**

- The above theorem is proved in [**10**] under the assumption that $k(0+) < \infty$, but all crucial estimates depend only on the integrability and monotonicity properties of the kernel $k$.
- The same technique can be used to prove the assertion of the Theorem for the Equation (1.2) with $K$ satisfying (K).
- When $a' > 0$ and $k \equiv 0$, a detailed description of the structure and convergence to the mean value for periodic data follows from the explicit representation obtained in [**16**] and [**19**]. Moreover, the hypothesis may be relaxed, assuming that the set of zero points of $a$ have no accumulation point on the real line, by means of the method of generalized characteristics due to [**2**]. Further generalizations as well as similar results for a two-dimensional conservation law were obtained in [**7**]. Nevertheless, it is clear that in this case the equation cannot be piecewise linear.

## 4 Asymptotic homogenization in the multidimensional case

The case $N > 1$ is surprisingly much more difficult. Positive results for $N = 2$, $k \equiv 0$, $a_i$, $i = 1, 2$ nonlinear were obtained in [**7**] by means of a sophisticated localization method. The general case for arbitrary $N$ and $k \equiv 0$ was treated in [**17**] employing the kinetic formulation of the problem. They found the *non-degeneracy* (LPT) condition concerning the velocity field $a_i$, $i = 1, ..., N$ which yields compactness estimates of bounded solutions. To state it, we set

$$\mathcal{M}(\xi) \equiv \{v \in R^1 \mid \xi_0 + \sum_{i=1}^N a_i(v)\xi_i = 0\},\ \xi = (\xi_0, ..., \xi_N) \in R^{N+1}.$$

The condition reads as follows,

$$(LPT) \qquad \text{meas}\mathcal{M}(\xi) = 0 \text{ for any } \xi \neq 0.$$

It means, roughly speaking, that the curve $v \mapsto (a_1(v), .., a_N(v))$ cannot remain in a given hyperplane.

To prove a multidimensional version of Theorem 3.1, we need compactness estimates of bounded solutions analogous to that of [**17**]. Such an estimate was indeed obtained in [**8**] by means of the kinetic method and forms the main ingredient of the proof of Theorem 4.1.

Owing to the presence of the memory term, we can relax the (LPT) condition. Of course, here the loss of compactness is compensated by the dissipative character of the kernel $k$.

$$(H)\quad \left\{\begin{array}{l} \text{If } y \in R^1 \text{ be a density point of } \mathcal{M}(\xi) \text{ for a certain } \xi \neq 0, i.e., \\ \qquad \lim_{\delta\to 0+} \frac{\text{meas}([y-\delta, y+\delta]\cap\mathcal{M}(\xi))}{2\delta} = 1, \\ \text{then there exists an open interval } I \subset R^1 \text{ such that } y \in I \\ \text{and all the functions } a_i,\ i = 1, ..., N \text{ are constant on } I \text{ and} \\ \sum_{i=1}^N a_i m_i \neq 0 \text{ for all N-tuples of integers } (m_1, ..., m_N) \neq 0. \end{array}\right.$$

**Theorem 4.1** *Let $u_0$, $k$ satisfy (1.4), (3.1) and, let $a_i \in C^2(R^1)$, $i = 1, ..., N$ satisfy the hypothesis (H).*

*Then the unique entropy solution $u$ of the problem (1.1), (1.3) tends to a constant as time goes to infinity, specifically,*

$$u(t) \to \overline{u} \text{ in } L^1(\mathcal{T}^N) \text{ as } t \to \infty$$

*where*

$$\overline{u} = \frac{\int_{\mathcal{T}^N} u_0(0,x)\, dx + \int_{-\infty}^0 k(-s) \int_{\mathcal{T}^N} u_0(x,s)\, dxds}{1 + \|k\|_{L^1(R^1)}}.$$

**Sketch of proof**

Similarly, as above, we use a localization method to reduce the problem to the linear and "strictly" nonlinear case.

If the equation is linear, we cannot expect convergence to a constant if $a_i$ are rationally dependent. On the other hand, if it is not the case, the multidimensional linear equation may be treated by the same methods as the one-dimensional. The most challenging part is the nonlinear problem. It was proved in [**11**] that compactness of the sequence

$$u^n(t,x) = u(nt, nx),\ n = 1, 2, ... \text{ in } L^1([0,T] \times \mathcal{T}^N), \tag{4.1}$$

where $u$ is a solution of (1.1), (1.3) yields the desired result.

The functions $u^n$ solve the equations

$$\frac{\partial}{\partial t}\left(u^n + \int_{-\infty}^t k^n(t-s)u^n(s)\, ds\right) + \sum_{i=1}^N a_i(u^n)\frac{\partial u^n}{\partial x_i} = 0 \tag{4.2}$$

with the data

$$u^n(t,x) = u_0(nt, nx) \text{ for } t \leq 0,$$

where

$$k^n(t) = nk(nt). \tag{4.3}$$

The crucial point in the proof is then to show compactness of the sequence $u^n$. This will be a particular case of the more general result given in the next section.

## 5 Compactness of bounded solutions

The main result of this section may be formulated as follows:

**Theorem 5.1** *Let $k$ satisfy the hypothesis (k). Let $u$ be an entropy solution of the equation (1.1) on the time interval $(0,T)$,*

$$-M \leq u(t,x) \leq M \text{ for a.e. } t \in (-\infty, T],\ x \in \mathcal{T}^N,$$

*and*

$$\|k\|_{L^1(0,\infty)} \leq \kappa.$$

*Moreover, let $a_i \in C^2[-M,M]$, $i = 1, ..., N$ and*

$$meas\{v \in [-M,M]\ ;\ \xi_0 + \Sigma_{i=1}^N a_i(v)\xi_i = 0\} = 0 \tag{5.1}$$

*for any fixed $\xi \neq 0$, $\xi \in R^{N+1}$.*

*Then $u|_{[0,T]}$ belongs to a compact set of $L^1\big((0,T)\times\mathcal{T}^N\big)$ depending on $M$, $\kappa$ only.*

### Sketch of proof

Following the ideas of [**17**] we introduce the kinetic formulation of the problem, which converts the Equation (1.1) to a linear equation at the expense of an additional variable.

To a solution $u$, we assign a function $f$,

$$f(t,x,v) = \begin{cases} 1 & \text{if } 0 < v \leq u(t,x) \\ -1 & \text{if } 0 > v \geq u(t,x) \\ 0 & \text{otherwise.} \end{cases} \tag{5.2}$$

Observe that

$$\int_{-\infty}^{\infty} f(t,x,v)\ dv = u(t,x) \text{ for all } t, x.$$

As $u$ satisfies the entropy inequality (2.1), it is possible to show that $f$ is a solution of a linear equation with a nonnegative distribution on the right-hand side.

**Proposition 5.2** *Under the hypotheses of Theorem 5.1, the function $f$ defined by (5.2) satisfies*

$$supp(f) \subset (-\infty, T] \times \mathcal{T}^N \times [-M,M],$$

*and*

$$\frac{\partial}{\partial t}\left(f(t,x,v) + \int_{-\infty}^{t} k(t-s)f(s,x,v)\ ds\right) + \sum_{i=1}^{N} a_i(v)\frac{\partial f(t,x,v)}{\partial x_i} = \frac{\partial m}{\partial v} \tag{5.3}$$

*on $(0,T)\times\mathcal{T}^N\times R^1$ in the sense of distributions where $m$ is a nonnegative bounded measure supported on the set $[0,T]\times\mathcal{T}^N\times[-M,M]$,*

$$\|m\|_{\mathcal{M}(R^{N+2})} \leq C(\kappa, M). \tag{5.4}$$

At this stage, we take advantage of the fact that (5.3) is linear for any fixed $v$ and, consequently, we can use the Fourier transform technique. The assumption (5.1) and the Lizorkin theorem on $L^p$- multipliers together with the compact embeding of the space of measures into $W^{-1,p}$ and (5.4) enables to obtain the desirable compactnesss result. See [**8**] for details.

**Remarks**

- For the application of the above Theorem to the convergence of solutions it is important that the compact set in which bounded solutions are contained depends solely on their $L^\infty$−norm and the $L^1$−norm of the kernel $k$. Indeed, the functions $u^n$, defined by (4.1) satisfy equations (4.2) with kernels (4.3) which are different but of the same $L^1$−norm.
- The conclusion of Theorem 5.1 remains true if we require (2.1) to hold only for $k_r \equiv 0$, $k_s \equiv k$. In this case, however, uniqueness of solutions is not guaranteed.

To prove similar result for the equation (1.2), the non-degeneracy hypotheses (5.1) may be weakened. It is sufficient to require

$$\text{meas}\left\{v \in [-M, M] \mid \sum_{i=1}^{N} a_i(v)\xi_i = 0\right\} = 0 \text{ for all } |\xi| \neq 0. \tag{5.5}$$

The entropy condition in this case reduces to

$$\frac{\partial}{\partial t}\left(\int_{-\infty}^{t} K(t-s)S(u(s))\, ds\right) + \text{div}\eta(u) \leq 0 \tag{5.6}$$

in the sense of distributions for any convex entropy $S$ and the corresponding flux $\eta$.

The following theorem is proved in [**9**].

**Theorem 5.3** *Let the hypotheses (K), (5.5) hold, $a_i \in C^2(R^1), i = 1, ..., N$. Then the set of all entropy solutions $u$ of the Equation (1.1) satisfying*

$$-M \leq u(t,x) \leq M \text{ for a.e. } t \in (-\infty, T),\ x \in R^N$$

*is compact in $L^1_{loc}([0,T] \times R^N)$.*

To see the difference between (5.1) and (5.5), consider the following simple example. If $N = 2$, then bounded solutions of the problem

$$(K * u)_t + au_x + b(u)u_y = 0$$

are locally compact in $L^1$ provided $K$ is as above, $a \neq 0$ and $b'(u)$ vanishes on at most countable set while such a condition obviously does not guarantee compactness for the "standard" conservation law

$$u_t + au_x + b(u)u_y = 0.$$

**Some open problems**

- Does the condition (5.6) guarantee uniqueness of solutions to the Equation (1.2)?
- Is it possible to prove convergence of solutions of the problem (1.1), (1.3) if the flux is linear in one direction and "genuinely nonlinear" in the others?

## References

[1] Cockburn, B., Gripenberg, G. and Londen, S.-O. [1996], *On convergence to entropy solutions of a single conservation law*, J. Differential Equations **128**, 206–251.

[2] Dafermos, C. M. [1985], *Regularity and large time behaviour of solutions of a conservation law without convexity*, Proc. Royal Soc. Edinburgh **99A**, 201–239.

[3] Dafermos, C. M [1988], *Solutions in $L^\infty$ for a conservation law with memory*, Analyse Mathématique et Applications, Gauthier- Villars, Paris, pages 117–128.

[4] DiPerna, R. J. [1983], *Convergence of approximate solutions to conservation laws*, Arch. Rat. Mech. Anal. **82**, 27–70.

[5] Engler, H. [1991], *Weak solutions of a class of quasilinear hyperbolic integro-differential equations describing viscoelastic materials*, Arch. Rat. Mech. Anal. **113**, 1–38.

[6] Engler, H. [1996], *Global smooth solutions for a class of parabolic integrodifferential equations*, Trans. Amer. Math. Soc. **348**, 267–290.

[7] Engquist, B. and Engquist, E. W. [1993], *Large time behavior and homogenization of solutions of two-dimensional conservation laws*, Commun. Pure Appl. Math. **46**, 1–26.

[8] Feireisl, E. and Petzeltová, H. [1998], *On compactness of bounded solutions to multidimensional conservation laws with memory*, NoDEA **5**, 193-204.

[9] Feireisl, E. and Petzeltová, H. [1998], *Singular kernels and compactness in nonlinear conservation laws*, J. Differential Equations **142**, 291-304.

[10] Feireisl, E. and Petzeltová, H. [1996], *On the long-time behaviour of solutions to a conservation law with memory*, Math. Meth. Appl. Sci. **20**, 569–581.

[11] Feireisl, E. and Petzeltová, H. [1997], *Asymptotic spatial homogeneity of solutions to conservation laws with memory in several space dimensions*, Math. Meth. Appl. Sci. **20**, 1459–1468.

[12] Greenberg, J. M. [1973], *The existence and qualitative properties of solutions of* $\frac{\partial u}{\partial t} + \frac{1}{2}\frac{\partial}{\partial x}[u^2 + \int_0^t c(s)u^2(x, t-s)\, ds] = 0$, J. Math. Anal. Appl. **42**, 205–220.

[13] Gripenberg, G. [1985], *Volterra integro-differential equations with accretive nonlinearity*, J. Differential Equations **60**, 57–79.

[14] Gripenberg, G. and Londen, S.-O. [1995], *Fractional derivatives and smoothing in nonlinear conservation laws*, Differential Integral Equations **8**, 1961–1976.

[15] Gripenberg, G., Londen, S.-O. and Staffans, O. [1990], *Volterra Integral and Functional Equations*, Cambridge University Press, Cambridge.

[16] Lax, P. D. [1957], *Hyperbolic systems of conservation laws, II*, Commun. Pure Appl. Math. **10**, 537–556.

[17] Lions, P.-L., Perthame, B. and Tadmor, E. [1994], *A kinetic formulation of multidimensional scalar conservation laws and related equations*, J. Amer. Math. Soc. **7**, 169–191.

[18] Malek-Madani, R. and Nohel, J. [1985], *Formation of singularities for a conservation law with memory*, SIAM J. Math. Anal. **16**, 530–540.

[19] Oleinik, O. A. [1975], *On discontinuous solutions of nonlinear differential equations* (in Russian) Usp. Mat. Nauk. **12**, 3–73.

[20] Prüss, J. [1993], *Evolutionary Integral Equations and Applications.* Birkhäuser, Basel, Boston, Berlin.

[21] Renardy, M., Hrusa, W. J. and Nohel, J. [1987], *Mathematical Problems in Viscoelasticity.* John Wiley, New York.

[22] Serre, D. [1986], *La compacité par compensation pour les systèmes hyperboliques non linéaires de deux équations a une dimension d'espace*, J. Math. Pures Appl. **65**, 423–468.

[23] Tartar, L. [1975], *Compensated compactness and applications to partial differential equations*, Nonlinear Anal. and Mech., Heriot-Watt Sympos. (L. J. Knopps, ed.), Research Notes in Math 39, Pitman, Boston, pages 136–211.

Fields Institute Communications
Volume **21**, 1999

# On Boundary Value Problems for Quasilinear Elliptic Equations

**Klaus Schmitt**
Department of Mathematics
University of Utah
Salt Lake City, UT 84112, USA
schmitt@math.utah.edu

**Abstract.** This paper provides a brief survey of results concerning the existence and nonexistence of positive solutions of a class of quasilinear elliptic differential equations subject to Dirichlet boundary conditions. The class contains as particular examples perturbations of the $p$–Laplacian and related operators.

## 1 The boundary value problem

Let $\Omega$ a bounded open set in $\mathbb{R}^N$ with smooth boundary and consider the boundary value problem

$$\begin{aligned} -\mathrm{div}\alpha(|\nabla u|^2)\nabla u &= f(\lambda, u), \ x \in \Omega \\ u &= \quad 0, \ x \in \partial\Omega, \end{aligned} \tag{1.1}$$

where the function

$$\phi(s) = \alpha(s^2)s \tag{1.2}$$

is an increasing homeomorphism of $\mathbb{R}$ and $f$ is a given nonlinear term. Such problems are derived from, for example, diffusion processes, where the diffusion coefficients are not necessarily constant and the reaction terms are nonlinear functions of the state variable; see e.g. [**26**].

Such problems have received much attention in recent years and it is the purpose of this article to report on some of the results obtained.

Particular examples that have been studied extensively (see, e.g. [**2**], [**7**], [**9**], [**10**], [**11**], [**21**], [**22**], [**23**], [**24**]) are the case of the Laplace operator, where $\phi(s) = s$, the $p$–Laplace operator, where

$$\phi(s) = |s|^{p-2}s, \ p > 1,$$

and the $p$–capillary surface operator, where

$$\phi(s) = \frac{s^{2p-2}}{\sqrt{1+s^{2p}}}s, \ p > 1$$

1991 *Mathematics Subject Classification.* Primary 35J25; Secondary 35J35, 35J70.

and $f$ satisfies various growth conditions.

We shall first consider the case that

$$f(\lambda, u) = \lambda f(u)$$

and $f$ is a function with similar behavior to $\phi$. In which case we discuss results about *principal eigenvalues* of such problems. We then become specific concerning the domain $\Omega$ and shall assume it to be a ball centered at the origin and discuss the existence and nonexistence of positive radial solutions. Here we distinguish between various growth conditions imposed on the nonlinearity $f$ (which will be assumed to be independent of the parameter $\lambda$, as it plays no role in these considerations), namely we shall distinguish between *subcritical* and *supercritical* growth. For the case of radial solutions on annular domains we refer to [**3**], [**5**], [**20**], [**24**].

## 2 Eigenvalue problems for general domains

One way to (formally) obtain the existence of eigenvalues and eigenfunctions of problem (1.1) is to regard the problem in a weak sense and to, hence, regard Equation (1.1) as an equation for the derivatives of certain functionals, i.e., one minimizes a certain functional subject to a constraint and employs Liusternik's theorem on Lagrange multipliers (see, e.g., [**25**], [**26**]).

**2.1 Theorem on Lagrange multipliers.** To apply this method we need the following set-up:

- $E$ is a reflexive Banach space with dual $E^*$,
- $$h_1 : E \to \mathbb{R}$$
  is a weakly lower semicontinuous, coercive functional,
- $$h_2 : E \to \mathbb{R}$$
  is such that
  $$S = \{u : h_2(u) = c = \text{constant}\}$$
  is weakly closed.

Whenever these conditions hold, $\exists u \in S$ such that

$$h_1(u) = \min_{v \in S} h_1(v).$$

- If it is also the case that
  $$h_1 : E \to \mathbb{R},\ h_2 : E \to \mathbb{R}$$
  are $C^1$ functionals and
  $$h_2'(u) \neq 0,\ \ u \in S,$$
  then $\exists \lambda \neq 0,\ \exists u \in S$ such that
  $$h_1'(u) = \lambda h_2'(u).$$

**2.2 The $p-$Laplacian.** As an application of the result just stated one obtains existence results for the existence of an eigenvalue for the $p-$Laplacian, see e.g. the results in the papers [**2**], [**8**], [**16**], [**23**]. We particularize the contents of subsection 2.1 as follows:

- $E = W_0^{1,p}(\Omega)$,
- $$h_1(u) = \frac{1}{p}\int_\Omega |\nabla u|^p,$$
$$h_2(u) = \frac{1}{q}\int_\Omega |u|^q,$$
$$1 < q < p^* = \begin{cases} \frac{Np}{N-p}, & p < N \\ \infty, & p \geq N \end{cases}$$

 and let
$$S = \{u : h_2(u) = c > 0\}.$$

It then follows from the restrictions on $q$ and from the Sobolev embedding theorem that $S$ is weakly closed in $E$.

We hence may conclude that $\exists\lambda \neq 0,\ \exists u \in S,\ u \geq 0$ such that

$$h_1'(u) = \lambda h_2'(u),$$

i.e.,

$$\int_\Omega |\nabla u|^{p-2}\nabla u \cdot \nabla v = \lambda \int_\Omega |u|^{q-2}uv,\ \forall v \in E$$

and hence $u$ is a weak solution of

$$-\mathrm{div}(|\nabla u|^{p-2}\nabla u) = \lambda |u|^{q-2}u.$$

To see that the eigenfunction $u$ may be chosen to satisfy $u \geq 0$ follows from the simple fact, that if $u$ is a minimizer then $|u|$ is also.

We also may draw the following conclusions:

- If $p \neq q$ any $\lambda > 0$ will be an eigenvalue for some $S_c$ (as follows by rescaling) and in particular
- the equation
$$-\mathrm{div}(|\nabla u|^{p-2}\nabla u) = |u|^{q-2}u$$
 has a nontrivial weak solution in $W_0^{1,p}(\Omega)$.
- If $p = q$, there exists a unique smallest eigenvalue which is simple (cf., [**2**], [**19**]).

**2.3 p-Laplacian-like operators.** The above example may be extended to the case where $\phi(s) = \alpha(|s|^2)s,\ \psi(s) = \beta(|s|^2)s$ are increasing homeomorphisms satisfying the growth conditions

$$c|s|^{p-1} \leq \phi(s), \psi(s) \leq d|s|^{p-1},$$

where $c$ and $d$ are positive constants (see, e.g., [**18**], [**22**], [**23**]). One concludes:

- For every constant $\mu > 0$ there exists $\lambda > 0$ and a positive (negative) solution $u$ with $\|u\|_{L^p} = \mu$.
- The set of such $\lambda$ is bounded away from 0 and bounded above.

To see this, one employs the considerations of the previous section by choosing the same Banach space $E$ and

$$\Phi(s) = \frac{1}{2}\int_0^{s^2} \alpha(\tau)d\tau,$$

$$\Psi(s) = \frac{1}{2}\int_0^{s^2} \beta(\tau)d\tau,$$

$$h_1(u) = \int_\Omega \Phi(|\nabla u|)dx,$$

$$h_2(u) = \int_\Omega \Psi(|u|)dx,$$

$$S_\mu = \{u \in W_0^{1,p}(\Omega) : h_2(u) = \mu\}.$$

In the case that the growth conditions imposed on the functions $\phi$ and $\psi$ are further relaxed, similar results have become recently available using Orlicz-Sobolev space considerations (cf., [**15**]) and in the case that $\Omega$ is a ball in [**12**]. Since, in general, Orlicz-Sobolev spaces are no longer reflexive (cf., [**17**]) and the corresponding functionals are no longer defined and finite everywhere, the arguments to establish such eigenvalues become considerably more involved. Also, a more general interpretation of the differential equation, using the concepts of subdifferentials and the theory of variational inequalities are needed in these cases (see, e.g., [**12**], [**18**]).

**2.4 $\Omega$ a ball.** In this section we consider the case that $\Omega$ is a ball of radius $R$ centered at the origin of $\mathbb{R}^N$ and we seek eigenvalues with corresponding eigenfunctions which are functions of the radius only (radial eigenfunctions), (see, e.g., [**7**], [**12**], [**13**], [**14**]). Such eigenfunctions are solutions of the boundary value problem

$$\begin{cases} -(r^{N-1}\phi(u'))' = r^{N-1}\lambda\psi(u) & r \in (0,R) \\ u'(0) = 0 = u(R). \end{cases} \tag{2.1}$$

**2.5 The radial $p$–Laplacian.** If it is the case that $\phi(s) = \phi_p(s) = |s|^{p-2}s = \psi(s)$, then (2.1) has an infinite sequence $\{\lambda_n(p)\}_n$ of positive eigenvalues such that $\lambda_n(p) \to \infty$ as $n \to \infty$, and associated with each $\lambda_n(p)$ there is a one-dimensional space spanned by a solution with exactly $n-1$ simple zeros in $(0, R)$ (see [**7**]).

**2.6 Equivalent integral equations.** We note that $u$ is a solution of

$$\begin{cases} -(r^{N-1}\phi(u'))' = r^{N-1}\lambda\psi(u) & r \in (0,R) \\ u'(0) = 0 = u(R). \end{cases} \tag{2.2}$$

if and only if $u$ is a fixed point of the completely continuous operator

$$T^\lambda_{\phi\psi} : C[0, R] \to C[0, R]$$

given by

$$T^\lambda_{\phi\psi}(u)(r) = \int_r^R \phi^{-1}\Big[\frac{1}{s^{N-1}}\int_0^s \xi^{N-1}\lambda\psi(u(\xi))d\xi\Big]ds. \tag{2.3}$$

This completely continuous operator has the following properties:

- If

$$\lim_{s\to 0}\frac{\phi(\sigma s)}{\psi(s)}=\sigma^{p-1},\quad \text{for all}\quad \sigma\in\mathbb{R}_+,$$

then

$$\deg_{LS}(I-T^{\lambda}_{\phi\psi},B(0,\varepsilon),0)=\begin{cases}1, & \text{if}\quad \lambda<\lambda_1(p),\\ (-1)^m, & \text{if}\quad \lambda\in(\lambda_m(p),\lambda_{m+1}(p)).\end{cases}$$

- If

$$\lim_{|s|\to\infty}\frac{\phi(\sigma s)}{\psi(s)}=\sigma^{q-1},\quad \text{for all}\quad \sigma\in\mathbb{R}_+,$$

then

$$\deg_{LS}(I-T^{\lambda}_{\phi\psi},B(0,R),0)=\begin{cases}1, & \text{if}\quad \lambda<\lambda_1(q)\\ \\ (-1)^l, & \text{if}\quad \lambda\in(\lambda_l(q),\lambda_{l+1}(q)).\end{cases}$$

The above results concerning the change of the Leray-Schauder degree, as $\lambda$ varies, therefore imply, via an application of global bifurcation theory (see [**13**]), that there exists a countably infinite number of solution continua $C_i\subset\mathbb{R}\times C[0,R]$ of (2.2) which bifurcate from the trivial solution and from infinity and which have the property that if $(\lambda,u)\in C_i,\ u\neq 0$, then $u$ has precisely $i-1$ simple zeros in $(0,R)$, $i=0,1,\cdots$, which is, of course, an extension of the, by now, classical theory of nonlinear boundary value problems of Sturm-Liouville type.

**2.7 The principal eigenvalue.** Using the integral operator introduced above one may show under the weaker assumptions $\forall x\neq 0,\ \forall\sigma>0$

$$a(\sigma)\le\frac{\phi(\sigma|x|)}{|\psi(x)|}\le b(\sigma),$$

where $a(\sigma)$ and $b(\sigma)$ are positive constants depending on $\sigma$ only, that:

- $\exists$ a principal eigenvalue for every level, i.e., for each $c>0$, there exists $\lambda(c)>0$, an eigenvalue with an associated positive eigenfunction $u$ and $\|u\|=u(0)=c$.
- This set of eigenvalues is bounded away from zero and bounded above (see [**14**]).

**2.8 Existence of eigenvalues.** Using the properties of the integral operator introduced above and the result in subsection 2.5 one obtains the following:

- Given any level $c>0$, $\exists$ a sequence $\{\lambda_n\}\to\infty$, $\exists$ a sequence $\{u_n\}$, $\|u_n\|=c$, such that $u_n$ has precisely $n$ simple zeros in $[0,R)$ and $(\lambda_n,u_n)$ solves

$$\begin{cases}-(r^{N-1}\phi(u'))'=r^{N-1}\lambda\psi(u) & r\in(0,R)\\ \\ u'(0)=0=u(R).\end{cases}$$

## 3 On *superlinear* problems

This section is devoted to the case where the nonlinear term $\psi$ is replaced by a nonlinearity whose growth may be faster than the growth of $\phi$ but other requirements are relaxed. We state both existence and nonexistence results and refer to [**4**] and [**14**] for the details.

We consider the boundary value problem

$$\begin{cases} \left[r^{N-1}\phi(u')\right]' + r^{N-1}f(\lambda,u) = 0, & r \in (0,R), \\ u'(0) = 0, \quad u(R) = 0, \end{cases} \tag{3.1}$$

where $\forall \sigma > 0$

$$\limsup_{x\to+\infty} \frac{\phi(\sigma x)}{\phi(x)} < +\infty \tag{3.2}$$

$$\limsup_{x\to 0} \frac{\phi(\sigma x)}{\phi(x)} < +\infty, \tag{3.3}$$

and

$$\Phi(x) = \int_0^x \phi(s)ds, \quad F(\lambda,x) = \int_0^x f(\lambda,s)ds, \quad \hat{F}(\lambda,x) = \int_0^x \hat{f}(\lambda,s)ds,$$

with $\hat{f}(\lambda,x) = \sup_{s\in[0,x]} f(\lambda,s)$. Further, let us put

$$\Gamma := \limsup_{x\to+\infty} \frac{\Phi(x)}{x\phi(x)},$$

and for $\theta \in (0,1)$,

$$\delta_\theta := \liminf_{x\to+\infty} \frac{F(\lambda,\theta x)}{x\hat{f}(x)}.$$

We shall be interested here in the existence of positive solutions of (3.1) under various assumptions placed on the function $f$, particularly concerning its behavior at 0 and at $\infty$. We note that $u$ will be a solution of the problem (3.1) if and only if it is a fixed point of the completely continuous operator $T$ given by

$$T(u)(r) = \int_r^R \phi^{-1}\Big[\frac{1}{s^{N-1}} \int_0^s \xi^{N-1} f(\lambda,u(\xi))d\xi\Big]ds. \tag{3.4}$$

It is this operator, which is being analyzed, using various tools from nonlinear analysis, to obtain the results described below.

**3.1 Existence and nonexistence.** In this section we shall be concerned with the case that $f$ is independent of $\lambda$ and

$$f(0) = 0, \ \ sf(s) \geq 0, \ \ s \in \mathbb{R}. \tag{3.5}$$

The following existence theorem is valid:

- Suppose that $\Gamma < 1$ and

  $$\lim_{x\to 0} \frac{\phi(x)}{f(x)} = +\infty, \qquad \lim_{x\to+\infty} \frac{f(x)}{\phi(x)} = +\infty. \tag{3.6}$$

  Further let there exist $\theta \in (0,1)$ such that $\delta_\theta > 0$ and

  $$N\delta_\theta > N\Gamma - 1. \tag{3.7}$$

  Then problem (3.1) has at least one positive solution.

In the case $\phi(s) = |s|^{p-2}s, \ p > 1$ and $f(s) = s^q$, the above requirements are satisfied, whenever

- $$p \geq N, \ q > p-1,$$

- $$1 < p < N,\ p-1 < q < \frac{Np}{N-p}$$

which are, what have been termed, *subcritical* growth conditions.

- If we assume that
$$\sup_{x\geq 0} \frac{NF(x)}{xf(x)} \leq \inf_{x\geq 0} \frac{N\Phi(x)}{x\phi(x)} - 1 \tag{3.8}$$
then no positive solutions of (3.1) will exist.

Again, for the special case that $\phi(s) = |s|^{p-2}s,\ p > 1$ and $f(s) = s^q$, (3.8) will follow provided that $1 < p < N$ and
$$\frac{Np}{N-p} \leq q + 1,$$
which is the *supercritical* growth condition.

The following examples will serve to further illustrate the above existence and nonexistence results (again see [**14**]). Consider the boundary value problem
$$\begin{aligned} &\operatorname{div}\left(\frac{|\nabla u|^{2p-2}\nabla u}{\sqrt{1+|\nabla u|^{2p}}}\right) + u^q = 0 \text{ in } B_R \\ &u = 0 \text{ on } \partial B_R. \end{aligned} \tag{3.9}$$
Then one obtains nonexistence of positive radial solutions, whenever
$$q \geq \frac{N(2p-1)+2p}{N-2p}$$
and existence of such, whenever
$$p - 1 < q < \frac{N(p-1)+p}{N-p}.$$
For the boundary value problem
$$\begin{aligned} &\operatorname{div}\left((1+|\nabla u|^2)^{\frac{m-1}{2}}\nabla u\right) + u^q = 0 \text{ in } B_R \\ &u = 0 \text{ on } \partial B_R \end{aligned}$$
the above conditions imply the existence of positive solutions when
$$m < q < \frac{Nm+m+1}{N-m-1}$$
and nonexistence provided
$$\begin{aligned} q &\geq \tfrac{Nm+m+1}{N-m-1}, \quad m > 1 \\ q &\geq \tfrac{N+2}{N-2}, \quad m \leq 1. \end{aligned}$$

**3.2 Positone problems etc.** Let us return to problem (3.1) and now assume that $f(\lambda, u)$ is of the form $\lambda f(u)$, where either
$$f(s) > 0,\ s \geq 0 \tag{3.10}$$
or
$$f(s) > 0,\ s \geq s_0 > 0,\ f(0) < 0. \tag{3.11}$$
Thus we consider the problem
$$\begin{cases} \left[r^{N-1}\phi(u')\right]' + \lambda r^{N-1} f(u) = 0, & r \in (0, R), \\ u'(0) = 0, \quad u(R) = 0. \end{cases} \tag{3.12}$$

(Such problems usually are referred to as *positone* or *semipositone* problems.) In addition, we shall require that $\phi$ is convex on $[0,\infty)$ and

$$\lim_{s\to+\infty}\frac{f(s)}{\phi(s)}=+\infty.$$

We have the following two results.

- Assume that (3.11) holds and that $f$ is eventually nondecreasing with
$$N\liminf_{s\to\infty}\frac{F(s)}{sf(s)}>N\limsup_{s\to\infty}\frac{\Phi(s)}{s\phi(s)}-1\geq 0. \tag{3.13}$$
Then there exists $\bar\lambda>0$, such that for $0<\lambda<\bar\lambda$, (3.12) has a positive solution $u_\lambda$ with
$$\lim_{\lambda\to 0}u_\lambda(r)=\infty$$
uniformly for $r$ in compact subsets of $[0,1)$.
- Assume that (3.10) and (3.13) hold and that $f$ is eventually nondecreasing. Then there exists $\bar\lambda>0$, such that for $0<\lambda<\bar\lambda$, (3.12) has at least two positive solutions, at least one for $\lambda=\bar\lambda$, and none for $\lambda>\bar\lambda$.

We note that in case $\phi(s)=|s|^{p-2}s,\ p\geq 2$, condition (3.13) becomes

$$\liminf_{s\to\infty}\frac{F(s)}{sf(s)}>\frac{N-p}{Np},$$

which is, of course, similar to what was required before. However, because of the convexity requirement placed upon $\phi$ the range of $p$ had to be further restricted. For details of proof we refer to [**4**].

## References


[1] Adams, R. [1975], *Sobolev Spaces*, Acad. Press, New York.

[2] Anane, A. [1987], *Simplicité et isolation de la première valeur propre du p-Laplacien*, C. R. Acad. Sci. Paris **305**, 725-728.

[3] Dang, H., Manásevich, R. and Schmitt, K. [1997], *Positive radial solutions of some nonlinear partial differential equations*, Math. Nachr. **186**, 101–113.

[4] Dang, H. and Schmitt, K. [1997], *On a class of quasilinear boundary value problems on a ball*, preprint.

[5] Dang, H., Schmitt, K. and Shivaji, R. [1996], *On the number of solutions for boundary value problems involving the p-Laplacian*, Electr. J. Diff. Equations **1996**, 1–9.

[6] Dang, H., Schmitt, K. and Shivaji, R. [1998], *Positive solutions of quasilinear boundary*, J. Math. Anal. Appl. **217**, 672-686.

[7] Del Pino, M. and Manásevich, R. [1992], *Global bifurcation from the eigenvalues of the p-Laplacian*, J. of Diff. Equa. **92**, 226-251.

[8] Coffman, C. [1973], *Lyusternik-Schnirelman theory and eigenvalue problems for monotone potential operators*, J. Funct. Anal. **14**, 237-252.

[9] Erbe, L. and Tang, M. [1997], *Uniqueness theorems for positive radial solutions of quasilinear elliptic equations in a ball*, J. Differential Equations **138**, 351–379.

[10] Erbe, L. and Tang, M. [1997], *Structure of positive radial solutions of semilinear elliptic equations*, J. Diff. Equations **133**, 179–202.

[11] Fukagai, N., Ito, M. and Narukawa, K. [1995], *Bifurcation of radially symmetric solutions of degenerate quasilinear elliptic equations*, Diff. Int. Equations **8**, 1709-1732.

[12] García-Huidobro, M., Manásevich, R. and Schmitt, K. [1996], *On principal eigenvalues of p-Laplacian like operators*, J. Diff. Equations **130**, 235–246.

[13] García-Huidobro, M., Manásevich, R. and Schmitt, K. [1997], *Some bifurcation results for a class of p-Laplacian like operators*, Diff. Int. Equations **10**, 51–66.

[14] García-Huidobro, M., Manásevich, R. and Schmitt, K., *Positive radial solutions of quasilinear elliptic partial differential equations on a ball*, Nonl. Anal., TMA, to appear.

[15] García-Huidobro, M., Manásevich, R., Le, V. and Schmitt, K. [1997], *On principal eigenvalues for quasilinear elliptic differential operators: An Orlicz-Sobolev space setting*, preprint.
[16] Idogawa, N. and Ôtani, M. [1995], *The first eigenvalue of some abstract elliptic operators*, Funk. Ekvacioy **38**, 1-9.
[17] Kufner, A., John, O. and Fućik, S. [1977], *Function Spaces*, Noordhoff, Leiden.
[18] Le, V. and Schmitt, K. [1997], *Global Bifurcation in Variational Inequalities: Applications to Obstacle and Unilateral Problems*, vol 137, Applied Math. Sciences, Springer Verlag, New York.
[19] Lindqvist, P. [1990], *On the equation* $\operatorname{div}(|\nabla u|^{p-2}\nabla u) + \lambda|u|^{p-2}u = 0$, Proc. A. M. S. **109**, 157–164.
[20] Manásevich, R. and Schmitt, K. [1996], *Boundary value problems for quasilinear second order differential equations*, in Nonlinear Analysis and Boundary Value Problems (F. Zanolin ed.), pp. 79–119, CISM Lecture Notes vol. 371, Springer Verlag, Wien.
[21] McGough, J. [1994], *On solution continua of supercritical quasilinear elliptic problems*, Diff. Int. Equations **7**, 1453–1472.
[22] Narukawa, K. and Suzuki, T. [1994], *Nonlinear eigenvalue problems for a modified capillary surface equation*, Funk. Ekvacioy **37**, 81-100.
[23] Ôtani, M. [1988], *Existence and nonexistence of nontrivial solutions of some nonlinear degenerate elliptic equations*, J. Funct. Anal. **76**, 140-159.
[24] Schmitt, K. [1995], *Positive solutions of semilinear elliptic equations*, in Topological Method in Differential Equations and Inclusions, (Granas/Frigon, eds.), pp. 447-500, Kluwer, Boston.
[25] Struwe, M. [1990], *Variational Methods*, Springer Verlag, New York.
[26] Zeidler, E. [1990], *Nonlinear Functional Analysis and Applications*, vol. 3, *Variational Methods and Optimization*, Springer Verlag, New York, 1985. vol. 2b, *Nonlinear Monotone Operators*, Springer Verlag, New York.

Fields Institute Communications
Volume **21**, 1999

# A Model for the Production of Ovarian Hormones during the Menstrual Cycle

**James F. Selgrade**
Department of Mathematics
North Carolina State University
Raleigh, NC 27695-8205, USA
selgrade@math.ncsu.edu

**Paul M. Schlosser**
Chemical Industry Institute of Toxicology
P.O. Box 12137
Research Triangle Park, NC 27709, USA
schlosser@ciit.org

**Abstract.** In this study, a mathematical model is developed for the production of the ovarian hormones (estradiol, progesterone, and inhibin) with input functions which represent blood levels of the gonadotropin hormones (luteinizing hormone and follicle stimulating hormone). A 9-dimensional system of linear, nonautonomous, ordinary differential equations tracks the capacities of 9 stages of the ovary to synthesize hormones. The model predicts blood levels of these ovarian hormones which match reasonably well data in the literature collected from normally cycling women. In addition, given input functions fitting the data for the gonadotropin hormones and of a fixed period, the system of differential equations is shown to have a globally asymptotically stable solution with the same period. This model will be merged with a similar model for the gonadotropin hormones to obtain a nonlinear system with delay describing the concentrations of five hormones important for regulation and maintenance of the menstrual cycle. Such a model may be useful for predicting the effects of hormonally active environmental substances on the menstrual cycle.

## 1 Introduction

Recent reports (e.g., [**21**] or [**9**]) have suggested that environmental substances which have estrogenic activity may disrupt the sexual endocrine systems in both humans and animals. In particular, such compounds may have contributed to the increase in breast cancer [**10**], to declines in sperm counts [**31**], and to developmental abnormalities [**20**]. At present, it is not clear what levels of exposure to estrogenic substances produce toxic effects in humans.

1991 *Mathematics Subject Classification.* Primary 92D99, 34C25.

In order to test the effects of exogenous compounds on the human menstrual cycle in a modelling environment, we intend to develop a mathematical model of the normal menstrual cycle which consists of a system of ordinary differential equations describing the concentrations of five hormones important for regulation and maintenance of the cycle. There are two primary components to this model. The first component is the pituitary's synthesis and release of luteinizing hormone ($LH$) and follicle stimulating hormone ($FSH$), which are controlled by the ovarian hormones, estradiol ($E_2$), progesterone ($P_4$), and inhibin ($Ih$). The derivation and rationale for the equations of the first component of the model are contained in [**29**] and will be reviewed below. This paper studies the second component of the model, which describes the effects of $LH$ and $FSH$ on the growth and development of the ovary and on the production of the ovarian hormones, $E_2$, $P_4$ and $Ih$. Various positive and negative feedback mechanisms that are supported by experimental data are incorporated into the model. A 9-dimensional system of linear, nonautonomous ordinary differential equations tracks the capacities of 9 stages of the ovary to synthesize hormones. This system of equations describes the qualitative features of the circulatory concentrations of ovarian hormones. Comparisons with data in [**22**] indicate that the peaks and valleys of our model simulations correspond precisely to those in the data. In addition, given input functions fitting the data for the gonadotropin hormones and of a fixed period, we show that the model has a unique, globally asymptotically stable solution with the same period. Hence, our model predicts that women with the same gonadotropin blood profiles will have the same ovarian hormone profiles.

The primary focus of this work is the structural aspects of the ovarian component of the model and its qualitative behavior. Integrating the two components into one system where all hormone concentrations are phase variables, estimating model parameters from clinical data, and validation of the model by comparing its predictions to clinical data will be the topics of future work.

## 2 Background and model structure

**2.1 Outline of the Menstrual Cycle.** The normal menstrual cycle (Figure 1) for an adult female is approximately 30 days in duration. Briefly, the first half of the cycle is called the follicular phase, during which the primary follicle develops and matures. Then ovulation occurs and the primary follicle is transformed into the corpus luteum. The luteal phase of the cycle follows, during which the corpus luteum produces hormones in preparation for pregnancy. If pregnancy does not occur, the corpus luteum becomes inactive and the cycle repeats.

Here we focus on five primary hormones produced during the month which are important for regulating and maintaining the cycle. The pituitary, in conjunction with the hypothalamus, synthesizes and releases the gonadotropin hormones, $FSH$ and $LH$. Although these hormones have a pulsatile secretion pattern, we assume that the ovary responds to average blood levels of $LH$ and $FSH$ [**23**], so we track their average concentrations in the blood. As part of its normal function, the ovary produces $E_2$, $P_4$, and $Ih$, which control the pituitary's synthesis and release of the gonadotropin hormones during the various stages of the cycle, as depicted in Figure 1.

The cycle begins with the first day of menstrual flow. During menstruation, the blood levels of $FSH$ are rising and promoting the growth of immature follicles. Usually 6 to 12 follicles develop into secondary follicles by adding layers of granulosa

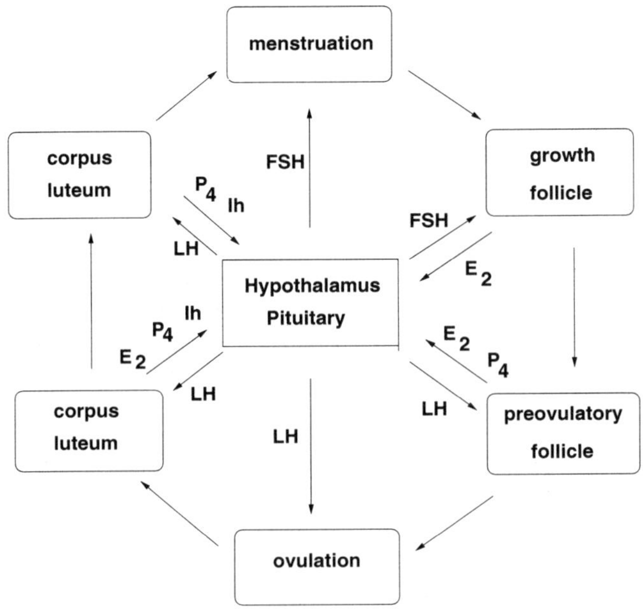


**Figure 1** Phases of the menstrual cycle with gonadotropin hormones, $LH$ and $FSH$, and ovarian hormones, $E_2$, $P_4$ and $Ih$.

cells [**23**]. During the second third of the follicular phase, the production of $E_2$ increases due to the growth of the secondary follicles and a primary follicle is selected to continue its development and ultimately to release its ovum. We refer to this time period as the growth follicle stage (see Figure 1). We do not try to model the follicle selection process, but models have been developed, e.g., see [**16**], [**15**], and [**6**]. During the growth follicle stage, $FSH$ reaches a maximum and secondary follicles begin to atrophy. As the cycle passes into the preovulatory follicular stage, the primary follicle grows more rapidly and produces $E_2$ in large amounts. During the first two-thirds of the follicular phase, $LH$ levels are constant. As $E_2$ increases toward a maximum, $E_2$ primes the pituitary for gonadotropin synthesis, so $LH$ and $FSH$ levels begin to rise. $E_2$ peaks and one day later $LH$ peaks at approximately 8 times its early follicular concentration. This rapid rise and fall of $LH$ over a period of 5 days is referred to as the $LH$ surge and is necessary for ovulation, which occurs within 16 to 36 hours after the surge. The day of the $LH$ peak is considered the midpoint of the menstrual cycle. After a significant decrease during the preovulatory follicular stage, $FSH$ also surges concurrently with $LH$.

The $LH$ surge followed by ovulation transforms the primary follicle into the corpus luteum. The corpus luteum ("yellow body") is characterized by increased fat storage in the theca and granulosa cells and increased secretion of $E_2$, $P_4$, and $Ih$. A low $P_4$ concentration during the follicular phase begins to increase approximately 2 days before the $LH$ surge and continues to increase to a maximum midway through the luteal phase. The $Ih$ profile is similar to that of $P_4$. $P_4$ and $Ih$ appear to inhibit the synthesis of $LH$ and $FSH$, respectively, during the luteal phase in spite of a second increase in $E_2$ (see [**17**], [**14**], [**22**], and [**27, 28**]). The corpus luteum grows during the first half of the luteal phase, reaching its maximal size roughly by the middle of the luteal phase. If fertilization does not occur then the corpus luteum decreases in size and hormone secretion and becomes inactive by the end of the menstrual cycle. The decline of the corpus luteum results in a decrease in $P_4$ and $Ih$ secretion and, consequently, the removal of the inhibition on $LH$ and $FSH$ synthesis. The resulting gradual rise in $LH$ and $FSH$ levels at the end of the month initiates the next cycle.

Previous models of the menstrual cycle or the estrus cycle in mice and rats (see [**30**], [**3, 4**], [**19**], and [**26**]) have useful components but also contain elements which are not based on biological mechanisms. For example, they may contain a switch to turn on the $LH$ surge, separate sources for tonic and surge $LH$ levels, or convolution integrals which weight the effects of hormone levels in past time. We make every effort to link the terms in our differential equations model to physiological mechanisms.

**2.2 Equations for the Pituitary Model.** In [**29**], we develop systems of differential equations for the concentrations of the gonadotropin hormones with the ovarian hormones considered as input functions.

The precise form of these inputs are obtained from data in [**22**] for 33 normally cycling women. A novel feature of this portion of our model is that we separate the processes of gonadotropin synthesis and gonadotropin release. Synthesis refers to the production of mature proteins and sequestering of these proteins into secretory vesicles, producing a releasable pool. The release of these proteins into circulation is treated as a separate step, whose regulation differs from the regulation of synthesis. The idea that gonadotropin synthesis and release are differentially regulated was described first in [**34**].

Two systems of 2-dimensional ordinary differential equations are used to model both the $LH$ and $FSH$ processes of synthesis, release, and clearance. Each system is a two compartment model consisting of the pituitary and the blood. Gonadotropin synthesis occurs in the pituitary where it is held in a reserve pool for release into the blood stream. $RP_{LH}(t)$ and $RP_{FSH}(t)$ denote the functions of time which represent the amounts of $LH$ and $FSH$, respectively, in the releasable pool. $LH(t)$ and $FSH(t)$ denote the concentrations in the blood. The differential equations for $RP_{LH}(t)$ and $RP_{FSH}(t)$ contain terms for synthesis and for release; the differential equations for $LH(t)$ and $FSH(t)$ contain terms for assimilation and for clearance.

Both these systems are linear in the phase variables, $RP_{LH}(t)$, $LH(t)$, $RP_{FSH}(t)$, and $FSH(t)$, with time-dependent coefficients which are functions of the ovarian hormones, i.e., $E_2$, $P_4$, and $Ih$. We use the following explicit functions of time $t$ days for $E_2(t)$, $P_4(t)$, and $Ih(t)$ which approximate the data in [**22**] reasonably well (see [**29**]):

$$E_2(t) = 300 - \frac{240\,(t+1)^2}{3+(t+1)^2} + 90 \exp\left(-\frac{(t-8)^2}{10}\right),$$

$$P_4(t) = 52 \exp\left(-\frac{(t-7)^2}{18}\right) \quad \text{and}$$

$$Ih(t) = 300 + 1330 \exp\left(-\frac{(t-7)^2}{19}\right).$$

Here "exp" denotes exponentiation base "$e$".

The $LH$ system is:

$$\frac{d}{dt}\,RP_{LH} = syn_{LH}(E_2,P_4)\; - \;rel_{LH}(E_2,P_4,RP_{LH}) \tag{2.1a}$$

$$\frac{d}{dt}\,LH = \frac{1}{v_{dis}}\,rel_{LH}(E_2,P_4,RP_{LH})\; - \;clear_{LH}(LH)\,, \tag{2.1b}$$

where

$$syn_{LH}(E_2,P_4) = \frac{1400 + \dfrac{95900\,[E_2(t-d_E)]^8}{[360]^8\, +\, [E_2(t-d_E)]^8}}{1 + P_4(t-d_P)/26}\,, \tag{2.1c}$$

$$rel_{LH}(E_2,P_4,RP_{LH}) = \frac{3\;[1 + 0.024\,P_4(t)]\;RP_{LH}}{1+0.008\,E_2(t)}\,, \qquad \text{and} \tag{2.1d}$$

$$clear_{LH}(LH) = 14\,LH\,. \tag{2.1e}$$

The 12 parameters in system (2) have been chosen so that the output of numerical simulation fits the data of [**22**] qualitatively. However, these parameters have not been optimized using that data. The time-delays $d_E = 0.42$ days and $d_P = 2.9$ days are parameters which describe the period between the time when changes in blood levels of the ovarian hormones occur and the time when subsequent changes in $LH$ synthesis rates occur. The volume of distribution, $v_{dis} = 2.5L$, appears in the second equation to account for $LH$ assimilation into the blood stream.

Experimental evidence from women [**33**] and from animals ([**32**] and [**7**]) indicate that the effect of estradiol on $LH$ synthesis is different than the effect on $LH$ release. For instance, experimental infusion of estradiol is followed by an immediate decrease in serum $LH$ levels, i.e., $LH$ release is inhibited by $E_2$. However, after a period of days, if the $E_2$ concentration is large enough, there is a substantial increase in serum $LH$ — much larger than a simple rebound effect would indicate. In addition, the absence of any presurge dip in $LH$ during the normal menstrual cycle indicates the the surge is much more than a rebound effect [**18**]. Both the size of the $LH$ surge and the lack of a presurge $LH$ dip provide evidence for a positive effect of $E_2$ on $LH$ synthesis; see [**29**] for a complete discussion. Hence, the synthesis term (2.1c) contains an increasing function of $E_2$ in the numerator. This fractional expression is called a Hill function and it increases rapidly as $E_2$ varies between 250 and 600 ng/L to reflect the positive effect of large estradiol concentration on the synthesis of $LH$. The denominator of the release term (2.1d) provides for the inhibitory effect of $E_2$ on $LH$ release. Progesterone has the opposite effects on $LH$ synthesis and release as indicated by the positions of the $P_4$ terms in (2.1c) and (2.1d). The term $clear_{LH}$ is a first order $LH$ clearance term.

The $FSH$ system has a similar form, i.e.,

$$\frac{d}{dt}\,RP_{FSH} = syn_{FSH}(Ih)\; - \;rel_{FSH}(E_2,P_4,RP_{FSH}) \tag{2.2a}$$

$$\frac{d}{dt}\,FSH = \frac{1}{v_{dis}}\,rel_{FSH}(E_2,P_4,RP_{FSH})\; - \;clear_{FSH}(FSH)\,, \tag{2.2b}$$

where

$$syn_{FSH}(Ih) = \frac{3800}{1 + Ih(t - d_{Ih})/1430}\,, \tag{2.2c}$$

$$rel_{FSH}(E_2, P_4, RP_{FSH}) = \frac{65\,[1 + 4\,P_4(t)]\,RP_{FSH}}{1 + 0.007\,[E_2(t)]^2}\,, \qquad \text{and} \tag{2.2d}$$

$$clear_{FSH}(FSH) = 8.21\,FSH\,. \tag{2.2e}$$

There does not seem to be strong evidence of a positive effect of $E_2$ on $FSH$ synthesis, only a negative effect on $FSH$ release [**33**]. In fact, the presurge dip in $FSH$ as $E_2$ increases suggests that the $FSH$ surge may be a rebound effect, and our release term (2.2d) captures this behavior with second order inhibition of $E_2$ on $FSH$ release. The $syn_{FSH}$ term (2.2c) reflects the negative effect of inhibin $Ih(t)$ on $FSH$ synthesis, (see [**12**] and [**1**]), which has a time-delay $d_{Ih} = 2$ days.

## 3 System for the ovary

In our model for the production of ovarian hormones, we use a 9-dimensional system of differential equations with explicit functions of $t$ as inputs for the gonadotropin hormones. The forms of these input functions are determined by the data in [**22**] which are depicted in Figures 2 and 3. To obtain the $FSH$ function, which exhibits two distinctly different rising profiles during the month, we superpose a rational function and two negative exponential functions (Figure 2),

$$\begin{aligned} FSH(t) = 250 - &\frac{250\,(t-15)^4}{1+(t-15)^4} \\ &+ 175\,\exp\left(-\frac{(t-5)^2}{120}\right) + 150\,\exp\left(-\frac{(t-35)^2}{160}\right). \end{aligned} \tag{3.1}$$

For $LH(t)$, we capture the surge with a rational function centered at day 15 (Figure 3),

$$LH(t) = 380 - \frac{352\,(t-15)^4}{1+(t-15)^4}\,. \tag{3.2}$$

Functions in this form are handled easily by the numerical software package we use (*XPP*, the UNIX version of PhasePlane, [**11**]). In addition, explicit input functions allow us to calculate certain integral conditions derived in Section 4 for the existence and stability of periodic solutions of our system of differential equations.

Our model for the ovary divides the follicular phase and the luteal phase into 9 distinct stages based on the capacity of each stage to produce hormones. We use ordinary differential equations to track this capacity as it is transferred from one stage to the next stage during the cycle. We assume the capacity at each stage is proportional to the mass of that stage. The follicular phase consists of the menstrual stage, $MsF(t)$, the growth follicle stage, $GrF(t)$, and the preovulatory stage, $PrF(t)$. We divide the transitional period between the follicular phase and the luteal phase into two stages referred to as ovulatory scar, $Sc_1(t)$ and $Sc_2(t)$. The luteal phase consists of four stages, $Lut_i$ for $i = 1,..,4$. The system of differential equations we obtain is:

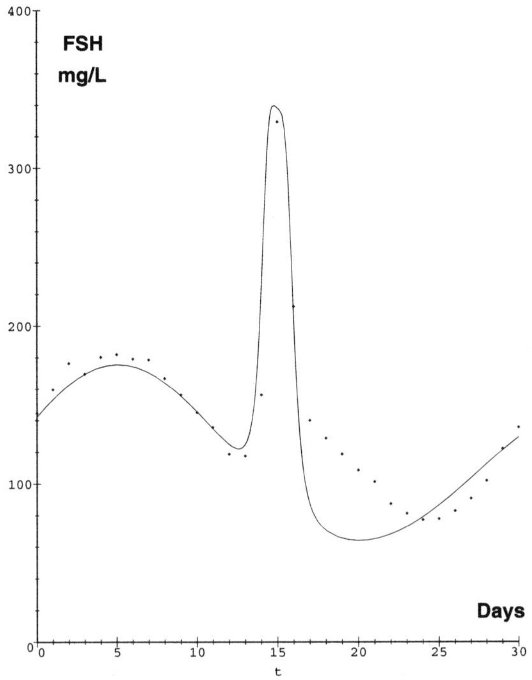


**Figure 2** Graph of the $FSH(t)$ input function (3.1) and data $(\cdots)$ from [**22**] for comparison.

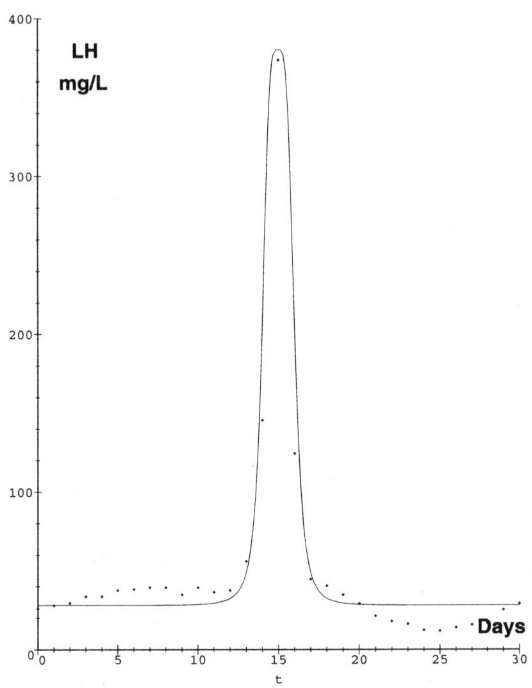


**Figure 3** Graph of the $FSH(t)$ input function (3.2) and data $(\cdots)$ from [**22**] for comparison.

$$
\begin{aligned}
\frac{d}{dt} MsF &= b\, FSH(t) + \{c_1\, FSH(t) - c_2\, [LH(t)]^a\}\, MsF \\
\frac{d}{dt} GrF &= c_2\, [LH(t)]^a\, MsF + \{c_3\, [LH(t)]^a - c_4\, LH(t)\}\, GrF \\
\frac{d}{dt} PrF &= c_4\, LH(t)\, GrF - c_5\, [LH(t)]^a\, PrF \\
\frac{d}{dt} Sc_1 &= c_5\, [LH(t)]^a\, PrF - d_1\, Sc_1 \\
\frac{d}{dt} Sc_2 &= d_1\, Sc_1 - d_2\, Sc_2 \\
\frac{d}{dt} Lut_1 &= d_2\, Sc_2 - k_1\, Lut_1 \\
\frac{d}{dt} Lut_2 &= k_1\, Lut_1 - k_2\, Lut_2 \\
\frac{d}{dt} Lut_3 &= k_2\, Lut_2 - k_3\, Lut_3 \\
\frac{d}{dt} Lut_4 &= k_3\, Lut_3 - k_4\, Lut_4\, .
\end{aligned}
\tag{3.3a}
$$

Since clearance from the blood of the ovarian hormones is on a fast time scale, we assume that the blood levels of $E_2$, $P_4$, and $Ih$ are essentially at steady-state as in [**3**]. Hence, we take these concentrations to be proportional to the hormone capacities at the appropriate stages of the cycle, i.e., we take:

$$E_2(t) = e_0 + e_1\, GrF(t) + e_2\, PrF(t) + e_3\, Lut_4(t)\,, \tag{3.3b}$$

$$P_4(t) = p_1\, Lut_3(t) + p_2\, Lut_4(t) \qquad \text{and} \tag{3.3c}$$

$$Ih(t) = h_0 + h_1\, PrF(t) + h_2\, Lut_2(t) + h_3\, Lut_3(t) + h_4\, Lut_4(t)\,. \tag{3.3d}$$

The first term of the equation for $MsF$ in (3.3a) represents the $FSH$ initiation of the development of premature follicles. This input from the pituitary begins the cyclic changes within the ovary. The follicle growth rate during this stage is assumed proportional to the $FSH$ blood levels [**23**]. The transition into the stage of development of the secondary follicles is promoted by $LH$ via the term $c_2\, [LH(t)]^a\, MsF$. We allow for a parameter $a$, $0 < a \leq 1$, to weaken slightly the $LH$ effect. As the secondary follicles grow, $E_2$ synthesis increases and therefore a $GrF$ term is included in (3.3b). The transition from the growth follicle stage to the preovulatory follicle stage corresponds to the selection of the primary follicle and depends on the full effect of $LH$. During the preovulatory stage, the primary follicle is secreting large amounts of $E_2$, [**2**], as reflected in (3.3b). Ovulation and luteinization are not instantaneous events [**23**] but occur over a period of 16 to 36 hours after the $LH$ surge. We represent this transition with two stages referred to as ovulatory scars. We assume that there is little hormone synthesis during this period. The transition is promoted by $LH$ as reflected by the first term in the equation for $Sc_1$. Then, in our model the capacity to produce hormones cascades through four luteal stages. Odell in [**23**] suggested that low concentrations of $LH$ are required to maintain normal corpus luteum function as depicted in Figure 1.

However, our $LH$ input function (Figure 3) is essentially constant during the luteal phase, so we assume constant growth rates in (3.3a) for our luteal stages. The choice of four stages allows us to position capacity peaks at times which correspond to the data in [**22**] for the luteal phase without using delays in the differential equations. Although delays pose no problem for the mathematical analysis of the present ovarian model, when the ovarian and the pituitary models are merged to yield a nonlinear system of differential equations eliminating delays will simplify that nonlinear system. The primary source of $P_4$ and $Ih$ is the corpus luteum as indicated in (3.3c) and (3.3d).

In Section 4, we obtain complicated analytic expressions for the solutions to (3.3a). However, these formulas do not provide insight into the qualitative behavior of the hormone profiles nor do they lend themselves to parameter estimation. Hence, we examine solution profiles using the UNIX simulation package of [**11**]. Some numerical experimentation was used to obtain the following tables of parameter values:

| Eq. (3.3a) | |
|---|---|
| $a$ | 0.6 |
| $b$ | 0.004 |
| $c_1$ | 0.0045 |
| $c_2$ | 0.077 |
| $c_3$ | 0.006 |
| $c_4$ | 0.008 |
| $c_5$ | 0.045 |
| $d_1$ | 0.5 |
| $d_2$ | 0.8 |
| $k_1$ | 0.6 |
| $k_2$ | 0.5 |
| $k_3$ | 0.6 |
| $k_4$ | 0.65 |

| Eqs. (3.3b,c,d) | |
|---|---|
| $e_0$ | 48 |
| $e_1$ | 0.7 |
| $e_2$ | 2.1 |
| $e_3$ | 1.7 |
| $p_1$ | 0.55 |
| $p_2$ | 0.45 |
| $h_0$ | 270 |
| $h_1$ | 2.5 |
| $h_2$ | 2 |
| $h_3$ | 10 |
| $h_4$ | 14 |

Table 1: Parameter values for the ovarian model (3.3).

The parameters of Table 1 are used to produce the profiles of ovarian hormones in Figure 4 below. As in [**3**], we assume that a basal level for $E_2$ is produced by the adrenal gland and the ovary, and we take it to be 48 ng/L because that is the minimum of the data values in [**22**]. The majority of the $E_2$ produced during a normal cycle is contributed by the growth and preovulatory stages in the follicular phase and the fourth luteal stage, with the largest portion produced by the primary follicle. Notice that the behavior of the simulated solution $E_2(t)$ in Figure 4 is not as sharp as the data in [**22**], but the peaks and valleys occur at precisely the same times during the cycle. The follicular $E_2$ peak occurs on day 14 and the luteal peaks for the ovarian hormones occur on days 22 and 23. In order to approximate data in [**22**], which is not periodic, we choose initial conditions which result in simulated solutions which are not periodic. The profiles of all three hormones are higher on day 30 than at the beginning of the cycle. In our model, $P_4$ is produced by the third and fourth luteal stages in approximately equal proportions. We assume a basal level for $Ih$ and take the majority of $Ih$ to be synthesized during the second, third, and fourth luteal stages. Also, notice that the $Ih$ data have a small spike on day 15 which we attribute to the primary follicle (see (3.3d)).

## 4 Mathematical analysis

Let $X(t) = (x_1(t), x_2(t).\dots, x_n(t))$ denote a $n$-dimensional, differentiable vector function of time $t \geq 0$. If $\delta$ is a vector of delays then let $W(t,\delta)$ be a continuous vector function of time and $\delta$. Assume that $A(t)$ represents a $n \times n$-matrix of continuous functions of time. The systems of differential equations for $LH$ (2.1), $FSH$ (2.2), and the ovary (3.3) are linear and nonautonomous with the general form:

$$X'(t) = A(t)\,X(t) + W(t,\delta), \tag{4.1}$$

where $'$ denotes the derivative with respect to time. For a given initial condition, a solution to (4.1) exists and is unique. If $\Phi(t)$ is a fundamental solution matrix of the homogeneous linear system $X'(t) = A(t)\,X(t)$ then the solution to (4.1) satisfying the initial condition $X(0)$ is given by

$$X(t) = \Phi(t)\,\Phi(0)^{-1}\,X(0) + \Phi(t)\int_0^t \Phi(s)^{-1}\,W(s,\delta)\,ds\,. \tag{4.2}$$

Equation (4.2) is the variation of constants formula which may be found in any standard text on ordinary differential equations, e.g., [**5**], [**8**], or [**13**].

If $A(t)$ and $W(t,\delta)$ are periodic functions of period $T$ then (4.1) has a solution of period $T$ if and only if $X(T) = X(0)$, (e.g., see [**5**]). Using this fact in (4.2) yields

$$\left[I - \Phi(T)\,\Phi(0)^{-1}\right]\,X(0) = \Phi(T)\int_0^T \Phi(s)^{-1}\,W(s,\delta)\,ds\,, \tag{4.3}$$

where $I$ is the $n \times n$ identity matrix. If $\left[I - \Phi(T)\,\Phi(0)^{-1}\right]$ has an inverse then the unique $T$-periodic solution to (4.1) has initial condition given by

$$X(0) = \left[I - \Phi(T)\,\Phi(0)^{-1}\right]^{-1}\,\Phi(T)\int_0^T \Phi(s)^{-1}\,W(s,\delta)\,ds\,. \tag{4.4}$$

If $\Phi(0)^{-1} = I$, which is the case in our models, then the properties of $\Phi(t)$ determine the existence, the uniqueness, and the stability of periodic solutions.

Because of the form of (2.1), (2.2), and (3.3), we assume that $A(t)$ has the bidiagonal form:

$$A(t) = \begin{bmatrix} a_{11}(t) & 0 & 0 & \cdots \\ a_{21}(t) & a_{22}(t) & 0 & \cdots \\ 0 & \ddots & \ddots & \\ \cdots & 0 & a_{n,n-1}(t) & a_{nn}(t) \end{bmatrix}. \tag{4.5}$$

A fundamental matrix for $X'(t) = A(t)\,X(t)$ may be obtained by repeated integration as outlined in the following proposition.

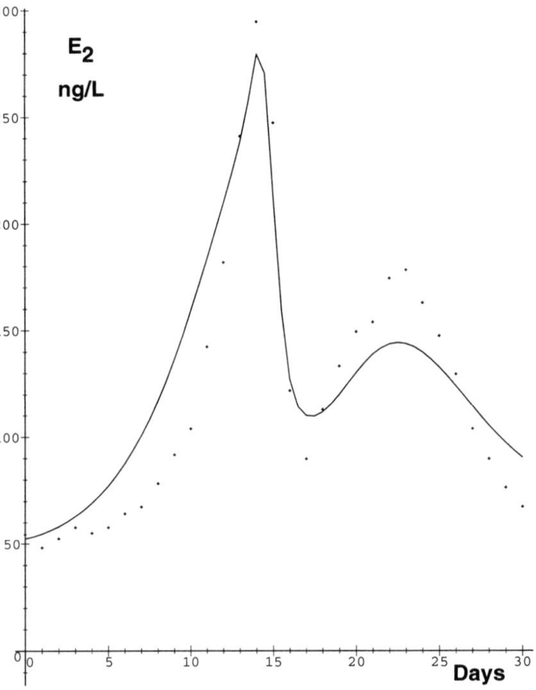


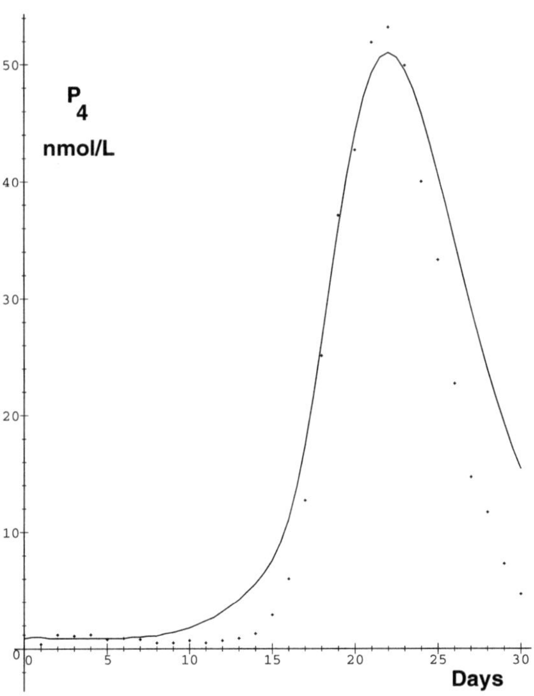


**Figure 4** continued on next page

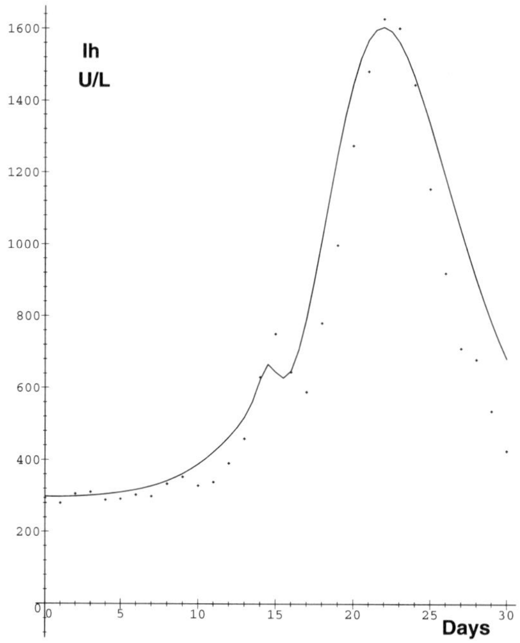


**Figure 4** Graphs of $E_2(t)$, $P_4(t)$, and $Ih(t)$ simulations of (3.3) and data $(\cdots)$ from [**22**] for comparison.

**Lemma 1** *Let $A(t)$ have the bidiagonal form given by (4.5). The fundamental matrix $\Phi(t) = (\phi_{ij}(t))$ for $X'(t) = A(t)\,X(t)$ with $\Phi(0) = I$ is lower triangular with diagonal entries, for $i = 1, ..., n$, defined by*

$$\phi_{ii}(t) = e^{\int_0^t a_{ii}(s)\,ds}\,, \tag{4.6}$$

*and the remaining nonzero entries recursively defined in terms of the previous row by*

$$\phi_{ij}(t) = \phi_{ii}(t) \int_0^t a_{i,i-1}(s)\,\phi_{i-1,j}(s)\,\phi_{ii}^{-1}(s)\,ds \qquad \text{for } j < i\,. \tag{4.7}$$

**Proof** First integrate the equation $x_1' = a_{11}(t)\,x_1$ to get $x_1(t) = x_1(0)e^{\int_0^t a_{11}(s)ds}$. Hence $\phi_{11}(t)$ is given by (4.6). Proceed by induction on the row index $i$.

For $i = 2$, the second row of the homogeneous system is

$$x_2' = a_{21}(t)\,x_1 + a_{22}(t)\,x_2\,,$$

which may be solved for $x_2(t)$ in terms of $x_1(t)$ giving

$$x_2(t) = x_2(0)\,e^{\int_0^t a_{22}(s)\,ds} + e^{\int_0^t a_{22}(s)\,ds} \int_0^t a_{21}(s)\,x_1(s)\,e^{-\int_0^s a_{22}(u)\,du}\,ds\,. \tag{4.8}$$

After substituting $x_1(s) = x_1(0)\,\phi_{11}(s)$ into (4.8), we conclude that $\phi_{22}(t)$ is given by (4.6) and $\phi_{21}(t)$ is given by (4.7).

By induction, assume that the $(i-1)$ row of the fundamental matrix has the desired form, i.e., the solution to the $(i-1)$ equation of the homogeneous system is

$$x_{i-1}(t) = \sum_{j=1}^{i-1} x_j(0)\,\phi_{i-1,j}(t), \tag{4.9}$$

where the $\phi_{i-1,j}$ is given by (4.6) or (4.7). As with the second equation, the solution to the $i$-th equation may be written as

$$x_i(t) = x_i(0)\,e^{\int_0^t a_{ii}(s)\,ds} + e^{\int_0^t a_{ii}(s)\,ds}\int_0^t a_{i,i-1}(s)\,x_{i-1}(s)\,e^{-\int_0^s a_{ii}(u)\,du}\,ds\,. \tag{4.10}$$

After substituting the left side of (4.9) into (4.10) for $x_{i-1}(s)$, we see that the $i$-th row of $\Phi(t)$ is given by (4.6) and (4.7). □

In the case of coefficients in (4.1) of period $T$, from (4.3) we see that there is a $T$-periodic solution if the inverse of $[I - \Phi(T)]$ exists. From Lemma 1, this is the case if none of the diagonal entries of $\Phi(T)$ are equal to 1. Hence, we obtain

**Proposition 2** *Take $T > 0$ and assume that $A(t)$ and $W(t,\delta)$ have period $T$ where $A(t)$ has the form of (4.5). Then (4.1) has a unique $T$-periodic solution given by (4.2) and (4.4) if for each $i = 1,\dots,n$*

$$\int_0^T a_{ii}(s)\,ds \neq 0\,.$$

If the $T$-periodic solution of Proposition 2 is globally asymptotically stable then, roughly speaking, women would have the same cycle if they produced the same periodic amounts of gonadotropin hormones during the cycle. Asymptotic stability holds if

$$\lim_{t\to\infty} \Phi(t) = 0\,. \tag{4.11}$$

To obtain (4.11) we assume that

$$\int_0^T a_{ii}(s)\,ds < 0\,. \tag{4.12}$$

From a biological perspective, (4.12) asserts that the average growth rate over the cycle for each phase of the ovary is negative. This permits the transition from one phase to the next, i.e., the decline of one phase as the next phase grows. So (4.12) precludes the continued growth of one phase throughout the cycle, which might be considered a cancerous situation. In order to see that (4.12) implies (4.11), we need bounds on the long-term behavior of the entries of $\Phi(t)$.

**Lemma 3** *Take $T > 0$ and assume that $a(t)$ is a continuous $T$-periodic function such that $\int_0^T a(s)\,ds < 0$. Then there are positive constants $m$, $M$, and $B$ such that for all $t \geq T$*

$$0 \leq m\,e^{-Bt} \leq e^{\int_0^t a(s)\,ds} \leq M\,e^{-Bt}. \tag{4.13}$$

**Proof** Let $A = -\int_0^T a(s)\,ds > 0$ and let $K = \int_0^T |a(s)|\,ds > 0$. Then for all $0 \le t \le T$

$$-K \le \int_0^t a(s)\,ds < \int_0^t |a(s)|\,ds \le K .$$

Fix $t \ge T$. There is a positive integer $n$ so that $0 \le t - nT < T$. Hence

$$e^{-(n+1)\,A} \le e^{-A\,t/T} \le e^{-n\,A},$$

which implies that

$$e^{-n\,A} \le e^{A}\,e^{-A\,t/T} \le e^{A}\,e^{-n\,A}.$$

Using the periodicity of $a(t)$, we write

$$e^{\int_0^t a(s)\,ds} = e^{\int_0^{nT} a(s)\,ds}\,e^{\int_{nT}^t a(s)\,ds} = e^{-nA}\,e^{\int_0^{t-nT} a(s)\,ds}. \tag{4.14}$$

This gives the upper bound

$$e^{\int_0^t a(s)\,ds} \le e^{-nA}\,e^{K} \le e^{A+K}\,e^{-A\,t/T}.$$

Choose $B = \frac{A}{T}$ and $M = e^{A+K}$. For the lower bound, from (4.14) we have

$$e^{\int_0^t a(s)\,ds} \ge e^{-nA}\,e^{-K} \ge e^{-K}\,e^{-A\,t/T}.$$

So we take $m = e^{-K}$ to obtain (4.13). □

From Lemma 1 and Lemma 3, we obtain the following bounds for the diagonal entries of $\Phi(t)$.

**Corollary 4** *Take $T > 0$ and assume that $A(t)$ has the form of (4.5) and has period $T$ such that $\int_0^T a_{ii}(s)\,ds < 0$ for each $i$. Then there are positive constants $m_i$, $M_i$, and $B_i$ such that for all $t \ge T$*

$$0 \le m_i\,e^{-B_i t} \le \phi_{ii}(t) \le M_i\,e^{-B_i t}, \tag{4.15}$$

*and*

$$\phi_{ii}^{-1}(t) \le \frac{1}{m_i}\,e^{B_i t}. \tag{4.16}$$

Bounds for the off-diagonal terms of $\Phi(t)$ may be recursively derived from the bounds of Corollary 4.

**Lemma 5** *With the same assumptions as Corollary 4, for $i = 1, \dots, n$ and for each $j < i$, there are positive constants $K_{ij}$ and $B_{ij}$ such that for all $t \ge T$*

$$|\phi_{ij}(t)| \le K_{ij}\,e^{-B_{ij} t}. \tag{4.17}$$

**Proof** We argue by induction on the row index $i$. There are positive constants $L_{ij}$ such that for all $t \ge 0$

$$|a_{ij}(t)| \le L_{ij}.$$

Fix $t \geq T$. For $i = 2$ and $j = 1$, using (4.7), (4.15), and (4.16), we have

$$\begin{aligned} |\phi_{21}(t)| &\leq \phi_{22}(t) \int_0^t |a_{21}(s)|\, \phi_{11}(s)\, \phi_{22}^{-1}(s)\, ds \\ &\leq L_{21} \frac{M_1 M_2}{m_2}\, e^{-B_2 t} \int_0^t e^{(B_2 - B_1)\, s}\, ds\,. \end{aligned} \tag{4.18}$$

The exponential integral in (4.18) is $[e^{(B_2-B_1)\,t} - 1]/(B_2 - B_1)$ if $B_2 \neq B_1$ but is $t$ if $B_2 = B_1$. In either case, by choosing $B_{21}$ positive and slightly less than the minimum of $B_1$ and $B_2$ and choosing $K_{21}$ large enough, we obtain

$$|\phi_{21}(t)| \leq K_{21}\, e^{-B_{21} t}\,.$$

By induction, assume the bound of (4.17) for $\phi_{i-1,j}(t)$ and consider $\phi_{ij}(t)$. Then from (4.7)

$$\begin{aligned} |\phi_{ij}(t)| &\leq \phi_{ii}(t) \int_0^t |a_{i,i-1}(s)|\, |\phi_{i-1,j}(s)|\, \phi_{ii}^{-1}(s)\, ds \\ &\leq L_{i,i-1}\, K_{i-1,j} \frac{M_i}{m_i}\, e^{-B_i t} \int_0^t e^{(B_i - B_{i-1,j})\, s}\, ds\,. \end{aligned} \tag{4.19}$$

The appropriate bound for $|\phi_{ij}(t)|$ is obtained from (4.19). □

Clearly, the bounds (4.15) and (4.17) imply that the fundamental matrix $\Phi(t)$ approaches 0 as $t \to \infty$. Hence we have the following result:

**Theorem 6** *Take $T > 0$ and assume that $A(t)$ and $W(t, \delta)$ have period $T$ where $A(t)$ has the form of (4.5). Then (4.1) has a unique $T$-periodic solution given by (4.2) and (4.4) which is globally asymptotically stable if for each $i = 1, ..., n$*

$$\int_0^T a_{ii}(s)\, ds < 0\,. \tag{4.20}$$

We apply Theorem 6 to (3.3a) with $T$=30 days. We assume that the input functions $FSH(t)$ and $LH(t)$ have periods of 30 days and show that the integrals over a period of the diagonal coefficients in (3.3a) are negative, i.e., we verify (4.20). Most of the diagonal coefficients in (3.3a) are either negative functions or negative constants. However, the coefficients of $MsF(t)$ and $GrF(t)$ are not. In fact, for the first third of the cycle, the coefficient of $MsF(t)$ is positive. We use the software *Maple* to evaluate the integrals of these two coefficient functions and obtain:

$$\begin{aligned} &\int_0^{30} \{0.0045\, FSH(s) \;-\; 0.077\, [LH(s)]^{0.6}\}\, ds \;\approx\; -4.3257 \quad \text{and} \\ &\int_0^{30} \{0.006\, [LH(s)]^{0.6} \;-\; 0.008\, LH(s)\}\, ds \;\approx\; -11.2218\,. \end{aligned} \tag{4.21}$$

Hence our result for (3.3a) may be stated simply as:

**Theorem 7** *If the gonadotropin input functions $FSH(t)$ and $LH(t)$ are given by (3.1) and (3.2) and have a period of 30 days then system (3.3a), which models* *the ovary, has a globally asymptotically stable solution and this solution has a period of 30 days.*

## 5 Discussion

In this study we develop a mathematical model for the synthesis of the ovarian hormones $E_2$, $P_4$, and $Ih$ with input functions which represent the gonadotropin hormones $FSH$ and $LH$. A 9-dimensional system (3.3a) of ordinary differential equations tracks the capacities of 9 stages of the ovary to synthesize hormones — 3 stages for the follicular phase, 2 stages for ovulation and the transitional period from the follicular phase to the luteal phase, and 4 stages for the luteal phase. Since clearance from the blood is so rapid for the ovarian hormones, we assume that their blood levels are at steady-state and take these concentrations to be proportional to the hormone capacities at the appropriate stages during the cycle, see (3.3b), (3.3c) and (3.3d).

This model describes the qualitative features of the circulatory concentrations of ovarian hormones. Comparisons with data in [**22**] in Figure 4 indicate that the peaks and valleys of our simulations correspond to those in the data although the growth and decay profiles of the data are somewhat sharper. In addition, given input functions fitting the data for the gonadotropin hormones and of period 30 days, we show that the model has a unique, globally asymptotically stable solution with a period of 30 days. Hence, our model predicts that women with the same gonadotropin blood profiles will have the same ovarian hormones profiles.

In numerical experiments with our ovarian model, we have noticed that delaying the effect of $LH$ on the luteal stages may reduce the number of stages needed to produce correctly timed luteal hormone peaks. In particular, delay may be used to eliminate the scar stages and still get hormone peaks at 22 and 23 days. However, we have not incorporated such delay into the model because we do not have a clear physiological basis for the delay.

In order to improve model behavior, we will optimize the parameters using the data of [**22**] for the 9-dimensional ovarian system and the systems for the synthesis and release of the gonadotropin hormones. Then we will merge the two components into one, nonlinear, autonomous system of ordinary differential equations. Because of the presence of delays in the gonadotropin component, the merged system will have delays. We intend to validate the model by seeing if it will reproduce responses to exogenous hormone administration which have been observed in clinical studies, e.g., [**33**], [**24**], and [**7**]. If this validation process is successful, our model should be useful in predicting the effects of environmental substances on the menstrual cycle.

## 6 Acknowledgments

The authors would like to thank Charlotte Lane McGlohon for providing the data of [**22**] in a form that our software could handle easily. Paul Schlosser's research is supported by the member companies of the Chemical Industry Institute of Toxicology.

## References

[1] Anderson, L. D. and DePaolo, L.V. [1981], *Control of inhibin secretion from the ovary*, Intragonadal Regulation of Reproduction, (Franchimont, P. and Channing, C. P. eds.), Academic Press, London, 343-363.

[2] Baird, D. T. [1976], *Ovarian steroid secretion and metabolism in women*, The Endocrine Function of the Human Ovary (James, V. H. T., Serio, M. and Giusti, G., eds.), Academic Press, London, 125-133.

[3] Bogumil, R. J., Ferin, M., Rootenberg, J., Speroff, L., and Vande Wiele, R. L. [1972], *Mathematical studies of the human menstrual cycle. I: Formulation of a mathematical model*, J. Clin. Endocrinol. Metab., **35**, 126-143.

[4] Bogumil, R. J., Ferin, M. and Vande Wiele, R. L. [1972], *Mathematical studies of the human menstrual cycle. II: Simulation performance of a model of the human menstrual cycle*, J. Clin. Endocrinol. Metab. **35**, 144-156.

[5] Brauer, F., and Nohel, J.A. [1969], *The Qualitative Theory of Ordinary Differential Equations*, W. A. Benjamin, New York.

[6] Chavez-Ross, A., Franks, S., Mason, H. D., Hardy, K. and Stark, J. [1997], *Modelling the control of ovulation and polycystic ovary syndrome*, J. Math. Biol. **36**, 95-118.

[7] Clarke, I. J. and Cummins, J. T. [1984], *Direct pituitary effects of estrogen and progesterone on gonadotropin secretion in the ovariectomized ewe*, Neuroendocrinology, **39**, 267-274.

[8] Coddington, E. A. and Levinson, N. [1955], *Theory of Ordinary Differential Equations*, McGraw-Hill, New York.

[9] Daston, G. P., Gooch, J. W., Breslin, W. J., Shuey, D. L., Nikiforov, A. I., Fico, T. A. and Gorsuch. J. W. [1997], *Environmental estrogens and reproductive health: A discussion of the human and environmental data*, Reproductive Toxicology **11**, 465-481.

[10] Davis, D. L., Bradlow, H. L., Wolff, M., Woodruff, T., Hoel, D. G. and Anton-Culver, H. [1993], *Medical hypothesis: Xenoestrogens as preventable causes of breast cancer*, Environ. Health Perspect. **101**, 372-377.

[11] Ermentrout, B. [1990], *PhasePlane*, Brooks/Cole, Pacific Grove.

[12] Franchimont, P., Henderson, K., Verhoeven, G., Hazee-Hagelstein, M., Charlet-Renard, C., Demoulin, A., Bourguignon, J., and Lecomte-Yerna, M. [1981], *Inhibin: Mechanisms of action and secretion*, Intragonadal Regulation of Reproduction, (Franchimont, P., and Channing, C.P. eds.), Academic Press, London, 167-191.

[13] Hale, J. [1969], *Ordinary Differential Equations*, John Wiley & Sons, New York.

[14] Krey, L. C., and Kamel, F. [1990], *Progesterone modulation of gonadotropin secretion by dispersed rat pituitary cells in culture, I. Basal and gonadotropin-releasing hormone-stimulated luteinizing hormone release*, Mol. Cell Endocrinol. **68**, 85-94.

[15] Lacker, H. M., Beers, W., Mueli, L. E. and Akin, E. [1987], *A theory of follicle selection: I and II*, Biology of Reproduction **37**, 570-580.

[16] Lacker, H. M. and Peskin, C. S. [1981], *Control of ovulation number in a model of ovarian follicular maturation*, Lectures on Mathematics in the Life Sciences **14**, 21-58.

[17] Legacé, L., Massicotte, J. and Labrie, F. [1980], *Acute stimulatory effects of progesterone on luteinizing hormone and follicle-stimulating hormone release in rat anterior pituitary cells in culture*, Endocrinology **106**, 684-689.

[18] Liu, J. H. and Yen, S. S. C. [1983], *Induction of midcycle gonadotropin surge by ovarian steroids in women: A critical evaluation*, J. Clin. Endocrinol. Metab. **57**, 797-802.

[19] McIntosh, J. E. A. and McIntosh, R. P. [1980], *Mathematical Modeling and Computers in Endocrinology*, Springer-Verlag, Berlin.

[20] McLachlan, J. A. [1985], *Estrogens in the Environment II: Influences on Development*, Elsevier North Holland, New York.

[21] McLachlan, J. A. and Korach, K. S. [1995], *Symposium on Estrogens in the Environment, III*, Environ. Health Perspect. **103 (Suppl. 7)**, 3-4.

[22] McLachlan, R. I., Cohen, N. L., Dahl, K. D., Bremner, W. J. and Soules, M. R. [1990], *Serum inhibin levels during the periovulatory interval in normal women: Relationships with sex steroid and gonadotrophin levels*, Clin. Endocrinol. **32**, 39-48.

[23] Odell, W. D. [1979], *The reproductive system in women*, Endocrinology (DeGroot, L.J. ed.), Grune & Stratton, New York, 1383-1400.

[24] Odell, W. D., and Molitch, M. E. [1974], *The pharmacology of contraceptive agents*, Ann. Rev. Pharmacol., **14**, 413-434.

[25] Odell, W. D., and Swerdloff, R. S. [1968], *Progestogen-induced luteinizing and follicle-stimulating hormone surge in postmenopausal women: A simulated ovulatory peak*, Proc. Natl. Acad. Sci. **61**, 529-536.

[26] Plouffe, Jr., L. and Luxenberg, S. N. [1992], *Biological modeling on a microcomputer using standard spreadsheet and equation solver programs: The hypothalamic-pituitary-ovarian axis as an example*, Comput. Biomed. Res. **25**, 117-130.

[27] Rivier, C., Meunier, H., Roberts, V. and Vale, W. [1990], *Inhibin: Role and secretion in the rat*, Recent Prog. Horm. Res. **46**, 231-259.

[28] Rivier, C., Schwall, R., Mason, A., Burton, L. and Vale, W. [1991], *Effect of recombinant inhibin on gonadotropin secretion during proestrus and estrus in the rat*, Endocrinology, **128**, 2223-2228.

[29] Schlosser, P. M. and Selgrade, J. F. [1997], *A model of gonadotropin regulation during the menstrual cycle in women: Qualitative features*, submitted.

[30] Schwartz, N. B. [1970], *Cybernetics of mammalian reproduction*, Mammalian Reproduction, (H. Gibian and E. J. Plotz, eds.), Springer-Verlag, Berlin, 97-111.

[31] Sharpe, R. M. and Skakkebaek, N. E. [1993], *Are oestrogens involved in falling sperm counts and disorders of the male reproductive tract?*, Lancet **341**, 1392-1395.

[32] Swerdloff, R. S., Jacobs, H. S. and Odell, W. D. [1972], *Synergistic role of progestogens in estrogen induction of LH and FSH surge*, Endocrinology **90**, 1529-1536.

[33] Tsai, C. C. and Yen, S. S. C. [1971], *The effect of ethinyl estradiol administration during early follicular phase of the cycle on the gonadotropin levels and ovarian function*, J. Clin. Endocr. **33**, 917-923.

[34] Yen, S. S. C. [1980], *Neuroendocrine regulation of the menstrual cycle*, Neuroendocrinology, (D. T. Krieger and J. C. Hughes, eds.), Sinauer Associates, Inc., Sunder-land, Mass.

Fields Institute Communications
Volume **21**, 1999

# Global Stability for a General Population Model with Time Delays

**Joseph W.–H. So**
Department of Mathematical Sciences
University of Alberta
Edmonton, Alberta, Canada T6G 2G1
joseph.so@ualberta.ca

**J. S. Yu**
Department of Applied Mathematics
Hunan University
Changsha, Hunan 410082, China
jsyu@dns.hunu.edu.cn

*This paper is dedicated to Professors Lynn Erbe and Herbert Freedman on the occasion of their retirements.*

**Abstract.** In this paper we obtain some new sufficient conditions for the global stability of the zero solution of a general delay population model

$$x'(t) = -[1 + x(t)] \sum_{i=1}^{n} \int_{t-r(t)}^{t} f_i(t, x(s))\, d\mu_i(t, s).$$

As applications we improve some known results in the literature.

## 1 Introduction

The scalar delay-differential equation

$$N'(t) = N(t)[a - bN(t) - N(t-1)], \quad t \geq 0 \tag{1.1}$$

is a special case of the so-called Volterra logistic delay equation which arises in population models of a single species where the death rate depends not only on the population at time $t$, but also on the population one unit of time earlier. Here $a$ and $b$ are positive constants. It was shown in [**5**, **17**] and [**21**] that if $b > 1$, then every positive solution of (1.1) tends to the equilibrium $N^* = a/(b + 1)$.

---

1991 *Mathematics Subject Classification.* Primary 34K15; Secondary 92D25.
Research partially supported by NSERC, grant number OGP36475.

Let $x(t) = \frac{b+1}{a}N(t) - 1$. Then (1.1) becomes

$$x'(t) = -\frac{a}{b+1}[1 + x(t)][bx(t) + x(t-1)]. \tag{1.2}$$

When $b = 0$, (1.2) reduces to the so-called Wright's equation (cf. [**25**]):

$$x'(t) = -a[1 + x(t)]x(t-1). \tag{1.3}$$

It was shown in [**25**] that if $0 < a \le \frac{3}{2}$, then every solution $x(t)$ of (1.3), which satisfies $x(t) > -1$ for $t \ge 0$, tends to zero as $t \to \infty$. In [**23**], we extended Wright's result to the nonautonomous equation

$$x'(t) = -a(t)[1 + x(t)]x(t-1) \tag{1.4}$$

where $a \in C([0,\infty),(0,\infty))$ and it was shown that if

$$\int_{t-1}^{t} a(s)\,ds \le \frac{3}{2} \quad \text{for large} \quad t \tag{1.5}$$

and

$$\int_{0}^{\infty} a(s)\,ds = \infty, \tag{1.6}$$

then every solution $x(t)$ of (1.4), satisfying $x(t) > -1$ for $t \ge 0$, converges to zero as $t \to \infty$.

A more general equation

$$x'(t) = -[1 + x(t)]\sum_{i=1}^{n}\int_{t-r(t)}^{t} f_i(t, x(s))\,d\mu_i(t,s) \tag{1.7}$$

involving possibly unbounded time delays, is discussed in [**12, 13**]. Here $f_i(t,x)$ and $r(t)$ are continuous with respect to their arguments, $\mu_i(t,s)$ is continuous with respect to $t$, nondecreasing with respect to $s$ and is defined for all $(t,s) \in \mathbb{R}^2$. In addition, it is assumed

($H_1$) for $xf_i(t,x) \ge 0$ for $x > -1$, and for all $t \ge 0, \sum_{i=1}^{n} f_i(t,x) = 0$ if and only if $x = 0$;
($H_2$) $r(t) > 0, t \mapsto t - r(t)$ is nondecreasing and $\lim_{t\to\infty}(t - r(t)) = \infty$;
($H_3$) $\mu_i(t,t) > \mu_i(t, t - r(t))$;
($H_4$) there exist continuous functions $a_i(t) \ge 0, b_i(|c|) \ge 0$ with $b_i(|c|) = 0$ if and only if $c = 0$, such that $|f_i(t,c)| \ge a_i(t)b_i(|c|)$ and

$$\lim_{t\to\infty}\int_0^t \left(\sum_{i=1}^{n}\int_{\tau-r(\tau)}^{\tau} a_i(\tau)\,d\mu_i(\tau,s)\right)d\tau = \infty.$$

Let $r = r(0)$. The initial value problem for (1.7) takes the form

$$x(\theta) = \phi(\theta), \quad \theta \in [-r, 0], \tag{1.8}$$

where $\phi \in C([-r,0],[-1,\infty))$ and $\phi(0) > -1$.

By ($H_1$) and ($H_3$), we see that (1.7) has exactly two steady states: $x(t) \equiv 0$ and $x(t) \equiv -1$. We say a function $x(t)$ defined on $[-r,\infty)$ is *oscillatory* if it is not eventually zero and if it has arbitrarily large zeros; otherwise, it is called *nonoscillatory*.

Equation (1.7) has attracted the attention and interest of many researchers because of its practical importance. Indeed, many models from population dynamics (c.f. [**1, 2, 3, 4, 5, 6, 7, 9, 10, 11, 12, 13, 17, 18, 19, 20, 21, 22, 23, 24, 25, 26**]) are special cases of (1.7) and the long term behavior of the solutions is of great interest. We refer the reader to [**12, 13**] for some general results on (1.7).

These results are normally obtained by constructing an appropriate Lyapunov functional or a Razumikhin-type function (cf. [**4, 5, 6, 8, 14, 15, 16**]).

Our aim in this paper is to apply the idea employed in [**23**] to obtain some new sufficient conditions for the global stability of the zero solution of (1.7). Our main result will be used to improve some known results in the literature (see Section 4). The rest of this paper is organized as follows: In Section 2, we establish a result on the boundedness of solutions of (1.7). In Section 3, the global stability of the zero solution of (1.7) is considered. Finally, in Section 4 we apply the results in Section 3 to some special models of (1.7). Throughout this paper, we assume that $(H_1)$–$(H_4)$ are satisfied.

## 2 Boundedness results

For the purpose of later reference, we state the following two results whose proofs are similar to Lemma 3.1 and Theorem 3.1 on p.127–129 of [**13**].

**Lemma 2.1** *Assume that there exists a continuous function $R : [0,\infty) \to [0,\infty)$ such that for $x \geq -1$,*

$$f_i(t,x) \geq -R(t), \quad i = 1, \cdots, n. \tag{2.1}$$

*Then the solution $x(t)$ of IVP (1.7)-(1.8) exists and $x(t) > -1$ for all $t \geq 0$. Moreover, if $x(t)$ is nonoscillatory, then $\lim_{t\to\infty} x(t) = 0$.*

**Theorem 2.2** *Assume that there exists a constant $N > 0$ and a nonnegative continuous function $R(t)$ such that*

$$f_i(t,x) \geq -R(t) \quad \text{for} \quad x \geq -1 \quad \text{and} \quad i = 1, \cdots, n \tag{2.2}$$

*and*

$$\int_{t-r(t)}^{t} R(\tau) \sum_{i=1}^{n} [\mu_i(\tau,\tau) - \mu_i(\tau, \tau - r(\tau))]\, d\tau \leq N \quad \text{for large} \quad t. \tag{2.3}$$

*Then the solution $x(t)$ of IVP (1.7)-(1.8) satisfies*

$$x(t) \leq e^N - 1 \quad \text{for large} \quad t.$$

*If, in addition, there exists a constant $\bar{N} > 0$ and a nonnegative continuous function $\bar{R}(t)$ such that*

$$f_i(t,x) \leq \bar{R}(t) \quad \text{for} \quad -1 \leq x \leq e^N - 1 \tag{2.4}$$

*and*

$$\int_{t-r(t)}^{t} \bar{R}(\tau) \sum_{i=1}^{n} [\mu_i(\tau,\tau) - \mu_i(\tau, \tau - r(\tau))]\, d\tau \leq \bar{N} \quad \text{for large} \quad t, \tag{2.5}$$

*then*

$$x(t) \geq -1 + e^{-\bar{N}} \quad \text{for large} \quad t.$$

## 3 Global stability results

In this section, we study the global stability of the zero solution of (1.7). The main result in this section is the following Theorem 3.1, which is an essential improvement of Theorem 3.2 in [**12**]; (see also Theorem 3.2 on p.129 of [**13**]).

**Theorem 3.1** *In addition to the assumptions in Theorem 2.2, assume further that there exist continuous functions $\alpha_i(t)$, such that*

$$|f_i(t,c)| \le \alpha_i(t)|c|, \quad i = 1, \cdots, n, \tag{3.1}$$

*and*

$$\int_{t-r(t)}^{t} \left( \sum_{i=1}^{n} \int_{\tau-r(\tau)}^{\tau} \alpha_i(\tau)\, d\mu_i(\tau,s) \right) d\tau \le \frac{3}{2} \quad \text{for large} \quad t. \tag{3.2}$$

*Then the solution $x(t)$ of IVP (1.7)–(1.8) satisfies*

$$\lim_{t\to\infty} x(t) = 0.$$

The following lemma is taken from [**23**], which will be used in the proof of Theorem 3.1.

**Lemma 3.2** *Let $0 < \alpha < \frac{1}{2}$. The following system of inequalities*

$$\begin{cases} \ln(1+x) \le y - \alpha y^2 \\ -\ln(1-y) \le x + \alpha x^2 \end{cases}$$

*has the unique solution: $(x,y) = (0,0)$ in the non–negative quadrant $\{(x,y) : x \ge 0, y \ge 0\}$.*

**Proof of Theorem 3.1** Since the proof is a straight–forward modification of the one we gave in [**23**] for a simpler situation, we will only highlight the main steps. The reader will be well–served by reading that proof first. By Lemma 2.1, we only need to consider the case when $x(t)$ is oscillatory. Set

$$u = \limsup_{t\to\infty} x(t), \quad v = -\liminf_{t\to\infty} x(t). \tag{3.3}$$

By Theorem 2.2, we see that $0 \le u < \infty, 0 \le v < 1$. It suffices to prove that $u = v = 0$. Set

$$A(t) = \sum_{i=1}^{n} \int_{t-r(t)}^{t} \alpha_i(t)\, d\mu_i(t,s).$$

For any $0 < \epsilon < 1 - v$, by (3.2) there exists $t_0 = t_0(\epsilon) > 0$ such that

$$\int_{t-r(t)}^{t} A(\tau)\, d\tau \le \frac{3}{2}, \quad \text{for} \quad t \ge t_0 - r(t_0) \tag{3.4}$$

and

$$-v_1 \equiv -v - \epsilon < x(t) < u + \epsilon \equiv u_1, \quad \text{for} \quad t \ge t_0 - r(t_0). \tag{3.5}$$

By $(H_1)$ and (3.1), we have

$$-\alpha_i(\tau)u_1 \le -f_i(\tau, x(s)) \le \alpha_i(\tau)v_1, \quad \text{for} \quad \tau \ge t_0 - r(t_0). \tag{3.6}$$

Substituting this into (1.7), we obtain

$$x'(t) \le [1+x(t)]\sum_{i=1}^{n}\int_{t-r(t)}^{t} \alpha_i(t)\, d\mu_i(t,s)v_1 = [1+x(t)]A(t)v_1, \qquad t \ge t_0 \quad (3.7)$$

and

$$x'(t) \ge -[1+x(t)]\sum_{i=1}^{n}\int_{t-r(t)}^{t} \alpha_i(t)\, d\mu_i(t,s)u_1 = -[1+x(t)]A(t)u_1, \qquad t \ge t_0. \quad (3.8)$$

Let $\{t_m^*\}$ be an increasing sequence such that $t_m^* \ge t_0, x'(t_m^*) = 0, \lim_{n\to\infty} x(t_m^*) = u$ and $t_m^* \to \infty$ as $m \to \infty$. By (1.7), we have

$$\sum_{i=1}^{n}\int_{t_m^*-r(t_m^*)}^{t_m^*} f_i(t_m^*, x(s))\, d\mu_i(t_m^*, s) = 0$$

which in turn implies that there exists $\xi_m \in (t_m^* - r(t_m^*), t_m^*)$ such that $x(\xi_m) = 0$ and $x(s) > 0$ for $s \in (\xi_m, t_m^*]$.

Now integrating (3.7) from $s$ to $\xi_m (\ge s)$, we have

$$x(s) \ge -1 + \exp\left(-v_1\int_s^{\xi_m} A(\tau)\, d\tau\right) \quad \text{for} \quad s \le \xi_m,$$

which, together with $(H_1)$ and (3.1), yields

$$-f_i(t, x(s)) \le \alpha_i(t)\left(1 - \exp\left(-v_1\int_s^{\xi_m} A(\tau)\, d\tau\right)\right) \quad \text{for} \quad s \le \xi_m.$$

Thus, for $\xi_m \le t \le t_m^*$, we have $t - r(t) < \xi_m$ and

$$\begin{aligned} x'(t) &\le -(1+x(t))\sum_{i=1}^{n}\int_{t-r(t)}^{\xi_m} f_i(t, x(s))\, d\mu_i(t,s) \\ &\le (1+x(t))\sum_{i=1}^{n}\int_{t-r(t)}^{\xi_m} \alpha_i(t)\left(1 - \exp\left(-v_1\int_s^{\xi_m} A(\tau)\, d\tau\right)\right) d\mu_i(t,s) \\ &\le (1+x(t))\left(1 - \exp\left(-v_1\int_{t-r(t)}^{\xi_m} A(\tau)\, d\tau\right)\right) A(t). \end{aligned}$$

Hence for $\xi_m \le t \le t_m^*$, we have

$$[\ln(1+x(t))]' \le \min\left\{A(t)v_1, \quad A(t)\left(1 - \exp\left(-v_1\int_{t-r(t)}^{\xi_m} A(\tau)\, d\tau\right)\right)\right\}. \quad (3.9)$$

We will prove that

$$\ln[1 + x(t_m^*)] \le v_1 - \frac{1}{6}v_1^2. \quad (3.10)$$

There are two possibilities:

**Case 1** $\int_{\xi_m}^{t_m^*} A(\tau)\, d\tau \le -\frac{\ln(1-v_1)}{v_1}$.

Then by (3.9),

$$\begin{aligned}
\ln[1+x(t_m^*)] &\le \int_{\xi_m}^{t_m^*} A(t)\left(1-\exp\left(1-v_1\int_{t-r(t)}^{\xi_m} A(\tau)\,d\tau\right)\right)dt \\
&\le \int_{\xi_m}^{t_m^*} A(t)\left(1-\exp\left(-\frac{3}{2}v_1+v_1\int_{\xi_m}^{t} A(\tau)\,d\tau\right)\right)dt \quad \text{by (3.4)} \\
&= \int_{\xi_m}^{t_m^*} A(t)\,dt \\
&\quad -\frac{1}{v_1}\exp\left(-v_1\left(\frac{3}{2}-\int_{\xi_m}^{t_m^*} A(\tau)\,d\tau\right)\right)\left(1-\exp\left(-v_1\int_{\xi_m}^{t_m^*} A(\tau)\,d\tau\right)\right).
\end{aligned}$$

Note that the function $\phi(x)=x-\frac{1}{v_1}e^{-v_1(\frac{3}{2}-x)}\left(1-e^{-v_1x}\right)$ is increasing for $0\le x\le \frac{3}{2}$. Thus if

$$\int_{\xi_m}^{t_m^*} A(\tau)\,d\tau \le -\frac{\ln(1-v_1)}{v_1} \le \frac{3}{2},$$

we have

$$\begin{aligned}
&\ln(1+x(t_m^*)) \\
&\le -\frac{\ln(1-v_1)}{v_1}-1+v_1\left(\frac{3}{2}+\frac{\ln(1-v_1)}{v_1}\right) \qquad (\text{since } e^{-x}>1-x \quad \text{for } x>0) \\
&\le v_1-\frac{1}{6}v_1^2
\end{aligned}$$

according to (2.21) on p.123 of [**13**]. On the other hand, if

$$\int_{\xi_m}^{t_m^*} A(\tau)\,d\tau \le \frac{3}{2} < -\frac{\ln(1-v_1)}{v_1},$$

then

$$\ln(1+x(t_m^*)) \le \frac{3}{2}-\frac{1}{v_1}\left(1-e^{-\frac{3}{2}v_1}\right) \le v_1-\frac{1}{6}v_1^2$$

according to (2.19) on p.123 of [**13**].

**Case 2** $\qquad -\frac{\ln(1-v_1)}{v_1} < \int_{\xi_m}^{t_m^*} A(\tau)\,d\tau \le \frac{3}{2}$.

Choose $\eta_m \in [\xi_m, t_m^*]$ such that

$$\int_{\eta_m}^{t_m^*} A(\tau)\,d\tau \equiv -\frac{\ln(1-v_1)}{v_1}.$$

Then

$$\begin{aligned}
&\ln(1+x(t_m^*)) \\
&\le \int_{\xi_m}^{\eta_m} A(t)v_1\,dt+\int_{\eta_m}^{t_m^*} A(t)\,dt-\int_{\eta_m}^{t_m^*} A(t)\exp\left(-\frac{3}{2}v_1+v_1\int_{\xi_m}^{t} A(\tau)\,d\tau\right)dt \\
&= v_1\int_{\xi_m}^{t_m^*} A(t)\,dt-\frac{(1-v_1)\ln(1-v_1)}{v_1}-\exp\left(-v_1\left(\frac{3}{2}-\int_{\xi_m}^{t_m^*} A(t)\,dt\right)\right) \\
&\le \frac{3}{2}v_1-\frac{(1-v_1)\ln(1-v_1)}{v_1}-1,
\end{aligned}$$

since $\phi(x) = v_1 x - e^{-v_1(\frac{3}{2}-x)}$ is increasing for $0 \le x \le \frac{3}{2}$. Thus, by (2.21) on p.123 of [**13**], we get

$$\ln(1 + x(t_m^*)) \le v_1 - \frac{1}{6}v_1^2.$$

Letting $m \to \infty$ and $\epsilon \to 0$ in (3.10), we have

$$\ln(1 + u) \le v - \frac{1}{6}v^2. \tag{3.11}$$

Next, let $\{s_m^*\}$ be an increasing sequence such that $s_m^* \ge t_0$, $\lim_{m\to\infty} s_m^* = \infty$, $x'(s_m^*) = 0$ and $\lim_{m\to\infty} x(s_m^*) = -v$. One can prove that

$$-\ln(1 + x(s_m^*)) \le u_1 + \frac{1}{6}u_1^2.$$

Letting $m \to \infty$ and $\epsilon \to 0$, we get

$$-\ln(1 - v) \le u + \frac{1}{6}u^2. \tag{3.12}$$

In view of Lemma 3.2, we see that (3.11) and (3.12) imply that $u = v = 0$. The proof is now complete. □

## 4 Some applications

In this section, we apply Theorem 3.1 to some concrete delay equations modelling single-species growth.

First we consider the well-known nonautonomous delay logistic equation

$$\frac{dN(t)}{dt} = R(t)N(t)\left[1 - \frac{N(t - r(t))}{k}\right], \quad t \ge 0, \tag{4.1}$$

where

$$k \in (0, \infty),\ R, r \in C([0, \infty), (0, \infty)) \quad \text{and} \quad \lim_{t\to\infty}(t - r(t)) = \infty. \tag{4.2}$$

Let $N(t) = k(1 + x(t))$. Then (4.1) reduces to

$$\frac{dx(t)}{dt} = -R(t)[1 + x(t)]x(t - r(t)). \tag{4.3}$$

Also, in this section, we make the same assumption on the initial function for (4.3) as that specified in Section 1, that is

$$x(\theta) = \phi(\theta) \ge -1 \quad \text{for} \quad \theta \in [-r, 0] \quad \text{and} \quad \phi(0) > -1, \tag{4.4}$$

where $\phi \in C([-r, 0], [-1, \infty))$ and $r = r(0)$. The global stability of the zero solution of (4.3) has been studied in [**1, 5, 12, 13, 26**]. In the following we list some of the known results:

**Theorem 4.1** ([**26**]) *Assume that $r(t)$ is bounded for $t \ge 0$ and that*

$$\int_{t-r(t)}^{t} R(s)\,ds \to 0 \quad as \quad t \to \infty \tag{4.5}$$

*and*

$$\int_0^\infty R(s)\,ds = \infty. \tag{4.6}$$

*Then the solution of IVP (4.3)–(4.4) tends to zero as $t \to \infty$.*

**Theorem 4.2** ([**5**]) *Assume that*

$$\sup_{t\geq 0} r(t) = \tau_0, \quad where \quad 0 < \tau_0 < \infty \tag{4.7}$$

*and*

$$R(t) \equiv r > 0 \quad and \quad r\tau_0 e^{2r\tau_0} < e^{-r\tau_0(e^{r\tau_0}-1)}. \tag{4.8}$$

*Then the solution of IVP (4.3)–(4.4) tends to zero as* $t \to \infty$.

In ([**12**], Theorem 4.2) was improved and the following result was shown.

**Theorem 4.3** ([**12, 13**]) *Assume that (4.6) holds and that there exists a constant* $M > 0$ *such that*

$$0 < R(t) < M \quad for \quad t \geq 0. \tag{4.9}$$

*If*

$$\int_{t-r(t)}^{t} R(s)\, ds \leq 1 \quad for\ large \quad t, \tag{4.10}$$

*then the solution of IVP (4.3)–(4.4) tends to zero as* $t \to \infty$.

Clearly, Theorem 4.3 also improves Theorem 4.1 under the additional assumption (4.9). In the recent work, [**1**], the boundedness condition (4.9) is removed.

If we apply Theorem 3.1 to Equation (4.3), we immediately obtain the following improvement of the above Theorems 4.1–4.3.

**Theorem 4.4** *Assume that (4.6) holds and that*

$$\int_{t-r(t)}^{t} R(s)\, ds \leq \frac{3}{2} \quad for\ large \quad t. \tag{4.11}$$

*Then the solution of IVP (4.3)-(4.4) tends to zero as* $t \to \infty$.

Next we consider the equation with several delays

$$x'(t) = -[1 + x(t)] \sum_{i=1}^{n} a_i(t) x(t - r_i(t)), \quad t \geq 0, \tag{4.12}$$

where $a_i(t)$ and $r_i(t)$ are nonnegative continuous functions, $r(t) \equiv \max\{r_i(t) : i = 1, 2, \cdots, n\} > 0$, and $\sum_{i=1}^{n} a_i(t) > 0$. This equation was studied in [**12, 13**] and it was shown that

**Theorem 4.5** ([**12, 13**]) *Assume that there exists a constant* $M > 0$ *such that*

$$0 < \sum_{i=1}^{n} a_i(t) < M, \quad for \quad t \geq 0. \tag{4.13}$$

*If*

$$\int_0^\infty \sum_{i=1}^{n} a_i(t)\, dt = \infty \tag{4.14}$$

*and*

$$\int_{t-r(t)}^{t} \sum_{i=1}^{n} a_i(\tau)\, d\tau \leq 1 \quad for\ large \quad t, \tag{4.15}$$

*then the solution of IVP (4.12)–(4.4) tends to zero as* $t \to \infty$.

When we apply Theorem 3.1 to Equation (4.12), we obtain the following improvement of Theorem 4.5.

**Theorem 4.6** *Assume that (4.14) holds and that*

$$\int_{t-r(t)}^{t} \sum_{i=1}^{n} a_i(\tau)\, d\tau \leq \frac{3}{2} \quad \text{for large} \quad t. \tag{4.16}$$

*Then the solution of IVP (4.12), (4.4) goes to zero as $t \to \infty$.*

When $a_i(t) = a_i > 0, r_i(t) = r_i \geq 0, r(t) = r > 0$, (4.12) becomes the autonomous logistic equation with several constant discrete delays

$$x'(t) = -[1 + x(t)] \sum_{i=1}^{n} a_i x(t - r_i). \tag{4.17}$$

This equation has been investigated in [**4, 5**]. In this case, Theorem 4.6 reduces to

**Corollary 4.7** *If*

$$r \sum_{i=1}^{n} a_i \leq \frac{3}{2}, \tag{4.18}$$

*then the solution of IVP (4.17), (4.4) tends to zero as $t \to \infty$.*

**Remark** Condition (4.18) greatly improves the following condition (for the special case $n = 2$)

$$(a_1 + a_2)(r_1 + r_2)e^{(a_1+a_2)(r_1+r_2)} < 1 \tag{4.19}$$

obtained in [**4, 5**]. At the same time, Corollary 4.7 also proves the conjecture proposed in [**4**] that (4.19) can be replaced by

$$(a_1 + a_2)(r_1 + r_2)e^{(a_1+a_2)(r_1+r_2)} < \frac{3}{2}. \tag{4.20}$$

Clearly, (4.18) is weaker than (4.20).

Next, we consider the so-called Michaelis–Menton single–species growth equation with several discrete delays

$$y'(t) = \gamma y(t) \left[ 1 - \sum_{i=1}^{n} \frac{a_i y(t - r_i)}{1 + c_i y(t - r_i)} \right], \tag{4.21}$$

where $\gamma > 0, a_i > 0, c_i > 0, r_i \geq 0, r > 0$, and $\sum_{i=1}^{n} a_i(1 + c_i)^{-1} = 1$. Let $x(t) = y(t) - 1$. Then (4.21) reduces to

$$x'(t) = -\gamma[1 + x(t)] \sum_{i=1}^{n} \frac{a_i x(t - r_i)}{(1 + c_i)[1 + c_i + c_i x(t - r_i)]}. \tag{4.22}$$

By applying Theorem 3.1 to (4.22), we immediately have the following improvement of Corollary 4.2 in [**12**].

**Theorem 4.8** *If*

$$\gamma r \leq \frac{3}{2}, \tag{4.23}$$

*then the solution of IVP (4.22), (4.4) tends to zero as $t \to \infty$.*

In [**6**], the authors studied the following so-called "food–limited" population model with time delay

$$\dot{N}(t) = rN(t)\frac{k - N(t-\tau)}{k + crN(t-\tau)}, \quad t \geq 0. \tag{4.24}$$

Letting $N(t) = k[1 + x(t)]$, (4.24) reduces to

$$\dot{x}(t) = -r[1 + x(t)]\frac{x(t-\tau)}{1 + cr(1 + x(t-\tau))} \tag{4.25}$$

where $k, r, \tau \in (0, \infty)$. It was shown in [**6**] that if

$$r\tau e^{r\tau} < 1, \tag{4.26}$$

then all solutions of (4.25), which are greater than $-1$, tends to zero as $t \to \infty$. Now if we apply Theorem 3.1 to (4.25), we can weaken (4.26) to

$$r\tau \leq \frac{3}{2}. \tag{4.27}$$

Finally, we consider an autonomous single-species growth equation with distributed delays

$$y'(t) = \gamma y(t)\left(1 - \alpha_1 \int_{-r}^{0} y(t+s)\, d\mu_1(s) - \alpha_2 \int_{-r}^{0} \frac{y(t+s)}{1 + cy(t+s)}\, d\mu_2(s)\right) \tag{4.28}$$

where $\int_{-r}^{0} d\mu_1(s) = \int_{-r}^{0} d\mu_2(s) = 1, \mu_1$ and $\mu_2$ are nondecreasing, $\gamma, r, \alpha_1, \alpha_2, c > 0$ are constants and $\alpha_1 + \frac{\alpha_2}{1+c} = 1$. By the change of variable $x(t) = y(t) - 1$, (4.28) can be reduced to

$$\begin{aligned} x'(t) = -\gamma[1 + x(t)]\Bigg(\alpha_1 \int_{-r}^{0} x(t+s)\, d\mu_1(s) \\ + \alpha_2 \int_{-r}^{0} \frac{x(t+s)}{(1+c)(1+c+cx(t+s))}\, d\mu_2(s)\Bigg). \end{aligned} \tag{4.29}$$

Again, an application of Theorem 3.1 leads to the following improvement of Corollary 4.3 in [**12**].

**Theorem 4.9** *If*

$$\gamma r \leq \frac{3}{2}, \tag{4.30}$$

*then the solution of IVP (4.29), (4.4) tends to zero as* $t \to \infty$.

## Acknowledgement

We would like to thank the anonymous referee whose comments allowed us to substantially trim down the paper.

## References


[1] Chen, M.-P. and Yu, J. S. [to appear] *Global attractivity in a nonautonomous delay logistic equation,* Dynamic Systems and Applications.

[2] Freedman, H. I. [1980] *Deterministic Mathematical Models in Population Ecology,*, Marcel Dekker, New York.

[3] Gilpin, M. E. and Ayala, F. J. [1973] *Global models of growth and competition,* Proceedings of the National Academy of Sciences, USA. **70**, 3590–3593.

[4] Gopalsamy, K. [1990] *Global stability in the delay-logistic equationwith discrete delays,* Houston Journal of Mathematics **16**, 347–356.

[5] Gopalsamy, K. [1992] *Stability and Oscillations in Delay Differential Equations of Population Dynamics,*, Kluwer Academic Publishers, Boston.

[6] Gopalsamy, K., Kulenovic, M. R. S. and Ladas, G. [1988] *Time lags in a "food-limited" population model,* Applicable Analysis **31**, 225–237.

[7] Haddock, J. R. and Kuang, Y. [1992] *Asymptotic theory for a class of nonautonomous delay differential equations,* J. Math. Anal. Appl. **168**, 147–162.

[8] Haddock, J. R. and Terjeki, J. [1983] *Liapunov–Razumikhin functions and invariance principle for functional differential equations,* J. Differential Equations, 95–122.

[9] Hale, J. K. [1977] *Theory of Functional Differential Equations*, Springer–Verlag, New York.

[10] Hutchinson, G. E. [1948] *Circular causal systems in ecology,* Annals of the New York Academy Sciences **50**, 221–246.

[11] Jones, G. S. [1962] *The existence of periodic solutions of* $f'(x) = -\alpha f(x-1)[1+f(x)]$, J. Math. Anal. Appl. **5**, 435–450.

[12] Kuang, Y. [1991] *Global stability for a class of nonlinear nonautonomous delay equations,* Nonlinear Analysis **17**, 627–634.

[13] Kuang, Y. [1993] *Delay Differential Equations with Applications in Population Dynamics,* Academic Press, Boston.

[14] Kuang, Y. and Smith, H. L. [1991] *Global stability in diffusive delay Lotka–Volterra systems,* Differential and Integral Equations **4**, 117–128.

[15] Kuang, Y. and Smith, H.L. [1993] *Global stability for infinite delay Lotka–Volterra type systems,* J. Differential Equations **103**, 221–246.

[16] Kuang, Y., Martin, R. H. and Smith, H. L. [1991] *Global stability for infinite delay, dispersive Lotka–Volterra systems: weakly interacting populations in nearly identical patches,* J. Dyn. Diff. Equs. **3**, 339–360.

[17] Lenhart, S. M. and Travis, C. C. [1986] *Global stability of a biological model with time delay,* Proc. Amer. Math. Soc. **96**, 75–78.

[18] May, R. M. [1973] *Time delay versus stability in population models with two and three trophic level,* Ecology **54**, 315–325.

[19] Miller, R. K. [1966] *On Volterra's population equation,* SIAM J. Appl. Math. **14**, 446–452.

[20] Nicholson, A. J. [1954] *An outline of the dynamics of animal populations,* Australian Journal of Zoology **2**, 651–663.

[21] Seifert, G. [1987] *On a delay–differential equation for single species population variations,* Nonlinear Analysis **11**, 1051–1059.

[22] Smith, F. E. [1963] *Population dynamics in Daphnia magna,* Ecology **44**, 651–663.

[23] So, J. W.-H. and Yu, J. S. [1995] *Global attractivity for a population model with time delay,* Proc. Amer. Math. Soc. **123**, 2687–2694.

[24] Sugie, J. [1992] *On the stability for a population growth equation with time delay,* Proc. Royal. Soc. Edinburgh **120A**, 179–184.

[25] Wright, E. M. [1955] *A nonlinear difference-differential equation,* J. Reine Angew. Math. **194**, 66–87.

[26] Zhang, B. G. and Gopalsamy, K. [1990] *Global attractivity in the delay logistic equation with variable parameters,* Math. Proc. Camb. Phil. Soc. **107**, 579–590.

Fields Institute Communications
Volume **21**, 1999

# Global Stability in Cyclic Epidemic Models with Disease Fatalities

**Horst R. Thieme**
Department of Mathematics
Arizona State University
Tempe, AZ 85787-1804, USA
thieme@math.la.asu.edu

**P. van den Driessche**
Department of Mathematics and Statistics
University of Victoria
Victoria, B.C., Canada V8W 3P4
pvdd@smart.math.uvic.ca

**Abstract.** A general disease transmission model of SIRS type is formulated and analyzed. There is constant recruitment into the susceptible class, mass action incidence, and stage age in the removed class. Disease fatalities occur in the infective class and the removed stages. Sufficient conditions are given for the endemic equilibrium to be globally asymptotically stable. For example, if the disease confers permanent immunity, then global stability of the endemic equilibrium (for parameter values such that it exists) is assured. Other examples are given that exhibit this same assured behavior.

## 1 Introduction

In most classical epidemic models, the total population is assumed to be constant. Anderson and May in [**1**] formulated one of the first o.d.e. epidemic models that includes population dynamics, thus the total population varies. They modeled the spread of *Pasteurella muris* in laboratory mice. The recruitment rate of mice was constant, and the rate at which mice acquired the infection was found to be proportional to the number of encounters between susceptible and infective mice (the law of mass action). This disease causes some fatalities, but other infective mice recover with temporary immunity. Thus the model is of SIRS type. Anderson and May in [**1**] gave a condition for the disease to remain endemic, and showed that in this case the disease decreases the population. This model was further analyzed by Mena-Lorca and Hethcote ([**6**], Section 2) and using a Lyapunov function they

1991 *Mathematics Subject Classification.* 92D30, 45J05, 45M10.

The research of HT was partially supported by NSF grant DMS-9403884.
The research of PD was partially supported by an NSERC research grant and the University of Victoria Committee on Faculty Research and Travel.

proved that the endemic equilibrium (when it exists) is globally asymptotically stable if there is no disease fatality or if immunity to the disease is permanent.

The SIRS model introduced by Anderson and May assumes that removed (recovered) individuals lose immunity at a rate proportional to their number, corresponding to an exponentially distributed time in the removed class. In other epidemic models a more general probability of remaining removed is assumed. For example, Hethcote et al. in [**4**] derived two SIRS models in a constant population that can exhibit Hopf bifurcation giving rise to periodic solutions for some parameter values. One model, ([**4**], Sections 2, 3) has a fixed period of temporary immunity, which corresponds to a step function probability. The other model ([**4**], Section 4) has a chain of $n \geq 3$ removed stages corresponding to a gamma distribution. For $n = 1, 2$, the endemic equilibrium is known to be globally asymptotically stable [**3, 9**].

In a complementary study on the constant population SIRS model, Stech and Williams in [**9**] used a result of Londen [**5**] to find conditions on the probability function that imply global stability of the endemic equilibrium. This result was used in ([**2**], Section 3.5) in a variable population SIRS model with a distributed length of the removed period and no disease fatality. This latter assumption means that the system has the same asymptotic behavior as that of an easier limit system.

Motivated by HIV/AIDS studies, Simon and Jacquez in [**7**] analyzed o.d.e. SI models with subpopulations and a chain of infective stages. Under the assumption that each subpopulation is isolated (restricted mixing), Simon and Jacquez ([**7**], Section 6) used a Lyapunov function to find sufficient conditions for the endemic equilibrium to be globally asymptotically stable. These conditions are more readily satisfied when the natural death rate is large compared with the rate of passing from one infective stage to the next.

Our aim is to use the method of [**9**] to prove global convergence to the endemic equilibrium (for parameter values such that it exists) in a variable population SIRS model with disease fatalities. We base our model on that of [**1**], but we include stage age for those individuals in the removed class. Our model system is formulated in Section 2, where the notation and assumptions are given. Further, a standard threshold condition is derived for stability of the disease-free equilibrium versus existence of an endemic equilibrium. In Section 3, the system is transformed into a scalar integro-differential equation and the result of [**9**] is used to give sufficient conditions for convergence to the endemic equilibrium. Section 4 continues with conditions for arbitrarily distributed time in the removed class. Each of Sections 5 and 6 considers a special case of the model system with several exponentially distributed removed stages. The model in Section 5 has one primary removed stage through which all individuals pass, and then an individual passes into one of several secondary removed stages that may be permanent or transient to the susceptible class. A limiting case of this model reduces to the SIRS model analyzed in [**1**] and ([**6**], Section 2). The model in Section 6 assumes that all individuals pass through a chain of two removed stages, then either return to the susceptible class or become permanently immune. Global asymptotic stability of the endemic equilibrium (whenever it exists) is proved for each of these o.d.e. models as well as for a chain of removed stages with permanent immunity (Section 4). Thus, there is a sharp threshold, with the disease dying out below the threshold, and tending to the endemic level above the threshold. This is in contrast to the model with a chain of at least three removed stages through which all individuals pass before losing immunity, which can have solutions that oscillate about the endemic level

([**4**], Section 4). The models that we consider in Sections 5 and 6 need no extra parameter conditions for global convergence to the endemic equilibrium, this is in contrast to the models analyzed in [**7**] with multiple infective stages. As kind of a trade-off, the method we present does not work if there is more than one infective stage, and it also does not work if an exposed class is included in the model (for diseases with a latent period). However, global stability of equilibria is preserved under small perturbations [**8**]; so global stability of the endemic equilibrium still holds when the latency period is sufficiently small, although this is difficult to quantify.

## 2 The model

A population of size $N$ is divided into susceptible, infective, and removed (or recovered) classes, with the number of individuals in each class given by $S$, $I$, $R$, respectively, and

$$N(t) = S(t) + I(t) + R(t).$$

Infection is modeled according to the law of mass action, thus,

$$I'(t) = \kappa S(t)I(t) - (\mu + \epsilon + \gamma)I(t).$$

Here $\kappa$ denotes the transmission coefficient, $\mu$ denotes the *per capita* natural death rate constant, $\epsilon$ denotes the extra *per capita* disease mortality rate constant for infectives, and $\gamma$ denotes the *per capita* removal (or recovery) rate constant of infectives.

Substituting the first into the second equation gives

$$I'(t) = \kappa\left(N(t) - I(t) - R(t)\right)I(t) - (\mu + \epsilon + \gamma)I(t). \tag{2.1}$$

In order to get a handle on $N$ we formulate the following differential equation

$$N'(t) = \Lambda - \mu N(t) - \epsilon I(t) - M_R(t). \tag{2.2}$$

Here $\Lambda$ is the constant recruitment rate into the population (actually into its susceptible class since vertical transmission is ignored). The remaining expressions on the right-hand side are various components of the population mortality rate; $\mu N$ is the mortality rate due to natural causes, $\epsilon I$ is the extra mortality rate for infectives due to the disease, and $M_R(t)$ denotes the extra mortality rate of removed (recovered) individuals due to the disease.

In the following we illustrate that it is possible to express $R$ and $M_R$ in terms of $I$. To this end we stratify the removed individuals according to their stage age,

$$R(t) = \int_0^\infty \rho(t,a)da, \tag{2.3}$$

with the stage age $a$ being the time since removal and $\rho(t,\cdot)$ the stage-age density at time $t$.

The stage age density $\rho$ for the removed individuals relates to the stage input (which here is $\gamma I$) and to the initial density $\rho_0$ in the usual way

$$\rho(t,a) = \begin{cases} \gamma I(t-a)\mathcal{F}(a), & t > a, \\ \rho_0(a-t)\frac{\mathcal{F}_(a)}{\mathcal{F}(a-t)}, & t < a, \end{cases} \tag{2.4}$$

where $\mathcal{F}(a)$ is the probability to be still in the removed class $a$ time units after removal from the infective class. This probability factorizes into three parts:

$$\mathcal{F}(a) = e^{-\mu a}\mathcal{F}_R(a)Q(a). \tag{2.5}$$

Here $e^{-\mu a}$ is the probability of not having died from natural causes in the time interval $[0, a]$, $\mathcal{F}_R$ is the survivorship of removed individuals with respect to the disease, i.e., $\mathcal{F}_R(a)$ is the probability of not having died from disease-related causes in the time from the moment of removal until $a$ time units later, and $Q$ describes the duration of the removed class, i.e., $Q(a)$ is the probability of not having returned to the susceptible class $a$ time units after removal from the infective class. If the removed period has a finite maximum length $L$, then $Q(a) = 0$ for $a > L$. If the disease confers permanent immunity, then $Q(a) \equiv 1$.

It follows from the interpretation that both $\mathcal{F}_R$ and $Q$ are non-negative, non-increasing functions on $[0, \infty)$ that have the value 1 at $a = 0$. For any function of this type, e.g., for $\mathcal{F}$, we make the convention that

$$\frac{\mathcal{F}(a)}{\mathcal{F}(a-t)} = 0, \qquad \text{whenever } a > t \text{ and } \mathcal{F}(a) = 0.$$

Another natural assumption states that the number of removed individuals is finite in the beginning and their density is non-negative, so $\rho_0 \in L^1[0, \infty)$, i.e.,

$$\int_0^\infty |\rho_0(a)| da < \infty.$$

In order to close our model, we have for the extra death rate of removed individuals, $M_R$, that

$$M_R(t) = -\int_0^\infty \frac{\rho(t,a)}{\mathcal{F}_R(a)} \mathcal{F}_R(da) \tag{2.6}$$

with the last integral being a Stieltjes integral.

The fastest way to understand the Stieltjes integral is to assume for a moment that $\mathcal{F}_R$ is differentiable. Then

$$\delta(a) = -\frac{\mathcal{F}_R'(a)}{\mathcal{F}_R(a)}$$

is the per capita extra mortality rate at which removed individuals die from the disease, and the integral

$$\int_0^\infty \rho(t,a)\delta(a) da$$

gives the associated population rate.

We substitute (2.4) and (2.5) into (2.3) and obtain that

$$R(t) = \gamma \int_0^t I(t-a) e^{-\mu a} g(a) da + e^{-\mu t} u(t)$$

with

$$g(a) = \mathcal{F}_R(a) Q(a), \quad \text{thus} \quad g(0) = 1, \tag{2.7}$$

and

$$u(t) = \int_t^\infty \rho_0(a-t) \frac{\mathcal{F}_R(a) Q(a)}{\mathcal{F}_R(a-t) Q(a-t)} da.$$

We have

$$u(t) \le \int_0^\infty \rho_0(a) da = u_0 < \infty.$$

We also substitute (2.4) and (2.5) into (2.6) obtaining

$$M_R(t) = \gamma \int_0^t I(t-a)e^{-\mu a} f(da) + e^{-\mu t}V(t),$$

with

$$f(a) = -\int_0^a Q(r)\mathcal{F}_R(dr), \quad a > 0, \qquad f(0) = 0, \tag{2.8}$$

and

$$V(t) = -\int_t^\infty \rho_0(a-t)\frac{Q(a)}{Q(a-t)\mathcal{F}_R(a-t)}\mathcal{F}_R(da).$$

We claim that $V \in L^1[0,\infty)$. Indeed,

$$\begin{aligned}
\int_0^t V(s)ds &\leq -\int_0^t \int_s^\infty \rho_0(a-s)\frac{1}{\mathcal{F}_R(a-s)}\mathcal{F}_R(da)ds \\
&\leq -\int_0^t \int_0^a \rho_0(a-s)\frac{1}{\mathcal{F}_R(a-s)}ds\mathcal{F}_R(da) \\
&= -\int_0^t \int_0^a \rho_0(s)\frac{1}{\mathcal{F}_R(s)}ds\mathcal{F}_R(da) \\
&\leq \int_0^t \rho_0(a)da.
\end{aligned}$$

In the last step we have used integration by parts for Stieltjes integrals. Remember that we have assumed that $\rho_0 \in L^1[0,\infty)$, hence $V \in L^1[0,\infty)$.

Summarizing, we have obtained the following model:

$$\begin{aligned}
I'(t) &= \kappa\,(N(t) - I(t) - R(t))\,I(t) - (\mu+\epsilon+\gamma)I(t), \qquad (2.9)\\
N'(t) &= \Lambda - \mu N(t) - \epsilon I(t) - M_R(t), \\
R(t) &= \gamma\int_0^t I(t-a)e^{-\mu a}g(a)da + e^{-\mu t}u(t), \\
M_R(t) &= \gamma\int_0^t I(t-a)e^{-\mu a}f(da) + e^{-\mu t}V(t).
\end{aligned}$$

In view of our derivation (recall (2.7) and (2.8)) we make the following assumptions.

**Assumptions 2.1.** All parameters are non-negative, $\mu > 0$.

- $g : [0,\infty) \to [0,1]$ is non-increasing.
- $f : [0,\infty) \to [0,1]$ is non-decreasing, and $f(0) = 0$.
- $u : [0,\infty) \to [0,\infty)$ is continuous and bounded.
- $V : [0,\infty) \to [0,\infty)$, $V \in L^1[0,\infty)$.

With the above assumptions, the model given by (2.9) is well posed, and solutions with non-negative initial conditions remain non-negative, while $S = N - I - R$ remains non-negative as well if it is non-negative initially. The system always has the disease free equilibrium, namely $N = S = \Lambda/\mu$, $I = R = 0$. The basic reproduction number, defined as

$$\mathcal{R}_0 = \frac{\Lambda\,\kappa}{\mu(\mu+\epsilon+\gamma)}, \tag{2.10}$$

is the product of the population size at the disease-free equilibrium, the transmission coefficient, and the mean infective period. The number $\mathcal{R}_0$ acts as a threshold parameter in the familiar way.

When $\mathcal{R}_0 \leq 1$, the disease free equilibrium is the only non-negative equilibrium. Applying Proposition 2.2 from [**10**] to the first equation in (2.9) shows that the disease-free equilibrium is globally asymptotically stable when $\mathcal{R}_0 \leq 1$. When $\mathcal{R}_0 > 1$, system (2.9) has a unique endemic equilibrium with $I_e > 0$ and $S_e = N_e - I_e - R_e = \Lambda/(\mu\mathcal{R}_0)$ from the first equation. Solving the second equation for $N_e$ and equating to $S_e + I_e + R_e$, gives the number of infectives at the endemic equilibrium as

$$I_e \left[\mu + \epsilon + \gamma\mu \int_0^\infty e^{-\mu a}\, g(a)\, da + \gamma \int_0^\infty e^{-\mu a}\, f(da)\right] = \Lambda(\mathcal{R}_0 - 1)/\mathcal{R}_0. \quad (2.11)$$

We focus on the endemic equilibrium, and our aim is to give sufficient conditions so that all solutions of (2.9) with $\mathcal{R}_0 > 1$ and having a positive number of initial infectives are attracted to this endemic equilibrium.

## 3 Reduction to a scalar integro-differential equation

Our strategy consists in transforming the system (2.9) into a scalar integro-differential equation for $I$ and in proceeding as in [**9**]. In order to find $N$ we integrate the second equation in (2.9) and obtain

$$N(t) = N_0 e^{-\mu t} + \frac{\Lambda}{\mu}\left(1 - e^{-\mu t}\right) - \epsilon \int_0^t I(t-s)e^{-\mu s} ds - \int_0^t M_R(s) e^{-\mu(t-s)} ds.$$

From the last equation in (2.9),

$$\begin{aligned}\int_0^t M_R(s)e^{-\mu(t-s)}ds &= \gamma \int_0^t \int_0^s I(s-a)e^{-\mu a} f(da) e^{-\mu(t-s)} ds \\ &\quad + e^{-\mu t}\int_0^t V(s)ds.\end{aligned}$$

Thus by changing the order of integration

$$\begin{aligned}\int_0^t M_R(s)e^{-\mu(t-s)}ds - e^{-\mu t}\int_0^t V(s)ds &= \gamma \int_0^t \int_a^t I(s-a)e^{-\mu(t+a-s)} ds f(da) \\ &= \gamma \int_0^t \int_0^{t-a} I(s)e^{-\mu(t-s)} ds f(da) \\ &= \gamma \int_0^t I(t-a)e^{-\mu a} f(a) da.\end{aligned}$$

The last equality follows from integration by parts, recall $f(0) = 0$. Summarizing, we have

$$N(t) = \frac{\Lambda}{\mu} - \int_0^t I(t-s)e^{-\mu s}\left(\epsilon + \gamma f(s)\right) ds - e^{-\mu t} v(t)$$

with a bounded continuous function $v$.

We substitute this equation and the third equation of (2.9) into the first equation of (2.9):

$$I'(t) = -(\mu + \epsilon + \gamma)I(t) + \kappa I(t)\left(\frac{\Lambda}{\mu} - I(t) - e^{-\mu t} w(t) - \int_0^t I(t-s)P(s)ds\right)$$

with a bounded continuous function $w = v + u$ and

$$P(s) = e^{-\mu s}(\epsilon + \gamma f(s) + \gamma g(s)). \quad (3.1)$$

To bring this equation to the same form as Equation (1.4) in [**9**], we introduce

$$I(t) = \frac{\Lambda}{\mu} x(t)$$

and obtain

$$x'(t) = -(\mu + \epsilon + \gamma)x(t) + \beta x(t)\left(1 - x(t) - e^{-\mu t}z(t) - \int_0^t x(t-s)P(s)ds\right) \tag{3.2}$$

with

$$\beta = \kappa \frac{\Lambda}{\mu}$$

and a bounded function $z(t)$.

We call the endemic equilibrium *globally asymptotically stable* if it is locally asymptotically stable and attracts all solutions of (3.2) with $x(0) > 0$. Here local asymptotic stability is formulated as in ([**9**], Lemma 4.1); see Sections 4, 5. Before we formulate our first global stability result, recall the definition of the Laplace transform of $P$, namely

$$\hat{P}(\lambda) = \int_0^\infty e^{-\lambda s} P(s) ds.$$

**Proposition 3.1** *Let* $-\hat{P}'(0) < \infty$ *and*

$$1 + \Re \hat{P}(\imath\nu) > 0 \quad \forall \nu > 0.$$

*Then the endemic equilibrium of (3.2), whenever it exists, is globally asymptotically stable.*

**Proof** It follows from ([**9**], Theorem 2.1) that the endemic equilibrium attracts all solutions of (3.2) with $x(0) > 0$. Local asymptotic stability follows from ([**9**], Lemma 4.1) and a similar continuation argument as in the proof of ([**2**], Theorem 5.2). □

Notice that this condition does not involve the parameter $\beta$ (the parameters $\gamma$ and $\epsilon$ are hidden in $P$), although $\beta > \mu + \epsilon + \gamma$ is needed for the endemic equilibrium to exist, see (2.10). This indicates that this condition may not be optimal. ([**9**], Theorem 3.1) shows that it is sharp in the following sense. If $-\hat{P}'(0) < \infty$, and

$$1 + \Re \hat{P}(\imath\nu) < 0 \quad \text{for some} \quad \nu > 0,$$

then there exists some $\beta > 0$ such that the endemic equilibrium exists and is unstable.

By (3.1),

$$\hat{P}(\lambda) = \frac{\epsilon}{\lambda + \mu} + \gamma \int_0^\infty e^{-(\lambda+\mu)a}(f(a) + g(a))da.$$

Hence

$$-\hat{P}'(0) = \frac{\epsilon}{\mu^2} + \gamma \int_0^\infty e^{-\mu a} a(f(a) + g(a))da.$$

Since $f$ and $g$ take values between 0 and 1, the condition $-\hat{P}'(0) < \infty$ is satisfied, and so Proposition 3.1 translates into the following result.

**Theorem 3.1** *Let the Assumptions 2.1 be satisfied and*

$$1 + \frac{\epsilon\mu}{\mu^2+\nu^2} + \gamma\Re\widehat{(f+g)}(\mu+\imath\nu) > 0 \quad \forall \nu > 0.$$

*Then the endemic equilibrium of (2.9), whenever it exists, is globally asymptotically stable.*

The following corollaries of the above result are useful for applications.

**Corollary 3.1** *Let the Assumptions 2.1 be satisfied and*

$$\int_0^\infty e^{-\mu a}(f(a)+g(a))da \le \frac{1}{\gamma}.$$

*Then the endemic equilibrium of (2.9), whenever it exists, attracts all solutions with* $I(0) > 0$.

**Corollary 3.2** *Let the Assumptions 2.1 be satisfied and the function* $f+g$ *be non-increasing and convex on* $[0,\infty)$. *Then the endemic equilibrium of (2.9), whenever it exists, attracts all solutions with* $I(0) > 0$.

**Proof**

$$\Re\widehat{(f+g)}(\mu+\imath\nu) = \int_0^\infty (f+g)(s)e^{-\mu s}\cos(\nu s)ds.$$

Since the product of two non-negative, non-increasing, convex functions has the same properties, then

$$h(s) = (f+g)(s)e^{-\mu s}$$

is non-increasing and convex. Without restricting the generality, it can be assumed that $h$ is differentiable. Otherwise it is possible to find a sequence of differentiable non-increasing convex functions $h_j$ such that

$$\int_0^\infty |h_j(s) - h(s)|ds \to 0, \quad j \to \infty.$$

Integrating by parts,

$$\Re\widehat{(f+g)}(\mu+\imath\nu) = -\frac{1}{\nu}\int_0^\infty \sin(\nu s)h'(s)ds.$$

Since $-h'$ is non-negative and non-increasing, the last integral is non-negative as can see be seen by splitting up the integral into appropriate pieces and comparing them. □

## 4 Some general results

In this section we look at the general case that the sojourn in the removed class is arbitrarily distributed. At the endemic equilibrium, the stage age distribution of the removed class is given by (2.4),

$$\rho_e(a) = \gamma I_e \mathcal{F}(a).$$

The endemic equilibrium is called *locally stable* if for every $\eta > 0$ there exists a $\tilde\eta > 0$ such that, for any solution of (2.1)–(2.6)

$$|N(t) - N_e| + |I(t) - I_e| + \int_0^\infty |\rho(t,a) - \rho_e(a)|da < \eta \quad \forall t \ge 0$$

whenever

$$|N(0) - N_e| + |I(0) - I_e| + \int_0^\infty |\rho(0,a) - \rho_e(a)| da < \tilde{\eta}.$$

The endemic equilibrium is globally asymptotically stable if it is locally stable and attracts all solutions with $I(0) > 0$. This notion of global asymptotic stability is equivalent to the one used in the previous section.

From (2.7) and (2.8), since $Q$ and $\mathcal{F}_R$ are monotone non-increasing,

$$\begin{aligned} f(a) + g(a) &= \mathcal{F}_R(a)Q(a) - \int_0^a Q(r)\mathcal{F}_R(dr) \\ &\geq \mathcal{F}_R(a)Q(a) - \int_0^a Q(a)\mathcal{F}_R(dr) \\ &= Q(a). \end{aligned}$$

From Corollary 3.1 we obtain the following result.

**Theorem 4.1** *Assume that*

$$\int_0^\infty e^{-\mu a} Q(a) da \leq \frac{1}{\gamma}.$$

*Then the endemic equilibrium, whenever it exists, is globally asymptotically stable.*

Recall that

$$D_R = \int_0^\infty Q(a) da$$

is the average duration of the removed class, while $1/\gamma$ is the average duration of the infective class. Hence the endemic equilibrium attracts all nontrivial solutions if the average duration of the removed class does not exceed the average duration of the infective class.

In order to apply Corollary 3.4, we integrate by parts and obtain

$$f(a) + g(a) = 1 + \int_0^a \mathcal{F}_R(r) Q(dr),$$

which shows that $f + g$ is non-increasing.

**Theorem 4.2** *Let*

$$\int_0^a \mathcal{F}_R(r) Q(dr)$$

*be a convex function of $a$. Then the endemic equilibrium of (2.9), whenever it exists, is globally asymptotically stable.*

If $Q$ is differentiable, then the condition in Theorem 4.2 is satisfied if $Q'$ has a representative such that $\mathcal{F}_R(a)Q'(a)$ is a non-decreasing function of $a$. Since $\mathcal{F}_R$ is non-increasing and $Q'$ is non-positive, it is sufficient that $Q'$ is non-decreasing. If $Q$ is convex, but not necessarily differentiable, a sequence $Q_j$ of convex differentiable functions can be chosen such that

$$\int_0^a \mathcal{F}_R(r) Q(dr) = \lim_{j\to\infty} \int_0^a \mathcal{F}_R(r) Q_j(dr)$$

pointwise in $a$. By the preceding considerations the assumptions of Theorem 4.2 are satisfied if $Q$ is convex.

**Corollary 4.1** *Let $Q$ be convex. Then the endemic equilibrium of (2.9), whenever it exists, is globally asymptotically stable.*

The conditions of Corollary 4.1 are satisfied, for example, if the length of the removed period is exponentially distributed. In particular, the endemic equilibrium attracts all nontrivial disease dynamics if the disease induces permanent immunity ($Q \equiv 1$) even if disease-related deaths are arbitrarily distributed over the removed class. This includes a model with a chain of exponentially distributed removed stages with no return to the susceptible class.

## 5 One primary and several secondary exponentially distributed removed stages

We now consider a cyclic model of

$$\begin{array}{ccccc} & & R_1 & & \\ & \nearrow & & \searrow & \\ \to S \to I \to R_0 & \to & \vdots & \to & S \\ & \searrow & & \nearrow & \\ & & R_n & & \end{array}$$

type where a removed individual first enters a primary removed stage where it is still prone to disease-related death and later one of several secondary removed stages where it is safe from disease-related death. The various secondary stages may have different durations; in particular it is possible that there is a stage where permanent immunity is acquired, but individuals in other secondary stages lose immunity and return to the susceptible class. The primary removed stage could be a quarantine period where the individual is still infective, but prevented from transmitting the disease. The total population size $N$ splits into

$$N = S + I + R, \quad \text{with} \quad R = R_0 + R_1 + \cdots + R_n, \tag{5.1}$$

where $R_0, \ldots, R_n$ are the number of individuals in the various stages of the removed class. In terms of $S, I, R_0, \ldots, R_n$ the model takes the following form:

$$\begin{aligned} S' &= \Lambda - \mu S - \kappa SI + \sigma_1 R_1 + \cdots + \sigma_n R_n, \\ I' &= \kappa SI - (\mu + \epsilon + \gamma) I, \\ R_0' &= \gamma I - (\mu + \delta + \rho) R_0, \qquad \rho = \rho_1 + \cdots + \rho_n. \\ R_j' &= \rho_j R_0 - (\mu + \sigma_j) R_j, \qquad j = 1, \ldots, n. \end{aligned} \tag{5.2}$$

Here $1/\rho$ and $1/\sigma_j$ are the mean durations of the primary removed stage and of the secondary removed stages; the $j$th secondary removed stage is associated with permanent immunity if $\sigma_j = 0$. The extra *per capita* mortality rate constant in the primary removed stage due to the disease is denoted by $\delta$.

For this o.d.e. system, local stability of the endemic equilibrium is defined in the usual sense, and its global asymptotic stability means that it is locally stable and attracts all solutions with $I(0) > 0$. It is possible, but cumbersome to rewrite this model in the form of Section 2, in particular to determine $\mathcal{F}_{\mathcal{R}}$. This is why we determine the functions $f$ and $g$ instead.

Adding the differential equations in (5.2) yields

$$N' = \Lambda - \mu N - \epsilon I - \delta R_0. \tag{5.3}$$

Set

$$g_0(a) = e^{-(\delta+\rho)a}.$$

Then

$$R_0(t) = R_0(0)e^{-\mu t}g_0(t) + \gamma \int_0^t I(t-s)e^{-\mu s}g_0(s)ds.$$

We identify from (5.3) and the last equation in (2.9) that

$$f(a) = \frac{\delta}{\delta+\rho}(1-g_0(a)). \tag{5.4}$$

Set

$$g_j(t) = \rho_j e^{-\sigma_j t}, \quad j = 1, \dots, n. \tag{5.5}$$

For convenience we recall the convolution notation

$$(g_0 * g_j)(t) = \int_0^t g_0(s)g_j(t-s)ds.$$

Solving the differential equations for $R_j$, $j = 1, \dots, n$, we obtain

$$R_j(t) = \frac{R_j(0)}{\rho_j}e^{-\mu t}g_j(t) + R_0(0)e^{-\mu t}(g_0 * g_j)(t) + \gamma \int_0^t I(t-s)e^{-\mu s}(g_0 * g_j)(s)ds.$$

Since $R = R_0 + \cdots + R_n$ we identify from the third equation in (2.9) that

$$g(a) = (g_0 + g_0 * (g_1 + \cdots + g_n))(a).$$

By (5.4),

$$\begin{aligned} f(a) + g(a) &= \frac{\delta}{\delta+\rho}(1-g_0(a)) + g_0(a) + \left(g_0 * \left(\sum_{j=1}^n g_j\right)\right)(a) \\ &= \frac{\delta}{\delta+\rho} + \frac{\rho}{\delta+\rho}g_0(a) + \left(g_0 * \left(\sum_{j=1}^n g_j\right)\right)(a). \end{aligned}$$

By (5.5) and the convolution theorem for Laplace transforms,

$$\begin{aligned} \widehat{(f+g)}(\mu+\imath\nu) &= \frac{\delta}{\delta+\rho}\,\frac{1}{\mu+\imath\nu} + \frac{\rho}{\delta+\rho}\,\frac{1}{\mu+\imath\nu+\delta+\rho} \\ &\quad + \frac{1}{\mu+\imath\nu+\delta+\rho}\left(\sum_{j=1}^n \frac{\rho_j}{\mu+\imath\nu+\sigma_j}\right). \end{aligned}$$

Multiplying by conjugates,

$$\begin{aligned} \widehat{(f+g)}(\mu+\imath\nu) &= \frac{\delta}{\delta+\rho}\,\frac{\mu-\imath\nu}{\mu^2+\nu^2} + \frac{\rho}{\delta+\rho}\,\frac{\mu+\delta+\rho-\imath\nu}{(\mu+\delta+\rho)^2+\nu^2} \\ &\quad + \frac{\mu+\delta+\rho-\imath\nu}{(\mu+\delta+\rho)^2+\nu^2}\left(\sum_{j=1}^n \rho_j \frac{\mu+\sigma_j-\imath\nu}{(\mu+\sigma_j)^2+\nu^2}\right). \end{aligned}$$

Taking real parts,

$$\begin{aligned}
&\Re\widehat{(f+g)}(\mu+\imath\nu)\\
&\quad=\frac{\delta}{\delta+\rho}\frac{\mu}{\mu^2+\nu^2}+\frac{\rho}{\delta+\rho}\frac{\mu+\delta+\rho}{(\mu+\delta+\rho)^2+\nu^2}\\
&\quad+\frac{\mu+\delta+\rho}{(\mu+\delta+\rho)^2+\nu^2}\left(\sum_{j=1}^{n}\rho_j\frac{\mu+\sigma_j}{(\mu+\sigma_j)^2+\nu^2}\right)\\
&\quad-\frac{\nu}{(\mu+\delta+\rho)^2+\nu^2}\left(\sum_{j=1}^{n}\rho_j\frac{\nu}{(\mu+\sigma_j)^2+\nu^2}\right)\\
&\quad\geq\frac{1}{(\mu+\delta+\rho)^2+\nu^2}\left(\rho\frac{\mu+\delta+\rho}{\delta+\rho}-\left(\sum_{j=1}^{n}\rho_j\frac{\nu^2}{(\mu+\sigma_j)^2+\nu^2}\right)\right).
\end{aligned}$$

Recalling that $\rho$ is the sum of the $\rho_j$, $\Re\widehat{(f+g)}(\mu+\imath\nu)\geq 0$, and Theorem 3.1 gives the following result.

**Theorem 5.1** *The endemic equilibrium of model (5.2), whenever it exists, is globally asymptotically stable.*

The SIRS model formulated in [**1**] and analyzed in [**6**] as described in the Introduction is easily seen as a limiting case of the model of this section (with the mean durations of all secondary removed stages being 0). There is a unique removed stage with no disease mortality, thus stages $R_1,\cdots,R_n$ are absent, $\sum_{j=1}^{n}\sigma_j R_j$ is replaced by $\rho R_0$ where $\rho$ is the loss of immunity rate constant, and $\delta=0$. The endemic equilibrium is then given by

$$(S_e, I_e, R_{0e}) = (\Lambda/(\mu\mathcal{R}_0),\ I_e,\ \gamma I_e/(\mu+\rho))$$

where $\mathcal{R}_0$ is defined in (2.10) and

$$I_e=\frac{\Lambda-\mu\,S_e}{\epsilon+\mu\,(1+\gamma/(\mu+\rho))}$$

from (2.11). Theorem 5.1 shows that if $\mathcal{R}_0>1$, then this endemic equilibrium is globally asymptotically stable. This completes the global stability analysis of the endemic equilibrium in ([**6**], Section 2).

## 6 Three subsequent removed stages with the last being permanent

In this section we consider a model of

$$\begin{array}{ccc}
 & & R_2\\
 & & \nearrow\\
\to S\to I\to R_0\to R_1 & & \\
 & & \searrow\\
 & & S
\end{array}$$

type where some removed individuals return into the susceptible class, but others become permanently immune. The total population size $N$ splits into

$$N = S + I + R, \quad \text{with} \quad R = R_0 + R_1 + R_2, \tag{6.1}$$

where $R_0, R_1, R_2$ are the number of individuals in the various stages of the removed class. In these terms, the model takes the following form:

$$\begin{aligned} S' &= \Lambda - \mu S - \kappa S I + \sigma_1 R_1, \\ I' &= \kappa S I - (\mu + \epsilon + \gamma) I, \\ R_0' &= \gamma I - (\mu + \delta_0 + \rho_0) R_0, \\ R_1' &= \rho_0 R_0 - (\mu + \delta_1 + \rho_1 + \sigma_1) R_1, \\ R_2' &= \rho_1 R_1 - (\mu + \delta_2) R_2. \end{aligned} \tag{6.2}$$

The parameters $\delta_j \geq 0$ are the extra *per capita* mortality rate constants in the $j$th removed stage, $1/\rho_j$ are the mean durations in the $j$th removed stage, and $\sigma_1$ is the return rate constant into the susceptible class. We define

$$g_0(s) = e^{-(\delta_0 + \rho_0)s}, \quad g_1(s) = \rho_0 e^{-(\delta_1 + \rho_1 + \sigma_1)s}, \quad g_2(s) = \rho_1 e^{-\delta_2 s}.$$

Integration of the equations in (6.2) and comparison with (2.9) yields

$$g = g_0 + g_0 * g_1 + g_0 * g_1 * g_2$$

and

$$f'(s) = \delta_0 g_0(s) + \delta_1 (g_0 * g_1)(s) + \delta_2 (g_0 * g_1 * g_2)(s).$$

Now

$$g'(s) = -\delta_0 g_0(s) - (\delta_1 + \sigma_1)(g_0 * g_1)(s) - \delta_2 (g_0 * g_1 * g_2)(s).$$

Hence

$$f'(s) + g'(s) = -\sigma_1 (g_0 * g_1)(s).$$

This implies that

$$f(s) + g(s) = 1 - \sigma_1 \int_0^s (g_0 * g_1)(s) ds.$$

By the convolution theorem for the Laplace transform,

$$\widehat{(f + g)}(\mu + \imath\nu) = \frac{1}{\mu + \imath\nu}(1 - \alpha)$$

with

$$\alpha = \frac{\sigma_1}{\delta_0 + \rho_0 + \mu + \imath\nu} \frac{\rho_0}{\delta_1 + \rho_1 + \sigma_1 + \mu + \imath\nu}.$$

Thus $\Re\widehat{(f + g)}(\mu + \imath\nu) \geq 0$ if

$$1 - \Re\alpha \geq 0, \quad \Im\alpha \leq 0.$$

Now

$$\alpha = \sigma_1 \rho_0 \frac{\delta_0 + \rho_0 + \mu - \imath\nu}{(\delta_0 + \rho_0 + \mu)^2 + \nu^2} \frac{\delta_1 + \rho_1 + \sigma_1 + \mu - \imath\nu}{(\delta_1 + \rho_1 + \sigma_1 + \mu)^2 + \nu^2}.$$

Obviously $\Im\alpha \leq 0$. Further

$$\begin{aligned}\Re\alpha &= \sigma_1\rho_0\frac{(\delta_0+\rho_0+\mu)(\delta_1+\rho_1+\sigma_1+\mu)-\nu^2}{[(\delta_0+\rho_0+\mu)^2+\nu^2][(\delta_1+\rho_1+\sigma_1+\mu)^2+\nu^2]} \\ &\leq \rho_0\frac{(\delta_0+\rho_0+\mu)(\delta_1+\rho_1+\sigma_1+\mu)}{(\delta_0+\rho_0+\mu)^2(\delta_1+\rho_1+\sigma_1+\mu)} \\ &\leq 1.\end{aligned}$$

From Theorem 3.1 we obtain the following result.

**Theorem 6.1** *Whenever the endemic equilibrium of (6.2) exists, it is globally asymptotically stable.*

## References

[1] Anderson, R. M. and May, R. M. [1979], *Population Biology of Infectious Diseases I*, Nature, **180**, 361–367.

[2] Castillo-Chavez, C. and Thieme, H. R. [1995], *Asymptotically autonomous epidemic models*, in Mathematical Population Dynamics: Analysis of Heterogeneity I. Theory of Epidemics (O. Arino et al. eds.), Wuerz, 33–50.

[3] Hethcote, H. W. [1976], *Qualitative Analysis for Communicable Disease Models*, Math. Biosci. **28**, 335–356.

[4] Hethcote, H. W., Stech, H. W. and van den Driessche, P. [1981], *Nonlinear Oscillations in Epidemic Models*, SIAM J. Appl. Math. **40**, 1–9.

[5] Londen, S.-O. [1975], *On the Variation of the Solution of a Nonlinear Integral Equation*, J. Math. Anal. Appl. **52**, 430–449.

[6] Mena-Lorca, J. and Hethcote, H. W. [1992], *Dynamic Models of Infectious Diseases as Regulators of Population Sizes*, J. Math. Biol. **30**, 693–716.

[7] Simon, C. P. and Jacquez, J. A. [1992], *Reproduction Numbers and the Stability of Equilibria of SI Models for Heterogeneous Populations*, SIAM J. Appl. Math. **52**, 541–576.

[8] Smith, H. L. and Waltman, P., *Perturbation of a Globally Stable Steady State*, Proc. Amer. Math. Soc., to appear.

[9] Stech, H. and Williams, M. [1981], *Stability in a Class of Cyclic Epidemic Models with Delay*, J. Math. Biol. **11**, 95–103.

[10] Thieme, H. R. [1993], *Persistence Under Relaxed Point-Dissipativity (With Application to an Endemic Model)*, SIAM J. Math. Anal. **24**, 407-435.

Fields Institute Communications
Volume **21**, 1999

# Feedback Loops Analysis for Chaotic Dynamics with an Application to Lorenz System

**B. Toni**
Departamento de Matematicas
Facultad de Ciencias
Universidad Autonoma del Estado de Morelos
Cuernavaca 62210, Morelos, Mexico
toni@servm.fc.uaem.mx

**D. Thieffry**
Max Planck Institute for History of Science
Wilhelmstrasse 44
10117 Berlin, Germany
denis@mpiwg-berlin.mpg.de

**R. Bulajich**
Departamento de Matematicas
Facultad de Ciencias
Universidad Autonoma del Estado de Morelos
Cuernavaca 62210, Morelos, Mexico
bulajich@servm.fc.uaem.mx

**Abstract.** The method of feedback loops as developed by Réné Thomas et al. is used to analyze chaotic dynamics. We confirm some of the requirements for a system to be chaotic, namely at least one 2−element negative loop and one positive loop in the system's loop structure. Application to Lorenz system and comparison with Rössler's system and its variants, lead to the following conjecture: *A union of negative loops of at least the dimension of the system is necessary for a chaotic behavior.*

## 1 Introduction

Various authors have addressed the role of *feedback loops* (or *feedback circuits*) in dynamical systems [**7, 4, 16, 9, 10**]. Here we refer mainly to the methodology and terminologies introduced by Thomas et al. ([**12, 14, 15**] and references therein). Given a differential system, a feedback loop is described by a combination of terms of the Jacobian matrix under a circular permutation of the indices. The loop is positive (resp. negative) for an even (resp. odd) number of negative terms whenever these terms have fixed signs.

1991 *Mathematics Subject Classification.* 34C25, 58F13, 93D15.

Previous work has shown that at least one positive loop is necessary to have multistationarity, whereas at least one negative loop is necessary to have sustained oscillations [**12, 11, 8, 5**]. Even more complex behavior including multiple periodicity and chaos might be understood as resulting from the combined effects of several feedback loops [**13**]. However, more systems need to be investigated to fully understand the role of feedback loops in the generation of complex dynamical behavior. Here we show how a chaotic dynamic can be uncovered by the analysis of the corresponding feedback loops structure. As an application we revisit the Lorenz system.

In Section 2 we give the settings and working definitions of our approach, encompassing some earlier work such as [**8, 11, 4**]. Section 3 describes the feedback loops analysis of the Lorenz system. The implications of this work are discussed in Section 4.

## 2 Preliminaries

### 2.1 Definitions

Consider a dynamical system described by

$$\dot{x} = F(x, \lambda), \quad x = (x_1, \cdots, x_n) \in \mathbb{R}^n, \quad \lambda = (\lambda_1, \cdots, \lambda_M) \in \mathbb{R}^M. \tag{2.1}$$

The component functions $F_i$, $i = 1, 2, \cdots, n$ are assumed to be at least $C^1(U)$, $U$ an open convex set of $\mathbb{R}^n$. The partial order relation

$$x \leq y \Longleftrightarrow x_i \leq y_i, \quad i = 1, ..., n \tag{$\mathcal{O}$}$$

defines the vector order in $\mathbb{R}^n$. Solving $F(x, \lambda_0) = 0$ yields the steady states for a parameter value $\lambda_0$. Their respective stabilities can be obtained from linear stability analysis. The general Jacobian matrix at $x = (x_1, \ldots, x_n)$ is given by

$$A = \left[\frac{\partial F_i}{\partial x_j}(x)\right] = [a_{ij}], \quad 1 \leq i, j \leq n, \tag{2.2}$$

and in general depends on the state variables, except for linear systems. The dynamical system in (2.1) said to be *monotone inscreasing* if it preserves the order $(\mathcal{O})$ for positive time, i.e.,

$$x \leq y \Rightarrow \varphi_t(x) \leq \varphi_t(y), \quad \text{for } t \geq 0 \text{ and with } \varphi \text{ denoting the dynamic.} \tag{2.3}$$

We further assume a region $\mathcal{R}$ of the phase space where the terms $a_{ij}$ have constant sign $s_{ij}$, not necessarily a neighborhood of a steady state. For $\Lambda$ the set of indices $1, \cdots, n$ we denote by $I_k = \{i_1, \cdots, i_k\}$ an ordered subset of $k$ different elements of $\Lambda$ and by $J_k = \sigma(I_k) = \{j_1, \cdots, j_k\}$, with $\sigma$ a permutation of $I_k$.

A nonzero product

$$P_k = P(I_k, J_k) = \prod_{l=1}^{l=k} a_{i_l \sigma(i_l)} = a_{i_1 \sigma(i_1)} a_{i_2 \sigma(i_2)} \cdots a_{i_k \sigma(i_k)} \tag{2.4}$$

is called a *loop product.* The set of nonzero terms $a_{ij}$, $i \in I_k$, and $j \in J_k$, describes a $k-$*element (or* $k-$*dimensional) simple feedback loop* for a circular permutation; its sign is that of the loop product. Hence, it is positive (resp. negative) for an even (resp. odd) number of negative loop factors in $P_k$. A non-circular permutation yields a *union* of simple feedback loops, with dimension given by the number of distinct elements involved.

A $k-$dimensional feedback loop $L_k$ is represented by an interaction graph $G_k$ of $k$ vertices and edges $(x_i, x_j, s_{ij})$. We derive from $A$ a *qualitative Jacobian* $A_Q$ containing exclusively the terms $s_{ij}$. The *loop structure (or qualitative structure)*, denoted $\mathbb{L}_{\mathcal{R}}$, corresponding to the region $\mathcal{R}$, is the set of all feedback loops found in the Jacobian matrix evaluated in $\mathcal{R}$. Determining and analyzing this qualitative structure anywhere in the phase space, not only around steady states, if any, yields some understanding of the local and global dynamics of the system.

### 2.2 Dynamical properties of Feedback loops

We describe here how the dynamical properties of a system are related to its qualitative structure. See also [**8, 7, 11, 12**]. The characteristic equation

$$\mathcal{P}_\lambda(A) = |\lambda I - A| = 0 \tag{2.5}$$

can be written in the form

$$\lambda^n + c_1\lambda^{n-1} + \cdots + c_k\lambda^{n-k} + \cdots + c_{n-1}\lambda + c_n = 0, \tag{2.6}$$

where $(-1)^n c_n = det(A)$, $-c_1 = trace(A)$, and the general coefficients $c_k$, $k = 1, \ldots, n$ are (up to the multiple constant (-1)) the sum of the diagonal minors of order $k$ of the matrix $A$. See, e.g., [**3**]. On the other hand, upon expanding the Jacobian determinant $A$, any $k-$dimensional principal subdeterminants with indices in $I_k$ may be displayed in the form

$$D(I_k) = \sum_{J_k} (-1)^\epsilon P(I_k, J_k), \tag{2.7}$$

over all permutations $J_k$, where $\epsilon$ is the number of even-dimensional components of $P_k$. Letting

$$S_k = \sum_{I_k} D(I_k), \tag{2.8}$$

summed over all $k-$principal subdeterminants in $A$, it readily appears that

$$(-1)^k S_k = c_k, \quad k = 1, 2, \cdots, n. \tag{2.9}$$

Further analysis yields the following:

1. The loop factors $a_{i\sigma(j)}$ are the only Jacobian entries contributing to the characteristic equation, and therefore, influence directly the eigenvalues of the matrix.
2. Assuming all coefficients $c_k$ positive, i.e., no positive real eigenvalue, implies no positive feedback loop.

Routh-Hurwitz stability criteria, that is, negative real parts of the eigenvalues as necessary and sufficient conditions of stability, readily imply

$$\begin{aligned} &(-1)^k S_k > 0 \\ &S_3 > S_1 \times S_2, \quad (\text{for } n = 3). \end{aligned} \tag{2.10}$$

Moreover, if $S_1 = 0$ then $S_{2r+1} = 0$, $r = 1, \ldots, (n-1)$. Consequently, we obtain the following, (see also [**8**]).

1. Positive loops, having a "negative" contribution to $S_k$, always tend to destabilize a steady state. Hence their presence promotes instability in the system.
2. A steady state with $S_k \neq 0$ but $S_{k+1} = 0$ cannot be asymptotically stable.
3. $S_1 < 0$ requires at least one 1−element negative loop.
4. The loop structure $\mathbb{L}_{\mathcal{R}}$ must contain at least one negative loop of dimension 2 or higher to satisfy $S_3 < 0$.

From the above relations were derived the essential dynamical features of the qualitative structure, namely multistationarity and oscillations.

*2.2.1 Multistationarity*

It has been established that a positive feedback loop is a necessary condition for multistationarity, in the sense of the appearance of multiple steady states or multiple limit cycles. See, e.g., [**13, 11, 8**]. In fact, a positive feedback loop ensures instability in the system by causing the real part of the eigenvalues to become positive. Multistationarity corresponds to a positive loop around a repulsive separatrix.

*2.2.2 Periodic behavior*

Assume that the vector field $F$ in $\mathbb{R}^n$ is *positive quasi-monotone (or cooperative),* that is, $\frac{\partial F_i}{\partial x_j} \geq 0$ whenever $i \neq j$; in other words, the off-diagonal terms of the Jacobian matrix are positive. This implies that all feedback loops are positive except possibly the 1−element ones. It has been proved in [**1**] that under this assumption the vector field $F$ is monotone inscreasing.

Moreover, every attractor has an orbitally stable equilibrium (nearby states have their trajectories remaining nearby for all future times). In particular, there are no attracting periodic orbits. See details in [**6**]. An immediate important consequence is that, in the absence of negative loops of at least two elements, the system cannot have any stable periodic behavior such as stable limit cycles. Similar conclusions have also been drawn by Snoussi in [**11**].

In summary, negative feedback loops promote oscillations, whereas multistationarity is promoted by positive feedback loops. Whenever these loops fulfill this dynamical role, they are said to be *functional.*

In passing, we briefly indicate how some of the dynamical properties outlined above are translated in the two and three dimensional cases, that is, the Jacobian is respectively

$$A_2 = \begin{pmatrix} a_{11} & a_{12} \\ a_{21} & a_{22} \end{pmatrix} \quad \text{and } A_3 = \begin{pmatrix} a_{11} & a_{12} & a_{13} \\ a_{21} & a_{22} & a_{23} \\ a_{31} & a_{32} & a_{33} \end{pmatrix}, \tag{2.11}$$

where the entries have constant sign and depend possibly on the parameters. A complete analysis goes beyond the scope of this paper.

*2.2.3 Two Dimensional Case*

The stability criteria are given by

$$\begin{aligned} S_1 &= a_{11} + a_{22} \leq 0 \\ S_2 &= a_{11} \times a_{22} - a_{12} \times a_{21} \geq 0. \end{aligned} \tag{2.12}$$

From linear stability analysis, there is a parabolic boundary line $S_1^2 - 4S_2 = 0$ between real and complex eigenvalues. Complex eigenvalues, i.e., oscillations, are possible only when $a_{12} \times a_{21} < 0$; this corresponds to a 2−element negative loop. Crossing the boundaries of the second quadrant is the fundamental way in which stability can be lost. It corresponds to the appearance of a positive feedback loop to render positive the real part of the eigenvalues, thereby destabilizing the stable periodicity into a limit cycle for complex eigenvalues, or promoting multistationarity, i.e., saddle point for real eigenvalues.

The necessary condition for oscillations, $S_1^2 - 4S_2 < 0$, amounts to

$$(a_{11} - a_{22})^2 < -4a_{12}a_{21}. \tag{2.13}$$

Thus a 2−element negative loop $L_{xy}^-$ is necessary for any periodic behavior (center, stable focus, or limit cycles).

### *2.2.4 Three Dimensional Case*

In a 3−variable system, a *saddle-focus* may be described as follows:

1. A repulsive periodicity in plane $xy$, attractive along the z-axis has the loop structure
$$\mathbb{L} = L_{xy}^- \cup L_{x/y}^+ \cup L_z^-, \tag{2.14}$$
where $L_{x/y}^+$ stands for a 1−element positive loop at either $x$ or $y$.
2. A steady state periodically attractive in plane $xz$ and repulsive along the y-axis requires
$$\mathbb{L} = L_{xz}^- \cup L_{x/z}^- \cup L_y^+. \tag{2.15}$$

These periodicities might be destabilized in the presence of a 3−element loop $L_{xyz}^{\pm}$. The combined effects of the previous structures are qualitative ingredients for chaotic behavior, as shown in the Lorenz case below. Note also that there are only two 3−element loops in $A_3$, given respectively by $a_{12}a_{23}a_{31}$ and $a_{13}a_{32}a_{21}$. Moreover, wherever they exist together, they are coupled to three 2−element loops given by $a_{12}a_{21}$, $a_{13}a_{31}$, and $a_{23}a_{32}$. Such a strong coupling might preclude the appearence of a chaotic attractor. Indeed, we are aware of no 3−dimensional chaotic systems consisting of two coupled 3−element loops.

## 3. Feedback loops in the Lorenz system

To illustrate the fundamental ideas about the role of feedback loops in dynamical analysis, particularly chaotic dynamics, we choose to study the Lorenz system. It is one of the paradigms for chaotic dynamics, and has been studied extensively. The equations are

$$\begin{aligned} \dot{x} &= \sigma(y - x) \\ \dot{y} &= \rho x - y - xz \\ \dot{z} &= -\beta z + xy, \end{aligned} \tag{$\mathcal{L}$}$$

where $(x, y, z) \in \mathbb{R}^3$, and $\sigma, \rho, \beta > 0$. The Jacobian at any point $M_0 = (x_0, y_0, z_0)$ is

$$A = \begin{pmatrix} -\sigma & \sigma & 0 \\ \rho - z_0 & -1 & -x_0 \\ y_0 & x_0 & -\beta \end{pmatrix}. \tag{3.1}$$

And the corresponding qualitative Jacobian is

$$A_Q = \begin{pmatrix} - & + & 0 \\ sign(\rho - z_0) & - & -sign(x_0) \\ sign(y_0) & sign(x_0) & - \end{pmatrix}, \tag{3.2}$$

where $sign(a_{ij})$ refers to the sign of the term $a_{ij}$. It will take us too far afield to go into details about this system. Note that much of the understanding of the Lorenz attractor comes from numerical experiments. Its main characteristics are:

1. The system is invariant under reflection in the $z-$axis, orbit attracted to the origin.
2. It has a steady state $O = (0,0,0)$ at the origin for all $\rho$, and two other steady states $ss_{\pm} = (\pm\sqrt{\beta(\rho-1)}, \pm\sqrt{\beta(\rho-1)}, \rho - 1)$ appearing at $\rho > 1$, and symmetrically placed with respect to the $z-$axis.

Here, we give a new description of this celebrated system based on the analysis of the corresponding loop structure.

First, the stability parameters $S_k$, $k = 1,2,3$, are obtained as

$$\begin{aligned} S_1 &= -(\sigma + \beta + 1) < 0 \\ S_2 &= \beta + \sigma(1+\beta) - \sigma\rho + \sigma z_0 + x_0^2 \\ S_3 &= -\sigma(\beta(1-\rho) + \beta z_0 + x_0 y_0 + x_0^2). \end{aligned} \tag{3.3}$$

Hence

1. $S_1$ is always negative, ensuring the constant presence of at least one $1-$element negative loop $L^-_{x/y/z}$.
2. $S_2 = 0$ for $\rho = \rho_c = \frac{\beta+\sigma(1+\beta)+\sigma z_0 + x_0^2}{\sigma}$. It follows that $S_2 > 0$ for $\rho < \rho_c$ (to ensure stability) and $S_2 < 0$ for $\rho > \rho_c$ (allowing instability).
3. $S_3 = 0$ for $\rho = \rho_1 = -\frac{-\sigma(\beta(1+z_0)+x_0 y_0 + x_0^2)}{\sigma\beta}$.

### 3.1 Qualitative structure

The qualitative Jacobian in (3.2) depends on the state of the system; this allows us to distingish three main regions in the variable space. We analyze their corresponding loop structures to uncover the underlying dynamics in view of the previous sections. First, one finds in $A_Q$, three persistent $1-$element negative loops $L^-_{x/y/z}$.

1. Under the plane $\mathcal{P}_\rho :\ z = \rho$, $A_Q$ has the expression

$$A_Q^1 = \begin{pmatrix} - & + & 0 \\ + & - & -sign(x_0) \\ sign(y_0) & sign(x_0) & - \end{pmatrix}. \tag{3.4}$$

Thus the corresponding loop structure $\mathbb{L}$ contains an additional $2-$element positive loop $L^+_{xy}$.

2. In the plane $\mathcal{P}_\rho$, we have

$$A_Q^2 = \begin{pmatrix} - & + & 0 \\ 0 & - & -sign(x_0) \\ sign(y_0) & sign(x_0) & - \end{pmatrix}. \tag{3.5}$$

The previous positive loop has disappeared in the whole plane.

3. Above the plane $\mathcal{P}_\rho$, we get

$$A_Q^3 = \begin{pmatrix} - & + & 0 \\ - & - & -sign(x_0) \\ sign(y_0) & sign(x_0) & - \end{pmatrix}. \tag{3.6}$$

A 2−element negative loop $L_{xy}^-$ emerges.

The symmetry of the Lorenz with respect to the z-axis imposes an analysis along the z-axis, and only in the region $\mathcal{R}^\pm$ with $x$ and $y$ of respectively the same sign, and opposite signs.

1. The z-axis has the qualitative Jacobian

$$A_Q^z = \begin{pmatrix} - & + & 0 \\ s_{21} & - & 0 \\ 0 & 0 & - \end{pmatrix}, \tag{3.7}$$

where $s_{21}$ is respectively $(+)$, $(0)$, and $(-)$ under, in, and above the plane $\mathcal{P}_\rho$. The loop structure consists only of the previous persistant ones.

2. In $\mathcal{R}^+$, we obtain

$$A_Q^+ = \begin{pmatrix} - & + & 0 \\ s_{21} & - & \mp \\ \pm & \pm & - \end{pmatrix}, \tag{3.8}$$

with here again $s_{21}$ taking respectively the signs $(+)$, $(0)$, and $(-)$ as we move vertically. A 3−element negative feedback loop $L_{xyz}^-$ persists throughout $\mathcal{R}^+$, coupled successively with a 2−element positive loop $L_{xy}^+$ under the plane $z = \rho$, with a 2−element negative loop $L_{xy}^-$ above the plane, and with a 2−element negative loop $L_{yz}^-$.

3. In the region $\mathcal{R}^-$, $A_Q$ becomes

$$A_Q^- = \begin{pmatrix} - & + & 0 \\ s_{21} & - & \mp \\ \mp & \pm & - \end{pmatrix}. \tag{3.9}$$

The previous 2−element loops are now coupled with a 3−element positive loop $L_{xyz}^+$.

### *3.1.1 Steady states loop structure*

We now describe the qualitative structure in the neighborhood of the steady states of the system.

1. Around the origin, $A_Q$ is

$$A_Q^O = \begin{pmatrix} - & + & 0 \\ + & - & 0 \\ 0 & 0 & - \end{pmatrix}. \tag{3.10}$$

In addition to the 1−element loops, a 2−element positive loop $L^+_{xy}$ appears. The corresponding stability parameters $S_k$ are given by

$$\begin{aligned} S_1 &= -(\sigma+\beta+1) < 0 \\ S_2 &= \beta(1+\sigma) + (1-\rho)\sigma \\ S_3 &= -\beta\sigma(1-\rho). \end{aligned} \tag{3.11}$$

And the stability breaking points are at $\rho = 1$, and, for $\rho > 1$, at $\rho = 1 + \frac{\beta(1+\sigma)}{\sigma}$.

2. Around the steady states $ss_\pm = (\pm\sqrt{\beta(\rho-1)}, \pm\sqrt{\beta(\rho-1)}, \rho-1)$ for $\rho > 1$, $A_Q$ becomes

$$A^{ss}_Q = \begin{pmatrix} - & + & 0 \\ + & - & \mp \\ \pm & \pm & - \end{pmatrix}. \tag{3.12}$$

Here there appears a 3−element negative loop $L^-_{xyz}$, coupled with $L^+_{xy}$ and $L^-_{yz}$. The stability parameters are now

$$\begin{aligned} S_1 &= -(\sigma+\beta+1) < 0 \\ S_2 &= \beta(\sigma+\rho) > 0 \\ S_3 &= -2\sigma\beta(\rho-1) < 0. \end{aligned} \tag{3.13}$$

The stability criteria are therefore met. But as $\rho$ approaches 1 from above, stability is lost. Also, we know from weakly nonlinear theory that there is a subcritical Hopf bifurcation at $ss_\pm$ when $1 < \rho = \rho_c = \frac{\sigma(\sigma+\beta+3)}{\sigma-\beta-1}$, for a range of values of $\sigma$ and $\beta$. In fact, $ss_\pm$ are stable either for $\sigma < \beta+1$ or $1 < \rho < \rho_c$ and $\sigma > \beta+1$.

### 3.2 Feedback loops analysis and strange attractor

First of all, independent of the state of the system, and for any value of $\rho > 0$, the loop structure has a 2−element positive loop $L^+_{xy}$, promoting multistationarity under the plane $z = \rho$, and becoming a generator of periodicity $L^-_{xy}$ above the plane. The corresponding oscillations are normal to the plane $xy$ due to $L^-_x$ and damped by $L^-_y$.

Around the origin, stable for low values of $\rho < 1$, the previous 2−element positive loop $L^+_{xy}$ is actually functional. It ensures multistationarity with the emergence of the steady $ss_\pm$ as stability is broken at $\rho = 1$. These steady states are symmetrical foci generated by the 2−element negative loop $L^-_{yz}$.

In the region $\mathcal{R}^+$, particularly at $ss_\pm$, we have a periodicity $L^-_{yz}$, damped by the 1−element loops $L^-_y$ and $L^-_z$ in the plane $yz$, attractive along the x-axis by $L^-_x$, and made transient by $L^-_{xyz}$. Moreover, under the plane, this periodicity is destabilized by the positive loop $L^+_{xy}$ into limit cycles.

However, the trajectories are forced by $L^-_{xyz}$ to switch from one attractor to another, crossing the z-axis periodically.

Upon crossing the plane $z = \rho$, the loop positive $L^+_{xy}$ disappears and reappears negative above the plane; thus the existence of two generators of periodicity, distinct yet coupled at $y$. The oscillations are damped by $L^-_y$. Here also, the 3−element loop $L^-_{xyz}$ is still active, destabilizing the periodic behavior and, sooner or later, forcing the trajectories to cross the z-axis, periodically due to the appearance of the negative $L_{xy}^{-}$ on the z-axis. But the normal attractors are absent in this region. Thus the combination of these local periodic movements, together with the global destabilizing influence generated by the 3−element negative loop, indicates the presence of the *strange attractor.*

Moreover, we have seen above the presence of the 3−element positive loop $L_{xyz}^{+}$ in the region $\mathcal{R}^{-}$ with $x$ and $y$ of opposite signs. This part of the space is therefore destabilized, making the appearance of the attractors impossible under the plane $\mathcal{P}_{\rho}$ as well as that of the "strange attractor" above the plane, with the additional effect of directing the orbits toward the attractors or the "strange attractors" in the region $\mathcal{R}^{+}$.

## 4 Discussion and conclusions

From Thomas's analysis, chaotic dynamics require the following:

1. A generator of periodicity, i.e., a negative feedback loop with at least two elements.
2. A generator of multistationarity, i.e., a positive feedback loop, which may have a single element.
3. An appropriate nonlinearity for nontrivial behavior and confinement of the system.

In the case of the Lorenz system, these conditions are met by the presence of the negative loop $L_{yz}^{-}$, the positive loop $L_{xy}^{+}$, and the nonlinear terms $-xz$ and $+xy$. However, here, a unique generator of periodicity, the negative loop $L_{yz}^{-}$, generates two symmetrically periodic attractors, because both loop terms depend on the sign of the variable $x$. Moreover, the 3−element negative loop $L_{xyz}^{-}$,

1. contributes to confining the system to the bounded regions of the variable space;
2. forces the system to switch from one periodic attractor to the other; and
3. allows the periodic recurrence of the whole process.

Given its fundamental qualitative nature, feedback loops analysis only allows necessary conditions, not sufficient ones. But such necessary conditions are powerful enough to predict the effect of different structural or parameter changes. Indeed,

1. Changes in the equations leading to a violation of any of the previous necessary conditions should yield the loss of the chaotic behavior.
2. Changes in the equations which affect neither the feedback loop structure of the Jacobian, nor the nonlinearities of the system, should preserve the possibility of chaotic behavior.

For example, inverting the signs of the terms $-xz$ and $+xy$ in the Lorenz equations gives the Jacobian

$$A = \begin{pmatrix} -\sigma & \sigma & 0 \\ \rho + z_0 & -1 & +x_0 \\ -y_0 & -x_0 & -\beta \end{pmatrix}.$$

All feedback loops are preserved. However,

1. The signs of the Jacobian terms defining the negative loop $L_{yz}^{-}$ are inverted.
2. The signs of two of the terms in the 3−element loop $L_{xyz}^{-}$ are also inverted.

3. The term $a_{21}$ is modified, but still depends on $\rho$ and can still be positive or negative.

The changes are symmetric in the Jacobian. For identical parameter values, this modified Lorenz system presents a chaotic attractor symmetrical to the original.

Such feedback loop analysis allowed Thomas to interpret/predict various changes in the Rössler system and its variants, as well as in other chaotic systems. Reanalyzing these systems and comparing them with the Lorenz case lead us to the following conjecture that we intend to settle in the near future: *For a given differential system, the generation of chaotic behavior requires a union of negative loops of the dimension of the system.* For Lorenz system, this criterion is fulfilled by either

1. The 3–element negative loop $L^-_{xyz}$, or
2. The union of the negative loops $L^-_x$ and $L^-_{yz}$, or
3. The union of the three 1–element negative loops $L^-_x$, $L^-_y$, and $L^-_z$.

In the Rössler system, it is fulfilled by either

1. The union of the negative loops $L^-_{xy}$ and $L^-_z$, or
2. The union of the coupled negative loops $L^-_{xy}$ and $L^-_{xz}$.

Usually, using stability analysis, one can only attain the local dynamics of the system and hence need to use, for instance, numerical integration in conjunction to obtain the global dynamics. However, in this article we were able to predict the global dynamics of the Lorenz system by means of the qualitative structure found in its Jacobian matrix analysed throughout the phase space. Thus feedback loops and their combinations provide valuable information about the dynamics of a system (e.g., number and types of steady states, periodicity, chaotic behavior) even when only the signs, not the magnitudes of the Jacobian terms, are known. This is a great advantage over the classical analysis. In other words, this qualitative structure allows one to assert whether sustained oscillations, multistationarity, or chaotic dynamics are possible.

The Feedback loops technique as described in this paper is easy to implement and can quickly demarcate both parameter and phase spaces into exciting (oscillations, multistationarity, chaotic behavior) and non-exciting (single stable fixed point) regions. Hence it could prove useful in surveying dynamical response of models simulating physical, chemical and biological systems. An article in preparation will apply this method to a 3–dimensional model of electrochemical corrosion.

## References

[1] Coppel, W. A. [1965] *Stability and Asymptotic Behavior of Differential Equations,* Heath, Boston.

[2] Drazin, P. G. [1992] *Nonlinear Systems*, Cambridge Texts in Applied Mathematics, Cambridge University Press.

[3] Demidovich, B. and Maron, I. [1979] *Eléments de calcul numérique (Translated from Russian)*, Mir. Moscow.

[4] Eisenfeld, J. and De Lisi, C. [1994] *On conditions for qualitative instability of regulatory loops with applications to immunological control loops,* Mathematics and Computers in Biomathematical Applications, Elsevier, New York, 39-53.

[5] Gouzé, J. L. [1998] *Positive and negative circuits in dynamical systems,* J. Biol. Syst. (In press).

[6] Hirsch, M. W. [1985] *Systems of differential equations that are competitive or cooperative,* SIAM J. Math. Anal. **16**, 432-439.

[7] Levins, R. [1974] *The qualitative analysis of partially specified systems,* Ann. N. Y. Acad. Sci. **231**, 123-138.

[8] Plahte, E. Mestl, T. and Omholt, W. [1995] *Feedback loop, stability and multistationarity in dynamical systems,* J. Biol. Syst. **3(2)**, 409-413.

[9] Quirk, R. and Ruppert, R. [1965] *Qualitative economics and the stability of equilibrium,* Rev. Econ. Studies **32**, 311-326.

[10] Rosen, R. [1968] *Recent developments in theory of control and regulation of cellular processes,* Int. Rev. of Cytology **23**, 25-88.

[11] Snoussi, E. H. [1998] *Necessary condition for multistationarity and stable periodicity,* J. Biol. Syst. (In press)..

[12] Thomas, R. [1994] *The role of feedback circuits: positive feedback circuits are a necessary condition for positive eigenvalues in the Jacobian matrix,* Ber. Bunzenges. Phys. Chem. **98**, 1148-1151.

[13] Thomas, R. [1996] *Analyse et synthèse de systèmes à dynamique chaotique en terme de loops de rétroaction,* Académie Royale de Belgique 6$^e$ **série, Tome VII**, 101-124.

[14] Thomas, R. and D'Ari, R. [1990] *Biological Feedback*, CRC Press Inc., Boca Raton (316 pp).

[15] Thomas, R., Thieffry, D. and Kaufman, M. [1995] *Dynamical behaviour of biological regulatory networks. I. Biological role of feedback loops and practical use of the concept of the loop-characteristic state,* Bull. Math. Biol. **57**, 247-276.

[16] Tyson, J. [1975] *Classification of instabilities in chemical reaction systems,* J. Chem. Phys. **62**, 1010-1015.

Fields Institute Communications
Volume **21**, 1999

# Absolute Stability of the Second Canonical Form of Control Systems

**Weihua Wu**
Central China Normal University
Wuhan 430079, Hubei, P. R. China
whwu@ccnu-1.ccnu.edu.cn

**Zongqi Deng**
Central China Normal University
Wuhan 430079, Hubei, P. R. China

**Abstract.** Making use of Popov's frequency method, this paper gives some new criteria of absolute stability for the second canonical form of control systems. Some previous results are improved.

## 1 Introduction

Consider the second canonical form of control systems

$$\begin{cases} \dot{x}_j &= -\rho_j x_j + \sigma, \quad j = 1, 2, \ldots, n, \\ \dot{\sigma} &= \sum\limits_{j=1}^{n} \beta_j x_j - p\sigma - rf(\sigma), \end{cases} \tag{1.1}$$

where $p > 0, r > 0, \rho_j > 0, j = 1, 2, \ldots, n, f(\sigma)$ is a real continuous function on $(-\infty, +\infty)$ and satisfies $f(0) = 0, f(\sigma)\sigma > 0$ for $\sigma \neq 0$.

Without loss of generality, we suppose

$$\left.\begin{array}{ll} \beta_j > 0, & j = 1, \ldots, s; \\ \beta_j < 0, & j = s+1, \ldots, m; \\ \beta_j = 0, & j = m+1, \ldots, n \end{array}\right\}. \tag{1.2}$$

Using the Liapunov method, Liao in [**1**] gave the following criterion of absolute stability of the zero solution of (1.1):

$$p - \sum_{j=1}^{s} \frac{\beta_j}{\rho_j} > 0. \tag{1.3}$$

Later, Qiu and Shu in [**4**] gave the following two improved criteria:

$$p - \min_{s+1 \leq j \leq m} \ \min_{\lambda \in \Lambda_j} \sum_{i=1}^{s} \frac{\beta_i \rho_i}{(\rho_i + \lambda_i \rho_j)^2} > 0, \tag{1.4}$$

---

1991 *Mathematics Subject Classification.* 34D20, 93D10.

where $\Lambda_j = \{\lambda | \lambda = (\lambda_1, \dots, \lambda_s), \lambda_i = 0$ or $1$ such that $\sum\limits_{i=1}^{s} \frac{\lambda_i \beta_i}{\rho_i} + \frac{\beta_j}{\rho_j} \leq 0\}$; and

$$p - \sum_{i=1}^{s} \frac{\beta_i}{\rho_i} - \min_{1 \leq i \leq s} \min_{\mu \in M_i} \sum_{j=s+1}^{m} \frac{\mu_j \beta_j (2\rho_i + \rho_j)}{(\rho_i + \rho_j)^2} > 0, \tag{1.5}$$

where $M_i = \{\mu | \mu = (\mu_{s+1}, \dots, \mu_m), \mu_j = 0$ or $1$ such that $\sum\limits_{j=s+1}^{m} \frac{\mu_j \beta_j}{\rho_j} + \frac{\beta_i}{\rho_i} \geq 0\}$.

In this paper we study the absolute stability of (1.1) by using Popov's frequency method and obtain some new criteria.

When $\beta_j, j = 1, \dots, n$ have the same sign, necessary and sufficient conditions for absolute stability of (1.1) are already known (see [**4**]) and so we will not discuss this case.

## 2 Several lemmas

Consider the direct control system

$$\dot{x} = Ax + bf(c^{\mathrm{T}} x), \tag{2.1}$$

where $A \in \mathbf{R}^{n \times n}, b, c \in \mathbf{R}^n, f(\sigma)$ is a real continuous function on $(-\infty, +\infty)$ and satisfies $f(0) = 0, f(\sigma)\sigma > 0$ for $\sigma \neq 0$.

For system (2.1), Meyer in [**3**] gave the following result.

**Lemma 2.1** *Suppose the matrix $A$ is stable, i.e., $\mathrm{Re}\lambda(A) < 0$, where $\lambda(A)$ denotes the eigenvalues of $A$. If there exists a real number $q \geq 0$ such that for $-\infty < \omega < +\infty$,*

$$\mathrm{Re}\{(1 + \mathrm{i}\omega q) W(\mathrm{i}\omega)\} \geq 0, \tag{2.2}$$

*where $W(\mathrm{i}\omega) = -c^{\mathrm{T}}(\mathrm{i}\omega I - A)^{-1} b$, then the zero solution of (2.1) is absolutely stable.*

In particular, if $q = 0$, then (2.2) simplifies to

$$\mathrm{Re}\{c^{\mathrm{T}}(\mathrm{i}\omega I - A)^{-1} b\} \leq 0. \tag{2.3}$$

We write Mihailov's criterion as the following lemma (see [**2**]):

**Lemma 2.2** *Assume that the real polynomial*

$$F(z) = \det(zI - A) = a_0 + a_1 z + \cdots + z^n$$

*has no pure imaginary roots. Then a necessary and sufficient condition for $A \in \mathbf{R}^{n \times n}$ to be stable is that the vector $F(\mathrm{i}\omega)$ rotates through the angle $\varphi = \frac{\pi}{2} n$ anti-clockwise when $\omega$ changes from $0$ to $+\infty$.*

**Lemma 2.3** *The following formula holds:*

$$\begin{aligned} D_{n+1} &= \begin{vmatrix} z+\rho_1 & 0 & \cdots & 0 & -1 \\ 0 & z+\rho_2 & \cdots & 0 & -1 \\ \vdots & \vdots & \ddots & \vdots & \vdots \\ 0 & 0 & \cdots & z+\rho_n & -1 \\ -\beta_1 & -\beta_2 & \cdots & -\beta_n & z+p \end{vmatrix} \\ &= (z+p)\prod_{j=1}^{n}(z+\rho_j) - \sum_{j=1}^{n}\beta_j \prod_{\substack{k=1 \\ k \neq j}}^{n}(z+\rho_k). \end{aligned} \tag{2.4}$$

Lemma 2.3 can be proved by mathematical induction. The following lemma is obvious.

**Lemma 2.4** *Let $z$ be a complex number. Then* $\mathrm{Re}z > 0$ *is equivalent to* $\mathrm{Re}\frac{1}{z} > 0$.

## 3 New criteria of absolute stability

**Theorem 3.1** *If for* $-\infty < \omega < +\infty$,

$$p - \sum_{j=1}^{n} \frac{\beta_j \rho_j}{\omega^2 + \rho_j^2} > 0, \tag{3.1}$$

*then the zero solution of (1.1) is absolutely stable.*

**Proof** Obviously, system (1.1) is a particular case of system (2.1), where

$$A = \begin{pmatrix} -\rho_1 & 0 & \cdots & 0 & 1 \\ 0 & -\rho_2 & \cdots & 0 & 1 \\ \vdots & \vdots & \ddots & \vdots & \vdots \\ 0 & 0 & \cdots & -\rho_n & 1 \\ \beta_1 & \beta_2 & \cdots & \beta_n & -p \end{pmatrix}, \quad b = \begin{pmatrix} 0 \\ 0 \\ \vdots \\ 0 \\ -r \end{pmatrix}, \quad c = \begin{pmatrix} 0 \\ 0 \\ \vdots \\ 0 \\ 1 \end{pmatrix}.$$

By Lemma 2.3, we can see that

$$F(z) = \det(zI - A) = (z + p) \prod_{j=1}^{n} (z + \rho_j) - \sum_{j=1}^{n} \beta_j \prod_{\substack{k=1 \\ k \neq j}}^{n} (z + \rho_k).$$

Let

$$G(z) = \prod_{j=1}^{n} (z + \rho_j).$$

By (3.1), for $-\infty < \omega < +\infty$, we have

$$\begin{aligned} \mathrm{Re}\frac{F(\mathrm{i}\omega)}{G(\mathrm{i}\omega)} &= \mathrm{Re}\left\{ \frac{(\mathrm{i}\omega + p) \prod_{j=1}^{n} (\mathrm{i}\omega + \rho_j) - \sum_{j=1}^{n} \beta_j \prod_{\substack{k=1 \\ k \neq j}}^{n} (\mathrm{i}\omega + \rho_k)}{\prod_{j=1}^{n} (\mathrm{i}\omega + \rho_j)} \right\} \\ &= \mathrm{Re}\left\{ (\mathrm{i}\omega + p) - \sum_{j=1}^{n} \frac{\beta_j}{\mathrm{i}\omega + \rho_j} \right\} \\ &= p - \sum_{j=1}^{n} \frac{\beta_j \rho_j}{\omega^2 + \rho_j^2} \\ &> 0. \end{aligned} \tag{3.2}$$

Since the $n$th-order polynomial $G(z)$ only has negative real roots, by Lemma 2.2 we obtain

$$\Delta_0^{+\infty} \arg G(\mathrm{i}\omega) = \arg G(\mathrm{i}\omega)|_{\omega=0}^{\omega=+\infty} = \frac{\pi}{2} n. \tag{3.3}$$

Because $\mathrm{Re}\frac{F(\mathrm{i}\omega)}{G(\mathrm{i}\omega)} > 0$ for $-\infty < \omega < +\infty, F(z)$ has no pure imaginary roots. Noticing the definition of $F(z)$ and $G(z)$, we can deduce that

$$\lim_{\omega \to +\infty} [\arg F(\mathrm{i}\omega) - \arg G(\mathrm{i}\omega)] = \frac{\pi}{2}.$$

Therefore

$$\lim_{\omega\to+\infty} \arg F(\mathrm{i}\omega) = \lim_{\omega\to+\infty} \arg G(\mathrm{i}\omega) + \frac{\pi}{2}. \tag{3.4}$$

Since $\arg F(0) = \arg G(0) = 0$, by (3.3) and (3.4) we obtain

$$\Delta_0^{+\infty} \arg F(\mathrm{i}\omega) = \Delta_0^{+\infty} \arg G(\mathrm{i}\omega) + \frac{\pi}{2} = \frac{\pi}{2}(n+1).$$

Thus, by Lemma 2.2, $A$ is stable, i.e., $\mathrm{Re}\lambda(A) < 0$.

In the following, we will prove that (2.3) holds for $-\infty < \omega < +\infty$. In fact

$$c^{\mathrm{T}}(\mathrm{i}\omega I - A)^{-1}b = \frac{-r\prod\limits_{j=1}^{n}(\mathrm{i}\omega + \rho_j)}{\det(\mathrm{i}\omega I - A)} = -\frac{rG(\mathrm{i}\omega)}{F(\mathrm{i}\omega)}.$$

By (3.2) and Lemma 2.4, for $-\infty < \omega < +\infty$ we have

$$\mathrm{Re}\{c^{\mathrm{T}}(\mathrm{i}\omega I - A)^{-1}b\} = -r\mathrm{Re}\frac{G(\mathrm{i}\omega)}{F(\mathrm{i}\omega)} < 0.$$

This implies that the zero solution of (1.1) is absolutely stable. The proof is complete. □

Assuming condition (1.2), let

$$\begin{aligned} \nu_1 &= \max\{\rho_{s+1},\dots,\rho_m\}, & \nu_2 &= \min\{\rho_1,\dots,\rho_s\},\\ A_1 &= \sum_{j=s+1}^{m} \beta_j\rho_j < 0, & A_2 &= \sum_{j=1}^{s} \beta_j\rho_j > 0. \end{aligned}$$

**Theorem 3.2** *If system (1.1) satisfies*

$$p > \max\left\{\frac{A_1 + A_2}{\nu_1^2 + \nu_2^2}, \frac{A_1}{\nu_1^2} + \frac{A_2}{\nu_2^2}\right\}, \tag{3.5}$$

*then the zero solution of (1.1) is absolutely stable.*

**Proof** Under condition (3.5), we verify that (3.1) holds for $-\infty < \omega < +\infty$. Since

$$\sum_{j=1}^{n}\frac{\beta_j\rho_j}{\omega^2+\rho_j^2} = \sum_{j=1}^{s}\frac{\beta_j\rho_j}{\omega^2+\rho_j^2} + \sum_{j=s+1}^{m}\frac{\beta_j\rho_j}{\omega^2+\rho_j^2} < \frac{A_2}{\omega^2+\nu_2^2} + \frac{A_1}{\omega^2+\nu_1^2},$$

it is enough to prove that

$$\frac{A_2}{\omega^2+\nu_2^2} + \frac{A_1}{\omega^2+\nu_1^2} < p,$$

which is equivalent to

$$p\omega^4 + [p(\nu_1^2+\nu_2^2) - (A_1+A_2)]\omega^2 + [p\nu_1^2\nu_2^2 - (A_1\nu_2^2 + A_2\nu_1^2)] > 0.$$

In fact, condition (3.5) ensures that the above inequality holds for $-\infty < \omega < +\infty$. The proof is complete. □

**Theorem 3.3** *Under condition (1.2), if*

$$p - \min_{s+1\le j\le m}\ \min_{\lambda\in\Lambda_j} \sum_{i=1}^{s}\frac{\beta_i(|\rho_i - \lambda_i\rho_j| + \rho_i - \lambda_i\rho_j)}{2\rho_i(\rho_i + \lambda_i\rho_j)} > 0, \tag{3.6}$$

*where* $\Lambda_j = \{\lambda | \lambda = (\lambda_1,\dots,\lambda_s), \lambda_i = 0$ *or* $1$ *such that* $\sum\limits_{i=1}^{s}\frac{\lambda_i\beta_i}{\rho_i} + \frac{\beta_j}{\rho_j} \le 0\}$, *then the zero solution of (1.1) is absolutely stable.*

**Proof** First, for $-\infty < \omega < +\infty$, we prove the inequality

$$\frac{1}{\omega^2+\rho_i^2} - \frac{\lambda_i \rho_j^2}{\rho_i^2(\omega^2+\rho_j^2)} \leq \frac{|\rho_i - \lambda_i \rho_j| + \rho_i - \lambda_i \rho_j}{2\rho_i^2(\rho_i + \lambda_i \rho_j)}, \tag{3.7}$$

where $\lambda_i = 0$ or 1. If $\lambda_i = 0$, (3.7) holds obviously. If $\lambda_i = 1$, then (3.7) is equivalent to

$$\frac{(\rho_i^2 - \rho_j^2)\omega^2}{\rho_i^2(\omega^2+\rho_i^2)(\omega^2+\rho_j^2)} \leq \frac{|\rho_i - \rho_j| + \rho_i - \rho_j}{2\rho_i^2(\rho_i+\rho_j)}. \tag{3.8}$$

If $\rho_i \leq \rho_j$, the left side of (3.8) is not more than zero, while the right side is zero. Clearly, (3.8) holds. If $\rho_i > \rho_j$, (3.8) is equivalent to

$$2\rho_i^2(\rho_i+\rho_j)(\rho_i^2-\rho_j^2)\omega^2 \leq 2\rho_i^2(\omega^2+\rho_i^2)(\omega^2+\rho_j^2)(\rho_i-\rho_j),$$

i.e.,

$$(\omega^2 - \rho_i\rho_j)^2 \geq 0. \tag{3.9}$$

Obviously, (3.9) holds for $-\infty < \omega < +\infty$. This implies that (3.7) holds for $-\infty < \omega < +\infty$.

Secondly, because of (3.6), there exist some $j(s+1 \leq j \leq m)$ and $\lambda = (\lambda_1, \ldots, \lambda_s) \in \Lambda_j$ such that

$$p - \sum_{i=1}^{s} \frac{\beta_i(|\rho_i - \lambda_i\rho_j| + \rho_i - \lambda_i\rho_j)}{2\rho_i(\rho_i+\lambda_i\rho_j)} > 0. \tag{3.10}$$

Since $\lambda \in \Lambda_j$, we have

$$\sum_{i=1}^{s} \frac{\lambda_i\beta_i}{\rho_i} + \frac{\beta_j}{\rho_j} \leq 0.$$

Therefore

$$\frac{\beta_j\rho_j}{\omega^2+\rho_j^2} \leq -\sum_{i=1}^{s} \frac{\beta_i\rho_i\lambda_i\rho_j^2}{\rho_i^2(\omega^2+\rho_j^2)}. \tag{3.11}$$

By (3.11), (3.7) and (3.10), we obtain

$$\begin{aligned}
\sum_{i=1}^{n} \frac{\beta_i\rho_i}{\omega^2+\rho_i^2} &\leq \sum_{i=1}^{s} \frac{\beta_i\rho_i}{\omega^2+\rho_i^2} + \frac{\beta_j\rho_j}{\omega^2+\rho_j^2} \\
&\leq \sum_{i=1}^{s} \frac{\beta_i\rho_i}{\omega^2+\rho_i^2} - \sum_{i=1}^{s} \frac{\beta_i\rho_i\lambda_i\rho_j^2}{\rho_i^2(\omega^2+\rho_j^2)} \\
&= \sum_{i=1}^{s} \beta_i\rho_i \left[\frac{1}{\omega^2+\rho_i^2} - \frac{\lambda_i\rho_j^2}{\rho_i^2(\omega^2+\rho_j^2)}\right] \\
&\leq \sum_{i=1}^{s} \frac{\beta_i(|\rho_i-\lambda_i\rho_j| + \rho_i - \lambda_i\rho_j)}{2\rho_i(\rho_i+\lambda_i\rho_j)} \\
&< p.
\end{aligned}$$

This implies that the hypothesis of Theorem 3.1 is satisfied. So the zero solution of (1.1) is absolutely stable. The proof is complete. □

**Theorem 3.4** *Under condition (1.2), if*

$$p - \sum_{i=1}^{s} \frac{\beta_i}{\rho_i} - \min_{1\leq i\leq s} \min_{\mu\in M_i} \sum_{j=s+1}^{m} \frac{4\mu_j\beta_j}{\rho_i + 3\rho_j + |\rho_j - \rho_i|} > 0, \tag{3.12}$$

*where* $M_i = \{\mu | \mu = (\mu_{s+1}, \cdots, \mu_m), \mu_j = 0$ *or* $1$ *such that* $\sum_{j=s+1}^{m} \frac{\mu_j \beta_j}{\rho_j} + \frac{\beta_i}{\rho_i} \geq 0\}$, *then the zero solution of (1.1) is absolutely stable.*

**Proof** Let $k_{ij} = \frac{\rho_i + 3\rho_j + |\rho_j - \rho_i|}{4}$. Obviously, $k_{ij} = \max\left\{\rho_j, \frac{\rho_i + \rho_j}{2}\right\}$. We first prove the inequality

$$\frac{\rho_j}{\omega^2 + \rho_j^2} - \frac{1}{k_{ij}} \geq -\frac{\omega^2}{\rho_j(\omega^2 + \rho_i^2)}, \tag{3.13}$$

which is equivalent to

$$(k_{ij} - \rho_j)\omega^4 + [2k_{ij}\rho_j^2 - \rho_j(\rho_i^2 + \rho_j^2)]\omega^2 + (k_{ij} - \rho_j)\rho_i^2\rho_j^2 \geq 0. \tag{3.14}$$

If $\rho_j \geq \rho_i$, then $k_{ij} = \rho_j$. Obviously, (3.14) holds. If $\rho_j < \rho_i$, then $k_{ij} = \frac{\rho_i + \rho_j}{2}$. Thus, the left side of (3.14) is

$$\left(\frac{\rho_i - \rho_j}{2}\right)\omega^4 + \rho_i\rho_j(\rho_j - \rho_i)\omega^2 + \frac{\rho_i - \rho_j}{2}\rho_i^2\rho_j^2 = \left(\frac{\rho_i - \rho_j}{2}\right)(\omega^2 - \rho_i\rho_j)^2.$$

This implies that (3.14) holds. Therefore (3.13) holds.

Secondly, because of (3.12) there exist some $i(1 \leq i \leq s)$ and $\mu = (\mu_{s+1}, \cdots, \mu_m) \in M_i$ such that

$$p - \sum_{i=1}^{s} \frac{\beta_i}{\rho_i} - \sum_{j=s+1}^{m} \frac{4\mu_j\beta_j}{\rho_i + 3\rho_j + |\rho_j - \rho_i|} = p - \sum_{i=1}^{s} \frac{\beta_i}{\rho_i} - \sum_{j=s+1}^{m} \frac{\mu_j\beta_j}{k_{ij}} > 0. \tag{3.15}$$

Since $\mu \in M_i$, we obtain

$$\sum_{j=s+1}^{m} \frac{\mu_j\beta_j}{\rho_j} + \frac{\beta_i}{\rho_i} \geq 0. \tag{3.16}$$

From (3.13) and (3.16), it follows that

$$\sum_{j=s+1}^{m} \mu_j\beta_j \left(\frac{\rho_j}{\omega^2 + \rho_j^2} - \frac{1}{k_{ij}}\right) \leq \sum_{j=s+1}^{m} \mu_j\beta_j \left(\frac{-\omega^2}{\rho_j(\omega^2 + \rho_i^2)}\right) \leq \frac{\beta_i\omega^2}{\rho_i(\omega^2 + \rho_i^2)},$$

thus

$$\sum_{j=s+1}^{m} \frac{\mu_j\beta_j\rho_j}{\omega^2 + \rho_j^2} + \frac{\beta_i\rho_i}{\omega^2 + \rho_i^2} \leq \sum_{j=s+1}^{m} \frac{\mu_j\beta_j}{k_{ij}} + \frac{\beta_i}{\rho_i}. \tag{3.17}$$

By (3.17) and (3.15), for $-\infty < \omega < +\infty$, we can deduce

$$\begin{aligned}
\sum_{j=1}^{n} \frac{\beta_j\rho_j}{\omega^2 + \rho_j^2} &= \sum_{j=1}^{s} \frac{\beta_j\rho_j}{\omega^2 + \rho_j^2} + \sum_{j=s+1}^{m} \frac{\beta_j\rho_j}{\omega^2 + \rho_j^2} \\
&\leq \sum_{\substack{j=1 \\ j \neq i}}^{s} \frac{\beta_j\rho_j}{\omega^2 + \rho_j^2} + \frac{\beta_i\rho_i}{\omega^2 + \rho_i^2} + \sum_{j=s+1}^{m} \frac{\mu_j\beta_j\rho_j}{\omega^2 + \rho_j^2} \\
&\leq \sum_{\substack{j=1 \\ j \neq i}}^{s} \frac{\beta_j}{\rho_j} + \frac{\beta_i}{\rho_i} + \sum_{j=s+1}^{m} \frac{\mu_j\beta_j}{k_{ij}} \\
&< p.
\end{aligned}$$

This implies that the hypothesis of Theorem 3.1 is satisfied. Hence, the zero solution of (1.1) is absolutely stable. This completes the proof. □

In the following, using a particular case of (1.1), we compare the criteria obtained here with some previous criteria.

In system (1.1), suppose that

$$\begin{aligned}&\rho_i = \rho > 0, \quad i = 1, 2, \dots, n,\\ &\beta_1 \geq \beta_2 \geq \cdots \geq \beta_s > 0,\\ &0 > \beta_{s+1} \geq \beta_{s+2} \geq \cdots \geq \beta_m,\\ &\beta_{m+1} = \cdots = \beta_n = 0.\end{aligned}$$

Consider the following two cases.

(i) $\sum_{i=1}^{s} \beta_i + \beta_m \leq 0$. In this case, condition (1.4) is equivalent to

$$p > \sum_{i=1}^{s} \frac{\beta_i}{4\rho},$$

while all of conditions (3.1), (3.5) and (3.6) hold when $p > 0$.

(ii) $\sum_{j=s+1}^{m} \beta_j + \beta_1 \geq 0$. In this case, condition (1.5) becomes

$$p > \sum_{i=1}^{s} \frac{\beta_i}{\rho} + \sum_{j=s+1}^{m} \frac{3\beta_j}{4\rho},$$

but either (3.1) or (3.12) is equivalent to

$$p > \sum_{i=1}^{m} \frac{\beta_i}{\rho}.$$

The above example shows that the criteria given in this paper are more effective in some cases.

## Acknowledgements

The authors would like to thank Professor Gail S. K. Wolkowicz for her careful reading and correcting the manuscript. Thanks are also due to Professor Shigui Ruan for his invitation and help.

## References

[1] Liao, Xiaoxin [1982], *Criteria for the absolute stability of control system*, Appl. Math. and Mech. **3**, 267-279.

[2] Liao, Xiaoxin [1988], *Mathematical Theory and Application of Stability*, Huazhong Normal University Press, Wuhan (in Chinese).

[3] Meyer, K. R. [1965], *On the existence of Lyapunov functions for the problem of Lurie*, J. SIAM Control, Ser. A, **3**, 373-383.

[4] Qiu, Xiaogang and Shu, Zhongzhou [1986], *Several criterions of absolute stability for the second canonical form of control system*, Appl. Math. and Mech. **7**, 785-800.

[5] Zhang, Wei [1989], *The criterions for absolute stability of direct control systems of Lurie-type*, Appl. Math. and Mech. **10**, 955-967.

Fields Institute Communications
Volume **21**, 1999

# Bogdanov-Takens Bifurcations in Predator-Prey Systems with Constant Rate Harvesting

**Dongmei Xiao**
Department of Mathematics
Central China Normal University
Wuhan, Hubei 430070, China
dmxiao@ccnu.edu.cn

**Shigui Ruan**
Department of Mathematics, Statistics and Computing Science
Dalhousie University
Halifax, Nova Scotia, Canada B3H 3J5
ruan@mscs.dal.ca

**Abstract.** In this paper, we consider a predator-prey system with constant rate predator harvesting. The system has a cusp of codimension 2 and exhibits the Bogdanov-Takens bifurcation. By choosing the death rate and the harvesting rate of the predator as the bifurcation parameters, we show that the system undergoes a sequence of bifurcations including the Hopf bifurcation, saddle-node bifurcations, and the homoclinic bifurcation. Global bifurcation diagrams and phase portraits in a small neighborhood of the cusp are sketched. Numerical simulations are given to illustrate the results.

## 1 Introduction

The study of population dynamics with harvesting is a very interesting subject in mathematical bioeconomics. It is related to the optimal management of renewable resources (see [**9**]). The exploitation of biological resources and the harvest of population species are commonly practiced in fishery, forestry, and wildlife management. The basic model, usually in the form of differential equations, is the generalized Gause-type model for two species,

$$\begin{aligned} \dot{x} &= xg(x) - p(x)y - h_1, \\ \dot{y} &= q(x)y - \delta y - h_2. \end{aligned} \tag{1.1}$$

1991 *Mathematics Subject Classification.* Primary 34C25, 92A15; Secondary 58F14.
Research of the first author was supported by the National Science Foundations of China.
Research of the second author was supported by NSERC of Canada and the Petro-Canada Young Innovator Award.

Here $x$ and $y$, are functions of time, representing population densities of prey and predator respectively, and $h_1, h_2$, and $\delta$ are positive constants. The biological setting requires that all functions be continuous. $g(0)$ is positive and $g(x)$ is decreasing in $x$. Both $p(x)$ and $q(x)$ are positive for $x > 0$ and vanish at 0.

System (1.1) is interesting for the biological implications as well as mathematically. In a sequence of papers by Brauer and Soudack [**4, 5, 6**], and the recent work by Dai and Tang in [**10**] and Dai and Xu in [**11**], the global behavior of system (1.1) for some special functions $g(x)$, $p(x)$ and $q(x)$ was analyzed by using qualitative theory and numerical techniques.

Brauer and Soudack in [**4**] discussed the following model

$$\begin{aligned}\dot{x} &= rx\left(1 - \frac{x}{k}\right) - \frac{yx}{a+x},\\ \dot{y} &= y\left(-d + \frac{x}{a+x}\right) - h,\end{aligned} \tag{1.2}$$

where $k$ is the carrying capacity of the prey population, $d$ is the death rate of the predator, $r$ is the intrinsic growth rate of the prey population, and $h$ is the harvesting rate. The function $\frac{x}{a+x}$ is often called the functional response of Holling type II. $k$, $d$, $r$, $a$ and $h$ are positive constants. Numerical studies in [**4**] indicated the existence of a homoclinic loop and a periodic orbit for some parameter values in (1.2).

In this paper, we do a bifurcation analysis of model (1.2). In particular we are interested in codimension 2 bifurcations that occur in a two-dimensional parameter region. Under some conditions we prove that system (1.2) undergoes the Bogdanov-Takens bifurcation, i.e., the bifurcation of a cusp of codimension 2 (see [**8**] and [**12**]). An example is given to show the existence of the Bogdanov-Takens bifurcation. In this framework, we demonstrate that system (1.2) can exhibit qualitatively different dynamical behavior, including Hopf bifurcations, saddle-node bifurcations as well as homoclinic bifurcations. In particular, for certain parameter values, the system can have a unique limit cycle or even a homoclinic loop. The corresponding bifurcation diagrams and phase portraits are sketched. Our analysis supports the numerical simulation in [**4**].

The paper is organized as follows. The general analysis and conditions appear in the next section. In Section 3 we reduce system (1.2) to the canonical family and obtain the main theorem. Finally, we choose two parameters, $d$ and $h$, as the bifurcation parameters for system (1.2) and show that system (1.2) exhibits the Bogdanov-Takens bifurcation. The global bifurcation diagram and all possible phase portraits are sketched when the bifurcation parameters vary in a small neighborhood of the origin in the two dimensional parameter plane. Numerical simulations are given to illustrate the obtained results.

## 2 Preliminaries

From system (1.2), we can see that the $y$-axis is invariant under the flow. However, this is not the case on the $x$-axis. All solutions touching the $x$-axis cross out of the first quadrant. Thus, the first quadrant is no longer positively invariant under the flow generated by the harvested system. From the standpoint of biology, we are only interested in the dynamics of system (1.2) in the first quadrant.

First, we consider the existence of equilibria of system (1.2). The prey isocline of system (1.2) is

$$y = r(1 - \frac{x}{k})(a + x).$$

This curve is convex and passes through the points $(0, ar)$ and $(k, 0)$. The predator isocline is

$$y = \frac{h(a + x)}{x - d(a + x)},$$

which passes through the point $(k, \frac{h(a+k)}{k-d(a+k)})$ and if $d < 1$ then $y \to \infty$ as $x$ approaches the straight line $x = \frac{ad}{1-d}$ from the right. We sketch the predator isocline curve and prey isocline curve in the first quadrant (see Figure 2.1).

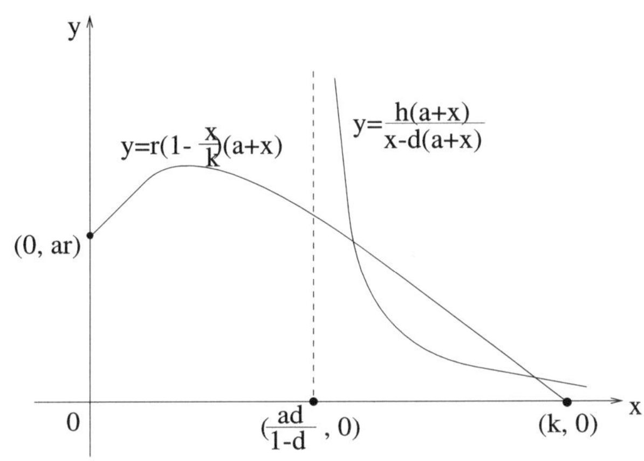


**Figure 2.1** The predator and prey isoclines

We assume throughout that $0 < d < 1$. Then it follows from the equations that the $x$ component of any equilibrium is less than $k$. If we fixed the parameters $r$, $k$, $a$, and $d$, then we can easily see that the number of equilibria of system (1.2) depends on the harvesting rate $h$. When $h$ is sufficiently large, system (1.2) has no equilibrium, and the species will be driven to extinction. From the point of view of the optimal management of renewable resources, we would like to determine the maximum sustainable yield (abbreviated as MSY) of the harvesting rates to ensure that the predator population can sustain itself. Straightforward calculations show that system (1.2) has a unique positive equilibrium if and only if the parameters $(r, k, d, a, h)$ satisfy

$$\left(d - 1 + \frac{ad}{k}\right)^2 - 4\frac{(1-d)(adr + h)}{kr} = 0 \quad \text{and} \quad k > \frac{ad}{1-d}. \tag{2.1}$$

The unique positive equilibrium is given by

$$(x_0, y_0) = \left(\frac{k(1-d) - ad}{2(1-d)}, \; r(1 - \frac{x_0}{k})(a + x_0)\right).$$

Mathematically, the surface, represented by the equation in (2.1), is called the *saddle-node bifurcation surface.* System (1.2) undergoes a saddle-node bifurcation when the parameters $(r,k,d,a,h)$ vary from one side to the other side of the surface. When the parameters $(r,k,d,a,h)$ satisfy the following inequality

$$\left(d-1+\frac{ad}{k}\right)^2-4\frac{(1-d)(adr+h)}{kr}>0 \text{ (or } <0) \quad \text{and} \quad k>\frac{ad}{1-d},$$

system (1.2) has two positive equilibria (or no equilibrium respectively).

In the following we will concentrate on the case when there exist parameter values $(r_0,k_0,d_0,a_0,h_0)$, denoted by $\lambda_0$, such that the unique equilibrium $(x_0,y_0)$ of system (1.2) is degenerate, and the variational matrix of system (1.2) at $(x_0,y_0)$

$$\begin{pmatrix} r-\dfrac{2rx}{k}-\dfrac{ay}{(a+x)^2} & -\dfrac{x}{a+x} \\ \dfrac{ay}{(a+x)^2} & -d+\dfrac{x}{a+x} \end{pmatrix}_{(x_0,y_0)}$$

has two zero eigenvalues. This is equivalent to the following conditions:

$$r-\frac{2rx_0}{k}-\frac{ay_0}{(a+x_0)^2}-d+\frac{x_0}{a+x_0}=0,$$
$$-rd+\frac{2rdx_0}{k}+\frac{adyy_0}{(a+x_0)^2}+\frac{rx_0}{a+x_0}-\frac{2rx_0^2}{k(a+x_0)}=0.$$

According to bifurcation theory ([**8**] and [**12**]), we know that under certain nondegeneracy conditions, the equilibrium $(x_0,y_0)$ is a cusp of codimension 2. If we choose suitable bifurcation parameters, then the system undergoes the Bogdanov-Takens bifurcation. In the next section we will introduce the nondegeneracy conditions and show how to choose the bifurcation parameters so that the system exhibits the Bogdanov-Takens bifurcation.

## 3 Reduction to canonical Bogdanov-Takens family

We consider system (1.2) as the parameters take value $\lambda_0$ and rewrite the system in the generic form

$$\begin{aligned} \dot{x}&=f(x,y,\lambda_0),\\ \dot{y}&=g(x,y,\lambda_0). \end{aligned} \tag{3.1}$$

According to the assumptions in Section 2, we know that system (3.1) has an isolated degenerate singular point $(x_0,y_0)$, $x_0>0$, $y_0>0$, the trace and the determinant of the variational matrix of system (3.1) at $(x_0,y_0)$ are zero but the variational matrix is not a zero matrix. System (3.1) is $C^\infty$ smooth with respect to the variables $x,y$ and $\lambda$ in a small neighborhood of $(x_0,y_0,\lambda_0)$. Next, we reduce system (3.1) to canonical form by using normal form theory.

Let $x_1=x-x_0$, $x_2=y-y_0$. Then system (3.1) can be transformed into

$$\begin{pmatrix} \dot{x}_1 \\ \dot{x}_2 \end{pmatrix}=Lx+\begin{pmatrix} \langle Px,x\rangle \\ \langle Qx,x\rangle \end{pmatrix}+O(|x|^3), \tag{3.2}$$

where $\langle\cdot,\cdot\rangle$ is a Cartesian product in $R^2$, the term $O(|x|^3)$ is $C^\infty$ in all variables, at least to the third order with respect to $x=col(x_1,x_2)$, and the matrices $L,P$ and

$Q$ are defined as follows

$$L = \begin{pmatrix} \dfrac{\partial f}{\partial x} & \dfrac{\partial f}{\partial y} \\ \\ \dfrac{\partial g}{\partial x} & \dfrac{\partial g}{\partial y} \end{pmatrix}_{(x_0,y_0,\lambda_0)} \triangleq \begin{pmatrix} a & b \\ c & d \end{pmatrix},$$

$$P = \begin{pmatrix} \dfrac{\partial^2 f}{\partial x^2} & \dfrac{\partial^2 f}{\partial x\partial y} \\ \\ \dfrac{\partial^2 f}{\partial x\partial y} & \dfrac{\partial^2 f}{\partial y^2} \end{pmatrix}_{(x_0,y_0,\lambda_0)} \triangleq \begin{pmatrix} p_{11} & p_{12} \\ p_{21} & p_{22} \end{pmatrix},$$

$$Q = \begin{pmatrix} \dfrac{\partial^2 g}{\partial x^2} & \dfrac{\partial^2 g}{\partial x\partial y} \\ \\ \dfrac{\partial^2 g}{\partial x\partial y} & \dfrac{\partial^2 g}{\partial y^2} \end{pmatrix}_{(x_0,y_0,\lambda_0)} \triangleq \begin{pmatrix} q_{11} & q_{12} \\ q_{21} & q_{22} \end{pmatrix}.$$

By the above assumptions, the matrix $L$ is similar to the Jordan block form $\begin{pmatrix} 0 & 1 \\ 0 & 0 \end{pmatrix}$. Hence $b^2 + c^2 \neq 0$. Assume that $b \neq 0$. Making the linear change of variables $y = Mx$, where $M = \begin{pmatrix} 1 & 0 \\ a & b \end{pmatrix}$ (if $b = 0$ and $c \neq 0$, then set $M = \begin{pmatrix} 0 & 1 \\ c & -d \end{pmatrix}$), system (3.2) becomes

$$\begin{pmatrix} \dot{y}_1 \\ \dot{y}_2 \end{pmatrix} = MLM^{-1}\begin{pmatrix} y_1 \\ y_2 \end{pmatrix} + M\begin{pmatrix} \langle (M^{-1})^T PM^{-1}y, y\rangle \\ \langle (M^{-1})^T QM^{-1}y, y\rangle \end{pmatrix} + O(|y|^3),$$

or in an abstract form

$$\dot{y} = Ay + h^2(y) + O(|y|^3), \tag{3.3}$$

where

$$A = \begin{pmatrix} 0 & 1 \\ bc - ad & a + d \end{pmatrix} = \begin{pmatrix} 0 & 1 \\ 0 & 0 \end{pmatrix},$$

the components of the vector function $h^2(y) \in C^\infty(R^2, R^2)$ are the homogeneous polynomials of degree 2 given by

$$h^2(y) = \begin{pmatrix} (p_{11} - \frac{2a}{b}p_{12} + \frac{a^2}{b^2}p_{22})y_1^2 + (\frac{2}{b}p_{12} - \frac{2a}{b^2}p_{22})y_1y_2 + \frac{1}{b^2}p_{22}y_2^2 \\ \\ d_1y_1^2 + \bar{d}_1y_1y_2 + (\frac{a}{b^2}p_{22} + \frac{1}{b}q_{22})y_2^2 \end{pmatrix},$$

$d_1 = ap_{11} - \frac{2a^2}{b}p_{12} + \frac{a^3}{b^2}p_{22} + bq_{11} - 2aq_{12} + \frac{a^2}{b}q_{22}$, and $\bar{d}_1 = \frac{2a}{b}p_{12} - \frac{2a^2}{b^2}p_{22} + 2q_{12} - \frac{2a}{b}q_{22}$. By the theory of normal forms, there exists a $C^\infty$ change of variables of system (2.3) in a small neighborhood of $(0,0)$,

$$\begin{pmatrix} z_1 \\ z_2 \end{pmatrix} = \begin{pmatrix} y_1 \\ y_2 \end{pmatrix} + \begin{pmatrix} -\frac{1}{2}(\frac{2}{b}p_{12} - \frac{a}{b^2}p_{22} + \frac{1}{b}q_{22})y_1^2 - \frac{1}{b^2}p_{22}y_1y_2 \\ \\ (p_{11} - \frac{2a}{b}p_{12} + \frac{a^2}{b^2})y_1^2 - (\frac{a}{b^2}p_{22} + \frac{1}{b}q_{22})y_1y_2 \end{pmatrix},$$

i.e.,

$$\begin{pmatrix} y_1 \\ y_2 \end{pmatrix} = \begin{pmatrix} z_1 \\ z_2 \end{pmatrix} + \begin{pmatrix} \frac{1}{2}(\frac{2}{b}p_{12} - \frac{a}{b^2}p_{22} + \frac{1}{b}q_{22})z_1^2 + \frac{1}{b^2}p_{22}z_1z_2 \\ -(p_{11} - \frac{2a}{b}p_{12} + \frac{a^2}{b^2})z_1^2 + (\frac{a}{b^2}p_{22} + \frac{1}{b}q_{22})z_1z_2 \end{pmatrix}$$

such that system (3.3) can be written as

$$\begin{pmatrix} \dot{z}_1 \\ \dot{z}_2 \end{pmatrix} = A \begin{pmatrix} z_1 \\ z_2 \end{pmatrix} + \begin{pmatrix} 0 \\ d_1z_1^2 + d_2z_1z_2 \end{pmatrix} + O(|z|^3), \tag{3.4}$$

where $d_2 = 2q_{12} + 2p_{11} - (p_{12} + p_{22})\frac{2a}{b}$. If $d_1d_2 \neq 0$, then the singular point $(x_0, y_0)$ of system (3.1) is a cusp of codim 2 by the qualitative theory of ordinary differential equations and the theory of differential manifolds. The condition $d_1d_2 \neq 0$ is called the *nondegeneracy condition of the cusp type of codimension 2* for the equilibrium $(x_0, y_0)$ of system (3.1). Moreover, we might as well assume that $d_1 > 0$.

From the above analysis, we obtain the following lemma.

**Lemma 3.1** *If $d_1d_2 \neq 0$, then in a small neighborhood of $(x_0, y_0, \lambda_0)$, system (1.2) is $C^\infty$ equivalent to*

$$\begin{aligned} \dot{x}_1 &= x_2 + w_1(x, \lambda), \\ \dot{x}_2 &= d_1x_1^2 + d_2x_1x_2 + w_2(x, \lambda), \end{aligned} \tag{3.5}$$

*where $\lambda = (r, k, d, a, h)$, $w_1, w_2 \in C^\infty(R^2 \times R^5, R^2)$ and $w_1(x, \lambda_0), w_2(x, \lambda_0)$ are power series in $(x_1, x_2)$ of powers $x_1^ix_2^j$ satisfying $i + j \geq 3$.*

**Lemma 3.2** *In a small neighborhood of $(0, 0, \lambda_0)$, system (3.5) is $C^\infty$ equivalent to*

$$\begin{aligned} \dot{y}_1 &= y_2, \\ \dot{y}_2 &= \phi_1(\lambda) + \phi_2(\lambda)y_1 + y_1^2 + y_2(\psi(\lambda) + \Phi(y_1, \lambda)) + y_2^2\Psi(y, \lambda), \end{aligned} \tag{3.6}$$

*where $\Phi, \Psi \in C^\infty$ and $\Phi(y_1, \lambda_0) = \frac{d_2}{\sqrt{d_1}}y_1 \neq 0$, $\Psi(y, \lambda_0) = 0$, $\phi_1(\lambda_0) = \phi_2(\lambda_0) = \psi(\lambda_0) = 0$.*

**Proof** Consider the $\lambda$-dependent change of coordinates

$$z_1 = x_1, \quad z_2 = x_2 + w_1(x, \lambda).$$

Then system (3.5) is transformed into

$$\begin{aligned} \dot{z}_1 &= z_2, \\ \dot{z}_2 &= P_1(z_1, \lambda) + z_2\Phi_1(z_1, \lambda) + z_2^2\Psi_1(z, \lambda), \end{aligned} \tag{3.7}$$

where $P_1, \Phi_1, \Psi_1 \in C^\infty$ and

$$P_1(0, \lambda_0) = \frac{\partial P_1(0, \lambda_0)}{\partial z_1} = 0, \quad \frac{\partial^2 P_1(0, \lambda_0)}{\partial z_1^2} = 2d_1 \neq 0,$$

$$\Phi_1(0, \lambda_0) = 0, \quad \frac{\partial \Phi_1(0, \lambda_0)}{\partial z_1} = d_2 \neq 0.$$

Applying the Malgrange Preparation Theorem (see [**7**], pp. 43) to the function $P_1(z_1, \lambda)$, we have

$$P_1(z_1, \lambda) = (\phi_1(\lambda) + \phi_2(\lambda)z_1 + z_1^2)B(z_1, \lambda),$$

where $\phi_1, \phi_2, B \in C^\infty$ and $B(0, \lambda_0) = d_1 \neq 0, \phi_i(\lambda_0) = 0, i = 1, 2$.

Hence, system (3.7) becomes

$$\begin{aligned}
\dot{z}_1 &= z_2,\\
\dot{z}_2 &= \left(\phi_1(\lambda)+\phi_2(\lambda)z_1+z_1^2+z_2\frac{\Phi_1(z_1,\lambda)}{B(z_1,\lambda)}+z_2^2\frac{\Psi_1(z,\lambda)}{B(z_1,\lambda)}\right)B(z_1,\lambda).
\end{aligned} \tag{3.8}$$

Since $\Phi_1(0,\lambda_0)=0$ and $\frac{\partial\Phi_1(0,\lambda_0)}{\partial z_1}=d_2\neq 0$, there exists a function $\psi(\lambda)$ such that $\frac{\Phi_1(0,\lambda)}{\sqrt{B(0,\lambda)}}=\psi(\lambda)$. Set

$$y_1=z_1,\ y_2=\frac{z_2}{\sqrt{B(z_1,\lambda)}},\ \tau=\int_0^t\sqrt{B(z_1(s),\lambda)}ds,$$

$$\Phi(z_1,\lambda)=\frac{\Phi_1(z_1,\lambda)}{\sqrt{B(z_1,\lambda)}}-\psi(\lambda),\ \Psi(z,\lambda)=\frac{\Psi_1(z,\lambda)}{B(z_1,\lambda)}.$$

We can transform system (3.8) into system (3.6). This completes the proof. □

In the following theorem we show how to choose the bifurcation parameters such that the system exhibits the Bogdanov-Takens bifurcation.

**Theorem 3.3** *If the rank of the matrix*

$$\begin{pmatrix}
\dfrac{\partial(\phi_1(\lambda)-\frac{1}{4}\phi_2^2(\lambda))}{\partial\lambda_1} & \cdots & \dfrac{\partial(\phi_1(\lambda)-\frac{1}{4}\phi_2^2(\lambda))}{\partial\lambda_5}\\[2ex]
\dfrac{\partial(\psi(\lambda)-\frac{d_2}{2\sqrt{d_1}}\phi_2(\lambda))}{\partial\lambda_1} & \cdots & \dfrac{\partial(\psi(\lambda)-\frac{d_2}{2\sqrt{d_1}}\phi_2(\lambda))}{\partial\lambda_5}
\end{pmatrix}_{\lambda_0}$$

*is two, then we can choose two bifurcation parameters such that system (1.2) undergoes the Bogdanov-Takens bifurcation.*

**Proof** By using the $C^\infty$ equivalences in Lemmas 3.1 and 3.2, we have transformed system (3.1) into the parameter dependent system (3.6). Let $x_1=y_1+\frac{1}{2}\phi_2(\lambda)$, $x_2=y_2$. System (3.6) can be written as

$$\begin{aligned}
\dot{x}_1 &= x_2,\\
\dot{x}_2 &= \phi_1(\lambda)-\frac{1}{4}\phi_2^2(\lambda)+\left(\psi(\lambda)-\frac{d_2}{2\sqrt{d_1}}\phi_2(\lambda)\right)x_2+x_1^2+\frac{d_2}{\sqrt{d_1}}x_1x_2+Q(x,\lambda),
\end{aligned} \tag{3.9}$$

where $Q(x,\lambda)$ is a power series in $(x_1,x_2)$ with powers $x_1^ix_2^j$ satisfying $i+j\geq 3$ and coefficients depending on $(\lambda_1,\dots,\lambda_5)$.

Without loss of generality, we assume that the determinant of the matrix

$$\begin{pmatrix}
\dfrac{\partial(\phi_1(\lambda)-\frac{1}{4}\phi_2^2(\lambda))}{\partial\lambda_1} & \dfrac{\partial(\phi_1(\lambda)-\frac{1}{4}\phi_2^2(\lambda))}{\partial\lambda_2}\\[2ex]
\dfrac{\partial(\psi(\lambda)-\frac{d_2}{2\sqrt{d_1}}\phi_2(\lambda))}{\partial\lambda_1} & \dfrac{\partial(\psi(\lambda)-\frac{d_2}{2\sqrt{d_1}}\phi_2(\lambda))}{\partial\lambda_2}
\end{pmatrix}_{\lambda_0}$$

is not zero. Denote

$$\mu_1=\phi_1(\lambda)-\frac{1}{4}\phi_2^2(\lambda),\ \mu_2=\psi(\lambda)-\frac{d_2}{2\sqrt{d_1}}\phi_2(\lambda),\ \mu_3=\lambda_3,\ \mu_4=\lambda_4,\ \mu_5=\lambda_5.$$

Obviously, the parameter transformation is nonsingular. System (3.9) can be rewritten as

$$\begin{aligned}\dot{x}_1 &= x_2,\\ \dot{x}_2 &= \mu_1 + \mu_2 x_2 + x_1^2 + \frac{d_2}{\sqrt{d_1}} x_1 x_2 + Q(x, \mu).\end{aligned} \tag{3.10}$$

By the theorems in [**2, 3**] and [**13**], we know that system (3.10) is strongly topologically equivalent to

$$\begin{aligned}\dot{x}_1 &= x_2,\\ \dot{x}_2 &= \mu_1 + \mu_2 x_2 + x_1^2 \pm x_1 x_2,\end{aligned} \tag{3.11}$$

where $\pm = sign(\frac{d_2}{d_1})$. Therefore, we can choose $\mu_1$ and $\mu_2$ as the bifurcation parameters such that system (1.1) undergoes the Bogdanov-Takens bifurcation (i.e., bifurcation of cusp singularity of codim 2). □

## 4 An example

In this section, as an example we consider system (1.2) with fixed $r$, $k$ and $a$, that is, we consider the following model

$$\begin{aligned}\dot{x} &= x\left(1 - \frac{x}{2}\right) - \frac{xy}{1+x},\\ \dot{y} &= y\left(-d + \frac{x}{1+x}\right) - h,\end{aligned} \tag{4.1}$$

where $d$ and $h$ are parameters. We will show that $d$ and $h$ in fact are the bifurcation parameters and system (4.1) exhibits the Bogdanov-Takens bifurcation.

From the analysis in Section 2 and Section 3, we know that the positive equilibrium $(x_0, y_0) = (1.12415, 0.930219)$ is a cusp of codimension 2 when $(d_0, h_0) = (0.19891, 0.30726)$.

Now we study the dynamics of system (4.1) when the parameters $d$ and $h$ vary in a small neighborhood of $(d_0, h_0)$. We use a series of changes of coordinates to transform system (4.1) into the canonical family in Section 3. Consider

$$\begin{aligned}\dot{x} &= x\left(1 - \frac{x}{2}\right) - \frac{xy}{1+x},\\ \dot{y} &= y\left(-d_0 + \lambda_1 + \frac{x}{1+x}\right) - h_0 + \lambda_2,\end{aligned} \tag{4.2}$$

where $\lambda_1$ and $\lambda_2$ are small parameters.

Let $x_1 = x - x_0$, $x_2 = y - y_0$, then system (4.2) becomes

$$\begin{aligned}\dot{x}_1 = &\left(1 - x_0 - \frac{y_0}{(1+x_0)^2}\right) x_1 - \frac{x_0}{1+x_0} x_2 + \left(-\frac{1}{2} + \frac{2 - x_0}{2(1+x_0)^2}\right) x_1^2\\ &- \frac{1}{(1+x_0)^2} x_1 x_2 + P_1(x_1, x_2),\\ \dot{x}_2 = &\, y_0 \lambda_1 + \lambda_2 + \frac{y_0}{(1+x_0)^2} x_1 + \left(-d_0 + \lambda_1 + \frac{x_0}{1+x_0}\right) x_2\\ &- \frac{y_0}{(1+x_0)^3} x_1^2 + \frac{1}{(1+x_0)^2} x_1 x_2 + P_2(x_1, x_2),\end{aligned} \tag{4.3}$$

where $P_1$ and $P_2$ are power series in $(x_1, x_2)$ with powers $x_1^i x_2^j$ satisfying $i + j \geq 3$. Denote

$$a = 1 - x_0 - \frac{y_0}{(1+x_0)^2}, \quad b = -\frac{x_0}{1+x_0}, \quad c = \frac{y_0}{(1+x_0)^2}, \quad d = -d_0 + \lambda_1 + \frac{x_0}{1+x_0};$$

$$p_{11} = -\frac{1}{2} + \frac{2-x_0}{2(1+x_0)^2}, \quad p_{12} = -\frac{1}{2(1+x_0)^2}, \quad q_{11} = -\frac{y_0}{(1+x_0)^3}, \quad q_{12} = \frac{1}{2(1+x_0)^2}.$$

Making the affine transformation

$$y_1 = x_1, \quad y_2 = ax_1 + bx_2,$$

we see that system (4.3) becomes

$$\begin{aligned} \dot{y}_1 &= y_2 + \left(p_{11} - \frac{2ap_{12}}{b}\right) y_1^2 + \frac{2p_{12}}{b} y_1 y_2 + Q_1(y_1, y_2), \\ \dot{y}_2 &= b(y_0\lambda_1 + \lambda_2) + a\lambda_1 y_1 + \lambda_1 y_2 + \left(ap_{11} - \frac{2a^2 p_{12}}{b} + bq_{11} - 2aq_{12}\right) y_1^2 \\ &+ \left(\frac{2ap_{12}}{b} + 2q_{12}\right) y_1 y_2 + Q_2(y_1, y_2). \end{aligned} \tag{4.4}$$

Here $Q_1$ and $Q_2$ are power series in $(y_1, y_2)$ with powers $y_1^i y_2^j$ satisfying $i + j \geq 3$.

Consider the $C^\infty$ change of coordinates in a small neighborhood of $(0, 0)$

$$z_1 = y_1 - \frac{p_{12}}{b} y_1^2, \quad z_2 = y_2 + \left(p_{11} - \frac{2ap_{12}}{b}\right) y_1^2.$$

Then system (4.4) is transformed into

$$\begin{aligned} \dot{z}_1 &= z_2 + R_1(z_1, z_2), \\ \dot{z}_2 &= b(y_0\lambda_1 + \lambda_2) + a\lambda_1 z_1 + \lambda_1 z_2 + \left(2p_{11} - \frac{2ap_{12}}{b} + 2q_{12}\right) z_1 z_2 \\ &+ \left(ap_{11} - \frac{2a^2 p_{12}}{b} + bq_{11} - 2aq_{12} + \frac{p_{12}a}{b}\lambda_1 - \lambda_1(p_{11} - \frac{2ap_{12}}{b})\right) z_1^2 \\ &+ R_2(z_1, z_2). \end{aligned} \tag{4.5}$$

Here $R_1$ and $R_2$ are power series in $(z_1, z_2)$ with powers $z_1^i z_2^j$ satisfying $i + j \geq 3$.

We choose the $C^\infty$ change of coordinates in a small neighborhood of $(0, 0)$

$$x_1 = z_1, \quad x_2 = z_2 + R_1(z_1, z_2),$$

so that system (4.5) becomes

$$\begin{aligned} \dot{x}_1 &= x_2, \\ \dot{x}_2 &= b(y_0\lambda_1 + \lambda_2) + a\lambda_1 x_1 + \lambda_1 x_2 + \left(2p_{11} - \frac{2ap_{12}}{b} + 2q_{12}\right) x_1 x_2 + F_1(x_1) \\ &+ \left(ap_{11} - \frac{2a^2 p_{12}}{b} + bq_{11} - 2aq_{12} + \frac{p_{12}a}{b}\lambda_1 - \lambda_1(p_{11} - \frac{2ap_{12}}{b})\right) x_1^2 \\ &+ x_2 F_2(x_1) + x_2^2 F_3(x_1, x_2). \end{aligned} \tag{4.6}$$

Here $F_1$, $F_2$ and $F_3$ are power series in $x_1$ and $(x_1, x_2)$ with powers $x_1^{k_1}$, $x_1^{k_2}$ and $x_1^i x_2^j$ satisfy $k_1 \geq 3$, $k_2 \geq 2$ and $i + j \geq 1$, respectively.

We substitute values of $a, b, c, d, p_{11}, p_{12}, q_{11}$ and $q_{12}$ for the above system and obtain the following equations

$$\begin{aligned}\dot{x}_1 &= x_2,\\ \dot{x}_2 &= 0.52922(0.93022\lambda_1+\lambda_2) - 0.33032\lambda_1 x_1 + \lambda_1 x_2 - 0.44592 x_1 x_2\\ &\quad + F_1(x_1) + (0.211978 - 0.19545\lambda_1)x_1^2 + x_2 F_2(x_1) + x_2^2 F_3(x_1,x_2).\end{aligned} \tag{4.7}$$

Applying the Malgrange Preparation Theorem, we have

$$\begin{aligned}&0.52922(0.93022\lambda_1+\lambda_2) - 0.33032\lambda_1 x_1 + (0.211978 - 0.19545\lambda_1)x_1^2 + F_1(x_1)\\ &= \left(x_1^2 - \frac{0.33032\lambda_1}{0.211978-0.19545\lambda_1}x_1 + \frac{0.52922(0.93022\lambda_1+\lambda_2)}{0.211978-0.19545\lambda_1}\right) B_1(x_1,\lambda),\end{aligned}$$

where $B_1(0,\lambda) = 0.211978 - 0.19545\lambda_1$ and $B_1$ is a power series in $x_1$ whose coefficients depend on parameters $\lambda = (\lambda_1, \lambda_2)$.

Let $y_1 = x_1,\ y_2 = \frac{x_2}{\sqrt{B_1(x_1,\lambda)}},\ \tau = \int_0^t \sqrt{B_1(x_1(s),\lambda)}ds$. Then system (4.7) becomes

$$\begin{aligned}\dot{y}_1 &= y_2,\\ \dot{y}_2 &= \frac{0.52922(0.93022\lambda_1+\lambda_2)}{0.211978-0.19545\lambda_1} - \frac{0.33032\lambda_1}{0.211978-0.19545\lambda_1}y_1\\ &\quad + y_1^2 + \frac{\lambda_1}{\sqrt{B_1(y_1,\lambda)}}y_2 - \frac{0.44592}{\sqrt{B_1(y_1,\lambda)}}y_1y_2 + G(y_1,y_2,\lambda),\end{aligned} \tag{4.8}$$

where $\dot{y}_1 = \frac{dy_1}{d\tau}$, $\dot{y}_2 = \frac{dy_2}{d\tau}$, $G(y_1,y_2,0)$ is a power series in $(y_1,y_2)$ with power $y_1^i y_2^j$ satisfying $i+j\geq 3$ and $j \geq 2$.

Making the parameter dependent affine transformation in (4.8)

$$x_1 = y_1 - \frac{0.33032\lambda_1}{2(0.211978-0.19545\lambda_1)}, \quad x_2 = y_2,$$

we have

$$\begin{aligned}\dot{x}_1 &= x_2,\\ \dot{x}_2 &= \frac{0.52922(0.93022\lambda_1+\lambda_2)}{0.211978-0.19545\lambda_1} - \left(\frac{0.33032\lambda_1}{2(0.211978-0.19545\lambda_1)}\right)^2 + x_1^2\\ &\quad + \frac{0.65257\lambda_1+\alpha_1(\lambda)}{\sqrt{B_1(\frac{0.33032\lambda_1}{2(0.211978-0.19545\lambda_1)},\lambda)}}x_2 - (0.96853+\alpha_2(\lambda))x_1x_2\\ &\quad + R(x_1,x_2,\lambda),\end{aligned} \tag{4.9}$$

where $\alpha_1(0,\lambda_2) = 0$, $\frac{\partial\alpha_1(0,\lambda_2)}{\partial\lambda_1} = 0$, $\alpha_2(0) = 0$, and $R(x_1,x_2,0)$ is a power series in $(x_1,x_2)$ with powers $x_1^i x_2^j$ satisfying $i+j\geq 3$ and $j\geq 2$.

For the sake of convenience, we assume that

$$\begin{aligned}\mu_1(\lambda_1,\lambda_2) &= \frac{0.52922(0.93022\lambda_1+\lambda_2)}{0.211978-0.19545\lambda_1} - \left(\frac{0.33032\lambda_1}{2(0.211978-0.19545\lambda_1)}\right)^2,\\ \mu_2(\lambda_1,\lambda_2) &= \frac{0.65257\lambda_1+\alpha_1(\lambda)}{\sqrt{B_1(\frac{0.33032\lambda_1}{2(0.211978-0.19545\lambda_1)},\lambda)}},\end{aligned}$$

which is a nonsingular parameter transformation. System (4.9) can be rewritten as

$$\begin{aligned}\dot{x}_1 &= x_2,\\ \dot{x}_2 &= \mu_1(\lambda_1,\lambda_2) + \mu_2(\lambda_1,\lambda_2)x_2 + x_1^2 - 0.96853x_1x_2 + S(x_1,x_2,\mu),\end{aligned} \tag{4.10}$$

where $S(x_1, x_2, \mu)$ is a power series in $(x_1, x_2, \mu_1, \mu_2)$ with powers $x_1^i x_2^j \mu_1^k \mu_2^l$ satisfying $i + j + k + l \geq 4$ and $i + j \geq 3$.

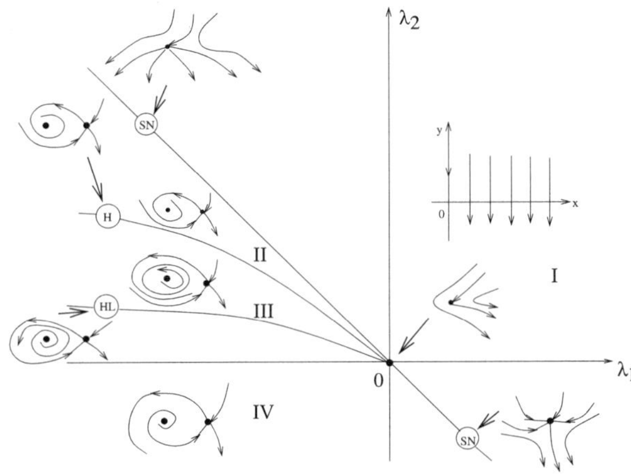


**Figure 4.1** The bifurcation set and the corresponding phase portraits for system (4.2)

By Theorem 3.3, we know that system (4.10) undergoes the Bogdanov-Takens bifurcation when the parameters vary in a small neighborhood of the origin. The local representations of the bifurcation curves in the small neighborhood of the origin are as follows:

(a) The saddle-node bifurcation curve

$$SN = \{(\lambda_1, \lambda_2) : \mu_1(\lambda_1, \lambda_2) = 0\}.$$

From the expression for $\mu_1(\lambda_1, \lambda_2)$, we can see that $SN$ is approximately a straight line when $\lambda_1$ is small. On the curve $SN$, system (4.10) has only a unique equilibrium which is a saddle-node. Some solutions of system (4.10) tend to the saddle-node and others either leave the first quadrant or increase without bound depending on the initial conditions (see Figure 4.1). If the parameter $\lambda_1$ and $\lambda_2$ satisfy $\mu_1(\lambda_1, \lambda_2) > 0$, then system (4.10) has no equilibrium. Hence all predator will tend to extinction. From this we can see that the harvesting rate must be limited, since otherwise the population cannot sustain itself. If the parameter $\lambda_1$ and $\lambda_2$ satisfy $\mu_1(\lambda_1, \lambda_2) < 0$, then system (4.10) has two positive equilibria. One is a hyperbolic saddle and the other is a focus. The trajectories of system (4.10) can either tend to an equilibrium or oscillate depending on the parameters.

(b) The Hopf bifurcation curve

$$H = \{(\lambda_1, \lambda_2) : \mu_2(\lambda_1, \lambda_2) = -0.96853\sqrt{-\mu_1(\lambda_1, \lambda_2)}, \mu_1(\lambda_1, \lambda_2) < 0\}.$$

When the parameters lie between the curve $SN$ and the curve $H$, system (4.10) has a hyperbolic saddle and a stable hyperbolic focus and no periodic orbit. On the curve $H$ system (4.10) has a hyperbolic saddle and a weak focus of degree one. The focus is stable and there are no periodic orbits.

(c) The homoclinic bifurcation curve

$$HL = \{(\lambda_1, \lambda_2) : \mu_2(\lambda_1, \lambda_2) = -0.6918\sqrt{-\mu_1(\lambda_1, \lambda_2)}, \mu_1(\lambda_1, \lambda_2) < 0\}.$$

When the parameters lie between the curve $H$ and the curve $HL$, system (4.10) has a unique stable limit cycle. However, system (4.10) has a hyperbolic saddle, an unstable focus, and no periodic orbits when the parameters lie between the curve $HL$ and the curve $SN$ (see Figure 4.1). On the $HL$ curve, system (4.10) has an unstable focus and a stable homoclinic loop.

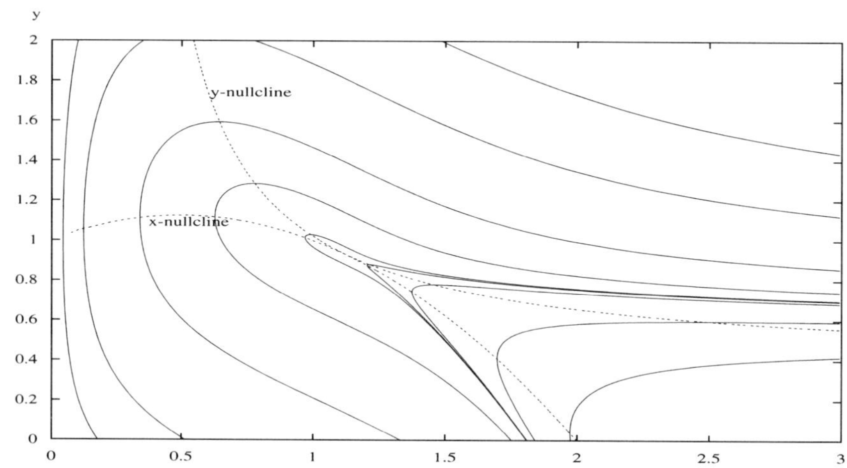


**Figure 4.2** When $(\lambda_1, \lambda_2) = (0, 0)$, the unique positive equilibrium is a cusp of codimension 2.

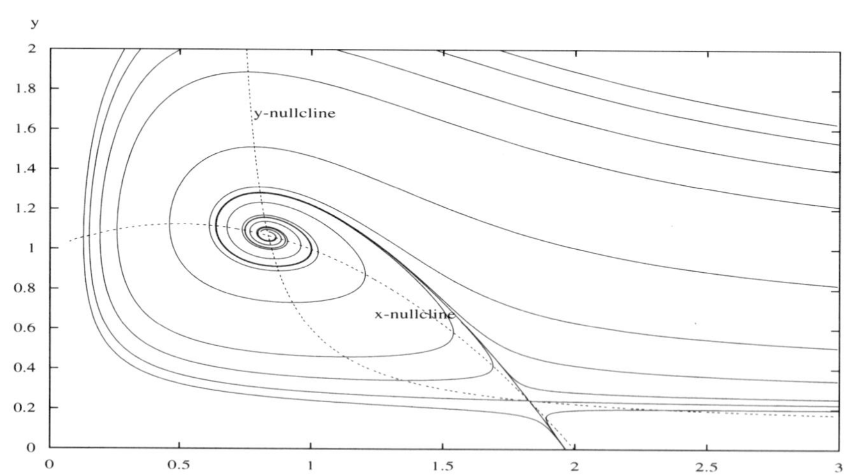


**Figure 4.3** When $(\lambda_1, \lambda_2) = (-0.20109, 0.24726)$ lies in the region II, there are two positive equilibria, a saddle and a stable focus.

Numerical simulations (using XPP) of system (4.2) are depicted in Figures 4.2 - 4.6. When $(\lambda_1, \lambda_2) = (0, 0)$, that is, when $(d_0, h_0) = (0.19891, 0.30726)$, there is a unique positive equilibrium $(x_0, y_0) = (1.12415, 0.930219)$, which is a cusp of

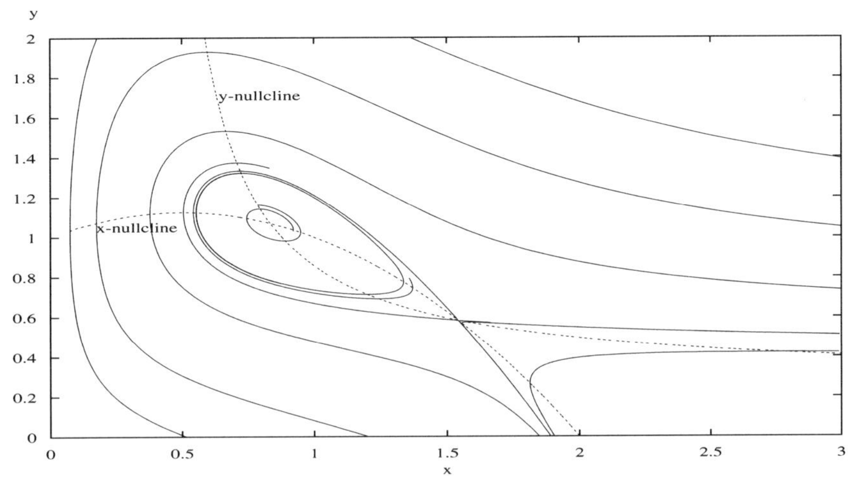


**Figure 4.4** When $(\lambda_1, \lambda_2) = (-0.07442, 0.11323$ lies in the region III, the stable focus becomes unstable and there is a stable periodic orbit.

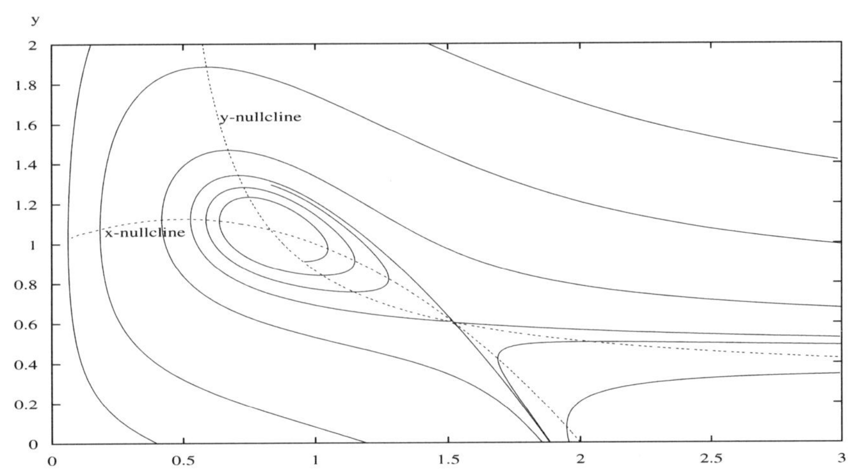


**Figure 4.5** When $(\lambda_1, \lambda_2) = (-0.06224, 0.10088)$ lies on the HL curve, there is a homoclinic loop.

codimension 2 (Figure 4.2). When $(\lambda_1, \lambda_2)$ varies, there is a saddle-node bifurcation. When $(\lambda_1, \lambda_2) = (-0.20109, 0.24726)$ lies in the region II, two positive equilibria bifurcate from the saddle-node, one is a saddle and the other is a stable focus (Figure 4.3). As $(\lambda_1, \lambda_2)$ keeps varying, there is a Hopf bifurcation. When $(\lambda_1, \lambda_2) = (-0.07442, 0.11323)$ lies in the region III, the stable focus becomes unstable and a stable periodic orbit bifurcates from the focus (Figure 4.4). When $(\lambda_1, \lambda_2)$ changes to $(-0.06224, 0.10088)$, the periodic orbit expands and reaches the stable and unstable manifolds of the saddle to create a homoclinic loop (Figure 4.5). When $(\lambda_1, \lambda_2) = (-0.04224, 0.07668)$ lies in the region IV, the homoclinic loop is broken, there are an unstable focus and a saddle (Figure 4.6).

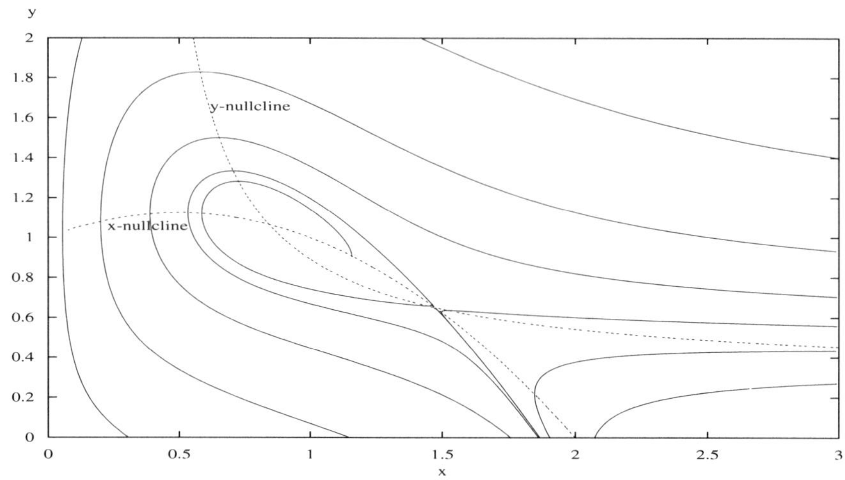


**Figure 4.6** When $(\lambda_1, \lambda_2) = (-0.04224, 0.07668)$ lies in the region IV, there is an unstable focus and a saddle.


## Acknowledgements

We would like to thank the referee for helpful comments and suggestions and Gail Wolkowicz for careful reading and corrections.


## References


[1] Andronov, A., Leontovich, E. A., Gordon, I. I. and Maier, A. G. [1971], *Theory of Bifurcations of Dynamical Systems on a Plane*, Israel Program for Scientific Translations, Jerusalem.

[2] Bogdanov, R. [1981], *Bifurcations of a limit cycle for a family of vector fields on the plane*, Selecta Math. Soviet. **1**, 373-388.

[3] Bogdanov, R. [1981], *Versal deformations of a singular point on the plane in the case of zero eigenvalues*, Selecta Math. Soviet **1**, 389-421.

[4] Brauer, F. and Soudack, A. C. [1979], *Stability regions and transition phenomena for harvested predator-prey systems*, J. Math. Biol. **7**, 319-337.

[5] Brauer, F. and Soudack, A. C. [1979], *Stability regions in predator-prey systems with constant rate prey harvesting*, J. Math. Biol. **8**, 55-71.

[6] Brauer, F. and Soudack, A. C. [1981], *Coexistence properties of some predator-prey systems under constant rate harvesting and stocking*, J. Math. Biol. **12**, 101-114.

[7] Chow, S.-N. and Hale, J. K. [1982], *Methods of Bifurcation Theory*, Springer-Verlag, New York - Heidelbeg - Berlin.

[8] Chow, S.-N., Li, C. Z. and Wang, D. [1994], *Normal Forms and Bifurcation of Planar Vector Fields*, Cambridge University Press, Cambridge.

[9] Clark, C. W. [1990], *Mathematical Bioeconomics, The Optimal management of Renewable Resources*, 2nd Ed., John Wiley & Sons, New York-Toronto.

[10] Dai, G. and Tang, M. [1998], *Coexistence region and global dynamics of a harvested predator-prey system*, SIAM J. Appl. Math. **58**, 193-210.

[11] Dai, G. and Xu, C. X. [1994], *Constant rate predator harvested predator-prey system with Holling-type I functional response*, Acta. Math. Scientia (in Chinese) **14**, 134-144.

[12] Kuznetsov, Y. A. [1995], *Elements of Applied Bifurcation Theory*, Appl. Math. Sci. **112**, Springer-Verlag, New York.

[13] Takens, F. [1974], *Forced oscillations and bifurcation*, in Applications of Global Analysis I, Comm. Math. Inst. Rijksuniversitat Utrecht, No. 3, pp. 1-59.

# Appendix

## List of Participants of the International Conference on Differential Equations with Applications to Biology

Abolghasem, Hossein, Dalhousie University, Halifax, Nova Scotia, Canada
Allegretto, Walter, University of Alberta, Edmonton, Alberta, Canada
Allen, Linda, Texas Tech. University, Lubbock, Texas, USA
Anderson, Erik, St. Francis Xavier University, Antigonish, Nova Scotia, Canada
Benoit, Patricia, Dalhousie University, Halifax, Nova Scotia, Canada
Brauer, Fred, University of Wisconsin, Madison, Wisconsin, USA
Brunner, Hermann, Memorial University of Newfoundland, St. John's, Newfoundland, Canada
Buchanan, Robert, Millersville University, Millersville, PA, USA
Campbell, Sue Ann, University of Waterloo, Waterloo, Ontario, Canada
Cantrell, Steve, University of Miami, Coral Gables, FL, USA
Carvalho, Luiz, Instituto de Ciências Matemáticas de São Carlos, São Carlos, Brazil
Chan, David, North Carolina State University, Raleigh, NC USA
Clements, Clyde, Dalhousie University, Halifax, Nova Scotia, Canada
Clements, John, Dalhousie University, Halifax, Nova Scotia, Canada
Coley, Alan, Dalhousie University, Halifax, Nova Scotia, Canada
Cosner, Chris, University of Miami, Coral Gables, FL USA
Cressman, Ross, Wilfrid Laurier University, Waterloo, Ontario, Canada
Culshaw, Rebecca Veronica, Dalhousie University, Halifax, Nova Scotia, Canada
Dormayer, Peter, Justus Leibig University, Giessen, Germany
Erbe, Lynn, University of Alberta, Edmonton, Alberta, Canada
Faria, Teresa, Universidade de Lisboa, Lisboa, Portugal
Farkas, Miklós, School of Mathematics, Budapest University of Technology, Budapest, Hungary
Feng, Zhilian, Purdue University, West Lafayette, IN USA
Fernandes, Rui L., Instituto Superior Técnico, Lisboa, Portugal
Foster, Andy, Memorial University of Newfoundland, St. John's, Newfoundland, Canada
Francois, Frederique, Centre d'Oceanologie de Marseille, Marseille, France
Franke, John, North Carolina State University, Raleigh, NC USA
Freedman, Herbert I., University of Alberta, Edmonton, Alberta, Canada
Freitas, Pedro, Instituto Superior Técnico, Lisboa, Portugal
Fusco, Domenico, Universitá di Messina, Messina, Italy
Gagnon, Linda, 6630 Sherbrooke West, Montreal, Quebec Canada
Gard, Thomas, The University of Georgia, Athens, GA USA
Garey, L. E., University of New Brunswick, Saint John, New Brunswick, Canada
Gedeon, Tomas, Montana State University, Bozeman, MT USA
Genik, Lynne, University of Victoria, Victoria, B.C. Canada

Glass, Leon, McGill University, Montreal, Quebec, Canada
Gopalsamy, K., Flinders University, Adelaide, Australia
Guillamon, Antoni, Universitat Politècnica de Catalunya, Barcelona, Spain
Hadeler, Karl-Peter, Universität Tübingen, Tübingen, Germany
Hamaya, Yoshihiro, Okayama Univ. of Science, Okayama, Japan
an der Heiden, Uwe, University of Witten, Witten, Germany
Hino, Yoshiyuki, Chiba University, Chiba, Japan
Hirsch, Morris, University of California, Berkeley, CA USA
Hofbauer, Josef, University of Vienna, Vienna, Austria
Horacek, B. Milan, Dalhousie University, Halifax, Nova Scotia, Canada
Hsu, Sze-Bi, Tsing-Hua University, Hsinchu, Taiwan
Huang, Ying Sue, Pace University, Pleasantville, NY USA
Infeld, Eric, Soltan Institute, Warsaw, Poland
Ingalls, Brian, Dalhousie University, Halifax, Nova Scotia, Canada
Irwin, Andrew, Queen's University, Kingston, Ontario, Canada
Johnson, Keith, Dalhousie University, Halifax, Nova Scotia, Canada
Kan-on, Yukio, Ehime University, Matsuyama, Japan
Kato, Junji, Tohoku University, Sendai, Japan
Ko, Bongsoo, Cheju National University, Cheju City, South Korea
Kong, Qingkai, Northern Illinois University, DeKalb, IL USA
Kuang, Yang, Arizona State University, Tempe, AZ USA
Ladeira, Luiz, Instituto de Ciências Matemáticas de São Carlos, São Carlos, Brazil
Lampreia, José Paulo, Quinta Da Torre, Monte Da Caparica, Portugal
Langford, William F., The Fields Institute, Toronto, Ontario, Canada
Lara-Aparicio, Miguel, Universidad Nacional Autonoma de Mexico, Mexico
Levin, Simon, Princeton University, Princeton, NJ USA
Lewis, Mark, University of Utah, Salt Lake City, UT USA
Li, Michael, Mississippi State University, Mississippi State, MS USA
Liu, Xinzhi, University of Waterloo, Waterloo, Ontario, Canada
Mackey, Michael C., McGill University, Montreal, Quebec, Canada
Macki, Jack, University of Alberta, Edmonton, Alberta, Canada
Mahaffy, Joseph M., San Diego State University, San Diego, CA USA
Mallet-Paret, John, Brown University, Providence, RI USA
Mangum, Wyatt, North Carolina State University, Raleigh, NC USA
McCluskey, Connell, University of Alberta, Edmonton, Alberta, Canada
Melrose, Gordon, Old Dominion University, Norfolk, VA USA
Milota, Jarovslav, Charles University, Praha, Czech Republic
Miyazaki, Rinko, Osaka Prefecture University, Osaka, Japan
Morita, Yoshihisa, Ryukoku University, Seta Ohtsu, Japan
Moson, Peter, Technical University of Budapest, Budapest, Hungary
Myers, Ransom A., Dalhousie University, Halifax, Nova Scotia, Canada
Peng, Qiu Liang, The Fields Institute, Toronto, Ontario, Canada
Petzeltova, Hana, Institute of Mathematics, Praha, Czech Republic
Pianigiani, Giulio, Universitá di Firenze, Firenze, Italy
Poggiale, Jean-Christophe, Centre d'Oceanologie de Marseille, Marseille, France
Ranger, Keith, University of Toronto, Toronto, Ontario, Canada
Rebelo, Carlota, Centro de Matematica e Aplicacoes Fundamentais, Lisboa, Portugal
Robinson, Bencion, BNEI - ATAROT, 60991, Israel

Ross, R. A., University of Toronto, Toronto, Ontario, Canada
Ruan, Shigui, Dalhousie University, Halifax, Nova Scotia, Canada
Ruan, Weihua, Purdue University Calumet, Hammond, IN USA
Ruess, Wolfgang M., Universität Essen, Essen, Germany
Sastri, Chelluri C. A., Dalhousie University, Halifax, Nova Scotia, Canada
Schmitt, Klaus, University of Utah, Salt Lake City, UT USA
Segel, Lee, Weizmann Institute, Rehovot, Israel
Selgrade, James, North Carolina State University, Raleigh, NC USA
Shaw, Ruth E., University of New Brunswick, Saint John, New Brunswick, Canada
Shayganmanesh, Ahmad, Iran University of Science and Technology, Narmak-Tehran, Iran
Singh, Manmohan, Swinburne University of Technology, Hawthorn, Victoria, Australia
Sleeman, Brian D., The University of Leeds, Leeds, United Kingdom
Smith, Bruce, Dalhousie University, Halifax, Nova Scotia, Canada
Smith, Hal L., Arizona State University, Tempe, AZ USA
So, Joseph, University of Alberta, Edmonton, Alberta, Canada
Sumner, Suzanne, Mary Washington College, Fredericskburg, VA USA
Swaminathan, S., Dalhousie University, Halifax, Nova Scotia, Canada
Takeuchi, Yasuhiro, Shizuoka University, Hamamatsu, Japan
Tang, Moxun, University of Minnesota, Minneapolis, MN USA
Thieme, Horst R., Arizona State University, Tempe, AZ USA
Toni, Bourama, Facultad de Ciencias,U.A.E.M., Cuernavaca, Mexico
Tweed, John, Old Dominion University, Norfolk, VA USA
van den Driessche, Pauline, University of Victoria, Victoria, B.C., Canada
Walde, Sandra, Dalhousie University, Halifax, Nova Scotia, Canada
Waltman, Paul, Emory University, Atlanta, GA USA
Webb, Glenn, Vanderbilt University, Nashville, TN USA
Wolansky, Gershon, Technion, Haifa, Israel
Wolkowicz, Gail S. K., McMaster University, Hamilton, Ontario, Canada
Wu, Jianhong, York University, North York, Ontario, Canada
Wu, Weihua, Central China Normal University, Wuhan, Hubei, China
Xiao, Dongmei, Central China Normal University, Wuhan, Hubei, China
Xu, Yao-Huan, Grove City College, Grove City, PA USA
Yung, Joyce, University of Alabama at Birmingham, Birminghan, AL USA
Zanolin, Fabio, University of Udine, Udine, Italy
Zeeman, Mary Lou, University of Texas at San Antonio, San Antonio, TX USA
Zhao, Xiao-Qiang, Academia Sinica, Beijing, China